Computational Methods in Elasticity and Plasticity

A. Anandarajah

Computational Methods in Elasticity and Plasticity

Solids and Porous Media

A. Anandarajah
Department of Civil Engineering
Johns Hopkins University
3400 Charles Street
Baltimore, Maryland 21218
USA
Rajah@jhu.edu

ISBN 978-1-4899-8241-4 ISBN 978-1-4419-6379-6 (eBook)
DOI 10.1007/978-1-4419-6379-6
Springer New York Heidelberg Dordrecht London

Printed on acid-free paper

Springer is part of Springer Science+Business Media (www.springer.com)

Preface

The past few decades have seen a great upsurge in research in the area of elasto-plastic modeling of solids, porous materials and structures. While research continues on more challenging topics, a certain degree of maturity has been reached on the basic principles of elasto-plastic numerical modeling. Understanding the basics of elasto-plastic modeling requires knowledge in a number of rather diverse areas such as continuum mechanics, finite element method, theories of elasticity and plasticity and integration of rate constitutive equations. The objectives of this book are (a) to provide in a single volume a detailed coverage of all necessary topics, (b) to develop the theories from the basic-level mathematics, mechanics and physics, (c) to cover the topics thoroughly, (d) to provide opportunity for the reader to develop hands-on experience and (e) to present adequate amount of numerical examples for the reader to develop a feel for the theories and to verify their own analysis results. In achieving these ambitious set of objectives, certain topics had to be excluded. For example, the consideration of large deformation, bifurcation, shear localization, gradient plasticity, damage models and contact problems have been omitted.

The book is intended to be a text book. Each chapter accompanies a set of problems. The level of presentation is aimed at graduate students. However, the style of presentation should allow undergraduate students to grasp the material as well. The material in the book is suitable for a number of different graduate-level courses, including, for instance, (a) computational plasticity, (b) numerical modeling in engineering, (c) finite element modeling and (d) theory of plasticity. It is left to the course instructor to select the topics to serve the course at hand. The book should serve courses in various engineering disciplines such as mechanical engineering, aerospace engineering, structural engineering, geotechnical engineering and geological engineering.

Chapter 1 presents an introduction to modeling. The mathematical and mechanics principles are presented in Chap. 2. The governing equations concerning the load-deformation problems of solids are covered in Chap. 3. The elastic constitutive laws for isotropic and anisotropic solids are described in Chap. 4. The finite element method for solving the load-deformation behavior of solids and structures is the subject of Chap. 5. The governing equations and their finite element implementations for porous media are presented in Chaps. 6 and 7. Newton's method of nonlinear analysis is described in Chap. 8. A comprehensive treatment of the theory of plasticity as applied to constitutive modeling is the subject of Chap. 9. The remaining chapters are devoted to the systematic treatment of the integration of constitutive rate equations. The closest point projection method is emphasized. The cutting-plane method and the two-step forward Euler method are also presented. The theories are presented in a general sense in Chap. 10. The theories are then applied to the von Mises model, modified Cam-clay model, Drucker-Prager model and sliding-rolling model in Chaps. 11, 12, 13, and 14, respectively. The appendices complete the topics presented in the Chapters.

The completion of this book would not have been possible without the help of numerous colleagues, students, former teachers, mentors, family and friends. I wish to thank my former teachers and mentors from the University of California, Davis. I learned mechanics and plasticity from Yannis Dafalias, finite elements from Len Herrmann, dynamics from Karl Romstad and geomechanics from the late professor Arul Arulanandan. The research topics that I have tackled over the past three decades and the scientific rigor with which I have solved the research problems are largely due to the knowledge and inspiration that I acquired from these fine researchers and gentlemen. I wish to express my sincere appreciation for the colleagues who reviewed some of the chapters and provided invaluable comments. In particular, I wish to thank Yannis Dafalias of University of California, Davis and National Technical University of Athens, Victor Kaliakin of University of Delaware, Xiang Song Li of the Hong Kong University of Science and Technology, Ning Lu of Colorado School of Mines, Majid Manzari of George Washington University, and Muralee Muraleetharan of University of Oklahoma. Dara Male provided assistance with some of the figures. The support and encouragement of the family members, Luxmi, Vinod and Nivay Anandarajah have been instrumental in the completion of the book. A special appreciation goes to Nivay Anandarajah for patiently reading and editing the original chapters. Finally, I wish to thank the staff at Springer who kept the project moving smoothly end expeditiously to completion. In particular, I acknowledge the help of Steve Elliot and Andrew Leigh.

Supplementary materials including a solution manual will be placed on the course website. Please read the README.PDF for complete update on the materials available on the website.

Johns Hopkins University A. Anandarajah

Contents

Chapter 1
Introduction

Mathematical (analytical or numerical) modeling of continuum mechanics problems is an integral part of engineering design, scientific research, and technological inventions. Mechanical engineers are concerned with the design of structures of machines, automobiles, and robots. Aerospace engineers design airplanes and space shuttles. Biomedical engineers are pursuing mathematical modeling of human anatomy. Civil engineers are charged with the design of skyscrapers, bridges, foundations, dams, and roads. Geologists and earth scientists are interested in quantifying fluid flow through and deformation of rocks. In all of these problems, it is of interest to *predict* the behavior of the structure or body at hand. The crucial step in the process of prediction in these cases is mathematical modeling.

Often the behavior of structures loaded into the nonlinear and/or plastic range needs to be predicted. The nonlinearity typically stems from large deformations (geometric nonlinearity) and inelasticity (material nonlinearity); the book is concerned with the latter. More specifically, the rate-independent plasticity is emphasized. In regard to mathematical modeling, a numerical method known as the finite element method is emphasized. An attempt is made to present a comprehensive coverage of all relevant aspects of the finite element modeling of rate-independent elasto-plastic solids, particulates, and structures experiencing small strain and deformation. To this end, the chapters of the book cover the necessary mathematical and mechanics principles, governing equations, and their finite element formulations of the underlying physics of solids and porous media, details of elasticity and plasticity theories, principles of nonlinear analysis, and integration and finite element implementation of some commonly used elasto-plastic constitutive laws. In the present chapter, we introduce the reader to various elements of mathematical modeling.

A. Anandarajah, *Computational Methods in Elasticity and Plasticity: Solids and Porous Media*, DOI 10.1007/978-1-4419-6379-6_1,

1.1 Mathematical Modeling of Continuum Mechanics Problems

1.1.1 Elements of a Mathematical Model

Consider a problem of finding the vertical displacement of point A, u_A, on the vertical column shown in Fig. 1.1. The column is approximately cylindrical and has a variable cross-section along its height. The surface of the column has an irregular geometry. The first step in solving this problem is the development of a mathematical model, which essentially means *idealization* of the following elements:

- Physics of the problem
- Geometry of the domain
- Boundary conditions
- Material behavior

1.1.1.1 Physics of the Problem

The deformation of point A (Fig. 1.1) will depend on whether the material is *porous* such as powders, soil and rock or *solid* such as metals and plastics. If it is solid, there is a need to know whether there are phenomena such as thermal effects to be accounted for. If it is porous, the pores may be filled with one or more liquids.

The problems involving more than one material are known as *multiphase problems*. For example, the problem involving rock saturated with water and oil is a multiphase problem, and rock saturated with just water is a two-phase problem. Airplane wings and automobile crank shafts are made of solids. Water-saturated powders, rock, and soil are two-phase materials. Heart muscle is a multiphase material.

For two-phase or multiphase materials, the *drainage conditions* must be known since there could be *coupling between deformation and fluid flow*. In certain cases, the *secondary compression* (creep) also may have to be considered. The nature of the contact between the base of the column and the support (Fig. 1.1) may dictate consideration of complex *contact physics*.

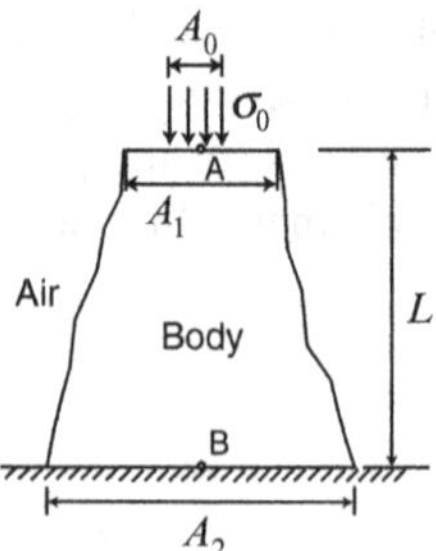

Fig. 1.1 Schematic of load-deformation problem

It is thus clear that the definition of the problem to be solved begins with recognizing the *underlying physics* and making certain *approximations*. For instance, the thermal effects may have a negligible influence on u_A and thus may be ignored. Whether the thermal effects are negligible or not depends on the required *accuracy* of the estimate of u_A.

For instance, the problem may be part of your research project, and you are seeking a high degree of accuracy on the estimate of u_A. In such a case, you may decide to consider the thermal effects even when they are small. On the contrary, if the problem is part of a design of a building support system, you may be able to neglect the thermal effects in estimating a value for u_A. Hence the selection of the physics to be considered for analysis is based on the context in which answers are sought.

Problems with more than one type of physics are referred to as *multiphysics* problems. We will refer to problems with single type of physics as *single-physics* problems.

Returning to the problem in Fig. 1.1, let us assume, for the sake of discussion, that the thermal and secondary compression effects are negligible. Let us further assume that the material is dry (i.e., with no fluid in the pore space), rendering drainage considerations irrelevant. These restrictions make the problem of interest a single-physics problem. Specifically, the problem shown in Fig. 1.1 is a single-physics problem known as the *load-deformation problem*.

In the analyses presented in the following sections, it is assumed that the structure is perfectly bonded to the support at the base in the vertical direction, making the consideration of contact modeling unnecessary.

Once the physics of the problem is identified, it then remains to develop *mathematical description of the physical laws*. In our case (for the load-deformation problem), the physical laws are *Newton's laws*.

1.1.1.2 Geometry of the Domain

The definition of the problem now requires the definition of the *domain* of interest. A domain consists of *volume* Ω and the *boundary* Γ surrounding it, as shown in Fig. 1.2a. The domain can be simply connected, as shown in Fig. 1.2a, or multiply connected, as shown in Fig. 1.2b. As far as the usage of general solution methods

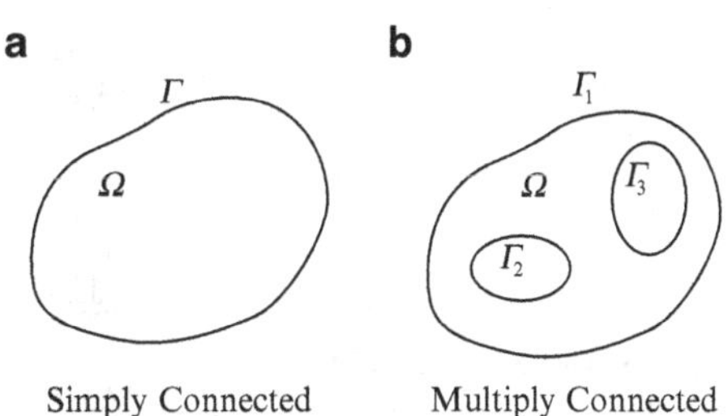

Fig. 1.2 Types of typical domains

such as the finite element method (FEM) is concerned, the treatment of simply connected and multiply connected domains boils down to just the specification of the conditions on the boundaries known as the boundary conditions (discussed next). Thus, there are no conceptual difficulties of dealing with a multiply connected domain.

The domain may further be subdivided into sub-regions based on various factors; for instance, subdivision based on physics, material properties, internal loading, etc. The domain may consist of a *homogeneous material* (spatially uniform properties) or *heterogeneous material* (spatially nonuniform properties). The properties can be directionally independent (*isotropic*) or directionally dependent (*anisotropic*).

In our example problem (Fig. 1.1), we need to decide on the nature of approximation to be adopted. The approximation leads to a *model of the geometry*. Figure 1.3a–c illustrates three different geometrical models that one may choose to adopt.

The first approximation (Model 1), where the column is approximated by a one-dimensional prismatic rod (Fig. 1.3a), simplifies the problem so that an analytical solution can be readily obtained. The model is one dimensional and the displacement of a point is approximated by one component in the vertical direction. The details of the surface irregularity are ignored, and the varying cross-sectional area is averaged into a single representative cross-sectional area A_a.

In the second model (Model 2), the surface irregularity is ignored, but the varying cross-section is considered on an average sense (Fig. 1.3b). The model is still one dimensional and the displacement of a point is approximated by one component in the vertical direction. The problem is intractable by hand calculations, but is amenable to analytic treatment; i.e., a differential equation governing the equilibrium of the rod can be developed and, in conjunction with the boundary conditions, solved to obtain an exact solution for u_{A} for the case of linear elastic material.

The third model (Model 3) represents almost an exact replica of the real domain. Nevertheless, this model of the domain is still an approximation since the surfaces are represented by approximate mathematical functions (be it continuous or piecewise). The model is at a minimum two dimensional, requiring at least two displacement components to model the deformation of a point. Depending on

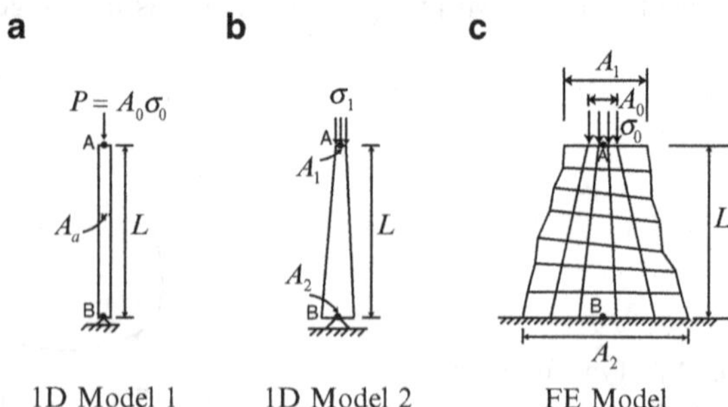

Fig. 1.3 Different idealized models

the irregularities of the geometry, nature of the boundary conditions, and spatial distribution of material properties, the problem could be three dimensional, requiring three displacement components to model the deformation of a point. Now the problem is intractable by analytical methods; more general numerical techniques such as the finite element method (Fig. 1.3c) is needed for obtaining the solution.

It is clear that idealization of the geometry is an important part of developing the model. The type of approximation depends on the *accuracy* of results sought and the *methods of solution* available at hand.

1.1.1.3 Boundary Conditions

Consider again the single-physics (load-deformation) idealization of the problem. In such problems, generally there are three types of boundary conditions: *displacement type*, *load type*, and *mixed type*.

Various approximations to the boundary conditions are possible. Unless the air that the body is immersed in exerts a surface traction, the irregular surfaces are free from any surface traction, and therefore are "free boundaries" (load boundary condition with zero traction). With this approximation, no external force acts on the side boundaries in any of the three idealized domains in Fig. 1.3a–c.

First, let us consider Model 1 (Fig. 1.3a). The boundary condition at point A of the rod is one of load type. In the actual problem, the surface traction acts only over a portion of the horizontal plane passing through point A (Fig. 1.1), but this detail cannot be considered in the one-dimensional rod in Fig. 1.3a. The resultant force $P = A_0\sigma_0$ is applied at point A as shown.

Now, consider Model 2 shown in Fig. 1.3b. If the details of the traction at point A are to be considered, then the problem becomes two or three dimensional, and an analytical solution cannot be easily obtained. Thus, an equivalent uniform stress $\sigma_1 = A_0\sigma_0/A_1$ is applied at point A, covering the entire area A_1 such that the resultant force is preserved. The *Saint Vernant principle* may be invoked to justify the assumptions concerning the development of the equivalent force P in Model 1 or the equivalent stress σ_1 in Model 2.

The details of the loading at point A can, however, be considered in the model shown in Fig. 1.3c; numerical techniques such as the finite element can easily handle the details.

At point B, the rods shown in Fig. 1.3a and b must be fixed, and this is the only type of boundary condition that can be considered here (for example, in a two- or three-dimensional modeling, a roller boundary condition can be considered). The boundary condition is a displacement type.

The model shown in Fig. 1.3c permits the simulation of the boundary condition as in the real structure, i.e., fixed in both vertical and horizontal directions. The boundary condition is a displacement type.

The problem does not involve a mixed type, which is generally encountered when, say, a spring is attached between a fixed point and a point on the structure.

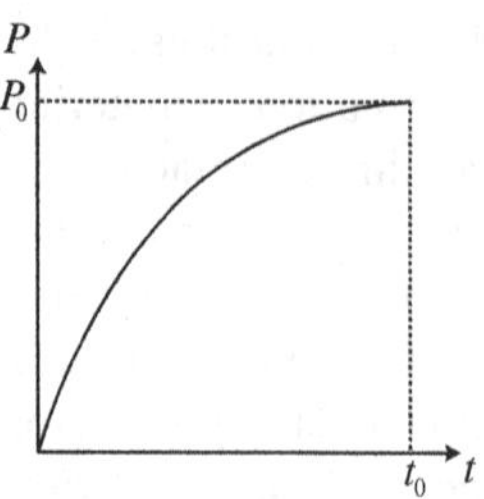

Fig. 1.4 Load-time history

The type of approximation on boundary conditions again depends on the *accuracy* of results sought and the *methods of solution* available at hand.

In addition to the recognition and idealization of the boundary conditions, the time history of loading (i.e., the load-time relation or the displacement-time relation) also must be characterized. No load in the real world can be applied at zero duration. When the loading is so slow as to make the inertial forces negligible compared to the boundary reaction forces, the loading is characterized as *static*, and when it is fast enough to produce nonnegligible inertial forces, it is characterized as *dynamic*. When it is dynamic, the precise nature of the variation of load with time (e.g., Fig. 1.4) is needed and must be mathematically characterized. Again, assumptions are made in approximating a real problem into either a static or a dynamic problem. If the problem is dynamic, the load-time relation is approximated by a mathematical function (e.g., sinusoidal, piecewise linear, etc.).

In the case of static loading, the variation of load with time is not needed for elastic bodies. However, even in static problems, the history of loading influences the behavior of a body consisting of history-dependent materials such as elasto-plastic materials.

For elastic bodies, the final load (i.e., the load at which the behavior of the body needs to be evaluated) is the only quantity that needs to be specified; the details of how the final load is reached are irrelevant. For instance, the displacement of the column in Fig. 1.1 at the end of the loading process where σ_0 varies as $\sigma_0 = 0 \rightarrow 10 \rightarrow 5$ units is the same as the displacement at the end of the loading process where σ_0 varies as $\sigma_0 = 0 \rightarrow 20 \rightarrow 5$ units, since the displacement depends only on the final value of σ_0, which is 5 units in both cases. In continuum mechanics, such bodies are called *conservative* bodies.

1.1.1.4 Material Behavior

Again let us limit our discussion to the single-physics (load-deformation) problem. To solve the problem, the relationships between stresses and strains, known as *constitutive laws*, are required. These relationships can take a variety of forms such as linear elastic (Fig. 1.5a), nonlinear elastic (Fig. 1.5b), elasto-plastic (Fig. 1.5c), etc. The constitutive relation is an inherent characteristic of the material, depending

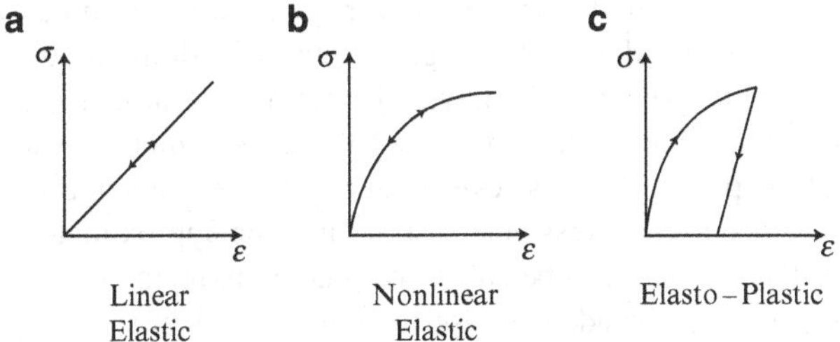

Fig. 1.5 Different types of stress-strain behavior

on its constitution. It is independent of the geometry or the boundary conditions of the problem.

From practical considerations, approximations are again made in various contexts. For example, all materials on earth are likely to be rate dependent, although for a given rate of loading, the degree of dependency varies from material to material and with such factors as temperature. Metals, powders, rocks, clays, and sands all show rate effects at very high rate of loading such as that involved during an explosive or impact loading. During an earthquake loading (0.1–20 Hz), cohesive soils (e.g., clay) are rate dependent, whereas cohesionless soils (e.g., sand) are rate independent. Hence the same material (e.g., sand) could be characterized as either rate dependent or rate independent depending on the rate of loading.

A related phenomenon is time dependency (creep or stress relaxation). Every material on earth is likely to exhibit time-dependent behavior, but if the duration of loading and the duration for which the solution is sought are short relative to the duration required for noticeable creep, the characterization of creep behavior of the material is unimportant. It is thus evident that the material behavior could be approximated as either time dependent or time independent.

Considering now a time-independent and rate-independent material and returning to the idealizations shown in Fig. 1.5a–c, we can choose a specific idealization or a model based on a number of considerations. That is, a *constitutive framework* is chosen (linear elastic, nonlinear elastic, or elasto-plastic) and a suitable mathematical model known as the *constitutive model* is developed and calibrated for describing the stress-strain relation of the material(s) under consideration. The real material behavior will not in general be in exact compliance with any model, but it may be closer to one model than the others. To achieve maximum accuracy of predictions, the model that matches the observations more closely should be the preferred choice of the model. However, adhering to this model may involve intricate mathematical complexities and difficulties in performing computations at affordable cost. Thus, approximations are again made.

As in the case with approximations on physics, geometry, and boundary conditions, the type of approximation on the constitutive behavior again depends on the *accuracy* of results sought and *methods of solution* available at hand. For instance, in the column problem shown in Fig. 1.1, the choice of a framework other than the linear elastic framework (Fig. 1.5a) would require some form of iterative solution strategy, making it difficult to obtain answers readily. The choice of, say, the nonlinear elastic behavior depicted in Fig. 1.5b to go with the model in Fig. 1.3b would make it difficult to obtain an analytical solution, defeating the purpose of the idealizations on the geometry and boundary conditions, which are meant to render analytical solution possible. Thus, even when the real behavior is close to the nonlinear elastic model, the stress versus strain may be approximated by a straight line just so that the problem can be solved by analytical methods. The accuracy of the solution may perhaps be adequate for the purpose at hand.

When a more accurate solution is needed, an iterative solution method may indeed be devised for the problem. Then a question arises as to whether or not it is worth using an elaborate nonlinear model to solve the problem using the simplified geometrical model in Fig. 1.3a. In general, it does not make sense to make gross approximations on geometry and to use highly accurate constitutive model for describing the stress-strain behavior. This is because the outcome, which in our case is the displacement of point A, depends on all input parameters, including geometrical and constitutive models. It makes sense to use a highly accurate constitutive model with a highly accurate geometrical model and realistic boundary conditions, such as those in the model in Fig. 1.3c. The degree of effort involved in dealing with the complex geometry by a numerical method such as the finite element method is compatible with the degree of effort involved in dealing with the mathematical complexity of an elasto-plastic model.

1.1.2 Problem Statement

Once a model is developed, the next step of the solution process is the development of a mathematical statement of the problem (the problem statement). Physical laws such as Newton's laws, thermodynamic laws, conservation of mass, etc., are used to develop a mathematical statement of the physics. In addition, the boundary and initial conditions, material behavior, and geometry must all be mathematically described.

Returning now to the problem at hand, the physics involved is the "motion of bodies due to mechanical forces" and the relevant physical laws are Newton's laws, in particular, Newton's second law. Let us derive the mathematical descriptions of the physics for each of the three models presented in Fig. 1.3 and complete the definitions of the problem statements.

1.1.2.1 Model 1: Prismatic Rod

Tensile stress and strain are taken positive. As the parameters of the problem (cross-sectional area, material properties, and boundary conditions) are assumed to vary only in one direction, the problem is one dimensional. With respect to the x axis, applying Newton's second law to the element of thickness $\mathrm{d}x$ (Fig. 1.6),

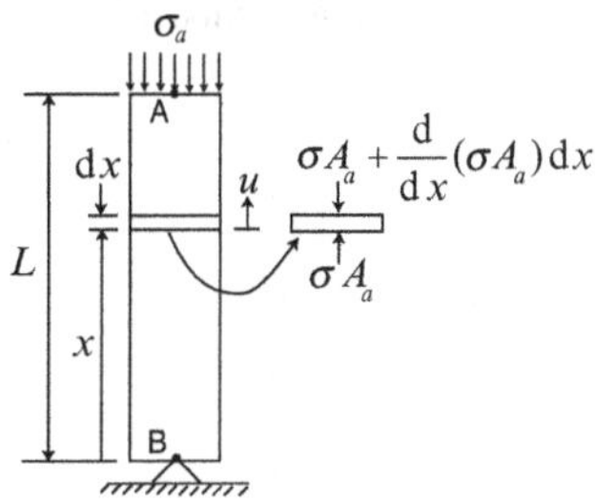

Fig. 1.6 Analysis of Model 1 (Fig. 1.3a)

$$\frac{\mathrm{d}}{\mathrm{d}x}(\sigma A_a)\mathrm{d}x = -(A_a\,\mathrm{d}x\rho)\ddot{u} \tag{1.1}$$

where σ is the normal stress (note that all other components of the stress tensor are zero), ρ is the mass density of the material, A_a is the average cross-sectional area, and $\ddot{u}$ is the acceleration. u denotes the displacement of the rod at x (in the direction of x axis). Here, we will neglect the inertia: $\ddot{u} = 0$. The problem is thus assumed to be static. Noting that A_a is constant with x, (1.1) becomes

$$\frac{\mathrm{d}\sigma}{\mathrm{d}x} = 0 \tag{1.2}$$

which is the differential equation governing the equilibrium of the rod.

The boundary conditions are

$$\text{At } x = 0, \quad u = 0,$$

and (1.3)

$$\text{At } x = L, \quad \sigma = \sigma_a = -\sigma_0 A_0/A_a; \quad A_a = \frac{A_1 + A_2}{2}$$

where σ_a is the average stress in the rod. Note that a negative sign is inserted with the stress boundary condition because of our sign convention that a tensile stress is positive. Assume that the material is *elastic*. The only relevant stress and strain variables are σ and ε, which are related to each other as

$$\sigma = E\varepsilon \tag{1.4}$$

where

$$\varepsilon = \frac{\mathrm{d}u}{\mathrm{d}x} \tag{1.5}$$

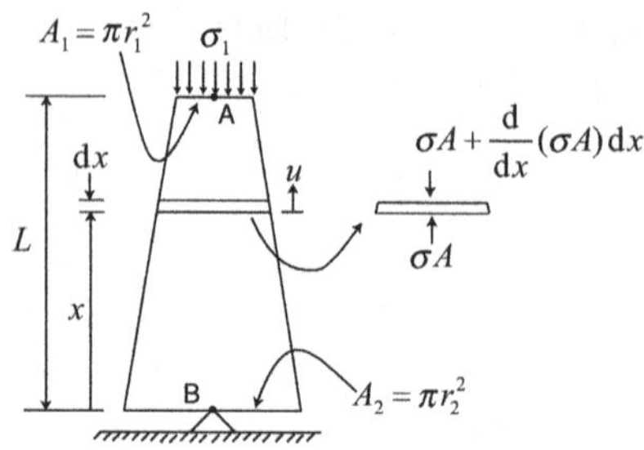

Fig. 1.7 Analysis of Model 2 (Fig. 1.3b)

and E is *Young's modulus*. ε is the normal strain in direction x (which could be called the uniaxial strain in the present case). Two- and three-dimensional strain-displacement relations are given in Chap. 3.

The geometry of the problem is trivially defined by

$$0 \leq x \leq L. \tag{1.6}$$

1.1.2.2 Model 2: Rod of Variable Cross-Section

Tensile stress and strain are taken positive. By following steps similar to those for Model 1, but recognizing here that the cross-sectional area varies with x (Fig. 1.7), the equation governing equilibrium of element of thickness $\mathrm{d}x$ is

$$\frac{\mathrm{d}}{\mathrm{d}x}(\sigma A)\mathrm{d}x = -\frac{1}{2}\left(A + A + \frac{\mathrm{d}A}{\mathrm{d}x}\mathrm{d}x\right)(\mathrm{d}x\,\rho)\ddot{u}. \tag{1.7}$$

Let us next define the geometry, so that the governing equation can be simplified a bit further. As in Model 1,

$$0 \leq x \leq L.$$

In addition, the radius of the circle representing the cross-section at any x varies linearly with x. Hence,

$$A = \pi r^2$$

$$r = r_1 - \left(\frac{r_1 - r_2}{L}\right)(L - x) = a + bx \quad \text{where } a = r_2 \text{ and } b = \frac{r_1 - r_2}{L} \tag{1.8a}$$

and thus

$$\frac{\mathrm{d}A}{\mathrm{d}x} = 2\pi r\frac{\mathrm{d}r}{\mathrm{d}x} = 2\pi r b. \tag{1.8b}$$

With the aid of these equations and the assumption $\ddot{u} = 0$ (static), (1.7) is simplified to

$$\begin{aligned} A\frac{\mathrm{d}\sigma}{\mathrm{d}x} + \sigma\frac{\mathrm{d}A}{\mathrm{d}x} &= 0 \\ \pi r^2\frac{\mathrm{d}\sigma}{\mathrm{d}x} + 2\pi r\sigma b &= 0 \\ (a+bx)\frac{\mathrm{d}\sigma}{\mathrm{d}x} &= -2\sigma b \end{aligned} \tag{1.9}$$

The boundary conditions for this problem are identical to those for Model 1, i.e.,

$$\text{At } x = 0, \quad u = 0,$$

and (1.10)

$$\text{At } x = L, \quad \sigma = \sigma_1 = -\sigma_0 A_0 / A_1$$

Assume that the material is elastic. Although the cross-section changes with x, there are only one stress and one strain components involved in the problem, which are interrelated as (with the assumption of elasticity)

$$\sigma = E\varepsilon = E\frac{\mathrm{d}u}{\mathrm{d}x}. \tag{1.11}$$

1.1.2.3 Model 3: Body with Arbitrarily Varying Cross-Section

In this case, the problem is two or three dimensional. Similar to the steps followed for Models 1 and 2, equilibrium equations are derived on the basis of Newton's law. The derivations are deferred to Chap. 3.

1.1.3 Solution Methods

We will now discuss methods of solution to the mathematical equations. Generally, the solution methods can be divided into two:

- Analytical methods
- Numerical methods

The *analytical methods* are those that yield *analytical or closed-form solutions* to the governing equations and *numerical methods* are those that yield *approximate solutions* (e.g., the finite element method). In some cases, the original governing equations may be further simplified for the purpose of obtaining *approximate closed-form solutions.*

Generally, the closed-form solutions are obtained by directly integrating the governing differential equations. This is possible, as demonstrated below, for Models 1 and 2 (Fig. 1.3a, b). It may be worth reminding ourselves that this does not mean the real problem is solved exactly, since the mathematical description is a "model" of the real problem, and approximations, although of varying degrees, have already been made.

Model 3 (Fig. 1.3c) is an example of a problem where the geometry is too complex for obtaining exact solutions. There are several numerical techniques available for solving such problems; for example,

- The method of undetermined parameters (e.g., collocation methods, subdomain collocation methods, least squares techniques, and Galerkin method)
- The energy methods (e.g., Ritz method and finite element method)
- The finite difference method

For introduction to these methods, the reader is referred to Reddy (1993). Among these different techniques, the two most widely used numerical methods are as follows:

- The finite element method
- The finite difference method.

Relative to the finite difference method, the finite element method permits straightforward treatment of complex boundary conditions, geometry, and constitutive models. In some limited cases, the finite difference method may prove to be computationally more efficient than the finite element method. The finite element method is chosen in this book, and is covered in detail in Chap. 5.

In certain problems, combined analytical-numerical methods (or semi-analytical methods) may prove to be the optimal approach, based on either computational efficiency or accuracy. For instance, in certain dynamic or quasi-static problems, while solving the spatial part numerically using a method such as the finite element method, the temporal part is solved exactly. The solution is first obtained in frequency domain and the time-domain solutions are then synthesized. Approximation is used in choosing the number of frequencies to be included in the solution.

In certain soil-foundation interaction problems, the domain to be analyzed could be divided into a core region that includes nonlinearity and boundary irregularity, and an outer region, which is simple enough to enable analytical treatment. The finite element method is used for the core region and analytical solutions are used for the outer region, and the two regions are coupled by a suitable method, such as the Lagrange multiplier method (Anandarajah 1990, 1993b).

Let us illustrate the use of analytical methods to solve simplified problems such as those represented by Models 1 and 2, and the use of finite element method to solve complex problems such as that represented by Model 3.

1.1.4 Solutions

1.1.4.1 Model 1: Analytical Solution

The problem is presented in Figs. 1.3a and 1.6. Equation (1.2) is directly integrated to obtain

$$\sigma = c_1, \tag{1.12}$$

where c_1 is an integration constant. Using the second of boundary conditions in (1.3) to evaluate c_1, the solution for the stress becomes

$$\sigma = \sigma_a. \tag{1.13}$$

Now from (1.13), (1.4), and (1.5), we have

$$\frac{\mathrm{d}u}{\mathrm{d}x} = \frac{\sigma_a}{E}, \tag{1.14}$$

which can be directly integrated to obtain

$$u = \frac{\sigma_a}{E}x + c_2. \tag{1.15}$$

The first of the boundary conditions listed in (1.3) is used to evaluate the integration constant c_2. The solution for the displacement then becomes

$$u = \frac{\sigma_a}{E}x. \tag{1.16}$$

The displacement varies linearly with x. The displacement of point A is obtained from (1.16) for $x = L$. One gets the expression of being familiar with the elementary statics:

$$u_{\mathrm{A}} = \frac{\sigma_a L}{E}; \quad \sigma_a = -\frac{\sigma_0 A_0}{A_a}; \quad A_a = \frac{A_1 + A_2}{2} \tag{1.17}$$

1.1.4.2 Model 2: Analytical Solution

The steps to be followed here are identical to those for Model 1. First, we directly integrate (1.9) as

$$\int_{\sigma_1}^{\sigma} \frac{d\sigma}{\sigma} = -\int_{L}^{x} \frac{2b}{a+bx} dx \tag{1.18}$$

$$\ln\left(\frac{\sigma}{\sigma_1}\right) = -2\ln\left(\frac{a+bx}{a+bL}\right) \tag{1.19}$$

$$\sigma = \sigma_1 \left(\frac{r_1}{a+bx}\right)^2 \tag{1.20}$$

$$\text{Recall the integral formula}: \int \frac{dx}{a+bx} = \frac{1}{b}\ln(a+bx)$$

Note that $a + bL = r_1$ (1.8a).
From (1.11) and (1.20),

$$E\frac{du}{dx} = \frac{\sigma_1 r_1^2}{(a+bx)^2}. \tag{1.21}$$

Integrating this with limits (first condition of (1.10))

$$\int_0^u du = \frac{\sigma_1 r_1^2}{E} \int_0^x \frac{1}{(a+bx)^2} dx \tag{1.22}$$

$$u = \frac{\sigma_1 r_1^2}{E}\left[\frac{1}{b(-1)(a+bx)}\right]_0^x$$

$$= \frac{\sigma_1 r_1^2}{aE}\left(\frac{x}{a+bx}\right) \tag{1.23}$$

$$\text{Recall the integral formula}: \int \frac{1}{(a+bx)^n} dx = \frac{1}{b(1-n)(a+bx)^{n-1}} \text{ if } n \neq 1$$

At $x = L$,

$$u_A = \frac{\sigma_1 r_1^2}{aE}\left(\frac{L}{a+bL}\right) = \frac{\sigma_1 L}{E}\left(\frac{r_1}{r_2}\right); \ \sigma_1 = -\frac{\sigma_0 A_0}{A_1} \tag{1.24}$$

For a prismatic rod, Model 2 is identical to Model 1 and the solutions coincide (compare (1.24) with (1.17)) for $r_1 = r_2$ and $\sigma_1 = \sigma_a = \sigma_0$.

Discussion

Consider a case where $A_0 = A_1$ (i.e., the load covers the whole area at the top). Then, denoting the displacement of point A estimated using Model 1 by u_A^1 and that estimated using Model 2 by u_A^2, we have from (1.17),

$$u_A^1 = \frac{\sigma_a L}{E}; \quad \sigma_a = -\frac{\sigma_0 A_0}{A_a} = -\frac{2\sigma_0 A_1}{(A_1 + A_2)}$$

from (1.24),

$$u_A^2 = \frac{\sigma_1 L}{E}\left(\frac{r_1}{r_2}\right) = -\frac{\sigma_0 L}{E}\left(\frac{r_1}{r_2}\right)$$

$$\frac{u_A^1}{u_A^2} = \frac{2A_1}{(A_1 + A_2)}\frac{r_2}{r_1} = \frac{2r_1^2}{(r_1^2 + r_2^2)}\frac{r_2}{r_1}$$

The error between the two solutions can be evaluated. For instance, when $r_1/r_2 = 0.5$, $\frac{u_A^1}{u_A^2} = 0.8$ and $e = \left|\frac{u_A^1 - u_A^2}{u_A^2}\right| \times 100\% = 20\%$. Intuition tells us that the error must depend on r_1/r_2, but the analysis presented here provides a quantitative means of evaluating the error, demonstrating the value of analytical solutions. (Note that the error will be different if u_A^1 is used in the denominator.)

1.1.4.3 Model 3: Finite Element Solution

The finite element method is treated systematically in Chap. 5. Here we present some results without describing the method. The domain to be analyzed is cylindrical with a total height of 1 unit. The radii at the base and top are 1 unit and 0.5 units, respectively.

The problem is axisymmetric, and hence only a two-dimensional plane needs to be analyzed. The domain is discretized using eight-noded isoparametric elements (Chap. 5) as shown in Fig. 1.8a (called the "mesh"). The cylinder is placed on rollers at the base and subjected to a compressive uniform pressure of 350 units at the top. The domain is assumed to be homogeneous and linearly elastic with Young's modulus of 1.0×10^6 units and Poisson's ratio of 0.3.

The problem is analyzed using the finite element method. Figure 1.8b compares the mesh with the deformed configuration. It is seen that the deformation is two dimensional, which cannot be captured using either Model 1 or Model 2. The vertical displacement u_A (at the center) is -0.289×10^{-3} units. The vertical displacement at the edge is -0.291×10^{-3} units. The displacements calculated by the simplified models are as follows:

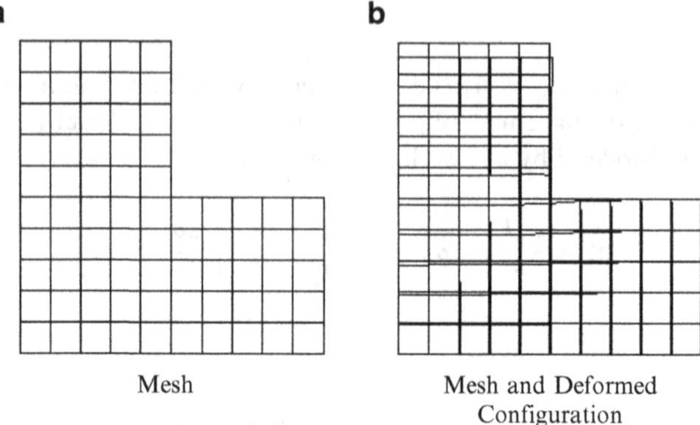

Fig. 1.8 Analysis of model shown in Fig. 1.3c by FEM

From (1.17),

$$\sigma^a = \frac{\sigma_0 A_0}{A_a} = \frac{2\sigma_0 A_1}{(A_1 + A_2)} = \frac{2 \times 1 \times 0.5^2 \sigma_0}{(1 + 0.5^2)} = 0.4\sigma_0;$$

$$u_A^1 = -\frac{\sigma^\alpha L}{E} = -0.140 \times 10^{-3} \text{ units}$$

From (1.24),

$$\sigma_1 = \sigma_0;$$

$$u_A^2 = -\frac{\sigma_1 L}{E}\left(\frac{r_1}{r_2}\right) = -\frac{\sigma_0 L}{E}\left(\frac{r_1}{r_2}\right) = -\frac{350 \times 1}{1.0 \times 10^6}\left(\frac{0.5}{1}\right) = -0.175 \times 10^{-3} \text{ units}$$

Both u_A^1 and u_A^2 are smaller than those predicted by the finite element method, emphasizing the need for exercising caution in using approximate models.

1.1.5 Consideration of Plasticity

In the analyses presented in the preceding sections, the material is assumed to be elastic. The true stress-strain behavior of most materials is elasto-plastic. Unless the elasto-plasticity is simulated, the true load-deformation behavior of a structure cannot be predicted. In particular, the failure of the structure cannot be predicted. Prediction of the elasto-plasticity is the major focus of this book, and the relevant aspects of this are treated in several of the subsequent chapters. First, a mathematical model capable of simulating the three-dimensional stress-strain behavior is required. Such a model is developed within the framework of the theory of plasticity, which is covered in Chap. 9. In constitutive models, the rate of stress is related to the rate of strain. However, the finite-element analysis requires the relation between the increments of stress and strain. To obtain this, the rate equations must be integrated. This task is nontrivial and involves iterative techniques. Several chapters beginning with Chap. 10 are devoted to this. Similarly, iterations are required at the global level. This is covered in Chap. 8.

Returning to the problem at hand, let us treat the material in Model 3 (Fig. 1.8a) as elasto-plastic and evaluate the behavior using the finite element method. A constitutive model known as the von Mises elasto-plastic model, covered in Chap. 11, is used to represent the stress-strain relation of the material. The elasto-plastic models also require an elastic relation; here we use isotropic linear elasticity with a Young's modulus of 1.0×10^6 units and a Poisson's ratio of 0.3. The von Mises model requires a uniaxial failure stress, which is taken as 350 units. In between the initial elastic part and the final failure stress, the material is modeled as an isotropic hardening material (Chap. 11). The remaining details are kept the same as in the elastic analysis presented in the preceding section.

The load-deformation relation predicted by the finite element method is shown in Fig. 1.9b. The deformed configuration is compared to the original mesh in Fig. 1.9a. Comparison of Fig. 1.8b (elastic deformation) and Fig. 1.9a (elasto-plastic deformation) reveals that the patterns of deformation are very different from each other. The material fails when the average vertical pressure is about 344 units, which is very close to the uniaxial failure strength used in the von Mises model (350 units).

Several key differences between the elastic and elasto-plastic responses may be noted. The major difference is that in elastic analysis, there is no limit to the vertical pressure and the structure never fails, whereas in the elasto-plastic analysis, the vertical pressure that can be applied to the rod is limited to the failure stress. The other differences are that the load-deformation relations are different from each other (linear versus nonlinear) and the deformed patterns are different from each other.

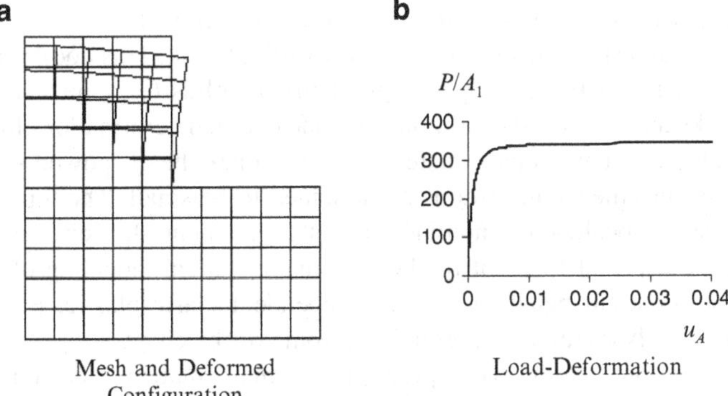

Fig. 1.9 Elasto-plastic behavior of Model 3

1.1.6 Challenges in Mathematical Modeling

Depending on the problem, challenges are encountered in all elements of mathematical modeling (physics, geometry, boundary conditions, and material behavior). The problems to be analyzed are in general of two types: *natural* and *man-made*. The problem of modeling the human heart falls in the former category and of modeling the behavior of an airplane falls in the latter category.

In general, the elements of modeling are more difficult to characterize for natural problems than for man-made problems. The physics associated with the functioning of a heart is numerous and complex. The in vivo heart is a "black box" and indirect methods are needed to quantify the geometry, boundary conditions, and material behavior. In man-made problems, as the designer chooses the geometry, boundary conditions, and material properties, they are more easily quantified. While there are these inherent differences between natural and man-made problems, either one can be very challenging to model owing to their shear sizes, details, and complexities.

There could be other challenges. For example, the physics could be too numerous and complex that idealization and mathematical description could be challenging. Elasto-plastic analyses can be computationally demanding that three-dimensional analyses involving a large number of unknowns in the finite element analysis can be difficult to perform in personal computers; parallel supercomputers may be needed.

1.2 Scope of the Book

An attempt is made in the preceding section to present a broad picture of mathematical modeling. The physics could be either single physics or multiphysics; the specific formulation depends on the specifics of the physics. The material behavior can be linear elastic, nonlinear elastic, elasto-plastic, visco-elastic, or visco-plastic. The deformation and strains can be small or large. Other aspects such as strain softening, shear localization, etc., may also need to be modeled.

Coverage of all aspects in a single book is a difficult task. This book is limited to continuum mechanics (single-physics) problems of solids (e.g., metals, powders, soils, and rocks under drained condition, etc.) and coupled deformation-flow (multiphysics) problems of two-phase materials (e.g., bones, heart, powders, soil, and rock). The displacements and strains are assumed to be small. The topics such as bifurcation, shear localization, nonlinear contacts, gradient plasticity, etc., are not considered. The special focus of the book is mathematical modeling of problems where the stress-strain behavior is rate-independent elasto-plastic. Examples of finite element analyses are also presented in some of the chapters.

Chapters 2–5 present the basic principles of mathematics, mechanics, elastic laws, and finite element method that are needed for all types of problems - problems involving solids, particulates, and structures. Chapters 6 and 7 present fully coupled formulations for saturated and partially saturated porous media. Chapters 8 and Chapters 9 present the principles of nonlinear analysis and the theory of plasticity that are needed for all types of problems. Chapters 10 and 11 treat the integration of constitutive models for solids and structures. Chapters 10–14 are relevant to the modeling of particulates, and geologic materials and structures. Several supplemental topics are covered in the Appendix.

1.3 Word of Caution

The preceding discussions present an overview of the mathematical modeling of continuum mechanics problems and the various choices of analysis methods available at the present time. It is clear that for a given problem, a range of analysis methods are normally available, depending on the solution accuracy sought. While in basic research and technology development, one may have a clear notion of accuracy, it is often blurred in design. In most engineering, a "factor of safety" is used to account for the "lack of accuracy" in estimates of design variables. While in principle the use of a sophisticated analysis method allows the estimation of design variables with high degree of accuracy, and hence the use of a lower factor of safety, the decision to use a lower factor of safety depends on factors such as the reliability and accuracy with which input parameters (material, boundary, and geometrical parameters) can be estimated, the degree of confidence on the validity of analytical or numerical method, the overall cost, etc. This reinforces the need for validation of the analysis techniques using adequate experimental evidence.

The transfer of knowledge of theoretical modeling to the design world then takes special efforts from the part of both the researcher who develops the analysis method and the designer who could potentially benefit a great deal from it. It is incumbent upon the researcher to strive toward simplicity, reliability, and robustness, and the designer to take special effort to give state-of-the-art methods a chance, first as a secondary method so that the method can be validated, and then as a primary design tool.

Regardless of the particular usage of the analysis method, there is a certain degree of danger in using an analysis method or a computer code as a "black box" that takes certain input and puts out certain output. Intelligent use of a given analysis method hinges on the understanding of the idealizations made in coming up with the model for the mathematical treatment, the model parameters involved and the sensitivity of output to these parameters, etc. For instance, if you did not know the analysis details behind the three models presented earlier, which one would you choose when you discover that you have three methods to choose from?

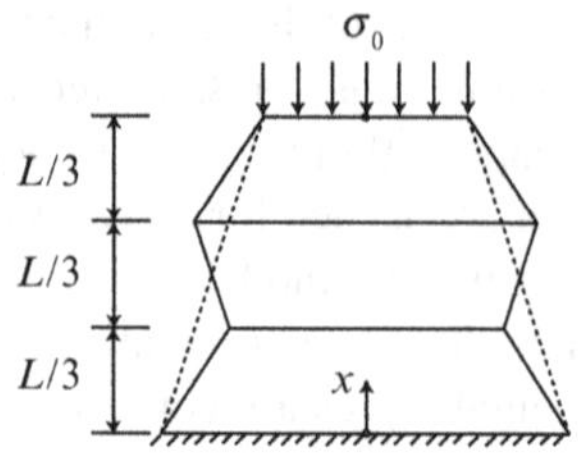

Fig. 1.10 Axisymmetric tower of variable cross-section

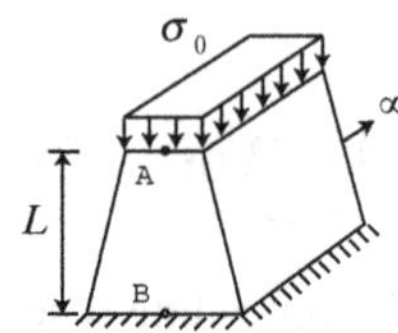

Fig. 1.11 Plane strain tower of variable cross-section

Problems

Problem 1.1 Describe (in about one page) an example of mechanics modeling (or something close to it) that you have done in your life. Clearly describe the physics, geometry, boundary conditions, and material model used in your modeling. (It could be a problem you solved in your Introduction to Engineering class or a similar one, or even your structural mechanics class. It could even be one of your high school projects!)

Problem 1.2 The solid tower shown in Fig. 1.10 is axisymmetric, having a variable cross-section. The radius of the cross-section varies linearly from $2r_1$ at $x = 0$ to $5r_1/3n$ at $x = L/3$ to $4r_1 n/3$ at $x = 2L/3$ to r_1 at $x = L$. Note that when $n = 1$, the tower's cross-section varies linearly from $x = 0$ to $x = L$ (like the column in Fig. 1.3b). Develop an equation for computing the displacement of point A, $u_A(n)$, by dividing the tower into three sections, each of length $L/3$, using (1.24) to compute the compression of each section and summing them up. Plot the ratio $u_A(n)/u_A(n = 1)$ as a function of n for $1 \leq n \leq 1.2$. Also, plot the ratio $u_A(n)/u_A^1$, where u_A^1 is the value of the displacement of A calculated using Model 1 (Fig. 1.3a).

Problem 1.3 Develop a variable cross-section model similar to Fig. 1.3b for a plane strain problem shown in Fig. 1.11. Derive an equation for calculating the vertical displacement of point A.

Chapter 2
Mathematical Foundations

In this chapter, we present certain mathematical and continuum mechanics principles that are relevant to constitutive modeling and boundary value analysis. One of the most important topics is the concept of tensors. Tensors are quantities that obey coordinate transformation rules. Examples of tensors include certain scalars (mass, area, and volume), vectors (displacements and forces) and certain quantities encountered in continuum mechanics (stresses and strains).

We will limit our discussions to rectangular Cartesian tensors, where tensors are defined with respect to a rectangular Cartesian (Euclidean) coordinate system. In a Cartesian coordinate system, the axes are orthogonal to each other. Tenors defined with respect to a curvilinear coordinate system require the consideration of the concepts of contravariant and covariant tensors. In a rectangular Cartesian system, there is no distinction between Cartesian, contravariant, and covariant tensors.

2.1 Transformation Rules

2.1.1 Scalars

Physical quantities such as the mass, volume, area, and length of a vector remain independent of the orientation of the coordinate axes. For example, denoting mass by m

$$(m)_{x_1-x_2-x_3} = (m)_{x'_1-x'_2-x'_3} \tag{2.1}$$

where $x_1 - x_2 - x_3$ is the original coordinate system and $x_1' - x_2' - x_3'$ is the rotated coordinate system.

A. Anandarajah, *Computational Methods in Elasticity and Plasticity: Solids and Porous Media*, DOI 10.1007/978-1-4419-6379-6_2,

2.1.2 Vectors

Examples of vectors include position, displacement, force, velocity, and acceleration. Such vectors can be quantified using a certain coordinate system. The directions of the axes are defined in terms of what are known as *basis vectors*. Denoting the basis vectors in three dimensions in the $x_1 - x_2 - x_3$ system as $\mathbf{e}_1$, $\mathbf{e}_2$, and $\mathbf{e}_3$ (Fig. 2.1), where

$$\mathbf{e}_1 = (1,0,0); \quad \mathbf{e}_2 = (0,1,0); \quad \mathbf{e}_3 = (0,0,1) \tag{2.2}$$

a vector such as a displacement vector $\mathbf{u}$ can be expressed as

$$\mathbf{u} = u_1\mathbf{e}_1 + u_2\mathbf{e}_2 + u_3\mathbf{e}_3 \tag{2.3}$$

where u_1, u_2, and u_3 are components of the vector $\mathbf{u}$.

The basis vectors must be independent of each other, i.e., $\alpha\mathbf{e}_1 + \beta\mathbf{e}_2 + \gamma\mathbf{e}_3 = 0$ only when $\alpha = 0$, $\beta = 0$ and $\gamma = 0$.

Definition 2.1: Indicial notation (or tensor notation). *A tensor may be expressed by appending indices to the letter representing the tensor. Hence, a vector* $\mathbf{x}$ *may be expressed as* x_i. *Similarly, higher order tensors that we will define later may be expressed as* α_{ij} *(second order),* $\beta_{ijk\ell}$ *(fourth order), etc.*

Definition 2.2: Summation convention. *Unless explicitly stated, repeated indices appearing as subscripts on quantities on the same side of an equation imply summation. The number of terms in the summation equals the dimension of the space. The convention is used regardless of whether the quantity is a tensor or not.*

For example, in a 3D Cartesian system,

$$\alpha = x_i x_i = x_1 x_1 + x_2 x_2 + x_3 x_3 = x_1^2 + x_2^2 + x_3^2$$

where α *is a scalar, and* x_1, x_2 *and* x_3 *are the components of the vector* $\mathbf{x}$. *The subscript i repeats on the right side of the equation. The subscript i used to indicate summation is known as the "dummy" index since any arbitrary letter can be used for the index. For example*

$$\alpha = x_i x_i = x_k x_k$$

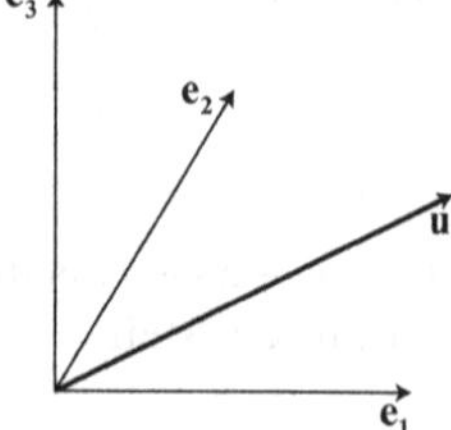

Fig. 2.1 Basis vectors for rectangular Cartesian coordinate system

Definition 2.3: The Kronecker delta. *The Kronecker delta δ_{ij} is defined as*

$$\begin{aligned}\delta_{ij} &= 1 \quad \text{for} \quad i=j \\ &= 0 \quad \text{for} \quad i \neq j\end{aligned}$$

Hence, the Kronecker delta can be thought of as a 3×3 identity matrix in three dimensions and a 2×2 identity matrix in two dimensions.

Using the summation convention (Definition 2.2), (2.3) may be written as

$$\mathbf{u} = u_i \mathbf{e}_i \tag{2.4a}$$

Note that the summation convention eliminates the need for using the customary summation sign as

$$\mathbf{u} = \sum_{i=1}^{i=3} u_i \mathbf{e}_i \tag{2.4b}$$

Let us first consider *coordinate transformation* of a *vector* in two dimensions. The $x_1 - x_2$ system is rotated through an angle θ in the counter clockwise direction to obtain the $x_1' - x_2'$ system as shown in Fig. 2.2.

Let the components of a vector $\bar{\mathbf{u}}$ in the two systems be

$$(u_1, u_2) \quad \text{and} \quad (u_1', u_2') \tag{2.5}$$

Projecting the vector $\bar{\mathbf{u}}$ onto the $x_1' - x_2'$ system results in

$$u_1' = u_1 \cos\theta + u_2 \sin\theta \tag{2.6a}$$

$$u_2' = -u_1 \sin\theta + u_2 \cos\theta \tag{2.6b}$$

Placing the components (u_1, u_2) and (u_1', u_2') in vectors as

$$\mathbf{u} = (u_1, u_2) \quad \text{and} \quad \mathbf{u}' = (u_1', u_2')$$

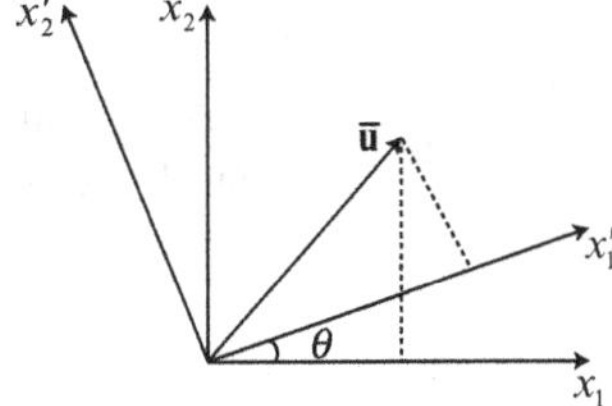

Fig. 2.2 Rotation of coordinate axes in two dimensions

Equation (2.6) can be expressed in a matrix-vector form as

$$\mathbf{u}' = \mathbf{a}\mathbf{u} \tag{2.7a}$$

where

$$\mathbf{a} = \begin{bmatrix} a_{11} & a_{12} \\ a_{21} & a_{22} \end{bmatrix} = \begin{bmatrix} \cos\theta & \sin\theta \\ -\sin\theta & \cos\theta \end{bmatrix} \tag{2.7b}$$

In indicial notation (Definition 2.1), (2.7a) is expressed as

$$u_i' = a_{ik}u_k \tag{2.7c}$$

Notice the use of summation convention in (2.7c).
By projecting the vector $\bar{\mathbf{u}}$ onto the $x_1 - x_2$ system

$$u_1 = u_1'\cos\theta - u_2'\sin\theta \tag{2.8a}$$

$$u_2 = u_1'\sin\theta + u_2'\cos\theta \tag{2.8b}$$

In a matrix-vector form

$$\mathbf{u} = \mathbf{b}\mathbf{u}' \tag{2.9a}$$

where

$$\mathbf{b} = \begin{bmatrix} b_{11} & b_{12} \\ b_{21} & b_{22} \end{bmatrix} = \begin{bmatrix} \cos\theta & -\sin\theta \\ \sin\theta & \cos\theta \end{bmatrix} \tag{2.9b}$$

In indicial notation

$$u_i = b_{ik}u_k' \tag{2.9c}$$

The matrices **a** and **b** are known as the *transformation* or *rotation* matrices.
The properties of the transformation matrices are

$$\mathbf{a} = \mathbf{b}^{\mathrm{T}}(\text{or } a_{ij} = b_{ji}) \quad \text{and} \quad \mathbf{b} = \mathbf{a}^{\mathrm{T}}(\text{or } b_{ij} = a_{ji}) \tag{2.10a}$$

$$\mathbf{a}^{\mathrm{T}}\mathbf{a} = \begin{bmatrix} \cos\theta & -\sin\theta \\ \sin\theta & \cos\theta \end{bmatrix} \begin{bmatrix} \cos\theta & \sin\theta \\ -\sin\theta & \cos\theta \end{bmatrix} = \begin{bmatrix} 1 & 0 \\ 0 & 1 \end{bmatrix} = \mathbf{I} \tag{2.10b}$$

$$\mathbf{a}^{-1} = \mathbf{a}^{\mathrm{T}} \tag{2.10c}$$

where **I** is the identity matrix. In indicial notation, (2.10b) is written as

$$a_{ki}a_{kj} = \delta_{ij} \tag{2.10d}$$

where δ_{ij} is the Kronecker delta (Definition 2.3).

Due to property (2.10c)

$$\mathbf{a}\mathbf{a}^{\mathrm{T}} = \mathbf{a}\mathbf{a}^{-1} = \mathbf{I}$$

or

$$a_{ik}a_{jk} = \delta_{ij} \tag{2.10e}$$

Hence, in general:

$$\mathbf{a}^{\mathrm{T}}\mathbf{a} = \mathbf{a}\mathbf{a}^{\mathrm{T}} = \mathbf{I} \tag{2.10f}$$

$$a_{ki}a_{kj} = a_{ik}a_{jk} = \delta_{ij} \tag{2.10g}$$

Definition 2.4: Linear transformation. *The transformation of vectors due to the rotation of coordinate system (2.7c)*

$$u'_i = a_{ik}u_k$$

is known as the linear transformation since there is a linear relationship between the original and rotated components of the vector.

Definition 2.5: Bilinear transformation and bilinear function. *A bilinear transformation assigns a scalar ϕ to two vectors* **u** *and* **v** *as*

$$\phi = S_{ij}u_iv_j$$

Note that ϕ is linear in the components of **u** *as well as in the components of* **v**. *Such functions are called the bilinear functions.*

Definition 2.6: Orthogonal matrices. *Matrices satisfying the property* $\mathbf{a}^{\mathrm{T}}\mathbf{a} = \mathbf{I}$ *are known as orthogonal matrices. Hence, the transformation matrix in the transformation of a vector in a Cartesian coordinate system is orthogonal.*

Definition 2.7: Orthogonal transformation. *When* **a** *appearing in the linear transformation $u'_i = a_{ik}u_k$ is orthogonal, the transformation is also called the orthogonal transformation.*

Examination of the components of **a** reveals that they are the direction cosines of the unit vectors along the axes. This may be formally shown as follows.

Let $\bar{\mathbf{u}}$ be a two-dimensional unit vector along the x_1–axis. Then in the $x_1 - x_2$ system,

$$\bar{\mathbf{u}} = \begin{bmatrix} 1 \\ 0 \end{bmatrix}$$

Defining $\cos(x_i', x_j)$ as the cosine of angle between x_i' and x_j axes, the components of $\bar{\mathbf{u}}$ in the $x_1' - x_2'$ system are

$$\bar{\mathbf{u}}' = \begin{bmatrix} \cos(x'_1, x_1) \\ \cos(x'_2, x_1) \end{bmatrix}$$

Now by the relation

$$\bar{\mathbf{u}}' = \mathbf{a}\bar{\mathbf{u}}$$

$$\begin{bmatrix} \cos(x'_1, x_1) \\ \cos(x'_2, x_1) \end{bmatrix} = \begin{bmatrix} a_{11} & a_{12} \\ a_{21} & a_{22} \end{bmatrix} \begin{bmatrix} 1 \\ 0 \end{bmatrix} = \begin{bmatrix} a_{11} \\ a_{21} \end{bmatrix} \tag{2.11a}$$

The above result will hold for other cases as well. Hence, in general

$$a_{ij} = cos(x_i', x_j) \tag{2.11b}$$

In two dimensions:

$$\mathbf{a} = \begin{bmatrix} \cos(x'_1, x_1) & \cos(x'_1, x_2) \\ \cos(x'_2, x_1) & \cos(x'_2, x_2) \end{bmatrix} \tag{2.11c}$$

By expressing the cosines in (2.11c) in θ shown in Fig. 2.2, we get (2.7b).

In three dimensions:

$$\mathbf{a} = \begin{bmatrix} \cos(x'_1, x_1) & \cos(x'_1, x_2) & \cos(x'_1, x_3) \\ \cos(x'_2, x_1) & \cos(x'_2, x_2) & \cos(x'_2, x_3) \\ \cos(x'_3, x_1) & \cos(x'_3, x_2) & \cos(x'_3, x_3) \end{bmatrix} \tag{2.11d}$$

According to Definition 2.4, the transformation $u_i' = a_{ik}u_k$ (2.7c), which relates the vector $\mathbf{u}$ to $\mathbf{u}'$ in a Cartesian coordinate system is a *linear transformation*. The function on the right-hand side of in (2.7c) is *linear* in the components of $\mathbf{u}$.

Consider a projection of the vector $\mathbf{u}$ onto an arbitrary direction $\boldsymbol{\eta} = (\eta_1, \eta_2)$

$$F_\eta = u_i \eta_i \tag{2.12a}$$

Let us show that the numerical value of F_η is independent of the orientation of the coordinate system:

$$\begin{aligned} F'_\eta &= u'_i \eta'_i \\ &= (a_{ik}u_k)(a_{i\ell}\eta_\ell) \text{ (using (2.7c))} \\ &= \delta_{k\ell}u_k\eta_\ell \text{ (using (2.10g))} \\ &= u_k\eta_k \text{(using contraction operation, defined below)} \\ &= F_\eta \end{aligned} \tag{2.12b}$$

Note that when $\boldsymbol{\eta}$=**u**, we have $F_\eta = u_i u_i = |\mathbf{u}|$. Hence, the length of a vector is independent of the orientation of the coordinate system in which the components of **u** are expressed.

While we will formally introduce the contraction operation for tensors later, the operation involved in the last step of (2.12b) may be easily understood by expanding the term $\delta_{k\ell}\eta_\ell$ and noting that a term in the summation is nonzero only when $\ell = k$ (from the definition of the Kronecker delta) as.

$$\delta_{k\ell}\eta_\ell = \delta_{k1}\eta_1 + \delta_{k2}\eta_2 + \delta_{k3}\eta_3$$

$$\delta_{k\ell}\eta_\ell = \delta_{11}\eta_1 + \delta_{12}\eta_2 + \delta_{13}\eta_3 = \eta_1 \quad \text{for} \quad k = 1$$

$$\delta_{k\ell}\eta_\ell = \delta_{21}\eta_1 + \delta_{22}\eta_2 + \delta_{23}\eta_3 = \eta_2 \quad \text{for} \quad k = 2$$

$$\delta_{k\ell}\eta_\ell = \delta_{31}\eta_1 + \delta_{32}\eta_2 + \delta_{33}\eta_3 = \eta_3 \quad \text{for} \quad k = 3$$

Hence,

$$\delta_{k\ell}\eta_\ell = \eta_k \tag{2.12c}$$

That is, replace the index of η (i.e., ℓ) with one of the indices of δ that is different from the index of η (i.e., k).

Equation (2.12b) proves that the component of the vector **u** in any arbitrary direction $\boldsymbol{\eta}$ is invariant with respect to the orientation of the coordinate system in which the components of **u** and $\boldsymbol{\eta}$ are expressed. Combining this with the fact that F_η is linear in η_i, we say that a vector **u** in an arbitrary direction $\boldsymbol{\eta}$ is an *invariant linear function of the components of the direction* $\boldsymbol{\eta}$ (Butkov 1968).

2.1.3 Stress

We discussed certain properties of a vector in the preceding section. We will see in the next section, after giving a formal definition of a tensor that vectors are first-order tensors. We will also see that stress is an example of a second-order tensor.

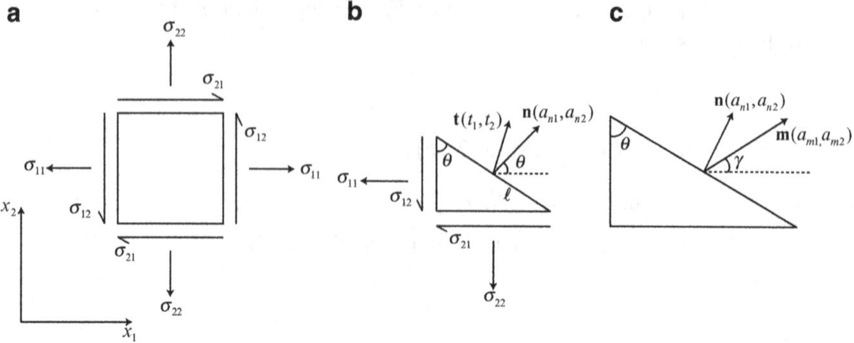

Fig. 2.3 Two-dimensional stresses

To facilitate the definition of tensors in general, we will first present some analysis of stress in this section.

First, consider a two-dimensional case. The stresses acting on a differential element is shown in Fig. 2.3a with respect to the $x_1 - x_2$ coordinate system. We are interested in finding the traction vector $\mathbf{t} = (t_1, t_2)$ on an inclined plane shown in Fig. 2.3b. Traction is defined as force per unit area; hence, traction is the resultant stress at a point on a plane.

The inclined plane has a unit normal $\mathbf{n}$=(a_{n1}, a_{n2}); hence

$$\mathbf{n} = (a_{n1}, a_{n2}) = [\cos(\mathbf{n},\mathbf{e}_1), \cos(\mathbf{n},\mathbf{e}_2)] = (\cos\theta, \sin\theta) \tag{2.13}$$

Let the length of the inclined side be ℓ. Now let us enforce equilibrium in x_1 – and x_2 – directions as

$$t_1\ell = \sigma_{11}\ell\cos\theta + \sigma_{21}\ell\sin\theta \Rightarrow t_1 = \sigma_{11}a_{n1} + \sigma_{21}a_{n2} \tag{2.14a}$$

$$t_2\ell = \sigma_{22}\ell\sin\theta + \sigma_{12}\ell\cos\theta \Rightarrow t_1 = \sigma_{12}a_{n1} + \sigma_{22}a_{n2} \tag{2.14b}$$

Hence,

$$\mathbf{t} = \boldsymbol{\sigma}^{\mathrm{T}}\mathbf{n}; \quad \boldsymbol{\sigma}^T = \begin{bmatrix} \sigma_{11} & \sigma_{21} \\ \sigma_{12} & \sigma_{22} \end{bmatrix} \tag{2.14c}$$

or

$$t_i = \sigma_{ji}a_{nj} \tag{2.14d}$$

It is assumed that the reader is familiar with the result that shear stresses acting on the faces parallel to the x_1 – and x_2 – axes are equal to each other (i.e., $\sigma_{12} = \sigma_{21}$) – these are called the complementary shear stresses. Utilizing this symmetry property, (2.14d) is written as

$$\mathbf{t} = \boldsymbol{\sigma}\mathbf{n} \quad \text{or} \quad t_i = \sigma_{ij}a_{nj} \tag{2.15}$$

The normal and shear stresses on the inclined plane may now be found by projecting **t** onto the normal **n** and shear directions.

But here we are interested in projecting onto an arbitrary direction (unit vector) **m** shown in Fig. 2.3c, where

$$\mathbf{m} = (a_{m1}, a_{m2}) = \cos(\mathbf{m},\mathbf{e}_1), \cos(\mathbf{m},\mathbf{e}_2) = (\cos\gamma, \sin\gamma) \tag{2.16}$$

We will call this σ_{nm} (stress component in direction **m** on a plane whose unit normal is **n**). Using (2.14d), the expression for σ_{nm} is

$$\sigma_{nm} = \mathbf{m}\cdot\mathbf{t} = \mathbf{m}^{\mathrm{T}}\boldsymbol{\sigma}\mathbf{n} = a_{mi}(\sigma_{ji}a_{nj}) = a_{nj}\sigma_{ji}a_{mi} \tag{2.17}$$

According to Definition 2.5, σ_{nm} given by (2.17) is a *bilinear function* in the components of the directions **m** and **n**.

Let the stress components with respect to the $x_1' - x_2'$ system be σ_{ij}'. When $\mathbf{n} = \mathbf{e}_1'$ and $\mathbf{m} = \mathbf{e}_1'$, then $\sigma_{nm} = \sigma_{11}'$. When $\mathbf{n} = \mathbf{e}_1'$ and $\mathbf{m} = \mathbf{e}_2'$, then $\sigma_{nm} = \sigma_{12}'$. Hence, it may be seen that, upon letting **n** and **m** coincide with $\mathbf{e}_1'$ and $\mathbf{e}_2'$ in an alternating manner, (2.17) generates the components of σ_{ij}'. Hence, (2.17) is a transformation rule that gives σ_{ij}' (i.e., components in the $x_1' - x_2'$ system) in terms of σ_{ij} (i.e., components in the $x_1 - x_2$ system), i.e.,

$$\boldsymbol{\sigma}' = \mathbf{a}\boldsymbol{\sigma}\mathbf{a}^{\mathrm{T}} \quad \text{or} \quad \sigma_{ij}' = a_{ik}\sigma_{k\ell}a_{j\ell} = \sigma_{k\ell}a_{ik}a_{j\ell} \tag{2.18a}$$

It is further noted that the transformation matrix **a** is the same transformation matrix that rotated a vector as $\mathbf{u}' = \mathbf{a}\mathbf{u}$ (2.7a).

Using the orthogonality property of **a** (2.10f), (2.18a) may be written (by pre- and post-multiplying (2.18a) with $\mathbf{a}^{\mathrm{T}}$ and **a**, respectively, and simplifying) as

$$\boldsymbol{\sigma} = \mathbf{a}^{\mathrm{T}}\boldsymbol{\sigma}'\mathbf{a} \quad \text{or} \quad \sigma_{ij} = a_{ki}\sigma_{k\ell}'a_{\ell j} = \sigma_{k\ell}'a_{ki}a_{\ell j} \tag{2.18b}$$

As in the case with vectors (2.12b), there exist an invariant property for stresses. Let us examine this below. In a Cartesian coordinate system, the numerical value of σ_{nm} given by (2.17) is independent of the orientation of the coordinate system in which the components σ_{ij} appearing on the right-hand side of (2.17) are expressed. While this is evident on physical grounds, it may be theoretically proved as follows. First to avoid confusion of notations, let us introduce the following notations

$$\boldsymbol{\mu} = (\mu_1, \mu_2) = \mathbf{m} = (a_{m1}, a_{m2}) \tag{2.19a}$$

$$\boldsymbol{\eta} = (\eta_1, \eta_2) = \mathbf{n} = (a_{n1}, a_{n2}) \tag{2.19b}$$

Now consider the term

$$
\begin{aligned}
a'_{mi}\sigma'_{ij}a'_{nj} &= \sigma'_{ij}\mu'_i\eta'_j \\
&= (\sigma_{k\ell}a_{ik}a_{j\ell})(a_{ip}\mu_p)(a_{jq}\eta_q) \text{ (using (2.7c) and (2.18a))} \\
&= \sigma_{k\ell}(a_{ik}a_{ip})(a_{j\ell}a_{jq})(\mu_p\eta_q) \text{ (rearranging)} \\
&= \sigma_{k\ell}\delta_{kp}\delta_{\ell q}\mu_p\eta_q \text{ (using the orthogonality property in (2.10g))} \\
&= \sigma_{k\ell}\mu_k\eta_\ell \text{ (using contraction operations (2.12c))}
\end{aligned}
\tag{2.20}
$$

Hence, (2.17) becomes

$$\sigma_{mn} = \sigma'_{ij}\mu'_i\eta'_j = \sigma_{ij}\mu_i\eta_j \tag{2.21}$$

which proves that σ_{mn} is independent (*invariant*) of the orientation of the coordinate system in which the components σ_{ij} appearing on the right-hand side of (2.17) are expressed.

It follows from (2.21) that stress with respect to two arbitrary directions $\boldsymbol{\mu}$ and $\boldsymbol{\eta}$ is an *invariant bilinear function of the components of the two directions* $\boldsymbol{\mu}$ and $\boldsymbol{\eta}$.

The relationships derived in this section in two dimensions are valid in three dimensions as well; the derivation is carried out by following the same approaches taken in this section. This is the subject of Problem 2.3.

2.2 Definition of a Cartesian Tensor

In Sect. 2.1, we examined the manner in which scalars, vectors, and stress transform when the coordinate system is rotated. In particular, we noted that (1) vectors and stress projected onto arbitrary directions remain invariant with regard to the orientation of the original coordinate system, and (2) the components of vectors and stress in a rotated coordinate system are obtained from the components in the original coordinate system through a coordinate transformation rule. Most scalars obey these rules (they remain unchanged when the coordinated system is rotated). These are the properties that define a tensor. Note, however, that the two properties listed above are equivalent (we will show this later).

One of the following two definitions may be used for a Cartesian tensor (Butkov 1968):

1. A tensor $T_{ij\,\ldots\,n}$ (with m indices) yields a multi-linear, invariant function in arbitrary directions η,μ, … ξ (m unit vectors) as

$$\alpha = T_{ij\ldots n}\eta_i\mu_j\ldots\xi_n \tag{2.22}$$

The summations are done from 1 to N, where N is the dimension of the space. $T_{ij...n}$ are called the component of the tensor, and are defined with respect to the same bases as those of $\eta, \mu, \ldots \xi$. The tensor $T_{ij...n}$ is known as the tensor of order (or rank) m

2. A tensor $T_{ij...n}$ (with m indices) is one that transforms under rotations of the coordinate system according to the equation

$$T'_{ij...n} = T_{pq...r} a_{ip} a_{jq} \ldots a_{nr} \tag{2.23}$$

where $\mathbf{a}$ is the transformation matrix that transforms a vector from the original coordinate system to a rotated coordinate system as $\mathbf{u}' = \mathbf{au}$. The tensor $T_{ij...n}$ is known as the tensor of order (or rank) m.

It may now be verified that vectors are first order tensors and stress is a second order tensor (by either the first or second definition). It is customary to consider scalars as tensors of zero order.

It is worth repeating the other examples mentioned earlier: velocity, acceleration, and force are vectors and hence are first order tensors. Strain is a second order tensor. In flow through porous media, one encounters a permeability tensor which is a second order tensor. The analogous quantity in heat flow is the heat conductivity tensor and in electric flow is the electric conductivity tensor.

It should also be noted that a quantity represented by a matrix is not always a tensor. A matrix is a mathematical concept and may or may not satisfy the requirements of a tensor as illustrated in Examples 2.1 and 2.2.

Example 2.1.

Question: Determine if the following matrix is a tensor or not.

$$\mathbf{A} = \begin{bmatrix} x_1^2 & x_1 x_2 \\ x_1 x_2 & x_2^2 \end{bmatrix}$$

where x_1 and x_2 are components of a vector $\mathbf{x}$ in two dimensions.

Answer: To examine this using the invariance property (2.22), consider

$$\alpha = A_{ij}\mu_i\eta_j = x_1^2\mu_1\eta_1 + x_1x_2\mu_1\eta_2 + x_1x_2\mu_2\eta_1 + x_2^2\mu_2\eta_2$$

Since $\boldsymbol{\mu}$ and $\boldsymbol{\eta}$ are arbitrary, let

$$\boldsymbol{\mu} = \boldsymbol{\eta} = \frac{\mathbf{x}}{|\mathbf{x}|} = \frac{1}{|\mathbf{x}|}(x_1 \quad x_2)$$

Then

$$\begin{aligned}\alpha &= \frac{1}{|\mathbf{x}|}(x_1^4 + x_1^2x_2^2 + x_1^2x_2^2 + x_2^4) \\ &= \frac{(x_1^2 + x_2^2)^2}{|\mathbf{x}|} \\ &= |\mathbf{x}|^3\end{aligned}$$

Since length of a vector is an invariant under rotation of coordinate systems, α is an invariant as well. Hence, the matrix $\mathbf{A}$ is a tensor.

To carry out the investigation by the transformation property (2.23), let us write $\mathbf{A}$ as

$$\mathbf{A}' = \begin{bmatrix} x_1'^2 & x_1'x_2' \\ x_1'x_2' & x_2'^2 \end{bmatrix} = \begin{bmatrix} x_1' \\ x_2' \end{bmatrix} [x_1' \quad x_2'] = \mathbf{x}'\mathbf{x}'^{\mathrm{T}} = \mathbf{a}\mathbf{x}\mathbf{x}^{\mathrm{T}}\mathbf{a}^{\mathrm{T}} = \mathbf{a}\mathbf{A}\mathbf{a}^{\mathrm{T}}$$

Hence, $\mathbf{A}$ is a tensor.

Example 2.2.

Question: Determine if the following matrix is a tensor or not.

$$\mathbf{B} = \begin{bmatrix} x_1'x_2' & x_2'^2 \\ x_1'^2 & -x_1'x_2' \end{bmatrix}$$

where x_1 and x_2 are components of a vector $\mathbf{x}$ in two dimensions.

Answer: To use the first definition of tensors (2.22), examine:

$$\alpha = B_{ij}\mu_i\eta_j = x_1x_2\mu_1\eta_1 + x_2^2\mu_1\eta_2 + x_1^2\mu_2\eta_1 - x_1x_2\mu_2\eta_2$$

Let

$$\boldsymbol{\mu} = \boldsymbol{\eta} = \frac{\mathbf{x}}{|\mathbf{x}|} = \frac{1}{|\mathbf{x}|}(x_1 \quad x_2)$$

Then

$$\begin{aligned}\alpha = C_{ij}\mu_i\eta_j &= \frac{1}{|\mathbf{x}|}(x_1^3x_2 + x_1x_2^3 + x_1^3x_2 - x_1x_2^3) \\ &= \frac{1}{|\mathbf{x}|}[2x_1^3x_2]\end{aligned}$$

which is not an invariant under rotations of coordinate systems. For example, when the coordinate system is rotated such that

$$\mathbf{x}' = |\mathbf{x}|(\cos 45°, \sin 45°) = \frac{|\mathbf{x}|}{\sqrt{2}}(1, 1), \text{ then } \alpha' = \frac{|\mathbf{x}|^3}{2}$$

When the coordinate system is rotated such that

$$\mathbf{x}' = |\mathbf{x}|(\cos 30°, \sin 30°) = \frac{|\mathbf{x}|}{2}(\sqrt{3}, 1), \text{ then } \alpha' = \frac{3\sqrt{3}|\mathbf{x}|^3}{8}$$

Example 2.3.

Question: Prove that the two definitions of a tensor (2.22) and (2.23) are equivalent.

Answer: It was proved earlier (2.21) that, for stress, which is a second order tensor, the transformation property (2.22) leads to the invariant property (2.23). Extension of the proof to higher order tensors is conceptually straightforward.

To prove the inverse (that the invariance property leads to the transformation property), let us consider a second order tensor again. By the invariance property

$$T'_{ij}\eta'_i\mu'_j = T_{k\ell}\eta_k\mu_\ell$$

By the inverse transformation (2.9c) and using (2.10a)

$$\eta_k = b_{kp}\eta'_p = a_{pk}\eta'_p$$

$$\mu_\ell = b_{\ell q}\mu'_q = a_{q\ell}\eta'_q$$

Hence,

$$T'_{ij}\eta'_i\mu'_j = T_{k\ell}(a_{pk}\eta'_p)(a_{q\ell}\mu'_q) = (T_{k\ell}a_{pk}a_{q\ell})\eta'_p\mu'_q$$

Let us consider two dimensions. Since the above relation must hold for any arbitrary directions $\boldsymbol{\eta}'$ and $\boldsymbol{\mu}'$, let us choose $\boldsymbol{\eta}' = (1, 0)$ and $\boldsymbol{\mu}' = (0, 1)$. Then

$$T'_{12} = T_{k\ell}a_{1k}a_{2\ell}$$

By repeating this process with appropriate choices for $\boldsymbol{\eta}'$ and $\boldsymbol{\mu}'$, it can be shown

$$T'_{ij} = T_{k\ell}a_{ik}a_{j\ell}$$

Hence, we have proven that the invariance property leads to the transformation property.

2.3 Operations with Tensors

2.3.1 Addition of Tensors

When a tenor is multiplied by a scalar constant, it remains a tensor with its order unchanged. Tensors of the same order may be added or subtracted to produce another tensor of the same order as

$$\alpha T_{ij\ldots n} \pm \beta S_{ij\ldots n} = U_{ij\ldots n} \tag{2.24}$$

where α and β are scalars.

2.3.2 Contraction Operation

In a contraction operation, two of the indices are set equal to each other. The summation convention applies over these repeated indices. For example, T_{ijkk} is a contraction of $T_{ijk\ell}$ over the last two indices, and expands as

$$T_{ijkk} = T_{ij11} + T_{ij22} + T_{ij33} \tag{2.25}$$

A contraction of a tensor of order m results in a tensor of order $m-2$. In the above example, a fourth order tensor becomes a second order tensor. To prove this, consider the contraction of a third order tensor A'_{iik} and simplify using the transformation property as

$$A'_{ijk} = A_{pqr} a_{ip} a_{jq} a_{kr}$$

$$A'_{iik} = A_{pqr} a_{ip} a_{iq} a_{kr} = A_{pqr} \delta_{pq} a_{kr} = A_{ppr} a_{kr} \text{ (using (2.10d))}$$

Hence, A_{iik} transforms as a first order tensor. The proof for tensors of other orders is similarly obtained.

Contraction of a second order tensor A_{ij} results in its *trace* $A_{ii} = A_{11} + A_{22} + A_{33}$.

Some examples of contraction operation are

- $B_{ij} = T_{ijkk}$
- $B_{ijk} = T_{ijk\ell}\, a_\ell$
- $B_{ij} = A_{ik} C_{kj}$

2.3.3 Outer Product of Tensors

An outer product (also called the tensor product) of two tensors of arbitrary orders is multiplication of its components with distinct indices for each as

$$C_{ij...kpq...r} = A_{ij...k}B_{pq...r} \tag{2.26}$$

If the order of **A** and **B** are m and n, respectively, then the order of **C** is $m+n$. Some examples of outer products are

- $C_{ijk\ell} = A_{ij}B_{k\ell}$
- $C_{ijpqrs} = A_{ij}B_{pqrs}$
- $C_{ij} = u_i v_j$

The last outer product $C_{ij} = u_i v_j$ is the outer product between two vectors. This is also known as the *dyad*.

2.3.4 Inner Product of Tensors

The inner product of two tensors is formed by contraction of their outer products. For example, contraction of the outer products of **A** and **B** in (2.26) in index j and q yields

$$C_{ij...kpj...r} = A_{ij...k}B_{pj...r} \tag{2.27}$$

It is seen that the inner product of two tensors of order m and n yields a tensor of order $m + n - 2$.

Some examples of inner products are

- $\alpha = u_i v_i$ (inner product between two vectors)
- $C_{ij} = A_{ik}B_{kj}$
- $C_i = A_{ik}B_k$
- $C_{i\ell mn} = A_{ik}B_{k\ell mn}$

2.3.5 The Permutation Symbol

The permutation symbol (also known as the *alternator symbol* or the *Levi-Civita symbol*) ε_{ijk} is a third order tensor defined as

- $\varepsilon_{ijk} = 1$ if ijk is an even permutation of 123 (i.e., if they appear in a sequence as in the arrangement 12312)
- $\varepsilon_{ijk} = -1$ if ijk is an odd permutation of 123 (i.e., if they appear in a sequence as in the arrangement 32132)
- $\varepsilon_{ijk} = 0$ if ijk is not a permutation of 123 (i.e., if two or more of its indices have the same value).

For example, $\varepsilon_{123} = 1$ (even); $\varepsilon_{213} = -1$ (odd); $\varepsilon_{231} = 1$ (even); $\varepsilon_{213} = -1$ (odd); $\varepsilon_{113} = 0$ (repeated indices).

2.3.6 Cross Product of Two Vectors

Referring to Fig. 2.4, the cross product of two vectors **x** and **y** is defined as

$$\mathbf{v} = \mathbf{x} \times \mathbf{y} = |\mathbf{x}||\mathbf{y}|(\sin\theta)\mathbf{e} \tag{2.28}$$

Hence, the cross product of two vectors **x** and **y** is vector **v**, whose direction **e** is normal to the plane in which **x** and **y** lie. The magnitude of the cross product is the area of the parallelogram formed by the vectors **x** and **y** as shown in Fig. 2.4.

The components of **v** may be expressed in two different ways:

1.
$$v_i = \varepsilon_{ik\ell} x_k y_\ell \tag{2.29a}$$

2. From the determinant of the following matrix

$$\mathbf{v} = \begin{vmatrix} \mathbf{e}_1 & \mathbf{e}_2 & \mathbf{e}_3 \\ x_1 & x_2 & x_3 \\ y_1 & y_2 & y_3 \end{vmatrix} = (x_2y_3 - x_3y_2)\mathbf{e}_1 + (x_3y_1 - x_1y_3)\mathbf{e}_2 + (x_1y_2 - x_2y_1)\mathbf{e}_3 \tag{2.29b}$$

It may be verified that both expressions are identical (Problem 2.7).

2.3.7 Symmetry of Tensors

A second order tensor T_{ij} is *symmetric* if

$$T_{ij} = T_{ji} \quad \text{or} \quad \mathbf{T} = \mathbf{T}^{\mathrm{T}} \tag{2.30a}$$

Examples of such tensors are the stress and strain tensors.

A second order tensor T_{ij} is skew-symmetric or antisymmetric if

$$T_{ij} = -T_{ji} \quad \text{or} \quad \mathbf{T} = -\mathbf{T}^{\mathrm{T}} \tag{2.30b}$$

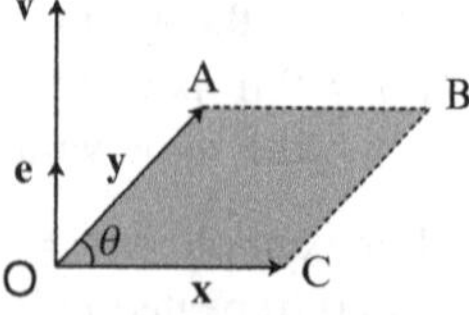

Fig. 2.4 Cross product of vectors

Any second order tensor can be uniquely decomposed into a symmetric and anti symmetric part as

$$T_{ij} = \frac{1}{2}(T_{ij} + T_{ji}) + \frac{1}{2}(T_{ij} - T_{ji}) \tag{2.31a}$$

Denoting

$$D_{ij} = \frac{1}{2}(T_{ij} + T_{ji}); \quad S_{ij} = \frac{1}{2}(T_{ij} - T_{ji}) \tag{2.31b}$$

Equation (2.31a) is written as

$$T_{ij} = D_{ij} + S_{ij} \tag{2.31c}$$

Noting that

$$D_{ji} = \frac{1}{2}(T_{ji} + T_{ij}) = D_{ij} \quad \text{and} \quad S_{ji} = \frac{1}{2}(T_{ji} - T_{ij}) = -S_{ij}$$

it is seen that D_{ij} is the symmetric part of T_{ij} and S_{ij} is the antisymmetric part of T_{ij}.

Note that

$$S_{ij} = 0 \ \text{ for } \ i = j \tag{2.32}$$

Hence, the diagonal elements of the antisymmetric part of second order tensors such as stress and strain are zero.

Extending these ideas to a tensor of any order, we define symmetric and antisymmetric tensors with respect to two of the indices of the tensor.

Some examples of such tensors are

- $T_{ijk\ell} = T_{kji\ell}$ (symmetric in indices i and k)
- $T_{ijk\ell} = T_{kji\ell}$ (antisymmetric in indices i and k)
- $T_{ijk\ell} = T_{k\ell ij}$ (symmetric in indices i and k, and in j and ℓ)
- $\varepsilon_{ijk} = \varepsilon_{jik}$ (antisymmetric in indices i and j)

Example 2.4.

Question: Given that A_{ij} and B_{ij} are symmetric and antisymmetric tensors, show that $A_{ij}B_{ij} = 0$.

Answer: To prove, let us expand $A_{ij}B_{ij}$ as

$$A_{ij}B_{ij} = -A_{ij}B_{ji} = -A_{ji}B_{ji} = -A_{pq}B_{pq}$$

$$2A_{ij}B_{ij} = 0$$

$$A_{ij}B_{ij} = 0$$

2.3.8 Further Operations Involving the Kronecker and Levi-Civita Symbols

First, let us prove that the Kronecker delta δ_{ij} is a tensor using the invariant property of tensors, i.e., consider

$$\begin{aligned}\alpha &= \delta_{ij}\eta_i\mu_j \\ &= \eta_i\mu_i\end{aligned}$$

Hence, α equals the inner product between the two unit vectors $\boldsymbol{\eta}$ and $\boldsymbol{\mu}$. This is equal to the cosine of the angle between $\boldsymbol{\eta}$ and $\boldsymbol{\mu}$, which is independent of the orientation of the coordinate system. Hence, the Kronecker delta δ_{ij} is a second order tensor.

Now let us use the transformation property

$$\delta'_{ij} = \delta_{k\ell}a_{ik}a_{j\ell} = a_{i\ell}a_{j\ell} = \delta_{ij}\text{ (using (2.10g))}$$

Hence, the components of δ_{ij} are independent of the orientation of the coordinate system. Such tensors (i.e., tensors whose components remain unchanged upon rotation of the coordinate system) are known as *isotropic tensors*.

Recall that, it was shown earlier $\delta_{k\ell}\eta_\ell = \eta_k$ ((2.12c)). Using this result, the following results are obtained:

- $\delta_{ik}\delta_{kj} = \delta_{ij}$
- $\delta_{ij}\delta_{ij} = \delta_{ii} = 3$
- $\delta_{ik}\delta_{k\ell}\delta_{\ell i} = \delta_{ii} = 3$
- $T_{ijk\ell}\delta_{ij} = T_{iik\ell} = B_{k\ell}$ (second order tensor)

It can be shown that

$$\varepsilon_{ijk}\varepsilon_{rst} = \begin{vmatrix} \delta_{ir} & \delta_{is} & \delta_{it} \\ \delta_{jr} & \delta_{js} & \delta_{jt} \\ \delta_{kr} & \delta_{ks} & \delta_{kt} \end{vmatrix} \tag{2.33}$$

To show this, consider the determinant

$$\begin{vmatrix} \delta_{11} & \delta_{12} & \delta_{13} \\ \delta_{21} & \delta_{22} & \delta_{23} \\ \delta_{31} & \delta_{32} & \delta_{33} \end{vmatrix} = 1$$

The magnitude is 1 since the off diagonal terms of the matrix are all 0's and the diagonal terms of the matrix are all 1's.

We recall from the definition of determinants that the magnitude of the determinant of a matrix changes sign when two rows or columns are switched. For instance, switching columns,

$$\begin{vmatrix} \delta_{11} & \delta_{12} & \delta_{13} \\ \delta_{21} & \delta_{22} & \delta_{23} \\ \delta_{31} & \delta_{32} & \delta_{33} \end{vmatrix} = -\begin{vmatrix} \delta_{12} & \delta_{11} & \delta_{13} \\ \delta_{22} & \delta_{21} & \delta_{23} \\ \delta_{32} & \delta_{31} & \delta_{33} \end{vmatrix} \text{ (switching the \textit{first} and \textit{second} columns)}$$

Based on the definition of the permutation symbol, this is achieved for any arbitrary column switches as

$$\begin{vmatrix} \delta_{1r} & \delta_{1s} & \delta_{1t} \\ \delta_{2r} & \delta_{2s} & \delta_{2t} \\ \delta_{3r} & \delta_{3s} & \delta_{3t} \end{vmatrix} = \varepsilon_{rst}\begin{vmatrix} \delta_{11} & \delta_{12} & \delta_{13} \\ \delta_{21} & \delta_{22} & \delta_{23} \\ \delta_{31} & \delta_{32} & \delta_{33} \end{vmatrix} = \varepsilon_{rst} \times 1 = \varepsilon_{rst}$$

By the same logic, the sign change due to an arbitrary number of row switches is represented by

$$\begin{vmatrix} \delta_{ir} & \delta_{is} & \delta_{it} \\ \delta_{jr} & \delta_{js} & \delta_{jt} \\ \delta_{kr} & \delta_{ks} & \delta_{kt} \end{vmatrix} = \varepsilon_{ijk}\varepsilon_{rst}$$

which is the required result.

Using this result, the following results may be obtained (Problem 2.9):

- $\varepsilon_{ijk}\varepsilon_{ist} = \delta_{js}\delta_{kt} - \delta_{jt}\delta_{ks}$
- $\varepsilon_{ijk}\varepsilon_{ijt} = 2\delta_{kt}$
- $\varepsilon_{ijk}\varepsilon_{ijk} = 6$

2.4 Tensor Calculus

2.4.1 Tensor Field

Denoting the position in space and time as $\mathbf{x}$ and t, a tensor expressed as a function of $\mathbf{x}$ and t is known as the *tensor field*. In general

$$\mathbf{T} = \mathbf{T}(\mathbf{x}, t) \quad \text{or} \quad T_{ij...n} = T_{ij...n}(x_i, t) \tag{2.34}$$

2.4.2 The del Operator

The *del* operator (or the differential operator) is defined as

$$\nabla = \mathbf{e}_i \frac{\partial}{\partial x_i} \tag{2.35a}$$

Hence, the *del* operator is a vector

$$\nabla = \mathbf{e}_1 \frac{\partial}{\partial x_1} + \mathbf{e}_2 \frac{\partial}{\partial x_2} + \mathbf{e}_3 \frac{\partial}{\partial x_3} \tag{2.35b}$$

When the *del* operator operates on a scalar $\phi(\mathbf{x}, \boldsymbol{t})$, we get a vector as

$$\text{grad}\,\phi = \nabla\phi = \mathbf{e}_1 \frac{\partial \phi}{\partial x_1} + \mathbf{e}_2 \frac{\partial \phi}{\partial x_2} + \mathbf{e}_3 \frac{\partial \phi}{\partial x_3} \,(\text{vector}) \tag{2.36a}$$

This is the gradient (or "grad" for short) of ϕ. The components of the gradient vector are

$$\frac{\partial \phi}{\partial x_i} \quad \text{or} \quad \phi_{,i} \,(\text{another common notation}) \tag{2.36b}$$

Consider a surface in space mathematically defined as

$$\phi(\mathbf{x}, t) = c \tag{2.37}$$

where c is a constant. Figure 2.5a shows a schematic of the surface. Then $\nabla\phi$ is a vector normal to the surface at $\mathbf{x}$ for a fixed t.

To prove this, consider two points A and B shown in Fig. 2.5b. As these points are near each other and lie on the surface, as you go from A to B:

$$\begin{aligned} \mathrm{d}\phi &= 0 \\ \phi_{,1}\,\mathrm{d}x_1 + \phi_{,2}\,\mathrm{d}x_2 + \phi_{,3}\,\mathrm{d}x_3 &= 0 \\ \nabla\phi \cdot \mathrm{d}\mathbf{x} &= 0 \end{aligned}$$

Hence, $\nabla\phi$ and $\mathrm{d}\mathbf{x}$ are orthogonal, proving that $\nabla\phi$ is normal to the surface. The gradient of a general tensor is given by

$$T_{ij\ldots n,x_k} = \frac{\partial}{\partial x_k} T_{ij\ldots n}$$

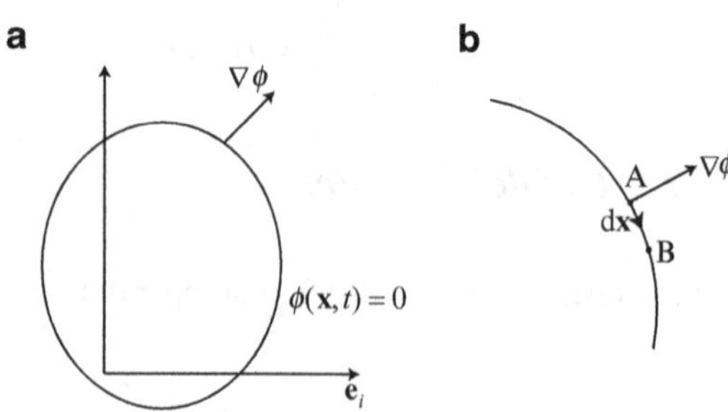

Fig. 2.5 Vector normal to a surface

If the original tensor $T_{ij\,\ldots\,n}$ is an mth order tensor, then the gradient defined above is an $(m+1)$th order tensor. We saw this to be the case above in the case of a scalar (i.e., a scalar became a vector, (2.36a)). To prove this, let us consider the gradient of a second order tensor T_{ij}

$$T'_{ij,x_k} = \frac{\partial}{\partial x'_k} T'_{ij} = \frac{\partial}{\partial x_\ell}(T_{pq} a_{ip} a_{jq}) \frac{\partial x_\ell}{\partial x'_k}$$

(using the transformation law, (2.23), and the chain rule)

From (2.9c) and (2.10a)

$$x_i = b_{ik} x'_k = a_{ki} x'_k \Rightarrow \frac{\partial x_i}{\partial x'_r} = a_{ki} \frac{\partial x'_k}{\partial x'_r} = a_{ki} \delta_{kr} = a_{ri}$$

Hence,

$$T'_{ij,x_k} = \frac{\partial}{\partial x_\ell}(T_{pq} a_{ip} a_{jq} a_{k\ell}) = T_{pq,x_\ell} a_{ip} a_{jq} a_{kl}$$

which proves that T'_{ij,x_k} transforms as a third order tensor. Hence, the derivative of a second order tensor is a third order tensor.

The other common operations involving the *del* operator are

The divergence operator:

$$\text{div}\mathbf{u} = \nabla \cdot \mathbf{u} = u_{i,i}\ (\text{scalar}),\ \text{where}\ \mathbf{u}\ \text{is a vector} \tag{2.38a}$$

The Laplace operator:

$$\nabla^2 \phi = \nabla \cdot \nabla \phi = \phi_{,ii} = \frac{\partial^2 \phi}{\partial x_1^2} + \frac{\partial^2 \phi}{\partial x_2^2} + \frac{\partial^2 \phi}{\partial x_3^2}\ (\text{scalar}) \tag{2.38b}$$

The Curl operator:

$$\mathbf{v} = \text{Curl}\,\mathbf{u} = \nabla \times \mathbf{u}; \quad \text{or} \quad v_i = \varepsilon_{ijk} \partial_j u_k = \varepsilon_{ijk} u_{k,j} \tag{2.38c}$$

2.4.3 *Stokes' Theorem*

Stokes' theorem relates a line integral of a vector **F** around a closed curve to a surface integral over a two-sided surface as shown in Fig. 2.6. The theorem states

$$\oint_C \mathbf{F}.\text{d}\mathbf{x} = \int_S \mathbf{n}.(\nabla \times \mathbf{F})\text{d}S \tag{2.39a}$$

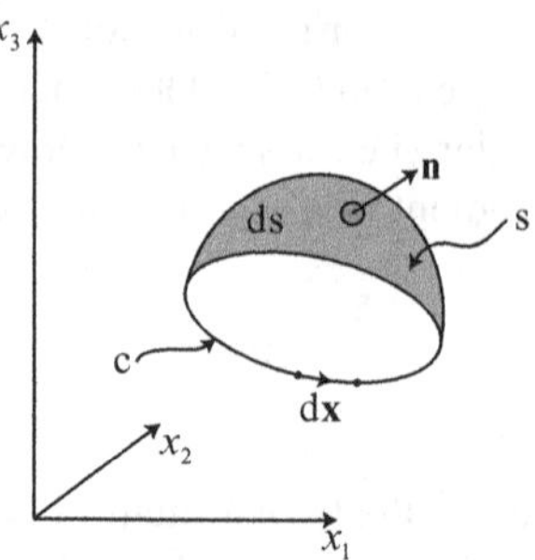

Fig. 2.6 Line–surface integral

In indicial notation:

$$\oint_C F_i \mathrm{d}x_i = \int_S n_i \varepsilon_{ijk} F_{k,j} \mathrm{d}S \tag{2.39b}$$

2.4.4 Divergence Theorem of Gauss

The divergence theorem of Gauss relates a surface integral to a volume integral (Fig. 2.7). Considering first a vector **u**, the Gauss theorem in indicial notation is

$$\int_S u_i n_i \mathrm{d}S = \int_\Omega u_{i,i} \mathrm{d}V \tag{2.40a}$$

and in symbolic notation is

$$\int_S \mathbf{u}.\mathbf{n}\, \mathrm{d}S = \int_\Omega \operatorname{div} \mathbf{u}\, \mathrm{d}V \tag{2.40b}$$

u.n is the flux of the vector field **u** through the surface S.

The theorem is generalized to a tensor as

$$\int_S T_{ij\ldots k} n_\ell \, \mathrm{d}S = \int_\Omega T_{ij\ldots k, x_\ell} \, \mathrm{d}V \tag{2.40c}$$

2.4.5 Differentiation of a Tensor with Respect to another Tensor

Let us define a tensor **T(A,B,...)** of arbitrary order that is a function of other tensors **A**, **B**, etc., of arbitrary orders. The partial derivative of **T(A,B,...)** with respect to one of its argument tensors (say, **B**) is

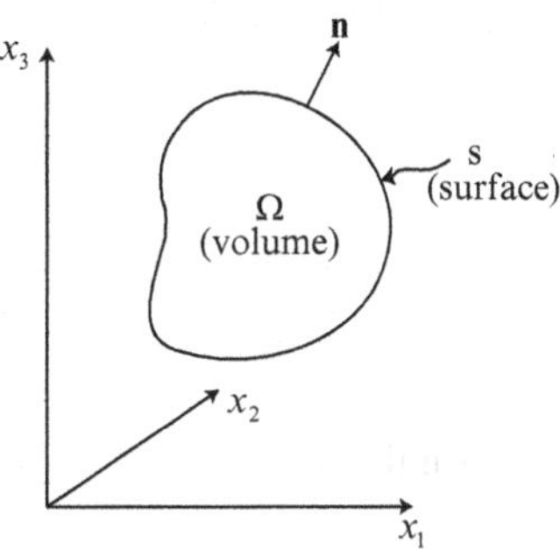

Fig. 2.7 Surface–volume integral

$$\mathbf{G} = \frac{\partial \mathbf{T}}{\partial \mathbf{B}} \tag{2.41}$$

If the orders of tensors **T** and **B** are m and n, respectively, the order of the derivative G is $m + n$.

Some examples are

- $G_{ij} = \frac{\partial \phi}{\partial T_{ij}}$ (derivative of a scalar with respect to a second order tensor)
- $G_{ijk\ell} = \frac{\partial A_{ij}}{\partial T_{k\ell}}$ (derivative of a second order tensor with respect to a another second order tensor)

Let us consider the derivative of a tensor by itself as (for example, for a second order tensor)

$$C_{ijk\ell} = \frac{\partial A_{ij}}{\partial A_{k\ell}}$$

Noting that $C_{ijkl} = 0$ except when the indices coincide as $i = k$ and $j = \ell$, it follows

$$C_{ijk\ell} = \frac{\partial A_{ij}}{\partial A_{k\ell}} = \delta_{ik}\delta_{j\ell} \tag{2.42}$$

To prove that $G_{ij} = \frac{\partial \phi}{\partial T_{ij}}$ is a second order tensor, let us consider

$$\begin{aligned} G'_{ij} &= \frac{\partial \phi'(\mathbf{T}')}{\partial T'_{ij}} \\ &= \frac{\partial \phi'(\mathbf{T}')}{\partial T_{k\ell}} \frac{\partial T_{k\ell}}{\partial T'_{ij}} \end{aligned}$$

From (2.18b):

$$
\begin{aligned}
T_{k\ell} &= T'_{pq} a_{pk} a_{q\ell} \\
\frac{\partial T_{k\ell}}{\partial T'_{ij}} &= \delta_{pi} \delta_{qj} a_{pk} a_{q\ell} \\
&= a_{ik} a_{j\ell}
\end{aligned}
$$

Assuming $\phi' = \phi$,

$$
\begin{aligned}
G'_{ij} &= \frac{\partial \phi}{\partial T_{k\ell}} a_{ik} a_{j\ell} \\
&= G_{k\ell} a_{ik} a_{j\ell}
\end{aligned}
$$

Hence, G_{ij} is a second order tensor ((2.23)).

The usual chain rule applies to tensor derivatives as well. For example, let ϕ be a function of a scalar α as $\phi(\alpha)$, where $\alpha = (\sigma_{k\ell}\sigma_{k\ell})^{1/2}$, then

$$
C_{ij} = \frac{\partial \phi}{\partial \sigma_{ij}} = \frac{\partial \phi}{\partial \alpha} \frac{\partial \alpha}{\partial \sigma_{ij}} \tag{2.43}
$$

Now differentiate α as

$$
\begin{aligned}
\alpha^2 &= \sigma_{k\ell}\sigma_{k\ell} \\
2\alpha \frac{\partial \alpha}{\partial \sigma_{ij}} &= 2\delta_{ki}\delta_{\ell j}\sigma_{k\ell} \\
\frac{\partial \alpha}{\partial \sigma_{ij}} &= \frac{\sigma_{ij}}{\alpha}
\end{aligned}
$$

Then

$$
C_{ij} = \frac{\partial \phi}{\partial \sigma_{ij}} = \frac{\partial \phi}{\partial \alpha} \frac{\sigma_{ij}}{\alpha}
$$

A more comprehensive application of this is addressed in Problem 2.12.

2.5 Invariants of Stresses and Strains

In most problems that the finite element method is used to solve (i.e., two- and three-dimensional problems), more than one component of the stress tensor varies with loading. Furthermore, the variation can be quite arbitrary. For instance, in a plane strain problem, all three of the nonzero components of the stress tensor

(σ_{xx}, σ_{yy}, σ_{xy}) can vary with the applied load, and the interrelationship among the three components also can be quite arbitrary. Thus, the three stress components (σ_{xx}, σ_{yy}, σ_{xy}) and the corresponding three strain components (ε_{xx}, ε_{yy}, ε_{xy}) can vary in a complex manner. In a truly three-dimensional problem, all six stress components (σ_{xx}, σ_{yy}, σ_{zz}, σ_{xy}, σ_{xz}, σ_{yz}) and the corresponding six strain components (ε_{xx}, ε_{yy}, ε_{zz}, ε_{xy}, ε_{xz}, ε_{yz}) can vary with the interrelationships being very complex. Hence, there is a need to seek simpler and/or more rational methods of relating the stresses and strains.

For these reasons, it is of interest to seek other measures of stresses and strains. Examples of such measures are

- Principal stresses and principal strains
- Volumetric and deviatoric components of the stress and strain tensors
- Invariants of stress and strain tensors

In the following section, we will define these quantities and present geometrical meaning of them.

2.5.1 Principal Stresses and Strains

The principal planes are those planes where the shear stresses are zero. Considering a Cartesian coordinate system, there are three perpendicular planes where this occurs. The normal stresses on the principal planes are called the principal stresses (also principal values). The principal strains are similarly defined.

It is assumed that the reader is familiar with the Mohr-circle method of finding the principal stresses, principal strains and their directions. Here we describe a general method of finding the principal values and directions that is useful in three-dimensional analyses.

It was shown earlier (2.14c) that the traction **t** on a plane having a unit normal **n** is related to the stress tensor at that point as

$$\mathbf{t} = \boldsymbol{\sigma}\mathbf{n} \tag{2.44}$$

If the shear stress on this plane is to be zero, then the direction of **t** must coincide with the direction of **n**, i.e.,

$$\mathbf{t} = \lambda\mathbf{n} \tag{2.45}$$

where λ is a scalar. Combining (2.44) and (2.45)

$$\boldsymbol{\sigma}\mathbf{n} = \lambda\mathbf{n} \tag{2.46a}$$

which the reader can recognize as the standard eigenvalue problem. λ is the eigenvalue of $\boldsymbol{\sigma}$ and **n** is the corresponding eigenvector. Since the size of $\boldsymbol{\sigma}$ is

3×3, there are three eigenvalues (not necessarily distinct) and eigenvectors, which are, respectively, the principal stresses and their directions. Equation (2.46a) may be written as

$$(\boldsymbol{\sigma} - \lambda \mathbf{I})\mathbf{n} = 0 \tag{2.46b}$$

where **I** is identity matrix. For (2.46b) to be true for a nontrivial **n**, the determinant of the matrix must be zero, i.e.,

$$|\boldsymbol{\sigma} - \lambda \mathbf{I}| = 0 \tag{2.46c}$$

Expanding (2.46c), one obtains a cubic equation in λ, with three solutions, which are the principal stresses associated with $\boldsymbol{\sigma}$. We illustrate this for a specific problem in Example 2.5.

Example 2.5.

Question: Find the principal stresses and principal directions for the following stress tensor:

$$\boldsymbol{\sigma} = \begin{bmatrix} 50 & 20 & 0 \\ 20 & 60 & 0 \\ 0 & 0 & 40 \end{bmatrix}$$

Answer: From (2.46c),

$$\begin{vmatrix} 50-\lambda & 20 & 0 \\ 20 & 60-\lambda & 0 \\ 0 & 0 & 40-\lambda \end{vmatrix} = 0$$

$$(50-\lambda)(60-\lambda)(40-\lambda) - 20 \times 20 \times (40-\lambda) = 0$$

$$(40-\lambda)[(60-\lambda)(50-\lambda) - 400] = 0$$

$$\lambda_1 = 40, \; \lambda_2 = 75.61, \; \lambda_3 = 34.39$$

The principal stress tensor is

$$\boldsymbol{\sigma} = \begin{bmatrix} 40.00 & 0 & 0 \\ 0 & 75.61 & 0 \\ 0 & 0 & 34.39 \end{bmatrix}$$

In finding the principal directions, it must be remembered that the three equations in (2.46b) are dependent, and there are two independent equations. A common procedure is to assume one of the nonzero components of the vector **n** to be 1.0 and find the remaining two.

In this specific problem, it is seen that direction 3 is already a principal direction (notice that the off diagonal terms on the third column and row are zero). This is why one of the principal values coincided with the diagonal term of 40.0 of the stress tensor. The analysis will ascertain this as well.

Let us evaluate the principal directions. First, consider the three equations (2.46b) for $\lambda = 40$:

$$(50 - 40)n_1 + 20n_2 + 0 \times n_3 = 0 \Rightarrow 10n_1 + 20n_2 = 0$$

$$20n_1 + (60 - 40)n_2 + 0 \times n_3 = 0 \Rightarrow 20n_1 + 20n_2 = 0$$

$$0 \times n_1 + 0 \times n_2 + (40 - 40)n_3 = 0$$

It is clear that n_3 cannot be solved from the first two equations. The third equation suggests that n_3 can assume any value. Let us assume $n_3 = 1.0$. From the first two equations, we get: $n_1 = 0$ and $n_2 = 0$. Thus, $\mathbf{n}_1 = \{0, 0, 1\}$, which is the direction along the x_3-axis. As expected, the principal direction coincides with the x_3-axis.

Now, let us consider the three equations for $\lambda = 75.61$:

$$(50 - 75.61)n_1 + 20n_2 + 0 \times n_3 = 0 \Rightarrow \; -25.61n_1 + 20n_2 = 0$$

$$20n_1 + (60 - 75.61)n_2 + 0 \times n_3 = 0 \Rightarrow 20n_1 - 15.61n_2 = 0$$

$$0 \times n_1 + 0 \times n_2 + (40 - 75.61)n_3 = 0 \Rightarrow \; -35.61n_3 = 0$$

This time, the third equation leads to $n_3 = 0$. It is seen that the first two equations are identical. With $n_1 = 1$, these equations yield $n_2 = 1.28$. Thus, the eigenvector is $\mathbf{n}_2 = \{1, 1.28, 0\}$. When a unit vector is desired, $\mathbf{n}_2 = \{0.616, 0.788, 0\}$.

By following the same procedure, it can be shown that the eigenvector corresponding to $\lambda = 34.39$ is $\mathbf{n}_3 = \{0.788, -0.616, 0\}$.

Also note that vectors $-\mathbf{n}_1$, $-\mathbf{n}_2$ and $-\mathbf{n}_3$ will also satisfy (2.46b), and therefore are legitimate eigenvectors. Noting that each principal plane has two sides, the meaning of $\mathbf{n}_i$ and $-\mathbf{n}_i$ for $i = 1$–3 are easily understood.

2.5.2 Spherical and Deviatoric Components

Any second order tensor can be split up into spherical and deviatoric components. Let us first consider the stress tensor σ_{ij}. A coordinate invariant I is defined as

$$I = \sigma_{kk} = \sigma_{k\ell}\delta_{k\ell} = \sigma_{11} + \sigma_{22} + \sigma_{33} \tag{2.47a}$$

I is, therefore, the trace of the stress tensor (sum of the diagonals). I is related to the mean normal pressure p as

$$I = 3p \tag{2.47b}$$

The spherical (also called the hydrostatic) part of σ_{ij} is

$$\frac{1}{3} I \delta_{ij} \tag{2.47c}$$

The deviatoric component of the stress tensor is obtained by subtracting the spherical part from the stress tensor as

$$s_{ij} = \sigma_{ij} - \frac{1}{3} I \delta_{ij} \tag{2.48}$$

Denoting the strain tensor by ε_{ij}, the corresponding quantities for the strain tensor are similarly defined as follows:

$$\varepsilon_v = \varepsilon_{kk} = \varepsilon_{k\ell}\delta_{k\ell} = \varepsilon_{11} + \varepsilon_{22} + \varepsilon_{33} \tag{2.49}$$

$$e_{ij} = \varepsilon_{ij} - \frac{1}{3}\varepsilon_v \delta_{ij} \tag{2.50}$$

where $(1/3)\varepsilon_v\delta_{ij}$ is the spherical (also called as the volumetric) part of ε_{ij}, and e_{ij} is the deviatoric part of ε_{ij}. ε_v is the volumetric strain.

2.5.3 *Invariants of Stress and Strain Tensors*

Invariants are scalar quantities which remain unchanged upon rotation of the coordinate system. Many physical scalars, which are 0th order tensors, are invariants. For example, mass of a body is invariant. For a vector, the linear function defined as part of the first definition of a tensor ((2.22) and (2.12a)) is an invariant. When the arbitrary vector $\boldsymbol{\eta}$ in (2.12a) is along the direction of the vector itself, the linear invariant function is the length of the vector; hence, the length of a vector is an invariant, which is easy to understand from physical grounds.

For a second order tensor, such as the stress and strain tensors, three scalar invariants can be defined. These invariants can be defined in a number of different ways. For example, when you expand the determinant in (2.46c), the following cubic equation in λ, known as the characteristic equation, is obtained.

$$\lambda^3 - I_1\lambda^2 + I_2\lambda - I_3 = 0$$

where $I_1 = \sigma_{kk}$ (trace of σ_{ij})

$$I_2 = \frac{1}{2}(\sigma_{ii}\sigma_{jj} - \sigma_{ij}\sigma_{ij})$$
$$I_3 = \det(\boldsymbol{\sigma})$$

I_1, I_2, and I_3 are invariants.

Here, we will define the invariants in a different way. Our first invariant is I and is taken to be the same as I_1 defined above. That is,

$$I = \sigma_{kk} = \sigma_{k\ell}\delta_{k\ell} = \sigma_{11} + \sigma_{22} + \sigma_{33} \tag{2.51a}$$

Note that I was defined earlier in (2.47a) as part of the definition of the spherical part of the stress tensor. The second (J) and the third (S) invariants are defined on the basis of the deviatoric stress tensor s_{ij} (2.48) as follows:

$$J = \left(\frac{1}{2}s_{ij}s_{ij}\right)^{1/2} \tag{2.51b}$$

$$S = \left(\frac{1}{3}s_{ik}s_{k\ell}s_{\ell i}\right)^{1/3} \tag{2.51c}$$

In addition to these, we define a fourth invariant, known as the "Lode" angle, by combining J and S as follows:

$$-\frac{\pi}{6} \le \alpha = \frac{1}{3}\sin^{-1}\left[\frac{3\sqrt{3}}{2}\left(\frac{S}{J}\right)^3\right] \le \frac{\pi}{6} \tag{2.52}$$

The invariants I, J and α have geometrical meaning as we will see later, and have special significance for constitutive modeling. These are the reasons for these particular choices of invariants for the discussion here.

The first invariant of the strain tensor is the volumetric strain ε_v defined in (2.49), which is repeated below for convenience:

$$\varepsilon_v = \varepsilon_{kk} = \varepsilon_{k\ell}\delta_{k\ell} = \varepsilon_{11} + \varepsilon_{22} + \varepsilon_{33} \tag{2.53a}$$

The second and third invariants of the strain tensor (i.e., the counterparts of J and S) are defined from the deviatoric strain tensor e_{ij} (2.50) as follows:

$$J_\varepsilon = \left(\frac{1}{2}e_{ij}e_{ij}\right)^{1/2} \tag{2.53b}$$

$$S_\varepsilon = \left(\frac{1}{3}e_{ik}e_{k\ell}e_{\ell i}\right)^{1/3} \tag{2.53c}$$

Example 2.6.

Question: Determine the volumetric and deviatoric components and the invariants (I, J, S, α) of the following stress tensor:

$$\boldsymbol{\sigma} = \begin{bmatrix} 50 & 20 & 0 \\ 20 & 60 & 0 \\ 0 & 0 & 40 \end{bmatrix}$$

Answer:

$$I = 50 + 60 + 40 = 150$$

$$\mathbf{s} = \begin{bmatrix} 50 & 20 & 0 \\ 20 & 60 & 0 \\ 0 & 0 & 40 \end{bmatrix} - \frac{150}{3}\begin{bmatrix} 1 & 0 & 0 \\ 0 & 1 & 0 \\ 0 & 0 & 1 \end{bmatrix} = \begin{bmatrix} 0 & 20 & 0 \\ 20 & 10 & 0 \\ 0 & 0 & -10 \end{bmatrix}$$

$$\begin{aligned} J &= \frac{1}{\sqrt{2}}\left(s_{11}^2 + s_{12}^2 + \cdots\right)^{1/2} \\ &= \frac{1}{\sqrt{2}}(0 + 400 + 0 + 400 + 100 + 0 + 0 + 0 + 100)^{1/2} = 22.36 \end{aligned}$$

$$\begin{aligned} S &= \frac{1}{\sqrt{3}}\begin{bmatrix} s_{11}s_{11}s_{11} + s_{11}s_{12}s_{21} + s_{11}s_{13}s_{31} & \text{for} \quad i = 1, k = 1, \ell = 1:3 \\ +s_{12}s_{21}s_{11} + s_{12}s_{22}s_{21} + s_{12}s_{23}s_{31} & \text{for} \quad i = 1, k = 2, \ell = 1:3 \\ + \cdots, \text{etc} & \text{for a total of 27 terms} \end{bmatrix} \\ &= 15.87 \end{aligned}$$

$$\alpha = \frac{1}{3}\sin^{-1}\left[\frac{3\sqrt{3}}{2}\left(\frac{15.87}{22.36}\right)^3\right] = 22.75^\circ \leq \left(\frac{\pi}{6} = 30^\circ\right)$$

2.5.4 Analysis of Stress Invariants in the Principal Stress Space

Let us now consider the principal stress space (where the stress tensor only has the diagonal terms), simplify the equations presented in the preceding section, and examine the geometrical meaning of the deviatoric tensor and the stress invariants. In the principal stress space, the deviatoric stress tensor is diagonal as well. This property allows us to treat the stress tensor as a vector and represent the state of stress by a point in the principal stress space, as shown in Fig. 2.8. The stress and the deviatoric stress tensors are expressed in vectors as:

$$\boldsymbol{\sigma}^* = \begin{bmatrix} \sigma_1 \\ \sigma_2 \\ \sigma_3 \end{bmatrix}; \quad \mathbf{s}^* = \begin{bmatrix} s_1 \\ s_2 \\ s_3 \end{bmatrix} = \begin{bmatrix} \sigma_1 - I/3 \\ \sigma_2 - I/3 \\ \sigma_3 - I/3 \end{bmatrix} = \begin{bmatrix} \dfrac{2\sigma_1 - \sigma_2 - \sigma_3}{3} \\ \dfrac{2\sigma_2 - \sigma_3 - \sigma_1}{3} \\ \dfrac{2\sigma_3 - \sigma_1 - \sigma_2}{3} \end{bmatrix} \tag{2.54a}$$

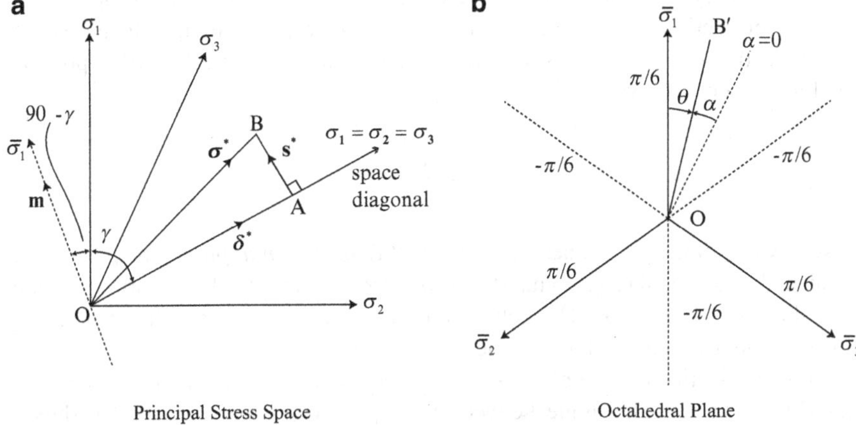

Fig. 2.8 Representation of stress in principal stress space and invariant space

$$s_i^* = \sigma_i - p; \quad p = \frac{1}{3}(\sigma_1 + \sigma_2 + \sigma_3) \tag{2.54b}$$

$$J = \frac{1}{\sqrt{2}}(s_1^2 + s_2^2 + s_3^2)^{1/2} \quad \text{and} \quad S = \frac{1}{\sqrt{3}}(s_1^3 + s_2^3 + s_3^3)^{1/3} \tag{2.54c}$$

p is the mean normal pressure. The Kronecker delta is expressed in a vector as:

$$\boldsymbol{\delta}^* = \begin{bmatrix} 1 \\ 1 \\ 1 \end{bmatrix} \tag{2.55}$$

Let us provide geometrical interpretation of these quantities. The *space diagonal* is the line in the (σ_1,σ_2,σ_3) space that is inclined at equal angle γ to σ_1-, σ_2-, and σ_3- axes.

The direction cosine of the space diagonal with each of the axes is cos γ. From the property

$$\cos^2\gamma + \cos^2\gamma + \cos^2\gamma = 1 \Rightarrow \cos\gamma = \frac{1}{\sqrt{3}} \tag{2.56}$$

A unit vector along the space diagonal is $\delta^*/\sqrt{3}$. Then, referring to Fig. 2.8a,

$$\left|\overrightarrow{OA}\right| = \frac{1}{\sqrt{3}}\boldsymbol{\sigma}^*.\,\boldsymbol{\delta}^* = \frac{I}{\sqrt{3}} \tag{2.57a}$$

$$\overrightarrow{AB} = \boldsymbol{\sigma}^* - \overrightarrow{OA} = \boldsymbol{\sigma}^* - \frac{I}{\sqrt{3}}\frac{\boldsymbol{\delta}^*}{\sqrt{3}} = \mathbf{s}^* \tag{2.57b}$$

Thus, the projection of the stress vector onto the space diagonal is proportional to I. The projection of the stress vector onto a plane perpendicular to the space diagonal is the deviatoric stress tensor. The magnitude of the vector AB is proportional to the second invariant J as

$$J = \frac{\overrightarrow{AB}}{\sqrt{2}} \tag{2.58}$$

Now we define two planes as follows: (a) *meridional plane*: A plane that contains the space diagonal and (b) *octahedral plane*: A plane normal to the space diagonal (Fig. 2.8b). The octahedral plane that passes through the origin is known as *the* π*-plane*. On the π-plane, $p = I/3 = 0$.

Let the projections of the σ_1-, σ_2-, and σ_3- axes onto the π-plane be $\bar{\sigma}_1$-, $\bar{\sigma}_2$-, $\bar{\sigma}_3$- axes (Fig. 2.8a, b). The angle between the σ_1- and $\bar{\sigma}_1$- axes is 90–γ as shown in Fig. 2.8a; hence the direction cosine of $\bar{\sigma}_1$- axis with the σ_1- axis is cos (90–γ) = sin γ. The $\bar{\sigma}_1$- axis is inclined at equal angles to σ_2- and σ_3- axes and hence have equal direction cosines with σ_2- and σ_3- axes; let this be x. Then

$$\sin^2\gamma + 2x^2 = 1 \Rightarrow \frac{2}{3} + 2x^2 = 1 \Rightarrow x = \frac{1}{\sqrt{6}}$$

Noting that $\bar{\sigma}_1$-axis is in the negative direction of σ_2- and σ_3-axes, the unit normal along the $\bar{\sigma}_1$-axis is

$$\mathbf{m} = \frac{1}{\sqrt{6}}\{2, -1, -1\} \tag{2.59}$$

Now equating the projection of vector $\overrightarrow{AB}$ onto the $\bar{\sigma}_1$–axis, and using the property $s_1 + s_2 + s_3 = 0 \Rightarrow s_2 + s_3 = -s_1$, and (2.58),

$$\mathbf{s}^*.\mathbf{m} = (OB')\cos\theta$$

$$\frac{1}{\sqrt{6}}(2s_1 - s_2 - s_3) = J\sqrt{2}\cos\theta$$

$$\frac{3}{\sqrt{6}}s_1 = J\sqrt{2}\cos\theta$$

$$s_1 = \frac{2}{\sqrt{3}}J\cos\theta$$

Now from (2.48)

$$\sigma_1 = \frac{2}{\sqrt{3}}J\cos\theta + p$$

Similarly, by finding the projection of $\overrightarrow{AB}$ onto $\bar{\sigma}_1$- and $\bar{\sigma}_2$-axes, it can be shown (Problem 2.18)

$$\sigma_1 = r_0 \cos\theta + p \tag{2.60a}$$

$$\sigma_2 = -r_0 \cos(\theta - 60^\circ) + p \tag{2.60b}$$

$$\sigma_3 = -r_0 \cos(\theta + 60^\circ) + p \tag{2.60c}$$

where

$$r_0 = \frac{2}{\sqrt{3}} J \tag{2.60d}$$

We will discuss more about the angle θ and α in the next section. Let us now present some other commonly used expressions. In the principle stress space, the octahedral normal stress σ_{oct} is given by (with the aid of using (2.14c))

$$\sigma_{\text{oct}} = \mathbf{n}^{\mathrm{T}} \boldsymbol{\sigma} \mathbf{n} = \frac{1}{3} \boldsymbol{\delta}^{*\mathrm{T}} \boldsymbol{\sigma} \boldsymbol{\delta} = \frac{1}{3}(\sigma_1 + \sigma_2 + \sigma_3) = p \tag{2.61}$$

where $\mathbf{n} = (1/\sqrt{3})\boldsymbol{\delta}^{*\mathrm{T}}$ is the unit normal to the octahedral plane, which is identical to the unit vector along the space diagonal. The octahedral shear stress τ_{oct} is

$$\tau_{\text{oct}} = \sqrt{|\boldsymbol{\sigma}\mathbf{n}|^2 - \sigma_{\text{oct}}^2} \tag{2.62}$$

which, after some manipulations, yields the following in the principal stress space

$$\tau_{\text{oct}} = \frac{1}{3}\left[(\sigma_1 - \sigma_2)^2 + (\sigma_2 - \sigma_3)^2 + (\sigma_1 - \sigma_3)^2\right]^{1/2} \tag{2.63}$$

In the general stress space,

$$\tau_{\text{oct}} = \frac{1}{3}\left[(\sigma_{11} - \sigma_{22})^2 + (\sigma_{22} - \sigma_{33})^2 + (\sigma_{11} - \sigma_{33})^2 + 6\sigma_{12}^2 + 6\sigma_{23}^2 + 6\sigma_{13}^2\right]^{1/2} \tag{2.64}$$

It can be further shown

$$\tau_{\text{oct}} = J\sqrt{\frac{2}{3}} \tag{2.65}$$

2.5.5 *Analysis of Invariants in the Equibiaxial (Triaxial) Stress Space*

This is a special case of the principal stress space where two of the principal stresses are equal to each other. Taking 2 and 3 to be the directions in which the stresses are equal to each other, an equibiaxial (also known as “triaxial” in disciplines that deal with soil, rock, and concrete) state of stress is then formally defined as: $\sigma_1 \neq \sigma_2 = \sigma_3$. The two variables that are used in constitutive theories are the *mean normal pressure*,

$$p = \frac{1}{3}(\sigma_1 + 2\sigma_3) \tag{2.66a}$$

and the *deviatoric stress*,

$$q = \sigma_1 - \sigma_3 \tag{2.66b}$$

An isotropic compression is one where $\sigma_1 = \sigma_2 = \sigma_3$. An equibiaxial compression loading is one where $\sigma_1 > \sigma_2 = \sigma_3$ and equibiaxial extension loading is one where $\sigma_1 < \sigma_2 = \sigma_3$. It then follows that

$$q > 0 \quad \text{for equibiaxial compression} \tag{2.67a}$$

$$q < 0 \quad \text{for equibiaxial extension} \tag{2.67b}$$

$$q = 0 \quad \text{for isotropic (or hydrostatic) compression loading} \tag{2.67c}$$

For equibiaxial state of stress:

$$\boldsymbol{\sigma}^* = \begin{bmatrix} \sigma_1 \\ \sigma_3 \\ \sigma_3 \end{bmatrix};$$

$$\mathbf{s}^* = \begin{bmatrix} s_1 \\ s_2 \\ s_3 \end{bmatrix} = \begin{bmatrix} \sigma_1 - I/3 \\ \sigma_3 - I/3 \\ \sigma_3 - I/3 \end{bmatrix} = \begin{bmatrix} \dfrac{2\sigma_1 - \sigma_3 - \sigma_3}{3} \\ \dfrac{2\sigma_3 - \sigma_3 - \sigma_1}{3} \\ \dfrac{2\sigma_3 - \sigma_1 - \sigma_3}{3} \end{bmatrix} = \begin{bmatrix} \dfrac{2}{3}q \\ -\dfrac{1}{3}q \\ -\dfrac{1}{3}q \end{bmatrix} \tag{2.68a}$$

$$\begin{aligned} J &= \tfrac{1}{\sqrt{2}}\left(s_1^2 + s_2^2 + s_3^2\right)^{1/2} \quad \text{and} \quad & S &= \tfrac{1}{\sqrt{3}}\left(s_1^3 + s_2^3 + s_3^3\right)^{1/3} \\ &= \tfrac{q}{\sqrt{3}} & &= \tfrac{2^{1/3}q}{3} \end{aligned} \tag{2.68b}$$

Now from (2.52),

$$\alpha = \frac{1}{3}\sin^{-1}\left[\frac{3\sqrt{3}}{2}(\pm)\frac{6\sqrt{3}}{27}\right] \tag{2.69a}$$

$$= \pm\frac{\pi}{6} \text{ with } + \text{ for equibiaxial compression and}$$

$$- \text{ for equibiaxial extension} \tag{2.69b}$$

Referring to Fig. 2.8b, the "Lode" angle α is an angular measure in this plane, with $\alpha = \pi/6$ on the positive σ_1-, σ_2-, and σ_3-axes, and $\alpha = -(\pi/6)$ on the negative σ_1-, σ_2-, and σ_1-axes. α takes on a value between $\pi/6$ and $-(\pi/6)$ at other points on the octahedral plane. J is a radial measure of the magnitude of stress on the π-plane.

Instead of using $(\sigma_1, \sigma_2, \sigma_3)$, an alternate way to represent the state of stress is using the invariants (I, J, α).

There are two advantages of this representation:

1. The stress is split up into spherical and deviatoric parts. This allows the constitutive models to be developed on the basis of the volumetric and deviatoric stress–strain relationships of the material.
2. In three-dimensional analyses, the stress tensor has six independent stress components. However, for isotropic materials, it suffices to develop the stress–strain behavior of the material in terms of the three invariants (I, J, α).

Problems

Problem 2.1 Determine the transformation matrix **a** for the original and rotated coordinate systems shown in Fig. 2.9a and b. Verify that the transformation matrices are orthogonal.

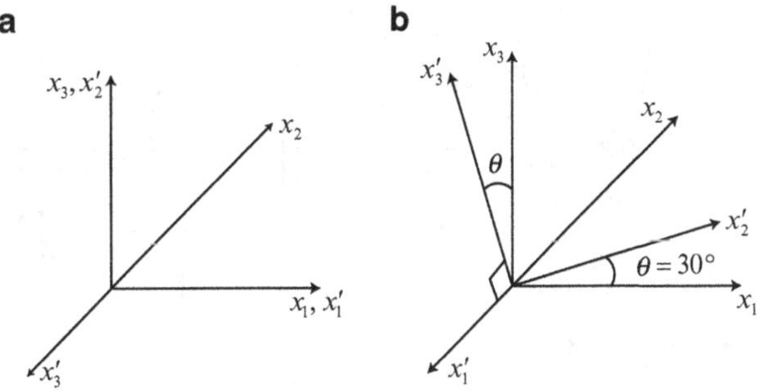

Fig. 2.9 Rotated coordinate systems

Problem 2.2 Determine the components of the vector **u** = (1, –1, 1) in rotated coordinate systems shown in Fig. 2.9a and b using the transformation rule in (2.7a) and the transformation matrices computed in Problem 2.1. Verify your answers by directly projecting **u** onto the rotated system.

Problem 2.3 An inclined plane ABC is defined in three-dimensional space in Fig. 2.10a. The vectors **n**, **m**, and **t** denote, respectively, the unit vector normal to the plane, an arbitrary unit vector and traction on the plane (refer to Fig. 2.3b and c for their counter parts in two dimensions).
Derive the following equations:

1. $t_i = \sigma_{ik} a_{nk}$
2. $\sigma_{mn} = \sigma'_{ij}\mu'_i\eta'_j = \sigma_{ij}\mu_i\eta_j$
3. $\sigma'_{ij} = \sigma_{k\ell}a_{ik}a_{j\ell}$

where **a** is the transformation matrix, and $\boldsymbol{\mu} = \mathbf{m}$ and $\boldsymbol{\eta} = \mathbf{n}$ (two arbitrary directions).

Hint: (area OAC) = (area ABC)(**n.e**$_2$) = (area ABC)a_{n2}. Hence, $(1/2)\mathrm{d}x_1\mathrm{d}x_3 = \mathrm{d}s_\mathrm{n}a_{n2}$, where $\mathrm{d}s_n$ = (area ABC).

Problem 2.4 The stress tensor in the x_1–x_2–x_3 system is

$$\boldsymbol{\sigma} = \begin{bmatrix} 10 & 5 & -6 \\ 5 & 20 & 15 \\ -6 & 15 & 8 \end{bmatrix}$$

Determine the components in the rotated systems shown in Fig. 2.9a and b.

Problem 2.5 Determine if the vector outer product

$$B_{ij} = u_i v_j$$

is a tensor or not. u_i and v_i are components of vectors **u** and **v**, respectively.

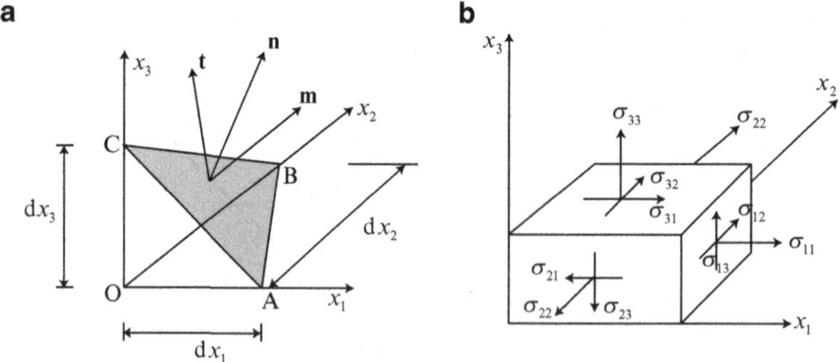

Fig. 2.10 Stresses in three dimensions

Problem 2.6 Assuming that x_1 and x_2 transform as components of a vector, determine which of the following matrices are tensors (Butkov 1968)

$$\mathbf{C} = \begin{bmatrix} x_2^2 & x_1x_2 \\ x_1x_2 & x_1^2 \end{bmatrix}; \quad \mathbf{D} = \begin{bmatrix} -x_1x_2 & x_1^2 \\ -x_2^2 & x_1x_2 \end{bmatrix}$$

Problem 2.7 Verify that the expressions given by (2.29a) and (2.29b) are identical.

Problem 2.8 Prove that $T_{ij}u_iu_j = D_{ij}u_iu_j$, where D_{ij} is the symmetric part of T_{ij}

Problem 2.9 Using the result of (2.33), show

- $\varepsilon_{ijk}\varepsilon_{ist} = \delta_{js}\delta_{kt} - \delta_{jt}\delta_{ks}$
- $\varepsilon_{ijk}\varepsilon_{ijt} = 2\delta_{kt}$
- $\varepsilon_{ijk}\varepsilon_{ijk} = 6$

Problem 2.10 Prove that $(P_{ijk} + P_{jki} + P_{jik})x_ix_jx_k = 3P_{ijk}x_ix_jx_k$ (from Mase 1970)

Problem 2.11 Prove that the following quantities derived from a stress tensor σ_{ij} are invariants under rotations of the coordinate system:

(a) $I = \sigma_{ii}$
(b) $J = \left(\frac{1}{2}s_{ij}s_{ij}\right)^{1/2}$
(c) $\varepsilon_{ijk}\varepsilon_{kjp}T_{ip}$ (from Mase 1970)

where $s_{ij} = \sigma_{ij} - (1/3)I\delta_{ij}$.

Problem 2.12 An yield surface is given by the following function

$$\phi(I, J, \alpha, \zeta) = 0$$

where

$$I = \sigma_{k\ell}\delta_{k\ell}$$

$$s_{ij} = \sigma_{ij} - \frac{1}{3}I\delta_{ij}; \quad J = \left(\frac{1}{2}s_{k\ell}s_{k\ell}\right)^{1/2}$$

$$S = \left(\frac{1}{3}s_{ij}s_{jk}s_{ki}\right)^{1/3}$$

$$-\frac{\pi}{6} \le \alpha = \frac{1}{3}\sin^{-1}\left[\frac{3\sqrt{3}}{2}\left(\frac{S}{J}\right)^3\right] \le \frac{\pi}{6}$$

Show that the gradient of the function (normal to the surface) in the stress space is given by

$$\frac{\partial \phi}{\partial \sigma_{ij}} = \frac{\partial \phi}{\partial I}\delta_{ij} + \frac{\partial \phi}{\partial J}\frac{s_{ij}}{2J} + \frac{\partial \phi}{\partial \alpha}\frac{\sqrt{3}}{2J\cos 3\alpha}\left[\frac{s_{ir}s_{rj}}{J^2} - \frac{2}{3}\delta_{ij} - \frac{3}{2}\left(\frac{S}{J}\right)^3\frac{s_{ij}}{J}\right]$$

Problem 2.13 For the function $\lambda = A_{ij}x_ix_j$ where A_{ij} is a constant, show that

- $\frac{\partial \lambda}{\partial x_i} = (A_{ij} + A_{ji})x_j$
- $\frac{\partial^2 \lambda}{\partial x_i \partial x_j} = A_{ij} + A_{ji}$

Simplify these derivatives for the case $A_{ij} = A_{ij}$ (from Mase 1970).

Problem 2.14 Show that the eigenvalues of the tensor

$$\boldsymbol{\sigma} = \begin{bmatrix} 30 & -10 & 0 \\ -10 & 30 & 0 \\ 0 & 0 & 10 \end{bmatrix}$$

are 10, 20, and 40. Also show that the eigen-vectors corresponding to these eigenvalues are

$$\mathbf{n}_1 = \{0, 0, \pm 1\},\ \mathbf{n}_2 = \left\{\pm 1/\sqrt{2}, \pm 1/\sqrt{2}, 0\right\} \text{ and}$$
$$\mathbf{n}_3 = \left\{\mp 1/\sqrt{2}, \pm 1/\sqrt{2}, 0\right\} \text{ respectively}$$

Problem 2.15 Show that the principal stresses are real.

Problem 2.16 Show that for a symmetric second order tensor, when $\lambda_i \neq \lambda_j$, the eigen-vectors are orthogonal to each other.

Problem 2.17 Form a matrix **T** by placing the eigenvectors in the columns of **T** as:

$$\mathbf{T} = \{\mathbf{n}_1 \quad \mathbf{n}_2 \quad \mathbf{n}_3\}$$

Then show:

$$\mathbf{T}^T \boldsymbol{\sigma}\, \mathbf{T} = \begin{bmatrix} \lambda_1 & 0 & 0 \\ 0 & \lambda_2 & 0 \\ 0 & 0 & \lambda_3 \end{bmatrix}$$

Remark: It is clear from the above equation, that **T** is the transformation matrix that transforms the stress tensor into the principal stress tensor.

Problem 2.18 By following the procedure presented in Sect. 2.5.4, show

$$\sigma_2 = -r_0 \cos(\theta - 60^\circ) + p$$

$$\sigma_3 = -r_0 \cos(\theta + 60^\circ) + p$$

where

$$r_0 = \frac{2}{\sqrt{3}} J$$

Chapter 3
Governing Equations in Solid Mechanics

The first step in finite element analysis of a given problem is the development of an appropriate model of the problem. As described in Chap. 1, the development of a model involves making approximations of the physics, geometry, boundary conditions and material behavior. Once a model is developed, (a) the mathematical equations governing the physics are developed, (b) the boundary value problem is formally defined, and (c) an appropriate technique is used to solve the problem. In this chapter, we focus on the derivation of the governing equations and the definition of the boundary/initial value problems for linear, single-physics, solid mechanics (load-deformation) problems.

3.1 Coordinate Systems

While the rectangular Cartesian coordinate system (Fig. 3.1a) is convenient to use in some problems, the polar coordinates (Fig. 3.1b, c) are convenient to use in the others. For instance, the geometry, loading and boundary conditions involved in a problem may be such that the deformation of the domain may be axisymmetric. In such a case, it is theoretically and computationally efficient to express the problem in cylindrical coordinates (Fig. 3.1b). Similarly, spherical coordinates (Fig. 3.1c) may be convenient in some cases. A point in space is represented by (x, y, z) in the Cartesian system, (r, θ, z) in the cylindrical system, and (r, θ, ϕ) in the spherical system.

In this chapter, *tensile stresses and strains are taken positive*. The notation for an ordinary derivative of a quantity α, say, with respect to x is $d\alpha/dx$. The corresponding partial derivative is $\partial\alpha/\partial x$. Another compact notation for the derivative is $\alpha_{,x}$. Thus, $\alpha_{,x}$, $\alpha_{,y}$, and $\alpha_{,z}$ are derivatives of α with respect to x, y, and z, respectively. $\alpha_{,1}$, $\alpha_{,2}$, and $\alpha_{,3}$ are derivatives of α with respect to x_1, x_2, and x_3, respectively.

A. Anandarajah, *Computational Methods in Elasticity and Plasticity: Solids and Porous Media*, DOI 10.1007/978-1-4419-6379-6_3,

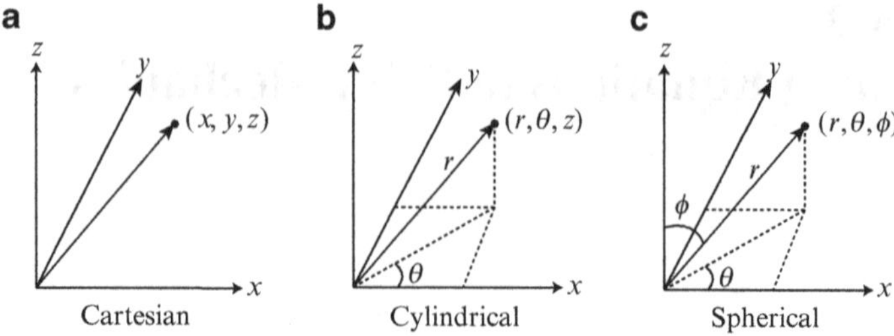

Fig. 3.1 Commonly used coordinate systems

3.2 Governing Equations for One-Dimensional Rod

Consider an isotropic, homogeneous, linear elastic, one-dimensional rod shown in Fig. 3.2a. We assume that all stress components are zero except σ_{xx}, which we simply refer to as σ. Generally, the strains in the directions perpendicular to the x-axis (ε_{yy} and ε_{zz}) are nonzero. For an elastic material (Chap. 4), once ε_{xx} (which we would simply refer to as ε) is determined, ε_{yy} and ε_{zz} can be calculated with the knowledge of Poisson's ratio.

The rod is held tightly ("fixed") at point A, and subjected to a tensile force $P_0(t)$ at point B and a distributed surface force $f_0(x,t)$ per unit length. The finer details of the support at point A and the details of the load application at B would make this problem two dimensional. However, we invoke the Saint Vernant principle and assume that, except very near points A and B, the load-deformation behavior of the rod can be assumed to be one dimensional. Derivation of the governing equations is achieved using Newton's second law.

While for emphasis the functional forms of f_0 (i.e., $f_0(x,t)$) and P_0 (i.e., $P_0(t)$) are indicated, it must be noted that other variables of the problem (i.e., u, ε, and σ) are also functions of the spatial coordinate x and time t (i.e., $u(x,t)$, $\varepsilon(x,t)$, and $\sigma(x,t)$). For simplicity of notations, the functional form will not always be shown in the equations below.

Let u be the displacement of the point at x (point C). Consider the free-body diagram associated with an element of length dx, shown in Fig. 3.2b, where dm is the mass of the element and $\ddot{u}$ is the (absolute) acceleration of the point at x. Let ρ be the mass density of the material. The equation governing the equilibrium of the rod for the domain defined by $0 < x < L$ (i.e., $x \in [0, L]$) is derived by summing up the forces in the x-direction and setting the sum to mass times the acceleration

$$(\sigma + \mathrm{d}\sigma)a - \sigma a + f_0 \mathrm{d}x = (\mathrm{d}m)\ddot{x} = a(\mathrm{d}x)\rho\ddot{x} \tag{3.1}$$

This leads to the following differential equation governing the equilibrium

$$\frac{\mathrm{d}\sigma}{\mathrm{d}x} + \frac{f_0}{a} = \rho\ddot{x} \quad \text{for} \quad x \in [0, L] \tag{3.2}$$

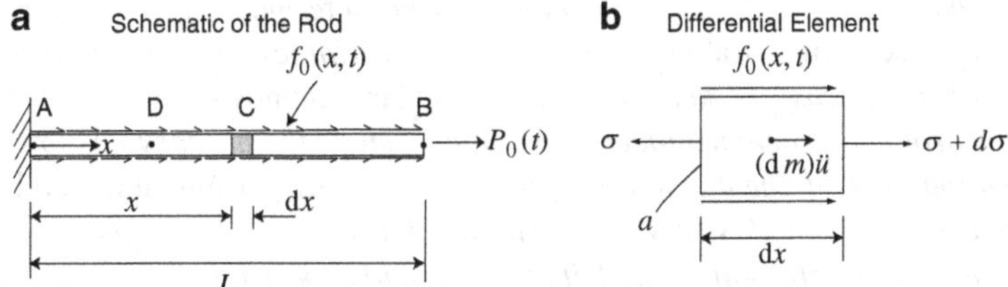

Fig. 3.2 One-dimensional rod

or

$$\sigma_{,x} + \frac{f_0}{a} = \rho \ddot{x} \quad \text{for} \quad x \in [0, L] \tag{3.3}$$

The boundary conditions of the problem are

$$u = 0 \quad \text{at} \quad x = 0 \quad \text{and} \tag{3.4a}$$

$$\sigma = \frac{P_0}{a} \quad \text{at} \quad x = L \tag{3.4b}$$

The condition given by (3.4a) is a displacement boundary condition, whereas the condition given by (3.4b) is a load boundary condition.

For completeness, let us present the constitutive relation as well. The relevant constitutive equation is (Chap. 4)

$$\sigma = E\varepsilon \tag{3.5a}$$

where ε is the strain and E is Young's modulus. ε is related to the displacement by the strain-displacement relation

$$\varepsilon = \frac{\mathrm{d}u}{\mathrm{d}X} \tag{3.5b}$$

Typically, answers are sought for the following questions:

- Determine the displacement, velocity, and acceleration at a certain point such as point D
- Determine the stress at a point such as point D
- Determine the reaction at the support at point A

Definition 3.1. *In the one-dimensional problem, there are three unknown variables: (u, ε, σ). Among these, there is only one independent variable, and the remaining two can be derived from this independent variable. For instance, if u is taken as the independent variable, (3.5b) is used to find ε and (3.5a) is used to find σ. The* *variables that are chosen as the unknowns to be determined first are known as the* primary unknown variables. *The dependent variables that are derived from the primary unknowns are known as the* secondary unknown variables. *In solid mechanics problem, when displacements are chosen as the primary unknowns in a given solution method, the method is referred to as the* displacement method. *Thus, in a displacement method, stress and strain are the secondary unknowns.*

While in the finite element method, it is common to choose the displacement as the primary unknown variable type, mixed methods in which both displacement and stress are used as primary unknown variable types have also been developed. In the mixed formulations, in general, by suitable manipulation of the problem, the number of unknown variable types to be solved for can be reduced. During this manipulation, a point is reached where the number of unknown variable types can not be further reduced. The corresponding formulation is known as the irreducible formulation. *The formulation that is not irreducible is called the* mixed formulation.

The boundary value problem is formally defined as:

(a) The governing equation:

$$\sigma_{,x} + \frac{f_0}{a} = \rho\ddot{x} \quad \text{for} \quad x \in [0, L] \tag{3.6a}$$

(b) The boundary conditions:

$$u = 0 \quad \text{at} \quad x = 0 \text{ at any time } t \text{ and} \tag{3.6b}$$

$$\sigma = \frac{P_0}{a} \quad \text{at} \quad x = L \quad \text{at any time} \quad t \tag{3.6c}$$

(c) The initial conditions:

$$u = 0 \quad \text{at} \quad t = 0 \quad \text{at any time } x \tag{3.6d}$$

(d) The constitutive law:

$$\sigma = E\varepsilon \tag{3.6e}$$

(e) The strain-displacement relation:

$$\varepsilon = \frac{\mathrm{d}u}{\mathrm{d}X} \tag{3.6f}$$

Definition 3.2. *When the problem involves both the boundary and initial conditions, the problem is known as a* mixed boundary/initial value problem.

3.3 Governing Equations for Two- and Three-Dimensional Solids in the Cartesian Coordinate System

Notations for the stresses: The first subscript denotes the axis that is normal to the plane, and the second subscript denotes the direction of the stress. Hence, σ_{xx} is the stress in the x-direction on the plane whose normal is x-axis. σ_{xy} is the stress in the y-direction on the plane whose normal is x-axis.

Let us start with the two-dimensional problem. Referring to Fig. 3.3, the stresses on an element of area d$x$dy are shown. Notice that the normal stress on the x-plane (plane whose normal is the x-axis) at x is σ_{xx}. This becomes

$$\sigma_{xx} + \mathrm{d}\sigma_{xx} = \sigma_{xx} + \frac{\partial \sigma_{xx}}{\partial x}\mathrm{d}x \quad \text{at} \quad x + \mathrm{d}x$$

The same procedure applies to the other components. The body is subjected to a body force of components b_x and b_y in the x- and y-directions, respectively. The inertial forces in the x- and y-directions are $(\mathrm{d}m)\ddot{u}$ and $(\mathrm{d}m)\ddot{v}$, respectively.

First, we sum up the moments (in the clockwise direction) about the center of the element and set it to zero:

$$\sigma_{yx}\mathrm{d}x\frac{\mathrm{d}y}{2} + \left(\sigma_{yx} + \frac{\partial \sigma_{yx}}{\partial y}\mathrm{d}y\right)\mathrm{d}x\frac{\mathrm{d}y}{2} - \sigma_{xy}\mathrm{d}y\frac{\mathrm{d}x}{2} - \left(\sigma_{xy} + \frac{\partial \sigma_{xy}}{\partial x}\mathrm{d}x\right)\mathrm{d}y\frac{\mathrm{d}x}{2} = 0 \quad (3.7)$$

After neglecting O(dy^3) terms in relation to the O(dy^2) terms and simplifying, we obtain

$$\sigma_{xy} = \sigma_{yx} \quad (3.8)$$

σ_{xy} and σ_{yx} are called the complementary shear stresses. Equation (3.8) proves that the stress tensor is symmetric.

Denote the mass density of the material by ρ. Now, summing up the forces in the x- and y-directions and simplifying each equation as above, we arrive at two partial differential equations governing the equilibrium of the body:

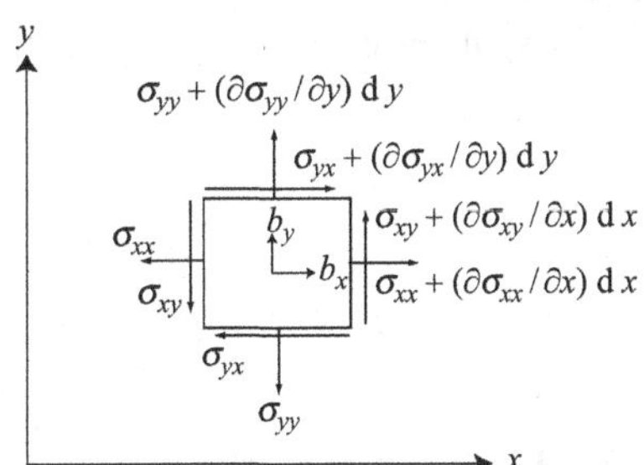

Fig. 3.3 Stresses acting on a two-dimensional element at a point

$$\frac{\partial \sigma_{xx}}{\partial x} + \frac{\partial \sigma_{yx}}{\partial y} + b_x = \rho \ddot{u} \quad (\text{the equation in the } x\text{-}direction) \tag{3.9a}$$

$$\frac{\partial \sigma_{xy}}{\partial x} + \frac{\partial \sigma_{yy}}{\partial y} + b_y = \rho \ddot{v} \quad (\text{the equation in the } y\text{-}direction) \tag{3.9b}$$

We define the boundary conditions in a general manner as

$$\mathbf{u} = \hat{\mathbf{u}}(t) \quad \text{on} \quad \Gamma_u \tag{3.9c}$$

$$\mathbf{t} = \hat{\mathbf{t}}(t) \quad \text{on} \quad \Gamma_t \tag{3.9d}$$

where $\mathbf{u} = (u, v)$ is the displacement vector and $\hat{\mathbf{u}}$ is its prescribed value over the boundary Γ_u. $\mathbf{t} = (t_x, t_y)$ is the boundary traction and $\hat{\mathbf{t}}$ is its prescribed value over the boundary Γ_t.

Similarly, initial conditions are specified as

$$\mathbf{u} = \mathbf{u}_0(x) \text{ on } \Gamma_u \tag{3.9e}$$

$$\dot{\mathbf{u}} = \dot{\mathbf{u}}_0(x) \text{ on } \Gamma_u \tag{3.9f}$$

It was shown in Chap. 2 that the traction $\mathbf{t}$ on a given plane (Fig. 3.4) is related to the stress tensor $\boldsymbol{\sigma}$ by the following equation:

$$\mathbf{t} = \boldsymbol{\sigma}\mathbf{n} \tag{3.10}$$

where $\mathbf{n}$ is the unit normal to the plane of interest, as shown in Fig. 3.4.

For small strains, the strain–displacement relations are

$$\varepsilon_{xx} = \frac{\partial u}{\partial x}; \quad \varepsilon_{yy} = \frac{\partial v}{\partial y}; \quad \varepsilon_{zz} = \frac{\partial w}{\partial z}; \quad \varepsilon_{xy} = \frac{1}{2}\left(\frac{\partial v}{\partial x} + \frac{\partial u}{\partial y}\right) \tag{3.11}$$

where (u, v, w) are the components of the displacements in the x-, y-, and z-directions.

Three types of two-dimensional idealizations are generally used:

1. Plane stress
2. Plane strain
3. Axisymmetric

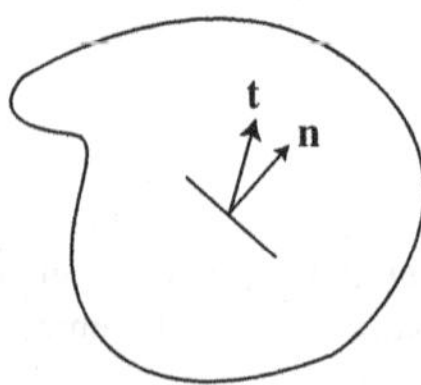

Fig. 3.4. Traction on a plane

We will return to axisymmetric idealization later after presenting the axisymmetric equilibrium equations. The plane stress and plane strain idealizations are discussed below.

While the out-of-plane normal stress (σ_{zz} in the present case) is assumed to remain constant in the plane stress problem, the out-of-plane normal strain (ε_{zz} in the present case) is assumed to remain constant in the plane strain problem. Thus, in the plane stress problem, ε_{zz} can vary during the loading, whereas in the plane strain problem, σ_{zz} can vary. However, given one, the other can be calculated using the constitutive equations. $\sigma_{zz} = 0$ is a special case of the plane stress problem, and $\varepsilon_{zz} = 0$ is a special case of the plane strain problem.

In both the plane stress and plane strain problems, the only components that must be treated as varying unknowns are $(\sigma_{xx}, \sigma_{yy}, \sigma_{xy})$ and $(\varepsilon_{xx}, \varepsilon_{yy}, \gamma_{xy})$, since σ_{zz} or ε_{zz} can be calculated from these unknowns and the constitutive relations. Placing these in vectors as: $\boldsymbol{\sigma} = (\sigma_{xx}, \sigma_{yy}, \sigma_{xy})$ and $\boldsymbol{\varepsilon} = (\varepsilon_{xx}, \varepsilon_{yy}, \gamma_{xy})$, the relationship between $\boldsymbol{\sigma}$ and $\boldsymbol{\varepsilon}$ is needed to complete the definition of the two-dimensional boundary value problem.

In general, the stress tensor is related to the strain tensor through a fourth order tensor as

$$\sigma_{ij} = D_{ijk\ell}\varepsilon_{k\ell}$$

where $D_{ijk\ell}$ is the fourth order stiffness modular tensor. When the material is elastic, $D_{ijk\ell}$ is equal to the elastic stiffness tensor. Various forms of elastic relations (isotropic and anisotropic) are presented in Chap. 4. When the stress-strain behavior is elasto-plastic, the above relation is expressed as a rate relation (i.e., $\dot{\boldsymbol{\sigma}}$ versus $\dot{\boldsymbol{\varepsilon}}$). $D_{ijk\ell}$ is the combined elasto-plastic stiffness tensor. A suitable elasto-plastic model is needed to define $D_{ijk\ell}$; this is discussed in Chaps. 9–14.

The symmetry of stress and strain tensors allows the stress-strain relation to be written in a matrix-vector form. In the rest of the chapter, we will limit our discussions to elastic stress-strain relations.

For a three-dimensional solid

$$\boldsymbol{\sigma} = \mathbf{C}\boldsymbol{\varepsilon} \tag{3.12}$$

$$\begin{Bmatrix} \sigma_{xx} \\ \sigma_{yy} \\ \sigma_{zz} \\ \sigma_{xy} \\ \sigma_{yz} \\ \sigma_{xz} \end{Bmatrix} = \begin{bmatrix} C_{11} & C_{12} & C_{13} & C_{14} & C_{15} & C_{16} \\ C_{21} & C_{22} & C_{23} & C_{24} & C_{25} & C_{26} \\ C_{31} & C_{32} & C_{33} & C_{34} & C_{35} & C_{36} \\ C_{41} & C_{42} & C_{43} & C_{44} & C_{45} & C_{46} \\ C_{51} & C_{52} & C_{53} & C_{54} & C_{55} & C_{56} \\ C_{61} & C_{62} & C_{63} & C_{64} & C_{65} & C_{66} \end{bmatrix} \begin{Bmatrix} \varepsilon_{xx} \\ \varepsilon_{yy} \\ \varepsilon_{zz} \\ \gamma_{xy} \\ \gamma_{yz} \\ \gamma_{xz} \end{Bmatrix} \tag{3.13}$$

The above 6×6 three-dimensional stress-strain relation simplifies to a 4×4 relation for plane stress and plane strain problems (Problem 3.1). The relationship is in the form

$$\begin{Bmatrix} \sigma_{xx} \\ \sigma_{yy} \\ \sigma_{zz} \\ \sigma_{xy} \end{Bmatrix} = \begin{bmatrix} C_{11} & C_{12} & C_{13} & C_{14} \\ C_{21} & C_{22} & C_{23} & C_{24} \\ C_{31} & C_{32} & C_{33} & C_{34} \\ C_{41} & C_{42} & C_{43} & C_{44} \end{bmatrix} \begin{Bmatrix} \varepsilon_{xx} \\ \varepsilon_{yy} \\ \varepsilon_{zz} \\ \gamma_{xy} \end{Bmatrix} \tag{3.14}$$

In a plane stress analysis

$$\sigma_{zz} = \sigma_{z0}\,(\text{constant}) \tag{3.15}$$

In a plane strain analysis

$$\varepsilon_{zz} = \varepsilon_{z0}\,(\text{constant}) \tag{3.16}$$

Hence, as pointed out earlier, the only components that vary during the loading are $\boldsymbol{\sigma} = (\sigma_{xx}, \sigma_{yy}, \sigma_{xy})$ and $\boldsymbol{\varepsilon} = (\varepsilon_{xx}, \varepsilon_{yy}, \gamma_{xy})$. Equation (3.14) is modified to obtain the following 3×3 matrix required for the analysis

$$\begin{Bmatrix} \sigma_{xx} \\ \sigma_{yy} \\ \sigma_{xy} \end{Bmatrix} = \begin{bmatrix} C^*_{11} & C^*_{12} & C^*_{13} \\ C^*_{21} & C^*_{22} & C^*_{23} \\ C^*_{31} & C^*_{32} & C^*_{33} \end{bmatrix} \begin{Bmatrix} \varepsilon_{xx} \\ \varepsilon_{yy} \\ \gamma_{xy} \end{Bmatrix} + \begin{Bmatrix} \sigma_{x0} \\ \sigma_{y0} \\ \sigma_{xy0} \end{Bmatrix} \tag{3.17}$$

For example, in the *plane stress* analysis, combining the third equation of (3.14) with (3.15), we have

$$\varepsilon_{zz} = \frac{1}{C_{33}}(\sigma_{z0} - C_{31}\varepsilon_{xx} - C_{32}\varepsilon_{yy} - C_{34}\gamma_{xy}) \tag{3.18}$$

which leads to $C^*_{11} = C_{11} - (C_{13}C_{31}/C_{33})$, etc., and $\sigma_{x0} = (C_{13}\sigma_{z0}/C_{33})$, etc.

Similarly, for the *plane strain* analysis, combining the third equation of (3.14) with (3.16), we have

$$\sigma_{zz} = C_{31}\varepsilon_{xx} + C_{32}\varepsilon_{yy} + C_{33}\varepsilon_{z0} + C_{34}\gamma_{xy} \tag{3.19}$$

which leads to $C^*_{11} = C_{11}$, etc., and $\sigma_{x0} = C_{13}\varepsilon_{z0}$, etc. (In Problem 3.1, you are asked to derive the expressions for these coefficients.)

The elements of the boundary value problem (governing equations, boundary conditions, constitutive laws and the strain-displacement equations) are now available for the two-dimensional case. Extension of the process for the three-dimensional case is straightforward. Only the final equations are presented here; the reader is referred to standard texts on solid mechanics for the details of derivations (see, for example, Boresi et al. 1993).

The three equations of equilibrium in the Cartesian coordinate system are

$$\frac{\partial \sigma_{xx}}{\partial x}+\frac{\partial \sigma_{yx}}{\partial y}+\frac{\partial \sigma_{zx}}{\partial z}+b_x=\rho\ddot{u} \tag{3.20a}$$

$$\frac{\partial \sigma_{xy}}{\partial x}+\frac{\partial \sigma_{yy}}{\partial y}+\frac{\partial \sigma_{zy}}{\partial z}+b_y=\rho\ddot{v} \tag{3.20b}$$

$$\frac{\partial \sigma_{xz}}{\partial x}+\frac{\partial \sigma_{yz}}{\partial y}+\frac{\partial \sigma_{zz}}{\partial z}+b_z=\rho\ddot{w} \tag{3.20c}$$

The boundary conditions are

$$\mathbf{u}=\hat{\mathbf{u}}(t) \quad \text{on} \quad \Gamma_u \tag{3.21a}$$

$$\mathbf{t}=\hat{\mathbf{t}}(t) \quad \text{on} \quad \Gamma_t \tag{3.21b}$$

The initial conditions are

$$\mathbf{u}=\mathbf{u}_0(x) \quad \text{on} \quad \Gamma_u \tag{3.21c}$$

$$\dot{\mathbf{u}}=\dot{\mathbf{u}}_0(x) \quad \text{on} \quad \Gamma_u \tag{3.21d}$$

The constitutive relation is given by (3.13).
With the assumption of small strains, the strain-displacement relations are

$$\begin{aligned} &\varepsilon_{xx}=\frac{\partial u}{\partial x}, \quad \varepsilon_{yy}=\frac{\partial v}{\partial y}, \quad \varepsilon_{zz}=\frac{\partial w}{\partial z}, \quad \varepsilon_{xy}=\frac{1}{2}\left(\frac{\partial v}{\partial x}+\frac{\partial u}{\partial y}\right), \\ &\varepsilon_{xz}=\frac{1}{2}\left(\frac{\partial w}{\partial x}+\frac{\partial u}{\partial z}\right), \quad \varepsilon_{yz}=\frac{1}{2}\left(\frac{\partial w}{\partial y}+\frac{\partial v}{\partial z}\right) \end{aligned} \tag{3.22}$$

3.4 Formal Definition of Boundary/Initial Value Problems in the Cartesian Coordinate System

It is seen that the definition of one-, two-, and three-dimensional boundary value problems involve the same elements (equilibrium equations, boundary conditions, etc.), permitting the generalization of the description. For the solid mechanics problem, referring to Fig. 3.5, and employing compact tensorial notations (and noting that due to symmetry of the stress tensor, $\sigma_{ji,j}=\sigma_{ij,j}$), the boundary/initial value problem definition involves the following:

The equilibrium equation is

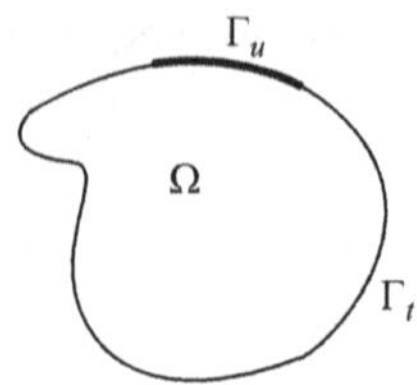

Fig. 3.5 Domain definition

$$\sigma_{ij,j} + b_i = \rho \ddot{u}_i \quad \text{in} \quad \Omega \tag{3.23a}$$

The boundary conditions are

$$\mathbf{u} = \hat{\mathbf{u}}(t) \quad \text{on} \quad \Gamma_u \tag{3.23b}$$

$$\mathbf{t} = \hat{\mathbf{t}}(t) \quad \text{on} \quad \Gamma_t \tag{3.23c}$$

The initial conditions are

$$\mathbf{u} = \mathbf{u}_0(x) \quad \text{on} \quad \Gamma_u \tag{3.23d}$$

$$\dot{\mathbf{u}} = \dot{\mathbf{u}}_0(x) \quad \text{on} \quad \Gamma_u \tag{3.23e}$$

The constitutive equation is

$$\boldsymbol{\sigma} = \mathbf{D}\boldsymbol{\varepsilon} \tag{3.23f}$$

Assuming that the strains are small, the strain-displacement relation is

$$\varepsilon_{ij} = \frac{1}{2}(u_{i,j} + u_{j,i}) \tag{3.23g}$$

3.5 Three-Dimensional Equations in the Cylindrical Coordinate System

For completeness, we present the relevant equations in the cylindrical coordinate system here; the reader is referred to any book on advanced solid mechanics for the derivations.

Noting that the stress tensor is symmetric (which allows replacing $\sigma_{\theta r}$ with $\sigma_{r\theta}$, etc.), the equilibrium equations are

$$\frac{\partial \sigma_{rr}}{\partial r} + \frac{1}{r}\frac{\partial \sigma_{r\theta}}{\partial \theta} + \frac{\partial \sigma_{rz}}{\partial z} + \frac{\sigma_{rr} - \sigma_{\theta\theta}}{r} + b_r = \rho \ddot{u} \tag{3.24a}$$

$$\frac{\partial \sigma_{\theta r}}{\partial r} + \frac{1}{r}\frac{\partial \sigma_{\theta\theta}}{\partial \theta} + \frac{\partial \sigma_{\theta z}}{\partial z} + \frac{2\sigma_{\theta r}}{r} + b_\theta = \rho \ddot{v} \tag{3.24b}$$

$$\frac{\partial \sigma_{zr}}{\partial r} + \frac{1}{r}\frac{\partial \sigma_{z\theta}}{\partial \theta} + \frac{\partial \sigma_{zz}}{\partial z} + \frac{\sigma_{zr}}{r} + b_z = \rho \ddot{w} \tag{3.24c}$$

The boundary and initial conditions are given by (3.23b)–(3.23e). The constitutive relation is given by (3.23f) or (3.13). The strain-displacement relations are

$$\begin{aligned} &\varepsilon_{rr} = \frac{\partial u}{\partial r}, \quad \varepsilon_{\theta\theta} = \frac{u}{r} + \frac{1}{r}\frac{\partial v}{\partial \theta}, \quad \varepsilon_{zz} = \frac{\partial w}{\partial z}, \quad \varepsilon_{r\theta} = \frac{1}{2}\left(\frac{1}{r}\frac{\partial u}{\partial \theta} + \frac{\partial v}{\partial r} - \frac{v}{r}\right), \\ &\varepsilon_{rz} = \frac{1}{2}\left(\frac{\partial w}{\partial r} + \frac{\partial u}{\partial z}\right), \quad \varepsilon_{\theta z} = \frac{1}{2}\left(\frac{1}{r}\frac{\partial w}{\partial \theta} + \frac{\partial v}{\partial z}\right) \end{aligned} \tag{3.25}$$

It can be shown (Problem 3.2) that for the axisymmetric problem, the three-dimensional equilibrium equations in (3.24a)–(3.24c) simplify to

$$\frac{\partial \sigma_{rr}}{\partial r} + \frac{\partial \sigma_{rz}}{\partial z} + \frac{\sigma_{rr} - \sigma_{\theta\theta}}{r} + b_r = \rho \ddot{u} \tag{3.26}$$

$$\frac{\partial \sigma_{zr}}{\partial r} + \frac{\partial \sigma_{zz}}{\partial z} + \frac{\sigma_{zr}}{r} + b_z = \rho \ddot{v} \tag{3.27}$$

The relevant constitutive relations are

$$\begin{Bmatrix} \sigma_{rr} \\ \sigma_{zz} \\ \sigma_{\theta\theta} \\ \sigma_{rz} \end{Bmatrix} = \begin{bmatrix} C_{11} & C_{12} & C_{13} & C_{14} \\ C_{21} & C_{22} & C_{23} & C_{24} \\ C_{31} & C_{32} & C_{33} & C_{34} \\ C_{41} & C_{42} & C_{43} & C_{44} \end{bmatrix} \begin{Bmatrix} \varepsilon_{rr} \\ \varepsilon_{zz} \\ \varepsilon_{\theta\theta} \\ \gamma_{rz} \end{Bmatrix} \tag{3.28}$$

where the elements C_{11}, etc., are identical to those defined in (3.14).

3.6 Three-Dimensional Equations in the Spherical Coordinate System

Noting that the stress tensor is symmetric (which allows replacing $\sigma_{\theta r}$ with $\sigma_{r\theta}$, etc.), the equilibriums equations are

$$\frac{\partial \sigma_{rr}}{\partial r} + \frac{1}{r}\frac{\partial \sigma_{r\theta}}{\partial \theta} + \frac{1}{r\sin\theta}\frac{\partial \sigma_{r\phi}}{\partial \phi} + \frac{1}{r}\left[2\sigma_{rr} - \sigma_{\theta\theta} - \sigma_{\phi\phi} + \sigma_{r\theta}\cot\theta\right] + b_r = \rho \ddot{u} \tag{3.29a}$$

$$\frac{\partial \sigma_{\theta r}}{\partial r} + \frac{1}{r}\frac{\partial \sigma_{\theta\theta}}{\partial \theta} + \frac{1}{r\sin\theta}\frac{\partial \sigma_{\theta\phi}}{\partial \phi} + \frac{1}{r}\left[(\sigma_{\theta\theta} - \sigma_{\phi\phi})\cot\theta + 3\sigma_{\theta r}\right] + b_\theta = \rho\ddot{v} \quad (3.29b)$$

$$\frac{\partial \sigma_{\phi r}}{\partial r} + \frac{1}{r}\frac{\partial \sigma_{\phi\theta}}{\partial \theta} + \frac{1}{r\sin\theta}\frac{\partial \sigma_{\phi\phi}}{\partial \phi} + \frac{1}{r}\left[3\sigma_{\phi r} + 2\sigma_{\phi\theta}\cot\theta\right] + b_\phi = \rho\ddot{w} \quad (3.29c)$$

The boundary and initial conditions are given by (3.23b)–(3.23e). The constitutive relation is given by (3.23f) or (3.13). The strain-displacement relations are

$$\varepsilon_{rr} = \frac{\partial u}{\partial r}, \quad \varepsilon_{\theta\theta} = \frac{u}{r} + \frac{1}{r}\frac{\partial v}{\partial \theta}, \quad \varepsilon_{\phi\phi} = \frac{u}{r} + \frac{v}{r}\cot\theta + \frac{1}{r\sin\theta}\frac{\partial w}{\partial \phi} \quad (3.30)$$

$$\varepsilon_{r\theta} = \frac{1}{2}\left(\frac{1}{r}\frac{\partial u}{\partial \theta} + \frac{\partial v}{\partial r} - \frac{v}{r}\right), \quad \varepsilon_{r\phi} = \frac{1}{2}\left(\frac{\partial w}{\partial r} + \frac{1}{r\sin\theta}\frac{\partial u}{\partial \phi} - \frac{w}{r}\right),$$
$$\varepsilon_{\theta\phi} = \frac{1}{2}\left(\frac{1}{r}\frac{\partial w}{\partial \theta} + \frac{1}{r\sin\theta}\frac{\partial v}{\partial \phi} - \frac{w}{r}\cot\theta\right) \quad (3.31)$$

3.7 Flexural Behavior of Beams

Certain problems may be efficiently solved by combining solid finite elements and bending elements. Considering a linear elastic beam (Fig. 3.6), the governing equations are summarized below.

For simplicity, consider a static problem. In the *Euler-Bernoulli beam theory*, the plane sections normal to the beam axis are assumed to remain planes during the deformation. With this assumption, denoting the transverse displacement of the beam by $w(x)$, the governing differential equation is given by

$$\frac{\mathrm{d}^2}{\mathrm{d}x^2}\left(EI\frac{\mathrm{d}^2 w}{\mathrm{d}x^2}\right) = \mathrm{q} \text{ in } 0 < x < \ell \quad (3.32)$$

Boundary conditions (specific to the problem shown in Fig. 3.6) are

$$w(x = 0) = 0 \,(\text{deflection}) \quad (3.33a)$$

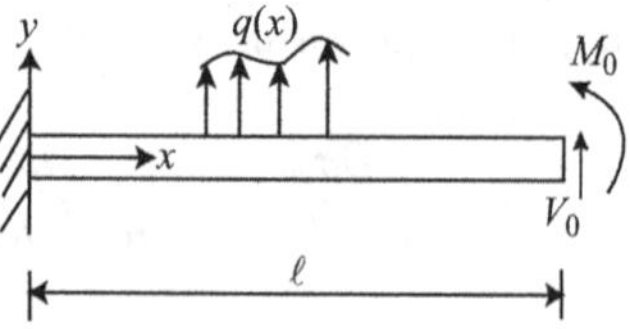

Fig. 3.6 Flexural behavior of a beam

$$\theta(x=0) = \frac{\mathrm{d}w}{\mathrm{d}x}(x=0) = 0\,(\text{slope}) \tag{3.33b}$$

$$M(x=\ell) = EI\frac{\mathrm{d}^2y}{\mathrm{d}x^2}(x=\ell) = M_0\,(\text{bending moment}) \tag{3.33c}$$

$$V(x=\ell) = -\frac{\mathrm{d}}{\mathrm{d}x}\left(EI\frac{\mathrm{d}^2y}{\mathrm{d}x^2}\right)(x=\ell) = V_0\,(\text{shear force}) \tag{3.33d}$$

The number of each of the four types of boundary conditions stated above (i.e., deflection, slope, bending moment and shear force) could vary from problem to problem.

The variables of the problem are (w,θ,M,V). Depending on the specific formulation used in the solution method, one or more of (w,θ,M,V) may be treated as the primary dependent variable(s) and the remaining ones become the secondary variables.

For equations associated with plates and/or shells, the reader is encouraged to consult a finite element book (e.g., Zienkiewicz and Taylor 1991).

Problems

Problem 3.1 Derive the relations between the coefficients of the matrix in (3.17) (i.e., C_{11}^*, etc.) and the coefficients of the elasticity stiffness matrix in (3.14) (C_{11}, etc.) for the cases of

(a) Plane stress
(b) Plane strain
(c) Axisymmetry

Problem 3.2 Derive the two-dimensional equilibrium equations in cylindrical coordinate system given by (3.26) and (3.27) from (3.24a)–(3.24c).

Problem 3.3 Deformation of a thin plate is given by $u = 0.0025xy$, $v = 0.0015xy$, and $w = 0$. Determine the components of the strain at $(x, y) = (0.5,1)$.

Problem 3.4 Determine if the following stress field satisfies the equilibrium equation in the Cartesian coordinate system

$$\sigma_{11} = 5x_1^2 + 8x_2^3, \quad \sigma_{22} = 5x_2^3; \quad \sigma_{33} = -5x_3^2$$

$$\sigma_{12} = -10x_1x_2, \quad \sigma_{23} = 10x_2x_3; \quad \sigma_{13} = 0$$

$$b_1 = b_2 = b_3 = 0; \quad \ddot{u}_1 = \ddot{u}_2 = \ddot{u}_3 = 0$$

Chapter 4
Elastic Constitutive Laws

The number of equations available through the principles of continuum mechanics (conservation of mass and energy, and principles of linear and angular momentums) is inadequate to solve boundary value problems. The additional required equations are provided through what are known as the *constitutive relations*. In this chapter, we will discuss the details of the *linear-elastic constitutive relations*.

4.1 Fundamentals

An *elastic material* is defined as one where the stress $\boldsymbol{\sigma}$ depends only on the strain $\boldsymbol{\varepsilon}$, and not on the past thermodynamic history (Eringen 1967). Limiting our discussion to processes that are adiabatic (where the heat loss or gain is absent) and isothermal (where the temperature remains constant), the first law of thermodynamics results in the following:

$$\sigma_{ij} = \frac{\partial u}{\partial \varepsilon_{ij}}; \quad u = u(\boldsymbol{\varepsilon}) \tag{4.1}$$

where σ_{ij} and ε_{ij} are the stress and strain tensors, respectively, and u is the *strain energy density* (energy per unit volume). The derivation is presented in Appendix 2.

The symmetry of the stress tensor is assured by writing (4.1) as

$$\sigma_{ij} = \frac{1}{2}\left(\frac{\partial u}{\partial \varepsilon_{ij}} + \frac{\partial u}{\partial \varepsilon_{ji}}\right) \tag{4.2}$$

Since u is a function of strain $\boldsymbol{\varepsilon}$ (4.1), (4.2) is a *constitutive law* for elastic solids when thermal effects are neglected. The models represented by (4.2) are called the *Green-elastic* or *hyperelastic* models. Equation (4.2) is valid for *isotropic* as well as *anisotropic solids*, and *linear* as well as *nonlinear* elastic solids.

A. Anandarajah, *Computational Methods in Elasticity and Plasticity: Solids and Porous Media*, DOI 10.1007/978-1-4419-6379-6_4,

On the other hand, in *Cauchy elastic* models, the stress is expressed as a direct function of the strain as

$$\boldsymbol{\sigma} = \phi(\boldsymbol{\varepsilon})$$

We will limit the discussion in this chapter to Green-elastic models.

Let us use a polynomial expression for u as

$$u = A + B_{k\ell}\varepsilon_{k\ell} + \frac{1}{2}\bar{C}_{k\ell mn}\varepsilon_{k\ell}\varepsilon_{mn} + \cdots \tag{4.3a}$$

We require that u be invariant under coordinate axis rotations. Based on the coordinate transformation rules (Chap. 2) for second order tensors, the strain tensor transforms as

$$\boldsymbol{\varepsilon}' = \mathbf{a}^{\mathrm{T}}\boldsymbol{\varepsilon}\mathbf{a} \quad \text{or} \quad \varepsilon'_{k\ell} = a_{ki}\varepsilon_{ij}a_{\ell j}$$

The strain energy density in the rotated coordinate system becomes

$$\begin{aligned} u' &= A' + B'_{k\ell}\varepsilon'_{k\ell} + \frac{1}{2}\bar{C}'_{k\ell mn}\varepsilon'_{k\ell}\varepsilon'_{mn} + \cdots \\ &= A' + B'_{k\ell}a_{ki}a_{\ell j}\varepsilon_{ij} + \frac{1}{2}\bar{C}'_{k\ell mn}(a_{ki}a_{\ell j}\varepsilon_{ij})(a_{mp}a_{nq}\varepsilon_{pq}) + \cdots \\ &= A' + (B'_{k\ell}a_{ki}a_{\ell j})\varepsilon_{ij} + \frac{1}{2}(\bar{C}'_{k\ell mn}a_{ki}a_{\ell j}a_{mp}a_{nq})\varepsilon_{ij}\varepsilon_{pq} + \cdots \end{aligned} \tag{4.3b}$$

where $\mathbf{a}$ is the transformation matrix appearing in the transformation of vectors $\mathbf{x}' = \mathbf{a}\mathbf{x}$ (Chap. 2). Noting that $u = u'$, comparison of (4.3a) and (4.3b) reveals that

$$A = A'; \qquad B_{ij} = B'_{k\ell}a_{ki}a_{\ell j}; \qquad \overline{C}_{ijpq} = \overline{C}'_{k\ell mn}a_{ki}a_{\ell j}a_{mp}a_{nq} \tag{4.3c}$$

The constants of the polynomial are, therefore, tensors themselves since they follow the transformation rules for tensors (see Chap. 2 for the definition of tensors).

Now, taking $u = 0$ when $\varepsilon_{ij} = 0$ in (4.3a), we obtain $A = 0$. Since we are only interested in *linear elasticity* in this chapter, we only need to retain up to the quadratic term in (4.3a) as

$$u = B_{k\ell}\varepsilon_{k\ell} + \frac{1}{2}\bar{C}_{k\ell mn}\varepsilon_{k\ell}\varepsilon_{mn} \tag{4.3d}$$

Differentiating u with respect to the strain tensor

$$\begin{aligned} \frac{\partial u}{\partial \varepsilon_{ij}} &= B_{k\ell}\delta_{ki}\delta_{\ell j} + \frac{1}{2}\bar{C}_{k\ell mn}\delta_{ki}\delta_{\ell j}\varepsilon_{mn} + \frac{1}{2}\bar{C}_{k\ell mn}\varepsilon_{k\ell}\delta_{mi}\delta_{nj} \\ &= B_{ij} + \frac{1}{2}\left(\bar{C}_{ijmn}\varepsilon_{mn} + \bar{C}_{k\ell ij}\varepsilon_{k\ell}\right) \end{aligned} \tag{4.4}$$

In view of (4.1), requiring that $\sigma_{ij} = 0$ for $\varepsilon_{ij} = 0$, we have $B_{ij} = 0$. This leads to

$$\sigma_{ij} = \frac{1}{2}\left(\bar{C}_{ijmn}\varepsilon_{mn} + \bar{C}_{k\ell ij}\varepsilon_{k\ell}\right) = \frac{1}{2}\left(\bar{C}_{ijk\ell} + \bar{C}_{k\ell ij}\right)\varepsilon_{k\ell} = C_{ijk\ell}\varepsilon_{k\ell} \tag{4.5}$$

where

$$C_{ijk\ell} = \frac{1}{2}\left(\bar{C}_{ijk\ell} + \bar{C}_{k\ell ij}\right) \tag{4.6}$$

$$C_{ijk\ell} = C_{k\ell ij} \tag{4.7}$$

Hence, $C_{ijk\ell}$ is symmetric both in i and k (i.e., $C_{ijk\ell}$ remains the same when ik is changed to ki) and in j and ℓ.

Remark. In general, an assumption of the relationship between the stress and strain of the form $\sigma_{ij} = C_{ijk\ell}\varepsilon_{k\ell}$ and the thermodynamic relation (4.1) is sufficient to show the symmetry in (4.7) as follows:

$$\sigma_{ij} = \frac{\partial u}{\partial \varepsilon_{ij}} = C_{ijpq}\varepsilon_{pq} \Rightarrow \frac{\partial^2 u}{\partial \varepsilon_{ij}\partial \varepsilon_{k\ell}} = C_{ijk\ell}$$

$$\sigma_{k\ell} = \frac{\partial u}{\partial \varepsilon_{k\ell}} = C_{k\ell pq}\varepsilon_{pq} \Rightarrow \frac{\partial^2 u}{\partial \varepsilon_{k\ell}\partial \varepsilon_{ij}} = C_{k\ell ij}$$

Since

$$\frac{\partial^2 u}{\partial \varepsilon_{ij}\partial \varepsilon_{k\ell}} = \frac{\partial^2 u}{\partial \varepsilon_{k\ell}\partial \varepsilon_{ij}} \Rightarrow C_{ijk\ell} = C_{k\ell ij}$$

Now referring back to (4.5), the symmetry of the stress and strain tensors requires

$$C_{ijk\ell} = C_{jik\ell} = C_{ij\ell k} \tag{4.8a}$$

Thus,

$$\sigma_{ij} = C_{ijk\ell}\varepsilon_{k\ell} \tag{4.8b}$$

with $C_{ijk\ell}$ possessing the symmetries stated in (4.7) and (4.8a).

Using Voigt's notation, let us arrange the six independent components of σ_{ij} and ε_{ij} in vectors as

$$\bar{\boldsymbol{\sigma}} = \{\, \sigma_{11} \quad \sigma_{22} \quad \sigma_{33} \quad \sigma_{12} \quad \sigma_{23} \quad \sigma_{13} \,\} \tag{4.9a}$$

$$\bar{\boldsymbol{\varepsilon}} = \{ \varepsilon_{11} \quad \varepsilon_{22} \quad \varepsilon_{33} \quad \gamma_{12} \quad \gamma_{23} \quad \gamma_{13} \} \tag{4.9b}$$

where $\gamma_{12} = (1/2)(\varepsilon_{12} + \varepsilon_{21})$, $\gamma_{23} = (1/2)(\varepsilon_{23} + \varepsilon_{32})$, and $\gamma_{13} = (1/2)(\varepsilon_{13} + \varepsilon_{31})$ are engineering measures of shear strains. We will thus use the engineering shear strains instead of the tensor shear strains ε_{12}, etc., to be compatible with the relation needed in the finite element analysis (Chap. 5).

In a matrix-vector form, (4.8b) can be written as

$$\begin{Bmatrix} \sigma_{11} \\ \sigma_{22} \\ \sigma_{33} \\ \sigma_{12} \\ \sigma_{23} \\ \sigma_{13} \end{Bmatrix} = \begin{bmatrix} C_{1111} & C_{1122} & C_{1133} & C_{1112} & C_{1123} & C_{1113} \\ C_{2211} & C_{2222} & C_{2233} & C_{2212} & C_{2223} & C_{2213} \\ C_{3311} & C_{3322} & C_{3333} & C_{3312} & C_{3323} & C_{3313} \\ C_{1211} & C_{1222} & C_{1233} & C_{1212} & C_{1223} & C_{1213} \\ C_{2311} & C_{2322} & C_{2333} & C_{2312} & C_{2323} & C_{2313} \\ C_{1311} & C_{1322} & C_{1333} & C_{1312} & C_{1323} & C_{1313} \end{bmatrix} \begin{Bmatrix} \varepsilon_{11} \\ \varepsilon_{22} \\ \varepsilon_{33} \\ \gamma_{12} \\ \gamma_{23} \\ \gamma_{13} \end{Bmatrix} \tag{4.10a}$$

$$= \begin{bmatrix} \bar{C}_{11} & \bar{C}_{12} & \bar{C}_{13} & \bar{C}_{14} & \bar{C}_{15} & \bar{C}_{16} \\ \bar{C}_{21} & \bar{C}_{22} & \bar{C}_{23} & \bar{C}_{24} & \bar{C}_{25} & \bar{C}_{26} \\ \bar{C}_{31} & \bar{C}_{32} & \bar{C}_{33} & \bar{C}_{34} & \bar{C}_{35} & \bar{C}_{36} \\ \bar{C}_{41} & \bar{C}_{42} & \bar{C}_{43} & \bar{C}_{44} & \bar{C}_{45} & \bar{C}_{46} \\ \bar{C}_{51} & \bar{C}_{52} & \bar{C}_{53} & \bar{C}_{54} & \bar{C}_{55} & \bar{C}_{56} \\ \bar{C}_{61} & \bar{C}_{62} & \bar{C}_{63} & \bar{C}_{64} & \bar{C}_{65} & \bar{C}_{66} \end{bmatrix} \begin{Bmatrix} \varepsilon_{11} \\ \varepsilon_{22} \\ \varepsilon_{33} \\ \gamma_{12} \\ \gamma_{23} \\ \gamma_{13} \end{Bmatrix} \tag{4.10b}$$

$$\bar{\boldsymbol{\sigma}} = \bar{\mathbf{C}}\bar{\boldsymbol{\varepsilon}} \tag{4.10c}$$

where the symmetry with respect to the last two subscripts (4.8a) has already been utilized in the simplifications $(1/2)(C_{1112} + C_{1121}) = C_{1112}$, etc. Thus, there are at most 36 independent constants in (4.10b). However, making use of the symmetry $C_{ijk\ell} = C_{k\ell ij}$ (4.7), it follows that $\bar{C}$ is symmetric. Hence, there are only 21 independent constants in (4.10b).

Remark. We recall Hooke's law for an isotropic, elastic solid in uniaxial loading as

$$\sigma = E\varepsilon \tag{4.11}$$

where σ and ε are the uniaxial stress and strain, respectively, and E is the Young's modulus. Extending this to the multiaxial case, we define the *generalized Hooke's law* as

$$\sigma_{ij} = C_{ijk\ell}\varepsilon_{k\ell} \tag{4.12}$$

where $C_{ijk\ell}$ is the fourth order elastic material tensor. Before considering symmetries, it is seen that 81 constants are needed to define $C_{ijk\ell}$. This is because a relationship between nine components of σ_{ij} and nine components of ε_{ij} involves $9 \times 9 = 81$ constants. Now the symmetry of σ_{ij} and ε_{ij} reduces the number of constants to 36 ($= 6 \times 6$). Further, we saw above that the use of Green-elasticity (4.1) reduces the number of constants from 36 to 21. Further reductions are possible based on certain other material symmetries, which are discussed in the subsequent sections.

4.2 Linear-Elastic Relations for Anisotropic Solids with Symmetries

The general linear-elastic stress-strain relation for an anisotropic solid is given by (4.10b). This requires the specification of 21 independent constants (when Green-elasticity is used). However, most materials possess certain material symmetries that can be exploited to further reduce the number of constants needed for (4.10b).

To understand material symmetries, first consider an isotropic solid shown in Fig. 4.1. The microstructure of the material has random orientation at the appropriate scale to render the elastic properties directionally independent. (Note that while a metal may be anisotropic at a nanoscale due to anisotropic orientation of atoms within crystals, it may still be isotropic at the macroscopic scale due to random orientation of grains.)

Let us call the coordinate system with respect to which the loading is applied as "loading directions." For an isotropic material, when the loading directions are rotated from $x_1 - x_2 - x_3$ to $x_1' - x_2' - x_3'$, the stress-strain behavior will remain unchanged.

For example, consider two specimens, A and B, shown in Fig. 4.1. The specimens are cut from the parent material. The difference is that the faces of specimen

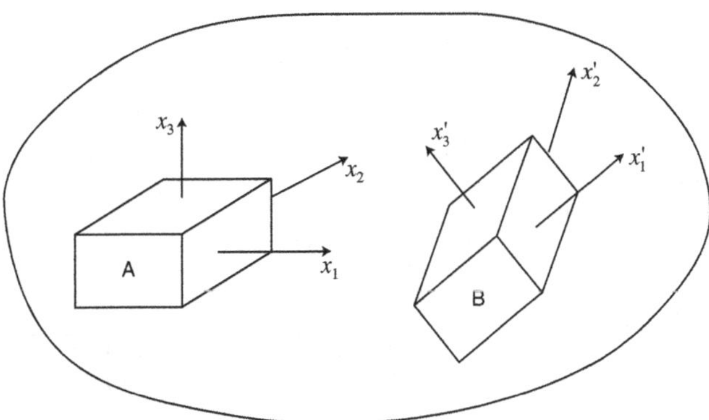

Fig. 4.1 Specimens cut from isotropic material

A are aligned with the $x_1 - x_2 - x_3$ directions, and those of specimen B are aligned with the $x'_1 - x'_2 - x'_3$ directions.

Suppose that when specimen A is subjected to the principal strains (0.01, 0.02, 0.03)% in the directions parallel to its faces, the stresses developed in these directions are (10, 20, 30) MPa. Then when specimen B is subjected to the same principal strain loading in the directions parallel to its faces, the stresses developed in these directions must still be (10, 20, 30) MPa.

This will not be the case when the material has different properties in different directions. However, when certain symmetries exist, the behavior in some specific loading directions remains unchanged. Three such symmetries are considered in this section (Figs. 4.2–4.4).

In Fig. 4.2, we have a material for which the behavior of specimens A and B is identical. Specimen B is obtained as a mirror-image (or reflection) of specimen A about the $x_{30} = 0$ plane. It is implied that the loading is applied in the $x_1 - x_2 - x_3$

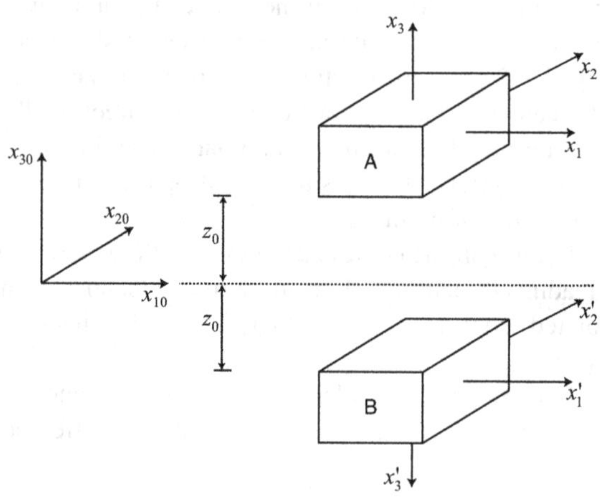

Fig. 4.2 Reflection about the $x_{30} = 0$ plane

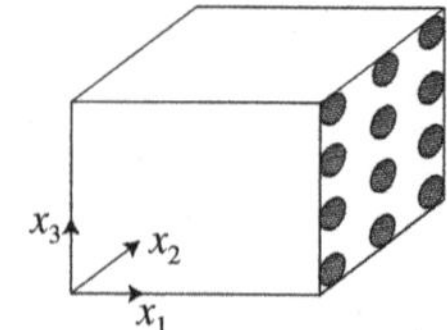

Fig. 4.3 Orthotropic material

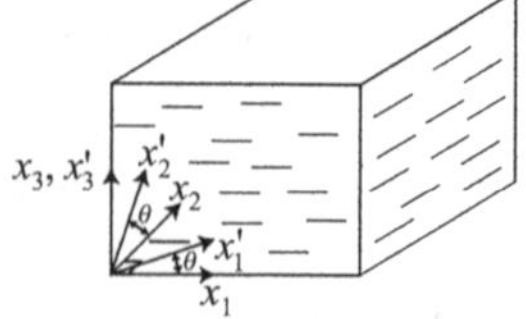

Fig. 4.4 Transversely isotropic material

directions for specimen A. An identical loading is applied to specimen B in the $x_1' - x_2' - x_3'$ directions.

Examples of materials that possess such symmetry are *monoclinic crystals* (gypsum, talc, etc.) and a rock mass with a set of parallel fractures arranged in a certain manner. The behavior is identical with respect to the $x_1 - x_2 - x_3$ and $x_1' - x_2' - x_3'$ coordinate systems. The material is said to have one plane of symmetry – the plane having the x_3 axis as its normal. Such materials are called *monoclinic* materials.

In Fig. 4.3, we have a material that has three orthogonal planes of symmetry; the planes of symmetry are those normal to the $x_1 -$, $x_2 -$ and $x_3 -$ axes. Such materials are called *orthotropic* materials. Isotropic materials reinforced with cylindrical fibers as shown in Fig. 4.3 are examples of orthotropic materials. Another good example of an orthotropic material is wood. (It may be pointed out that traditionally, such materials have been idealized as transversely isotropic ones, see, Simácek et al. 1993.)

In Fig. 4.4, we have a material for which the properties on the $x_3 = 0$ plane are directionally independent. Such materials are called the transversely isotropic materials. The axis of symmetry is the $x_3 -$ axis. Natural soils are transversely isotropic with the axis of symmetry being the original depositional direction (usually vertical).

The material symmetries reduce the number of constants from 21. The reduction is determined as follows. Let the stress-strain relation in the $x_1 - x_2 - x_3$ and $x_1' - x_2' - x_3'$ systems be

$$\sigma_{ij} = C_{ijk\ell}\varepsilon_{k\ell} \tag{4.13a}$$

and

$$\sigma'_{ij} = C'_{ijk\ell}\varepsilon'_{k\ell} \tag{4.13b}$$

respectively. Recall the discussion we had in the case of the isotropic material (Fig. 4.1). In a similar manner, when the symmetry dictates that the stress-strain relation be the same in the $x_1 - x_2 - x_3$ and $x_1' - x_2' - x_3'$ directions, we require that $\sigma'_{ij} = \sigma_{ij}$ when $\varepsilon'_{ij} = \varepsilon_{ij}$ and vice versa. This requires that

$$C'_{ijk\ell} = C_{ijk\ell} \tag{4.14}$$

Based on (4.3c), (4.14) becomes

$$C_{ijk\ell} - C_{pqrs}a_{pi}a_{qj}a_{rk}a_{s\ell} \tag{4.15}$$

In the following sections, we apply (4.15) for each of the three symmetries discussed in the preceding paragraphs and determine the reduction in the number of constants.

4.2.1 Monoclinic Materials (One Plane of Symmetry, Fig. 4.2)

Referring to Fig. 4.2, the transformation matrix appearing in $\mathbf{x}' = \mathbf{ax}$ (see Chap. 2 for the definition of $\mathbf{a}$) is

$$\mathbf{a} = \begin{bmatrix} 1 & 0 & 0 \\ 0 & 1 & 0 \\ 0 & 0 & -1 \end{bmatrix}$$

Noting the zeroes in $\mathbf{a}$, let us evaluate two typical constants in $\bar{\mathbf{C}}$ (4.10a) and (4.10b):

$$\bar{C}'_{44} = C_{1212} = C_{pqrs}a_{p1}a_{q2}a_{r1}a_{s2} = C_{1212}a_{11}a_{22}a_{11}a_{22} = C_{1212} \Rightarrow C_{1212} \neq 0$$

That is, C_{1212} is not necessarily zero. It may be noted that a similar result is obtained when

$$\text{sign}(a_{pi}) \times \text{sign}(a_{qj}) \times \text{sign}(a_{rk}) \times \text{sign}(a_{s\ell}) = 1$$

This leads to the conclusion that the following 13 constants are not necessarily zero (considering only the upper diagonal coefficients in (4.10a):

$$C_{1111}; C_{1122}; C_{1133}; C_{1112}; C_{2222}; C_{2233}; C_{2212}; C_{3333}; C_{3312}; C_{1212}; C_{2323}; C_{2313}; C_{1313}$$

Now consider:

$$\bar{C}'_{15} = C_{1123} = C_{pqrs}a_{p1}a_{q1}a_{r2}a_{s3} = C_{1123}a_{11}a_{11}a_{22}a_{33} = -C_{1123} \Rightarrow C_{1123} = 0$$

Noting that this occurs when

$$\text{sign}(a_{pi}) \times \text{sign}(a_{qj}) \times \text{sign}(a_{rk}) \times \text{sign}(a_{s\ell}) = -1$$

the following eight constants (among the upper diagonal coefficients) are found to fall in this group:

$$C_{1123};\ C_{1113};\ C_{2223};\ C_{2213};\ C_{3323};\ C_{3313};\ C_{1223};\ C_{1213}$$

The stiffness matrix in (4.10a) now becomes:

$$\bar{\mathbf{C}} = \begin{bmatrix} C_{1111} & C_{1122} & C_{1133} & C_{1112} & 0 & 0 \\ C_{1122} & C_{2222} & C_{2233} & C_{2212} & 0 & 0 \\ C_{1133} & C_{2233} & C_{3333} & C_{3312} & 0 & 0 \\ C_{1112} & C_{2212} & C_{3312} & C_{1212} & 0 & 0 \\ 0 & 0 & 0 & 0 & C_{2323} & C_{2313} \\ 0 & 0 & 0 & 0 & C_{2313} & C_{1313} \end{bmatrix} \tag{4.16}$$

The number of independent constants is thus 13.

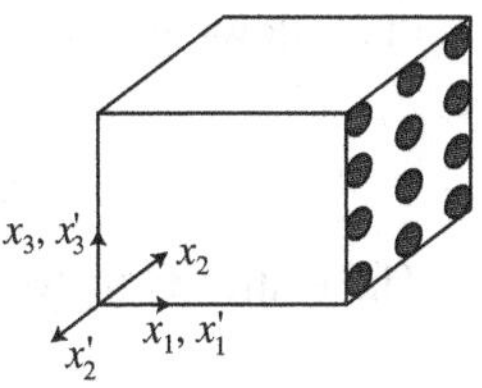

Fig. 4.5 Reflection about the $x_2 = 0$ plane

4.2.2 *Orthotropic Materials (Fig. 4.3)*

The symmetry shown in Fig. 4.2, which represents reflection about the $x_3 = 0$ plane, gave us the matrix in (4.16). In addition, we now use the symmetry represented by the $x_1' - x_2' - x_3'$ axes shown in Fig. 4.5. The $x_1' - x_2' - x_3'$ axes are obtained by reflecting the $x_1 - x_2 - x_3$ axes about the $x_2 = 0$ plane. It turns out that the number of independent constants obtained by using two orthogonal planes of symmetry is the same as that using three orthogonal planes of symmetry. This implies that the third symmetry is dependent on the first two.

The transformation matrix is

$$\mathbf{a} = \begin{bmatrix} 1 & 0 & 0 \\ 0 & -1 & 0 \\ 0 & 0 & 1 \end{bmatrix}$$

We will start with (4.16) and see how the matrix $\bar{\mathbf{C}}$ further simplifies. Let us evaluate a typical constant C_{1122}:

$$\bar{C}'_{12} = C_{1122} = C_{pqrs}a_{p1}a_{q1}a_{r2}a_{s2} = C_{1122}a_{11}a_{11}a_{22}a_{22} = C_{1122} \Rightarrow C_{1122} \neq 0$$

That is, C_{1122} is not necessarily zero. A similar result is obtained when

$$\text{sign}(a_{pi}) \times \text{sign}(a_{qj}) \times \text{sign}(a_{rk}) \times \text{sign}(a_{s\ell}) = 1$$

This leads to the conclusion that the following nine constants (among the upper diagonal coefficients) are not necessarily zero:

$$C_{1111};\ C_{1122};\ C_{1133};\ C_{2222};\ C_{2233};\ C_{3333};\ C_{1212};\ C_{2323};\ C_{1313}$$

Now consider:

$$\bar{C}'_{14} = C_{1112} = C_{pqrs}a_{p1}a_{q1}a_{r1}a_{s2} = C_{1112}a_{11}a_{11}a_{11}a_{22} = -C_{1112} \Rightarrow C_{1112} = 0$$

Noting that this occurs when

$$\text{sign}(a_{pi}) \times \text{sign}(a_{qj}) \times \text{sign}(a_{rk}) \times \text{sign}(a_{s\ell}) = -1$$

the following four constants are found to fall in this group:

$$C_{1112};\ C_{2212};\ C_{3312};\ C_{2313}$$

The matrix $\bar{\mathbf{C}}$ (4.16) now becomes:

$$\bar{\mathbf{C}} = \begin{bmatrix} C_{1111} & C_{1122} & C_{1133} & 0 & 0 & 0 \\ C_{1122} & C_{2222} & C_{2233} & 0 & 0 & 0 \\ C_{1133} & C_{2233} & C_{3333} & 0 & 0 & 0 \\ 0 & 0 & 0 & C_{1212} & 0 & 0 \\ 0 & 0 & 0 & 0 & C_{2323} & 0 \\ 0 & 0 & 0 & 0 & 0 & C_{1313} \end{bmatrix} \tag{4.17}$$

The number of independent constants is thus 9.

The reader may now verify that the reflection about the $x_1 = 0$ plane, which corresponds to the transformation matrix,

$$\mathbf{a} = \begin{bmatrix} -1 & 0 & 0 \\ 0 & 1 & 0 \\ 0 & 0 & 1 \end{bmatrix}$$

leaves (4.17) unchanged (Problem 4.2).

By writing down the six equations in (4.10) with the $\bar{\mathbf{C}}$ matrix given by 4.17, it may be seen that the normal stress-strain relations are uncoupled from the shear stress-strain relations. A shear strain component does not produce a normal stress component and a normal strain component does not produce a shear stress component. Equation (4.17) may be inverted to find the inverse relation, which is of the form:

$$\begin{Bmatrix} \varepsilon_{11} \\ \varepsilon_{22} \\ \varepsilon_{33} \\ \gamma_{12} \\ \gamma_{23} \\ \gamma_{13} \end{Bmatrix} = \begin{bmatrix} \bar{S}_{11} & \bar{S}_{12} & \bar{S}_{13} & 0 & 0 & 0 \\ \bar{S}_{21} & \bar{S}_{22} & \bar{S}_{23} & 0 & 0 & 0 \\ \bar{S}_{31} & \bar{S}_{32} & \bar{S}_{33} & 0 & 0 & 0 \\ 0 & 0 & 0 & \bar{S}_{44} & 0 & 0 \\ 0 & 0 & 0 & 0 & \bar{S}_{55} & 0 \\ 0 & 0 & 0 & 0 & 0 & \bar{S}_{66} \end{bmatrix} \begin{Bmatrix} \sigma_{11} \\ \sigma_{22} \\ \sigma_{33} \\ \sigma_{12} \\ \sigma_{23} \\ \sigma_{13} \end{Bmatrix} \tag{4.18a}$$

$$\boldsymbol{\varepsilon} = \bar{\mathbf{S}}\boldsymbol{\sigma} \tag{4.18b}$$

where $\bar{\mathbf{S}}$ is the compliance matrix. $\bar{\mathbf{S}}$ is symmetric, i.e.,

$$\bar{S}_{21} = \bar{S}_{12};\quad \bar{S}_{32} = \bar{S}_{23};\quad \text{and}\quad \bar{S}_{31} = \bar{S}_{13} \tag{4.18c}$$

It is useful to interpret the coefficients in (4.18a) in terms of the familiar engineering constants E and ν. Consider a uniaxial stress loading in the x_1 – direction. For this loading, for an orthotropic material

$$\sigma_{22} = 0 \quad \text{and} \quad \sigma_{33} = 0$$

$$\varepsilon_{11} = \bar{S}_{11}\sigma_{11}; \quad \varepsilon_{22} = \bar{S}_{21}\sigma_{11} = \frac{\bar{S}_{21}}{\bar{S}_{11}}\varepsilon_{11}; \quad \varepsilon_{33} = \bar{S}_{31}\sigma_{11} = \frac{\bar{S}_{31}}{\bar{S}_{11}}\varepsilon_{11}$$

Compare these with the corresponding relations for an isotropic material (discussed in the next section)

$$\varepsilon_{11} = \frac{1}{E}\sigma_{11}; \quad \varepsilon_{22} = -\nu\varepsilon_{11}; \quad \varepsilon_{33} = -\nu\varepsilon_{11}$$

where E and ν are Young's modulus and Poison's ratio, respectively, of the material. Based on the similarities seen, the following orthotropic constants are introduced:

$$\bar{S}_{11} = \frac{1}{E_1}; \quad \frac{\bar{S}_{21}}{\bar{S}_{11}} = -\nu_{21} \Rightarrow \bar{S}_{21} = -\frac{\nu_{21}}{E_1}; \quad \frac{\bar{S}_{31}}{\bar{S}_{11}} = -\nu_{31} \Rightarrow \bar{S}_{31} = -\frac{\nu_{31}}{E_1}$$

We follow a similar logic for the remaining constants involved in the normal stress-strain relations. E_1 is the Young's modulus in direction 1. ν_{21} is negative of the normal strain in direction 2 due to a unit normal strain in direction 1. (Note that ν_{21} is just a scalar, not a component of a second order tensor.)

The orthotropic shear stress-strain relations are

$$\gamma_{12} = \bar{S}_{44}\sigma_{12}; \quad \gamma_{23} = \bar{S}_{55}\sigma_{23}; \quad \gamma_{13} = \bar{S}_{66}\sigma_{13}$$

Comparing these with the corresponding relations for an isotropic material

$$\gamma_{12} = G\sigma_{12}; \quad \gamma_{23} = G\sigma_{23}; \quad \gamma_{13} = G\sigma_{13}$$

where G is the shear modulus of an isotropic material, we introduce the following orthotropic shear moduli as

$$S_{44} = G_{12}; \quad S_{55} = G_{23}; \quad S_{66} = G_{13}$$

In terms of these constants, the compliance matrix for an orthotropic material $\bar{\mathrm{S}}$ (4.18a) becomes

$$\begin{Bmatrix} \varepsilon_{11} \\ \varepsilon_{22} \\ \varepsilon_{33} \\ \gamma_{12} \\ \gamma_{23} \\ \gamma_{13} \end{Bmatrix} = \begin{bmatrix} 1/E_1 & -\nu_{12}/E_2 & -\nu_{13}/E_3 & 0 & 0 & 0 \\ -\nu_{21}/E_1 & 1/E_2 & -\nu_{23}/E_3 & 0 & 0 & 0 \\ -\nu_{31}/E_1 & -\nu_{32}/E_2 & 1/E_3 & 0 & 0 & 0 \\ 0 & 0 & 0 & 1/G_{12} & 0 & 0 \\ 0 & 0 & 0 & 0 & 1/G_{23} & 0 \\ 0 & 0 & 0 & 0 & 0 & 1/G_{13} \end{bmatrix} \begin{Bmatrix} \sigma_{11} \\ \sigma_{22} \\ \sigma_{33} \\ \sigma_{12} \\ \sigma_{23} \\ \sigma_{13} \end{Bmatrix} \tag{4.19}$$

There are 12 independent constants in the above equation: $E_1, E_2, E_3, G_{12}, G_{23}, G_{13}, \nu_{12}, \nu_{13}, \nu_{23}, \nu_{21}, \nu_{31}$, and ν_{32}.

However, the compliance matrix in the above equation must be symmetric. This requires

$$\frac{\nu_{21}}{E_1} = \frac{\nu_{12}}{E_2}; \quad \frac{\nu_{31}}{E_1} = \frac{\nu_{13}}{E_3}; \quad \frac{\nu_{32}}{E_2} = \frac{\nu_{23}}{E_3} \tag{4.20}$$

These relations reduce the number of independent constants from 12 to 9. It is to be noted that the coefficients of Poisson's ratios themselves are not symmetric (i.e., $\nu_{12} \neq \nu_{21}$, etc.), but rather Poisson's ratios normalized with the stiffness coefficients as in (4.20) are symmetric.

The $\bar{\mathbf{S}}$ matrix in (4.19) may be inverted to find the $\bar{\mathbf{C}}$ matrix. Skipping the details, the final results are

$$\begin{aligned} &\bar{C}_{11} = (1 - \nu_{23}\nu_{32})/(E_2E_3d); \quad \bar{C}_{12} = (\nu_{12} + \nu_{13}\nu_{32})/(E_2E_3d); \\ &\bar{C}_{13} = (\nu_{13} + \nu_{12}\nu_{23})/(E_2E_3d) \end{aligned} \tag{4.21a}$$

$$\begin{aligned} &\bar{C}_{21} = (\nu_{21} + \nu_{23}\nu_{31})/(E_1E_3d); \quad \bar{C}_{22} = (1 - \nu_{13}\nu_{31})/(E_1E_3d); \\ &\bar{C}_{23} = (\nu_{23} + \nu_{21}\nu_{13})/(E_1E_3d) \end{aligned} \tag{4.21b}$$

$$\begin{aligned} &\bar{C}_{31} = (\nu_{31} + \nu_{21}\nu_{32})/(E_1E_2d); \quad \bar{C}_{32} = (\nu_{32} + \nu_{12}\nu_{31})/(E_1E_2d); \\ &\bar{C}_{33} = (1 - \nu_{12}\nu_{21})/(E_1E_2d) \end{aligned} \tag{4.21c}$$

where d is the determinant of the top-left 3×3 sub-matrix in (4.19), given by

$$d = \frac{1}{E_1E_2E_3}(1 - \nu_{12}\nu_{21} - \nu_{23}\nu_{32} - \nu_{13}\nu_{31} - 2\nu_{12}\nu_{23}\nu_{31}) \tag{4.21d}$$

$$\bar{C}_{44} = G_{12}; \quad \bar{C}_{55} = G_{23}, \quad \text{and} \quad \bar{C}_{55} = G_{13}. \tag{4.21e}$$

$\bar{\mathbf{C}}$ is given by (4.17) and involves only nine independent constants. With (4.20), it is shown that $\bar{C}_{21} = \bar{C}_{12}$, $\bar{C}_{31} = \bar{C}_{13}$, and $\bar{C}_{32} = \bar{C}_{23}$ as follows:

$$\begin{aligned} \bar{C}_{21} &= \frac{1}{E_1E_3d}(\nu_{21} + \nu_{23}\nu_{31}) = \frac{1}{E_1E_3d}\left(\frac{E_1\nu_{12}}{E_2} + \frac{E_3\nu_{32}}{E_2}\frac{E_1\nu_{13}}{E_3}\right) \\ &= \frac{1}{E_2E_3d}(\nu_{12} + \nu_{13}\nu_{32}) = \bar{C}_{12} \end{aligned}$$

$$\begin{aligned} \bar{C}_{31} &= \frac{1}{E_1E_2d}(\nu_{31} + \nu_{21}\nu_{32}) = \frac{1}{E_1E_2d}\left(\frac{E_1\nu_{13}}{E_3} + \frac{E_1\nu_{12}}{E_2}\frac{E_2\nu_{23}}{E_3}\right) \\ &= \frac{1}{E_2E_3d}(\nu_{13} + \nu_{12}\nu_{23}) = \bar{C}_{13} \end{aligned}$$

$$\bar{C}_{32} = \frac{1}{E_1 E_2 d}(\nu_{32} + \nu_{12}\nu_{31}) = \frac{1}{E_1 E_2 d}\left(\frac{E_2 \nu_{23}}{E_3} + \frac{E_2 \nu_{21}}{E_1}\frac{E_1 \nu_{13}}{E_3}\right)$$
$$= \frac{1}{E_1 E_3 d}(\nu_{23} + \nu_{21}\nu_{13}) = \bar{C}_{23}$$

For isotropic materials, a value of Poisson's ratio larger than 0.5 is thermodynamically inadmissible. However, for orthotropic materials, Poisson's ratio has been found to be larger than 1 in many cases. Lempriere (1968) provides an analysis of this and establishes ranges of values for orthotropic coefficients of Poisson's ratio. He shows that all coefficients of Poisson's ratio cannot be larger than 0.5 at the same time.

4.2.3 Transversely Isotropic Material (Fig. 4.4)

In this form of symmetry, the behavior must be the same for any value of θ shown in Fig. 4.4. It turns out that by considering three different values for θ, we can derive the final stiffness matrix for this case. We will consider $\theta = 180°$, $90°$, and $45°$ in this order. The key steps are summarized below, but the derivations are largely left for the reader to carry out.

For $\theta = 180°$:

$$\mathbf{a} = \begin{bmatrix} -1 & 0 & 0 \\ 0 & -1 & 0 \\ 0 & 0 & 1 \end{bmatrix} \tag{4.22a}$$

By following an approach similar to that used in the previous sections (e.g., Sect. 4.2.2), it can be shown that

$$\bar{\mathbf{C}} = \begin{bmatrix} C_{1111} & C_{1122} & C_{1133} & C_{1112} & 0 & 0 \\ C_{2211} & C_{2222} & C_{2233} & C_{2212} & 0 & 0 \\ C_{3311} & C_{3322} & C_{3333} & C_{3312} & 0 & 0 \\ C_{1211} & C_{1222} & C_{1233} & C_{1212} & 0 & 0 \\ 0 & 0 & 0 & 0 & C_{2323} & C_{2313} \\ 0 & 0 & 0 & 0 & C_{1323} & C_{1313} \end{bmatrix} \tag{4.22b}$$

For $\theta = 90°$:

$$\mathbf{a} = \begin{bmatrix} 0 & 1 & 0 \\ -1 & 0 & 0 \\ 0 & 0 & 1 \end{bmatrix} \tag{4.23a}$$

By following a similar procedure as above, we get

$$\bar{\mathbf{C}} = \begin{bmatrix} C_{1111} & C_{1122} & C_{1133} & C_{1112} & 0 & 0 \\ C_{1122} & C_{1111} & C_{1133} & -C_{1112} & 0 & 0 \\ C_{1133} & C_{1133} & C_{3333} & 0 & 0 & 0 \\ C_{1112} & -C_{1112} & 0 & C_{1212} & 0 & 0 \\ 0 & 0 & 0 & 0 & C_{2323} & 0 \\ 0 & 0 & 0 & 0 & 0 & C_{2323} \end{bmatrix} \tag{4.23b}$$

For $\theta = 45°$:

$$\mathbf{a} = \begin{bmatrix} 1/\sqrt{2} & 1/\sqrt{2} & 0 \\ -1/\sqrt{2} & 1/\sqrt{2} & 0 \\ 0 & 0 & 1 \end{bmatrix} \tag{4.24a}$$

$\bar{\mathbf{C}}$ becomes

$$\bar{\mathbf{C}} = \begin{bmatrix} C_{1111} & C_{1122} & C_{1133} & 0 & 0 & 0 \\ C_{1122} & C_{1111} & C_{1133} & 0 & 0 & 0 \\ C_{1133} & C_{1133} & C_{3333} & 0 & 0 & 0 \\ 0 & 0 & 0 & (C_{1111} - C_{1122})/2 & 0 & 0 \\ 0 & 0 & 0 & 0 & C_{2323} & 0 \\ 0 & 0 & 0 & 0 & 0 & C_{2323} \end{bmatrix} \tag{4.24b}$$

It may be verified that (4.24b) remains valid for any other value of θ. The number of independent constants is thus 5.

$\bar{\mathbf{C}}$ given in (4.24b) may be inverted to obtain the corresponding compliance matrix, which will have the following form:

$$\bar{\mathbf{S}} = \begin{bmatrix} S_{11} & S_{12} & S_{13} & 0 & 0 & 0 \\ S_{12} & S_{11} & S_{13} & 0 & 0 & 0 \\ S_{13} & S_{13} & S_{33} & 0 & 0 & 0 \\ 0 & 0 & 0 & S_{44} & 0 & 0 \\ 0 & 0 & 0 & 0 & S_{55} & 0 \\ 0 & 0 & 0 & 0 & 0 & S_{66} \end{bmatrix} \tag{4.25}$$

Note that the pattern seen in $\bar{\mathbf{C}}$ given in (4.24b) (i.e., $C_{13} = C_{23}$, etc.) is still preserved in $\bar{\mathbf{S}}$ given above. We will obtain a special relation for S_{44} later. First, following an approach similar to that for the orthotropic materials (Sect. 4.2.2), we develop the following representation for $\bar{\mathbf{S}}$ in terms of the familiar engineering constants:

$$\begin{Bmatrix} \varepsilon_{11} \\ \varepsilon_{22} \\ \varepsilon_{33} \\ \gamma_{12} \\ \gamma_{23} \\ \gamma_{13} \end{Bmatrix} = \begin{bmatrix} 1/E_1 & -\nu_{12}/E_1 & -\nu_{13}/E_3 & 0 & 0 & 0 \\ -\nu_{12}/E_1 & 1/E_1 & -\nu_{13}/E_3 & 0 & 0 & 0 \\ -\nu_{31}/E_1 & -\nu_{31}/E_1 & 1/E_3 & 0 & 0 & 0 \\ 0 & 0 & 0 & 1/G_{12} & 0 & 0 \\ 0 & 0 & 0 & 0 & 1/G_{23} & 0 \\ 0 & 0 & 0 & 0 & 0 & 1/G_{23} \end{bmatrix} \begin{Bmatrix} \sigma_{11} \\ \sigma_{22} \\ \sigma_{33} \\ \sigma_{12} \\ \sigma_{23} \\ \sigma_{13} \end{Bmatrix} \tag{4.26a}$$

There are seven independent constants in the above equation: E_1, E_3, G_{12}, G_{23}, ν_{12}, ν_{13}, and ν_{31}. Direction 3 is the axis of symmetry and directions 1 and 2 are in the plane of symmetry. There are two interrelations that reduce the number of independent constants from 7 to 5. One of them comes from the symmetry of the compliance matrix in (4.26a). This leads to

$$\frac{\nu_{31}}{E_1} = \frac{\nu_{13}}{E_3} \tag{4.26b}$$

We will discuss the other later in this section. The $\bar{\mathbf{S}}$ matrix in (4.26a) may be inverted to find the $\bar{\mathbf{C}}$ matrix. Skipping the details, the final results are

$$\begin{aligned} &\bar{C}_{11} = (1 - \nu_{13}\nu_{31})/(E_1E_3d); \quad \bar{C}_{12} = (\nu_{12} + \nu_{13}\nu_{31})/(E_1E_3d); \\ &\bar{C}_{13} = \nu_{13}(1 + \nu_{12})/(E_1E_3d) \end{aligned} \tag{4.27a}$$

$$\bar{C}_{21} = (\nu_{12} + \nu_{13}\nu_{31})/(E_1E_3d) = \bar{C}_{12}; \quad \bar{C}_{22} = \bar{C}_{11}; \quad \bar{C}_{23} = \bar{C}_{13} \tag{4.27b}$$

$$\bar{C}_{31} = \nu_{31}(1 + \nu_{12})/(E_1^2 d) = \bar{C}_{13}; \quad \bar{C}_{32} = \bar{C}_{31}; \quad \bar{C}_{33} = (1 - \nu_{12}^2)/(E_1^2 d) \tag{4.27c}$$

where

$$d = \frac{1}{E_1^2 E_3}(1 + \nu_{12})(1 - \nu_{12} - 2\nu_{13}\nu_{31}) \tag{4.27d}$$

and

$$\bar{C}_{44} = G_{12}; \quad \bar{C}_{55} = G_{13}, \quad \text{and} \quad \bar{C}_{66} = G_{13} \tag{4.27e}$$

From (4.24b), we have

$$\bar{C}_{44} = (C_{1111} - C_{1122})/2 = (\bar{C}_{11} - \bar{C}_{12})/2$$

This leads to

$$G_{12} = \frac{(1 - \nu_{12} - 2\nu_{13}^2)}{2E_1E_3d} = \frac{E_1}{2(1 + \nu_{12})} \tag{4.28a}$$

Hence,

$$S_{44} = \frac{2(1+\nu_{12})}{E_1} \tag{4.28b}$$

Equation (4.28a) is the second interrelation needed to reduce the number of independent constants from 7 to 5.

4.3 Linear-Elastic Relations for Isotropic Solids

In isotropic solids, the material behavior is identical with respect to loading in any arbitrary directions. Isotropy could be viewed as transverse isotropy with respect to two orthogonal axes.

We derived (4.24b) by considering transverse isotropy with respect to symmetry about the x_3-axis (Fig. 4.4). By repeating the process with respect to one other axis of symmetry, it can be shown:

$$\bar{\mathbf{C}} = \begin{bmatrix} C_{1111} & C_{1122} & C_{1122} & 0 & 0 & 0 \\ C_{1122} & C_{1111} & C_{1122} & 0 & 0 & 0 \\ C_{1122} & C_{1122} & C_{1111} & 0 & 0 & 0 \\ 0 & 0 & 0 & (C_{1111}-C_{1122})/2 & 0 & 0 \\ 0 & 0 & 0 & 0 & (C_{1111}-C_{1122})/2 & 0 \\ 0 & 0 & 0 & 0 & 0 & (C_{1111}-C_{1122})/2 \end{bmatrix} \tag{4.29}$$

There are only two independent constants (C_{1111} and C_{1122}) in the above equation for $\bar{\mathbf{C}}$. $\bar{\mathbf{C}}$ may be inverted to find the compliance matrix $\bar{\mathbf{S}}$.

In terms of Young's modulus E and Poisson's ratio ν, the inverse relation (obtained by simplifying (4.26)) becomes

$$\begin{Bmatrix} \varepsilon_{11} \\ \varepsilon_{22} \\ \varepsilon_{33} \\ \gamma_{12} \\ \gamma_{23} \\ \gamma_{13} \end{Bmatrix} = \begin{bmatrix} 1/E & -\nu/E & -\nu/E & 0 & 0 & 0 \\ -\nu/E & 1/E & -\nu/E & 0 & 0 & 0 \\ -\nu/E & -\nu/E & 1/E & 0 & 0 & 0 \\ 0 & 0 & 0 & 1/G & 0 & 0 \\ 0 & 0 & 0 & 0 & 1/G & 0 \\ 0 & 0 & 0 & 0 & 0 & 1/G \end{bmatrix} \begin{Bmatrix} \sigma_{11} \\ \sigma_{22} \\ \sigma_{33} \\ \sigma_{12} \\ \sigma_{23} \\ \sigma_{13} \end{Bmatrix} \tag{4.30a}$$

The coefficients of the stiffness matrix are obtained by simplifying (4.27) and (4.28). This leads to

$$\bar{\mathbf{C}} = \frac{E}{(1+v)(1-2v)} \begin{bmatrix} 1-v & v & v & 0 & 0 & 0 \\ v & 1-v & v & 0 & 0 & 0 \\ v & v & 1-v & 0 & 0 & 0 \\ 0 & 0 & 0 & (1-2v)/2 & 0 & 0 \\ 0 & 0 & 0 & 0 & (1-2v)/2 & 0 \\ 0 & 0 & 0 & 0 & 0 & (1-2v)/2 \end{bmatrix} \tag{4.30b}$$

$\bar{\mathbf{C}}$ can be expressed in terms of the Lamé constants λ and μ, where in (4.29),

$$C_{1122} = \lambda \text{ and } \frac{1}{2}(C_{1111} - C_{1122}) = \mu \tag{4.30c}$$

This leads to

$$\bar{\mathbf{C}} = \begin{bmatrix} \lambda+2\mu & \lambda & \lambda & 0 & 0 & 0 \\ \lambda & \lambda+2\mu & \lambda & 0 & 0 & 0 \\ \lambda & \lambda & \lambda+2\mu & 0 & 0 & 0 \\ 0 & 0 & 0 & \mu & 0 & 0 \\ 0 & 0 & 0 & 0 & \mu & 0 \\ 0 & 0 & 0 & 0 & 0 & \mu \end{bmatrix} \tag{4.30d}$$

The coefficients of $\bar{\mathbf{C}}$ may also be generated using the following tensorial equation:

$$C_{ijk\ell} = \lambda\delta_{ij}\delta_{k\ell} + \mu(\delta_{ik}\delta_{j\ell} + \delta_{i\ell}\delta_{jk}) \tag{4.31}$$

For example,

$$\begin{aligned} \bar{C}_{11} &= C_{1111} = \lambda\delta_{11}\delta_{11} + \mu(\delta_{11}\delta_{11} + \delta_{11}\delta_{11}) = \lambda + 2\mu \\ \bar{C}_{13} &= C_{1122} = \lambda\delta_{11}\delta_{22} + \mu(\delta_{12}\delta_{12} + \delta_{12}\delta_{12}) = \lambda \\ \bar{C}_{44} &= (C_{1111} - C_{1122})/2 = \mu \end{aligned}$$

The stress-strain relationship can be split up into spherical and deviatoric parts as follows:

$$\begin{aligned} \sigma_{ij} &= C_{ijk\ell}\varepsilon_{k\ell} \\ \sigma_{ij} &= \left[\lambda\delta_{ij}\delta_{k\ell} + \mu(\delta_{ik}\delta_{j\ell} + \delta_{i\ell}\delta_{jk})\right]\varepsilon_{k\ell} \\ \sigma_{ii} &= \sigma_{ij}\delta_{ij} = \left[\lambda\delta_{ij}\delta_{k\ell}\delta_{ij} + \mu(\delta_{ik}\delta_{j\ell} + \delta_{i\ell}\delta_{jk})\delta_{ij}\right]\varepsilon_{k\ell} \\ &= [3\lambda + 2\mu]\delta_{k\ell}\varepsilon_{k\ell} \\ &= [3\lambda + 2\mu]\varepsilon_{kk} \end{aligned}$$

$$\sigma_{ii} = 3\left(\lambda + \frac{2}{3}\mu\right)\varepsilon_{kk} = 3K\varepsilon_{kk}; \quad K = \lambda + \frac{2}{3}\mu \tag{4.32}$$

Equation (4.32) is the spherical part of the stress-strain relation. This relates the mean normal pressure $p = \sigma_{ii}/3$ to the volumetric strain $\varepsilon_v = \varepsilon_{kk}$. K is the *bulk modulus*.

The deviatoric stress-strain relation is obtained as follows:

$$\begin{aligned} s_{ij} &= \sigma_{ij} - \frac{\sigma_{kk}}{3}\delta_{ij} \\ &= \left[\lambda\delta_{ij}\delta_{k\ell} + \mu(\delta_{ik}\delta_{j\ell} + \delta_{i\ell}\delta_{jk})\right]\varepsilon_{k\ell} - \left(\lambda + \frac{2}{3}\mu\right)\varepsilon_{kk}\delta_{ij} \\ &= 2\mu\left\{\varepsilon_{ij} - \frac{1}{3}\varepsilon_{kk}\delta_{ij}\right\} \\ &= 2\mu e_{ij} \end{aligned} \tag{4.33}$$

e_{ij} is the deviatoric strain. Equation (4.33) is the deviatoric stress-strain relation. It may be recognized that the Lamé constant μ is the shear modulus G, i.e.,

$$G = \mu \tag{4.34}$$

By combining (4.32) and (4.33), the inverse stress-strain relation is derived as

$$\begin{aligned} \varepsilon_{ij} &= e_{ij} + \frac{1}{3}\varepsilon_{kk}\delta_{ij} \\ &= \frac{1}{2\mu}s_{ij} + \frac{1}{3(3\lambda + 2\mu)}\sigma_{kk}\delta_{ij} \\ &= \frac{1}{2\mu}(\sigma_{ij} - \frac{1}{3}\sigma_{kk}\delta_{ij}) + \frac{1}{3(3\lambda + 2\mu)}\sigma_{kk}\delta_{ij} \\ &= \frac{1}{2\mu}\sigma_{ij} + \left[\frac{6\mu - 9\lambda - 6\mu}{18(3\lambda + 2\mu)\mu}\right]\sigma_{kk}\delta_{ij} \end{aligned}$$

$$\varepsilon_{ij} = \frac{1}{2\mu}\sigma_{ij} - \frac{\lambda}{2\mu(3\lambda + 2\mu)}\sigma_{kk}\delta_{ij} \tag{4.35a}$$

or

$$\varepsilon_{ij} = \left[-\frac{\lambda}{2\mu(3\lambda + 2\mu)}\delta_{ij}\delta_{k\ell} + \frac{1}{\mu}(\delta_{ik}\delta_{j\ell} + \delta_{i\ell}\delta_{jk})\right]\sigma_{k\ell} \tag{4.35b}$$

$$\varepsilon_{ij} = S_{ijk\ell}\sigma_{k\ell}; \quad S_{ijk\ell} = -\frac{\lambda}{2\mu(3\lambda + 2\mu)}\delta_{ij}\delta_{k\ell} + \frac{1}{\mu}(\delta_{ik}\delta_{j\ell} + \delta_{i\ell}\delta_{jk}) \tag{4.35c}$$

where $S_{ijk\ell}$ is the compliance tensor.

In a matrix-vector form:

$$\begin{Bmatrix} \varepsilon_{11} \\ \varepsilon_{22} \\ \varepsilon_{33} \\ \varepsilon_{12} \\ \varepsilon_{23} \\ \varepsilon_{13} \end{Bmatrix} = \begin{bmatrix} \alpha & \beta & \beta & 0 & 0 & 0 \\ \beta & \alpha & \beta & 0 & 0 & 0 \\ \beta & \beta & \alpha & 0 & 0 & 0 \\ 0 & 0 & 0 & \gamma & 0 & 0 \\ 0 & 0 & 0 & 0 & \gamma & 0 \\ 0 & 0 & 0 & 0 & 0 & \gamma \end{bmatrix} \begin{Bmatrix} \sigma_{11} \\ \sigma_{22} \\ \sigma_{33} \\ \sigma_{12} \\ \sigma_{23} \\ \sigma_{13} \end{Bmatrix} \tag{4.35d}$$

where

$$\alpha = \frac{\lambda+\mu}{\mu(3\lambda+2\mu)}; \quad \beta = -\frac{\lambda}{2\mu(3\lambda+2\mu)}; \quad \gamma = \frac{1}{2\mu} \tag{4.35e}$$

As pointed out earlier, only two independent constants are needed to describe the stress-strain relations of an isotropic elastic material. Young's modulus E and Poison's ratio v are related to the bulk modulus and shear modulus by

$$K = \frac{E}{3(1-2v)}; \quad G = \frac{E}{3(1+v)} \tag{4.36}$$

4.4 Limits on Elastic Parameter Values

We will consider only isotropic materials in this section. The bounds for the values of elastic parameters may be established from the principles of thermodynamics or from physical observations. We use the latter approach here. For finite (i.e., not infinite) strains, the stress is finite, and for finite stress, the strain is finite. This leads to:

$$E, G, K < \infty; \quad E, G, K \neq 0 \tag{4.37}$$

When a material is subjected to (a) a compressive linear stress, the material experiences a compressive linear strain; (b) a compressive hydrostatic stress, the material experiences a compressive volumetric strain, and (c) a shear stress, the material experiences a shear strain of the same sign. These observations yield

$$E, G, K > 0 \tag{4.38}$$

Combining (4.37) and (4.38):

$$0 < E, G, K < \infty \tag{4.39}$$

Table 4.1 Examples of the values of elastic parameters

Material	E (GPa)	ν
Rubber	0.003	~ 0.5
Timber	4	0.5
Concrete	10	0.2
Glass	72.1	0.23
Aluminum	70.3	0.345
Brass	103.5	0.33
Iron	211	0.29
Tungsten	400	0.27
Tungsten carbide	534.4	0.22

The combination of the condition (4.39) and the relations in (4.36), one obtains the following restriction for the Poisson's ratio:

$$-1 < \nu < \frac{1}{2}$$

Limits on the properties for orthotropic materials may be found in Lempriere (1968).

Some examples of the values of E and ν are presented in Table 4.1 (Rees 2000). Note that the values listed for some materials (e.g., timber) are typical values. Poisson's ratio of rubber is close to 0.5. The Poison's ratio of cork (not presented here) is approximately 0. Orthotropic properties of various types of wood may be found in Green et al. (1999).

4.5 Loading in Non-principal Material Directions

The simple relations derived in Sect. 4.2 for anisotropic materials are valid only when the loading axes coincide with the principal material directions. For example, the orthotropic relations (4.17)–(4.19) and (4.21) are valid only when the loading directions coincide with the $x_1 - x_2 - x_3$ axes shown in Fig. 4.3. Otherwise, a coordinate transformation is needed to obtain the relevant relations. Let:

$$\sigma_{ij} = C_{ijk\ell}\varepsilon_{k\ell} \quad \text{in the principal material directions}$$

$$\sigma'_{ij} = C'_{ijk\ell}\varepsilon'_{k\ell} \quad \text{in the loading directions}$$

Since $C_{ijk\ell}$ is a tensor (4.3c), the following inverse relation holds:

$$C'_{ijpq} = C_{k\ell mn}a_{ik}a_{j\ell}a_{pm}a_{qn} \tag{4.40}$$

Equation (4.40) can be used to calculate the stiffness tensor in the loading directions from the elastic parameters defined in the principal material directions.

The coefficients of the 6 × 6 matrix $\bar{\mathbf{C}}$ can then be generated as before. The matrix will still be symmetric.

Problems

Problem 4.1 Determine the form of $\bar{\mathbf{C}}$ (4.10b) when there is one plane of symmetry with respect to the $x_1 = 0$ plane. Ans:

$$\bar{\mathbf{C}} = \begin{bmatrix} C_{1111} & C_{1122} & C_{1133} & 0 & C_{1123} & 0 \\ C_{1122} & C_{2222} & C_{2233} & 0 & C_{2223} & 0 \\ C_{1133} & C_{2233} & C_{3333} & 0 & C_{3323} & 0 \\ 0 & 0 & 0 & C_{1212} & 0 & C_{1213} \\ C_{2311} & C_{2322} & C_{2333} & 0 & C_{2323} & 0 \\ 0 & 0 & 0 & C_{1312} & 0 & C_{1313} \end{bmatrix}$$

Problem 4.2 Equation (4.17) was obtained for an orthotropic material by imposing symmetries about the $x_3 = 0$ and $x_2 = 0$ planes. Show that a further consideration of symmetry with respect to the $x_1 = 0$ plane leaves (4.17) unchanged.

Problem 4.3 The properties of an orthotropic material are: $E_1 = 200$ GPa, $E_2 = 150$ GPa, $E_3 = 100$ GPa, $\nu_{12} = 0.25$, $\nu_{23} = 0.30$, $\nu_{13} = 0.35$, $G_{12} = 100$ GPa, $G_{23} = 80$ GPa, and $G_{13} = 60$ GPa. Determine the stresses corresponding to the following strains:

$$\boldsymbol{\varepsilon} = \begin{bmatrix} 0.01 & 0.005 & 0 \\ 0.005 & 0.02 & 0 \\ 0 & 0 & 0.03 \end{bmatrix}$$

What are the values of the mean normal pressure and the deviatoric stress invariant (i.e., J defined in Chap. 2)?

Problem 4.4 Verify that with the transformation matrix given by (4.23a), the stiffness matrix given by (4.22b) reduces to the one given by (4.23b). Ans:

$$\begin{aligned} &\bar{C}'_{11} = C_{1111} = +C_{2222}, \bar{C}'_{12} = C_{1122} = +C_{2211}, \bar{C}'_{13} = C_{1133} = +C_{2233}, \\ &\bar{C}'_{14} = C_{1112} = -C_{2221}, \bar{C}'_{22} = C_{2222} = +C_{1111}, \bar{C}'_{23} = C_{2233} = +C_{1133}, \\ &\bar{C}'_{24} = C_{2212} = -C_{1121}, \bar{C}'_{33} = C_{3333} = +C_{3333}, \bar{C}'_{34} = C_{3312} = -C_{3321}, \\ &\bar{C}'_{44} = C_{1212} = +C_{2121}, \bar{C}'_{55} = C_{2323} = +C_{1313}, \bar{C}'_{56} = C_{2313} = -C_{1323}, \\ &\bar{C}'_{66} = C_{1313} = +C_{2323} \end{aligned}$$

These results, in conjunction with the symmetries such as $C_{1323} = C_{2313}$, lead to the solution.

Problem 4.5 Assume that the stiffness matrix $\bar{\mathbf{C}}$ of Problem 4.3 is defined in the $x_1 - x_2 - x_3$ coordinate system. Find the values of its coefficients in a rotated coordinate system $x_1' - x_2' - x_3'$ obtained by rotating $x_1 - x_2 - x_3$ system about the x_2-axis through an angle of 45° clockwise.

If the specimen is loaded in the $x_1' - x_2' - x_3'$ with the following strain tensor, determine the induced stresses:

$$\boldsymbol{\varepsilon}' = \begin{bmatrix} 0.01 & 0 & 0 \\ 0 & 0.02 & 0 \\ 0 & 0 & 0.03 \end{bmatrix}$$

The following algorithm may be used for computer implementation:

1. Define and initialize the following arrays
   (a) Stress array S(6)
   (b) Strain array E(6)
   (c) Stiffness tensor array (in $x_1 - x_2 - x_3$ system) C(3,3,3,3)
   (d) Transformation matrix array A(3, 3)
   (e) An integer array II(6) = (11,22,33,12,23,13)
2. Read in or type in the values of E(6), C(3,3,3,3) and A(3, 3)
3. Calculate $\bar{\mathbf{C}}'$ using the following algorithm
   (a) DO 1000 M = 1,6
   (b) I = II(M)/10
   (c) J = MOD(II(M),10)
   (d) DO 1000 N = 1,6
   (e) K = II(N)/10
   (f) L = MOD(II(N), 10)
   (g) CBAR(M,N) = 0.0
   (h) DO 1000 NP = 1,3
   (i) DO 1000 NQ = 1,3
   (j) DO 1000 NR = 1,3
   (k) DO 1000 NS = 1,3
   (l) CBAR(M,N) = CBAR(M,N) + C(NP, NQ, NR, NS)*A(NP, I)*A(NQ, J) *A(NR, K)*A(NS, L)
   (m) 1000 CONTINUE
4. Calculate the stress vector in the $x_1' - x_2' - x_3'$ system as
   (a) DO 2000 I = 1,6
   (b) S(I) = 0.0
   (c) DO 2000 K = 1,6
   (d) S(I) = S(I) + CBAR(I, K)*E(K)
   (e) 2000 CONTINUE
5. Print results

Chapter 5
Finite Element Analysis of Solids and Structures

In Chap. 3, the equations governing the single-physics, load-deformation problems are presented. The boundary/initial value problem is formally defined. In the present chapter, the finite element method of solving this boundary/initial value problem is described. The notations used in this chapter are similar to those in Chap. 3. While the finite element method is presented in a way that is suitable for both the linear and nonlinear problems, the specific formulations and the application examples in this chapter are restricted to elastic solids. The theory of elasticity for isotropic and anisotropic solids is presented in Chap. 4. Applications to elasto-plastic solids are addressed in Chaps. 8 and 10–13.

The finite element method is a numerical method for obtaining approximate solutions for boundary value problems that are defined in terms of a set of governing mathematical equations, boundary and/or initial conditions, constitutive laws and other secondary relations such as the strain–displacement relations. In general, any one of a *number of different principles* (e.g., virtual work principle, minimum potential energy principle, variational principle, etc.) may be used as the *basis* for developing the finite element equations for a given boundary value problem. All these methods share some common characteristics.

As illustrated in Fig. 5.1 for a typical two-dimensional problem, the domain is divided into a number of *elements*, interconnected at a set of *nodes*. The *primary unknowns* at a given point within the element are expressed in terms of their *nodal values* with the aid of piece-wise shape functions. For example, in a displacement-based finite element method, the displacement vector at any point within an element is expressed as a function of its nodal values as

$$\mathbf{u} = \mathbf{N}\hat{\mathbf{u}} \tag{5.1a}$$

where $\mathbf{u}$ is the vector containing the displacement unknowns at any point within the element, $\hat{\mathbf{u}}$ is a vector containing the nodal unknowns surrounding that element and $\mathbf{N}$ is a matrix containing the shape functions. The size of $\mathbf{N}$ depends on the number of displacement components (one, two or three) and the order of the shape functions (linear, quadratic, cubic, etc.).

A. Anandarajah, *Computational Methods in Elasticity and Plasticity: Solids and Porous Media*, DOI 10.1007/978-1-4419-6379-6_5,

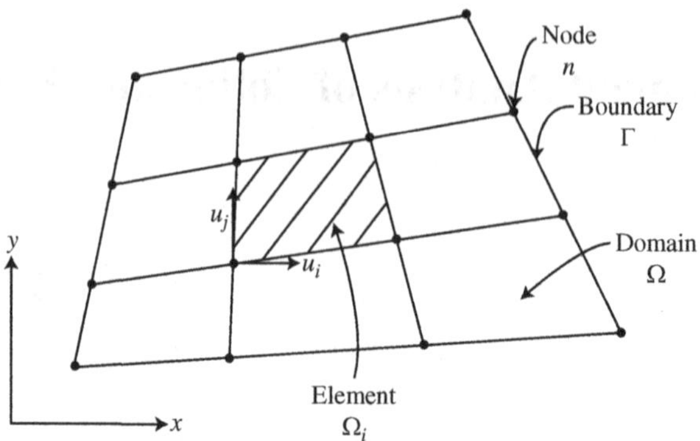

Fig. 5.1 Typical finite element mesh

The *secondary unknowns* are then derived from (5.1a). They are expressed in terms of $\hat{\mathbf{u}}$ as well. For example, in a solid mechanics problem, the strain vector is expressed in terms of $\hat{\mathbf{u}}$ as

$$\boldsymbol{\varepsilon} = \mathbf{B}\hat{\mathbf{u}} \tag{5.1b}$$

Using a suitable principle as the basis, the nodal unknowns $\hat{\mathbf{u}}$ are first determined. The secondary variables are then evaluated using appropriate equations. For example, in the solid mechanics problems, the strain is calculated using (5.1b). The stress is then calculated using the stress–strain (i.e., the constitutive) relations.

The assumed displacement field is continuous within the elements and across the interelement boundaries (as dictated by the governing equations and the method of solution). The stresses and strains are continuous within elements, but discontinuous across interelement boundaries. Special interpolation procedures may be used to determine the best estimates of stresses and strains on the interelement boundaries.

With this cursory introduction to the finite element method, we now illustrate some of the principles that are suitable for the finite element modeling of solid mechanics problems. For additional information, the reader may refer to any standard text on the finite element method or approximate methods (e.g., Zienkiewicz and Taylor 1991; Reddy 1993; Kaliakin 2001; Bathe 1982).

5.1 Basis for Finite Element Equations: Static Problems

5.1.1 Virtual Work Method

5.1.1.1 Principle

A principle that is widely used in solving solid mechanics problems is the virtual work principle. The virtual work principle is a variation of the equilibrium equations and boundary conditions (Chap. 3) involved in the definition of boundary value

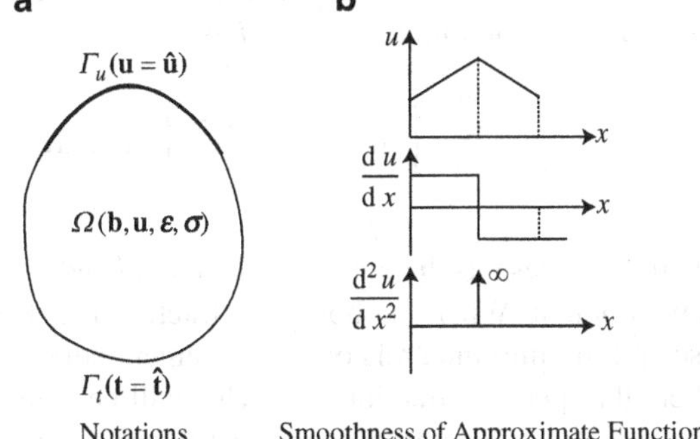

Fig. 5.2 Notations used for variables and smoothness of approximate functions

problem in solid mechanics. The connection is described in Appendix 1. Also, it will be shown in Sect. 5.1.2 that the virtual work principle is identical to what is known as the "weak" form of the boundary value problem definition. The generalized method of deriving the "weak" form is presented in Sect. 5.1.2. The notations used for the variables are presented in Fig. 5.2a.

Specifically, we will apply the *principle of virtual displacements*, which is a variation of the virtual work principle. The virtual displacement is taken as the variation of the actual displacement, i.e., $\delta\mathbf{u}$, where $\mathbf{u}$ is the actual displacement. As stated in Appendix 1, $\delta\mathbf{u}$ must be continuous. The consistent virtual strain $\delta\boldsymbol{\varepsilon}$ is derived from δu through the strain–displacement relation. Since $\mathbf{u} = \hat{\mathbf{u}}$ on Γ_u, its variation must vanish on Γ_u.

On the basis of the above, we define a "pertinent kinematically admissible field" ($\delta\mathbf{u}$, $\delta\boldsymbol{\varepsilon}$) as one that satisfies

- $\delta\mathbf{u} = 0$ on Γ_u
- Continuous in Ω and smooth enough for $\delta\boldsymbol{\varepsilon}$ to be derived from it

Applying the virtual work principle:
The internal virtual work – external virtual work = 0

$$\int_{\Omega} \delta\boldsymbol{\varepsilon}^{\mathrm{T}}\boldsymbol{\sigma}\,\mathrm{d}v - \int_{\Omega} \delta\mathbf{u}^{\mathrm{T}}\mathbf{b}\,\mathrm{d}v - \int_{\Gamma_t} \delta\mathbf{u}^{\mathrm{T}}\mathbf{t}\,\mathrm{d}s = 0 \tag{5.2}$$

Since $\delta\mathbf{u} = 0$ over Γ_u, the boundary term includes only the integral over Γ_t. $\mathbf{b}$ is the body force vector. The first term represents the internal virtual work and the remaining two terms represent the external virtual work.

Equation (5.2) provides us with a basis for determining the unknowns of the problem. Taking the displacements as the primary unknown variables, (a) we approximate the displacement field within the domain in terms of the primary unknowns, (b) express all other secondary variables of the problem ($\boldsymbol{\sigma}$ and $\boldsymbol{\varepsilon}$ in the present case) in terms of the primary unknowns, and (c) use (5.2) to determine the primary unknowns. It is understood that a constitutive law is used to relate $\boldsymbol{\sigma}$ and $\boldsymbol{\varepsilon}$.

In the finite element analysis, the volume integrals in (5.2) are divided up into integrals over each finite element i as

$$\sum_{i=1}^{N_E}\int_{\Omega_i}\delta\boldsymbol{\varepsilon}^{\mathrm{T}}\boldsymbol{\sigma}\,\mathrm{d}v-\sum_{i=1}^{N_E}\int_{\Omega_i}\delta\mathbf{u}^{\mathrm{T}}\mathbf{b}\,\mathrm{d}v-\int_{\Gamma_t}\delta\mathbf{u}^{\mathrm{T}}\mathbf{t}\,\mathrm{d}s=0 \tag{5.3}$$

where Ω_i represents the volume of element i and N_E is the number of finite elements in the domain. When the boundary traction $\mathbf{t}$ is distributed, the surface integral is also split up into integrals over the element edges that coincide with the boundary where the specified traction acts. This will be elaborated later.

Consider a typical term in the first integral, associated with a typical element Ω_i (Fig. 5.1). From (5.1a) and (5.1b), by denoting the vector containing the nodal unknowns surrounding element i as $\hat{\mathbf{u}}_{\mathrm{i}}$, we have

$$\mathbf{u}=\mathbf{N}\hat{\mathbf{u}}_{\mathrm{i}} \tag{5.4a}$$

$$\boldsymbol{\varepsilon}=\mathbf{B}\hat{\mathbf{u}}_{\mathrm{i}} \tag{5.4b}$$

and

$$\delta\mathbf{u}=\mathbf{N}\delta\hat{\mathbf{u}}_{\mathrm{i}} \tag{5.4c}$$

$$\delta\boldsymbol{\varepsilon}=\mathbf{B}\delta\hat{\mathbf{u}}_{\mathrm{i}} \tag{5.4d}$$

Equation (5.4a) implies a certain assumption concerning the variation of the displacement within a given element and on its boundary (and hence on the inter-element boundaries). The "order" of the shape function is chosen so that the integrals in (5.2) will not become infinite. Equation (5.2) involves stresses and strains. In linear elastic materials, stresses are linearly related to strains. The strains are derived from the first derivatives of the displacements (Chap. 3). Thus, the first derivative of the displacements must not become infinite within the element. Further, the function must be continuous on interelement boundaries, as shown in Fig. 5.2b. Such functions are called C_0 continuous functions This can be generalized to state that if the highest order of derivatives in the integral is n, the shape function must be at least $C_{n\text{-}1}$ *continuous.*

For an elastic material, the constitutive law is of the form

$$\boldsymbol{\sigma}=\mathbf{D}\boldsymbol{\varepsilon} \tag{5.4e}$$

where $\mathbf{D}$ is the symmetric material stiffness matrix (6×6 in size in three-dimensional analysis). The symmetry of $\mathbf{D}$ is proved in Chap. 4.

The first integral in (5.3) for element i becomes

$$I_{\mathrm{i}}^{1}=\int_{\Omega_i}\delta\boldsymbol{\varepsilon}^{\mathrm{T}}\boldsymbol{\sigma}\,\mathrm{d}v=\delta\hat{\mathbf{u}}_{\mathrm{i}}^{\mathrm{T}}\int_{\Omega_i}\mathbf{B}^{\mathrm{T}}\boldsymbol{\sigma}\,\mathrm{d}v=\delta\hat{\mathbf{u}}_{\mathrm{i}}^{\mathrm{T}}\left[\int_{\Omega_i}\mathbf{B}^{\mathrm{T}}\mathbf{D}\mathbf{B}\,\mathrm{d}v\right]\hat{\mathbf{u}}_{\mathrm{i}}=\delta\hat{\mathbf{u}}_{\mathrm{i}}^{\mathrm{T}}\mathbf{k}_{\mathrm{i}}\hat{\mathbf{u}}_{\mathrm{i}} \tag{5.5a}$$

where

$$\mathbf{k}_i = \int_{\Omega_i} \mathbf{B}^{\mathrm{T}}\mathbf{D}\mathbf{B}\mathrm{d}v \tag{5.5b}$$

and is referred to as the *element stiffness matrix*. The second integral for element i takes the form

$$\mathbf{I}_i^2 = \int_{\Omega_i} \delta\mathbf{u}_i^{\mathrm{T}}\mathbf{b}\,\mathrm{d}v = \delta\hat{\mathbf{u}}_i^{\mathrm{T}} \int_{\Omega_i} \mathbf{N}^{\mathrm{T}}\mathbf{b}\,\mathrm{d}v = \delta\hat{\mathbf{u}}_i^{\mathrm{T}}\mathbf{f}_i^b \tag{5.6a}$$

where

$$\mathbf{f}_i^b = \int_{\Omega_i} \mathbf{N}^{\mathrm{T}}\mathbf{b}\,\mathrm{d}v \tag{5.6b}$$

and is referred to as the *element force vector* due to body force. Even the third term in (5.3) can be split up into elemental components. Each elemental term leads to element force vector $\mathbf{f}_\mathrm{i}^\mathrm{t}$ due to boundary traction as

$$\mathbf{I}_i^3 = \int_{\Gamma_{ti}} \delta\mathbf{u}_i^{\mathrm{T}}\mathbf{t}\,\mathrm{d}s = \delta\hat{\mathbf{u}}_i^{\mathrm{T}} \int_{\Gamma_{ti}} \mathbf{N}^{\mathrm{T}}\mathbf{t}\,\mathrm{d}s = \delta\hat{\mathbf{u}}_i^{\mathrm{T}}\mathbf{f}_i^t \tag{5.7a}$$

where

$$\mathbf{f}_i^t = \int_{\Gamma_{ti}} \mathbf{N}^{\mathrm{T}}\mathbf{t}\,\mathrm{d}s \tag{5.7b}$$

Note that the boundary integral in (5.7b) is performed only over that part of the boundary that coincides with the boundaries of element i. Equation (5.3) now becomes

$$\sum_{i=1}^{N_\mathrm{E}} I_i^1 - \sum_{i=1}^{N_\mathrm{E}} I_i^2 - \sum_{i=1}^{N_\mathrm{E}} I_i^3 = 0 \tag{5.8a}$$

$$\sum_{i=1}^{N_\mathrm{E}} \delta\hat{\mathbf{u}}_i^{\mathrm{T}}\mathbf{k}_i\hat{\mathbf{u}}_i - \sum_{i=1}^{N_\mathrm{E}} \delta\hat{\mathbf{u}}_i^{\mathrm{T}}\mathbf{f}_i^b - \sum_{i=1}^{N_\mathrm{E}} \delta\hat{\mathbf{u}}_i^{\mathrm{T}}\mathbf{f}_i^t = 0 \tag{5.8b}$$

There is a systematic way of performing the summations involved in (5.8b), which we will discuss in detail later. Briefly, a global primary unknown nodal displacement vector $\hat{\mathbf{u}}$ is defined. Consistent with the definition of $\hat{\mathbf{u}}$, a global stiffness matrix $\mathbf{K}$ and a global load vector $\mathbf{P}$ are defined. $\mathbf{K}$ is formed by assembling $\mathbf{k}_i$ for $i = 1, N_E$. $\mathbf{P}$ is formed by assembling $\mathbf{f}_i^b$ and $\mathbf{f}_i^t$ for $i = 1, N_E$. In addition, lumped nodal forces are added into $\mathbf{P}$. Equation (5.8b) then takes the form:

$$\delta\hat{\mathbf{u}}^T(\mathbf{K}\hat{\mathbf{u}} - \mathbf{P}) = 0 \tag{5.9}$$

If (5.9) is to be true for any arbitrary and nontrivial (nonzero) virtual nodal displacement $\delta\hat{\mathbf{u}}$, then

$$\mathbf{K}\hat{\mathbf{u}} = \mathbf{P} \tag{5.10}$$

Thus, a set of linear simultaneous equations are obtained. At this stage, the displacement boundary conditions are imposed and a suitable solution method is used to solve for $\hat{\mathbf{u}}$.

$\mathbf{K}$ is obtained by assembling element $\mathbf{k}_i$ (5.5b). $\mathbf{K}$ is symmetric if $\mathbf{k}_i$ is symmetric. If follows from (5.5b) that $\mathbf{k}_i$ is symmetric if $\mathbf{D}$ is symmetric. In elastic analysis, $\mathbf{D}$ is indeed symmetric (Chap. 3) and hence $\mathbf{K}$ is symmetric.

The virtual work method presented here is applicable to both linear and nonlinear systems. For nonlinear systems, $\mathbf{D}$ is not in general symmetric, and hence $\mathbf{K}$ is not in general symmetric. As will be seen in Chap. 9, when $\mathbf{D}$ is derived from an elasto-plastic constitutive model that employs a nonassociated flow rule, $\mathbf{D}$ is nonsymmetric. Depending on the nonlinear method employed, $\mathbf{D}$ may either be what is known as the continuum tangent operator or the consistent tangent operator (Chaps. 8–10). When associated flow rule is employed, the former is symmetric, but the latter may or may not be.

It can be shown that in the displacement method, the structure is modeled to be stiffer than the actual structure, i.e., the calculated displacements are smaller than the actual displacements. As the number of elements are increased, the solution approaches the exact solution for $N_E \geq N_E^*$. N_E^* varies with the actual nonlinearity of the spatial variation of displacements and the order of shape functions employed in the development of the element stiffness matrices. For instance, in the problem of an axially loaded one-dimensional rod subjected to a point load at one end and fixed at the other end, the displacement varies linearly with x. In this case, one element based on linear shape function is adequate to provide exact results, i.e., $N_E^* = 1$.

5.1.1.2 Finite Element Equations for One-Dimensional Rod

In this section, we develop finite element equations by using the virtual work principle to solve the one-dimensional rod problem shown in Fig. 5.3a. The objective is to solve for the displacements at points B and C.

We begin by defining the matrices $\mathbf{N}$ and $\mathbf{B}$ for a typical element i. Since the problem is one-dimensional, we represent the displacement vector at any point x by a scalar $u(x)$. To develop the element equations, we introduce a *local coordinate*

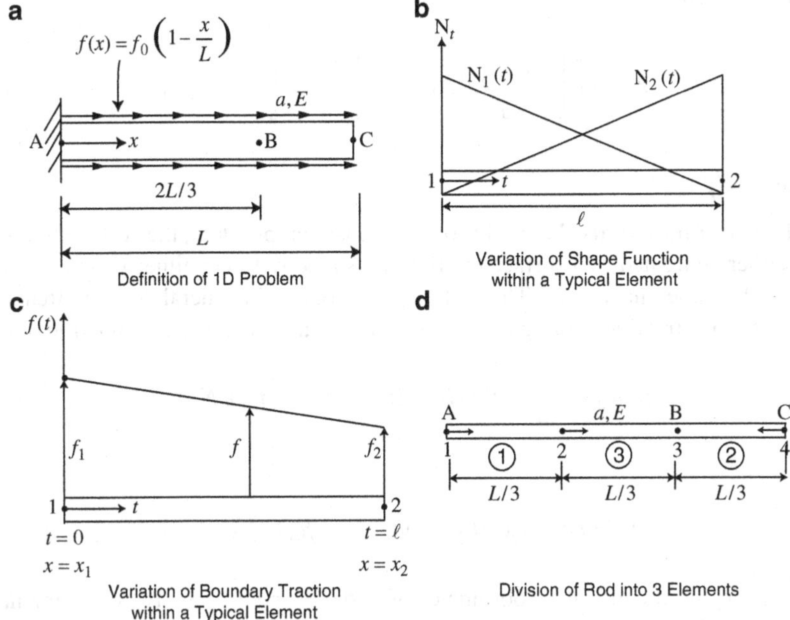

Fig. 5.3 Definition of a one-dimensional rod problem

system t as shown in Fig. 5.3b. Note that the positive direction of t is taken from local node 1 to local node 2. The next step is to choose a *shape function* of an *acceptable order*. The shape function must be a polynomial of at least order 0. Here we choose a linear function (order 1), as shown in Fig. 5.3b. The displacement unknown is expressed as

$$u = N_1\hat{u}_1 + N_2\hat{u}_2 = \mathbf{N}\hat{\mathbf{u}} \tag{5.11}$$

where $N_1 = 1 - t/\ell$ and $N_2 = t/\ell$, and $\underset{1\times 2}{\mathbf{N}} = [N_1\ N_2]$ is a 1×2 shape matrix (the N-matrix). Note the special characteristics of the shape functions: at local node 1, $N_1 = 1$ and $N_2 = 0$. At local node 2, $N_1 = 0$ and $N_2 = 1$. The result is that $u = \hat{u}_1$ at node 1, and $u = \hat{u}_2$ at node 2.

From the strain–displacement relation

$$\varepsilon = \frac{\mathrm{d}u}{\mathrm{d}x} = \frac{\mathrm{d}u}{\mathrm{d}t} = B_1\hat{u}_1 + B_2\hat{u}_2 = \mathbf{B}\hat{\mathbf{u}} \tag{5.12}$$

where $B_1 = -1/\ell$ and $B_2 = 1/\ell$, and $\underset{1\times 2}{\mathbf{B}} = [B_1\ B_2]$ is a 1×2 matrix (the B-matrix).

The relevant constitutive equation is

$$\sigma = E\varepsilon = D\varepsilon \tag{5.13}$$

where E is Young's modulus, and $\underset{1\times 1}{D} = [E]$ is the 1×1 material stiffness matrix.

From (5.5b):

$$\underset{2\times 2}{\mathbf{k}_i} = \int_0^{\ell} \left[\begin{bmatrix} B_1 \\ B_2 \end{bmatrix} [E][B_1 \quad B_2] \right] a\mathrm{d}t = \begin{bmatrix} k & -k \\ -k & k \end{bmatrix} \tag{5.14}$$

where $k = \frac{Ea}{\ell}$.

The reader may verify that in this one-dimensional problem, the surface traction may either be treated as a surface traction or as a body force; either way, one ends up with the same final results. This is, however, not true in general. We will treat the load as surface traction. Noting that $x = x_1 + t$, we can express f_x as a function of t as

$$f_x = f_t = (1 - t/\ell)f_1 + (t/\ell)f_2 = N_1 f_1 + N_2 f_2 \tag{5.15a}$$

where

$$f_1 = f_0(1 - x_1/L) \quad \text{and} \quad f_2 = f_0(1 - x_2/L) \tag{5.15b}$$

x_1 and x_2 are the global x-coordinates of nodes 1 and 2 of the typical element i. f_1 and f_2 are the values of f at nodes 1 and 2. Note that the particular form of (5.15a) in terms of N_1 and N_2 is unique to the specific loading (linear) and not a general result. Now, from (5.7b):

$$\mathbf{f}_i^t = \int_0^{\ell} \left[\begin{bmatrix} N_1 \\ N_2 \end{bmatrix} [f_x] \right] \mathrm{d}t = \int_0^{\ell} \left[\begin{bmatrix} N_1 \\ N_2 \end{bmatrix} [N_1 f_1 + N_2 f_2] \right] \mathrm{d}t = \frac{\ell}{6} \begin{bmatrix} 2f_1 + f_2 \\ f_1 + 2f_2 \end{bmatrix} \tag{5.16}$$

Equations (5.14) and (5.16) are generalized expressions for the element stiffness matrix and the load vector due to surface traction for a typical element i, and can be used to evaluate the respective quantities for any given element.

Let us divide the rod into three elements of equal length as shown in Fig. 5.3d. There is no theoretical restriction for the choice of local directions. Just for generality and to demonstrate the manner in which global matrices are assembled from element matrices for general cases, we use an irregular numbering system. Specifically, we take the positive local directions of elements 1, 2, and 3 to be from (global node) 1 to 2, 4 to 3, and 2 to 3, respectively. The element stiffness and load vectors for each of the three elements are

$$\mathrm{k}_1 = \mathrm{k}_2 = \mathrm{k}_3 = \begin{bmatrix} k & -k \\ -k & k \end{bmatrix} \quad \text{where} \quad k = \frac{3Ea}{L} \tag{5.17a}$$

$$\mathrm{f}_1^t = \bar{f}\begin{bmatrix} 8 \\ 7 \end{bmatrix}; \quad \mathrm{f}_2^t = \bar{f}\begin{bmatrix} 1 \\ 2 \end{bmatrix}; \quad \mathrm{f}_3^t = \bar{f}\begin{bmatrix} 5 \\ 4 \end{bmatrix} \quad \text{where} \quad \bar{f} = \frac{Lf_0}{54} \tag{5.17b}$$

Now the element matrices are assembled into the global matrices. To achieve this, we must find the correspondence between the two nodes surrounding each element and the global nodes. In this example, the local nodes 1 and 2 of element 1 go with the global nodes 1 and 2, respectively; those of element 2 go with global nodes 4 and 3, respectively; and those of element 3 go with global nodes 2 and 3, respectively. At this point, let us define a global displacement vector as

$$\hat{\mathbf{u}} = [\hat{u}_1, \hat{u}_2, \hat{u}_3, \hat{u}_4] \tag{5.18}$$

Denoting the stiffness matrix of element 1 by

$$\mathrm{k}_1 = \begin{bmatrix} k_{11}^1 & k_{12}^1 \\ k_{21}^1 & k_{22}^1 \end{bmatrix}, \text{ etc.} \tag{5.19}$$

the first summation in (5.8b) can be written as

$$[\delta\hat{u}_1, \delta\hat{u}_2, \delta\hat{u}_3, \delta\hat{u}_4] \left[\begin{bmatrix} k_{11}^1 & k_{12}^1 & 0 & 0 \\ k_{21}^1 & k_{22}^1 & 0 & 0 \\ 0 & 0 & 0 & 0 \\ 0 & 0 & 0 & 0 \end{bmatrix} + \begin{bmatrix} 0 & 0 & 0 & 0 \\ 0 & 0 & 0 & 0 \\ 0 & 0 & k_{22}^2 & k_{21}^2 \\ 0 & 0 & k_{12}^2 & k_{11}^2 \end{bmatrix} + \begin{bmatrix} 0 & 0 & 0 & 0 \\ 0 & k_{11}^3 & k_{12}^3 & 0 \\ 0 & k_{21}^3 & k_{22}^3 & 0 \\ 0 & 0 & 0 & 0 \end{bmatrix} \right] \begin{bmatrix} \hat{u}_1 \\ \hat{u}_2 \\ \hat{u}_3 \\ \hat{u}_4 \end{bmatrix} \tag{5.20a}$$

$$= [\delta\hat{u}_1, \delta\hat{u}_2, \delta\hat{u}_3, \delta\hat{u}_4] \begin{bmatrix} k_{11}^1 & k_{12}^1 & 0 & 0 \\ k_{21}^1 & k_{22}^1 + k_{11}^3 & k_{12}^3 & 0 \\ 0 & k_{21}^3 & k_{22}^2 + k_{22}^3 & k_{21}^2 \\ 0 & 0 & k_{12}^2 & k_{11}^2 \end{bmatrix} \begin{bmatrix} \hat{u}_1 \\ \hat{u}_2 \\ \hat{u}_3 \\ \hat{u}_4 \end{bmatrix} \tag{5.20b}$$

$$= [\delta\hat{u}_1, \delta\hat{u}_2, \delta\hat{u}_3, \delta\hat{u}_4] \begin{bmatrix} k & -k & 0 & 0 \\ -k & 2k & -k & 0 \\ 0 & -k & 2k & -k \\ 0 & 0 & -k & k \end{bmatrix} \begin{bmatrix} \hat{u}_1 \\ \hat{u}_2 \\ \hat{u}_3 \\ \hat{u}_4 \end{bmatrix} \tag{5.20c}$$

The 4 × 4 matrix in the middle of (5.20c) is the global stiffness matrix **K**.

5.1.1.3 Method of Assembling Global Stiffness Matrix and Load Vector

The coefficients of a given element matrix may be directly placed into the global stiffness matrix by following the rule illustrated on Fig. 5.4. Here, the local nodes 1 and 2 of element i coincide with global nodes ℓ and k, respectively. Then, the first and second rows of k_i get placed on the ℓ th and kth rows of the global matrix, respectively. The first and second columns of k_i get placed on the ℓ th and kth columns of the global matrix, respectively.

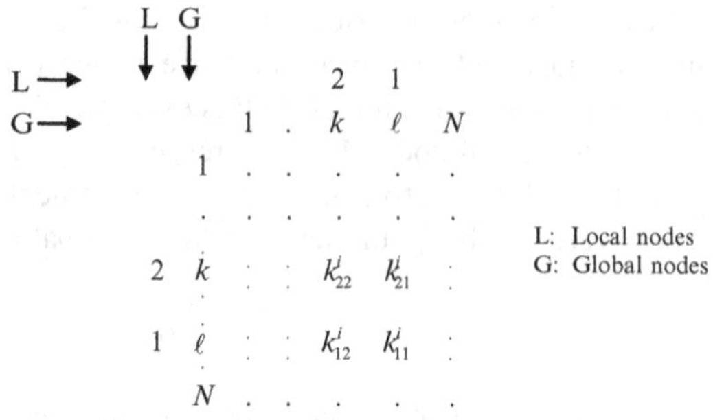

Fig. 5.4 The rule for assembling element stiffness matrices into the global stiffness matrix

$$
\begin{array}{ccccccc}
 & & & & 2 & 1 & \\
 & & 1 & . & k & \ell & N \\
 & 1 & . & . & . & . & . \\
 & . & . & . & . & . & . \\
2 & k & : & : & K^i_{22} & K^i_{21} & : \\
1 & \ell & : & : & K^i_{12} & K^i_{11} & : \\
 & N & . & . & . & . & .
\end{array}
$$

L: Local nodes
G: Global nodes

The global load vector is similarly assembled from the element load vectors. Denoting the load vector of element 1 by (dropping the superscript t to avoid confusion)

$$
\mathrm{f}_1 = \begin{bmatrix} f_1^1 \\ f_1^2 \end{bmatrix}, \text{ etc.} \tag{5.21}
$$

the third term in (5.8b) becomes

$$
[\delta\hat{u}_1, \delta\hat{u}_2, \delta\hat{u}_3, \delta\hat{u}_4] \left[\begin{bmatrix} f_1^1 \\ f_1^2 \\ 0 \\ 0 \end{bmatrix} + \begin{bmatrix} 0 \\ 0 \\ f_2^2 \\ f_2^1 \end{bmatrix} + \begin{bmatrix} 0 \\ f_3^1 \\ f_3^2 \\ 0 \end{bmatrix} \right] \tag{5.22a}
$$

$$
= [\delta\hat{u}_1, \delta\hat{u}_2, \delta\hat{u}_3, \delta\hat{u}_4] \begin{bmatrix} f_1^1 \\ f_1^2 + f_3^1 \\ f_2^2 + f_3^2 \\ f_2^1 \end{bmatrix} \tag{5.22b}
$$

$$
= [\delta\hat{u}_1, \delta\hat{u}_2, \delta\hat{u}_3, \delta\hat{u}_4] \begin{bmatrix} 8\bar{f} \\ 12\bar{f} \\ 6\bar{f} \\ \bar{f} \end{bmatrix} \tag{5.22c}
$$

The column vector in (5.22c) is the global load vector. Thus, the global equation becomes

$$
\begin{bmatrix} k & -k & 0 & 0 \\ -k & 2k & -k & 0 \\ 0 & -k & 2k & -k \\ 0 & 0 & -k & k \end{bmatrix} \begin{bmatrix} \hat{u}_1 \\ \hat{u}_2 \\ \hat{u}_3 \\ \hat{u}_4 \end{bmatrix} = \begin{bmatrix} 8\bar{f} \\ 12\bar{f} \\ 6\bar{f} \\ \bar{f} \end{bmatrix} \tag{5.23}
$$

The displacement boundary condition that

$$\hat{u}_1 = 0 \tag{5.24}$$

is imposed at this stage. This may be achieved by simply dropping the first column and first row from (5.23). We then get the following equation:

$$\begin{bmatrix} 2k & -k & 0 \\ -k & 2k & -k \\ 0 & -k & k \end{bmatrix} \begin{bmatrix} \hat{u}_2 \\ \hat{u}_3 \\ \hat{u}_4 \end{bmatrix} = \begin{bmatrix} 12\bar{f} \\ 6\bar{f} \\ 1\bar{f} \end{bmatrix} \tag{5.25a}$$

$$\begin{bmatrix} 2 & -1 & 0 \\ -1 & 2 & -1 \\ 0 & -1 & 1 \end{bmatrix} \begin{bmatrix} \hat{u}_2 \\ \hat{u}_3 \\ \hat{u}_4 \end{bmatrix} = \frac{\bar{f}}{k} \begin{bmatrix} 12 \\ 6 \\ 1 \end{bmatrix} \tag{5.25b}$$

which can be easily solved for the unknown displacements. The answer is $\{\hat{u}_2, \hat{u}_3, \hat{u}_4\} = \{19, 26, 27\}(\bar{f}/k)$.

Example 5.1.

Question. Referring to the rod with a variable cross-section shown in Fig. 5.5, determine the vertical deflection at Point 3 by the finite element method. Approximate the rod into a one-dimensional body. Represent the rod by two 2-noded linear finite elements. The radius of the rod varies linearly from 5 units at the fixed support (bottom) to 1 unit at the top. The Young's modulus of the rod is E. Compare your answers with the exact solution given in Chap. 1.

Answer. The element stiffness matrices of the bottom and top elements are

$$\mathbf{k}_1 = \frac{E\pi}{50}\begin{bmatrix} 17 & -17 \\ -17 & 17 \end{bmatrix} \quad \text{and} \quad \mathbf{k}_2 = \frac{E\pi}{50}\begin{bmatrix} 5 & -5 \\ -5 & 5 \end{bmatrix}$$

Local nodes 1 and 2 of element 1 coincide with the global nodes 1 and 2. Local nodes 1 and 2 of element 2 coincide with the global nodes 2 and 3.

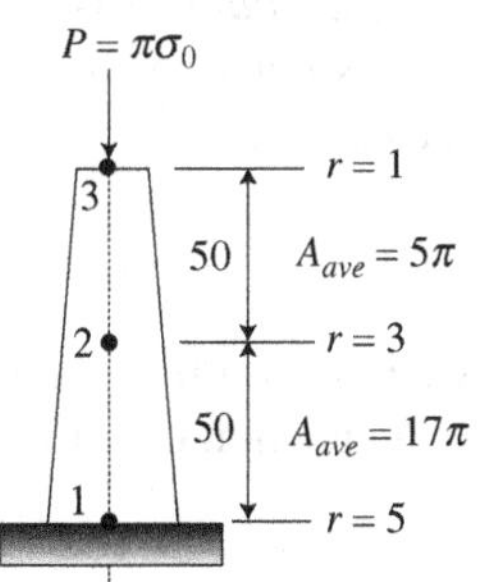

Fig. 5.5 Rod with variable cross-section

(These connectivities apparently do not make a difference in this problem.) The global stiffness matrix and the load vectors are

$$\mathbf{K} = \frac{E\pi}{50}\begin{bmatrix} 17 & -17 & 0 \\ -17 & 17+5 & -5 \\ 0 & -5 & 5 \end{bmatrix} \quad \text{and} \quad \mathbf{P} = \begin{bmatrix} 0 \\ 0 \\ -\pi\sigma_0 \end{bmatrix}$$

Applying the displacement boundary condition, the reduced system equation is

$$\frac{E\pi}{50}\begin{bmatrix} 22 & -5 \\ -5 & 5 \end{bmatrix}\begin{bmatrix} \hat{u}_2 \\ \hat{u}_3 \end{bmatrix} = \pi\sigma_0\begin{bmatrix} 0 \\ -1 \end{bmatrix}$$

Solving this

$$\hat{u}_3 = -\frac{220}{17}\frac{\sigma_0}{E} = -12.9\frac{\sigma_0}{E}$$

The analytical solution from Chap. 1 is

$$\hat{u}_3 = -\frac{\sigma_0 \times 100}{E}\left(\frac{1}{5}\right) = -20\frac{\sigma_0}{E}$$

5.1.2 *Variational Methods*

5.1.2.1 Definitions: Functional, Differential Operators

Let us begin by defining certain quantities that are central to the understanding of variational principles.

A *function* u is one that is represented in terms of independent variables, such as the spatial coordinates (x, y, z) and time t, e.g.,

$$u = \hat{u}(x, y, z, t) \tag{5.26}$$

A *functional* I is one that is represented in terms of independent variables, as well as functions of independent variables (i.e., "function of functions"), and is a scalar, e.g.,

$$I = \hat{I}(x, y, z, t, u, u_{,i}, u') \tag{5.27}$$

where u' is time derivative of u.

A functional $I(u)$ is *linear* if and only if it satisfies

$$I(\alpha u + \beta v) = \alpha I(u) + \beta I(v) \tag{5.28}$$

A functional $I(u, v)$ is bilinear if and only if it is linear in each of its arguments (u, v), i.e.,

$$I(\alpha u_1 + \beta u_2, v) = \alpha I(u_1, v) + \beta I(u_2, v) \tag{5.29}$$

and

$$I(u, \alpha v_1 + \beta v_2) = \alpha I(u, v_1) + \beta I(u, v_2) \tag{5.30}$$

A differential operator is one that operates on a variable when taking first derivative, second derivative, etc. For example,

$$\frac{\partial}{\partial x}\left(k_x \frac{\partial h}{\partial x}\right) = \frac{\partial}{\partial x}\left(k_x \frac{\partial}{\partial x}\right)(h) = L(h) \tag{5.31a}$$

where

$$L = \frac{\partial}{\partial x}\left(k_x \frac{\partial}{\partial x}\right) \tag{5.31b}$$

is the differential operator. With this compact notation, the equilibrium equations described in Chap. 3, which are linear differential equations, can be written in the following form:

$$\mathbf{A}(\mathbf{u}) = \mathbf{L}(\mathbf{u}) - \mathbf{f} = 0 \tag{5.32}$$

where $\mathbf{f}$ is in general a function of $\mathbf{u}$ and t.

In certain cases, $\mathbf{L}(\mathbf{u})$ is a scalar (i.e., involves a single differential equation) and in others, it is a vector (i.e., involves more than one differential equation). For instance, the steady-state heat flow or water flow is governed by a scalar equation, given by

$$A(h) = L(h) = 0; \quad L = k_x \frac{\partial^2}{\partial x^2} + k_y \frac{\partial^2}{\partial y^2} \tag{5.33a}$$

Where h is the hydraulic head, and k_x and k_y are permeabilities in the x- and y-directions, respectively. Whereas for the three-dimensional equilibrium equations given in Chap. 3 (where there are three equations), $\mathbf{L}(\mathbf{u})$ is a vector, given by

$$\mathbf{A}(\mathbf{u}) = \mathbf{L}(\mathbf{u}) - \mathbf{f} = 0 \tag{5.33b}$$

where

$$\mathbf{L}(\mathbf{u}) = [L_1(\mathbf{u}), L_2(\mathbf{u}), L_3(\mathbf{u})]; \quad \mathbf{A}(\mathbf{u}) = [A_1(\mathbf{u}), A_2(\mathbf{u}), A_3(\mathbf{u})] \quad \text{and } \mathbf{u} = (u, v, w) \tag{5.33c}$$

The governing equations and the boundary conditions for linear differential equations (e.g., for load-deformation problems) may then be defined as

$$\mathbf{A}(\mathbf{u}) = \mathbf{L}(\mathbf{u}) - \mathbf{f} = 0 \quad \text{in} \quad \Omega \tag{5.34a}$$

$$\mathbf{B}(\mathbf{u}) = 0 \quad \text{in} \quad \Gamma \tag{5.34b}$$

5.1.2.2 Variational Method

A variational method involves the determination of a functional corresponding to the governing equations and the boundary conditions in the general form

$$I(\mathbf{u}) = \int_{\Omega} F(\mathbf{u}, \mathbf{u}_{,x}, \mathbf{u}', \ldots) \mathrm{d}\boldsymbol{\Omega} + \int_{\Gamma} G(\mathbf{u}, \mathbf{u}_{,x}, \mathbf{u}', \ldots) \mathrm{d}s \tag{5.35}$$

such that the solution to the boundary value problem makes the functional $I(\mathbf{u})$ stationary.

It turns out that for some boundary value problems, a *natural variational statement* exists, and for the others, it has to be *contrived*. One of the conditions for the existence of a natural variational statement is that the differential operator be *self-adjoint*.

Definition. *Self-Adjoint Operator*

A differential operator is self adjoint if

$$\int_{\Omega} \mathbf{u}_1 L(\mathbf{u}_2) \mathrm{d}\Omega = \int_{\Omega} \mathbf{u}_2 L(\mathbf{u}_1) \mathrm{d}\Omega + \text{boundary terms} \tag{5.36}$$

Example 5.2.

Question. Show that the equation governing one-dimensional water flow

$$\frac{\mathrm{d}}{\mathrm{d}x}\left(k\frac{\mathrm{d}h}{\mathrm{d}x}\right) = 0 \tag{5.37}$$

is self adjoint.

Answer. Formula for Integration by Parts:

$$\int v\frac{\mathrm{d}u}{\mathrm{d}x}\mathrm{d}x = [uv]_{x=a}^{x=b} - \int u\frac{\mathrm{d}v}{\mathrm{d}x} \tag{5.38}$$

Integrating (5.37) by parts twice:

$$\int_{\Omega} \mathrm{u}_1 L(\mathrm{u}_2)\mathrm{d}\Omega = \int h_1 \frac{\mathrm{d}}{\mathrm{d}x}\left(k\frac{\mathrm{d}h_2}{\mathrm{d}x}\right)\mathrm{d}x \tag{5.39}$$

$$= -\int k\frac{\mathrm{d}h_2}{\mathrm{d}x}\frac{\mathrm{d}h_1}{\mathrm{d}x}\mathrm{d}x + \text{boundary terms}$$

$$= \int h_2 \frac{\mathrm{d}}{\mathrm{d}x}\left(k\frac{\mathrm{d}h_1}{\mathrm{d}x}\right)\mathrm{d}x + \text{boundary terms}$$

$$= \int \mathbf{u}_2 L(\mathbf{u}_1)\mathrm{d}\Omega + \text{boundary terms}$$

Thus, the differential operator involved in (5.37) is self adjoint.

According to the variational principle, the solution to the boundary value problem is obtained when

$$\delta I = 0 \tag{5.40}$$

where δI is the "variation" of I (5.35).

Consider a problem where the unknown variable is a scalar u (e.g., temperature), depending only on the spatial coordinates. Let the solution is approximated in terms of a set of global unknowns as

$$u \approx \sum N_i \hat{u}_i \tag{5.41}$$

where N_i are a set of global shape functions. A "hat" over the variable u_i is used to indicate that the variable is an approximation to u_i. (This type of approximation is referred to as the Rayleigh–Ritz approximation, and is different from the finite element approximation, where piece-wise shape functions are used.) The functional I given by (5.35) can then be expressed in terms of the global unknowns as

$$I \approx \hat{I}(u_1, u_2, \ldots u_n) \tag{5.42a}$$

Equation (5.40) becomes

$$\delta I \approx \delta\hat{I} = \frac{\partial I}{\partial \hat{u}_1}\delta\hat{u}_1 + \frac{\partial I}{\partial \hat{u}_2}\delta\hat{u}_2 + \cdots + \frac{\partial I}{\partial \hat{u}_n}\delta\hat{u}_n = 0 \tag{5.42b}$$

$$= \delta\hat{\mathbf{u}}^{\mathrm{T}}\frac{\partial \hat{I}}{\partial \hat{\mathbf{u}}} = 0 \tag{5.42c}$$

where

$$\delta\hat{\mathbf{u}}^{\mathrm{T}} = [\delta\hat{u}_1, \delta\hat{u}_2, \ldots.\delta\hat{u}_n] \tag{5.42d}$$

and

$$\left(\frac{\partial \hat{I}}{\partial \hat{\mathbf{u}}}\right)^{\mathrm{T}} = \left\{\frac{\partial \hat{I}}{\partial \hat{u}_1}, \frac{\partial \hat{I}}{\partial \hat{u}_2}, \ldots \frac{\partial \hat{I}}{\partial \hat{u}_n}\right\} \tag{5.42e}$$

If follows from (5.42c) that for an arbitrary (nontrivial) variation $\delta\hat{\mathbf{u}}$,

$$\frac{\partial \hat{I}}{\partial \hat{\mathbf{u}}} = 0 \tag{5.42f}$$

which leads to a set of n simultaneous equations in terms of the unknowns. The equation is solved by a suitable method. If the functional I is quadratic, then the system of equations are linear as

$$\frac{\partial \hat{I}}{\partial \hat{\mathbf{u}}} = \mathbf{K}\hat{\mathbf{u}} - \mathbf{f} = 0 \tag{5.42g}$$

and $\mathbf{K}$ is symmetric (5.43). Thus, the final outcome of this procedure is identical to that from the virtual work method described in Sect. 5.1.1 (5.10).

Remark. Consider a quadratic function $\alpha = \mathbf{x}^{\mathrm{T}}\mathbf{A}\mathbf{x}$. Taking its variation, we have

$$\delta\alpha = \delta\mathbf{x}^{\mathrm{T}}\mathbf{A}\mathbf{x} + \mathbf{x}^{\mathrm{T}}\mathbf{A}\delta\mathbf{x} = \delta\mathbf{x}^{\mathrm{T}}\mathbf{A}\mathbf{x} + \delta\mathbf{x}^{\mathrm{T}}\mathbf{A}^{\mathrm{T}}\mathbf{x} = \delta\mathbf{x}^{\mathrm{T}}\left[\mathbf{A} + \mathbf{A}^{\mathrm{T}}\right]\mathbf{x} = \delta\mathbf{x}^{\mathrm{T}}\mathbf{B}\mathbf{x} \tag{5.43}$$

where $\mathbf{B} = \frac{1}{2}\left[\mathbf{A} + \mathbf{A}^{\mathrm{T}}\right]$. It is seen that $\mathbf{B} = \mathbf{B}^{\mathrm{T}}$.

Then $\delta\alpha = 0 \Rightarrow \mathbf{B}\mathbf{x} = 0$. Hence, the resulting equation is linear in $\mathbf{x}$ and the matrix $\mathbf{B}$ is symmetric.

Example 5.3.

Question. Show that the following quadratic equation leads to a set of linear simultaneous equations involving a symmetric coefficient matrix:

$$\hat{I} = 2\hat{u}_1^2 - \hat{u}_1\hat{u}_2 + 4\hat{u}_2^2 - 3\hat{u}_1 + 5\hat{u}_2 = 0 \tag{5.44a}$$

Answer.

$$\frac{\partial \hat{I}}{\partial \hat{\mathbf{u}}} = \begin{bmatrix} \dfrac{\partial \hat{I}}{\partial \hat{u}_1} \\ \dfrac{\partial \hat{I}}{\partial \hat{u}_2} \end{bmatrix} = \begin{bmatrix} 4 & -1 \\ -1 & 8 \end{bmatrix}\begin{bmatrix} \hat{u}_1 \\ \hat{u}_2 \end{bmatrix} - \begin{bmatrix} 3 \\ -5 \end{bmatrix} = 0 \tag{5.44b}$$

Hence, a set of 2×2 linear equations are obtained and the coefficient matrix is symmetric.

Now, consider the second variation of I, $\delta^2\hat{I}$. From (5.42c):

$$\delta^2\hat{I} = \delta\hat{\mathbf{u}}^{\mathrm{T}} \begin{bmatrix} \dfrac{\partial}{\partial\hat{u}_1}\left(\dfrac{\partial\hat{I}}{\partial\hat{u}_1}\right)\delta\hat{u}_1 & \dfrac{\partial}{\partial\hat{u}_2}\left(\dfrac{\partial\hat{I}}{\partial\hat{u}_1}\right)\delta\hat{u}_2 & \cdot & \cdot \\ \dfrac{\partial}{\partial\hat{u}_1}\left(\dfrac{\partial\hat{I}}{\partial\hat{u}_2}\right)\delta\hat{u}_1 & \dfrac{\partial}{\partial\hat{u}_2}\left(\dfrac{\partial\hat{I}}{\partial\hat{u}_2}\right)\delta\hat{u}_2 & \cdot & \cdot \\ \cdot & \cdot & \cdot & \cdot \\ \cdot & \cdot & \cdot & \cdot \end{bmatrix} \tag{5.45a}$$

$$\delta^2\hat{I} = \delta\hat{\mathbf{u}}^{\mathrm{T}} \begin{bmatrix} \dfrac{\partial^2\hat{I}}{\partial\hat{u}_1^2} & \dfrac{\partial^2\hat{I}}{\partial\hat{u}_2\partial\hat{u}_1} & \cdot & \cdot \\ \dfrac{\partial^2\hat{I}}{\partial\hat{u}_1\partial\hat{u}_2} & \dfrac{\partial^2\hat{I}}{\partial\hat{u}_2^2} & \cdot & \cdot \\ \cdot & \cdot & \cdot & \cdot \\ \cdot & \cdot & \cdot & \cdot \end{bmatrix} \begin{bmatrix} \delta\hat{u}_1 \\ \delta\hat{u}_2 \\ \cdot \\ \cdot \end{bmatrix} = \delta\hat{\mathbf{u}}^{\mathrm{T}}\mathbf{K}^t\delta\hat{\mathbf{u}} \tag{5.45b}$$

where $\mathbf{K}^t$ is referred to as the *tangent stiffness matrix*. There are two aspects to note from (5.45b):

1. The tangent stiffness matrix is symmetric since

$$K_{ij}^t = \frac{\partial^2\hat{I}}{\partial\hat{u}_i\partial\hat{u}_j} = \frac{\partial^2\hat{I}}{\partial\hat{u}_j\partial\hat{u}_i} = K_{ji}^t \tag{5.46a}$$

   Note that $\hat{I}$ in the above equation is not necessarily quadratic.

2. When $\hat{I}$ is quadratic (which is the case in solid mechanics problems), $\mathbf{K}^t$ (appearing in (5.45b)) and $\mathbf{K}$ (appearing in (5.42g)) are the same. This is shown from (5.42g) as

$$\delta^2\hat{I} = \delta\hat{\mathbf{u}}^{\mathrm{T}}\delta\left[\frac{\partial\hat{I}}{\partial\hat{\mathbf{u}}}\right] = \delta\hat{\mathbf{u}}^{\mathrm{T}}\delta[\mathbf{K}\hat{\mathbf{u}} - \mathrm{f}] = \delta\hat{\mathbf{u}}^{\mathrm{T}}\mathbf{K}\delta\hat{\mathbf{u}} \tag{5.46b}$$

   Comparison of (5.45b) and (5.46b) indicates that $\mathbf{K} = \mathbf{K}^t$. The fact that $\mathbf{K}^t$ is symmetric proves that $\mathbf{K}$ is also symmetric in this case (where the functional is quadratic).

3. If the stiffness matrix $\mathbf{K}$ is *positive definite*, then $\delta^2\hat{I} > 0$, indicating that the *stationary point is a minimum*. Conversely, if it known that the stationary point is a minimum, then $\delta^2\hat{I} > 0$ and it follows from (5.46b) that $\mathbf{K}$ is *positive definite*. This is the case in solid mechanics problems. The stationary principle coincides with the well-known *minimum potential energy principle*, as we will show next

A task that still remains is the determination of the scalar functional corresponding to the boundary value problem of interest. There are again several methods of deriving the functional. We will discuss two such methods. The first of which is deriving directly from the virtual work principle that we have already discussed. The method is relevant only to solid mechanics problems. The second method to be discussed is very general and powerful in that a variational statement can be found for most, any boundary value problem that is defined in terms of a set of differential equations and a set of boundary conditions. Thus, this method can be applied to such problems as heat flow/water flow problems.

5.1.2.3 Variational Statement Derived from Virtual Work Principle

In the case of elastic solid mechanics problems, the functional needed in the variational statement can be directly derived from the virtual work equation presented in (5.2) (more precisely, the virtual displacement principle) as follows. For a linear elastic material, the stress and strain vectors are related through a symmetric elastic stiffness tensor **D**

$$\boldsymbol{\sigma} = \mathbf{D}\boldsymbol{\varepsilon} \tag{5.47}$$

Equation (5.2) then can be written as

$$V = \int_{\Omega} \delta\boldsymbol{\varepsilon}^{\mathrm{T}}\boldsymbol{\sigma}\,\mathrm{d}v - \int_{\Omega} \delta\mathbf{u}^{\mathrm{T}}\mathbf{b}\,\mathrm{d}v - \int_{\Gamma_t} \delta\mathbf{u}^{\mathrm{T}}\mathbf{t}\,\mathrm{d}s = 0 \tag{5.48a}$$

$$= \int_{\Omega} \delta\boldsymbol{\varepsilon}^{\mathrm{T}}\mathbf{D}\boldsymbol{\varepsilon}\,\mathrm{d}v - \int_{\Omega} \delta\mathbf{u}^{\mathrm{T}}\mathbf{b}\,\mathrm{d}v - \int_{\Gamma_t} \delta\mathbf{u}^{\mathrm{T}}\mathbf{t}\,ds = 0 \quad (\text{from (5.47)}) \tag{5.48b}$$

Noting that

$$= \delta\left[\int_{\Omega} \boldsymbol{\varepsilon}^{\mathrm{T}}\mathbf{D}\boldsymbol{\varepsilon}\mathrm{d}v\right] = \int_{\Omega} \left[\delta\boldsymbol{\varepsilon}^{\mathrm{T}}\mathbf{D}\boldsymbol{\varepsilon} + \boldsymbol{\varepsilon}^{\mathrm{T}}\mathbf{D}\,\delta\boldsymbol{\varepsilon}\right]\mathrm{d}v = \int_{\Omega} \left[\delta\boldsymbol{\varepsilon}^{\mathrm{T}}\mathbf{D}\boldsymbol{\varepsilon} + \delta\boldsymbol{\varepsilon}^{\mathrm{T}}\mathbf{D}^{\mathrm{T}}\boldsymbol{\varepsilon}\right]\mathrm{d}v$$

and using symmetry of **D**

$$\int_{\Omega} \delta\boldsymbol{\varepsilon}^{\mathrm{T}}\mathbf{D}\boldsymbol{\varepsilon}\,\mathrm{d}v = \delta\left[\int_{\Omega} \frac{\mathbf{1}}{\mathbf{2}}\boldsymbol{\varepsilon}^{\mathrm{T}}\mathbf{D}\boldsymbol{\varepsilon}\,\mathrm{d}v\right]$$

Then (5.48b) becomes

$$V = \delta\left[\frac{1}{2}\int_{\Omega} \boldsymbol{\varepsilon}^{\mathrm{T}}\mathbf{D}\boldsymbol{\varepsilon}\,\mathrm{d}v - \int_{\Omega} \mathbf{u}^{\mathrm{T}}\mathbf{b}\,\mathrm{d}v - \int_{\Gamma_t} \mathbf{u}^{\mathrm{T}}\mathbf{t}\,\mathrm{d}s\right] = 0 \tag{5.48c}$$

$$= \delta\pi = 0 \tag{5.48d}$$

where

$$\pi = \frac{1}{2}\int_{\Omega} \boldsymbol{\varepsilon}^{\mathrm{T}}\mathbf{D}\boldsymbol{\varepsilon}\,\mathrm{d}v - \int_{\Omega} \mathbf{u}^{\mathrm{T}}\mathbf{b}\,\mathrm{d}v - \int_{\Gamma_t} \mathbf{u}^{\mathrm{T}}\mathbf{t}\,\mathrm{d}s \tag{5.49a}$$

$$= \frac{1}{2}\int_{\Omega} \boldsymbol{\sigma}^{\mathrm{T}}\boldsymbol{\varepsilon}\,\mathrm{d}v - \int_{\Omega} \mathbf{u}^{\mathrm{T}}\mathbf{b}\,\mathrm{d}v - \int_{\Gamma_t} \mathbf{u}^{\mathrm{T}}\mathbf{t}\,\mathrm{d}s \quad \text{(using(5.47)and symmetry of } \mathbf{D}) \tag{5.49b}$$

$$= U - W \tag{5.49c}$$

π is the potential energy of the body, with

$$U = \frac{1}{2}\int_{\Omega} \boldsymbol{\sigma}^{\mathrm{T}}\boldsymbol{\varepsilon}\,\mathrm{d}v \tag{5.49d}$$

representing the internal energy of the body, and

$$W = \int_{\Omega} \mathbf{u}^{\mathrm{T}}\mathbf{b}\,\mathrm{d}v + \int_{\Gamma_t} \mathbf{u}^{\mathrm{T}}\mathbf{t}\,\mathrm{d}s \tag{5.49e}$$

representing the work done by the external forces.

Noting that (5.48d) is the mathematical condition for finding stationarity (minimum in the case of solid mechanics problems) of the potential energy π. It then follows that the *virtual work principle and the minimum potential energy principle are exactly the same*. The former states that (5.48a) recovers the equilibrium equations and the boundary conditions, and the latter states that minimization of the potential energy π recovers the equilibrium equations and the boundary conditions.

Note that even for a nonlinear elastic material the minimum potential energy principle is valid since (5.48a) can be written as follows.

Without using the linear relation $\boldsymbol{\sigma} = \mathbf{D}\boldsymbol{\varepsilon}$, the internal virtual work δU is written as

$$\delta U = \int_{\Omega} \delta\boldsymbol{\varepsilon}^{\mathrm{T}}\boldsymbol{\sigma}\,\mathrm{d}v$$

Equation (5.48a) becomes

$$\begin{aligned} V &= \delta U - \int_{\Omega} \delta \mathbf{u}^{\mathrm{T}} \mathbf{b}\, \mathrm{d}v - \int_{\Gamma_t} \delta \mathbf{u}^{\mathrm{T}} \mathbf{t}\, \mathrm{d}s = 0 \\ &= \delta U - \delta \left[\int_{\Omega} \mathbf{u}^{\mathrm{T}} \mathbf{b}\, \mathrm{d}v + \int_{\Gamma_t} \mathbf{u}^{\mathrm{T}} \mathbf{t}\, \mathrm{d}s \right] = 0 \\ &= \delta U - \delta \left[\int_{\Omega} \mathbf{u}^{\mathrm{T}} \mathbf{b}\, \mathrm{d}v + \int_{\Gamma_t} \mathbf{u}^{\mathrm{T}} \mathbf{t}\, \mathrm{d}s \right] = 0 \\ &= \delta U - \delta W = 0 \\ &= \delta (U - W) = 0 \\ V &= \delta \pi = 0 \end{aligned}$$

5.1.2.4 Generalized Method of Deriving Variational Statement

In this section, we describe a method known as the *weighted residual method*. This is a very powerful method for developing variational statement for any boundary value problem that is defined in terms of a set of differential equations given by (5.34a) and a set of boundary conditions given by (5.34b). When the primary unknowns are approximated by $\hat{\mathbf{u}}$, the quantities $\mathbf{A}(\hat{\mathbf{u}})$ and $\mathbf{B}(\hat{\mathbf{u}})$ (residuals) will not be exactly zero, and our objective in the numerical technique will be to find that function $\hat{\mathbf{u}}$ that will make the residuals $\mathbf{A}(\hat{\mathbf{u}})$ and $\mathbf{B}(\hat{\mathbf{u}})$ as close to zero as possible. This must be achieved on an average sense for the entire body. That is, what we want is for $\mathbf{A}(\hat{\mathbf{u}})$ and $\mathbf{B}(\hat{\mathbf{u}})$ to be as close to zero as possible over the entire domain and the boundary, rather than at one or two points. But of course, in some cases, when $\mathbf{A}(\hat{\mathbf{u}})$ and $\mathbf{B}(\hat{\mathbf{u}})$ are zero at sufficiently large number of points, then $\hat{\mathbf{u}}$ may be taken as a "satisfactory approximation" for $\mathbf{u}$ for the whole domain. Such techniques are known as method of collocation; we will not cover this in detail in this book.

First let us discuss the mathematical statement of the weighted residual technique. We seek an approximation such that for any arbitrary functions $\mathbf{v} = [v_1, v_2, \ldots v_n]$ and $\mathbf{w} = [w_1, w_2, \ldots w_n]$, the integral representation of the residuals will be zero as follows:

$$I = \int_{\Omega} \mathbf{v}^{\mathrm{T}} \mathbf{A}(\hat{\mathbf{u}}) \mathrm{d}v + \int_{\Gamma} \mathbf{w}^{\mathrm{T}} \mathbf{B}(\hat{\mathbf{u}}) \mathrm{d}s = 0 \tag{5.50a}$$

$$\begin{aligned} &= \int_{\Omega} v_1 A_1(\hat{\mathbf{u}}) + v_2 A_2(\hat{\mathbf{u}}) + \cdots + v_n A_n(\hat{\mathbf{u}}) \mathrm{d}v \\ &\quad + \int_{\Gamma} w_1 B_1(\hat{\mathbf{u}}) + w_2 B_2(\hat{\mathbf{u}}) + \cdots + w_n B_n(\hat{\mathbf{u}}) \mathrm{d}s = 0 \end{aligned} \tag{5.50b}$$

The admissibility requirements on $\hat{\mathbf{u}}$, $\mathbf{v}$ and $\mathbf{w}$ are the same as those in the virtual work method: (1) $\hat{\mathbf{u}}$ should be $C_{n\text{-}1}$ continuous, where n is the highest order of derivative involved in $\mathbf{A}(\hat{\mathbf{u}})$ or $\mathbf{B}(\hat{\mathbf{u}})$ (thus, C_0 continuous for solid mechanics, elasticity problems), and (2) $\mathbf{v}$ and $\mathbf{w}$ must be any functions such that the integrals in (5.50) do not become infinity.

To develop an appreciation for the general applicability of the generalized method discussed here, we will learn to develop variational statements for the following problems:

1. Two- and three-dimensional solid mechanics elasticity problem
2. Elastic beam bending problems
3. Two-dimensional water flow problem
4. One-dimensional diffusion-convection problem

Items 1 and 2 are done here (Sect. 5.3), and 3 and 4 are left for the reader to try in exercises (Problems 5.4 and 5.6).

The following mathematical theorems and some identities derived from them are useful in the ensuing mathematical developments.

Definitions. *Considering the Cartesian coordinate system*

Gradient operator:

$$\nabla = \mathbf{i}\frac{\partial}{\partial x} + \mathbf{j}\frac{\partial}{\partial y} + \mathbf{k}\frac{\partial}{\partial z} \text{ (vector)} \tag{5.51a}$$

Laplace operator:

$$\nabla^2 = \frac{\partial^2}{\partial x^2} + \frac{\partial^2}{\partial y^2} + \frac{\partial^2}{\partial z^2} \text{ (scalar)} \tag{5.51b}$$

Let $F(x,y,z)$ be any differentiable scalar function. Then the gradient (the grad) operator is given by

$$\text{grad}(F) = \nabla F = \left[\frac{\partial F}{\partial x}, \frac{\partial F}{\partial y}, \frac{\partial F}{\partial z}\right] = \mathbf{i}\frac{\partial F}{\partial x} + \mathbf{j}\frac{\partial F}{\partial y} + \mathbf{k}\frac{\partial F}{\partial z} \text{ (vector)} \tag{5.52}$$

Let $\mathbf{v} = [v_1, v_2, v_3]$ be a vector function. The divergence (the div) operator is given by

$$\text{div}(\text{v}) = \nabla \cdot \text{v} = \frac{\partial \text{v}_1}{\partial \text{x}} + \frac{\partial \text{v}_2}{\partial \text{y}} + \frac{\partial \text{v}_3}{\partial \text{z}} \text{ (scalar)} \tag{5.53}$$

Gradient Theorem (Green Theorem): The Green theorem allows the transformation of a volume integral into a surface integral. Referring to Fig. 5.6, the Green theorem states:

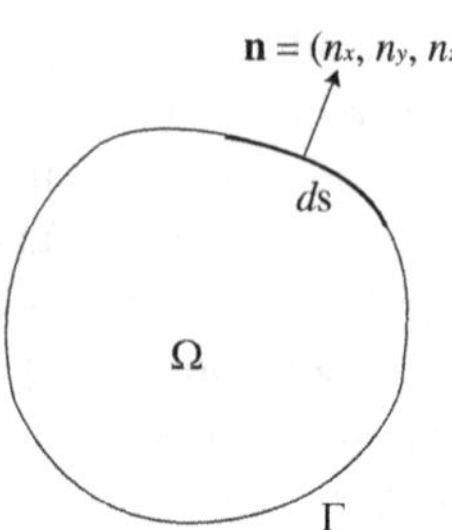

Fig. 5.6 Domain details

$$\int_{\Omega} \text{grad}\,(F)\mathrm{d}x\,\mathrm{d}y\,\mathrm{d}z = \int_{\Gamma} \mathbf{n}F\,\mathrm{d}s \tag{5.54a}$$

where

$$\int_{\Omega} \text{grad}(F)\mathrm{d}x\,\mathrm{d}y\,\mathrm{d}z = \int_{\Omega} \left[\mathbf{i}\frac{\partial F}{\partial x} + \mathbf{j}\frac{\partial F}{\partial y} + \mathbf{k}\frac{\partial F}{\partial z}\right]\mathrm{d}x\,\mathrm{d}y\,\mathrm{d}z \text{ and} \tag{5.54b}$$

$$\int_{\Gamma} \mathbf{n}F\,\mathrm{d}s = \int_{\Gamma} \left[\mathbf{i}n_x + \mathbf{j}n_y + \mathbf{k}n_z\right]\mathrm{d}s \tag{5.54c}$$

The Divergence Theorem: The divergence theorem permits the conversion of a volume integral into a surface integral of the divergence of a vectorial quantity, as follows:

$$\int_{\Omega} \text{div}(\mathbf{v})\mathrm{d}x\,\mathrm{d}y\,\mathrm{d}z = \int_{\Gamma} \mathbf{n}\cdot\mathbf{v}\,\mathrm{d}s \tag{5.55a}$$

where

$$\int_{\Omega} \text{div}(\mathbf{v})\mathrm{d}x\,\mathrm{d}y\,\mathrm{d}z = \int_{\Omega} \left[\frac{\partial v_x}{\partial x} + \frac{\partial v_y}{\partial y} + \frac{\partial v_z}{\partial z}\right]\mathrm{d}x\,\mathrm{d}y\,\mathrm{d}z \tag{5.55b}$$

and

$$\int_{\Gamma} \mathbf{n}\cdot\mathbf{v}\,\mathrm{d}s = \int_{\Gamma} \left[n_x v_x + n_y v_y + n_z v_z\right]\mathrm{d}s \tag{5.55c}$$

An Identity: Certain useful identities can be derived from the above theorems, which can be directly made use of in the development of variational statements; these are presented below.

Let F(x, y, z) and G(x, y, z) be two scalar functions. Consider

$$\int_{\Omega} (\nabla F) G \, dx \, dy \, dz$$

$$= \int_{\Omega} \left[\mathbf{i} \frac{\partial F}{\partial x} G + \mathbf{j} \frac{\partial F}{\partial y} G + \mathbf{k} \frac{\partial F}{\partial z} G \right] dx \, dy \, dz \text{ (writing it explicitly)} \tag{5.56a}$$

which can be written as

$$= \int_{\Omega} \left[\mathbf{i} \frac{\partial (FG)}{\partial x} + \mathbf{j} \frac{\partial (FG)}{\partial y} + \mathbf{k} \frac{\partial (FG)}{\partial z} \right] dx dy dz - \int_{\Omega} \left[\mathbf{i} \frac{\partial G}{\partial x} F + \mathbf{j} \frac{\partial G}{\partial y} F + \mathbf{k} \frac{\partial G}{\partial z} F \right] dx dy dz \tag{5.56b}$$

Let us use the gradient theorem (5.54a) to convert the first volume integral on the right-hand side into a surface integral as

$$\int_{\Omega} \left[\mathbf{i} \frac{\partial (FG)}{\partial x} + \mathbf{j} \frac{\partial (FG)}{\partial y} + \mathbf{k} \frac{\partial (FG)}{\partial z} \right] dx \, dy \, dz = \int_{\Gamma} \mathbf{n} FG \, ds \tag{5.56c}$$

Thus,

$$\int_{\Omega} (\nabla F) G \, dx \, dy \, dz = -\int_{\Omega} (\nabla G) F \, dx \, dy \, dz + \int_{\Gamma} \mathbf{n} FG \, ds \tag{5.57}$$

Basically, the variables F and G have been switched as you go from the left-hand side to the right-hand side.

5.1.2.5 Variational Statement for Three-Dimensional Elastic Solid Mechanics Problems

The relevant equilibrium equations and the boundary conditions are listed in Chap. 3. In (5.50), the functions $\mathbf{v}$ and $\mathbf{w}$ are arbitrary, with the restriction that they should result in finite integrals. While such arbitrary functions are indeed employed in certain techniques such as the collocation methods, in deriving the variational statements, they are taken to be variations of functions. Applying (5.50b) to the equilibrium equations, we have

$$I = \int_{\Omega} v_1 A_1(\hat{\mathbf{u}}) + v_2 A_2(\hat{\mathbf{u}}) + v_3 A_3(\hat{\mathbf{u}}) \mathrm{dv} + \int_{\Gamma} w_1 B_1(\hat{\mathbf{u}}) + w_2 B_2(\hat{\mathbf{u}}) \mathrm{ds} = 0 \tag{5.58a}$$

$$I = \int_{\Omega} \left[\delta v_1 \left\{ \frac{\partial \sigma_{xx}}{\partial x} + \frac{\partial \sigma_{xy}}{\partial y} + \frac{\partial \sigma_{xz}}{\partial z} + b_x \right\} + \delta v_1 \left\{ \frac{\partial \sigma_{yx}}{\partial x} + \frac{\partial \sigma_{yy}}{\partial y} + \frac{\partial \sigma_{yz}}{\partial z} + b_y \right\} \right.$$
$$\left. + \delta v_3 \left\{ \frac{\partial \sigma_{zx}}{\partial x} + \frac{\partial \sigma_{zy}}{\partial y} + \frac{\partial \sigma_{zz}}{\partial z} + b_z \right\} \mathrm{d}v \right] + \int_{\Gamma_t} \delta \boldsymbol{w}^{\mathrm{T}} (\mathbf{t} - \bar{\mathbf{t}}) \mathrm{d}s + \int_{\Gamma_u} \delta \mathbf{s}^{\mathrm{T}} (\mathbf{u} - \bar{\mathbf{u}}) \mathrm{d}s = 0 \tag{5.58b}$$

where $\delta \mathbf{v} = \{\delta v_1, \delta v_2, \delta v_3\}$, $\delta \mathbf{w} = \{\delta w_1, \delta w_2, \delta w_3\}$, and $\delta s = \{\delta s_1, \delta s_2, \delta s_3\}$ are arbitrary variations with the requirement that the integrals are bounded. Now, we use the identity given by (5.57) and transfer the differentiation in the first volumetric integral terms from σ's to δv's as follows: With $F = \sigma_{xx}$ and $G = \delta v_1$, (5.57) becomes

$$\int_{\Omega} \left\{ \frac{\partial \sigma_{xx}}{\partial x} \mathbf{i} + \frac{\partial \sigma_{xx}}{\partial y} \mathbf{j} + \frac{\partial \sigma_{xx}}{\partial z} \mathbf{k} \right\} \delta v_1 \mathrm{d}v$$
$$= - \int_{\Omega} \left\{ \frac{\partial(\delta v_1)}{\partial x} \mathbf{i} + \frac{\partial(\delta v_1)}{\partial y} \mathbf{j} + \frac{\partial(\delta v_1)}{\partial z} \mathbf{k} \right\} \sigma_{xx} \, \mathrm{d}v + \int_{\Gamma} \{ n_x \mathbf{i} + n_y \mathbf{j} + n_z \mathbf{k} \} \sigma_{xx} \delta v_1 \, \mathrm{d}s$$

Taking just the first component of the vector equation above, we have

$$\int_{\Omega} \delta v_1 \frac{\partial \sigma_{xx}}{\partial x} \mathrm{d}v = - \int_{\Omega} \sigma_{xx} \frac{\partial(\delta v_1)}{\partial x} \mathrm{d}v + \int_{\Gamma} \delta v_1 \sigma_{xx} n_{xx} \, \mathrm{d}s \tag{5.59}$$

Performing the above operation on the other terms in the volumetric integral, except for the body force terms, and grouping the terms in a certain way, (5.58b) becomes

$$I = - \int_{\Omega} \left[\sigma_{xx} \frac{\partial(\delta v_1)}{\partial x} + \sigma_{yy} \frac{\partial(\delta v_2)}{\partial y} + \sigma_{zz} \frac{\partial(\delta v_3)}{\partial z} + \sigma_{xy} \left\{ \frac{\partial(\delta v_1)}{\partial y} + \frac{\partial(\delta v_2)}{\partial x} \right\} \right.$$
$$\left. + \sigma_{xz} \left\{ \frac{\partial(\delta v_1)}{\partial z} + \frac{\partial(\delta v_3)}{\partial x} \right\} + \sigma_{yz} \left\{ \frac{\partial(\delta v_2)}{\partial z} + \frac{\partial(\delta v_3)}{\partial y} \right\} \right] \mathrm{d}v$$
$$+ \int_{\Gamma} \delta v_1 \{ \sigma_{xx} n_x + \sigma_{xy} n_y + \sigma_{xz} n_z \} \mathrm{d}s + \int_{\Gamma} \delta v_2 \{ \sigma_{yx} n_x + \sigma_{yy} n_y + \sigma_{yz} n_z \} \mathrm{d}s$$
$$+ \int_{\Gamma} \delta v_3 \{ \sigma_{zx} n_x + \sigma_{zy} n_y + \sigma_{zz} n_z \} \mathrm{d}s + \int_{\Omega} \{ \delta v_1 b_x + \delta v_2 b_y + \delta v_3 b_z \} \mathrm{d}v$$
$$+ \int_{\Gamma_t} \delta \mathbf{w}^{\mathrm{T}} (\mathbf{t} - \bar{\mathbf{t}}) \mathrm{d}s + \int_{\Gamma_u} \delta \mathbf{s}^{\mathrm{T}} (\mathbf{u} - \bar{\mathbf{u}}) \mathrm{d}s = 0 \tag{5.60}$$

Equation (5.60) is further simplified with the aid of the following:

- Recall the relationship between traction and stress: $\mathbf{t} = \boldsymbol{\sigma}\mathbf{n}$ or

$$\begin{bmatrix} t_x \\ t_y \\ t_z \end{bmatrix} = \begin{bmatrix} \sigma_{xx}n_x + \sigma_{xy}n_y + \sigma_{xz}n_z \\ \sigma_{yx}n_x + \sigma_{yy}n_y + \sigma_{yz}n_z \\ \sigma_{zx}n_x + \sigma_{zy}n_y + \sigma_{zz}n_z \end{bmatrix} \tag{5.61a}$$

- Introduce variations of strains as

$$\delta\varepsilon_{xx} = \frac{\partial(\delta v_1)}{\partial x}; \quad \delta\varepsilon_{xz} = \frac{\partial(\delta v_1)}{\partial z} + \frac{\partial(\delta v_3)}{\partial x}, \text{ etc.} \tag{5.61b}$$

- Arrange the stress and strain tensor components in vectors as

$$\sigma = \left\{\sigma_{xx}, \sigma_{yy}, \sigma_{zz}, \sigma_{xy}, \sigma_{xz}, \sigma_{yz}\right\} \text{ and } \delta\varepsilon = \left\{\delta\varepsilon_{xx}, \delta\varepsilon_{yy}, \delta\varepsilon_{zz}, \delta\varepsilon_{xy}, \delta\varepsilon_{xz}, \delta\varepsilon_{yz}\right\} \tag{5.61c}$$

Equation (5.60a) can then be written in a more compact manner as

$$I = -\int_{\Omega} \delta\boldsymbol{\varepsilon}^{\mathrm{T}}\boldsymbol{\sigma}\,\mathrm{d}v + \int_{\Omega} \delta\mathbf{v}^{\mathrm{T}}\mathbf{b}\,\mathrm{d}v + \int_{\Gamma} \delta\mathbf{v}^{\mathrm{T}}\mathbf{t}\,\mathrm{d}s + \int_{\Gamma_t} \delta\mathbf{w}^{\mathrm{T}}(\mathbf{t}-\bar{\mathbf{t}})\mathrm{d}s + \int_{\Gamma_u} \delta\mathrm{s}^{\mathrm{T}}(\mathbf{u}-\bar{\mathbf{u}})\mathrm{d}s = 0$$

Let us split the third term into integrals over Γ_t and Γ_u separately:

$$\begin{aligned} I = -\int_{\Omega} \delta\boldsymbol{\varepsilon}^{\mathrm{T}}\boldsymbol{\sigma}\,\mathrm{d}v + \int_{\Omega} \delta\mathbf{v}^{\mathrm{T}}\mathbf{b}\,\mathrm{d}v + \int_{\Gamma_t} \delta\mathbf{v}^{\mathrm{T}}\mathbf{t}\,\mathrm{d}s + \int_{\Gamma_u} \delta\mathbf{v}^{\mathrm{T}}\mathbf{t}\,\mathrm{d}s \\ + \int_{\Gamma_t} \delta\mathbf{w}^{\mathrm{T}}(\mathbf{t}-\bar{\mathbf{t}})\mathrm{d}s + \int_{\Gamma_u} \delta\mathrm{s}^{\mathrm{T}}(\mathbf{u}-\bar{\mathbf{u}})\mathrm{d}s = 0 \end{aligned} \tag{5.62b}$$

Let us further simplify with the aid of the following:

- $\delta\mathbf{v} = \mathbf{0}$ on Γ_u, which makes the fourth term disappear.
- Without loss of generality, let us assume that $\delta\mathbf{w} = -\delta\mathbf{v}$, making the third term and the first part of the fifth terms to cancel each other. This causes $\mathbf{t}$ to disappear from the boundary integral over Γ_t, satisfying the traction boundary condition automatically. Such a boundary condition is known as the *natural boundary condition.*
- The last term involves the displacement boundary condition, and this boundary condition will be exactly satisfied only if $\mathbf{u} = \bar{\mathbf{u}}$, and this is therefore, enforced. Such a boundary condition is known as the *forced boundary condition* or *essential boundary condition.*
- Without loss of generality, multiply the entire equation by -1.

Equation (5.62b) then becomes

$$I = \int_{\Omega} \delta\boldsymbol{\varepsilon}^{\mathrm{T}}\boldsymbol{\sigma}\,\mathrm{d}v - \int_{\Omega} \delta\mathbf{v}^{\mathrm{T}}\mathbf{b}\,\mathrm{d}v - \int_{\Gamma_t} \delta\mathbf{v}^{\mathrm{T}}\bar{\mathbf{t}}\,\mathrm{d}s = 0 \tag{5.63}$$

which is identical to the virtual work equation given by (5.2). Recall that (5.63) lead to the variational principle (the minimum potential energy principle) presented earlier in this section (5.47–5.49e).

5.1.2.6 What is a "Weak Form"?

The original differential equations in solid mechanics problems (Chap. 3) involve terms such as $(\partial\sigma_{xx}/\partial x)$. After substitution of σ_{xx} in terms of strains (with the aid of the constitutive laws) and then the strains in terms of displacements, the differential equations will become functions of the second-order displacement derivatives such as $C_1(\partial^2 u_1/\partial x^2)$. Displacements are the primary unknowns in displacement-based methods. Thus, any approximate solutions must be at least twice differentiable or C_1 continuous (e.g., a quadratic approximation). A linear approximation is not permissible.

However, when (5.63) is used as the basis for the approximate solution, the order of required shape function is reduced from C_1 continuous to C_0 continuous, since (5.63) consists only of stresses and strains, not their derivatives. On this basis, a variational functional such as that given by (5.63) is called the "weak form." In many cases, such weak forms do exist, reducing the burden of having to come up with higher order shape functions in approximate solution methods. As we will subsequently note, such stringent requirements may be difficult to meet in problems such as the beam bending problem that has fourth-order derivatives in the original differential equation.

5.1.2.7 The Need for Self-Adjoint Property

The following analysis is performed for a scalar variable u, but is applicable to a vector simply with the substitution $u \rightarrow \mathbf{u}$. Consider the integral

$$\delta I = \frac{1}{2}\delta\left[\int_{\Omega} uL(u)\mathrm{d}v\right] \tag{5.64a}$$

$$\delta I = \frac{1}{2}\int_{\Omega} [\delta u L(u) + u\delta(L(u))]\mathrm{d}v \tag{5.64b}$$

When L is a linear operator:

$$\delta I = \frac{1}{2}\int_{\Omega} [\delta u L(u) + u L(\delta u)] \mathrm{d}v \tag{5.64c}$$

If the differential operator is self adjoint,

$$\int_{\Omega} \delta u L(u) \mathrm{d}v = \int_{\Omega} u L(\delta u) \mathrm{d}v + \text{boundary terms} \tag{5.64d}$$

Then the first and the second parts of the integral in (5.64c) are equal to each other, and (5.64c) becomes

$$\delta I = \int_{\Omega} u L(\delta u) \mathrm{d}v + \text{boundary terms} \tag{5.64e}$$

From (5.64a) and (5.64e),

$$\int_{\Omega} u L(\delta u) \mathrm{d}v = \frac{1}{2}\delta\left[\int_{\Omega} u L(u) \mathrm{d}v\right] + \text{boundary terms} \tag{5.65}$$

With the aid of this result, we will now show that a variational method exist for the differential operator *L(u)*. Let us expand the integral form of *L(u)* given by (5.50a) as

$$W = \int_{\Omega} \delta u L(u) \mathrm{d}v \tag{5.66a}$$

$$= \int_{\Omega} u L(\delta u) \mathrm{d}v + \text{boundary terms} \tag{5.66b}$$

$$= \frac{1}{2}\delta\left[\int_{\Omega} u L(u) \mathrm{d}v\right] + \text{boundary terms} \quad \text{from (5.65)} \tag{5.66c}$$

$$= \delta\left[\frac{1}{2}\int_{\Omega} u L(u) \mathrm{d}v + \text{modified boundary terms}\right] \tag{5.66d}$$

$$W = \delta I = 0 \tag{5.67}$$

where

$$I = \frac{1}{2}\int_{\Omega} uL(u)\mathrm{d}v + \text{modified boundary terms} \tag{5.68}$$

Hence, a variational method exists. It is clear that one cannot go from (5.64c) to (5.64e) if not for the self-adjoint property. Thus, in the absence of a self-adjoint differential operator, a natural variational principle does not exist. However, one can in most cases be contrived, as shown in (5.69)–(5.70) below.

5.1.2.8 Summary of Requirements on Shape Functions

In the finite element analysis, the primary unknowns are approximated by (5.4a)

$$\mathbf{u} = \mathbf{N}\hat{\mathbf{u}}_{\mathrm{i}}$$

The shape functions (the coefficients of the N-matrix in the above equation) must satisfy the following:

- The shape functions must be C_{n-1} continuous, where n is the highest order of derivatives in the variational functional.
- The polynomial used for the shape functions must be complete. (For example, a variable that remains constant within the domain cannot be modeled using a quadratic polynomial without the constant term.)
- The shape functions must be continuous within the elements and along the interelement boundaries. (Recall that continuity of displacements is a requirement in the virtual displacement method.) This requirement is relaxed in special elements called the nonconforming elements.
- When displacements and rotations are defined as nodal unknowns (as in bending problems), the approximations for the displacements and rotations must be such that they assume the nodal values at the nodes. This also guarantees that the essential boundary conditions can be enforced.

5.1.2.9 Variational Statement for One-Dimensional Diffusion–Convection Problem

You are asked to show in Problem 5.4 that the differential operator associated with the one-dimensional diffusion–convection problem

$$L(c) = \alpha\frac{\mathrm{d}}{\mathrm{d}x} - \frac{\mathrm{d}}{\mathrm{d}x}\left(\beta\frac{\mathrm{d}}{\mathrm{d}x}\right) + k \tag{5.69a}$$

is not self adjoint. Here we describe one method (Guymon et al. 1970) of transforming this into a self-adjoint operator. Let us define a modified operator $\overline{L}$ as

$$\overline{L} = sL \tag{5.69b}$$

where s is some function of x. We now investigate the conditions necessary to make $\overline{L}$ self adjoint (if at all possible). We follow the usual procedure and start examining the integral

$$I = \int_{\Omega} c_1 \overline{L}(c_2) \mathrm{d}v \tag{5.69c}$$

$$= \int_{\Omega} c_1 \left[s\alpha \frac{\mathrm{d}c_2}{\mathrm{d}x} - s \frac{\mathrm{d}}{\mathrm{d}x} \left(\beta \frac{\mathrm{d}c_2}{\mathrm{d}x} \right) + ksc_2 \right] \mathrm{d}v \tag{5.69d}$$

By employing integration by parts:

$$I = -\int_{\Omega} \left[\alpha c_1 s \frac{\mathrm{d}c_2}{\mathrm{d}x} - \beta \frac{\mathrm{d}c_2}{\mathrm{d}x} \frac{\mathrm{d}(c_1 s)}{\mathrm{d}x} + ksc_2 \right] \mathrm{d}v - \int_{\Gamma} \beta c_1 s \frac{\mathrm{d}c_2}{\mathrm{d}x} \mathrm{d}s \tag{5.70a}$$

$$I = -\int_{\Omega} \left[\left(\alpha s - \beta \frac{\mathrm{d}s}{\mathrm{d}x} \right) c_1 \frac{\mathrm{d}c_2}{\mathrm{d}x} - \beta \frac{\mathrm{d}c_2}{\mathrm{d}x} s \frac{\mathrm{d}c_1}{\mathrm{d}x} + ksc_2 \right] \mathrm{d}v - \int_{\Gamma} \beta c_1 s \frac{\mathrm{d}c_2}{\mathrm{d}x} \mathrm{d}s \tag{5.70b}$$

It is easily verified that the second and third terms in the volume integral possesses the self-adjoint property; it is the first term that is not. This term will be self adjoint if and only if the term is zero, i.e.,

$$\alpha s - \beta \frac{\mathrm{d}s}{\mathrm{d}x} = 0 \tag{5.70c}$$

which after integration leads to

$$s = e^{\frac{\alpha}{\beta} x} \tag{5.70d}$$

Remark. In the analysis of isotropic elastic solids, when Poisson's ratio becomes close to 0.5, the bulk modulus approaches ∞ (i.e., the material becomes incompressible). The standard formulation presented in the preceding sections breaks down. To overcome this problem, either the special incompressible formulation suggested by Herrmann (1965) may be used or a suitable method of reduced integration technique may be used (Zienkiewicz and Taylor 1991).

5.2 Basis for Finite Element Equations: Dynamic Problems

Including the acceleration term, the equation governing equilibrium of solids is (Chap. 3)

$$\sigma_{ij,j} + b_i = \rho \ddot{u}_i \text{ in } \Omega$$

or

$$\sigma_{ij,j} + \bar{b}_i = 0 \text{ in } \Omega, \tag{5.71a}$$

where

$$\bar{b}_i = b_i - \rho \ddot{u}_i \tag{5.71b}$$

Equation (5.71a) is the static equivalent of dynamic equations, where the inertial force is incorporated in to the body force, and a modified body force $\bar{b}_i$ is defined. This is the same as using D'Alembert's principle to incorporate the inertial force with a minus sign into the static forces before applying Newton's law. At any given time t, the finite element formulation presented in Sect. 5.1 for static problems can be applied to dynamic problems with the modified body force $\bar{b}_i$. Then, the expression for the element stiffness matrix remains unmodified (5.5b), but the load vector becomes modified as follows.

By taking the time rate of (5.4a) twice

$$\ddot{\mathbf{u}} = \mathbf{N}\ddot{\hat{\mathbf{u}}}_i \tag{5.72a}$$

From (5.6a),

$$I_i^2 = \int_{\Omega_i} \delta\hat{\mathbf{u}}_i^{\mathrm{T}} \bar{\mathbf{b}} \mathrm{d}v = \delta\hat{\mathbf{u}}_i^{\mathrm{T}} \int_{\Omega_i} \mathbf{N}^{\mathrm{T}} \left[\mathbf{b} - \rho \mathbf{N}\ddot{\hat{\mathbf{u}}}\right] \mathrm{d}v = \delta\hat{\mathbf{u}}_i^{\mathrm{T}} \left[\mathbf{f}_i^b - \mathbf{m}\ddot{\hat{\mathbf{u}}}\right] \tag{5.72b}$$

where f_i^b is the same as that in (5.6b) (i.e., element load vector due to the static body forces), $\mathbf{m}\ddot{\hat{\mathbf{u}}}$ is element inertial force. The quantity

$$\mathbf{m} = \int_{\Omega_i} \mathbf{N}^{\mathrm{T}} \rho \mathbf{N} \, \mathrm{d}v \tag{5.73}$$

is known as the element mass matrix.

The static global equation (5.10) becomes modified as

$$\mathbf{M}\ddot{\hat{\mathbf{u}}} + \mathbf{K}\hat{\mathbf{u}} = \mathbf{P} \tag{5.74}$$

where **M** is the global mass matrix, obtained by assembling the element mass matrices by following exactly the same rules as those used for the stiffness matrix (i.e., the rule presented in Fig. 5.4).

The mass matrix calculated using (5.73) is known as the consistent mass matrix (Archer 1965) since the interpolations used for the mass matrix are the same (i.e., consistent) as those used for the stiffness matrix. The consistent mass is in general fully populated with nonzero off diagonal terms, which couples the degrees of freedoms within elements. It is, however, more common to approximate **m** by a diagonal matrix, known as the "lumped" mass matrix. In lumping, masses are assigned only to translational degrees of freedoms and not to the rotational degrees of freedoms (Cook 1981). While lumping is intuitive for simple elements like the two-noded one-dimensional element presented Fig. 5.3 (lump half the total mass of the element at each mode), it is not so straightforward for more complicated elements – we will return to this in a subsequent section.

Most materials are dissipative, even though their stress–strain behavior may be approximated by an elastic constitutive law. The energy dissipation is generally due to plasticity of the material and hence the use of a suitable elasto-plastic model is the most accurate way to model the energy dissipation. However, elastic methods are far more computationally efficient than elasto-plastic methods. In addition, elastic constitutive models are much simpler than elasto-plastic models. For these reasons, when the degree of plasticity is small, it is still common to use elastic methods, and augment the equations with additional terms to capture the energy dissipation. The most popular method of modeling the energy dissipation is to employ the linear viscoelastic models to describe the stress–strain behavior, where the viscous forces are assumed to be proportional to the velocity. The body force is further modified as follows

$$\bar{b}_i = b_i - \rho\ddot{u}_i - \mu\dot{u}_i \tag{5.75a}$$

Equation (5.72b) is then modified as

$$I_i^2 = \int_{\Omega_i} \delta\hat{\mathbf{u}}_i^{\mathrm{T}}\bar{\mathbf{b}}dv = \delta\hat{\mathbf{u}}_i^{\mathrm{T}} \int_{\Omega_i} \mathbf{N}^{\mathrm{T}}\left[\mathbf{b} - \rho\mathbf{N}\ddot{\hat{\mathbf{u}}} - \mu\mathbf{N}\dot{\hat{\mathbf{u}}}\right]dv = \delta\hat{\mathbf{u}}_i^{\mathrm{T}}\left[\mathbf{f}_i^b - \mathbf{m}\ddot{\hat{\mathbf{u}}} - \mathbf{c}\dot{\hat{\mathbf{u}}}\right] \tag{5.75b}$$

where $\mathbf{c}\dot{\hat{\mathbf{u}}}$ is referred to as the element damping force, and

$$\mathbf{c} = \int_{\Omega_i} \mathbf{N}^{\mathrm{T}}\mu\mathbf{N}dv \tag{5.76}$$

is known as the element damping matrix. The global matrix equation (5.74) becomes further modified as

$$\mathbf{M}\ddot{\hat{\mathbf{u}}}+\mathbf{C}\dot{\hat{\mathbf{u}}}+\mathbf{K}\hat{\mathbf{u}}=\mathbf{P} \tag{5.77}$$

where $\mathbf{C}$ is the global damping matrix. The damping matrix given by (5.76) is known as the consistent damping matrix.

As it is typically difficult to experimentally determine μ, a different model (e.g., Rayleigh damping model) is used to calculate $\mathbf{C}$ – we will return to this in a subsequent section.

It may also be pointed out that even when an elasto-plastic model is used to describe the stress–strain relationship, the viscous damping term is sometimes added to the equation to compensate for the inadequacies of the constitutive law. For example, the classical plasticity models assume the material behavior to be purely elastic when the stress point is inside the yield surface, whereas for materials such as soils, the size of the elastic domain is almost negligible. Hence, when the stresses are small, the behavior may be modeled as elastic whereas it may actually be elasto-plastic. The use of viscous damping can compensate for this deficiency in the constitutive model.

5.3 Element Stiffness Matrix, Mass Matrix, and Load Vector

By following a procedure very similar to that involved in the development of the one-dimensional rod element presented in Sect. 5.1, elements of various shapes (such as triangular, quadrilateral, cubical, etc.) can easily be developed. Triangular elements are some of the first to have been developed and are still widely used.

An element type that is even more widely used is the so-called *isoparametric element*. This element allows domains with curved boundaries to be discretized in a more natural manner than elements that are restricted to have straight edges as, for example, the three-noded triangular elements.

In this section, the following four element types are covered in detail:

1. Linear two-dimensional triangle (2D3)
2. Eight-noded two-dimensional isoparametric element (2D8)
3. Elastic beam element
4. Elastic space frame element (with reference to books on structural analysis)

Details needed to follow the formulations of some of the other elements are also presented.

5.3.1 Linear Triangular Element for Plane-Strain Elasticity Problems

The domain Ω is spatially discretized into a number of triangular elements as shown in Fig. 5.7a. In two-dimensional analyses, the displacement vector has two degrees of freedom as: $\mathbf{u} = \{u, v\}$. We will develop the necessary equations for a typical element e. Each component is assumed to vary linearly within e, as shown in

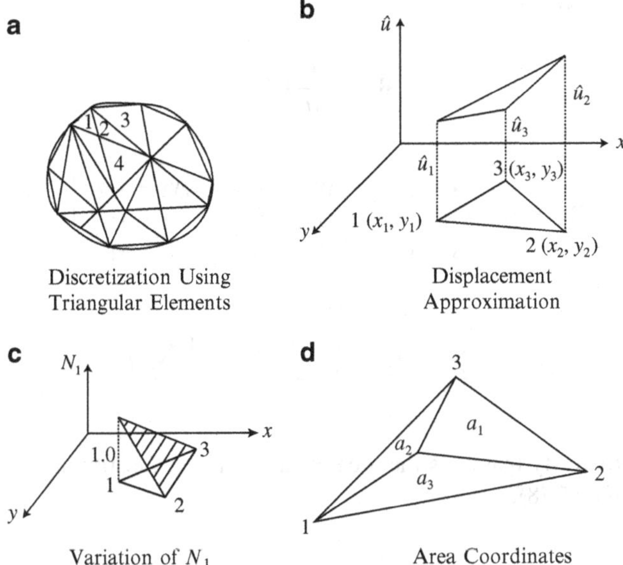

Fig. 5.7 Details concerning linear two-dimensional triangular element

Fig. 5.7b for component u. Let $\mathbf{x} = \{x, y\}$ be the position vector of any point within e with respect to the global x- and y-axes. Let us first consider the approximation $\hat{u}$ to the component u. The spatial variation of $\hat{u}$ can be written as

$$\hat{u} = c_1 + c_2 x + c_3 y \tag{5.78}$$

Let $\hat{u}_1$, $\hat{u}_2$, and $\hat{u}_3$ be the values of u at the three nodes 1, 2, and 3 of element e. The global coordinates of the three nodes are (x_1, y_1), (x_2, y_2), and (x_3, y_3), respectively. Then,

$$\begin{bmatrix} \hat{u}_1 \\ \hat{u}_2 \\ \hat{u}_3 \end{bmatrix} = \begin{bmatrix} 1 & x_1 & y_1 \\ 1 & x_2 & y_2 \\ 1 & x_3 & y_3 \end{bmatrix} \begin{bmatrix} c_1 \\ c_2 \\ c_3 \end{bmatrix} \quad \text{or} \quad \hat{\mathbf{u}}^* = \mathbf{A}\mathbf{c} \tag{5.79}$$

where c_1, c_2, and c_3 are constants. By inverting (5.79)

$$\mathbf{c} = \mathbf{A}^{-1}\hat{\mathbf{u}}^* = \mathbf{B}^*\hat{\mathbf{u}}^* \tag{5.80}$$

It may be noted that for triangles where two of the nodes are very close to each other, then the corresponding rows of **A** are nearly the same, making **A** nearly singular. Numerical problems will be encountered during the inversion involved in (5.80). This occurs when all three nodes are nearly on a straight line as well.

It can be shown that

$$\mathbf{B}^{*} = \frac{1}{2a_e}\mathbf{B}^{**} \tag{5.81a}$$

$$\mathbf{B}^{**} = \begin{bmatrix} x_2y_3 - x_3y_2 & x_3y_1 - x_1y_3 & x_1y_2 - x_2y_1 \\ y_2 - y_3 & y_3 - y_1 & y_1 - y_2 \\ x_3 - x_2 & x_1 - x_3 & x_2 - x_1 \end{bmatrix} \tag{5.81b}$$

$$a_e = \frac{1}{2}\sum_{i=1}^{3} x_i B_{2i}^{**} = \frac{1}{2}\sum_{i=1}^{3} y_i B_{3i}^{**} = \frac{1}{2}\sum_{i=1}^{3} B_{1i}^{**} \tag{5.81c}$$

It is easy to verify that a_e is the area of the triangle. Substituting the constants from (5.80) into (5.78),

$$\begin{aligned} \hat{u} &= \sum_{j=1}^{3} B_{1j}^{*}\hat{u}_j + \left(\sum_{j=1}^{3} B_{2j}^{*}\hat{u}_j\right)x + \left(\sum_{j=1}^{3} B_{3j}^{*}\hat{u}_j\right)y \\ &= \sum_{j=1}^{3}\left(B_{1j}^{*} + B_{2j}^{*}x + B_{3j}^{*}y\right)\hat{u}_j = \sum_{j=1}^{3} N_j\hat{u}_j \end{aligned} \tag{5.82a}$$

where

$$N_j(x,y) = B_{1j}^{*} + B_{2j}^{*}x + B_{3j}^{*}y \tag{5.82b}$$

Remarks.

- $N_i(x_j, y_j) = \delta_{ij}$, (5.83a)

where δ_{ij} is the Kronecker delta (=1 for $i = j$ and 0 for $i \neq j$). As an example, the variation of N_1 is shown in Fig. 5.7c.

- It can be shown that (Zienkiewicz 1977)

  - $\int_{\Omega_e} N_i \mathrm{d}x\, \mathrm{d}y = (a_e/3) \quad \text{for } i = 1,2,3$ (5.83b)

  - Referring to Fig. 5.7d,

$$\begin{aligned} &a_1 = N_1a_e,\ a_2 = N_2a_e,\ \text{and } a_3 = N_3a_e, \\ &a_1 + a_2 + a_3 = a_e \text{ and } N_1 + N_2 + N_3 = 1 \end{aligned} \tag{5.83c}$$

On this basis, instead of defining a point within the triangle by (x, y), the set (a_1, a_2, a_3) could be used to define the point. a_1, a_2, and a_3 are called the area coordinates. Note that, consistent with the fact that only two independent coordinates are needed to define a point in a two-dimensional space, there are only two independent area coordinates since $a_1 + a_2 + a_3 = a_e$

- In performing integrations using area coordinates, the following formula is useful:

$$\int_{\Omega_e} a_1^a a_2^b a_3^c \mathrm{d}x\mathrm{d}y = \frac{a!b!c!}{(a+b+c+2)!}2a_e \tag{5.83d}$$

The N-matrix is then obtained as

$$\mathbf{u}_e = \mathbf{N}\hat{\mathbf{u}}_e \tag{5.84a}$$

where

$$\mathbf{u}_e = \{u(x,y), v(x,y)\} \tag{5.84b}$$

$$\hat{\mathbf{u}}_e = \{\hat{u}_1, \hat{v}_1, \hat{u}_2, \hat{v}_2, \hat{u}_3, \hat{v}_3\} \tag{5.84c}$$

$$\mathbf{N} = \begin{bmatrix} N_1 & 0 & N_2 & 0 & N_3 & 0 \\ 0 & N_1 & 0 & N_2 & 0 & N_3 \end{bmatrix} \tag{5.84d}$$

Defining

$$F_i = \frac{\partial N_i}{\partial x} \text{ and } G_i = \frac{\partial N_i}{\partial y} \tag{5.85a}$$

it follows from (5.82b)

$$F_i = B^*_{2i} \text{ and } G_i = B^*_{3i} \tag{5.85b}$$

It may be noted that F_i and G_i are constants (i.e., independent of x and/or y). This is expected for linear shape functions. The B-matrix is obtained as

$$\boldsymbol{\varepsilon} = \begin{bmatrix} \varepsilon_{xx} \\ \varepsilon_{yy} \\ \gamma_{xy} \end{bmatrix} = \begin{bmatrix} \dfrac{\partial u}{\partial x} \\ \dfrac{\partial v}{\partial y} \\ \dfrac{\partial u}{\partial y} + \dfrac{\partial v}{\partial x} \end{bmatrix} = \mathbf{B}\hat{\mathbf{u}}_e \tag{5.86a}$$

$$\mathbf{B} = \begin{bmatrix} F_1 & 0 & F_2 & 0 & F_3 & 0 \\ 0 & G_1 & 0 & G_2 & 0 & G_3 \\ G_1 & F_1 & G_2 & F_2 & G_3 & F_3 \end{bmatrix} \tag{5.86b}$$

The B-matrix is independent of (x, y) and hence the strains (and consequently the stresses) are modeled to be constants within an element. For this reason, the linear triangle is also referred to as constant strain triangle.

The element stiffness matrix is readily obtained from (5.5b) as

$$\mathbf{k_e} = \int_{\Omega_e} \mathbf{B}^{\mathrm{T}}\mathbf{D}\mathbf{B}\mathrm{d}v = \left[\mathbf{B}^{\mathrm{T}}\mathbf{D}\mathbf{B}\right] a_e t^* \tag{5.87}$$

where t^* is the thickness of the element. $\mathbf{D}$ is the elastic stiffness matrix appearing in $\boldsymbol{\sigma} = \mathbf{D}\varepsilon$.

Now let us evaluate the load vector due to a boundary traction. Let us assume that the traction is linear along the side 1–2 as shown in Fig. 5.8. The formula for computing the element load vector is (5.7b). It follows from (5.83a) that along the boundary 1–2, N_3=0, and N_1 and N_2 are linear. Defining a local coordinate t as shown in Fig. 5.8, N_1 and N_2 may be expressed as

$$N_1 = \bar{N}_1 = 1 - \frac{t}{\ell} \text{ and } N_2 = \bar{N}_2 = \frac{t}{\ell} \tag{5.88}$$

The traction may be written as

$$S(t) = S_1\left(1 - \frac{t}{\ell}\right) + S_2\frac{t}{\ell} = \bar{N}_1 S_1 + \bar{N}_2 S_2 \tag{5.89}$$

Then

$$\mathbf{f}_e^t = \int_{\Gamma_{ti}} \mathbf{N}^{\mathrm{T}}\mathbf{t}\,\mathrm{d}s = \int_{\Gamma_{1-2}} \begin{bmatrix} \bar{N}_1 \\ \bar{N}_2 \end{bmatrix} [\bar{N}_1 S_1 + \bar{N}_2 S_2]\mathrm{d}t = \frac{\ell}{6}\begin{bmatrix} 2S_1 + S_2 \\ S_1 + 2S_2 \end{bmatrix} \tag{5.90}$$

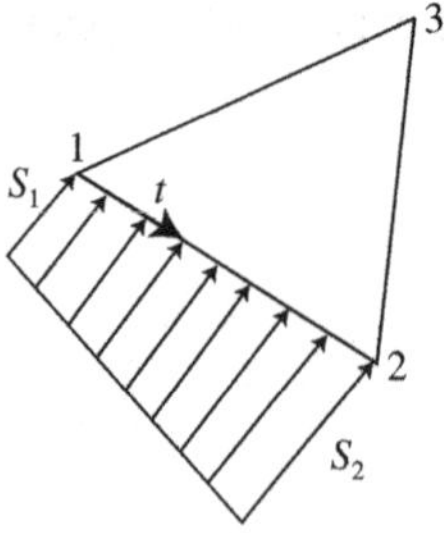

Fig. 5.8 Boundary traction on a triangular element

Elements of $\bar{\mathbf{f}}_e^t$ may then be easily placed into the 6×1 element load vector $\mathbf{f}_e^t$.

In view of (5.83b), the element load vector due to a body force vector $\mathrm{b} = \{b_x, b_y\}$ is computed as

$$\mathbf{f}_e^b = \int_{\Omega_e} \mathbf{N}^{\mathrm{T}} \mathbf{b}\, \mathrm{d}v = \int_{\Omega_e} \begin{bmatrix} N_1 & 0 \\ 0 & N_1 \\ N_2 & 0 \\ 0 & N_2 \\ N_3 & 0 \\ 0 & N_3 \end{bmatrix} \begin{bmatrix} b_x \\ b_y \end{bmatrix} \mathrm{d}v = \begin{bmatrix} a_e t^* b_x/3 \\ a_e t^* b_y/3 \\ a_e t^* b_x/3 \\ a_e t^* b_y/3 \\ a_e t^* b_x/3 \\ a_e t^* b_y/3 \end{bmatrix} \tag{5.91}$$

Example 5.4.

Question. For the one-element triangular problem shown in Fig. 5.9, determine the global stiffness matrix and the load vector. Assume plane strain condition with zero Poisson's ratio. The thickness of the element is t^*.

Answer. The global coordinates of the nodes are

$$(x_1, y_1) = (1, 1), \quad (x_2, y_2) = (3, 1), \quad (x_3, y_3) = (1, 3)$$

Using these coordinates, we have

$$\mathbf{B}^{**} = \begin{bmatrix} 8 & -2 & -2 \\ -2 & 2 & 0 \\ -2 & 0 & 2 \end{bmatrix}$$

$$\begin{aligned} a_e &= \frac{1}{2}\sum_{i=1}^{3} x_i B_{2i}^{**} = \frac{1}{2}[1 \times (-2) + 3 \times 2 + 1 \times 0] = 2 \\ &= \frac{1}{2}\sum_{i=1}^{3} y_i B_{3i}^{**} = \frac{1}{2}[1 \times (-2) + 1 \times 0 + 3 \times 2] = 2 \\ &= \frac{1}{2}\sum_{i=1}^{3} B_{1i}^{**} = \frac{1}{2}[8 - 2 - 2] = 2 \end{aligned}$$

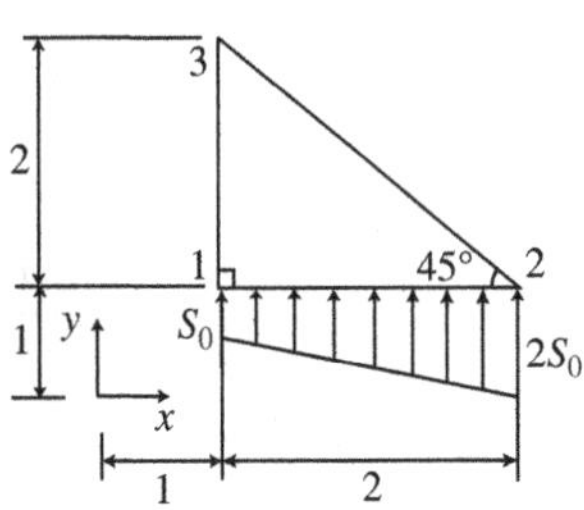

Fig. 5.9

$$N_1 = B_{11}^* + B_{21}^* x + B_{31}^* y = \frac{1}{4}[8 - 2x - 2y] = 2 - \frac{1}{2}x - \frac{1}{2}y$$

$$N_2 = B_{12}^* + B_{22}^* x + B_{32}^* y = \frac{1}{4}[-2 + 2x + 0 \times y] = -\frac{1}{2} + \frac{1}{2}x$$

$$N_3 = B_{13}^* + B_{23}^* x + B_{33}^* y = \frac{1}{4}[-2 + 0 \times x + 2y] = -\frac{1}{2} + \frac{1}{2}y$$

$$\mathbf{B}^* = \frac{1}{2a_e}\mathbf{B}^{**} = \frac{1}{4}\mathbf{B}^{**}$$

$$F_i = B_{2i}^* \Rightarrow [F_1, F_2, F_3] = \frac{1}{2}[-1, 1, 0]$$

$$G_i = B_{3i}^* \Rightarrow [G_1, G_2, G_3] = \frac{1}{2}[-1, 0, 1]$$

$$\mathbf{B} = \begin{bmatrix} F_1 & 0 & F_2 & 0 & F_3 & 0 \\ 0 & G_1 & 0 & G_2 & 0 & G_3 \\ G_1 & F_1 & G_2 & F_2 & G_3 & F_3 \end{bmatrix} = \frac{1}{2}\begin{bmatrix} -1 & 0 & 1 & 0 & 0 & 0 \\ 0 & -1 & 0 & 0 & 0 & 1 \\ -1 & -1 & 0 & 1 & 1 & 0 \end{bmatrix}$$

With $v = 0$, based on the equations in Chaps. 3 and 4,

$$\mathbf{D} = \frac{E}{(1+v)(1-2v)}\begin{bmatrix} 1-v & v & 0 \\ v & 1-v & 0 \\ 0 & 0 & (1-2v)/2 \end{bmatrix} = E\begin{bmatrix} 1 & 0 & 0 \\ 0 & 1 & 0 \\ 0 & 0 & 1/2 \end{bmatrix}$$

$$\mathbf{k} = \int_{\Omega_e} \mathbf{B}^{\mathrm{T}}\mathbf{D}\mathbf{B}\,\mathrm{d}v = \frac{a_e t^* E}{4}\begin{bmatrix} -1 & 0 & -1/2 \\ 0 & -1 & -1/2 \\ 1 & 0 & 0 \\ 0 & 0 & 1/2 \\ 0 & 0 & 1/2 \\ 0 & 1 & 0 \end{bmatrix}\begin{bmatrix} -1 & 0 & 1 & 0 & 0 & 0 \\ 0 & -1 & 0 & 0 & 0 & 1 \\ -1 & -1 & 0 & 1 & 1 & 0 \end{bmatrix}$$

$$= \frac{a_e t^* E}{8}\begin{bmatrix} 3 & 1 & -2 & -1 & -1 & 0 \\ 1 & 3 & 0 & -1 & -1 & -2 \\ -2 & 0 & 2 & 0 & 0 & 0 \\ -1 & -1 & 0 & 1 & 1 & 0 \\ -1 & -1 & 0 & 1 & 1 & 0 \\ 0 & -2 & 0 & 0 & 0 & 2 \end{bmatrix}$$

From (5.90)

$$\bar{\mathbf{f}}_e^t = \begin{bmatrix} 4S_0t^*/3 \\ 5S_0t^*/3 \end{bmatrix}$$

which is placed in the 6×1 element vector as

$$\mathbf{f}_e^t = \begin{bmatrix} 0 \\ 4S_0t^*/3 \\ 0 \\ 5S_0t^*/3 \\ 0 \\ 0 \end{bmatrix}$$

Example 5.5.

Question. Determine $\hat{u}_1$ and $\hat{u}_2$ corresponding to the global degrees of freedoms indicated by the arrows at the nodes where the rollers are (Fig. 5.10).

Answer. The rectangular domain is divided into two triangular elements 1 and 2. The numbers inside the elements are local element node numbers. It may be noted that the parameters associated with element 1 are exactly the same as those in Example 5.4. For the numbering shown for element 2, it can be shown that the element stiffness matrix is exactly the same as that of element 1. Also the 2×1 load vector associated with the traction is exactly the same as that in Example 5.4. The global stiffness matrix and the load vector are obtained by assembling these element quantities. There are only two nonzero degrees of freedom: $\hat{u}_1$ and $\hat{u}_2$. After imposing the fixed displacement boundary conditions, the reduced stiffness matrix has a size of 2×2 and the reduced load vector has a size of 2×1, which are

$$\mathbf{K} = \begin{bmatrix} k_{22}^1 & k_{24}^1 \\ k_{42}^1 & k_{44}^1 + k_{66}^2 \end{bmatrix} = \frac{Et^*}{4}\begin{bmatrix} 3 & -1 \\ -1 & 1+2 \end{bmatrix} = \frac{Et^*}{4}\begin{bmatrix} 3 & -1 \\ -1 & 3 \end{bmatrix}$$

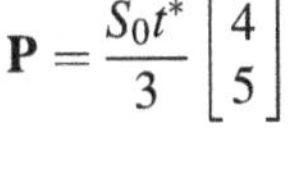

$$\mathbf{P} = \frac{S_0t^*}{3}\begin{bmatrix} 4 \\ 5 \end{bmatrix}$$

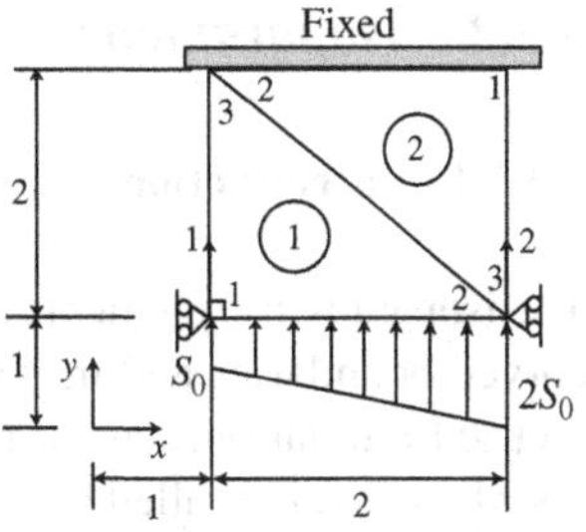

Fig. 5.10

The solution is

$$\hat{u}_1 = \frac{17S_0}{6E} \text{ and } \hat{u}_2 = \frac{19S_0}{6E}$$

Note that when the traction varies uniformly and has a value S_0

$$\mathbf{P} = S_0 \begin{bmatrix} 1 \\ 1 \end{bmatrix}$$

resulting in $\hat{u}_1 = \hat{u}_2 = (2S_0/E)$, which may be verified to be the solution for a axially loaded plane strain rod.

Using area coordinates (5.83c) and the integral formula in (5.83d), it can be shown

$$\int_{\Omega_e} N_i N_j \,\mathrm{d}x\,\mathrm{d}y\, t^* = \frac{a_e t^*}{6} \quad \text{for} \quad i = j \tag{5.92a}$$

and

$$= \frac{a_e t^*}{12} \quad \text{for} \quad i \neq j \tag{5.92b}$$

The consistent mass matrix may then be shown to be

$$\mathbf{m}_e = \int_{\Omega_i} \mathbf{N}^{\mathrm{T}} \rho \mathbf{N} \,\mathrm{d}v = \frac{a_e t^* \rho}{12} \begin{bmatrix} 2 & 0 & 1 & 0 & 1 & 0 \\ 0 & 2 & 0 & 1 & 0 & 1 \\ 1 & 0 & 2 & 0 & 1 & 0 \\ 0 & 1 & 0 & 2 & 0 & 1 \\ 1 & 0 & 1 & 0 & 2 & 0 \\ 0 & 1 & 0 & 1 & 0 & 2 \end{bmatrix} \tag{5.93}$$

5.3.2 *Isoparametric Elements*

5.3.2.1 Introduction

The isoparametric elements employ the technique of transformation in dealing with curved boundaries and irregular shapes of element volumes (or areas). In isoparametric formulation, standard Cartesian coordinates are transformed into curvilinear coordinates (also called natural or intrinsic coordinates). ξ and η shown in Fig. 5.11

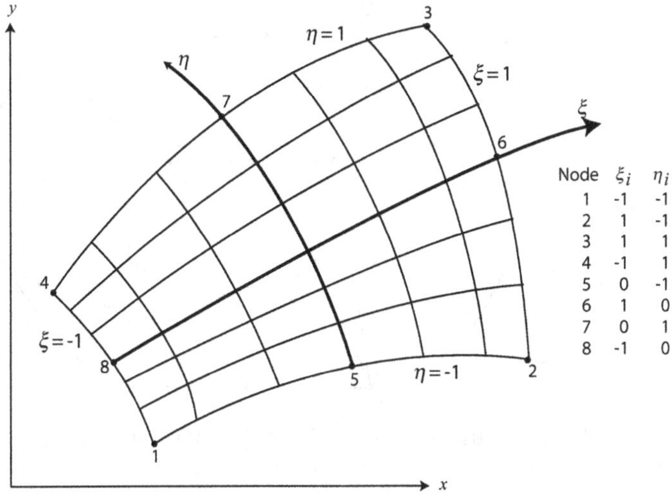

Fig. 5.11 Eight-noded, two-dimensional isoparametric element (2D8)

are curvilinear coordinates. They are normalized so that their values range between -1 and 1 as

$$-1 \leq \xi \leq 1 \tag{5.94a}$$

$$-1 \leq \eta \leq 1 \tag{5.94b}$$

Several types of isoparametric elements are available for modeling one-, two- and three-dimensional problems. Figure 5.12a is a schematic of a four-noded bilinear quadrilateral element that could be used to model two-dimensional problems. Figure 5.12b is a sketch of a 20-noded (incomplete quadratic) hexahedral element that could be used to model three-dimensional problems.

Equations associated with these different types of elements are similar. Once the fundamentals are understood for one type of element, it is easy to follow the details of the formulation for another. We will describe the details for an eight-noded, quadratic, quadrilateral, two-dimensional element, shown in Fig. 5.11, abbreviated here as 2D8. Many of the details are the same regardless of the number of unknowns at each node. For example, the number of unknowns at a node is (a) one for water flow problem, where the unknown is the hydraulic head h, (b) two for two-dimensional elasticity problems treated in this section, where the unknowns are the x- and y-components of the displacements u and v, and (c) three when modeling a fully coupled problems using two solid displacement variables and one pore pressure variable. We will develop the equations for the two-dimensional elasticity problems.

We will develop the equations for the plane strain problems, and point out the modifications needed to extend the formulation for plane stress and axisymmetric problems.

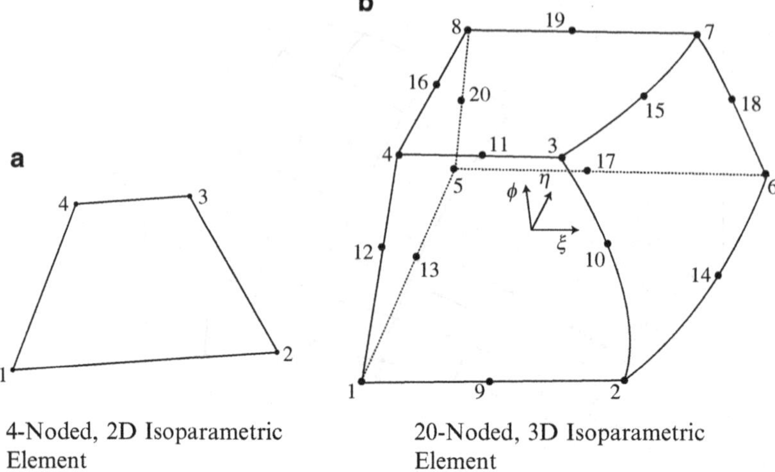

Fig. 5.12 Examples of other isoparametric elements

5.3.2.2 Element Stiffness Matrix for 2D8 Element

Note that 2D8 not only has nodes at the corners, but also the middle of the sides as shown in Fig. 5.11. 2D8 does not have nodes within the element (Note that there are elements that have intraelement nodes). The values of (ξ, η) at the eight nodes are listed in Fig. 5.11, i.e., (ξ_i, η_i) for $i = 1, 8$. The x- and y-coordinates of any point within the element is expressed in terms of the values of the coordinates at the nodes as

$$x = \sum_{1}^{8} N_i x_i \tag{5.95a}$$

$$y = \sum_{1}^{8} N_i y_i \tag{5.95b}$$

where (x_i, y_i) are nodal coordinates with respect to the global x- and y-coordinate axes, and N_i are isoparametric shape functions given by

$$N_i = \frac{1}{4}(1 + \xi_i \xi)(1 + \eta_i \eta)(\xi_i \xi + \eta_i \eta - 1) \quad \text{for} \quad i = 1, 2, 3, 4 \tag{5.96a}$$

$$N_i = \frac{1}{2}(1 - \xi^2)(1 + \eta_i \eta) \quad \text{for} \quad i = 5, 7 \tag{5.96b}$$

$$N_i = \frac{1}{2}(1 - \eta^2)(1 + \xi_i \xi) \quad \text{for} \quad i = 6, 8 \tag{5.96c}$$

The reader may verify that

$$\begin{aligned} N_i &= 1 \text{ at } (\xi_k, \eta_k); \quad k = i \\ &= 0 \text{ at } (\xi_k, \eta_k); \quad k = 1, 8 \quad \text{except when } k = i \end{aligned} \tag{5.96d}$$

In isoparametric formulation, the same set of shape functions are used for the interpolation of coordinates (5.95a) and (5.95b) and the displacements. There are two degrees of freedom (d.o.f.) at every point, $(u,\ v)$. It then follows that there are two d.o.f. at every node, $(\hat{u}_i,\ \hat{v}_i)$. The interpolation relations for the displacements are

$$u = \sum_{i=1}^{8} N_i \hat{u}_i \tag{5.97a}$$

$$v = \sum_{i=1}^{8} N_i \hat{v}_i \tag{5.97b}$$

In a matrix-vector form:

$$\underset{2\times 1}{\begin{bmatrix} u \\ v \end{bmatrix}} = \underset{2\times 16}{\begin{bmatrix} N_1 & 0 & N_2 & 0 & .. & .. & N_8 & 0 \\ 0 & N_1 & 0 & N_2 & .. & .. & 0 & N_8 \end{bmatrix}} \underset{16\times 1}{\begin{bmatrix} \hat{u}_1 \\ \hat{v}_1 \\ \hat{u}_2 \\ \hat{v}_2 \\ . \\ . \\ \hat{u}_8 \\ \hat{v}_8 \end{bmatrix}} \tag{5.98a}$$

In a compact format

$$\underset{2\times 1}{\mathbf{u}_e} = \underset{2\times 16}{\mathbf{N}} \ \underset{16\times 1}{\hat{\mathbf{u}}_e} \tag{5.98b}$$

It is clear that the chosen shape functions are continuous within an element. It can be verified that the displacement field is continuous along interelement boundaries as well. For instance, consider the boundary containing nodes 2, 6, and 3 (Fig. 5.11), where $\xi = 1$, and $-1 \le \eta \le 1$. Consider, for instance, the horizontal displacement u. Due to the property given by (5.96d), the variation of u along this boundary only involves $(\hat{u}_2, \hat{u}_3, \hat{u}_6)$. From (5.96a), (5.96c), and (5.97a), we have

$$u = -\frac{1}{2}(1-\eta)\eta\hat{u}_2 + \frac{1}{2}(1+\eta)\eta\hat{u}_3 + (1-\eta^2)\hat{u}_6 \tag{5.98c}$$

Thus, u varies quadratically with η on this boundary and is uniquely defined in terms of the values of $(\hat{u}_2, \hat{u}_3, \hat{u}_6)$. Thus, as long as the adjacent elements share the same set $(\hat{u}_2, \hat{u}_3, \hat{u}_6)$, then u is continuous at this interelement boundary. The fact that each element surrounding a given global node shares the same displacement unknown values, is a fundamental feature of the finite element method. (However, sometimes there are reasons for employing elements which violate the interelement continuity requirement. Such elements are known as nonconforming elements. Coverage of this is beyond the scope of the book.)

Defining

$$F_i = \frac{\partial N_i}{\partial x} \tag{5.99a}$$

and

$$G_i = \frac{\partial N_i}{\partial y} \tag{5.99b}$$

the strain vector can be written as

$$\underset{3\times1}{\begin{bmatrix} \varepsilon_{xx} \\ \varepsilon_{yy} \\ \gamma_{xy} \end{bmatrix}} = \begin{bmatrix} \dfrac{\partial u}{\partial x} \\ \dfrac{\partial v}{\partial y} \\ \dfrac{\partial u}{\partial y} + \dfrac{\partial v}{\partial x} \end{bmatrix} = \underset{3\times16}{\begin{bmatrix} F_1 & 0 & F_2 & 0 & .. & .. & F_8 & 0 \\ 0 & G_1 & 0 & G_2 & .. & .. & 0 & G_8 \\ G_1 & F_1 & G_2 & F_2 & .. & .. & G_8 & F_8 \end{bmatrix}} \underset{16\times1}{\begin{bmatrix} \hat{u}_1 \\ \hat{v}_1 \\ \hat{u}_2 \\ \hat{v}_2 \\ . \\ . \\ \hat{u}_8 \\ \hat{v}_8 \end{bmatrix}} \tag{5.99c}$$

In a more compact form:

$$\underset{3\times1}{\boldsymbol{\varepsilon}} = \underset{3\times16}{\mathbf{B}} \; \underset{16\times1}{\hat{\mathbf{u}}} \tag{5.99d}$$

The equations for F_i and G_i depend on (shown below) the derivatives of N_i with respect to ξ and η, which are

For $i = 1, 2, 3, 4$

$$\frac{\partial N_i}{\partial \xi} = \frac{\xi_i}{4}(1 + \eta\eta_i)(2\xi\xi_i + \eta\eta_i) \tag{5.100a}$$

$$\frac{\partial N_i}{\partial \eta} = \frac{\eta_i}{4}(1 + \xi\xi_i)(\xi\xi_i + 2\eta\eta_i) \tag{5.100b}$$

For $i = 5, 7$

$$\frac{\partial N_i}{\partial \xi} = -\xi(1 + \eta\eta_i) \tag{5.100c}$$

$$\frac{\partial N_i}{\partial \eta} = \frac{1}{2}(1 - \xi^2)\eta_i \tag{5.100d}$$

For $i = 6, 8$

$$\frac{\partial N_i}{\partial \xi} = \frac{1}{2}(1 - \eta^2)\xi_i \tag{5.100e}$$

$$\frac{\partial N_i}{\partial \eta} = -\eta(1 + \xi\xi_i) \tag{5.100f}$$

Remarks. Formulations for plane stress and axisymmetric problems

To see how the equations differ between plane strain and plane stress cases, let us for a moment go back to the potential energy expression given by (5.49b), and write the first term in a three-dimensional expanded form as

$$\begin{aligned} \pi_1 &= \int_\Omega \boldsymbol{\sigma}^{\mathrm{T}} \boldsymbol{\varepsilon} \, \mathrm{d}v \\ &= \int_\Omega \left\{ \left[\sigma_{xx}\varepsilon_{xx} + \sigma_{yy}\varepsilon_{yy} + \sigma_{xy}\gamma_{xy}\right] + \left[\sigma_{zz}\varepsilon_{zz} + \sigma_{xz}\gamma_{xz} + \sigma_{yz}\gamma_{yz}\right] \right\} \mathrm{d}v \end{aligned}$$

For a plane strain problem, since $\varepsilon_{zz} = \gamma_{xz} = \gamma_{yz} = 0$, the second groups of terms in (5.100g) is zero. The element stiffness matrix will be based on $\boldsymbol{\varepsilon} = \left[\varepsilon_{xx}, \varepsilon_{yy}, \gamma_{xy}\right]$, $\boldsymbol{\sigma} = \left[\sigma_{xx}, \sigma_{yy}, \sigma_{xy}\right]$, and the 3×3 plane strain **D**-matrix given in Chap. 3.

For a plane stress problem, $\gamma_{xz} = \gamma_{yz} = 0$, but $\varepsilon_{zz} \neq 0$. This leads to a modification of the constitutive law. The element stiffness matrix is based on $\boldsymbol{\varepsilon} = \left[\varepsilon_{xx}, \varepsilon_{yy}, \gamma_{xy}\right]$, $\boldsymbol{\sigma} = \left[\sigma_{xx}, \sigma_{yy}, \sigma_{xy}\right]$, and the 3×3 plane stress **D**-matrix.

In axisymmetric problems, all three normal components of the stresses and strains are nonzero, but two of the shear stress and strain components are zero. The element stiffness matrix is based on $\boldsymbol{\varepsilon} = [\varepsilon_{rr}, \varepsilon_{\theta\theta}, \varepsilon_{zz}, \gamma_{rz}]$, $\boldsymbol{\sigma} = [\sigma_{rr}, \sigma_{\theta\theta}, \sigma_{zz}, \sigma_{rz}]$, and the 4×4 axisymmetric **D**-matrix.

Returning now to the plane strain problem, the element stiffness matrix is given by (5.5b), and the element load vectors due to body force and boundary traction are given by (5.6b) and (5.7b), respectively. The consistent mass and damping matrices are given by (5.73) and (5.76), respectively. The integrals need to be evaluated with respect to x and y coordinates, whereas the **N** and **B** matrices are in intrinsic coordinates, requiring transformation between the two systems. The relevant transformation is derived directly from (5.95a) and (5.95b) as

$$\underset{2\times 1}{\begin{bmatrix} \mathrm{d}x \\ \mathrm{d}y \end{bmatrix}} = \begin{bmatrix} \dfrac{\partial x}{\partial \xi} & \dfrac{\partial x}{\partial \eta} \\ \dfrac{\partial y}{\partial \xi} & \dfrac{\partial y}{\partial \eta} \end{bmatrix} \begin{bmatrix} \mathrm{d}\xi \\ \mathrm{d}\eta \end{bmatrix} = \begin{bmatrix} J_{11} & J_{12} \\ J_{21} & J_{22} \end{bmatrix} \underset{2\times 1}{\begin{bmatrix} \mathrm{d}\xi \\ \mathrm{d}\eta \end{bmatrix}} = \underset{2\times 2}{\mathbf{J}} \underset{2\times 1}{\begin{bmatrix} \mathrm{d}\xi \\ \mathrm{d}\eta \end{bmatrix}} \tag{5.101a}$$

where the 2×2 matrix $\mathbf{J}$ is called the Jacobian matrix. Equation (5.5b) requires a volume element in three dimensions and an area in two dimensions (assuming a unit length in the direction perpendicular to the plane). The needed area element transforms as

$$\mathrm{d}x\,\mathrm{d}y = |\mathbf{J}|\mathrm{d}\xi\,\mathrm{d}\eta \tag{5.101b}$$

That is, the area integrals transform as

$$\int_{\Omega} (..)\mathrm{d}x\,\mathrm{d}y = \int_{\Omega} (..)|\mathbf{J}|\mathrm{d}\xi\,\mathrm{d}\eta \tag{5.101c}$$

Note that

$$\int_{\Omega} \mathrm{d}x\,\mathrm{d}y = \int_{\Omega} |\mathbf{J}|\mathrm{d}\xi\,\mathrm{d}\eta = \text{area of a plane element} \tag{5.101d}$$

The three-dimensional equivalent will give the volume and the one-dimensional equivalent will give the length. In the case of the axisymmetric approximation, the volume integrals involve:

$$\int_{\Omega} (..)r\,\mathrm{d}r\,\mathrm{d}\theta\,\mathrm{d}z = 2\pi \int_{\Omega} (..)\left\{\sum_{i=1}^{n} r_i N_i\right\}|\mathbf{J}|\mathrm{d}\xi\,\mathrm{d}\eta \tag{5.101e}$$

For the plane problems, by the chain rule of differentiation:

$$\begin{bmatrix} \dfrac{\partial N_i}{\partial \xi} \\ \dfrac{\partial N_i}{\partial \eta} \end{bmatrix} = \begin{bmatrix} \dfrac{\partial x}{\partial \xi} & \dfrac{\partial y}{\partial \xi} \\ \dfrac{\partial x}{\partial \eta} & \dfrac{\partial y}{\partial \eta} \end{bmatrix} \begin{bmatrix} \dfrac{\partial N_i}{\partial x} \\ \dfrac{\partial N_i}{\partial y} \end{bmatrix} \tag{5.102a}$$

From (5.95a) and (5.95b), we have

$$\frac{\partial x}{\partial \xi} = \sum_{1}^{8} \frac{\partial N_i}{\partial \xi} x_i \tag{5.102b}$$

$$\frac{\partial y}{\partial \xi} = \sum_{1}^{8} \frac{\partial N_i}{\partial \xi} y_i \tag{5.102c}$$

$$\frac{\partial x}{\partial \eta} = \sum_{1}^{8} \frac{\partial N_i}{\partial \eta} x_i \tag{5.102d}$$

$$\frac{\partial y}{\partial \eta} = \sum_{1}^{8} \frac{\partial N_i}{\partial \eta} y_i \tag{5.102e}$$

The derivatives of N_i with respect to the intrinsic coordinates are given in (5.100). The matrix on the right-hand side of (5.102a) can thus be evaluated. Equation (5.102a) is then inverted to find F_i and G_i (5.99a) and (5.99b).

Integrations need to be performed in evaluating the element stiffness matrix, mass matrix, and load vectors. For the plane strain problem of interest here, the expression for the element stiffness matrix given in (5.5b) becomes

$$\mathbf{k}_i = \int_{\Omega_i} \mathbf{B}^{\mathrm{T}} \mathbf{D} \mathbf{B} t^* \, \mathrm{d}x \, \mathrm{d}y \tag{5.103a}$$

$$= \int_{-1}^{1} \int_{-1}^{1} \mathbf{B}^{\mathrm{T}} \mathbf{D} \mathbf{B} t^* |\mathbf{J}| \mathrm{d}\xi \, \mathrm{d}\eta \tag{5.103b}$$

where t^* is the thickness of the element in the direction perpendicular to plane of loading, and is normally taken as unity.

5.3.2.3 Numerical Integration

A typical term in (5.103b) is of the form:

$$I = \int_{-1}^{1} \int_{-1}^{1} F(\xi, \eta) \mathrm{d}\xi \, \mathrm{d}\eta \tag{5.103c}$$

where $F(\xi, \eta)$ is a scalar function of certain order in ξ and η. Typically, both the numerator and the denominator of $F(\xi, \eta)$ have polynomial terms of complex nature, and as such, it is generally not possible to carry out the integration analytically. A numerical method needs to be employed. One of the most commonly used techniques is known as the Gauss method. The reader is referred to standard texts on numerical analysis (e.g., Kopal 1955; Stroud and Secrest 1966) or finite element books (e.g., Cook 1981) for theoretical background on this method.

In brief, the continuous integral in (5.103c) is replaced by discrete sum as

$$I = \sum_{i=1}^{n}\sum_{j=1}^{n} w_i w_j F(\xi_i, \eta_j) \tag{5.104}$$

where n is the order of integration, w_i and w_j are weights, and (ξ_i, η_j) are intrinsic coordinates at the sampling points (Gauss quadrature). In Gauss method, the sampling points and the weights are chosen so that for a given order of integration, the best results are obtained (i.e., error between the actual and numerical values of the integral is minimum). The sampling points for order 1, 2, and 3 integrations are shown in Figs. 5.13 and 5.14 for one- and two-dimensional elements, respectively. The values of ξ_i and w_i up to order 5 are presented in Table 5.1.

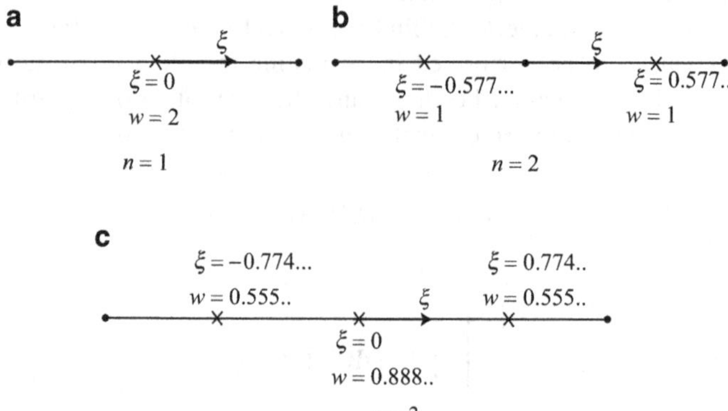

Fig. 5.13 Sampling points for gauss integration of one-dimensional elements

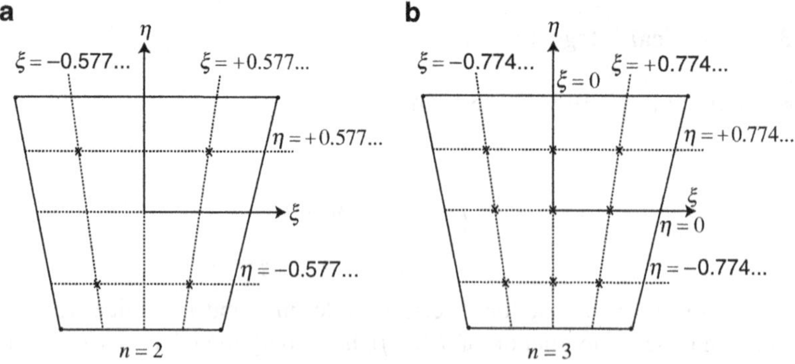

Fig. 5.14 Sampling points for gauss integration of two-dimensional elements

Table 5.1 Sampling points and weights for gauss quadrature

Order n	Coordinate ξ_i	Weight w_i
1	0.0	2.0
2	±0.57735 02691 89626	1.0
3	±0.77459 66692 41483	0.55555 55555 55556
	0.0	0.88888 88888 88889
4	±0.86113 63115 94053	0.34785 48451 37454
	±0.33998 10435 84856	0.65214 51548 62546
5	±0.90617 98459 38664	0.23692 68850 56189
	±0.53846 93101 05683	0.47862 86704 99366
	0.0	0.56888 88888 88889

The Gauss method is capable of integrating simple polynomials exactly: a polynomial of order $2n-1$ is integrated exactly by nth *order Gauss quadrature*. Thus, for instance, for a quadratic equation ($n = 2$), all we need is a third-order Gauss quadrature for exact integration. While $F(\xi, \eta)$ involved in element stiffness matrices are generally complicated to know exactly the lowest order needed for exact integration, experience exists with most standard elements.

A question arises as to whether it is necessary to perform the integrations exactly. Another question concerns whether there is a limit on the maximum order of integration to use in a given case. Obviously, the lower the order used, the higher the computational efficiency. Recall that in a finite element method, the structure is modeled to be "stiffer" than actual, resulting in smaller displacements than actual. It turns out that as the order of increases (which in turn increases the number of sampling points within the element), the structure becomes stiffer. Hence, an order that is higher than necessary will make the structure even stiffer, and lower than necessary will make the structure softer. Thus, from this point of view, it is good to use a low order Gauss quadrature. In fact, in Chap. 7, we will use lower order of integration for fluid pressure terms. Lower order is used for certain terms in plate/shell elements. Then the question is what the minimum required order is. The rule is that the order chosen must at least integrate the length (in one-dimensional) or area (in two-dimensional) or volume (in three-dimensional) exactly. That is, the quantity $\int |\mathbf{J}| \mathrm{d}\,\xi$ or $\int |\mathbf{J}| \mathrm{d}\xi\, \mathrm{d}\eta$ or $\int |\mathbf{J}| \mathrm{d}\xi\, \mathrm{d}\eta\, \mathrm{d}\,\zeta$ needs to be integrated exactly. Further discussion is beyond the scope of this book, and the reader is referred to standard texts.

The optimal order of integration for evaluating the stiffness for the 2D8 element is 3×3 (i.e., $n = 3$).

5.3.2.4 Element Load Vector for 2D8 Element

In evaluating the element load vector associated with the body force, the same order of integration as in the evaluation of the element stiffness matrix is used.

In the case of boundary traction, the integrations involved in the element load vectors can mostly be performed exactly. For instance, let us assume that over one of the element boundaries (say, the boundary containing nodes 2, 3, and 6 in Fig. 5.11), there is a uniform traction normal to the surface. The element load vector to be evaluated is given by (5.7b), which is associated with the virtual work terms in (5.7a).

Owing to the property given by (5.96d), the integral involved in (5.7a) can be simplified by recognizing the fact that the variation of the displacement over the boundary depends only on the degrees of freedom associated with the nodes on that boundary $(\hat{u}_2, \hat{u}_3, \hat{u}_6)$. Thus, we can define a new modified shape matrix in terms of only $(\hat{u}_2, \hat{u}_3, \hat{u}_6)$ and simplify the integral. You may recall that this was already done in the treatment of triangular elements (5.81)–(5.91).

Let us do this for a typical element for the case of the constant traction (Fig. 5.15a). For simplicity, we will call the nodes as 1, 2, and 3 as shown in the figure. The boundary is aligned with the global x-direction. Then, (5.7a) can be modified as

$$I_i^3 = \int_{\Gamma_{ti}} \delta \mathbf{u}_i^{\mathrm{T}} \mathbf{t} \, \mathrm{d}s \tag{5.105a}$$

$$= \int_{\Gamma_{ti}} \delta \mathbf{u}_n \mathbf{t}_n \, \mathrm{d}s \tag{5.105b}$$

$$= \delta \hat{\mathbf{u}}_n^{\mathrm{T}} \int_{\Gamma_{ti}} \mathbf{N}^{\mathrm{T}} \mathbf{t}_n(x) \mathrm{d}x \tag{5.105c}$$

a

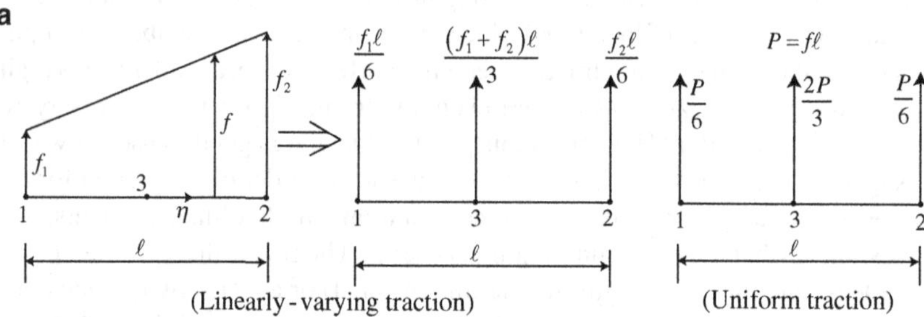

Equivalent Nodal Forces for Linearly Varying and Uniform Tractions on 8-Noded Plane Elements

b

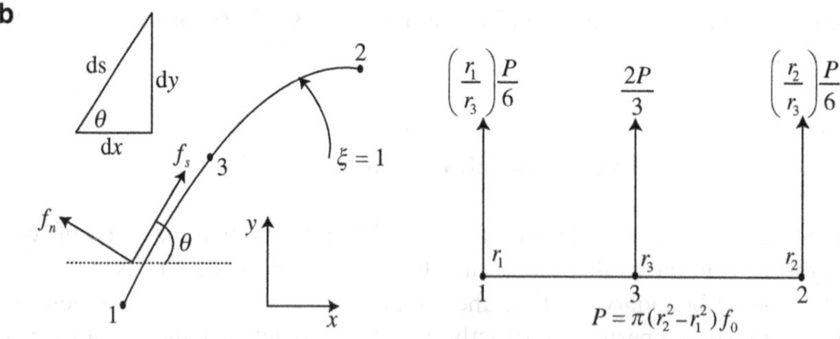

Equivalent Nodal Forces Due to Traction on Curved Boundaries

Equivalent Nodal Forces Due to Uniform Traction on an Axisymmetric Element

Fig. 5.15 Evaluation of equivalent nodal forces from distributed tractions

where

$$\delta u_n^{\mathrm{T}} = [\delta \hat{u}_{n1}, \delta \hat{u}_{n1}, \delta \hat{u}_{n1}], \quad \text{and from (5.98c)}, \tag{5.105d}$$

$$\mathbf{N} = [N_1, N_2, N_3] = \left[-\frac{1}{2}(1-\eta)\eta; \quad \frac{1}{2}(1+\eta)\eta; \quad (1-\eta^2) \right] \tag{5.105e}$$

$$\mathbf{t}_n(x) = [f_0] \tag{5.105f}$$

Note that the vectors $\delta \mathbf{u}_n$ and $\mathbf{t}_n$ are in the direction perpendicular to the boundary, which is the y-direction in this case. Along the boundary,

$$\begin{aligned} x &= N_1 x_1 + N_2 x_2 + N_3 x_3 \\ &= \left[-\frac{1}{2}(1-\eta)\eta \right] x_1 + \left[\frac{1}{2}(1+\eta)\eta \right] x_2 + \left[(1-\eta^2) \right] x_3 \end{aligned} \tag{5.106a}$$

From this, noting that node 3 is the midside node, we have

$$x_3 = \frac{x_1 + x_2}{2}; \quad \ell = x_2 - x_1 \quad \mathrm{d}x = \frac{\ell}{2} \mathrm{d}\eta \tag{5.106b}$$

Then, from (5.105c), the load vector is evaluated as

$$\begin{bmatrix} P_1 \\ P_2 \\ P_3 \end{bmatrix} = \int_{-1}^{+1} \begin{bmatrix} N_1 \\ N_2 \\ N_3 \end{bmatrix} [f_0] \frac{\ell}{2} \mathrm{d}\eta = \begin{bmatrix} \dfrac{f_0 \ell}{6} \\ \dfrac{f_0 \ell}{6} \\ \dfrac{2 f_0 \ell}{3} \end{bmatrix} = \begin{bmatrix} \dfrac{P}{6} \\ \dfrac{P}{6} \\ \dfrac{2P}{3} \end{bmatrix} \tag{5.107}$$

where $P = f_0 \ell$. Note that the forces in (5.107) are equivalent nodal forces perpendicular to the edge shown in Fig. 5.15a. If the edge is not parallel to either the x- or y-axis, the forces can be resolved in these directions before assembling into the global load vector. The nodal forces corresponding to shear force traction can be similarly handled.

It is left for the reader, as an exercise, to verify that the equivalent nodal forces for the case of the linearly varying traction are given by those shown in Fig. 5.15a (Problem 5.8). Note the similarities between this problem and the one addressed in Problem 5.2.

The details are somewhat involved if the element edge is curved. In this case, both the x- and y-components of the displacements and the traction along the edge must now be considered, even when the traction is normal or tangent to the boundary. Owing to the property given by (5.96d), as in the case of a straight boundary, the variation of the displacements along this boundary only involves the displacement unknowns associated with the nodes on the edge under consideration. Consequently, only these unknowns affect the work done in (5.7a). Thus, we can derive the relevant equations by considering a typical edge such as that shown in Fig. 5.15b. Equation (5.7b) can be written as

$$\begin{bmatrix} P_1 \\ Q_1 \\ P_2 \\ Q_2 \\ P_3 \\ Q_3 \end{bmatrix} = \int_{-1}^{+1} \begin{bmatrix} N_1 & 0 \\ 0 & N_1 \\ N_2 & 0 \\ 0 & N_2 \\ N_3 & 0 \\ 0 & N_3 \end{bmatrix} \begin{bmatrix} f_s \cos\theta - f_n \sin\theta \\ f_s \sin\theta + f_n \cos\theta \end{bmatrix} \mathrm{d}s \tag{5.108}$$

where P and Q represent the x- and y-components of the nodal load vectors, and $[N_1\ N_2\ N_3]$ are given by (5.105e). In performing the integration in (5.108), first note that

$$\mathrm{d}x = \mathrm{d}s \cos\theta \tag{5.109a}$$

and

$$\mathrm{d}y = \mathrm{d}s \sin\theta \tag{5.109b}$$

Noting that $\mathrm{d}\xi = 0$ on the edge under consideration, from (5.101a),

$$\mathrm{d}x = J_{12}\,\mathrm{d}\eta \tag{5.110a}$$

and

$$\mathrm{d}y = J_{22}\,\mathrm{d}\eta \tag{5.110b}$$

Equation (5.108) becomes

$$\begin{bmatrix} P_1 \\ Q_1 \\ P_2 \\ Q_2 \\ P_3 \\ Q_3 \end{bmatrix} = \int_{-1}^{1} \begin{bmatrix} N_1(f_sJ_{12} - f_nJ_{22}) \\ N_1(f_sJ_{22} + f_nJ_{12}) \\ N_2(f_sJ_{12} - f_nJ_{22}) \\ N_2(f_sJ_{22} + f_nJ_{12}) \\ N_3(f_sJ_{12} - f_nJ_{22}) \\ N_3(f_sJ_{22} + f_nJ_{12}) \end{bmatrix} \mathrm{d}\eta \tag{5.111}$$

Equation (5.111) can be evaluated numerically by the Gauss integration procedure just as the coefficients of the stiffness matrix are evaluated.

By following a similar procedure, equivalent nodal forces due boundary traction can be evaluated for axisymmetric elements (Fig. 5.15c). This is addressed in Problem 5.9.

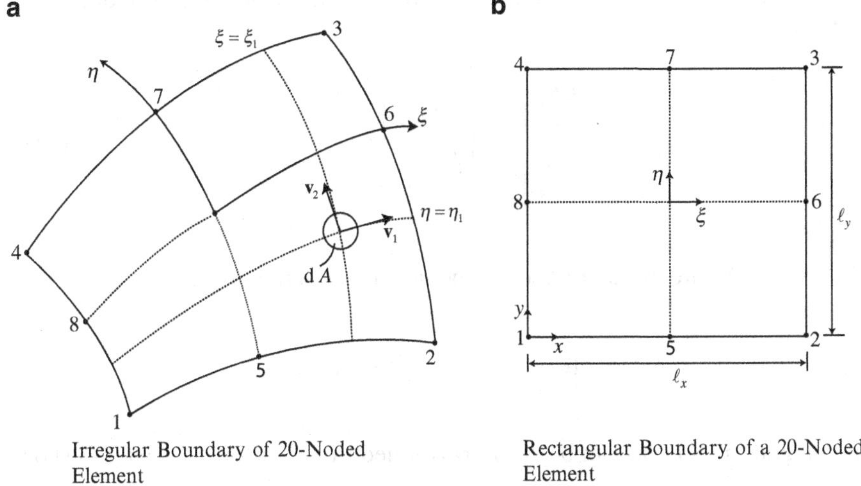

Fig. 5.16 Illustration of boundary integral to obtain equivalent nodal forces for three-dimensional elements

Now, we will present some details on the calculation of element load vectors for three-dimensional elements (Fig. 5.16). The calculation of element load vectors corresponding to body forces are performed using (5.6b) using a suitable numerical integration method.

The evaluation of element load vector due to boundary traction involves some further details. It was noted in the case of two-dimensional elements, the details of integration are somewhat simpler for straight boundaries ((5.105)–(5.107)) than for curved boundaries ((5.108)–(5.111)). The same is true for three-dimensional elements as well.

Referring to Fig. 5.12b, the three-dimensional shape functions are

$$
\begin{aligned}
N_i &= \frac{1}{8}(1+\xi_i\xi)(1+\eta_i\eta)(1+\phi_i\phi)(\xi_i\xi+\eta_i\eta+\phi_i\phi-2) \quad \text{for } i=1,8\\
N_i &= \frac{1}{4}(1-\xi^2)(1+\eta_i\eta)(1+\phi_i\phi) \quad \text{for } i=9,17,19,11\\
N_i &= \frac{1}{4}(1-\eta^2)(1+\xi_i\xi)(1+\phi_i\phi) \quad \text{for } i=13,14,15,16\\
N_i &= \frac{1}{4}(1-\phi^2)(1+\xi_i\xi)(1+\eta_i\eta) \quad \text{for } i=12,10,18,20
\end{aligned}
$$

When the element boundary has a rectangular shape as in Fig. 5.16b and is a plane, some simplifications are realized. For an eight-noded element boundary as shown in Fig. 5.16b, for instance, when a uniform traction f_0 acts normal to the plane, we obtain an equation similar to (5.107) for the element load vector normal to the plane as

$$\begin{bmatrix} P_1 \\ P_2 \\ \cdot \\ P_8 \end{bmatrix} = \iint \begin{bmatrix} N_1 \\ N_2 \\ \cdot \\ N_8 \end{bmatrix} f_0 \mathrm{d}x\,\mathrm{d}y \tag{5.112a}$$

where $N_1 \ldots N_8$ are the appropriate shape functions. Noting that

$$\mathrm{d}x = \frac{\ell_x}{2}\mathrm{d}\xi \text{ and } \mathrm{d}y = \frac{\ell_y}{2}\mathrm{d}\eta,$$

the integral in (5.112a) can be easily performed. It is left for the reader to show that

$$[P_1, P_2, \ldots P_8] = \left[-\frac{1}{12}, -\frac{1}{12}, -\frac{1}{12}, -\frac{1}{12}, \frac{1}{3}, \frac{1}{3}, \frac{1}{3}, \frac{1}{3}\right] f_0 \ell_x \ell_y \tag{5.112b}$$

The equivalent nodal forces are shown in Fig. 5.17 for this case.

When the element boundary is not rectangular as shown in Fig. 5.16a and/or curved, the formulation is somewhat involved. Let us illustrate the procedure for the case with the traction being normal to the plane. $\mathbf{v}_1$ and $\mathbf{v}_2$ are unit vectors at a given point on the boundary in the ξ- and η-directions, respectively. Let us consider $\mathbf{v}_1$ first, which can be expressed as

$$\mathbf{v}_1 = \mathbf{i}\,\mathrm{d}x + \mathbf{j}\,\mathrm{d}y + \mathbf{k}\,\mathrm{d}z \tag{5.113}$$

where (dx,dy,dz) is differential change along $\mathbf{v}_1$

Note that the boundary itself is normal to the third intrinsic coordinate ϕ. Since $\phi =$ constant on the boundary, we have $\mathrm{d}\phi = 0$ on the boundary. As $\mathbf{v}_1$ is along the

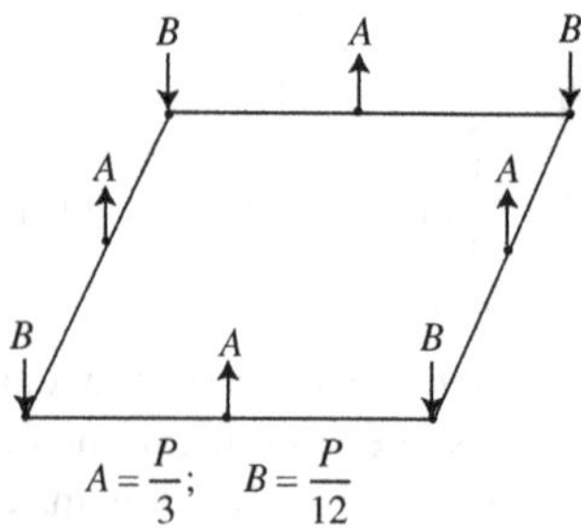

Fig. 5.17 Nodal forces corresponding to a uniform traction for 20-noded three-dimensional element with a planar rectangular side

ξ-axis, $\mathrm{d}\,\eta = 0$ along $\mathbf{v}_1$. That is, along the ξ-axis, $\mathrm{d}\,\eta = 0$ and $\mathrm{d}\,\phi = 0$. The three-dimensional Jacobian matrix takes the form:

$$\begin{bmatrix} \mathrm{d}x \\ \mathrm{d}y \\ \mathrm{d}z \end{bmatrix} = \begin{bmatrix} \frac{\partial x}{\partial \xi} & \frac{\partial x}{\partial \eta} & \frac{\partial x}{\partial \phi} \\ \frac{\partial y}{\partial \xi} & \frac{\partial y}{\partial \eta} & \frac{\partial y}{\partial \phi} \\ \frac{\partial z}{\partial \xi} & \frac{\partial z}{\partial \eta} & \frac{\partial z}{\partial \phi} \end{bmatrix} \begin{bmatrix} \mathrm{d}\xi \\ \mathrm{d}\eta \\ \mathrm{d}\phi \end{bmatrix} = \begin{bmatrix} J_{11} & J_{12} & J_{13} \\ J_{21} & J_{22} & J_{23} \\ J_{31} & J_{32} & J_{33} \end{bmatrix} \begin{bmatrix} \mathrm{d}\xi \\ \mathrm{d}\eta \\ \mathrm{d}\phi \end{bmatrix} \tag{5.114}$$

From this, $(\mathrm{d}x, \mathrm{d}y, \mathrm{d}z) = (J_{11}\,\mathrm{d}\xi, J_{21}\,\mathrm{d}\xi, J_{31}\,\mathrm{d}\xi)$ along the ξ-axis. By a similar logic, the components of $\mathbf{v}_2$ are evaluated. Then:

$$\mathbf{v}_1 = (\mathbf{i}J_{11} + \mathbf{j}J_{21} + \mathbf{k}J_{31})\mathrm{d}\xi \tag{5.115a}$$

and

$$\mathbf{v}_2 = (\mathbf{i}J_{12} + \mathbf{j}J_{22} + \mathbf{k}J_{32})\mathrm{d}\eta \tag{5.115b}$$

Let $\mathbf{v}_3$ be the unit vector perpendicular to the element boundary at the point under consideration. Noting that the magnitude of the cross product $\mathbf{v}_1 \times \mathbf{v}_2$ is the area element $\mathrm{d}A$ and the direction coincides with the needed direction $\mathbf{v}_3$, one has

$$\mathbf{v}_3\,\mathrm{d}A = \mathbf{v}_1 \times \mathbf{v}_2 = \begin{bmatrix} J_{21}J_{32} - J_{31}J_{22} \\ -J_{11}J_{32} + J_{31}J_{12} \\ J_{11}J_{22} - J_{21}J_{12} \end{bmatrix} \mathrm{d}\xi\,\mathrm{d}\eta \tag{5.116}$$

The element load vector corresponding to a uniform traction normal to the element boundary is now easily evaluated as

$$\begin{aligned} \begin{bmatrix} P_1 \\ Q_1 \\ R_1 \\ \cdot \\ \cdot \end{bmatrix}_{24\times 1} &= \iint \begin{bmatrix} N_1 & 0 & 0 \\ 0 & N_1 & 0 \\ 0 & 0 & N_1 \\ N_2 & 0 & 0 \\ \cdot & \cdot & \cdot \end{bmatrix}_{24\times 3} \mathbf{v}_3 f_n\,\mathrm{d}A \\ &= \int_{-1}^{+1}\int_{-1}^{+1} \begin{bmatrix} N_1 & 0 & 0 \\ 0 & N_1 & 0 \\ 0 & 0 & N_1 \\ N_2 & 0 & 0 \\ \cdot & \cdot & \cdot \end{bmatrix}_{24\times 3} \begin{bmatrix} J_{21}J_{32} - J_{31}J_{22} \\ -J_{11}J_{32} + J_{31}J_{12} \\ J_{11}J_{22} - J_{21}J_{12} \end{bmatrix}_{3\times 1} f_n\,\mathrm{d}\xi\,\mathrm{d}\eta \end{aligned} \tag{5.117}$$

Equation (5.117) can be evaluated numerically by the Gauss integration method.

Fig. 5.18

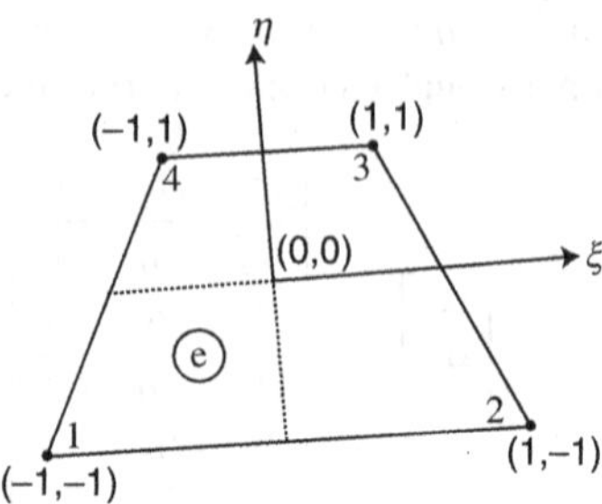

Example 5.6.

Question. For the four-noded linear two-dimensional isoparametric element shown above (Fig. 5.18), the shape functions are given as

$$N_i = \frac{1}{4}(1 + \xi_i\xi)(1 + \eta_i\eta)$$

Where (ξ_i, η_i) are the nodal (intrinsic) coordinates shown in the figure by the local node numbers 1–4. Considering the plane strain or plane stress case, develop the equations necessary to evaluate the element stiffness matrix and the load factor corresponding to boundary traction and body forces.

Answer. Similarly to (5.95) and (5.97), approximations for the coordinates and displacements are

$$x = \sum_1^4 N_i x_i; \quad y = \sum_1^4 N_i y_i$$

$$u = \sum_{i=1}^4 N_i \hat{u}_i; \quad v = \sum_{i=1}^4 N_i \hat{v}_i$$

Differentiating the shape functions

$$\frac{\partial N_i}{\partial \xi} = \frac{\xi_i}{4}(1 + \eta\eta_i); \quad \frac{\partial N_i}{\partial \eta} = \frac{\eta_i}{4}(1 + \xi\xi_i)$$

Defining

$$F_i = \frac{\partial N_i}{\partial x} \text{ and } G_i = \frac{\partial N_i}{\partial y}$$

the counterparts of (5.102a)–(5.102e) are

$$\begin{bmatrix} \dfrac{\partial N_i}{\partial \xi} \\ \dfrac{\partial N_i}{\partial \eta} \end{bmatrix} = \begin{bmatrix} \dfrac{\partial x}{\partial \xi} & \dfrac{\partial y}{\partial \xi} \\ \dfrac{\partial x}{\partial \eta} & \dfrac{\partial y}{\partial \eta} \end{bmatrix} \begin{bmatrix} F_i \\ G_i \end{bmatrix} = \begin{bmatrix} \sum_{i=1}^{4} \dfrac{\partial N_i}{\partial \xi} x_i & \sum_{i=1}^{4} \dfrac{\partial N_i}{\partial \xi} y_i \\ \sum_{i=1}^{4} \dfrac{\partial N_i}{\partial \eta} x_i & \sum_{i=1}^{4} \dfrac{\partial N_i}{\partial \eta} y_i \end{bmatrix} \begin{bmatrix} F_i \\ G_i \end{bmatrix} = [\mathbf{J}] \begin{bmatrix} F_i \\ G_i \end{bmatrix}$$

The above equation is solved for F_i and G_i.
The **B** matrix is given by

$$\mathbf{B} = \begin{bmatrix} F_1 & 0 & F_2 & 0 & F_3 & 0 & F_4 & 0 \\ 0 & G_1 & 0 & G_2 & 0 & G_3 & 0 & G_4 \\ G_1 & F_1 & G_2 & F_2 & G_3 & F_3 & G_4 & F_4 \end{bmatrix}$$

The appropriate form of the **D**-matrix (i.e., the constitutive matrix) is given in Chap. 3 for the plane strain and plane stress cases.

It is easy to verify that the displacements vary linearly along the element boundaries, just as the case in the linear triangles is. Hence, the expression for the element load vector due to boundary traction is exactly the same as those in the case of linear triangles (i.e., (5.88)–(5.91)). The element load vector due to body forces is, however, different, and is established using (5.6b).

Example 5.7.

Question. Referring to the problem shown above (Fig. 5.19), representing the body by one 4-noded isoparametric element, determine $\hat{u}_1$ and $\hat{u}_2$ corresponding to the global degrees of freedoms indicated by the arrows at the nodes where the rollers are. Assume plane strain condition with zero Poisson's ratio. The thickness of the element is t^*.

Answer. Noting that

$(x_1, y_1) = (1, 1), \quad (x_2, y_2) = (3, 1), \quad (x_3, y_3) = (3, 3), \quad (x_4, y_4) = (1, 3)$, and
$(\xi_1, \xi_2, \xi_3, \xi_4) = (-1, 1, 1, -1)$
$(\eta_1, \eta_2, \eta_3, \eta_4) = (-1, -1, 1, 1)$

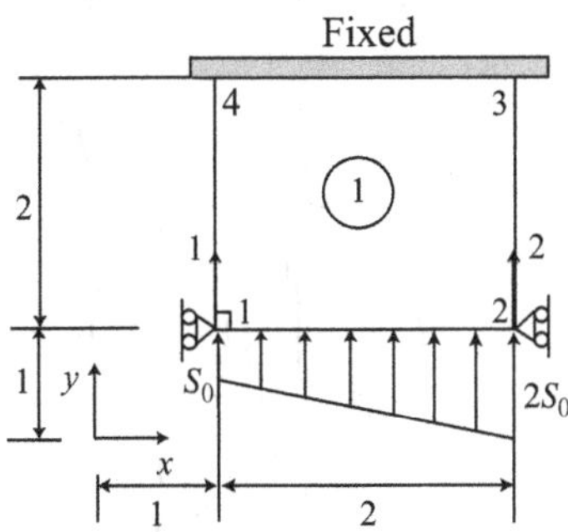

Fig. 5.19

Let us form a table as

	$i=1$	$i=2$	$i=3$	$i=4$	$\sum_{i=1}^{4}$
x_i	1	3	3	1	
y_i	1	1	3	3	
$\frac{\partial N_i}{\partial \xi}$	$-\frac{1}{4}(1-\eta)$	$\frac{1}{4}(1-\eta)$	$\frac{1}{4}(1+\eta)$	$-\frac{1}{4}(1+\eta)$	
$\frac{\partial N_i}{\partial \eta}$	$-\frac{1}{4}(1-\xi)$	$-\frac{1}{4}(1+\xi)$	$\frac{1}{4}(1+\xi)$	$\frac{1}{4}(1-\xi)$	
$\frac{\partial N_i}{\partial \xi}x_i$	$-\frac{1}{4}(1-\eta)$	$\frac{3}{4}(1-\eta)$	$\frac{3}{4}(1+\eta)$	$-\frac{1}{4}(1+\eta)$	1
$\frac{\partial N_i}{\partial \xi}y_i$	$-\frac{1}{4}(1-\eta)$	$\frac{1}{4}(1-\eta)$	$\frac{3}{4}(1+\eta)$	$-\frac{3}{4}(1+\eta)$	0
$\frac{\partial N_i}{\partial \eta}x_i$	$-\frac{1}{4}(1-\xi)$	$-\frac{3}{4}(1+\xi)$	$\frac{3}{4}(1+\xi)$	$\frac{1}{4}(1-\xi)$	0
$\frac{\partial N_i}{\partial \eta}y_i$	$-\frac{1}{4}(1-\xi)$	$-\frac{1}{4}(1+\xi)$	$\frac{3}{4}(1+\xi)$	$\frac{3}{4}(1-\xi)$	1

The Jacobian is calculated to be an identity matrix with $|\mathbf{J}| = 1$. Hence, the third and fourth rows of the above table gives F_i and G_i, respectively. For convenience, they are repeated below.

	$i=1$	$i=2$	$i=3$	$i=4$
F_i	$-\frac{1}{4}(1-\eta)$	$\frac{1}{4}(1-\eta)$	$\frac{1}{4}(1+\eta)$	$-\frac{1}{4}(1+\eta)$
G_i	$-\frac{1}{4}(1-\xi)$	$-\frac{1}{4}(1+\xi)$	$\frac{1}{4}(1+\xi)$	$\frac{1}{4}(1-\xi)$

Let us evaluate the following integrals ahead of time:

$$I_1 = \int_{\eta=-1}^{+1} \mathrm{d}\eta = [\eta]_{-1}^{+1} = 2$$

$$I_2 = \int_{\eta=-1}^{+1} (1-\eta)^2 \mathrm{d}\eta = \int_{\eta=-1}^{+1} (1-2\eta+\eta^2)\mathrm{d}\eta = \left[\eta - \eta^2 + \frac{\eta^3}{3}\right]_{-1}^{+1}$$

$$= \left[1 - 1 + \frac{1}{3} - \left\{-1 - 1 - \frac{1}{3}\right\}\right] = \frac{8}{3}$$

$$I_3 = \int_{\eta=-1}^{+1} (1+\eta)^2 \mathrm{d}\eta = \int_{\eta=-1}^{+1} (1+2\eta+\eta^2)\mathrm{d}\eta = \left[\eta + \eta^2 + \frac{\eta^3}{3}\right]_{-1}^{+1}$$

$$= \left[1 + 1 + \frac{1}{3} - \left\{-1 + 1 - \frac{1}{3}\right\}\right] = \frac{8}{3}$$

$$I_4 = \int_{\eta=-1}^{+1} (1-\eta)(1+\eta)\mathrm{d}\eta = \int_{\eta=-1}^{+1} (1-\eta^2)\mathrm{d}\eta = \left[\eta - \frac{\eta^3}{3}\right]_{-1}^{+1}$$

$$= \left[1 - \frac{1}{3} - \left\{-1 + \frac{1}{3}\right\}\right] = \frac{4}{3}$$

With the aid of these integrals, we will evaluate the element stiffness matrix. Now, to solve the problem at hand, we do not need to evaluate all of the coefficients of the stiffness matrix. Since we are using one element to represent the domain, the global stiffness and element stiffness matrices have the same size (8×8). However, after imposing the displacement boundary conditions, the size of the reduced matrix is only 2×2. The reduced equation to be solved is

$$\begin{bmatrix} k_{22}^1 & k_{24}^1 \\ k_{42}^1 & k_{44}^1 \end{bmatrix} \begin{bmatrix} \hat{u}_1 \\ \hat{u}_2 \end{bmatrix} = \begin{bmatrix} f_2^1 \\ f_4^1 \end{bmatrix}$$

where k_{22}^1, etc., are the coefficients of the element stiffness matrix, and f_2^1, etc., are the coefficients of the element load vector due to the boundary traction.

As in Examples 5.4 and 5.5, for $v = 0$, the constitutive matrix $\mathbf{D}$ becomes diagonal with the diagonal elements being equal to E, E and $0.5E$ (Example 5.4). With the aid of the integrals evaluated earlier and noting that $|\mathbf{J}| = 1$, the required coefficients of the element stiffness matrix are evaluated as follows:

$$\mathbf{k}_1 = t^* \int_{\eta=-1}^{+1}\int_{\xi=-1}^{+1} \begin{bmatrix} F_1 & 0 & G_1 \\ 0 & G_1 & F_1 \\ F_2 & 0 & G_2 \\ 0 & G_2 & F_2 \\ F_3 & 0 & G_3 \\ 0 & G_3 & F_3 \\ F_4 & 0 & G_4 \\ 0 & G_4 & F_4 \end{bmatrix} [\mathbf{D}] \begin{bmatrix} F_1 & 0 & F_2 & 0 & F_3 & 0 & F_4 & 0 \\ 0 & G_1 & 0 & G_2 & 0 & G_3 & 0 & G_4 \\ G_1 & F_1 & G_2 & F_2 & G_3 & F_3 & G_4 & F_4 \end{bmatrix} \mathrm{d}\eta\,\mathrm{d}\xi$$

$$= Et^* \int_{\eta=-1}^{+1}\int_{\xi=-1}^{+1} \begin{bmatrix} \cdot & \cdot & \cdot & \cdot & \cdot & \cdot & \cdot & \cdot \\ \cdot & G_1^2 + F_1^2/2 & \cdot & G_1G_2 + F_1F_2/2 & \cdot & \cdot & \cdot & \cdot \\ \cdot & \cdot & \cdot & \cdot & \cdot & \cdot & \cdot & \cdot \\ \cdot & G_2G_1 + F_2F_1/2 & \cdot & G_2^2 + F_2^2/2 & \cdot & \cdot & \cdot & \cdot \\ \cdot & \cdot & \cdot & \cdot & \cdot & \cdot & \cdot & \cdot \\ \cdot & \cdot & \cdot & \cdot & \cdot & \cdot & \cdot & \cdot \\ \cdot & \cdot & \cdot & \cdot & \cdot & \cdot & \cdot & \cdot \\ \cdot & \cdot & \cdot & \cdot & \cdot & \cdot & \cdot & \cdot \end{bmatrix} \mathrm{d}\eta\,\mathrm{d}\xi$$

$$= \frac{Et^*}{16}\begin{bmatrix} \cdot & \cdot & \cdot & \cdot & \cdot & \cdot & \cdot & \cdot \\ \cdot & 2\left(\frac{8}{3}+\frac{4}{3}\right) & \cdot & 2\left(\frac{4}{3}-\frac{4}{3}\right) & \cdot & \cdot & \cdot & \cdot \\ \cdot & \cdot & \cdot & \cdot & \cdot & \cdot & \cdot & \cdot \\ \cdot & 2\left(\frac{4}{3}-\frac{4}{3}\right) & \cdot & 2\left(\frac{8}{3}+\frac{4}{3}\right) & \cdot & \cdot & \cdot & \cdot \\ \cdot & \cdot & \cdot & \cdot & \cdot & \cdot & \cdot & \cdot \\ \cdot & \cdot & \cdot & \cdot & \cdot & \cdot & \cdot & \cdot \\ \cdot & \cdot & \cdot & \cdot & \cdot & \cdot & \cdot & \cdot \\ \cdot & \cdot & \cdot & \cdot & \cdot & \cdot & \cdot & \cdot \end{bmatrix}$$

The reduced equation to be solved is

$$\frac{Et^*}{16}\begin{bmatrix} \frac{24}{3} & 0 \\ 0 & \frac{24}{3} \end{bmatrix}\begin{bmatrix} \hat{u}_1 \\ \hat{u}_2 \end{bmatrix} = \begin{bmatrix} \frac{4S_0t^*}{3} \\ \frac{5S_0t^*}{3} \end{bmatrix} \Rightarrow \hat{u}_1 = \frac{8S_0}{3E}; \quad \hat{u}_2 = \frac{10S_0}{3E}$$

Note that when the traction varies uniformly and has a value S_0, the right-hand side of the above equation is

$$\mathbf{P} = S_0\begin{bmatrix} 1 \\ 1 \end{bmatrix}$$

resulting in $\hat{u}_1 = \hat{u}_2 = 2S_0/E$, which may be verified to be the solution for a axially loaded plane strain rod.

It may also be noted that the results computed using two triangular elements (Example 5.5) and one isoparametric element (this example) are the same for uniform traction ($\hat{u}_1 = \hat{u}_2 = 2S_0/E$ in both cases). They are slightly different from each other ($\hat{u}_1 = 2.83$ and $\hat{u}_2 = 3.17$ by using triangular elements and $\hat{u}_1 = 2.67$ and $\hat{u}_2 = 3.33$ by using isoparametric element) when the traction is linearly varying.

5.3.3 Elastic (Euler-Bernoulli) Beam Element

5.3.3.1 Element Stiffness Matrix and Load Vectors

The boundary value problem for the bending behavior of beams (i.e., one-dimensional bending members) has been defined in Chap. 3. In Problem 5.5, you are asked to derive the variational statement. The functional that needs to be minimized is

$$I(w) = \int_0^L \left[\frac{1}{2}EI\left(\frac{\mathrm{d}^2 w}{\mathrm{d}x^2}\right)^2 - wq\right]\mathrm{d}x - \int_{\Gamma_t}\left[wV_0 + \frac{\mathrm{d}w}{\mathrm{d}x}M_0\right]\mathrm{d}s \tag{5.118}$$

where w is the transverse displacement of the beam. E and I are Young's modulus and the second moment of inertia, respectively. The definition of the remaining parameters is given in Chap. 3.

Introducing two nodes for an element as shown in Fig. 5.20a, transverse displacements and rotations are selected as primary unknowns as shown in Fig. 5.20a, where $\hat{w}_1$ and $\hat{w}_2$ are the displacements at nodes 1 and 2, respectively, and $\hat{\theta}_1$ and $\hat{\theta}_2$ are the rotations at nodes 1 and 2, respectively.

The highest order of derivative in (5.118) is 2. Hence, the interpolation function must be at least C_1 continuous; for example, a linear polynomial. However, to satisfy the other two requirements discussed in Sect. 5.1.2 (i.e., the requirement for the polynomial to be complete and to satisfy the essential boundary conditions), a cubic polynomial is necessary. A cubic polynomial has four constants, which can be related to the four unknown indicated in Fig. 5.20a.

With respect to the local coordinate t, let the polynomial be

$$\hat{u} = c_0 + c_1 t + c_2 t^2 + c_3 t^3 \tag{5.119a}$$

Noting that $\mathrm{d}x = \mathrm{d}t$, the rotation is obtained as

$$\hat{\theta} = \frac{\mathrm{d}\hat{w}}{\mathrm{d}x} = \frac{\mathrm{d}\hat{w}}{\mathrm{d}t} = c_1 + 2c_2 t + 3c_3 t^2 \tag{5.119b}$$

Then by setting the displacements and rotations to the nodal unknowns at the nodes, four equations are obtained for constants $\{c_0, c_1, c_2, c_3\}$ in terms of the nodal unknowns $\hat{\mathrm{u}} = \left\{\hat{w}_1, \hat{\theta}_1, \hat{w}_2, \hat{\theta}_2\right\}$. Solving these equations, it can be shown that

$$\mathbf{u} = \mathbf{N}\hat{\mathbf{u}} \tag{5.120a}$$

where

$$\mathrm{u} = \{w(t)\}, \quad \mathbf{N} = \{N_1, N_2, N_3, N_4\}$$

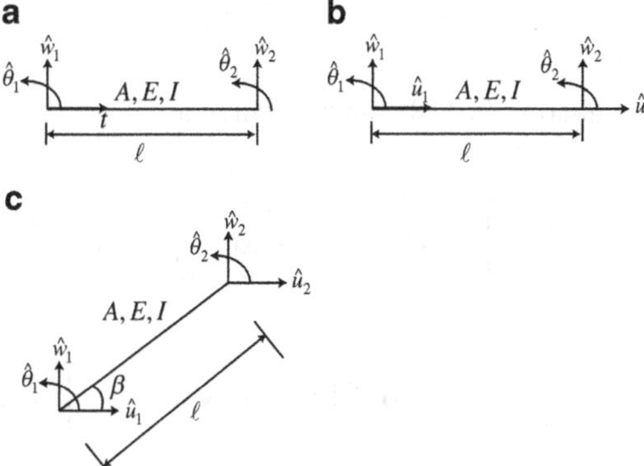

Fig. 5.20 Definition of variables for beam element

and

$$N_1 = 1 - \frac{3t^2}{\ell^2} + \frac{2t^3}{\ell^3};\quad N_2 = t - \frac{2t^2}{\ell} + \frac{t^3}{\ell^2};\quad N_3 = \frac{3t^2}{\ell^2} - \frac{2t^3}{\ell^3};\quad N_4 = -\frac{t^2}{\ell} + \frac{t^3}{\ell^2} \tag{5.120b}$$

$$\text{Define a curvature vector as } \underset{1\times 1}{\boldsymbol{\kappa}} = \left\{\frac{\mathrm{d}^2 w}{\mathrm{d}x^2}\right\} \tag{5.121a}$$

$$\text{The constitutive matrix as } \underset{1\times 1}{\mathbf{D}} = \{EI\} \tag{5.121b}$$

$$\text{A gradient vector as } \underset{1\times 1}{\mathbf{g}} = \left\{\frac{\mathrm{d}w}{\mathrm{d}x}\right\} \tag{5.121c}$$

Differentiating (5.120a)

$$\boldsymbol{\kappa} = \mathbf{B}\hat{\mathbf{u}};\quad \mathbf{g} = \bar{\mathbf{N}}\hat{\mathbf{u}} \tag{5.122a}$$

where

$$\bar{\mathbf{N}} = \{\bar{N}_1, \bar{N}_2, \bar{N}_3, \bar{N}_4\} \tag{5.122b}$$

$$\bar{N}_1 = -\frac{6t}{\ell^2} + \frac{6t^2}{\ell^3};\quad \bar{N}_2 = 1 - \frac{4t}{\ell} + \frac{3t^2}{\ell^2};\quad \bar{N}_3 = \frac{6t}{\ell^2} - \frac{6t^2}{\ell^3};\quad \bar{N}_4 = -\frac{2t}{\ell} + \frac{3t^2}{\ell^2} \tag{5.122c}$$

$$\mathrm{B} = \{B_1, B_2, B_3, B_4\} \tag{5.122d}$$

$$B_1 = -\frac{6}{\ell^2} + \frac{12t}{\ell^3};\quad B_2 = -\frac{4}{\ell} + \frac{6t}{\ell^2};\quad B_3 = \frac{6}{\ell^2} - \frac{12t}{\ell^3};\quad B_4 = -\frac{2}{\ell} + \frac{6t}{\ell^2} \tag{5.122e}$$

Equation (5.118) can be written as

$$\begin{aligned} I(w) &= \int_0^L \left[\frac{1}{2}\boldsymbol{\kappa}^{\mathrm{T}}\mathbf{D}\boldsymbol{\kappa} - wq\right]\mathrm{d}x - \int_{\Gamma_t}\left[wV_0 + \frac{\mathrm{d}w}{\mathrm{d}x}M_0\right]\mathrm{d}s \\ &= \int_0^L \left[\frac{1}{2}\hat{\mathbf{u}}^{\mathrm{T}}\mathbf{B}^{\mathrm{T}}\mathbf{D}\mathbf{B}\hat{\mathbf{u}} - \hat{\mathbf{u}}^{\mathrm{T}}\mathbf{N}^{\mathrm{T}}q\right]\mathrm{d}x - \int_{\Gamma_t}\left[\hat{\mathbf{u}}^{\mathrm{T}}\mathbf{N}^{\mathrm{T}}V_0 + \hat{\mathbf{u}}^{\mathrm{T}}\mathbf{N}^{\mathrm{T}}M_0\right]\mathrm{d}s \end{aligned} \tag{5.123}$$

Now following the steps in Sect. 5.1.1, the element stiffness matrix and load vectors are obtained as

$$\mathbf{k}_e = \int_0^L \mathbf{B}^{\mathrm{T}}\mathbf{D}\mathbf{B}\,\mathrm{d}x \tag{5.124a}$$

$$\mathbf{f}_e = \mathbf{f}_q + \mathbf{f}_v + \mathbf{f}_m \tag{5.124b}$$

$$\mathbf{f}_q = \int_{\Omega_e} \mathbf{N}^{\mathrm{T}} q\,\mathrm{d}x; \quad \mathbf{f}_v = \int_{\Gamma_t} \mathbf{N}^{\mathrm{T}} V_0\,\mathrm{d}s; \quad \mathrm{f}_m = \int_{\Gamma_t} \mathbf{N}^{\mathrm{T}} M_0\,\mathrm{d}s \tag{5.124c}$$

To consider the axial deformation along with the transverse displacements and rotations, two axial degrees of freedoms can be added as shown in Fig. 5.20b. The axial load-deformation behavior is completely uncoupled from the bending behavior. On this basis, the stiffness matrix for the system shown in Fig. 5.20b can be simply obtained by inserting the axial stiffness coefficients in the appropriate places. Arranging the element unknown variables as

$$\hat{\mathbf{u}} = \left\{\hat{u}_1, \hat{w}_1, \hat{\theta}_1, \hat{u}_2, \hat{w}_2, \hat{\theta}_2\right\} \tag{5.125a}$$

and defining

$$S = \frac{AE}{\ell}; \quad A = \frac{4EI}{\ell}; \quad B = \frac{2EI}{\ell}; \quad C = \frac{6EI}{\ell^2}; \quad \text{and } D = \frac{12EI}{\ell^3}, \tag{5.125b}$$

the stiffness matrix can be shown to be (Cook 1981)

$$k_e = \begin{bmatrix} S & 0 & 0 & -S & 0 & 0 \\ 0 & D & C & 0 & -D & C \\ 0 & C & A & 0 & -C & B \\ -S & 0 & 0 & S & 0 & 0 \\ 0 & -D & -C & 0 & D & -C \\ 0 & C & B & 0 & -C & A \end{bmatrix} \tag{5.126}$$

For the system of loads shown in Fig. 5.21, the equivalent nodal forces are (Cook 1981)

$$\mathbf{f}_q = q\begin{bmatrix} \ell/2 \\ \ell^2/12 \\ \ell/2 \\ -\ell^2/12 \end{bmatrix}; \quad \mathbf{f}_v = V\begin{bmatrix} 1/2 \\ \ell/8 \\ 1/2 \\ -\ell/8 \end{bmatrix}; \quad \mathbf{f}_m = M\begin{bmatrix} 3/2\ell \\ 1/4 \\ -3/2\ell \\ 1/4 \end{bmatrix} \tag{5.127}$$

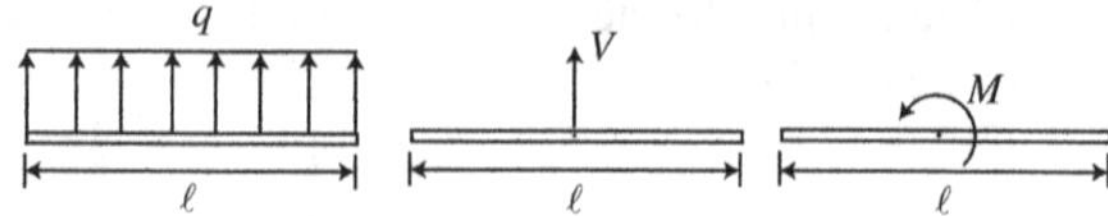

Fig. 5.21 Three types of forces acting on a beam

Noting that N_i's assume a value of either 0 or 1 at the nodes, it is easy to verify from (5.124c) that when the shear force V_0 or the bending moment M_0 act at the nodes, the corresponding nodal values become exactly equal to V_0 or M_0.

5.3.3.2 Coordinate Transformation

In the triangular element presented in Sect. 5.3.1 and the isoparametric elements presented in Sect. 5.3.2, the coordinate system used to define the element unknown variables coincides with the global coordinate system. On the other hand, they are different in the case of the beam element, and a coordinate transformation needs to be performed before the element quantities can be added into the global quantities. This is achieved as follows: Let $\hat{\mathbf{u}}_\ell$ and $\hat{\mathbf{u}}_g$ be unknown variable vectors with respect to local and global coordinate systems, respectively. The first step is to establish the coordinate transformation relationship

$$\mathbf{u}_\ell = \mathbf{T}\mathbf{u}_g \tag{5.128}$$

where $\mathbf{T}$ is the coordinate transformation matrix (Chap. 2). (Recall that $\mathbf{T}^{\mathrm{T}}\mathbf{T} = \mathbf{I}$, where $\mathbf{I}$ is the identity matrix.) The scalars given by (5.5a) and (5.6a) are defined in terms of $\hat{\mathbf{u}}_\ell$. They can be defined in terms of $\hat{\mathbf{u}}_g$ as

$$I_i^1 = \delta\hat{\mathbf{u}}_\ell^{\mathrm{T}}\mathbf{k}_\ell\hat{\mathbf{u}}_\ell = \delta\hat{\mathbf{u}}_g^{\mathrm{T}}\mathbf{T}^{\mathrm{T}}\mathbf{k}_\ell\mathbf{T}\hat{\mathbf{u}}_g = \delta\hat{\mathbf{u}}_g^{\mathrm{T}}\mathbf{k}_g\hat{\mathbf{u}}_g \tag{5.129a}$$

$$I_i^2 = \delta\hat{\mathbf{u}}_\ell^{\mathrm{T}}\mathbf{f}_\ell^b = \delta\hat{\mathbf{u}}_g^{\mathrm{T}}\mathbf{T}^{\mathrm{T}}\mathbf{f}_\ell^b = \delta\hat{\mathbf{u}}_g^{\mathrm{T}}\mathbf{f}_g^b \tag{5.129b}$$

$$\mathbf{k}_g = \mathbf{T}^{\mathrm{T}}\mathbf{k}_\ell\mathbf{T} \text{ and } \mathbf{f}_g^b = \mathbf{T}^{\mathrm{T}}\mathbf{f}_\ell^b. \tag{5.129c}$$

The scalar given by (5.7a) is in the same form as that of (5.6a) and hence transforms in exactly the same way. The global functional can now be written in terms of the global unknown variables defined with respect to the global coordinate system. The functional can then be minimized with respect to the global variables. The final global stiffness matrix and the load vector are assembled from $\mathbf{k}_g$ and $\mathbf{f}_g^b$ given by (5.129c).

Note that for the two-dimensional case considered here (i.e., the plane-frame case), the local and global coordinate systems coincide for the rotational degrees of freedoms. Hence, the transformation needs to be performed only on the linear degrees of freedoms (i.e., the axial and transverse degrees of freedoms). Referring to the notations in Fig. 5.21, it can be shown that (Chap. 2)

$$\begin{bmatrix} u_\ell \\ w_\ell \end{bmatrix} = \begin{bmatrix} \cos\beta & \sin\beta \\ -\sin\beta & \cos\beta \end{bmatrix} \begin{bmatrix} u_g \\ w_g \end{bmatrix} \tag{5.130a}$$

Let us define

$$\ell = \cos\beta \text{ and } m = \sin\beta, \quad \mathbf{T} = \begin{bmatrix} \ell & m \\ -m & \ell \end{bmatrix} \tag{5.130b}$$

Now by carrying out the operations involved in (5.129c), it can be shown that (Cook 1981)

$$\mathbf{k}_g = \begin{bmatrix} F & G & H & -F & -G & H \\ G & P & Q & -G & -P & Q \\ H & Q & A & -H & -Q & B \\ -F & -G & -H & F & G & -H \\ -G & -P & -Q & G & P & -Q \\ H & Q & B & -H & -Q & A \end{bmatrix} \tag{5.131}$$

where $F = S\ell^2 + Dm^2$; $G = Sm\ell - Dm\ell$; $H = -Cm$; $P = Sm^2 + D\ell^2$; and $Q = C\ell$, and S, A, B, C, and D are given by (5.125b).

5.3.4 Elastic Space Frame Element

In space frames, the orientation of the beams, linear loads and moments can be truly three-dimensional. While Euler-Bernoulli theory is used to determine the stiffness coefficients with respect to the local principal axes of inertia, a rather sophisticated coordinated transformation is necessary to express the element stiffness matrix and load vector with respect to the global coordinate axes. Two nodes are introduced on each element. At each node, six degrees of freedom, three linear and three rotational, are introduced. The size of the resulting element stiffness matrix is 12×12. The reader is referred to a standard textbook on structural analysis (e.g., Kassimali 1999) for the details.

5.4 Solution of Equations

5.4.1 Static Problems

The system of equations to be solved is (5.10)

$$\mathbf{K\hat{u}} = \mathbf{P}$$

As pointed out earlier, **K** is in general a banded matrix. However, there are circumstances where the banded nature may be disrupted even for a systematic node numbering. For example, in earthquake analysis of earth structures or soil-foundation systems, to enforce free-field conditions far from the domain being analyzed, it is customary to tie up the nodes on the left-most vertical boundary with those at the corresponding elevations on the right-most vertical boundary. This introduces a few very tall columns in **K**. The remaining columns will still remain short. In elastic analyses, **K** is positive definite and symmetric. In elasto-plastic analyses involving elasto-plastic constitutive models based on associated flow rules, depending on the algorithms used to integrate the rate constitutive equations, **K** may still remain symmetric. When a consistent tangent operator is employed, **K** is in general non symmetric even when a associated flow rule is employed in the constitutive model. The nonassociated flow rules always lead to a nonsymmetric **K**. The equation solvers used in the computer program must be selected based on these considerations.

The theories behind equation solving are very well established. The reader is referred to a suitable book on linear algebra or to some of the finite element books (e.g., Bathe 1982). The available methods can be classified into

- Direct methods
- Iterative methods

5.4.1.1 Comments on Direct Solution Methods

In direct methods, equations are directly manipulated in a fixed number of operations in order to obtain the solutions. The number of operations involved varies with the specific algorithm, but are predictable. There are several types of direct methods. The most popular methods are those based on the Gauss elimination technique. Specific algorithms are available for symmetric and nonsymmetric matrices, and for banded and full matrices. When **K** has zeros on the diagonals, the algorithms with pivoting are used. An algorithm known as the active column technique is particularly effective for matrices with variable column lengths. In this technique, the height of columns (skylines) are first established and only the coefficients of **K** within the skylines are stored. Operations are performed taking advantage of the skyline structure of the matrix and avoiding multiplications with zeros outside the skylines. There are variations of this for symmetric and nonsymmetric matrices.

The solution of the equation is obtained in two steps. In the first step, the matrix is reduced into an upper/lower triangular form. In the second step, the solution is obtained by back substitution. Hence, if the matrix does not change in successive analyses (e.g., dynamic analysis of elastic solids), the matrix has to be reduced only once. The back substitution is performed as many times as necessary.

For a symmetric matrix, the reduction step takes $nm^2/2$ operations and the back substitution takes $2nm$ operations, where n is the number of equations and m is half bandwidth. Hence, the reduction step involves $m/4$ times more operations than the back-substitution step, and hence the former is much more computationally intensive than the latter.

Three factor affecting the computational efficiency in elasto-plastic analyses may be noted. (1) As already pointed out, the use of consistent tangent operators generally render **K** nonsymmetric, increasing the required computational effort. (2) The analyses involve several incremental steps, and within each increments, several iterative steps. During each step, the matrix **K** generally changes depending on the method of iterations (e.g., the full Newton-Raphson method). This requires the reduction step to be performed every time, increasing the required computational effort. (3) At every iterative step of the analysis, given a strain increment, the corresponding stress increment needs to be evaluated. This step is rather involved. This is to be done at every Gauss integration point and hence increases proportionately to the number of elements in the mesh. Let us assume that the number of nodes in the mesh is approximately equal to the number of elements in the mesh. Then the ratio between the operations required for equation solving and stress computations is proportional to $nm^2/nk = m^2/k$, where k is the number of operations required for stress computations. If the bandwidth remains the same as n increases, the above ratio remains the same. As k increases, the stress computation tends to dominate the required computational effort, and hence the need for an efficient algorithm for computing stresses in elasto-plastic analyses.

5.4.1.2 Comments on Iterative Solution Methods

Iterative methods start with a trial solution and follow a particular iterative strategy in obtaining a solution with an acceptable error (Axelsson 1976). One of the oldest methods is the Gauss-Seidel iterative method. There are also methods based on minimization of energy, where mathematical program is used to obtain the solution (e.g., Mallett and Schmit 1967).

The conjugate gradient technique is considered to be one of the most efficient methods, provided the matrix is preconditioned (Axelsson 1996). Compared to some of the other techniques, the method does not involve method parameters (e.g., extreme eigenvalues), which makes it highly desirable. However, when applied without preconditioning, the rate of convergence was found to be very slow in most cases. In the preconditioned strategy, instead of solving $\mathbf{K}\hat{\mathbf{u}} = \mathbf{P}$, we solve $\mathbf{A}^{-1}\mathbf{K}\hat{\mathbf{u}} = \mathbf{A}^{-1}\mathbf{P}$, where $\mathbf{A}$ is the preconditioning matrix. The need for preconditioning arises due to the fact that the condition number (the ratio between the largest and smallest eigenvalues) of most matrices leads to slow convergence. Thus, the objective of preconditioning is reducing the condition number. Of course, when $\mathbf{A} = \mathbf{K}$ then the solution is obtained in one step, but nothing is gained since $\mathbf{K}^{-1}$ needs to be determined. Thus, preconditioning must be done so that the inverse can be found much more efficiently. This is where most of the research effort has been spent over the years, resulting in various methods. Iterative solution methods are also becoming popular in parallel computing; see Duff and van der Vorst (1999) for a comprehensive discussion on these.

5.4.2 Dynamic Problems

The system of equations to be solved is (5.77)

$$\mathbf{M}\ddot{\hat{\mathbf{u}}} + \mathbf{C}\dot{\hat{\mathbf{u}}} + \mathbf{K}\hat{\mathbf{u}} = \mathbf{P} \tag{5.132a}$$

In general, there are two parts to the solution of (5.132a): (1) The solution to the eigenvalue problem, and (2) the solution to (5.132a).

5.4.2.1 The Eigenvalue Problem

Considering the undamped ($\mathbf{C} = \mathbf{0}$) free vibration ($\mathbf{P} = \mathbf{0}$) problem, and assuming that $\hat{\mathbf{u}}$ vary harmonically as

$$\hat{\mathbf{u}} = \hat{\mathbf{u}}_0 e^{i\omega t} \tag{5.132b}$$

Equation (5.132a) becomes

$$\left[\mathbf{K} - \omega^2\mathbf{M}\right]\hat{\mathbf{u}}_0 e^{i\omega t} = 0 \Rightarrow \mathbf{K}\hat{\mathbf{u}}_0 = \omega^2\mathbf{M}\hat{\mathbf{u}}_0 \tag{5.132c}$$

Equation (5.132c) is a generalized eigenvalue problem. If the size of $\mathbf{K}$ is $n\times n$, (5.132c) will have n eigenvalues ω^2. ω_i, $i = 1, n$, are referred to as the natural frequencies of the system, and the corresponding eigenvectors $\hat{\mathbf{u}}_0$ are known as the mode shapes. The lowest natural frequency is known as the fundamental frequency of the system.

Since $\mathbf{K}$ is positive definite in elasticity problems, the eigenvalues are real and positive, and the eigenvectors form a complete set spanning C^n, and are orthogonal (i.e., $\mathbf{x}_k^{\mathrm{T}}\mathbf{x}_\ell \neq 0$ for $k = \ell$ and $\mathbf{x}_k^{\mathrm{T}}\mathbf{x}_\ell = 0$ for $k \neq \ell$) (see Stewart 1973). Any arbitrary vector may then be expressed as a linear combination of the eigenvectors.

It is clear that the natural frequencies of the system depend on the stiffness and mass properties of the domain and the displacement boundary conditions. Since the stiffness matrix changes with loading for nonlinear problems, the natural frequencies are properly defined only for linear problems.

The knowledge of natural frequencies is useful in some cases and required in others. For example, as described in a subsequent section, the damping matrix $\mathbf{C}$ is commonly defined based on the Rayleigh damping model, which requires some of the natural frequencies. As described in a subsequent section, one of the indirect methods of solving (5.132a) is known as the *mode superposition (or modal decomposition) method.* The method requires knowledge of the natural frequencies. Explicit integration of (5.132a) is conditionally stable and the condition is based on the knowledge of the largest natural frequency of the system. These are examples where the knowledge of one or more of the natural frequencies is required.

Understanding the nature of the dynamic response of the system requires the knowledge of the natural frequencies. For example, when a system is excited with a harmonic forcing function of a frequency whose value is close to the fundamental frequency of the system, the system may experience resonance. Hence, the knowledge of the values of the natural frequencies is useful in this case.

There are various methods for solving eigenvalue problems. The methods are quite well established and hence the theories for solving eigenvalue problems will not be covered here (see, e.g., Stewart 1973).

5.4.2.2 Solution of Dynamic Equilibrium Equation

Introduction

It may be recalled that (5.132a) is a static equilibrium at time t with the inertial force $\mathbf{M}\ddot{\hat{\mathbf{u}}}$ added to the system by D'Alembert's law and with the viscous force $\mathbf{C}\dot{\hat{\mathbf{u}}}$ added to the system in a similar manner. Thus, it is useful (especially in nonlinear systems) to consider each term in (5.132a) as forces

$$\mathbf{M}\ddot{\hat{\mathbf{u}}} + \mathbf{C}\dot{\hat{\mathbf{u}}} + \mathbf{K}\hat{\mathbf{u}} = \mathbf{P} \tag{5.133a}$$

$$\mathbf{f}_I + \mathbf{f}_C + \mathbf{f}_S = \mathbf{P} \tag{5.133b}$$

where $\mathbf{f}_I = \mathbf{M}\ddot{\hat{\mathbf{u}}}$ is the inertial force, $\mathbf{f}_C = \mathbf{C}\dot{\hat{\mathbf{u}}}$ is the damping force and $\mathbf{f}_S = \mathbf{K}\hat{\mathbf{u}}$ is the spring force.

In one point of view, the solution methods may be classified as

- Frequency-domain methods
- Time-domain methods

In frequency-domain methods, the response is obtained at different frequencies and superposed to obtain the time variations of quantities. The method is, therefore, applicable only to linear systems since the principle of superposition is applied. This is of marginal interest in this book and hence will not be covered.

In the time-domain methods, the time dimension is discretized at a number of points at an interval Δt (either constant or variable) and the solution is obtained by enforcing (5.133) at these discrete points. The methods are also called the time-marching methods. The methods can be applied to either linear or nonlinear problems; the methods are described in some detail below.

In another point of view, the solution methods may be grouped into

- Direct methods
- Indirect methods

In direct methods, (5.133a) is directly integrated in the time dimension to obtain the solution, whereas in indirect methods, the equation is first transformed into another form before solving. The frequency-domain method described above is an indirect method. The direct methods are the most common for nonlinear problems. We mentioned earlier that the *mode superposition (or modal decomposition) method* is very popular in the structural dynamics community. In this method, the $n \times n$ coupled set of equations given by (5.133a) is first uncoupled into n separate scalar equations (modal equations). The scalar equations are then solved separately to obtain modal solutions. The system solution is then obtained by synthesizing the modal responses. The method assumes that the frequencies remain constant during the solution period. Also, it uses the principle of superposition, and thus is valid only to linear systems. However, since the solution process underlying this method allows the fundamentals of dynamics to be understood, we will cover this method.

Mode Superposition Method

Let us first assume that $\mathbf{C} = 0$, and later discuss methods for considering $\mathbf{C}$. The solution is expressed as a linear combination of the eigenvectors as

$$\hat{\mathbf{u}} = \mathbf{X}\boldsymbol{\psi}, \tag{5.134a}$$

$$\text{where} \quad \underset{n\times n}{\mathbf{X}} = [\mathrm{x}_1, \mathrm{x}_2, \mathrm{x}_3, \ldots..\mathrm{x}_n] \quad \text{and} \quad \underset{n\times 1}{\boldsymbol{\psi}} = [\psi_1, \psi_2, \psi_3, \ldots.\psi_n] \tag{5.134b}$$

$\mathbf{x}_1, \mathbf{x}_2, \mathbf{x}_3$, etc., are eigenvectors (mode shapes). The square matrix $\mathbf{X}$ is referred to as the modal matrix, and ψ is the vector of generalized displacements, also known as *normal coordinates*. Substituting (5.134) into (5.133a) and premultiplying the equation by $\mathbf{X}^{\mathrm{T}}$,

$$[\mathbf{X}^{\mathrm{T}}\mathbf{M}\mathbf{X}]\ddot{\boldsymbol{\psi}} + [\mathbf{X}^{\mathrm{T}}\mathbf{K}\mathbf{X}]\boldsymbol{\psi} = \mathbf{X}^{\mathrm{T}}\mathbf{P} = \mathbf{L} \tag{5.135}$$

It can be shown (Problem 5.12) that

$$[\mathbf{X}^{\mathrm{T}}\mathbf{M}\mathbf{X}] = \mathbf{U} \text{ and } [\mathbf{X}^{\mathrm{T}}\mathbf{K}\mathbf{X}] = \mathbf{W} \tag{5.136}$$

where $\mathbf{U}$ and $\mathbf{W}$ diagonal matrices. The relations in (5.136) are called the orthogonality property. The ith diagonals of these matrices are

$$U_{ii} = m_{ii} \text{ and } W_{ii} = m_{ii}\omega_i^2$$

where $m_{ii} = \mathbf{x}_i^{\mathrm{T}}\mathbf{M}\mathbf{x}_i$. Equation (5.135) then uncouples into n separate scalar equations. The ith equation is in the form

$$m_{ii}\ddot{\psi}_i + m_{ii}\omega_i^2\psi_i = L_i \tag{5.137}$$

where

$$L_i = \mathbf{x}_i^{\mathrm{T}}\mathbf{P} \tag{5.138}$$

It is easy to see that the natural frequency of the single degree of freedom system represented by (5.137) is ω_i. Equation (5.137) is solved for ψ_i and substituted into (5.134) for obtaining the solution $\hat{\mathbf{u}}$.

Let us assume that $\mathbf{P}$ is harmonic with the frequency $\overline{\omega}$. If $\overline{\omega}$ is close to one of the natural frequencies, say ω_k, then ψ_k tends to become significant due to resonance. The effect is even higher for small values of k and becomes maximum when $k = 1$. That is, when the excitation frequency is close to the fundamental (i.e., the lowest) frequency, the system experiences the greatest amplification. The reason for this is the following: The signs of the components of $\mathbf{x}_i$ (mode shapes) alternate more and more as i increases. Consequently, the value of L_i is the largest for $i = 1$ and decreases as i increases. Hence, the resonance near the fundamental frequency has a significantly larger effect on the overall response than the resonance near a high frequency. In fact, it turns out that when the system has thousands of degrees of freedom, and hence thousands of mode shapes, the system response can be obtained simply by solving the first few hundred modal equations.

Rayleigh Damping Model

When the system damping is considered, the system equation (5.133a) may still be uncoupled if $\mathbf{C}$ has a certain structure. Rayleigh showed that $\mathbf{C}$ expressed in the form

$$\mathbf{C} = \alpha_1\mathbf{M} + \alpha_2\mathbf{K} \tag{5.139}$$

retains the orthogonality property that $\mathbf{K}$ and $\mathbf{M}$ have (5.136) and hence permits uncoupling. It can be seen that with $\mathbf{C}$ expressed by (5.139), (5.137) becomes

$$m_{ii}\ddot{\psi}_i + \left[\alpha_1 m_{ii} + \alpha_2 m_{ii}\omega_i^2\right]\dot{\psi}_i + m_{ii}\omega_i^2\psi_i = L_i \tag{5.140}$$

Comparing (5.140) with the standard damped single degree of freedom dynamic equation (i.e., $m\ddot{x} + 2\beta m\omega\dot{x} + m\omega^2 x = p(t)$), it is seen that the equivalent fraction of critical damping applied to the ith modal equation is

$$\beta_i = \frac{\alpha_1 + \alpha_2\omega_i^2}{2\omega_i} \tag{5.141}$$

The two constants α_1 and α_2 may be evaluated by fixing the damping to a specific value (say, β_0) at two frequencies as follows

$$\alpha_1 + \alpha_2\omega_k^2 = 2\omega_k\beta_0$$

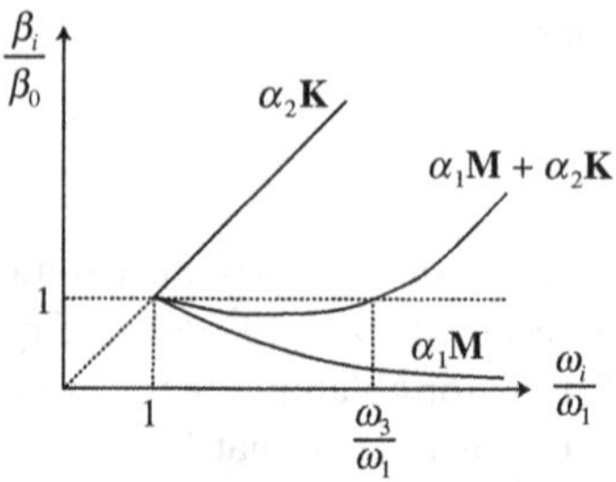

Fig. 5.22 Variation of damping applied to modal equations according to Rayleigh damping model

$$\alpha_1 + \alpha_2\omega_\ell^2 = 2\omega_\ell\beta_0$$

Solving the two equations

$$\alpha_1 = \frac{2\omega_k\omega_\ell}{\omega_k + \omega_\ell}\beta_0 \text{ and } \alpha_2 = \frac{2}{\omega_k + \omega_\ell}\beta_0 \tag{5.142}$$

Substituting (5.142) into (5.141)

$$\beta_i = \frac{1}{1 + \frac{\omega_k}{\omega_\ell}}\left[\frac{\omega_k}{\omega_i} + \frac{\omega_i}{\omega_\ell}\right]\beta_0 \tag{5.143}$$

It can be verified from (5.143) that $\beta_i = \beta$ at $i = k$ and at $i = \ell$. For an example with $k = 1$ and $\ell = 3$, the damping applied to other modal equations vary as shown in Fig. 5.22. Also shown in Fig. 5.22 are the schematics of the variation of damping applied to the modal equations for the models with $\alpha_1 = 0$ (stiffness proportional damping) and $\alpha_2 = 0$ (mass proportional damping). In both of these cases, the damping is fixed at β_0 for the first modal equation. It is seen that higher modal equations receive higher damping than β_0 with the stiffness proportional damping model, and smaller damping than β_0 with the mass proportional damping model.

It can be shown (Clough and Penzien 1975) that the Rayleigh damping model can be generalized to include more constants, and hence to have the ability to fix the damping at higher number of modal equations. The following expression is orthogonal in the sense of (5.136):

$$\mathbf{C} = \mathbf{M}\sum_k \alpha_k\left[\mathbf{M}^{-1}\mathbf{K}\right]^k; \quad -\infty < k < \infty \tag{5.144}$$

It can be verified that for $k = 0$, one obtains the mass proportional damping and for $k = 1$, one obtains the stiffness proportional damping. For $k = -1, 0$ and 1, we have

$$\mathbf{C} = \alpha_{-1}\mathbf{M}\mathbf{K}^{-1}\mathbf{M} + \alpha_0\mathbf{M} + \alpha_1\mathbf{K} \tag{5.145}$$

It is left to the reader to show that by fixing the damping at β_1, β_2, and β_3 at three frequencies ω_1, ω_2, and ω_3, the corresponding constants α_{-1}, α_0, and α_1 are related to β_1, β_2, and β_3 by

$$\begin{bmatrix} \beta_1 \\ \beta_2 \\ \beta_3 \end{bmatrix} = \frac{1}{2} \begin{bmatrix} 1/\omega_1^3 & 1/\omega_1 & \omega_1 \\ 1/\omega_2^3 & 1/\omega_2 & \omega_2 \\ 1/\omega_3^3 & 1/\omega_3 & \omega_3 \end{bmatrix} \begin{bmatrix} \alpha_{-1} \\ \alpha_0 \\ \alpha_1 \end{bmatrix} \tag{5.146}$$

Equation (5.146) can be solved to find the constants α_{-1}, α_0, and α_1.

It may be pointed out that, in general, that the damping matrix given by (5.144) becomes full (i.e., nonbanded) for $k > 2$, resulting in a significant increase in computational effort in solving the system equations.

In problems with composite structures and materials, the damping may vary drastically from region to region. In this case, the Rayleigh damping model may be applied at the element level with different fraction of critical damping for different elements. For example, in a soil-structure interaction problem, the damping associated with soil may be significantly higher than the damping associated with the structural elements. However, when this method is used, the damping matrix is neither proportional to the system stiffness matrix nor the system mass matrix. Hence, the system equation will not uncouple as in (5.140). The direct integration methods to be discussed in Sect. 5.4.2.3 can still be used.

Finally it may be noted that although the Rayleigh damping model is introduced to uncouple the equations in linear analyses, it serves as a method of calculating the damping matrix even in nonlinear analyses (even in elasto-plastic analyses). For example, in a soil-structure interaction analysis, one may use an elastic constitutive model for the superstructure and an elasto-plastic model for the soil. The Rayleigh damping model may then be used at the element level to apply damping for the structural part.

5.4.2.3 Time Integration Procedures

The objective of the procedures presented in this section is to, given the displacements, velocities and accelerations at time t, find the displacements, velocities and accelerations at time $t + \Delta t$, satisfying the system equilibrium equation (5.133) at $t + \Delta t$. It is assumed that (5.133) is satisfied at time t. The variation of the load with time is given either at discrete times or as continuous mathematical functions.

Symbolically, the problem is: given $\hat{\mathbf{u}}_n$, $\dot{\hat{\mathbf{u}}}_n$, $\ddot{\hat{\mathbf{u}}}_n$, and $\mathbf{P}(t)$, find $\hat{\mathbf{u}}_{n+1}$, $\dot{\hat{\mathbf{u}}}_{n+1}$, and $\ddot{\hat{\mathbf{u}}}_{n+1}$, where n represents the state at t and $n + 1$ the state at $t + \Delta t$.

At the start of the integration process, given the displacements and velocities, (5.133) may be used to establish the initial accelerations.

The procedures to be presented in this section are known as the time-marching techniques, and can be applied for integrating (5.133) directly or indirectly after uncoupling using the mode superposition technique presented in the preceding section. The procedures can be applied to linear as well as nonlinear problems, including elasto-plastic problems.

The methods may be broadly classified into

- Explicit methods
- Implicit methods

In the explicit methods, $\hat{\mathbf{u}}_{n+1}$, $\dot{\hat{\mathbf{u}}}_{n+1}$, and $\ddot{\hat{\mathbf{u}}}_{n+1}$ are obtained by satisfying (5.133) at the state n. In implicit method on the other hand, $\hat{\mathbf{u}}_{n+1}$, $\dot{\hat{\mathbf{u}}}_{n+1}$, and $\ddot{\hat{\mathbf{u}}}_{n+1}$ are obtained by satisfying (5.133) at the state $n+1$.

In both explicit and implicit methods, the variation of $\hat{\mathbf{u}}$, $\dot{\hat{\mathbf{u}}}$, and $\ddot{\hat{\mathbf{u}}}$ within a time step needs to be assumed. Different assumptions are made in different methods; for example, acceleration is assumed to remain constant in *the constant acceleration method*, and to vary linearly in *the linear acceleration method*. An example of the explicit methods is *the central difference method*. Examples of implicit methods are *the Houbolt method, the Wilson θ method and the Newmark method*. We will discuss the central difference method and the Newmark method in detail in this book. The reader is referred to a standard textbook on finite elements or structural dynamic for more details (e.g., Bathe 1982; Clough and Penzien 1975).

The explicit methods are conditionally stable, i.e., the time step must be smaller than a critical time step for stability. Some of the implicit methods are unconditionally stable; i.e., the time step can be as large as we want. Stability and accuracy are not synonymous. Accuracy still requires the time step to be small enough. The stability and accuracy issues are discussed in Sect. 5.4.2.6.

5.4.2.4 An Explicit Method: The Central Difference Method

The approximations for the acceleration and velocity used in this method are

$$\ddot{\hat{\mathbf{u}}}_n = \frac{1}{\Delta t^2}\left[\hat{\mathbf{u}}_{n-1} - 2\hat{\mathbf{u}}_n + \hat{\mathbf{u}}_{n+1}\right] \tag{5.147a}$$

$$\dot{\hat{\mathbf{u}}}_n = \frac{1}{2\Delta t}\left[\hat{\mathbf{u}}_{n+1} - \hat{\mathbf{u}}_{n-1}\right] \tag{5.147b}$$

Error in the above expansions is of order $(\Delta t)^2$. Substituting these expression in (5.133) at time t, and rearranging the equations, $\hat{\mathbf{u}}_{n+1}$ is obtained as

$$\mathbf{M}\ddot{\hat{\mathbf{u}}}_n + \mathbf{C}\dot{\hat{\mathbf{u}}}_n + \mathbf{K}\hat{\mathbf{u}}_n = \mathbf{P}_n \tag{5.148}$$

$$\mathbf{A}\hat{\mathbf{u}}_{n+1} = \bar{\mathbf{P}}_n \tag{5.149a}$$

where

$$\mathbf{A} = \frac{1}{\Delta t^2}\mathbf{M} + \frac{1}{2\Delta t}\mathbf{C} \tag{5.149b}$$

and

$$\bar{\mathbf{P}}_n = \mathbf{P}_n - \left[\mathbf{K} - \frac{2}{\Delta t^2}\mathbf{M}\right]\hat{\mathbf{u}}_n - \left[\frac{1}{\Delta t^2}\mathbf{M} - \frac{1}{2\Delta t}\mathbf{C}\right]\mathbf{u}_{n-1} \tag{5.149c}$$

Equation (5.149a) is solved for $\hat{\mathbf{u}}_{n+1}$. The procedure is repeated for finding $\hat{\mathbf{u}}_{n+1}$ for different values of n. Once $\hat{\mathbf{u}}_{n+1}$ is found, (5.147a) and (5.147b) may be used to find $\dot{\hat{\mathbf{u}}}_{n+1}$ and $\ddot{\hat{\mathbf{u}}}_{n+1}$.

At $t = 0$ (i.e., $n = 1$), the application of the above procedure requires $\hat{\mathbf{u}}_{n-1}$, which is obtained from (5.147a) as

$$\hat{\mathbf{u}}_{n-1} = \hat{\mathbf{u}}_n - \dot{\hat{\mathbf{u}}}_n\Delta t + \frac{1}{2}\ddot{\hat{\mathbf{u}}}_n\Delta t^2 \tag{5.150}$$

When **C** is null or diagonal, and **M** is diagonal, the solution of (5.149a) does not involve matrix inversion, and hence is trivial. However, the price to pay is that the time step must be very small. It will be shown in Sect. 5.4.2.6 that the time step must be

$$\Delta t \leq \Delta t_{\mathrm{cr}} = \frac{T_{\mathrm{s}}}{\pi} \tag{5.151}$$

where T_{s} is the smallest period of the system ($T_{\mathrm{s}} = 2\pi/\omega_\ell$, where ω_ℓ is the largest frequency of the system).

5.4.2.5 Implicit Methods

Linear Acceleration Method

The acceleration is assumed to vary linearly, and the velocity and the displacement are obtained by integration.

Defining $\bar{t} = t - t_n$

$$\ddot{\hat{\mathbf{u}}}(\bar{t}) = \ddot{\hat{\mathbf{u}}}_n + a\bar{t} \tag{5.152a}$$

$$\dot{\hat{\mathbf{u}}}(\bar{t}) = \dot{\hat{\mathbf{u}}}_n + \ddot{\hat{\mathbf{u}}}_n\bar{t} + a\frac{\bar{t}^2}{2} \tag{5.152b}$$

$$\hat{\mathbf{u}}(\bar{t}) = \hat{\mathbf{u}}_n + \dot{\hat{\mathbf{u}}}_n\bar{t} + \ddot{\hat{\mathbf{u}}}_n\frac{\bar{t}^2}{2} + a\frac{\bar{t}^3}{6} \tag{5.152c}$$

At $\bar{t} = \Delta t$

$$\ddot{\hat{\mathbf{u}}}_{n+1} = \ddot{\hat{\mathbf{u}}}_n + a\Delta t \tag{5.153a}$$

$$\dot{\hat{\mathbf{u}}}_{n+1} = \dot{\hat{\mathbf{u}}}_n + \ddot{\hat{\mathbf{u}}}_n\Delta t + a\frac{\Delta t^2}{2} \tag{5.153b}$$

$$\hat{\mathbf{u}}_{n+1} = \hat{\mathbf{u}}_n + \dot{\hat{\mathbf{u}}}_n\Delta t + \ddot{\hat{\mathbf{u}}}_n\frac{\Delta t^2}{2} + a\frac{\Delta t^3}{6} \tag{5.153c}$$

Solving for a from (5.153a) and substituting into (5.153b) and (5.153c)

$$a = \frac{1}{\Delta t}\left[\ddot{\hat{\mathbf{u}}}_{n+1} - \ddot{\hat{\mathbf{u}}}_n\right] \tag{5.154a}$$

$$\dot{\hat{\mathbf{u}}}_{n+1} = \dot{\hat{\mathbf{u}}}_n + \frac{\Delta t}{2}\left[\ddot{\hat{\mathbf{u}}}_n + \ddot{\hat{\mathbf{u}}}_{n+1}\right] \tag{5.154b}$$

$$\hat{\mathbf{u}}_{n+1} = \hat{\mathbf{u}}_n + \dot{\hat{\mathbf{u}}}_n\Delta t + \ddot{\hat{\mathbf{u}}}_n\frac{\Delta t^2}{3} + \ddot{\hat{\mathbf{u}}}_{n+1}\frac{\Delta t^2}{6} \tag{5.154c}$$

Solve for $\ddot{\hat{\mathbf{u}}}_{n+1}$ from (5.154c) and substitute into (5.154b)

$$\ddot{\hat{\mathbf{u}}}_{n+1} = \frac{6}{\Delta t^2}\left(\hat{\mathbf{u}}_{n+1} - \mathbf{u}_n\right) - \frac{6}{\Delta t}\dot{\mathbf{u}}_n - 2\ddot{\hat{\mathbf{u}}}_n \tag{5.155a}$$

$$\dot{\hat{\mathbf{u}}}_{n+1} = \frac{3}{\Delta t}\left(\hat{\mathbf{u}}_{n+1} - \hat{\mathbf{u}}_n\right) - 2\dot{\mathbf{u}}_n - \frac{\Delta t}{2}\ddot{\hat{\mathbf{u}}}_n \tag{5.155b}$$

Now substitute (5.155a) and (5.155b) into (5.133a) at $t + \Delta t$

$$\mathbf{M}\ddot{\hat{\mathbf{u}}}_{n+1} + \mathbf{C}\dot{\hat{\mathbf{u}}}_{n+1} + \mathbf{K}\hat{\mathbf{u}}_{n+1} = \mathbf{P}_{n+1} \tag{5.156}$$

$$\overline{\mathbf{K}}\hat{\mathbf{u}}_{n+1} = \overline{\mathbf{P}} \tag{5.157a}$$

where

$$\overline{\mathbf{K}} = \mathbf{K} + \frac{6}{\Delta t^2}\mathbf{M} + \frac{3}{\Delta t}\mathbf{C} \tag{5.157b}$$

and

$$\bar{\mathbf{P}} = \mathbf{P}_{n+1} + \mathbf{C}\left[\frac{3}{\Delta t}\hat{\mathbf{u}}_n + 2\dot{\hat{\mathbf{u}}}_n + \frac{\Delta t}{2}\ddot{\hat{\mathbf{u}}}_n\right] + \mathbf{M}\left[\frac{6}{\Delta t^2}\hat{\mathbf{u}}_n + \frac{6}{\Delta t}\dot{\hat{\mathbf{u}}}_n + 2\ddot{\hat{\mathbf{u}}}_n\right] \tag{5.157c}$$

Equation (5.157a) is solved for $\hat{\mathbf{u}}_{n+1}$, and (5.155a) and (5.155b) are used to find $\ddot{\hat{\mathbf{u}}}_{n+1}$ and $\dot{\hat{\mathbf{u}}}_{n+1}$. At $t = 0$ (i.e., for $n = 1$), (5.133a) is used to find $\ddot{\hat{\mathbf{u}}}_1$ from the knowledge of $\hat{\mathbf{u}}_1$ and $\dot{\hat{\mathbf{u}}}_1$.

The linear acceleration method is only *conditionally stable*. Hence, unless the time step is small enough, not only that the accuracy will be affected, but also the solution can diverge.

The Wilson θ Method

The Wilson θ method is a modified form of the linear acceleration method. The method is unconditionally stable for $\theta > 1.37$. The method assumes that the acceleration varies linearly over an extended period

$$\tau = \theta \Delta t$$

where $\theta \geq 1$. From a stability analysis, it was determined that for unconditional stability, $\theta > 1.37$. However, as the value of θ increases, the response period becomes elongated and the amplitude becomes attenuated. Hence, a value close to 1.37, which is the minimum value for unconditional stability, is recommended for use.

The relevant equations are derived by modifying those associated with the linear acceleration method as follows. The assumption that the acceleration varies linearly for an extended period τ is equivalent to using $\bar{t} = \tau = \theta \Delta t$ in (5.152). The remaining steps are exactly the same as those in (5.153)–(5.157), but the displacement solved by (5.157a) is the displacement at $t + \theta \Delta t$. The load vector $\mathbf{P}_{n+1}$ appearing in (5.157c) should be replaced with

$$\mathbf{P}_{n+\theta} = \mathbf{P}_n + \theta(\mathbf{P}_{n+1} - \mathbf{P}_n)$$

The acceleration found by (5.155a) is that at $t + \theta \Delta t$, say, $\ddot{\hat{\mathbf{u}}}_{n+\theta}$. The acceleration at $t + \Delta t$ is then obtained by linearly scaling back the incremental acceleration as

$$\ddot{\hat{\mathbf{u}}}_{n+1} = \ddot{\hat{\mathbf{u}}}_n + \frac{1}{\theta}\left[\ddot{\hat{\mathbf{u}}}_{n+\theta} - \ddot{\hat{\mathbf{u}}}_n\right] \tag{5.158}$$

Equation (5.157b) and (5.157c) are used as is (i.e., with Δt instead of $\theta \Delta t$) to find $\dot{\hat{\mathbf{u}}}_{n+1}$ and $\hat{\mathbf{u}}_{n+1}$, respectively.

The Newmark Method

The Newmark method is another variation of the linear acceleration method. The method involves two parameters for controlling the integration accuracy and stability. For specific choices of these parameters, the method degenerates into the linear acceleration method and the constant-average-acceleration method (also known as the trapezoidal rule). The latter is not discussed in this book.

The velocity and displacement are written as

$$\dot{\hat{\mathbf{u}}}_{n+1} = \dot{\hat{\mathbf{u}}}_n + \left[(1-\delta)\ddot{\hat{\mathbf{u}}}_n + \delta\ddot{\hat{\mathbf{u}}}_{n+1}\right]\Delta t \tag{5.159a}$$

$$\hat{\mathbf{u}}_{n+1} = \hat{\mathbf{u}}_n + \dot{\hat{\mathbf{u}}}_n\Delta t + \left[\left(\frac{1}{2}-\alpha\right)\ddot{\hat{\mathbf{u}}}_n + \alpha\ddot{\hat{\mathbf{u}}}_{n+1}\right]\Delta t^2 \tag{5.159b}$$

It can be verified that for $\delta = (1/2)$ and $\alpha = (1/6)$, (5.159a) and (5.159b) are identical to (5.154b) and (5.154c), respectively, rendering the Newmark method identical to the linear acceleration method.

From (5.159b):

$$\ddot{\hat{\mathbf{u}}}_{n+1} = \frac{1}{\alpha\Delta t^2}(\hat{\mathbf{u}}_{n+1} - \hat{\mathbf{u}}_n) - \frac{1}{\alpha\Delta t}\dot{\hat{\mathbf{u}}}_n - \left(\frac{1}{2\alpha} - 1\right)\ddot{\hat{\mathbf{u}}}_n \tag{5.159c}$$

Substituting (5.159c) into (5.159a), $\dot{\hat{\mathbf{u}}}_{n+1}$ can be expressed in terms of $\hat{\mathbf{u}}_{n+1}$.

By following the steps identical those in the linear acceleration method, it can be shown that (5.157) become

$$\overline{\mathbf{K}}\hat{\mathbf{u}}_{n+1} = \overline{\mathbf{P}}_{n+1} \tag{5.160a}$$

where

$$\overline{\mathbf{K}} = \mathbf{K} + \frac{1}{\alpha\Delta t^2}\mathbf{M} + \frac{\delta}{\alpha\Delta t}\mathbf{C} \tag{5.160b}$$

and

$$\begin{aligned}\overline{\mathbf{P}}_{n+1} = \mathbf{P}_{n+1} &+ \mathbf{C}\left[\frac{\delta}{\alpha\Delta t}\hat{\mathbf{u}}_n + \left(\frac{\delta}{\alpha} - 1\right)\dot{\hat{\mathbf{u}}}_n + \Delta t\left(\frac{\delta}{2\alpha} - 1\right)\ddot{\hat{\mathbf{u}}}_n\right]\\ &+ \mathbf{M}\left[\frac{1}{\alpha\Delta t^2}\hat{\mathbf{u}}_n + \frac{1}{\alpha\Delta t}\dot{\hat{\mathbf{u}}}_n + \left(\frac{1}{2\alpha} - 1\right)\ddot{\hat{\mathbf{u}}}_n\right]\end{aligned} \tag{5.160c}$$

Equation (5.160a) is used to find $\hat{\mathbf{u}}_{n+1}$, (5.159c) to find $\ddot{\hat{\mathbf{u}}}_{n+1}$ and (5.159a) to find $\dot{\hat{\mathbf{u}}}_{n+1}$.

For $\delta = (1/2)$ and $\alpha = (1/4)$, the method is called the constant-average-acceleration method. This method is Newmark's original method and is unconditionally stable. It may be verified that for $\delta = (1/2)$ and $\alpha = (1/6)$, (5.160) is identical to (5.157) (the linear acceleration method). The Newmark method is unconditionally stable for

$$\delta \geq \frac{1}{2} \text{ and } \alpha \geq \frac{1}{16}(2\delta + 1)^2$$

Hilber–Hughes–Taylor Method

Hilber et al. (1977) modified the Newmark method to introduce controllable numerical damping; the method is now known as the Hilber–Hughes–Taylor method. The Newmark's integration parameters are given as

$$\delta = \frac{1}{2} - \gamma \text{ and } \alpha = \frac{1}{4}(1 - \gamma)^2$$

where γ is the numerical damping parameter, lying in the range $-(1/3) \leq \gamma \leq 0$. When $\gamma = 0$, the method is the Newmark method with no numerical damping. When $\gamma < 0$, a numerical damping is applied with the magnitude depending on the value of γ.

5.4.2.6 Stability of Time Integration Methods

Introduction

From the section on mode superposition method, it is clear the overall system response depends on the modal responses. To integrate a given modal equation, the time step must not only be small enough to capture the variation of the load, but also must be small enough to capture its free vibration characteristics. To capture the latter, the time step must be smaller than about a tenth of the natural period, i.e.,

$$\Delta t < \frac{T_i}{10} = \frac{1}{10} \frac{2\pi}{\omega_i}$$

where T_i is the ith modal period and ω_i is the ith modal frequency. Hence, to integrate one of the higher modal equations, Δt must be really small. However, it was pointed out in the section on mode superposition method that the contribution of the higher modal responses to the overall system response is very small. Hence, the higher modal responses are not needed and thus the corresponding equations do not need to be integrated. If the mode superposition method is used, only the first few modal equations need to be integrated. Hence, the value of Δt can in principle be large.

However, when the system equations are directly integrated, a single time step needs to be chosen. The question is whether the choice of a large time step will lead to divergence. A stable algorithm must permit the solution to be continued without divergence regardless of the value chosen for Δt. An unstable algorithm can also lead to divergence due to errors introduced by numerical rounding on displacements, velocities and accelerations.

Stability Analysis

At any time t, the recursive relationships can be written in a matrix–vector form as

$$\mathbf{x}_{n+1} = \mathbf{A}\mathbf{x}_n + \mathbf{B}\mathbf{P}_{n+\theta} \tag{5.161}$$

Where **A** and **B** are known as the integration approximation and load operators (Bathe 1982). To see if the solution diverges for any given set of initial conditions (i.e., initial velocities and/or displacements), it suffices to consider (5.161) without the load term. From the start, if (5.161) is applied k times

$$\mathbf{x}_{k+1} = \mathbf{A}^k\mathbf{x}_k \tag{5.162}$$

For stability, as k increases, $\mathbf{A}^k$ must remain bounded.

It can be shown that if λ is an eigenvalue of **A** and **y** is the corresponding eigenvector, then the eigenvalue and eigenvector of $\mathbf{A}^k$ are λ^k and **y**, respectively. With this knowledge, spectral decomposition of $\mathbf{A}^k$ may be written as

$$\mathbf{A}^k = \mathbf{Q}\mathbf{J}\mathbf{Q}^{-1} \tag{5.163}$$

where **Q** contains the eigenvectors of **A** and the ith diagonal element of **J** is λ_i^k. The spectral radius of a matrix is defined as its maximum eigenvalue; i.e.,

$$\rho(\mathbf{A}) = \lambda_{\max} = \max\{\lambda_i\}, \quad i = 1, 2, \ldots n \tag{5.164}$$

As k increases, the matrix **J** is bounded if and only if $\lambda_{\max} \leq 1$. Note that if $\lambda_{\max} \leq 1$, then $\lambda_{\max}^k \to 0$ as $k \to \infty$.

Detailed analyses of different integration methods are beyond the scope of this book; the reader is referred to Bathe (1982). By considering the integration of the single degree of freedom equation

$$\ddot{x} + 2\xi\omega\dot{x} + \omega^2 x = 0 \tag{5.165a}$$

Bathe (1982) shows that for the Wilson θ method,

$$\mathbf{A} = \begin{bmatrix} 1 - \dfrac{\beta\theta^2}{3} - \dfrac{1}{\theta} - \kappa\theta & \dfrac{1}{\Delta t}(-\beta\theta - 2\kappa) & \dfrac{1}{\Delta t^2}(-\beta) \\ \Delta t\left(1 - \dfrac{1}{2\theta} - \dfrac{\beta\theta^2}{6} - \dfrac{\kappa\theta}{2}\right) & 1 - \dfrac{\beta\theta}{2} - \kappa & \dfrac{1}{\Delta t}\left(-\dfrac{\beta}{2}\right) \\ \Delta t^2\left(\dfrac{1}{2} - \dfrac{1}{6\theta} - \dfrac{\beta\theta^2}{18} - \dfrac{\kappa\theta}{6}\right) & \Delta t\left(1 - \dfrac{\beta\theta}{6} - \dfrac{\kappa}{3}\right) & 1 - \dfrac{\beta}{6} \end{bmatrix} \tag{5.165b}$$

where

$$\beta = \left(\frac{\theta}{\omega^2 \Delta t^2} + \frac{\xi \theta^2}{\omega \Delta t} + \frac{\theta^3}{6}\right)^{-1} \quad \text{and } \kappa = \frac{\xi \beta}{\omega \Delta t} \tag{5.165c}$$

By plotting $\lambda_{\max}$ of **A** for different values of θ, Bathe (1982) shows that $\lambda_{\max} \leq 1$ for $\theta \geq 1.37$ and $\lambda_{\max} > 1$ for $\theta < 1.37$, proving that for unconditional stability is realized for $\theta \geq 1.37$.

Similar analysis showed that the condition for the central different explicit method to be stable is $\Delta t \leq \Delta t_{cr} = (T/\pi)$, where T is the natural period of the system. The Newmark method is unconditionally stable when

$$\delta \geq \frac{1}{2} \text{ and } \alpha \geq \frac{1}{4}\left(\delta + \frac{1}{2}\right)^2.$$

5.4.2.7 Calculation of Lumped Mass Matrices

In general, only the translational degrees of freedom are assigned lumped masses. Conceptually, the lumped masses are based on the assumption that the translational inertial forces at a node depend only on the mass properties in its neighborhood. On this basis, for certain simple, regular elements such the four-noded rectangular element or the eight-noded brick element, it is intuitively straightforward. In the four-noded rectangular element, 1/4 of the total mass is placed at each node. In the eight-noded brick element, 1/8 of the total mass of placed at each node. However, in elements such as the eight-noded two-dimensional quadrilateral (2D8), it is not intuitive.

Several methods have been suggested for mass lumping in the past (see, for example, Cook 1981; Zienkiewicz 1977). A particularly convenient method is as follows (Hinton et al. 1976; Surana 1978): Determine the consistent mass matrix and scale the masses associated with the translational degrees of freedoms in each direction by the total mass of the element so that the total mass is preserved.

For example, the consistent mass for a linear triangle is given by (5.93). Denoting the total mass of the element by m ($m = a_e t^* \rho$), the sum of the diagonal elements in direction 1 is

$$\frac{m}{6} + \frac{m}{6} + \frac{m}{6} = \frac{m}{2}$$

The scale factor is total mass ($=m$) divided by the sum ($=m/2$), which is 2. Hence, the lumped mass for each of the three degrees of freedom in direction 1 is $2(m/6) = m/3$. By following the same procedure, the lumped mass for each of the three degrees of freedom in direction 2 is also $m/3$. Hence, the lumped mass matrix is obtained from the consistent mass matrix as

$$\mathbf{m}_e = \frac{m}{12}\begin{bmatrix} 2 & 0 & 1 & 0 & 1 & 0 \\ 0 & 2 & 0 & 1 & 0 & 1 \\ 1 & 0 & 2 & 0 & 1 & 0 \\ 0 & 1 & 0 & 2 & 0 & 1 \\ 1 & 0 & 1 & 0 & 2 & 0 \\ 0 & 1 & 0 & 1 & 0 & 2 \end{bmatrix} \Rightarrow \frac{m}{3}\begin{bmatrix} 1 & 0 & 0 & 0 & 0 & 0 \\ 0 & 1 & 0 & 0 & 0 & 0 \\ 0 & 0 & 1 & 0 & 0 & 0 \\ 0 & 0 & 0 & 1 & 0 & 0 \\ 0 & 0 & 0 & 0 & 1 & 0 \\ 0 & 0 & 0 & 0 & 0 & 1 \end{bmatrix}$$

The total mass is preserved in each translational direction. Hence, a third of the total mass gets placed at the nodes in each translational direction; the outcome in this case agrees with intuitive results.

Consider the two-noded beam element considered in Sect. 5.3.3 (Fig. 5.20a). The unknowns are arranged as $\hat{\mathbf{u}} = \left\{\hat{w}_1, \hat{\theta}_1, \hat{w}_2, \hat{\theta}_2\right\}$. The consistent mass is given below on the left in the matrix equation below, where $m = a_e \ell \rho$ is the total mass of the element. Sum of the diagonals associated with the translational degrees of freedom is

$$\frac{156m}{420} + \frac{156m}{420} = \frac{312m}{420}$$

The scale factor is (420/312). By multiplying each degree of freedom (including the rotational degree of freedom), the lumped mass matrix on the right is obtained.

$$\mathbf{m}_e = \frac{m}{420}\begin{bmatrix} 156 & 22\ell & 54 & -13\ell \\ 22\ell & 4\ell^2 & 13\ell & -3\ell^2 \\ 54 & 13\ell & 156 & -22\ell \\ -13\ell & -3\ell^2 & -22\ell & 4\ell^2 \end{bmatrix} \Rightarrow \frac{m}{78}\begin{bmatrix} 39 & 0 & 0 & 0 \\ 0 & \ell^2 & 0 & 0 \\ 0 & 0 & 39 & 0 \\ 0 & 0 & 0 & \ell^2 \end{bmatrix}$$

Example 5.8.

Question. For the discrete system shown in Fig. 5.23a,

1. Derive the system dynamic equilibrium equation.
2. Given in consistent units that $k_1 = k_2/4 = 1$ and $m_1 = m_2 = 1$, determine the (undamped) natural frequencies and mode shapes.
3. Determine the mass proportional damping matrix by fixing the fraction of damping to β for the modal equation at the fundamental frequency.

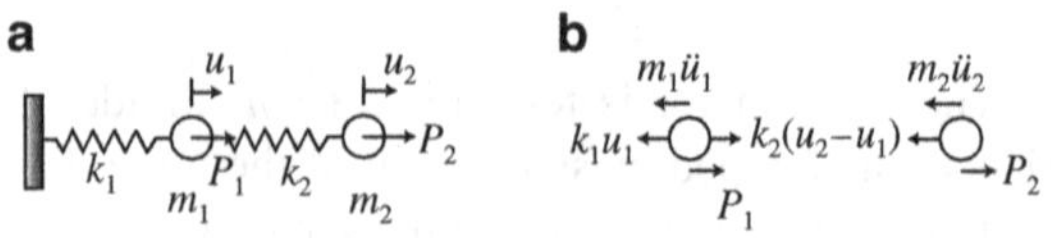

Fig. 5.23 Dynamics of two degree of freedom system

Answer.
Part(1): With u_1 and u_2 representing the displacements of the two masses from their static equilibrium positions, the forces acting on masses m_1 and m_2 are shown in Fig. 5.23b, where the inertial forces are introduced as static forces by D'Alembert's principle. Writing down the equilibrium equations in the horizontal direction for m_1 and m_2

$$\leftarrow \quad m_1\ddot{u}_1 + k_1u_1 - k_2(u_2 - u_1) - P_1 = 0$$

$$\leftarrow \quad m_2\ddot{u}_2 + k_2(u_2 - u_1) - P_2 = 0$$

Writing in a matrix-vector form

$$\begin{bmatrix} m_1 & 0 \\ 0 & m_2 \end{bmatrix}\begin{bmatrix} \ddot{u}_1 \\ \ddot{u}_2 \end{bmatrix} + \begin{bmatrix} k_1 + k_2 & -k_2 \\ -k_2 & k_2 \end{bmatrix}\begin{bmatrix} u_1 \\ u_2 \end{bmatrix} = \begin{bmatrix} P_1 \\ P_2 \end{bmatrix}$$

The first matrix is the mass matrix and the second matrix in the stiffness matrix.

Part(2):

$$\mathbf{M} = \begin{bmatrix} 1 & 0 \\ 0 & 1 \end{bmatrix}; \quad \mathbf{K} = \begin{bmatrix} 5 & -4 \\ -4 & 4 \end{bmatrix}$$

The natural frequencies are determined by solving the following equation (characteristic equation)

$$|\mathbf{K} - \omega^2\mathbf{M}| = 0$$

$$\begin{vmatrix} 5 - \omega^2 & -4 \\ -4 & 4 - \omega^2 \end{vmatrix} = 0 \Rightarrow \omega_1 = 0.685\,\text{rad/s and } \omega_2 = 2.92\,\text{rad/s}$$

The mode shapes are the eigenvectors, which are obtained by solving

$$[\mathbf{K} - \omega^2\mathbf{M}]\mathbf{x} = 0$$

For each of the two values for ω. As the above equation is singular, a fixed value is assumed for one of (x_1, x_2) and one of the two equations is used to solve the other. After solving for the eigenvectors, it may be normalized in any way desired. Here they are normalized such that the maximum positive value is unity. The results are

$$\mathrm{x}_1 = (0.882, 1.0) \text{ and } \mathrm{x}_2 = (1.0, -0.882)$$

Part(3): Referring to (5.139)–(5.142), for a mass proportional damping, using $\omega = \omega_1$, we have

$$\alpha_1 = 2\omega_1\beta = 2.0 \times 0.685 \times 0.05 = 0.0685$$

$$\mathbf{C} = \begin{bmatrix} 0.0685 & 0 \\ 0 & 0.0685 \end{bmatrix}$$

Example 5.9.

Question.

1. Referring to the problem in Example 5.8, determine the response of the masses for $0 < t < 10$ seconds by directly integrating the system equation by the central difference (explicit) method. The nodal loads are given as $P_1 = 0$ and $P_2 = \sin \omega_1 t$. Initial conditions are $\mathbf{u}(0) = \mathbf{0}$ and $\dot{\mathbf{u}}(0) = \mathbf{0}$. Use acceptable time step.
2. Repeat part(1) with a time step that is larger than the critical time step and investigate the convergence behavior. Carryout the analysis for $0 < t < 200$s.

Answer.

Part(1): As the method is explicit, it is conditionally stable, and $\Delta t \leq \Delta t_{cr} = (T_s/\pi)$. Let us take

$$\Delta t = \frac{T_2}{10} = \frac{1}{10}\frac{2\pi}{\omega_2} = \frac{2\pi}{10 \times 2.92} = 0.215 < \frac{T_2}{\pi} = 0.685\text{s}$$

Applying the equilibrium equation at $t = 0$, the acceleration is evaluated to be null, i.e., $\ddot{\mathbf{u}}(0) = \mathbf{0}$. Let $n = 1$ at $t = 0$. From (5.150), $\mathbf{u}_{-1} = \mathbf{0}$. The relevant equations to use from this point on are (5.147)–(5.149). The matrix $\mathbf{A}$ is evaluated as

$$\mathbf{A} = \begin{bmatrix} 21.79 & 0 \\ 0 & 21.79 \end{bmatrix}$$

Writing (5.149c) as

$$\overline{\mathbf{P}}_n = \mathbf{P}_n + \mathbf{B}_1\hat{\mathbf{u}}_n + \mathbf{B}_2\mathbf{u}_{n-1},$$

then

$$\mathbf{B}_1 = -\left[\mathbf{K} - \frac{2}{\Delta t^2}\mathbf{M}\right] = \begin{bmatrix} 38.27 & 4 \\ 4 & 39.27 \end{bmatrix}$$

$$\mathbf{B}_2 = -\left[\frac{1}{\Delta t^2}\mathbf{M} - \frac{1}{2\Delta t}\mathbf{C}\right] = \begin{bmatrix} -21.47 & 0 \\ 0 & -21.47 \end{bmatrix}$$

Equations involved in the recursive application are

$$\mathbf{A}\hat{\mathbf{u}}_{n+1} = \bar{\mathbf{P}}_n$$

$$\dot{\hat{\mathbf{u}}}_n = \frac{1}{2\Delta t}[\hat{\mathbf{u}}_{n+1} - \hat{\mathbf{u}}_n] = 2.326[\hat{\mathbf{u}}_{n+1} - \hat{\mathbf{u}}_n]$$

$$\ddot{\hat{\mathbf{u}}}_n = \frac{1}{\Delta t^2}[\hat{\mathbf{u}}_{n-1} - 2\hat{\mathbf{u}}_n + \hat{\mathbf{u}}_{n+1}] = 21.633[\hat{\mathbf{u}}_{n-1} - 2\hat{\mathbf{u}}_n + \hat{\mathbf{u}}_{n+1}]$$

Table 5.2 summarizes the relevant quantities for the first ten time steps

Part(2): The analysis is carried out to 200s with $\Delta t = 0.215 < 0.685$s and with $\Delta t = 0.7 > 0.685$s. The results are compared in Fig. 5.24. It can be seen that the response, which reaches steady state after some time, is captured well with a time step that is smaller than the critical time step. The calculated response diverges very quickly when the time step is larger than the critical value.

Example 5.10.

Question. Redo Example 5.9 using the Newmark method with $\delta = 1/2$ and $\alpha = 1/6$ (the linear acceleration method). Use $\Delta t = T_1/10 = 0.92$ s.

Also show that the linear acceleration method diverges when Δt is too large (e.g., $\Delta t = T_1/5 = 1.84$ s), whereas the original Newmark method (the constant-average-acceleration method), which is unconditionally stable, works.

Table 5.2 Response of the two-mass system by central difference method

t (s)	0.000	0.215	0.430	0.645	0.860	1.075	1.290	1.505	1.720	1.935
P	0.000	0.000	0.000	0.000	0.000	0.000	0.000	0.000	0.000	0.000
	0.000	0.147	0.290	0.427	0.555	0.671	0.773	0.858	0.924	0.970
$\bar{\mathbf{P}}$	0.000	0.000	0.027	0.149	0.471	1.113	2.176	5.711	5.709	8.096
	0.000	0.147	0.555	1.287	2.355	5.734	5.384	7.278	9.414	11.809
u	0.000	0.000	0.000	0.001	0.007	0.022	0.051	0.100	0.170	0.262
	0.000	0.000	0.007	0.025	0.059	0.108	0.171	0.247	0.334	0.432
$\dot{\mathbf{u}}$	0.000	0.000	0.003	0.016	0.047	0.103	0.182	0.277	0.377	0.468
	0.000	0.016	0.059	0.122	0.192	0.261	0.323	0.378	0.430	0.483
$\ddot{\mathbf{u}}$	0.000	0.000	0.027	0.095	0.199	0.317	0.417	0.470	0.459	0.386
	0.000	0.146	0.259	0.322	0.333	0.308	0.270	0.243	0.240	0.257

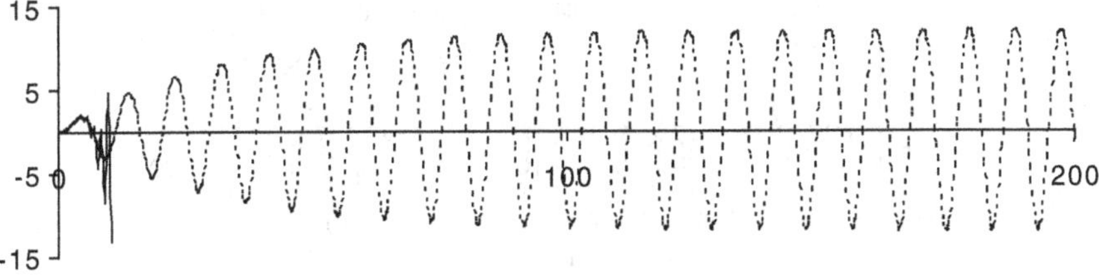

Fig. 5.24 Comparison of u_2 calculated with $\Delta t = 0.215$ s (*dotted line*) and $\Delta t = 0.7$ s (*solid line*)

Answer. Using $\Delta t = 0.92$, $\delta = 1/2$, and $\alpha = 1/6$, (5.160) becomes

$$\overline{\mathbf{K}}\hat{\mathbf{u}}_{n+1} = \overline{\mathbf{P}}_{n+1}$$

$$\overline{\mathbf{K}} = \begin{bmatrix} 12.31 & -4 \\ -4 & 11.31 \end{bmatrix}$$

$$\overline{\mathbf{P}}_{n+1} = \mathbf{P}_{n+1} + \mathbf{C}\left[3.26\hat{\mathbf{u}}_n + 2\dot{\hat{\mathbf{u}}}_n + 0.46\ddot{\hat{\mathbf{u}}}_n\right] + \mathbf{M}\left[7.09\hat{\mathbf{u}}_n + 6.52\dot{\hat{\mathbf{u}}}_n + 2\ddot{\hat{\mathbf{u}}}_n\right]$$

Equations (5.159c) and (5.159a) become

$$\ddot{\hat{\mathbf{u}}}_{n+1} = 7.09(\hat{\mathbf{u}}_{n+1} - \hat{\mathbf{u}}_n) - 6.52\dot{\hat{\mathbf{u}}}_n - 2\ddot{\hat{\mathbf{u}}}_n$$

$$\dot{\hat{\mathbf{u}}}_{n+1} = \dot{\hat{\mathbf{u}}}_n + 0.46\left[\ddot{\hat{\mathbf{u}}}_n + \ddot{\hat{\mathbf{u}}}_{n+1}\right]$$

The initial conditions are $\hat{\mathbf{u}}_1(0) = \mathbf{0}$, $\dot{\hat{\mathbf{u}}}_1(0) = \mathbf{0}$, and $\ddot{\hat{\mathbf{u}}}_1(0) = \mathbf{0}$.

The calculated variables for the first ten steps are listed in Table 5.3.

The responses calculated up to 200 s with different sets of parameters are plotted in Fig. 5.25. It is seen that the results obtained by the linear acceleration method with $\Delta t = 0.92$ s are very close to those obtained by the central difference method

Table 5.3 Response of the two-mass system by newmark method with $\Delta t = 0.92$, $\delta = 1/2$, and $\alpha = 1/6$

t (s)	0.920	1.840	2.760	5.680	4.600	5.520	6.440	7.360	8.280	9.200
P	0.000	0.000	0.000	0.000	0.000	0.000	0.000	0.000	0.000	0.000
	0.589	0.952	0.950	0.582	−0.008	−0.596	−0.955	−0.947	−0.576	0.016
$\bar{\mathbf{P}}$	0.000	0.830	4.938	9.236	11.232	9.974	2.756	−7.600	−16.641	−21.681
	0.589	5.507	7.136	11.358	15.368	9.570	1.326	−10.029	−20.634	−24.167
u	0.019	0.190	0.685	1.216	1.464	1.226	0.296	−1.023	−2.197	−2.774
	0.059	0.377	0.873	1.434	1.700	1.279	0.222	−1.248	−2.601	−5.117
$\dot{\mathbf{u}}$	0.062	0.370	0.627	0.466	0.051	−0.637	−1.313	−1.443	−1.043	−0.096
	0.192	0.463	0.612	0.534	−0.051	−0.833	−1.437	−1.663	−1.117	0.036
$\ddot{\mathbf{u}}$	0.136	0.534	0.026	−0.376	−0.527	−0.967	−0.502	0.220	0.651	1.406
	0.417	0.171	0.154	−0.326	−0.945	−0.754	−0.560	0.069	1.118	1.388

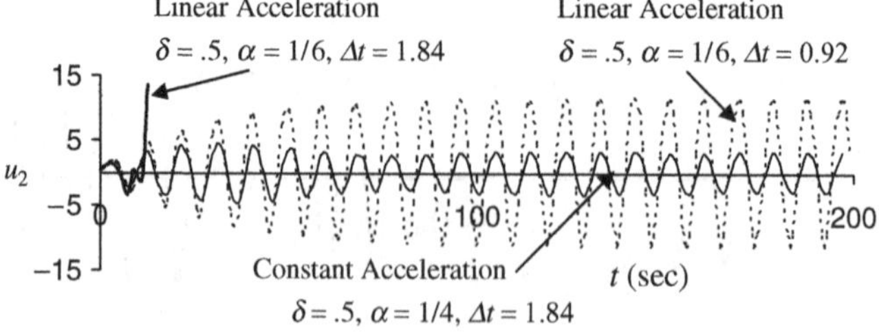

Fig. 5.25 Response calculated using newmark method

presented in Fig. 5.24. It may be noted that the central difference method with $\Delta t = 0.70$ s diverges, whereas the linear acceleration method converges even with a larger time step of 0.92 s.

When the time step is doubled, the linear acceleration method diverges and but the constant-average-acceleration method still converges. However, the results are inaccurate. Inaccurate representation of loading is one reason for the inaccuracy. Further, the time step is too large even to capture the first modal response $(T_1/10 = 2\pi/(10\omega_1) = 2\pi/(10 \times 0.685) = 0.917\,\text{s})$.

Problems

Problem 5.1. The one-dimensional rod shown in Fig. 5.26 has three sections, whose properties are given. The rod is subjected to point loads P and $2P$ at nodes 3 and 2, respectively. With respect to the global node numbers and the local directions shown in the figure, determine the element stiffness matrices for the three elements, the 4×4 global stiffness matrix and the 4×1 global load vector. Impose the boundary conditions and solve for the unknown displacements.

Ans.

$$u_3 = \frac{P}{9k}$$

and

$$u_2 = -\frac{5P}{9k},$$

where

$$k = \frac{AE}{\ell}.$$

Problem 5.2. Referring to Fig. 5.27, the displacement at any t is expressed in terms of quadratic shape functions as

$$\hat{u} = N_1\hat{u}_1 + N_2\hat{u}_2 + N_3\hat{u}_3$$

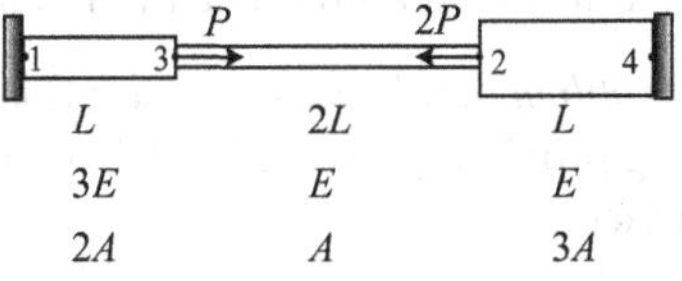

Fig. 5.26

Fig. 5.27

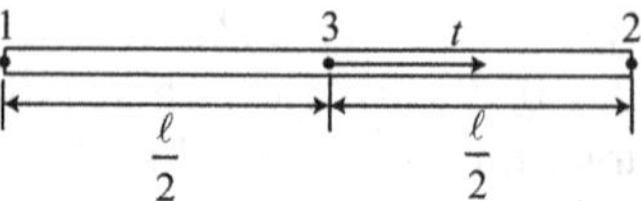

where

$$N_1 = -\frac{t}{\ell}\left(1 - \frac{2t}{\ell}\right); \quad N_2 = \frac{t}{\ell}\left(1 + \frac{2t}{\ell}\right); \quad N_3 = 1 - \frac{4t^2}{\ell^2}$$

with three nodes 1, 2, and 3, and the local axis t taken as shown in Fig. 5.27. Derive a general equation for the element stiffness matrix and traction load vector. Apply these equations to solve the problem presented in Fig. 5.3a. Divide the rod into two elements, one from point A to point B and another from point B to point C, and form the global stiffness and load vectors. Impose the displacement boundary condition and solve for the displacement at points B and C. Compare your results with those obtained using linear shape functions. Comment on the results.

Problem 5.3. Verify that

$$I(u) = -\int_0^L uf \, \mathrm{d}x + \frac{\mathrm{d}u}{\mathrm{d}x}(x = L)M_0$$

is a linear functional and

$$I(u, v) = \int_0^L b \frac{\mathrm{d}u}{\mathrm{d}x}\frac{\mathrm{d}v}{\mathrm{d}x} \mathrm{d}x$$

is bilinear.

Problem 5.4. The equation governing one-dimensional diffusion is

$$\alpha \frac{\mathrm{d}c}{\mathrm{d}x} - \frac{\mathrm{d}}{\mathrm{d}x}\left(\beta \frac{\mathrm{d}c}{\mathrm{d}x}\right) - Q + kc = 0$$

where α is the convection parameter, β is the dispersion parameter, Q is the source term, k is the sink term and c is the concentration of the quantity of interest. Show that this differential equation is not self adjoint.

Problem 5.5. Develop a variational statement for the one-dimensional beam bending problem presented in Chap. 3. Hint: Integration by parts needs to be performed twice.

Ans.

$$I(w) = \int_0^L \left[\frac{1}{2}EI\left(\frac{\mathrm{d}^2 w}{\mathrm{d}x^2}\right)^2 - wq\right]\mathrm{d}x - \int_{\Gamma_t}\left[wV_0 + \frac{\mathrm{d}w}{\mathrm{d}x}M_0\right]\mathrm{d}s$$

The boundary term will assume specific forms depending on the actual boundary conditions.

Problem 5.6. The boundary value problem associated with two-dimensional steady-state water flow through porous media is defined as follows:
The governing differential equation is

$$\frac{\partial v_x}{\partial x} + \frac{\partial v_y}{\partial y} = 0 \quad \text{in} \quad \Omega$$

where $\mathbf{v} = (v_x, v_y)$ is the fluid (Darcy's) velocity vector.

The boundary conditions are

$q_n = \mathbf{v}:\mathbf{n} = v_x n_x + v_y n_y = \hat{q}$ in Γ_q, where q_n is the flux across Γ_q, and
$h = \hat{h}$ on Γ_h

where h is the hydraulic head. The constitutive equation is

$$\mathbf{v} = \mathbf{kg}$$

or

$$\begin{bmatrix} v_x \\ v_y \end{bmatrix} = \begin{bmatrix} k_{xx} & k_{xy} \\ k_{yx} & k_{yy} \end{bmatrix} \begin{bmatrix} \dfrac{\partial h}{\partial x} \\ \dfrac{\partial h}{\partial y} \end{bmatrix}$$

Show that the variational functional is

$$I(h) = \frac{1}{2}\int_\Omega \mathbf{v}^\mathrm{T}\mathbf{g}\,\mathrm{d}v - \int_{\Gamma_q} h\bar{\mathbf{q}}\,\mathrm{d}s$$

Also, show that for the case of $k_{xy} = k_{yx} = 0$, the functional becomes

$$I(h) = \frac{1}{2}\int_\Omega \left[k_{xx}\left(\frac{\partial h}{\partial x}\right)^2 + k_{yy}\left(\frac{\partial h}{\partial y}\right)^2\right]\mathrm{d}v - \int_{\Gamma_q} h\bar{q}\,\mathrm{d}s$$

Problem 5.7. Referring to the plane strain body shown Fig. 5.28, modeling the problem with two 3-noded triangular elements shown in the figure, determine $\hat{u}_1$ and $\hat{u}_2$ corresponding to the global degrees of freedoms indicated by the arrows at

Fig. 5.28

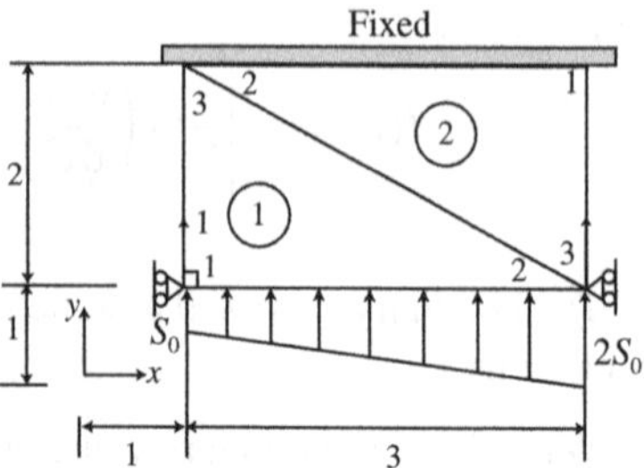

the nodes where the rollers are. The material properties are that Young's modulus is E and Poisson's ratio is zero. The thickness of the element is t^*. Perform the calculations by hand as in Examples 5.4 and 5.5.

Problem 5.8. Referring to Fig. 5.15a, show that for a linearly varying normal traction over a straight edge, (5.107) becomes

$$\begin{bmatrix} P_1 \\ P_2 \\ P_3 \end{bmatrix} = \begin{bmatrix} \dfrac{f_1 \ell}{6} \\ \dfrac{f_2 \ell}{6} \\ \dfrac{(f_1 + f_2)\ell}{3} \end{bmatrix}$$

where f_1 and f_2 are the values of the traction at nodes 1 and 2 respectively.

Problem 5.9. Referring to Fig. 5.15c, show that for a uniform normal traction over a straight edge (i.e., edge parallel to the radial direction) of an axisymmetric element, (5.107) becomes

$$\begin{bmatrix} P_1 \\ P_2 \\ P_3 \end{bmatrix} = \begin{bmatrix} \left(\dfrac{r_1}{r_3}\right)\dfrac{P}{6} \\ \left(\dfrac{r_2}{r_3}\right)\dfrac{P}{6} \\ \dfrac{2P}{3} \end{bmatrix}$$

where $P = \pi(r_2^2 - r_1^2)\, f_0$ and f_0 is the value of the traction.

Problem 5.10. Determine the nodal force vector in (5.108) for the element edge shown in Fig. 5.29, where the element edge is a circular arc. The traction acts normal to the edge.

Problem 5.11. A plane strain body is shown Fig. 5.30. By modeling the problem with one 4-noded isoparametric element shown in the figure, determine $\hat{u}_1$ and $\hat{u}_2$ corresponding to the global degrees of freedoms indicated by the arrows at the

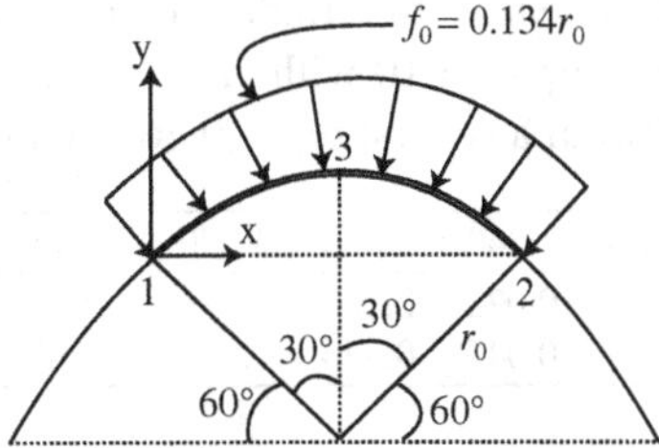

Fig. 5.29 Curved edge of an eight-noded isoparametric plane element

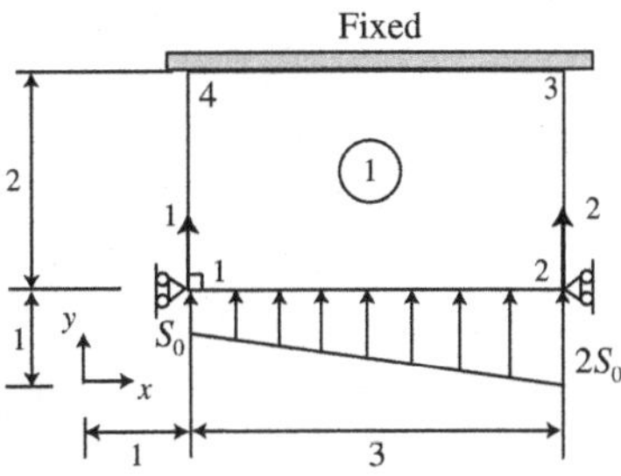

Fig. 5.30

nodes where the rollers are. The material properties are that Young's modulus is E and Poisson's ratio is zero. The thickness of the element is t^*. Perform the calculations by hand as in Examples 5.6 and 5.7.

Problem 5.12. Referring to (5.135) and assuming that the material is linear elastic, show that

$$\left[\mathbf{X}^{\mathrm{T}}\mathbf{M}\mathbf{X}\right] = \mathbf{U} \text{ and } \left[\mathbf{X}^{\mathrm{T}}\mathbf{K}\mathbf{X}\right] = \mathbf{W} \qquad \text{(From 5.136)}$$

where **U** and **W** diagonal matrices. The ith diagonals of these matrices are

$$U_{ii} = m_{ii} \text{ and } W_{ii} = m_{ii}\omega_i^2$$

where $m_{ii} = \mathbf{x}_i^{\mathrm{T}}\mathbf{M}\mathbf{x}_i$ and ω_i is the ith natural frequency. (Hint: Assume $\omega_i \neq \omega_j$ and use the fact that the stiffness and mass matrices are symmetric.)

Problem 5.13. Redo (using central difference method) the problem presented in Examples 5.8 and 5.9 with $m_1 = (m_2/3) = 1$.

Partial answers: $\omega_1 = 0.467$ and $\omega_2 = 2.473$ rad/s and $\Delta t_{cr} = 0.809$ s. Using $\Delta t = 0.3$s:

t (s)	0	0.3	0.6	0.9	1.2	1.5	1.8	2.1	2.4	2.7	5.0
u	0.000	0.000	0.000	0.001	0.008	0.024	0.056	0.104	0.170	0.250	0.342
	0.000	0.000	0.004	0.016	0.038	0.072	0.119	0.180	0.256	0.347	0.454

Problem 5.14. Redo (using Newmark method) the problem presented in Examples 5.10 with $m_1 = (m_2/3) = 1$.
Partial answers: $\omega_1 = 0.467$ and $\omega_2 = 2.473$ rad/s. Using $\Delta t = 1.34$ s:

t (s)	1.340	2.680	4.020	5.360	6.700	8.040	9.380	10.72	12.06	15.40	14.74
u	0.022	0.208	0.693	1.162	1.478	1.175	0.340	0.947	2.129	2.695	2.303
	0.047	0.326	0.851	1.458	1.734	1.402	0.331	1.175	2.580	5.234	2.700

Chapter 6
Governing Equations in Porous Media

The equations governing deformation of single-phase materials and the associated finite element formulations have been presented in Chaps. 3 and 5, respectively. While all materials are porous at some scale, they may be modeled as single-phase materials when the pores in the materials are macroscopically homogeneous and empty. In some cases, the stresses in the skeleton may be so much greater than those in the fluid that the effect of the fluid on the behavior of the skeleton may be neglected (e.g., dry concrete used in members supporting a bridge where the pores are filled with air). In a two-phase material (e.g., saturated soil), if the conditions (e.g., high permeability and/or slow loading) allow full drainage, the loading does not cause pressure build up and hence the fluid phase does not influence the behavior of the skeleton (i.e., the behavior of the skeleton under fully saturated and dry conditions are the same). At the other extreme, in a fully saturated material, if the conditions are such that relative movement of the fluid with respect to the skeleton is negligible (as in undrained behavior), the material may be modeled as a single-phase material. The theories presented in Chaps. 3 and 5 may then be used to model such problems.

While the situations presented above are exceptions, single-phase theories are in general inadequate to capture the behavior of multi-phase materials. Materials such as soils, rocks, powders, plants, bones, tissues, and polymers are examples of which the pores may be filled with one or more fluids. The consideration of various factors is necessary to mathematically model the behavior of such materials. The skeleton, the solid particles forming the skeleton (e.g., soil) and the fluids may be compressible or relatively incompressible. The fluids may be miscible or immiscible. Even when the problem at hand only concerns the prediction of behavior of skeleton, the above details will have an effect on the behavior, and hence must be modeled as an integral part. The analysis that considers the behavior of skeleton and fluids, and the interaction between them is known as the "fully coupled" analysis.

In this chapter, we consider problems where the pores are filled with one liquid and gas (e.g., unsaturated soil or rock). The cases of the dry and saturated materials become special cases of the general unsaturated case. The derivation of

A. Anandarajah, *Computational Methods in Elasticity and Plasticity: Solids and Porous Media*, DOI 10.1007/978-1-4419-6379-6_6,

the governing equations is presented in this chapter. The formulation of the finite element equations is presented in Chap. 7.

6.1 Preliminaries

6.1.1 Introduction

The problems that we are interested in this chapter generally involve deformation of the different phases comprising the materials and the flow of the fluid(s) through the pores in the skeleton. The consolidation behavior of a layer of soil under application of an external load (e.g., footing loading) is an example of such a problem. Solution of such problems requires not only the behavior of each phase, but also the consideration of the interactions between the phases. For example, when a fluid flows through a porous material, it imparts a viscous drag force on the solid skeleton. This must be considered in the equilibrium analysis of the skeleton. In addition, phase changes also may take place; for example, the solid phase may be converted into the liquid and/or the gas phase, the liquid phase may evaporate into the gas phase, and the gas phase may condense into the liquid phase.

Terzaghi (1925) devised the widely used one-dimensional consolidation theory. Biot (1941, 1955) developed a generalized theory of consolidation for poro-elastic solids. Biot (1941, 1955) took a *physical approach* to developing his theory. The *mixture theory* formulated by Truesdell and Toupin (1960) was later used by many to further generalize Biot's theory (e.g., Green and Naghdi 1965; Gurtin et al. 1972; Williams 1973; Bowen 1975, 1982; Atkin and Craine 1976a, b; Voyiadjis and Song 2006). Morland (1972) introduced the idea of volume fractions into the mixture theory. The modern porous media theories utilize this concept extensively. Most of these theories yield Biot's theory as a special case under the assumption of linearity. In mixture theories, balance laws are written down for each constituent at the macroscopic scale, with proper exchange laws between phases. The thermodynamic laws are used to establish the structure of constitutive laws.

There is another class of theories that uses "*the averaging theories*" to arrive at macroscopic field equations. The conservation and constitutive laws are established at the microscopic level. The corresponding equations at the macroscopic level are obtained by suitable averaging techniques (Slattery 1981; Whitaker 1986; Bear and Bachmat 1986). A brief comparison between the averaging and mixture theories may be found in Hassanizadeh and Gray (1990).

While the mixture theories provide practical methods in simulating observations, they do not allow transfer of microscopic information to macroscopic scales. Mixture theories also usually do not explicitly consider the interfaces between phases. For example, surface tension occurs at the interface between liquid and gas phases. A more accurate description of the capillary tension and the associated concepts (e.g., capillary pressure) would require explicit treatment of the interface

(Gray and Hassanizadeh 1989). On the other hand, the averaging theories usually require drastic assumptions before obtaining macroscopic equations.

In another circle, a form of a hybrid mixture theory, which combines aspects of the classical mixture theories and averaging theories, has been pursued (Drew 1971; Ishii 1975; Nigmatulin 1979; Marle 1982; Ahmadi and Farshad 1974; Kalaydjian 1987; Hassanizadeh and Gray 1979a, b, 1980, 1990; Gray and Hassanizadeh 1989, 1990a,b). In this approach, the balance laws are developed at the microscopic level and averaged to obtain the corresponding macroscopic equations. The constitutive theories are directly developed at the macroscopic level. In addition, Hassanizadeh and Gray (1979a, b, 1980, 1990) and Gray and Hassanizadeh (1989, 1990a,b) also explicitly consider interfaces. A recent application of the hybrid mixture theory with interfaces may be found in Muraleetharan and Wei (1999).

The modern theory of porous media is a result of contributions from numerous researchers spanning more than two centuries. An extensive historical note and a description of various approaches may be found in De Boer (1996).

Most fully coupled problems are too complex to solve analytically. The finite element method has been used as the solution method in many cases (e.g., Ghaboussi and Wilson 1972; Sandhu and Wilson 1969; Selvadurai 1996a, b). Application to elasto-plastic materials may be found in Lewis and Schrefler (1998), Prevost (1980), Schrefler and Simoni (1995), Zienkiewicz et al. (1990a, b, 1999). The fully coupled method has been applied extensively to elasto-plastic analysis of earthquake problems (e.g., Zienkiewicz et al. 1990a, b, 1999; Anandarajah 1993a, 2000; Anandarajah et al. 1995).

The physical approaches employ variables and processes with clear physical meaning. The mixture theories provide a systematic method for deriving governing equations. The descriptions in this book are limited to physical approaches and mixture theories. The averaging theories and hybrid mixture theories with interfaces are beyond the scope of this book.

6.1.2 Skeleton and Fluid Stresses

As in previous chapters, tensile normal stresses in the solid skeleton (effective stresses) and the mixture (total stresses) are taken positive, but the tensile liquid and gas pressures are taken negative.

In this section, we develop a physical interpretation of various stresses that are involved in poromechanics (total and effective stresses, and liquid and gas pressures). For further details on similar treatments, the reader may refer to recent publications such as Lu and Likos (2004) and Lu (2008).

6.1.2.1 Matric Suction

From basic physics, we know that a liquid rises in a capillary tube due to surface tension as in Fig. 6.1a. The free-body diagram showing the vertical forces acting on

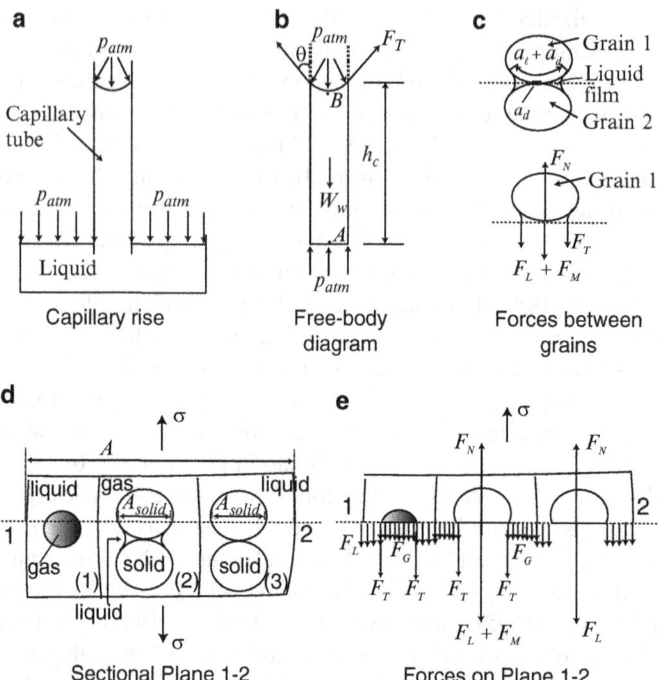

Fig. 6.1 Forces between grains and fluids in an unsaturated particulate

the liquid column is shown in Fig. 6.1b. $F_T = 2\pi r T_s$ is the force due to surface tension acting at an angle θ to vertical (contact angle), T_s is the surface tension force per unit length, $W_\ell = \pi r^2 h_c \rho_\ell g$ is the weight of the liquid column, h_c is the height of capillary rise, ρ_ℓ is the mass density of liquid, g is the gravitational acceleration, r is the radius of the tube and p_{atm} is the atmospheric pressure. By considering the equilibrium of forces in the vertical direction, it can be shown that

$$h_c = \frac{2T_s}{\rho_\ell g r} \cos\theta \tag{6.1}$$

For water/glass contact, $\theta \approx 0$. For example, when $r = 0.1\,\text{mm}$, noting that $T_s = 0.074\,\text{N/m}$, we have $h_c = 0.15\,\text{m}$. The absolute pressure at point A (Fig. 6.1b) is p_{atm}. The absolute pressure at point B (taken inside water) is

$$p_\ell = p_{atm} - h_c\gamma_w = p_{atm} - 0.15 \times 9.81\,\text{kPa} = (100 - 1.47)\,\text{kPa} = 98.53\,\text{kPa}$$

The quantity

$$\begin{aligned} p_c &= p_{atm} - p_\ell \\ &= p_{atm} - p_{atm} + h_c\gamma_w = h_c\gamma_w \end{aligned} \tag{6.2}$$

is known as the *capillary pressure* or the *matric suction*, which in this example is $0.15 \times 9.81 = 1.47\,\text{kPa}$. The radius at which p_ℓ will be zero is

$$p_\ell = p_{atm} - h_c\gamma_w = p_{atm} - \frac{2T_s}{r} = 0 \quad \Rightarrow \quad r = 0.0015\,\text{mm} = 1.5\,\mu\text{m}$$

The value of p_c at this radius is p_{atm}. In the capillary region (i.e., above the level where the absolute pressure is p_{atm}), $p_c \geq 0$. Equation (6.2) says that the value of p_c is independent of p_{atm}.

When $h_c\gamma_w = p_{atm}$, we have $p_\ell = 0$, which is the absolute pressure at point B. The relative pressure (i.e., relative to the atmospheric pressure) at this point is $p_\ell - p_{atm} = -p_{atm}$. In other words, the relative pressure in the liquid at point B is $-p_{atm}$, while the absolute pressure at this point is zero. When the tube's radius is less than 1.5 μm, the absolute pressure in the liquid becomes negative.

While some argue that water cannot carry tension (e.g., Koorevaar et al. 1983), others contend that water has a very high (several times the atmospheric pressure) tension-carrying capacity (e.g., Fisher 1948; Guan and Fredlund 1997). The matric suction is known to be very high in fine-grained soils like clays (e.g., $100\,p_{atm}$). Taylor (1948) leaves open the possibility of an unknown molecular adhesion being responsible for the high matric suction between clay particles. Lu and Likos (2004) use the concept of phase changes to explain the high tension-carrying capacity of water in contact with soil minerals. With this background on matric suction, let us now examine in detail the forces and stresses in an unsaturated particulate material such as soil.

6.1.2.2 Volume Fractions of Phases

Let us consider a case where the pores are filled with a liquid (e.g., water) and gas (e.g., air or liquid vapor). Let v_s, v_ℓ, and v_g be the volume fractions of the solid, liquid, and gas phases. Then

$$v_s + v_\ell + v_g = 1; \quad v_v = v_\ell + v_g \tag{6.3}$$

where v_v is the volume fraction of void. The parameters that are commonly used in soil mechanics are

$$S = \frac{v_\ell}{v_v}; \quad n = v_v; \quad e = \frac{v_v}{v_s};$$
$$n = \frac{e}{1+e}; \quad e = \frac{n}{1-n} \tag{6.4}$$

where S is degree of saturation, n is porosity, and e is void ratio. It can be verified that

$$v_s = 1 - n; \quad v_\ell = Sn; \quad \text{and} \quad v_g = (1 - S)n \tag{6.5}$$

6.1.2.3 Stresses

Let us first try and understand various forms of forces that act on an arbitrary plane running through a mixture and then define the corresponding stresses. In the absence of a fluid, the type of force that exists in the material is the grain-to-grain reaction force F_N (and forces such as the van der Waals attraction, which are not explicitly treated in this chapter). When the liquid is gradually introduced into the material, the nature of contacts between grains, liquid and gas changes. As the degree of (liquid) saturation increases, the gas phase changes from connected to disconnected, whereas the liquid phase changes from disconnected to connected (Wroth and Houlsby 1985). At an intermediate degree of saturation, both become connected.

When the liquid phase is disconnected (i.e., at low degree of saturation), the liquid film between the grains is separated from the gas phase by menisci (Fig. 6.1c) and consequently the grains are pulled together by surface tension and possibly other molecular forces. The liquid film between the grains may be capable of carrying very high tension. On the other hand, the tension-carrying capacity of free liquid is likely to be negligible. The typical forces between two grains are shown in Fig. 6.1c. The forces include the grain-to-grain reaction F_N (compression), surface tension force F_T (tension), molecular adhesion F_M (tension), and the free liquid force F_L. F_L is the force due to the liquid pressure at any point (e.g., points A and B shown in Fig. 6.1b) in the liquid. This could either be tensile or compressive.

At the other extreme, when the degree of saturation is high (say, $S > 70\%$), the grains will be completely surrounded by the liquid. The gas phase exists as occluded bubbles in the liquid phase. Pietruszczak and Pande (1996) considered such a system in their study on the development of a constitutive model for the system. In such a system, the relevant forces between the liquid and grains are the grain-to-grain reaction F_N and the free liquid force F_L. The relevant force between the occluded bubbles and liquid are the free liquid force F_L, gas force F_G (force due to the gas pressure) and liquid/gas surface tension force F_T.

In general, depending on the degree of saturation, the force system in an unsaturated particulate material may consist of one or more of the above forces. Here we present a qualitative analysis with the objective of providing physical interpretation for various components of the stresses in an unsaturated particulate material.

As shown in Fig. 6.1d, let us take an unsaturated material specimen and divide its domain into three parts: (1) gas-liquid part (occluded gas bubbles in connected liquid), (2) solid-gas-liquid part (one that occurs in disconnected liquid), and (3) solid-liquid part (one that occurs in connected liquid). Consider a sectional plane 1–2 passing through the specimen as shown in Fig. 6.1d. The specimen is subjected to an externally applied uniaxial tensile normal stress of σ. Considering the top half of the specimen, the free-body diagram of forces acting on this part is shown in Fig. 6.1e. By considering equilibrium of forces in the vertical direction, the external stress is related to the various internal forces by

$$\sigma = \frac{1}{A}\left[-\sum F_N + \sum F_G + \sum F_L + \sum F_T + \sum F_M\right] \tag{6.6}$$

A is the total cross sectional area of plane 1–2. A may be divided as $A = A_{solid} + A_{viod}$, where A_{solid} and A_{void} are the areas covering the solid phase and void, respectively. Noting that a portion of A_{void} is covered by liquid and the remaining portion is covered by gas, the corresponding area fractions may be related to the degree of saturation S as:

$$A_{void} = A_{viod}^{liquid} + A_{void}^{gas}; \quad A_{viod}^{liquid} = SA_{void}; \quad A_{void}^{gas} = (1-S)A_{void} \tag{6.7a}$$

In the above, the definition of the degree of saturation (which is a volumetric concept) is assumed to be applicable to the ratios of areas (Biot 1956; Schiffman1970).

The surface of each grain in general comprises three parts: (1) the part in contact with the liquid (a_ℓ indicated in Fig. 6.1c), (b) the part in contact with the gas (a_g, not shown in the figure), and (3) the part that makes the actual solid-solid to contact, where liquid or gas cannot enter (the “dry” area a_d indicated in Fig. 6.1c). Hence, the total surface area of a grain could be split up as

$$a = a_\ell + a_g + a_d = a_0 + a_d,$$

where

$$a_0 = a_\ell + a_g$$

When $a_d = 0$, following (6.7a), we can write

$$a_\ell = Sa_0; \quad a_g = (1-S)a_0$$

Let us introduce a reduction factor

$$0 \le \bar{\alpha} = \frac{a_\ell + a_g}{a} \le 1$$

to account for non-zero a_d. Then the above equations are modified to

$$a_\ell = \bar{\alpha}Sa_0; \quad a_g = \bar{\alpha}(1-S)a_0$$

Now referring to Fig. 6.1d, A_{solid} is sectional area of the grain in the mixture on plane 1–2. Let us assume that an identical decomposition can be used for A_{solid} as

$$A_{solid} = A_{solid}^{fluid} + A_{solid}^{dry}; \quad A_{solid}^{fluid} = A_{solid}^{liquid} + A_{solid}^{gas}$$

$$A_{solid}^{liquid} = \bar{\alpha} S A_{solid}; \quad A_{solid}^{gas} = \bar{\alpha}(1-S) A_{solid} \tag{6.7b}$$

Then on the sectional plane 1–2, the fractions covered by the liquid and gas phases may be written as

$$\frac{A^{gas}}{A} = \frac{A_{void}^{gas} + A_{solid}^{gas}}{A} = (1-S)\frac{A_{void} + \bar{\alpha} A_{solid}}{A} = \alpha(1-S);$$

$$\frac{A^{liquid}}{A} = \frac{A_{void}^{liquid} + A_{solid}^{liquid}}{A} = S\frac{A_{void} + \bar{\alpha} A_{solid}}{A} = \alpha S; \tag{6.7c}$$

where

$$0 \leq \alpha = \frac{A_{void} + \bar{\alpha} A_{solid}}{A} = n + \bar{\alpha}(1-n) \leq 1 \tag{6.8a}$$

and

$$\alpha = n + \bar{\alpha}(1-n); \quad 0 \leq \bar{\alpha} = \frac{\alpha - n}{1-n} \leq 1 \tag{6.8b}$$

Note that when $\bar{\alpha} = 1 \Rightarrow \alpha = 1$ and when $\bar{\alpha} < 1 \Rightarrow \alpha < 1$.

Rearranging (6.6):

$$\sigma = -\frac{\sum F_N}{A} + \frac{A^{gas}}{A}\frac{\sum F_{gas}}{A^{gas}} + \frac{A^{liquid}}{A}\left[\frac{\sum F_L}{A^{liquid}} + \frac{\sum F_M}{A^{liquid}} + \frac{\sum F_T}{A^{liquid}}\right] \tag{6.9}$$

Defining

$$\sigma' = -\frac{\sum F_N}{A}; \quad p_g = -\frac{\sum F_{gas}}{A^{gas}}; \quad p_\ell = -\left[\frac{\sum F_L}{A^{liquid}} + \frac{\sum F_M}{A^{liquid}} + \frac{\sum F_T}{A^{liquid}}\right] \tag{6.10}$$

Equation (6.9) is written as (making use of (6.7c)

$$\sigma = \sigma' - \alpha(1-S)p_g - \alpha S p_\ell$$

or

$$\sigma' = \sigma + \alpha p \tag{6.11a}$$

$$= \sigma + \alpha p_g - \alpha S(p_g - p_\ell) \tag{6.11b}$$

where

$$p = (1-S)p_g + S p_\ell \tag{6.11c}$$

The following observations are made:

1. For saturated materials, $S = 1, p_g = 0$, and hence (6.11b) becomes $\sigma' = \sigma + \alpha p_\ell$. This is the effective stress equation proposed by Terzaghi for saturated porous media (Terzaghi 1936) and became widely accepted over the years. α is called the effective area coefficient, which is found to be nearly unity for particulates such as soils. It can also be shown that $\alpha = 1 - (K_T/K_s)$ (Biot 1941; Biot and Willis 1957; Skempton 1960; Zienkiewicz 1982; Zienkiewicz et al. 1999), where K_T is the elastic bulk modulus of the solid skeleton, and K_s is the elastic bulk modulus of the solid phase. To show this, consider a specimen subjected to an external pressure Δp and an internal pore pressure of the same magnitude. Let $\Delta\varepsilon_v$ be the corresponding volumetric strain of the specimen. We express the volumetric strain in two different ways (using the skeleton elastic relation and using the solid phase elastic relation) and equate the two as follows:

   Using the skeleton elastic relation: with the aid of (6.11a), we have

$$\Delta\varepsilon_v = \frac{\Delta\sigma'_{kk}}{3K_T} = \frac{\Delta\sigma_{kk} + 3\alpha\Delta p}{3K_T} = \frac{3\Delta p + 3\alpha\Delta p}{3K_T} = (1+\alpha)\frac{\Delta p}{K_T}$$

   Using the solid phase elastic relation

$$\Delta\varepsilon_v = \frac{\Delta p}{K_s}$$

   Equating the two, we have

$$(1+\alpha)\frac{1}{K_T} = \frac{1}{K_s} \Rightarrow \alpha = 1 - \frac{K_T}{K_s} \tag{6.11d}$$

This expression for α is the same as the one used by Biot (1941). For materials such as soil, $K_T << K_s$ and hence $\alpha \approx 1$, whereas for materials such as rocks and concrete, the values of K_T and K_s are in the same order of magnitude, and α can be as low as 2/3 (Zienkiewicz et al. 1999).

2. Since $\alpha \approx 1$ for soils, for unsaturated soils, (6.11b) becomes,

$$\sigma' = \sigma + p_g - S(p_g - p_\ell)$$

This may be compared with the more general expression proposed by Bishop (1959):

$$\sigma' = \sigma + p_g - \chi(p_g - p_\ell) \tag{6.11e}$$

where χ is a function of S to be determined experimentally.

3. It follows from (6.10) that what is referred to as the liquid pressure is in general an average concept of pressure for the liquid incorporating the effects of free liquid pressure, capillary tension, and possible molecular adhesion. In other words, the liquid phase is highly heterogeneous. To model such heterogeneity, Hassanizadeh and Gray (Hassanizadeh and Gray 1979a, b, 1980, 1990) in a series of papers proposed theories that consider separately bulk phases and the interfaces between them. A detailed presentation of their theory is beyond the scope of this book.
4. When the degree of saturation is high (say greater than 70%), the solids are completely surrounded by liquid, and the gas phase exist in the form of disconnected bubbles. In this case, (6.10) is clearly an approximation since there is no basis for dividing $\sum F_T$ by A^{liquid}. Furthermore, $F_M \approx 0$ in this case, since the grains the completely surrounded by free liquid. The effective stress equation may be directly written from (6.9) in the form

$$\sigma' = \sigma + p_g - S(p_g - p_\ell) - p_T \tag{6.11f}$$

Where

$$p_g = -\frac{\sum F_{gas}}{A^{gas}},$$

$$p_\ell = \frac{A^{liquid}}{A}\frac{\sum F_L}{A^{liquid}}$$

and

$$p_T = -\frac{\sum F_T}{A}$$

p_ℓ is the pressure representing the bulk liquid and p_T is the pressure associated with the surface tension at the interface between the liquid phase and the gas bubbles. Pietruszczak and Pande (1996) showed that

$$p_T = \frac{2}{3}T_s\frac{\sqrt{1-S}}{r_v} \tag{6.11g}$$

where r_v is average pore size.

Also note that (6.6) may be modified to include the effects of physico-chemical forces such as the double-layer repulsive forces (Anandarajah and Lu 1992) and van der Waals attractive forces (Anandarajah and Chen 1997) that exist between cohesive soil minerals such as clay (Lu 2008).

In the analysis to be presented in the subsequent sections, we will use the following definition for the effective stress:

Uniaxial version:

$$\sigma' = \sigma + \alpha p; \quad p = x_g p_g + x_\ell p_\ell \tag{6.12a}$$

Multiaxial version:

$$\sigma'_{ij} = \sigma_{ij} + \alpha p \delta_{ij}; \quad p = x_g p_g + x_\ell p_\ell \tag{6.12b}$$

Generalizing Bishop's (1959) equation, the coefficients x_g and x_λ are:

$$x_g = 1 - \chi; \quad x_\ell = \chi \tag{6.12c}$$

Using the definition of Zienkiewicz et al. (1999)

$$\alpha = 1 - \frac{K_T}{K_s} \tag{6.12d}$$

The matric suction is defined as

$$p_c = p_g - p_\ell \tag{6.12e}$$

6.1.2.4 Soil–Water Retention Relationship

It has been experimentally shown (Fredlund and Rahardjo 1993) that p_c correlates with S reasonable well, although not uniquely. When a fully saturated soil is gradually dried ("drying") or drained, p_c increases with decreasing S, typically following a curve such as Curve A shown in Fig. 6.2. The $p_c - S$ relationship is known as the water-retention curve (also as soil-water characteristic curve or SWCC). When S is again increased ("wetting"), p_c decreases, but the $p_c - S$ relationship follows a different curve such as Curve B shown in Fig. 6.2. This is known as hysteresis. Factors such as heterogeneous pore size distribution, capillary condensation, occluded air bubbles, fabric changes, and contact angle changes have been cited as possible causes (Likos and Lu 2004; Li 2004). In view of this hysteresis, p_c is not a unique function of S. Assuming that appropriate measures can be taken to account for hysteresis, the following approximate relation is assumed:

$$p_c = \hat{p}_c(S) \tag{6.13}$$

6.1.3 Viscous Drag on Solid Skeleton

Let us first consider water flow through saturated soil and develop certain equations. The equations are applicable to any liquid flowing through a porous material. Let us

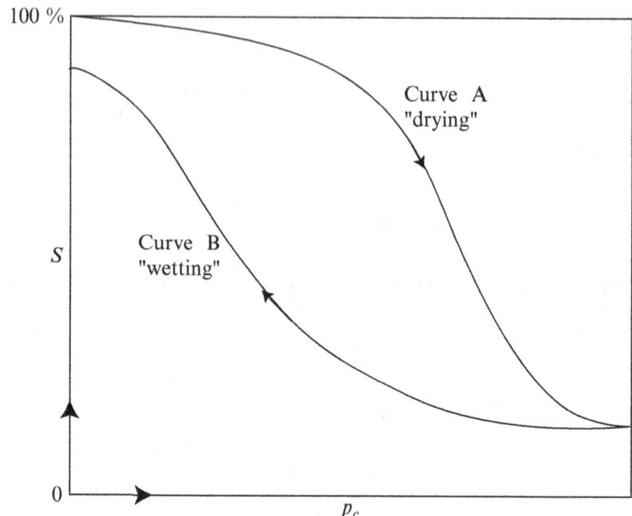

Fig. 6.2 Water-retention curve

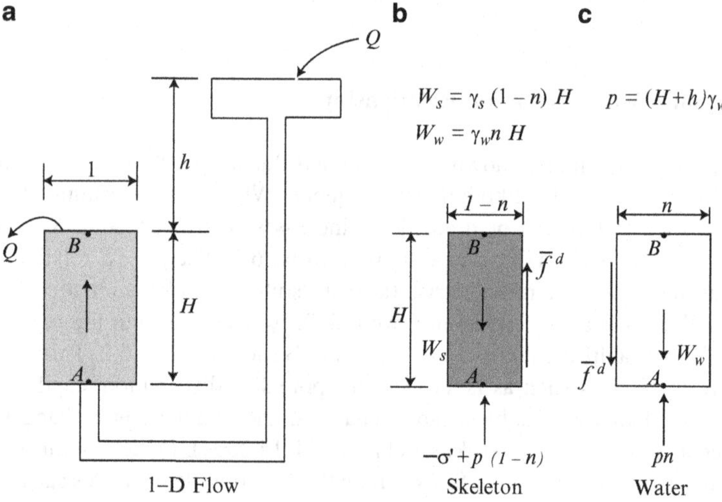

Fig. 6.3 One-dimensional flow of liquid through material

neglect the inertial effects. When the water flows through the soil specimen, it imparts a viscous drag force on the solid skeleton. This is known as the seepage force in the discipline of flow through porous media. To illustrate this, let us first consider a simple one-dimensional flow arrangement shown in Fig. 6.3a. The tube on the left is filled with saturated soil, with the lower end (point A) connected to a constant-head water supply line and the upper end (point B) is left open so that water can freely flow out of the tube. Given sufficient length of time, a steady-state flow will develop with water flowing from point A to point B.

Taking the datum through point A and neglecting the velocity head, the total hydraulic heads (pressure + elevation) at points A and B are $(H+h)\gamma_w$ and $H\gamma_w$, respectively, where $\gamma_w = \rho_w g$ is the density of water. As the head at point A is higher than the head at point B, water flows from point A to point B. The effective stress at point A is

$$\sigma' = \sigma + p = -[\gamma_{sat}H - (H+h)\gamma_w] = -[\gamma' H - \gamma_w h] \tag{6.14}$$

where γ_{sat} and $\gamma' = \gamma_{sat} - \gamma_w$ are the (total) saturated density and buoyant density of the soil, respectively. The effective compressive stress is positive at point A. (Note the use of + sign between σ and p in the above equation. This is due to the use of opposite sign conventions for the two stresses.) Hence, as h increases from zero, the effective compressive stress at point A decreases. This is true for any point in the soil column. The reason for the decrease in the effective compressive stress is that as the water flows around the soil grains, it exerts a viscous drag force on the grains and hence on the soil skeleton. This decreases the intergranular force (i.e., F_N shown in Fig. 6.1c, d).

To determine the magnitude of this drag force, let us develop free-body diagrams for the skeleton and water phase separately. These diagrams are schematically shown in Fig. 6.3b and c, respectively. The viscous drag force $\bar{f}^d$ is introduced as an interface force on both the skeleton and water phases. Noting that the cross sectional area of the column (Fig. 6.3a) is taken as unity, the areas covered by the solid and water phases are $1-n$ and n, respectively. At point A, the skeleton (Fig. 6.3b) is subjected to effective stress and water pressure. σ' is defined by dividing the inter-granular force by A and p is defined by dividing the fluid force by A^{liquid} (6.10). The forces acting on the skeleton due to effective stress and water pressure are then obtained by multiplying the corresponding stresses by 1 and $1-n$, respectively (Fig. 6.3b). The force on the water phase (Fig. 6.3c) is obtained by multiplying the water pressure by n. The expression for F_d can be obtained by either considering equilibrium of the skeleton (which is the question in Problem 6.2) or the water phase; the latter is done here as follows.

$$\bar{f}^d = pn - W_w = (H+h)\gamma_w n - Hn\gamma_w = h\gamma_w n = \left(\frac{h}{H}\gamma_w n\right)H = (i\gamma_w n)H \tag{6.15a}$$

where $i = h/H$ is the hydraulic gradient between points A and B. Now according to Darcy's (1856) law,

$$\dot{w}^{ave} = ik \tag{6.15b}$$

where $\dot{w}^{ave}$ is the average velocity of water flow (units: length/time) defined as

$$\dot{w}^{ave} = \frac{Q}{A} = \frac{Q}{1} = Q \tag{6.15c}$$

Q is the volumetric rate of water flow through the column, and k is the permeability (units: length/time). Using (6.15b), (6.15a) can be written as

$$\bar{f}^d = \frac{\dot{w}^{ave} nH}{(k/\gamma_w)} = n\bar{k}^{-1}\dot{w}^{ave}H \tag{6.15d}$$

where $\bar{k} = k/\gamma_w$. The drag force per unit volume is

$$f^d = \frac{\bar{f}^d}{H \times 1} = n\bar{k}^{-1}\dot{w}^{ave} \tag{6.15e}$$

The true velocity (also known as seepage velocity or intrinsic velocity) of the water particles $\dot{u}_w$ can be found by the principle of mass conservation. Assuming that the density of water does not change as the water flows through the soil:

$$Q = \dot{w}^{ave}A = \dot{u}_w A_w \quad \Rightarrow \quad \dot{w}^{ave} = \frac{A_w}{A}\dot{u}_w = n\dot{u}_w \tag{6.15f}$$

where A_w is the cross sectional area covered by the water phase. If the solid particles are also moving, then (6.15f) is modified as

$$\dot{w}^{ave} = n(\dot{u}_w - \dot{u}_s) = n\dot{w}; \quad \dot{w} = \dot{u}_w - \dot{u}_s \tag{6.15g}$$

where $\dot{u}_s$ is the intrinsic velocity of the solid particles and $\dot{w}$ is the intrinsic relative velocity between the liquid and solid phases.

The corresponding relations for an unsaturated soil are (Problem 6.3)

$$\dot{w}^{ave} = Sn(\dot{u}_w - \dot{u}_s) = Sn\dot{w} \tag{6.16a}$$

$$f^d = nS\bar{k}^{-1}\dot{w}^{ave} = n^2S^2\bar{k}^{-1}\dot{w} \tag{6.16b}$$

Equations (6.15b), (6.15g), (6.16a), and (6.16b) are generalized to a *multiaxial flow of any liquid* as

$$\dot{w}_i^{ave} = Sn(\dot{u}_i^\ell - \dot{u}_i^s) = Sn\dot{w}_i; \quad \dot{w}_i = \dot{u}_i^\ell - \dot{u}_i^s \tag{6.16c}$$

$$\dot{w}_i^{ave} = k_{ij}i_j \tag{6.16d}$$

$$f_i^d = nS\bar{k}_{ij}^{-1}\dot{w}_j^{ave} = n^2S^2\bar{k}_{ij}^{-1}\dot{w}_j; \quad \bar{k}_{ij} = \frac{k_{ij}}{\gamma_\ell} \tag{6.16e}$$

The superscripts ℓ and s denote "liquid" and "solid" phases, respectively. Equation (6.16d) is Darcy's law generalized to a multiaxial flow, with k_{ij} denoting the permeability tensor and i_j denoting the hydraulic gradient in direction j. i_j is dimensionless and hence the coefficients of k_{ij} have a unit of velocity. $\bar{k}_{ij}$ is obtained by normalizing k_{ij} with respect to the fluid density γ_ℓ.

6.1.4 Balance Principles: Single-Phase Materials

The key principles of continuum mechanics that are relevant to mechanics of all materials are the principles of the conservation of mass, balance of linear momentum, balance of moment of momentum (angular momentum), first law of thermodynamics (conservation of energy) and second law of thermodynamics (entropy production inequality).

The first law of thermodynamics encapsulates the principle of energy conservation and postulates interconvertibility of thermal and mechanical energies. The second law of thermodynamics provides a criterion for irreversible processes. The laws of thermodynamics provide a systematic means of identifying pertinent state variables and formulating constitutive laws. For example, the structure of Darcy's law can be derived by the application of the principles of thermodynamics (Coussy 1995). A detailed treatment of the topic is beyond the scope of this book.

In this section, we consider a single phase material and present the details of the principles of the conservation of mass, balance of linear momentum and balance of angular momentum. In Sect. 6.2, we will extend these principles to a porous media. To achieve these, some description of the kinematics of continua is necessary.

Referring to Fig. 6.4, a body having a volume V_0 at $t = 0$ (initial or undeformed configuration) deforms to one having a volume V at any time t (current or deformed configuration). A material point ("particle") at point A now occupies a point at point B. The vector representing the position of point A is $\mathbf{X}$ ("material coordinate") and that representing point B is $\mathbf{x}$ ("spatial coordinate"). (Note that the position of point B may be represented with respect to a moving coordinate system if desired).

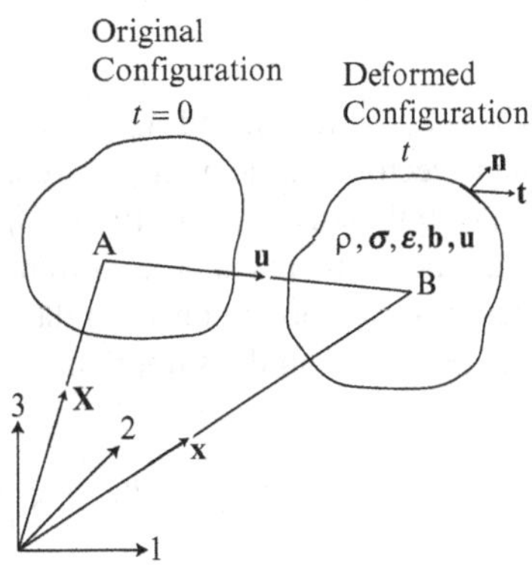

Fig. 6.4 Original and deformed configurations

The motion of any material point such as point A may then be represented either in terms of the material coordinate as

$$\mathbf{x} = \mathbf{x}(\mathbf{X}, t) \tag{6.17a}$$

or in terms of the spatial coordinate as

$$\mathbf{X} = \mathbf{X}(\mathbf{x}, t) \tag{6.17b}$$

The former (6.17a) is referred to as Lagrangian or material description and the latter (6.17b) is referred to as the Eulerian or spatial description. All other relevant quantities may similarly be given either a Lagrangian or an Eulerian description; for example, the mass density may be described as

$$\rho = \rho(\mathbf{X}, t) - \text{Lagrangian} \tag{6.18a}$$

or

$$\rho = \rho(\mathbf{x}, t) - \text{Eulerian} \tag{6.18b}$$

Let us now define a few related quantities:

$$F_{ij} = \frac{\partial x_i}{\partial X_j},\ \mathrm{d}x_i = F_{ik}\mathrm{d}X_k\ (\text{or}\ \mathrm{d}\mathbf{x} = \mathbf{F}\mathrm{d}\mathbf{X}) \tag{6.19a}$$

$$J = |F_{ij}| \tag{6.19b}$$

$$\mathbf{u} = \mathbf{x} - \mathbf{X} \tag{6.19c}$$

$$E_{ij} = \frac{1}{2}\left(F_{ki}F_{kj} - \delta_{ij}\right) \quad \text{or} \quad \mathbf{E} = \frac{1}{2}\left(\mathbf{F}^{\mathrm{T}}\mathbf{F} - \boldsymbol{\delta}\right) \tag{6.19d}$$

$$A_{ij} = \frac{1}{2}\left(\delta_{ij} - F_{ik}F_{jk}\right) \tag{6.19e}$$

where F_{ij}, J, $\mathbf{u}$, E_{ij}, and A_{ij} are known as deformation gradient, Jacobian, displacement vector, Lagrangian or Green finite strain tensor, and Eulerian or Almansi's finite strain tensor, respectively. δ_{ij} is the Kronecker delta. $\mathrm{d}\mathbf{x}$ and $\mathrm{d}\mathbf{X}$ are infinitesimal line elements in the current and original configurations, respectively. The condition for obtaining the relationship given by (6.17b) by inverting (6.17a) is that the Jacobian does not vanish. It can be shown (Problem 6.4) that

$$E_{ij} = \frac{1}{2}\left[\frac{\partial u_i}{\partial X_j} + \frac{\partial u_j}{\partial X_i} + \frac{\partial u_k}{\partial X_i}\frac{\partial u_k}{\partial X_j}\right] \tag{6.20a}$$

When $|F_{ij}|<<1$ (6.19a), the last term in (6.20a) is negligible compared to the first two terms, and

$$E_{ij} \approx \frac{1}{2}\left[\frac{\partial u_i}{\partial X_j}+\frac{\partial u_j}{\partial X_i}\right] = \varepsilon_{ij} \tag{6.20b}$$

where ε_{ij} is the infinitesimal strain tensor. When in addition to the elements of the deformation gradient, the displacements are also small, the Lagrangian, Eulerian, and the infinitesimal strain tensors are all approximately equal to each other. The corresponding theories are known as the small-deformation theories (the kind treated in this book).

Let us now define a few rate quantities that are needed for the understanding of the balance laws to be presented. The time derivative of a quantity $\boldsymbol{\phi}$ (tensor of any order) given in Lagrangian description, i.e., $\boldsymbol{\phi}$,($\mathbf{X}$, t), is straightforward as

$$\frac{\mathrm{d}\boldsymbol{\phi}}{\mathrm{d}t} = \frac{\partial \boldsymbol{\phi}(\mathbf{X},t)}{\partial t} \tag{6.21}$$

The derivative is obtained with $\mathbf{X}$ held constant, i.e., the derivative is defined for the same particle. The derivative is known as the *material derivative*. Applying (6.21) to $\mathbf{x}$:

$$\frac{\mathrm{d}\mathbf{x}}{\mathrm{d}t} = \frac{\partial \mathbf{x}(\mathbf{X},t)}{\partial t} = \dot{\mathbf{x}} = \mathbf{v} \tag{6.22a}$$

where $\mathbf{v}$ is the velocity of the particle. *Note that a dot over a variable will be used in this chapter to denote material derivatives*. The acceleration is obtained from

$$\frac{\mathrm{d}\mathbf{v}}{\mathrm{d}t} = \frac{\partial \mathbf{v}(\mathbf{X},t)}{\partial t} = \dot{\mathbf{v}} = \mathbf{a} \tag{6.22b}$$

When $\boldsymbol{\phi}$ is given in Eulerian description ($\boldsymbol{\phi} = \boldsymbol{\phi}(\mathbf{x}, t)$), the material derivative is

$$\dot{\boldsymbol{\phi}} = \frac{D\boldsymbol{\phi}(\mathbf{x},t)}{Dt} = \frac{\partial \boldsymbol{\phi}(\mathbf{x},t)}{\partial t} + \frac{\partial \boldsymbol{\phi}(\mathbf{x},t)}{\partial \mathbf{x}}\frac{\partial \mathbf{x}}{\partial t} \tag{6.23a}$$

Note that for emphasis, $D(.)/Dt$ is used to represent material derivatives. The first term is obtained by holding $\mathbf{x}$ constant and hence is known as the *local rate of change*. The second term represents the contribution due to the change of position of the particle in space and is known as the *convective rate of change*. When the flow is steady, the quantity $\boldsymbol{\phi}$ at a given $\mathbf{x}$ does not change, and hence the local rate of change is zero. The convective rate of change is zero for a uniform flow.

By making use of (6.22a), (6.23a) may be written as

$$\dot{\boldsymbol{\phi}} = \boldsymbol{\phi}_{,t} + \boldsymbol{\phi}_{,i} v_i \tag{6.23b}$$

where note that a compact notation is employed for derivatives ($\partial\phi/\partial t = \phi_{,t}$, etc.). As in previous chapters, repeated indices imply summation. The acceleration is the material time derivative of velocity and hence is (from (6.23b))

$$\mathbf{a} = \dot{\mathbf{v}} = \mathbf{v}_{,t} + v_{,i}v_i \tag{6.23c}$$

It is seen that even for a steady flow (for which $\mathbf{v}_{,t} = 0$), the acceleration may not be zero due to the presence of the convective term in (6.23c).

Now let us define a few other important rates as follows:

$$\dot{F}_{ij} = \frac{\partial}{\partial t}\left(\frac{\partial x_i}{\partial X_j}\right)\bigg|_{\mathbf{X}\ fixed} = \frac{\partial}{\partial X_j}\left(\frac{\partial x_i}{\partial t}\right)\bigg|_{\mathbf{X}\ fixed} = \frac{\partial v_i}{\partial X_j} = \frac{\partial v_i}{\partial x_k}\frac{\partial x_k}{\partial X_j} = v_{i,k}F_{kj} \tag{6.24a}$$

or

$$\dot{\mathbf{F}} = \mathbf{LF}; \quad L_{ij} = v_{i,j} \tag{6.24b}$$

where L_{ij} is known as the velocity gradient. $\mathbf{L}$ may be written as

$$\mathbf{L} = \frac{1}{2}(\mathbf{L} + \mathbf{L}^{\mathrm{T}}) + \frac{1}{2}(\mathbf{L} - \mathbf{L}^{\mathrm{T}}) = \mathbf{D} + \mathbf{W} \tag{6.24c}$$

$$\mathbf{D} = \frac{1}{2}(\mathbf{L} + \mathbf{L}^{\mathrm{T}}) \tag{6.24d}$$

$$\mathbf{W} = \frac{1}{2}(\mathbf{L} - \mathbf{L}^{\mathrm{T}}) \tag{6.24e}$$

$\mathbf{D}$ is known as the rate of deformation or stretching tensor and $\mathbf{W}$ is known as the vorticity or spin tensor. It can be verified that $\mathbf{D}$ and $\mathbf{W}$ are symmetric and anti-symmetric, respectively. Taking the rate of $\mathbf{E}$ (6.19d), it can be shown (Problem 6.5) that

$$\dot{\mathbf{E}} = \mathbf{F}^{\mathrm{T}}\mathbf{DF} \tag{6.25}$$

Using these results, it can be shown that the material derivative of the Jacobian (Problem 6.6) and the volume element in the current configuration (Problem 6.7) are given by

$$\dot{J} = Jv_{i,i} \tag{6.26a}$$

$$\frac{\mathrm{d}(\mathrm{d}v)}{\mathrm{d}t} = v_{i,i}\mathrm{d}v \tag{6.26b}$$

The kinematics necessary to establish the balance laws are now available. In the following, we will describe the balance of mass and momentum principles. It is assumed that

- The material is a single-phase material.
- The phase changes do not occur, i.e., the original particles remain as is, though they move and deform.

6.1.4.1 Conservation of Mass

The mass of a differential volume element at time t is $\mathrm{d}m=\rho\mathrm{d}v$, where ρ is the mass density and $\mathrm{d}v$ is the differential volume element in the current configuration. The principle of the conservation of mass states that mass cannot be destroyed. This is equivalent to stating that the rate of change of mass must be zero, i.e.,

$$\frac{\mathrm{d}m}{\mathrm{d}t}=\frac{\mathrm{d}}{\mathrm{d}t}\int \mathrm{d}m=\frac{\mathrm{d}}{\mathrm{d}t}\int \rho\mathrm{d}v=0 \tag{6.27a}$$

Expanding this:

$$\int\left[\dot{\rho}\mathrm{d}v+\rho\frac{\mathrm{d}(\mathrm{d}v)}{\mathrm{d}t}\right]=0 \tag{6.27b}$$

where $\dot{\rho}$ is the material derivative of ρ. Using the result in (6.26b)

$$\int\left[\dot{\rho}+\rho v_{i,i}\right]\mathrm{d}v=0 \tag{6.27c}$$

From which the local form of the mass conservation law is written as

$$\dot{\rho}+v_{i,i}\rho=0 \tag{6.28}$$

6.1.4.2 Balance of Linear Momentum

Based on Newton's laws (second and third laws), the principle of balance of linear momentum states that the rate of change of linear momentum must be equal to the resultant forces acting on any body. Applying this to the continuum shown in Fig. 6.4 at time t:

$$\int_{\Gamma}\mathbf{t}\,\mathrm{d}s+\int_{\Omega}\rho\mathbf{b}\,\mathrm{d}v=\frac{\mathrm{d}}{\mathrm{d}t}\int_{\Omega}\rho\dot{\mathbf{u}}\,\mathrm{d}v \tag{6.29a}$$

where $\mathbf{t}$ is the surface traction at any point where the unit normal to the boundary is $\mathbf{n}$, and $\mathbf{b}$ is the body force (force/mass). (Note that in some of the previous chapters, the body force was defined as force per unit volume.) $\boldsymbol{\Gamma}$ and $\boldsymbol{\Omega}$ represent the boundary and volume of the domain, respectively. The traction is related to the internal stress (more precisely, to the Cauchy stress, which is defined in the current configuration) σ_{ij} as

$$t_i = \sigma_{ij} n_j \tag{6.29b}$$

By making use of (6.29b) and applying the Gauss theorem to the first term and carrying out the differentiation on the right-hand side, (6.29a) becomes

$$\int_{\Omega} \left[\sigma_{ij,j} + \rho b_i\right] \mathrm{d}v = \int_{\Omega} \rho \ddot{u}_i \mathrm{d}v + \int_{\Omega} \dot{u}_i \frac{\mathrm{d}}{\mathrm{d}t}(\rho \mathrm{d}v) \tag{6.29c}$$

By the mass conservation principle (6.28), the last term in (6.29c) equals zero, resulting in

$$\int_{\Omega} \left[\sigma_{ij,j} + \rho b_i - \rho \ddot{u}_i\right] \mathrm{d}v = 0 \tag{6.29d}$$

This leads to the following local form for the balance of linear momentum

$$\sigma_{ij,j} + \rho b_i - \rho \ddot{u}_i = 0 \tag{6.30}$$

where $\ddot{u}_i$ is the acceleration of the material point. Equation (6.30) is also known as the differential equation governing equilibrium.

6.1.4.3 Balance of Angular Momentum

The angular momentum principle (also called the moment of momentum principle) states that for any body, the rate of change of angular momentum must be equal to the resultant moment of forces acting on the body about an arbitrary point. Referring to Fig. 6.4, the mathematical statement of the principle is

$$\int_{\Gamma} \mathbf{x} \times \mathbf{t}\, \mathrm{d}s + \int_{\Omega} \mathbf{x} \times \rho \mathbf{b}\, \mathrm{d}v = \frac{\mathrm{d}}{\mathrm{d}t} \int_{\Omega} \mathbf{x} \times \rho \dot{\mathbf{u}}\, \mathrm{d}v \tag{6.31a}$$

where "×" represents cross product. For convenience, the moment is taken about the origin. The last term is expanded as

$$\frac{\mathrm{d}}{\mathrm{d}t}\int_{\Omega}\mathbf{x}\times\rho\dot{\mathbf{u}}\mathrm{d}v=\int_{\Omega}\dot{\mathbf{x}}\times\rho\dot{\mathbf{u}}\mathrm{d}v+\int_{\Omega}\mathbf{x}\times\dot{\mathbf{u}}\frac{\mathrm{d}}{\mathrm{d}t}(\rho\mathrm{d}v)+\int_{\Omega}\mathbf{x}\times\rho\ddot{\mathbf{u}}\mathrm{d}v \tag{6.31b}$$

Noting that $\dot{\mathbf{x}}=\dot{\mathbf{u}}$, and $\dot{\mathbf{u}}\times\dot{\mathbf{u}}=0$, the first term is zero. The second term is also zero by the mass conservation principle (6.28). The cross product between two vectors can be expressed with the aid of the alternating tensor ε_{ijk} (refer to Chap. 2 and Problem 6.6). With the aid of (6.31b), (6.31a) becomes

$$\int_{\Gamma}\varepsilon_{ijk}x_jt_k\,\mathrm{d}s+\int_{\Omega}\varepsilon_{ijk}x_jb_k\rho\,\mathrm{d}v=\int_{\Omega}\varepsilon_{ijk}x_j\ddot{u}_k\rho\,\mathrm{d}v \tag{6.31c}$$

By applying the Gauss theorem (Chap. 2) and making use of (6.29b), the first term in (6.31c) becomes

$$\begin{aligned}\int_{\Gamma}\varepsilon_{ijk}x_jt_k\,\mathrm{d}s&=\int_{\Gamma}\varepsilon_{ijk}x_j\sigma_{k\ell}n_\ell\,\mathrm{d}s=\int_{\Omega}\left[\varepsilon_{ijk}x_j\sigma_{k\ell}\right]_{,\ell}\mathrm{d}v\\&=\int_{\Omega}\varepsilon_{ijk}x_j\sigma_{k\ell,\ell}\,\mathrm{d}v+\int_{\Omega}\varepsilon_{ijk}\delta_{j\ell}\sigma_{k\ell}\,\mathrm{d}v\\&=\int_{\Omega}\varepsilon_{ijk}x_j\sigma_{k\ell,\ell}\,\mathrm{d}v+\int_{\Omega}\varepsilon_{ijk}\sigma_{kj}\,\mathrm{d}v\end{aligned} \tag{6.31d}$$

Combining (6.31c) and (6.31d) and further simplifying:

$$\int_{\Omega}\varepsilon_{ijk}\left[\sigma_{k\ell,\ell}+\rho b_k-\rho\ddot{u}_k\right]x_j\,\mathrm{d}v-\int_{\Omega}\varepsilon_{ijk}\sigma_{kj}\,\mathrm{d}v=0 \tag{6.31e}$$

By (6.30), the first term is zero. The local form of the second term is

$$\varepsilon_{ijk}\sigma_{kj}=0 \tag{6.31f}$$

Now we make use of another property of ε_{ijk}, i.e., $\varepsilon_{ijk}=\varepsilon_{irs}\varepsilon_{ijk}=\delta_{rj}\delta_{sk}-\delta_{rk}\delta_{sj}$ (Chap. 2). Equation (6.31e) then leads to

$$\sigma_{sr}=\sigma_{rs} \tag{6.32}$$

The final outcome of the angular momentum principle is that the Cauchy stress is symmetric.

6.2 Balance Principles for Unsaturated Porous (Three-Phase) Materials: Full-Formulation

6.2.1 Introduction

We are interested in a three-phase material involving solid, liquid, and gas. The definitions of relevant parameters are first presented. Some parameters have already been defined, but they are again formally defined in this section. The mass conservation and the balance of linear momentum principles are applied individually for the three phases. The corresponding equations for the mixture are derived by summing up the individual equations for the separate phases. The angular momentum equations will not be derived; instead the stress tensor is assumed to be symmetric (6.32).

In the balance equations presented in Sect. 6.1, a control volume V_0 was considered (Fig. 6.4). The material points in this control volume move over to the control volume V at time t. The material points that are in V_0 are still contained in V. This is not true in porous media theories since the fluid keeps flowing relative to the solid skeleton. Thus, the control volume considered for the analysis only concerns the solid skeleton, and the fluid keeps flowing in and out of this control volume. In the absence of phase changes (e.g., conversion of solid phase into the liquid or gas phase), the material points representing the solid skeleton in control volume V_0 are still the same as that in V; this is not true for the fluid phases (De Boer 1996). The phase changes are allowed, but such changes must be appropriately accounted for.

Let ΔV be the smallest possible volume around a point such that an average of the values of a variable within ΔV can be considered to represent the variable at that point for the purpose of macroscopic analyses. Such a volume, known as Representative Elementary Volume (REV), is the definition of a material point. If the material, for instance, is soil, REV must be large enough to contain sufficient number of grains. If the material is concrete or rock, REV must be large enough to contain sufficient number of pore structure details. Let ΔV_s, ΔV_ℓ, and ΔV_g be volumes of solid, liquid, and gas phases in ΔV. The volume fractions are formally defined as

$$v_s = \frac{\Delta V_s}{\Delta V}; \quad v_\ell = \frac{\Delta V_\ell}{\Delta V}; \quad v_g = \frac{\Delta V_g}{\Delta V} \tag{6.33a}$$

where v_s, v_ℓ, and v_g are volume fractions of solid, liquid, and gas, respectively. It follows from (6.33a) that

$$v_s + v_\ell + v_g = 1 \tag{6.33b}$$

As stated earlier (6.3), the following relations hold

$$v_s = 1 - n; \quad v_\ell = Sn \quad \text{and} \quad v_g = (1 - S)n \tag{6.33c}$$

The definitions of S and n are defined in (6.4).

In the modern macroscopic theory of porous media, we ignore the exact details of the microscopic structure of the individual phases, and assume that all phases exist in the REV simultaneously in a homogeneous manner.

Besides S and n, the key variables involved in the problem are density, displacement, body force, stress (Fig. 6.4), and a few others that are related to these quantities (e.g., velocity, surface traction, etc.).

The body forces (force/mass) are denoted by $\mathbf{b}^s$, $\mathbf{b}^\ell$, and $\mathbf{b}^g$ for solid, liquid, and gas phases, respectively. Note that in most cases (e.g., gravity, centrifugal force, etc.) $\mathbf{b}^s = \mathbf{b}^\ell = \mathbf{b}^g = \mathbf{b}$, but we will keep the analysis general at this point.

The motion of a material point representing a given phase is defined in an average sense. For example, if we focus on the liquid in the REV, it is clear that the liquid particles (i.e., the liquid molecules) are moving in different directions and by different amounts at different points. The same is in general true for the solid phase as well. For example, in soil, the solid skeleton consists of an assemblage of individual grains. Each individual grains in general are moving in different directions and by different amounts at different points. In mixture theories, we ignore these details, and define an average motion for each material point (Truesdell 1965). For example, the solid phase material occupying the spatial position at $\mathbf{X}^s$ at time $t = 0$ (i.e., in the reference configuration) occupies a spatial position $\mathbf{x}^s$ at any time t (i.e., in the current configuration), where

$$\mathbf{x} = \mathbf{x}^s(\mathbf{X}^s, t)$$

As done in Sect. 6.1.4 for a single-phase material, the kinematical variables can then be defined based on the above equation. For example, the displacement can be defined by (6.19c) for the solid, liquid, and gas phases separately as

$$\mathbf{u}^s = \mathbf{x}^s(\mathbf{X}^s, t) - \mathbf{X}^s; \quad \mathbf{u}^\ell = \mathbf{x}^\ell(\mathbf{X}^\ell, t) - \mathbf{X}^\ell; \quad \mathbf{u}^g = \mathbf{x}^g(\mathbf{X}^g, t) - \mathbf{X}^g$$

For completeness, we repeat (6.16c):

$$\dot{\mathbf{w}}^{ave} = Sn(\dot{\mathbf{u}}^\ell - \dot{\mathbf{u}}^s) = Sn\dot{\mathbf{w}}; \quad \dot{\mathbf{w}} = \dot{\mathbf{u}}^\ell - \dot{\mathbf{u}}^s \tag{6.34}$$

where $\dot{\mathbf{w}}^{ave}$ is the macroscopic average velocity of liquid flow (also known as superficial velocity, Whitaker 1999) and $\dot{\mathbf{w}}$ is the relative velocity between the liquid and solid phases. Analogous quantities can be defined for the gas phase. $\dot{\mathbf{w}}$ is also known as the seepage velocity (Taylor 1948).

The definitions of the remaining variables are discussed below; a summary of the final results are presented in Table 6.1. For the mass densities and stresses, we will define two types: (a) *Intrinsic* and (b) *macroscopic average*. The former are microscopic averages defined for each of the phases and the latter are macroscopic averages. For example, considering a volume of soil consisting of quartz particles, whose specific gravity is approximately 2.7, the intrinsic mass density is 2.7 γ_w/g, where γ_w is the density of water and g is the gravitational acceleration. We will define the macroscopic average in the following.

Table 6.1 Intrinsic and macroscopic averages relevant to the three-phase problem

Var. type	Intrinsic/ average	Solid	Liquid	Gas	Mixture
Density	Intrinsic	ρ_s	ρ_l	ρ_g	
	Average	$\bar{\rho}_s = v_s \rho_s = (1-n)\rho_s$	$\bar{\rho}_\ell = v_\ell \rho_\ell = Sn\rho_\ell$	$\bar{\rho}_g = v_g \rho_g = (1-S)n\rho_g$	$\bar{\rho} = v_s\rho_s + v_s\rho_s + v_s\rho_s = (1-n)\rho_s + Sn\rho_\ell + (1-S)n\rho_g$
Stress	Intrinsic	$\boldsymbol{\sigma}^s = \frac{\sigma'}{(1-n)} - \frac{p(\alpha-n)}{(1-n)}$	$\boldsymbol{\sigma}^\ell = -p_\ell \boldsymbol{\delta}$	$\boldsymbol{\sigma}^g = -p_g \boldsymbol{\delta}$	
	Average	$\bar{\boldsymbol{\sigma}}^s = v_s^* \boldsymbol{\sigma}^s = \boldsymbol{\sigma}' - p(\alpha-n)\boldsymbol{\delta}$	$\bar{\boldsymbol{\sigma}}^\ell = v_\ell^* \boldsymbol{\sigma}^\ell = -\chi n p_\ell \boldsymbol{\delta}$	$\bar{\boldsymbol{\sigma}}^g = v_g^* \boldsymbol{\sigma}^g = -(1-\chi) n p_g \boldsymbol{\delta}$	$\bar{\boldsymbol{\sigma}} = v_s^* \bar{\boldsymbol{\sigma}}^s + v_\ell^* \bar{\boldsymbol{\sigma}}^\ell + v_g^* \bar{\boldsymbol{\sigma}}^g = \boldsymbol{\sigma}' - \alpha p \boldsymbol{\delta}$

Let us first consider the mass density. Within an REV, the intrinsic density itself could vary from point to point, and hence must be defined in an average sense. Let ρ_s, ρ_ℓ, and ρ_g be average intrinsic mass densities of solid, liquid, and gas phases within the REV. For example, if the material is a soil consisting of two types of grains, one with a mass density of ρ_{s1} and another with a mass density of ρ_{s2}, ρ_s will be calculated as

$$\rho_s = \frac{\rho_{s1}\Delta V_{s1} + \rho_{s2}\Delta V_{s2}}{\Delta V_s}$$

where ΔV_{s1} and ΔV_{s2} are volumes of grain type 1 and grain type 2, respectively, out of the total solid volume ΔV_s, and $\Delta V_s = \Delta V_{s1} + \Delta V_{s2}$. In the event ρ_s is statistically distributed, a suitable integration technique may be used. ρ_ℓ and ρ_g are similarly defined.

Now *macroscopic average densities* are defined as follows. Let us remove the liquid and gas phases out of the REV, leaving only the solid phase. The material left in the REV is what is defined as the solid *skeleton*, which will have a porous structure. Let us now define a density as the total mass of the skeleton in the REV divided by the volume of the REV ΔV, and simplify the resulting equation using (6.33a) and (6.33c) as

$$\bar{\rho}_s = \frac{\rho_s \Delta V_s}{\Delta V} = v_s \rho_s = (1-n)\rho_s \tag{6.35a}$$

Similarly, the macroscopic average densities of the liquid and gas phases are

$$\bar{\rho}_\ell = \frac{\rho_\ell \Delta V_\ell}{\Delta V} = v_\ell \rho_\ell = Sn\rho_\ell \tag{6.35b}$$

$$\bar{\rho}_g = \frac{\rho_g \Delta V_g}{\Delta V} = v_g \rho_g = (1-S)n\rho_g \tag{6.35c}$$

We also define the macroscopic average density of the mixture as

$$\bar{\rho} = \rho = \frac{\rho_s \Delta V_s + \rho_\ell \Delta V_\ell + \rho_g \Delta V_g}{\Delta V} = v_s \rho_s + v_\ell \rho_\ell + v_g \rho_s$$
$$\bar{\rho} = \rho = (1-n)\rho_s + Sn\rho_\ell + (1-S)n\rho_g \tag{6.35d}$$

The definition of stresses will be presented later. $\bar{\rho}_s$, $\bar{\rho}_\ell$, and $\bar{\rho}_g$ are also known as the partial densities (Morland 1972; De Boer 1996).

6.2.2 *Mass Conservation*

Now the principle of the conservation of mass is applied to each phase separately. We allow for the possibility of phase changes. For example, the liquid may evaporate and become gas, and the solid may become biodegraded into liquid and gas. Let r_ℓ and r_s be mass rates (buildup taken as positive) at which liquid and solid become converted. Let us further assume that (a) the liquid is converted into gas and (b) a portion of the solid converted becomes liquid and the other portion becomes gas. The mass conservation for the *solid phase* is:

$$\frac{\mathrm{d}}{\mathrm{d}t}\int_{\Delta V} \bar{\rho}_s \,\mathrm{d}v - r_s \int_{\Delta V} \bar{\rho}_s \,\mathrm{d}v = 0$$

Performing the material derivative in the above for a solid material particle (6.27c),

$$\dot{\bar{\rho}}_s + \dot{u}^s_{i,i}\bar{\rho}_s - r_s\bar{\rho}_s = 0$$

By using (6.35a)

$$(1-n)\dot{\rho}_s - \rho_s\dot{n} + \rho_s(1-n)\dot{u}^s_{i,i} - r_s\rho_s(1-n) = 0 \tag{6.36a}$$

Let $0 \le \beta_\ell \le 1$ be the fraction of the solid experiencing phase change that becomes liquid. The mass conservation for the *liquid phase* is:

$$\frac{\mathrm{d}}{\mathrm{d}t}\int_{\Delta V} \bar{\rho}_\ell \,\mathrm{d}v + \beta_\ell r_s \int_{\Delta V} \bar{\rho}_s \,\mathrm{d}v - r_\ell \int_{\Delta V} \bar{\rho}_\ell \,\mathrm{d}v = 0$$

By simplifying with the aid of (6.27c), (6.35a), and (6.35b), we have

$$\dot{\bar{\rho}}_\ell + \dot{u}^\ell_{i,i}\bar{\rho}_\ell + \beta_\ell r_s \bar{\rho}_s - r_\ell \bar{\rho}_\ell = 0$$

$$Sn\dot{\rho}_\ell + S\rho_\ell \dot{n} + \rho_\ell n\dot{S} + Sn\rho_\ell \dot{u}^\ell_{i,i} + \beta_\ell r_s \rho_s (1-n) - r_\ell Sn\rho_\ell = 0 \tag{6.36b}$$

The last term in (6.36b) represents evaporation. If desired, another term could be added to simulate condensation.

The equation for the *gas phase* is

$$\frac{\mathrm{d}}{\mathrm{d}t}\int\limits_{\Delta V} \bar{\rho}_g \,\mathrm{d}v + \beta_g r_s \int\limits_{\Delta V} \bar{\rho}_s \,\mathrm{d}v + r_\ell \int\limits_{\Delta V} \bar{\rho}_\ell \,\mathrm{d}v = 0$$

$$\dot{\bar{\rho}}_g + \dot{u}^g_{i,i}\bar{\rho}_g + \beta_g r_s \bar{\rho}_s + r_\ell \bar{\rho}_\ell = 0$$

$$(1-S)n\dot{\rho}_g + (1-S)\rho_g \dot{n} - \rho_g n\dot{S} + (1-S)n\rho_g \dot{u}^g_{i,i} + \beta_g r_s \rho_s (1-n) + r_\ell Sn\rho_\ell = 0 \tag{6.36c}$$

where $0 \le \beta_g \le 1$ is the fraction of the solid experiencing phase change that becomes gas. The term involving r_ℓ represents phase change from liquid to gas; note that this term is added in (6.36c) and subtracted in (6.36b). Also note that $\beta_\ell + \beta_g = 1$. Just as evaporation was added in (6.36b), condensation could be added in (6.36c) if desired. Equations (6.27c), (6.35a), (6.35b), and (6.35c) have been used in obtaining (6.35c).

By adding (6.36a), (6.36b), and (6.36c) and simplifying with the aid of (6.35a)–(6.35d), the mass conservation equation for the *mixture* is obtained as

$$\begin{aligned}
&(1-n)\dot{\rho}_s - \rho_s \dot{n} + \rho_s (1-n)\dot{u}^s_{i,i} - r_s \rho_s (1-n) \\
&\quad + Sn\dot{\rho}_\ell + S\rho_\ell \dot{n} + \rho_\ell n\dot{S} + Sn\rho_\ell \dot{u}^\ell_{i,i} + \beta_\ell r_s \rho_s (1-n) - r_\ell Sn\rho_\ell \\
&\quad + (1-S)n\dot{\rho}_g + (1-S)\rho_g \dot{n} - \rho_g n\dot{S} + (1-S)n\rho_g \dot{u}^g_{i,i} + \beta_g r_s \rho_s (1-n) + r_\ell Sn\rho_\ell = 0
\end{aligned}$$

$$\dot{\rho} + \dot{u}^s_{i,i}\bar{\rho}_s + \dot{u}^\ell_{i,i}\bar{\rho}_\ell + \dot{u}^g_{i,i}\bar{\rho}_g = 0 \tag{6.36d}$$

$$\dot{\rho} + \rho \dot{\bar{\bar{u}}}_{i,i} = 0 \tag{6.36e}$$

where

$$\dot{\bar{\bar{u}}}_{i,i} = \frac{\dot{u}^s_{i,i}\bar{\rho}_s + \dot{u}^\ell_{i,i}\bar{\rho}_\ell + \dot{u}^g_{i,i}\bar{\rho}_g}{\rho} \tag{6.36f}$$

$\dot{\bar{\mathbf{u}}}$ is the mass-averaged mean velocity of the mixture and ρ is the volume-averaged mixture density given by (6.35d).

6.2.3 *Balance of Linear Momentum*

It is now necessary to define the intrinsic and macroscopic average stresses. Since interfaces between phases (and hence quantities such as the capillary pressure) are not explicitly considered, certain approximations will have to be made. This is done here with the objective of reproducing the widely-used affective stress equation (6.12b):

$$\boldsymbol{\sigma}' = \boldsymbol{\sigma} + \alpha p \boldsymbol{\delta} \tag{6.37a}$$

$$p = x_\ell p_\ell + x_g p_g = \chi p_\ell + (1 - \chi) p_g \tag{6.37b}$$

When $\alpha = 1$, (6.37) coincides with that proposed by Bishop (1959) for unsaturated soils. When $\alpha = 1$ and $\chi = 1$, it coincides with the effective stress equation proposed by Terzaghi (Terzaghi 1936) for saturated soils. When $\chi = 1$, (6.37) coincides with that of Biot (1941), Biot and Willis (1957), and Skempton (1960) for general porous materials such as rocks and concrete, and extensively used by Zienkiewicz (1982) and Zienkiewicz et al. (1999). The particular form in (6.37) is also the one used by Zienkiewicz et al. (1999) for unsaturated soils. Some investigators have used the assumption $S = \chi$ (e.g., Gray and Hassanizadeh 1990a, b; Meroi et al. 1995; Gawin and Schrefler 1996) for unsaturated soils. Muraleetharan et al. (2009) assumed that $\chi = v_\ell$, where v_ℓ is the volume fraction of the liquid.

Equation (6.37) is still not general enough to include the effect of capillary tension for cases with a high degree of saturation (i.e., $S > 70\%$). In this case, the equation must include an addition term $-p_T$ (Pietruszczak and Pande 1996) as discussed in Sect. 2.1.1 (6.11f). In the following, for simplicity, $-p_T$ is ignored and (6.37) is adopted.

Referring back to the analysis presented in Sect. 2.1.1 and Fig. 6.1, let us define the intrinsic stresses. Each grain is subjected to an intergranular stress and fluid pressure. Considering first the fluid pressure, let us assume that the grains are subjected to an average fluid (liquid + gas) pressure p given by (6.37b). However, a fraction of the grain's surface is not covered by fluids and this must be accounted for.

Let us introduce an average intrinsic pressure p^s arising from the fluid pressure and calculate its value by

$$p^s A_{solid} = p A_{solid}^{fluid} \tag{6.38a}$$

where A_{solid} and A_{solid}^{fluid} are defined in (6.7b) (also see Fig. 6.1). Using the relationships in (6.7b) and (6.8b):

$$p^s = p\frac{A_{solid}^{fluid}}{A_{solid}} = p\frac{\bar{\alpha}A_{solid}}{A_{solid}} = p\bar{\alpha} = p\frac{\alpha - n}{1 - n} \tag{6.38b}$$

p^s is the contribution to the intrinsic solid stress from the fluids. Noting that $\boldsymbol{\sigma}'$ is already a macroscopic stress (recall that $\boldsymbol{\sigma}'$ was defined by dividing the intergranular force by the total area in (6.10)), the contribution of the effective stress to the intrinsic solid stress is $\boldsymbol{\sigma}'/(1 - n)$. The total intrinsic solid stress then is (Table 6.1)

$$\boldsymbol{\sigma}^s = \frac{\boldsymbol{\sigma}'}{1 - n} - \frac{\alpha - n}{1 - n}p\boldsymbol{\delta} \tag{6.39a}$$

Note that the negative sign in the above equation is due to the use of opposite sign conventions for $\boldsymbol{\sigma}'$ and p. The intrinsic liquid and gas pressures are p_ℓ and p_g, respectively. The spherical stress tensors corresponding to these are

$$\boldsymbol{\sigma}^\ell = p_\ell\boldsymbol{\delta} \tag{6.39b}$$

$$\boldsymbol{\sigma}^g = p_g\boldsymbol{\delta} \tag{6.39c}$$

In keeping up with the mixture theory assumptions, the intrinsic stresses are to be multiplied by the actual volume fractions to obtain the macroscopic average stresses. While this works for the solid stresses, modified volume fractions are to be employed when using the fluid pressure defined by (6.37b). The modified volume fractions are:

$$v_s^* = v_s = 1 - n \tag{6.40a}$$

$$v_\ell^* = \chi n \tag{6.40b}$$

$$v_g^* = (1 - \chi)n \tag{6.40c}$$

The definitions coincide with the standard definitions when $\chi = S$. The macroscopic average stresses are then obtained as products between intrinsic stresses and the corresponding volume fractions as

$$\bar{\boldsymbol{\sigma}}^s = v_s^*\boldsymbol{\sigma}^s = \boldsymbol{\sigma}' - p(\alpha - n)\boldsymbol{\delta} \tag{6.41a}$$

$$\bar{\boldsymbol{\sigma}}^\ell = v_\ell^*\boldsymbol{\sigma}^\ell = -\chi n p_\ell\boldsymbol{\delta} \tag{6.41b}$$

$$\bar{\boldsymbol{\sigma}}^g = v_g^*\boldsymbol{\sigma}^g = -(1 - \chi)np_g\boldsymbol{\delta} \tag{6.41c}$$

The total macroscopic average stress (i.e., the mixture stress) is obtained by adding the solid, liquid, and gas stresses as

$$\bar{\boldsymbol{\sigma}} = \bar{\boldsymbol{\sigma}}^s + \bar{\boldsymbol{\sigma}}^\ell + \bar{\boldsymbol{\sigma}}^g = \boldsymbol{\sigma}' - \alpha p \boldsymbol{\delta} \tag{6.41d}$$

This total stress coincide with the total stress obtained by rearranging (6.37a), i.e.,

$$\boldsymbol{\sigma} = \bar{\boldsymbol{\sigma}} = \boldsymbol{\sigma}' - \alpha p \boldsymbol{\delta} \tag{6.41e}$$

These results are summarized in Table 6.1. $\bar{\boldsymbol{\sigma}}^s$,$\bar{\boldsymbol{\sigma}}^\ell$, and $\bar{\boldsymbol{\sigma}}^g$ are also known as partial stresses (Morland 1972; De Boer 1996).

Now we are ready to write down the linear momentum equations for each of the three phases separately. The domain details, analogous to those for a single phase material shown in Fig. 6.4, are presented in Fig. 6.5 for each of the three phases of a three-phase mixture. The macroscopic average quantities, listed in Table 6.1, are used for the densities and stresses. *For simplicity, we will assume that the drag force between the gas phase and the other two phases (solid and liquid) is negligible*, and the only interface drag force involved is the viscous drag force between the solid and liquid phases (6.16e).

The linear momentum equation for the solid phase is

$$\int_\Gamma \bar{\mathbf{t}}^s \, \mathrm{d}s + \int_\Omega \bar{\rho}^s \mathbf{b}^s \, \mathrm{d}v + \int_\Omega \mathbf{f}^d \, \mathrm{d}v = \frac{\mathrm{d}}{\mathrm{d}t} \int_\Omega \bar{\rho}^s \dot{\mathbf{u}}^s \, \mathrm{d}v \tag{6.42a}$$

where $\bar{t}_i^s = \bar{\sigma}_{ij}^s n_j$ (Table 6.1). As done in (6.29), (6.42a) is simplified with the aid of (6.41a) and (6.35a) as

$$\int_\Omega \left[\sigma'_{ij,j} - (\alpha - n)p_{,i} + (1-n)\rho_s b_i^s - (1-n)\rho_s \ddot{u}_i^s + f_i^d\right] dv$$
$$= \int_\Omega \dot{u}_i^s \frac{\mathrm{d}}{\mathrm{d}t}(\bar{\rho}^s \, \mathrm{d}v) = \int_\Omega \dot{u}_i^s r_s (1-n)\rho_s \, \mathrm{d}v \tag{6.42b}$$

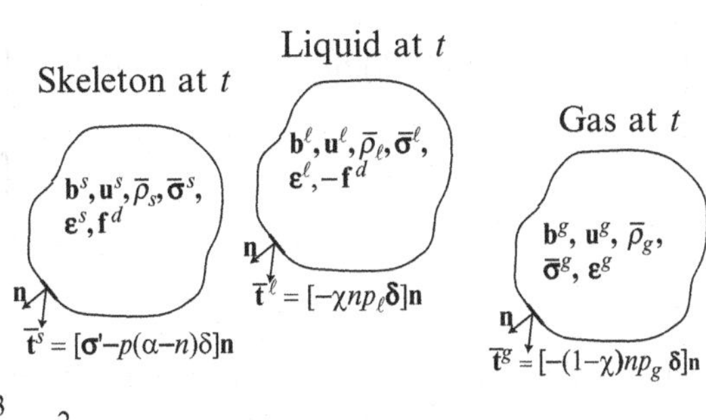

Fig. 6.5 Deformed configurations of the domain containing solid, liquid, and gas phases

$$\sigma'_{ij,j} - (\alpha - n)p_{,i} + (1 - n)\rho_s b_i^s - (1 - n)\rho_s \ddot{u}_i^s - (1 - n)\rho_s r_s \dot{u}_i^s + f_i^d = 0 \quad (6.42c)$$

Note that the viscous drag force $\mathbf{f}^d$ is included in the equation. The mass conservation of the solid phase (6.36a) is used in the last step of (6.42b).

The linear momentum equation for the liquid phase is

$$\int_\Gamma \bar{\mathbf{t}}^\ell \, ds + \int_\Omega \bar{\rho}^\ell \mathbf{b}^\ell \, dv - \int_\Omega \mathbf{f}^d \, dv = \frac{d}{dt} \int_\Omega \bar{\rho}^\ell \dot{\mathbf{u}}^\ell \, dv \quad (6.43a)$$

where $\bar{t}_i^\ell = \bar{\sigma}_{ij}^\ell n_j$ (Table 6.1). Note that the drag force term is introduced in the above equation with a sign that is opposite to that in (6.42a). Simplifying further with the aid of (6.41b) and (6.35b), we have

$$\begin{aligned} \int_\Omega \left[-\chi n p_{,i}^\ell + S n \rho_\ell b_i^\ell - S n \rho_\ell \ddot{u}_i^\ell - f_i^d \right] &= \int_\Omega \dot{u}_i^\ell \frac{d}{dt} (\bar{\rho}^\ell dv) \\ &= \int_\Omega \dot{u}_i^\ell [-\beta_\ell r_s (1 - n)\rho_s + r_\ell S n \rho_\ell] dv \end{aligned} \quad (6.43b)$$

$$-\chi n p_{,i}^\ell + S n \rho_\ell b_i^\ell - S n \rho_\ell \ddot{u}_i^\ell + (1 - n)\rho_s \beta_\ell r_s \dot{u}_i^\ell - S n \rho_\ell r_\ell \dot{u}_i^\ell - f_i^d = 0 \quad (6.43c)$$

The mass conservation of the liquid phase (6.36b) is used in the last step of (6.43b).

The linear momentum equation for the gas phase is

$$\int_\Gamma \bar{\mathbf{t}}^g ds + \int_\Omega \bar{\rho}^g \mathbf{b}^g dv = \frac{d}{dt} \int_\Omega \bar{\rho}^g \dot{\mathbf{u}}^g dv \quad (6.44a)$$

where $\bar{t}_i^g = \bar{\sigma}_{ij}^g n_j$ (Table 6.1). Simplifying further with the aid of (6.41c) and (6.35c), we have

$$\begin{aligned} \int_\Omega \left[-(1-\chi) n p_{,i}^g + (1-S) n \rho_g b_i^g - (1-S) n \rho_g \ddot{u}_i^g \right] &= \int_\Omega \dot{u}_i^g \frac{d}{dt} (\bar{\rho}^g dv) \\ &= \int_\Omega \dot{u}_i^g \left[-\beta_g r_s (1-n)\rho_s - r_\ell S n \rho_\ell \right] dv \end{aligned} \quad (6.44b)$$

$$-(1 - \chi) n p_{,i}^g + (1 - S) n \rho_g b_i^g - (1 - S) n \rho_g \ddot{u}_i^g + (1 - n)\rho_s \beta_g r_s \dot{u}_i^g + S n \rho_\ell r_\ell \dot{u}_i^g = 0 \quad (6.44c)$$

The mass conservation of the gas phase (6.36c) is used in the last step of (6.43b).

The linear momentum equation for the mixture can be obtained by adding (6.42c), (6.43c), and (6.44c) as:

$$\begin{aligned}&\sigma'_{ij,j} - (\alpha - n)p_{,i} + (1-n)\rho_s b_i^s - (1-n)\rho_s \ddot{u}_i^s - (1-n)\rho_s r_s \dot{u}_i^s + f_i^d \\ &- \chi n p_{,i}^\ell + Sn\rho_\ell b_i^\ell - Sn\rho_\ell \ddot{u}_i^\ell + (1-n)\rho_s \beta_\ell r_s \dot{u}_i^\ell - Sn\rho_\ell r_\ell \dot{u}_i^\ell - f_i^d - (1-\chi) n p_{,i}^g \\ &+ (1-S)n\rho_g b_i^g - (1-S)n\rho_g \ddot{u}_i^g + (1-n)\rho_s \beta_g r_s \dot{u}_i^g + Sn\rho_\ell r_\ell \dot{u}_i^g = 0\end{aligned}$$

After rearranging the terms and simplifying, the following equation is obtained:

$$\sigma'_{ij,j} - \alpha p_{,i} + \rho \bar{\bar{b}}_i - \rho \ddot{\bar{\bar{u}}}_i + a_s \dot{u}_i^s + a_\ell \dot{u}_i^\ell + a_g \dot{u}_i^g = 0 \tag{6.45a}$$

The above equation is in terms of the effective stress. The equation can be expressed in terms of the total stress (6.37) as

$$\sigma_{ij,j} + \rho \bar{\bar{b}}_i - \rho \ddot{\bar{\bar{u}}}_i + a_s \dot{u}_i^s + a_\ell \dot{u}_i^\ell + a_g \dot{u}_i^g = 0 \tag{6.45b}$$

In (6.45a) and (6.45b)

$$\begin{aligned} a_s &= -r_s \rho_s (1-n) \\ a_\ell &= (1-n)\rho_s \beta_\ell r_s - Sn\rho_\ell r_\ell \\ a_g &= \beta_g r_s (1-n)\rho_s + r_\ell Sn\rho_\ell \end{aligned} \tag{6.45c}$$

$\bar{\bar{\mathbf{b}}}$ and $\ddot{\bar{\bar{\mathbf{u}}}}$ are mass-averaged quantities defined (similarly to (6.36f)) as:

$$\bar{\bar{\mathbf{b}}} = \frac{\mathbf{b}^s \bar{\rho}_s + \mathbf{b}^\ell \bar{\rho}_\ell + \mathbf{b}^g \bar{\rho}_g}{\rho}; \quad \ddot{\bar{\bar{u}}}_i = \frac{\ddot{u}_i^s \bar{\rho}_s + \ddot{u}_i^\ell \bar{\rho}_\ell + \ddot{u}_i^g \bar{\rho}_g}{\rho} \tag{6.45d}$$

For the special case of $\mathbf{b}^s = \mathbf{b}^\ell = \mathbf{b}^g = \mathbf{b}$ (e.g., gravity), we have $\bar{\bar{\mathbf{b}}} = \mathbf{b}$.

6.2.4 *Summary of Balance Equations for the Full-Formulation*

It is assumed that the viscous drag forces from the flow of gas relative to the other phases can be neglected. The viscous drag force from the flow of liquid phase is assumed to be only between the solid and liquid phases.

The mass conservation equations for the solid, liquid, and gas phases, respectively, are ((6.36a), (6.36b), and (6.36c))

$$(1-n)\dot{\rho}_s - \rho_s \dot{n} + \rho_s (1-n)\dot{u}_{i,i}^s - r_s \rho_s (1-n) = 0 \tag{6.46a}$$

$$Sn\dot{\rho}_\ell + S\rho_\ell \dot{n} + \rho_\ell n\dot{S} + Sn\rho_\ell \dot{u}^\ell_{i,i} + \beta_\ell r_s \rho_s (1-n) - r_\ell S n \rho_\ell = 0 \tag{6.46b}$$

$$(1-S)n\dot{\rho}_g + (1-S)\rho_g \dot{n} - \rho_g n\dot{S} + (1-S)n\rho_g \dot{u}^g_{i,i} + \beta_g r_s \rho_s (1-n) + r_\ell S n \rho_\ell = 0 \tag{6.46c}$$

The mass conservation equation for the mixture is (6.36e)

$$\dot{\rho} + \rho \dot{u}_{i,i} = 0 \tag{6.46d}$$

The linear momentum equations for the solid, liquid, and gas phases, respectively, are ((6.42c), (6.43c), and (6.44c))

$$\sigma'_{ij,j} - (\alpha - n)p_{,i} + (1-n)\rho_s b^s_i - (1-n)\rho_s \ddot{u}^s_i - (1-n)\rho_s r_s \dot{u}^s_i + f^d_i = 0 \tag{6.46e}$$

$$-\chi n p^\ell_{,i} + Sn\rho_\ell b^\ell_i - Sn\rho_\ell \ddot{u}^\ell_i + (1-n)\rho_s \beta_\ell r_s \dot{u}^\ell_i - Sn\rho_\ell r_\ell \dot{u}^\ell_i - f^d_i = 0 \tag{6.46f}$$

$$-(1-\chi)np^g_{,i} + (1-S)n\rho_g b^g_i - (1-S)n\rho_g \ddot{u}^g_i + (1-n)\rho_s \beta_g r_s \dot{u}^g_i + Sn\rho_\ell r_\ell \dot{u}^g_i = 0 \tag{6.46g}$$

The linear momentum equation for the mixture is (6.45a)

$$\sigma'_{ij,j} - \alpha p_{,i} + \rho \bar{\bar{b}}_i - \rho \ddot{\bar{\bar{u}}}_i + a_s \dot{u}^s_i + a_\ell \dot{u}^\ell_i + a_g \dot{u}^g_i = 0 \tag{6.46i}$$

where the coefficients a_s, a_ℓ, and a_g are given in (6.45c) and $\bar{\bar{\mathbf{b}}}$ and $\ddot{\bar{\bar{\mathbf{u}}}}$ are mass-averaged body force and acceleration terms, respectively, given by (6.45d).

6.2.5 The Role of Constitutive Equations

The formulation presented in the preceding section (referred to here as the full-formulation) is very general. The following approximations have, however, been made: (1) The fluid pressure is approximated by (6.37b), which does not consider the influence of surface tension explicitly (which will entail inclusion of an additional pressure term P_T as in (6.11f)), and (2) the drag force arising from the flow of gas is neglected. Apart from these assumptions, the formulation is very general.

The unknown variables involved in the full-formulation are listed in Table 6.2. The total number of unknown is 25. As shown in Table 6.3, the number of equations developed thus far (i.e., those based on the principles of the balance of linear momentum and mass conservation) adds up to a total of 12. This requires an additional 13 equations for solving for all of the unknowns. These additional equations come from the specification of constitutive relations among variables.

Table 6.2 Unknowns in the full-formulation

Variable	$\boldsymbol{\sigma}'$	p_ℓ	p_g	S	n	ρ_s	ρ_ℓ	ρ_g	$\mathbf{u}^s$	$\mathbf{u}^\ell$	$\mathbf{u}^g$	$\mathbf{f}^d$	Total
No. unknowns	6	1	1	1	1	1	1	1	3	3	3	3	25

Table 6.3 Equations typically available in the full-formulation

Principle	Relation	No. equations
Linear momentum: solid	Equation (6.46e)	3
Linear momentum: liquid	Equation (6.46f)	3
Linear momentum: gas	Equation (6.46g)	3
Mass conservation: solid	Equation (6.46a)	1
Mass conservation: liquid	Equation (6.46b)	1
Mass conservation: gas	Equation (6.46c)	1
Constitutive equation	$\boldsymbol{\sigma}'\ vs\ \boldsymbol{\varepsilon}^s$	6
Constitutive equation	$p_s\ vs\ \rho_s$	1
Constitutive equation	$p_\ell\ vs\ \rho_\ell$	1
Constitutive equation	$p_g\ vs\ \rho_g$	1
Constitutive equation	$p_c = (p_g - p_\ell)\ vs\ S$	1
Darcy's law for fluid drag	$\mathbf{f}^d\ vs\ (\mathbf{u}^\ell - \mathbf{u}^s)$	3
Total		25

Table 6.3 lists the typical variables related by constitutive relations and the number of equations resulting from these relations. In the relation $\boldsymbol{\sigma}'\ vs\ \boldsymbol{\varepsilon}^s$, the strains derived from $\mathbf{u}^s$ appear; it is understood that the use of $\boldsymbol{\varepsilon}^s$ introduces six additional variables (noting that $\boldsymbol{\varepsilon}^s$ is symmetric), however, there are six additional equations relating $\boldsymbol{\varepsilon}^s$ to $\mathbf{u}^s$ (the strain-displacement relations). Also note that the suggested constitutive functional forms are not to be taken as universal. However, the number of constitutive relations listed in Table 6.3 is the minimum required for solving the problem. We will give some details of the constitutive equations in Sect. 6.3.

6.3 Balance Principles for Unsaturated Porous (Three-Phase) Materials: The Approximate *u-u-p* Formulation

6.3.1 Approximations

Now we introduce a number of additional assumptions before proceeding. The major assumptions made are that (a) the gas pressure is negligible in comparison to the liquid pressure and the solid stresses and (b) the gas density is negligible in comparison to the solid and fluid densities. Due to the first assumption, (6.37b) simplifies to

$$p = x_\ell p_\ell + x_g p_g = \chi p_\ell + (1-\chi)p_g \Rightarrow p = \chi p_\ell \qquad (6.47a)$$

Due to the second assumption, (6.35d) simplifies to

$$\bar{\rho} = (1-n)\rho_s + Sn\rho_\ell + (1-S)n\rho_g \Rightarrow \bar{\rho} = (1-n)\rho_s + Sn\rho_\ell \tag{6.47b}$$

In addition, we will not seek to determine $\mathbf{u}^g$, and $\dot{\mathbf{u}}^g$ and $\ddot{\mathbf{u}}^g$ are assumed to be negligible. The number of unknowns is reduced by five (dropping p_g, ρ_g, and $\mathbf{u}^g$, Table 6.2). We will drop the linear momentum equation for gas (6.46g), the mass conservation equation for gas (6.46c) and the constitutive law p_g *vs* ρ_g, a total of five equations in all.

6.3.2 Constitutive Equations

The following constitutive relations are assumed. The skeleton effective stresses are assumed to be uniquely related to the skeleton strains as

$$\dot{\boldsymbol{\sigma}}' = \mathbf{D}\dot{\boldsymbol{\varepsilon}}^s \tag{6.48a}$$

where $\mathbf{D}$ is a suitable tangent modulus matrix; e.g., elastic or elasto-plastic stiffness tensor.

Equation (6.36b) was obtained by applying the mass conservation principle for the fluid phase at the macroscopic scale. A constitutive equation of the form p_ℓ *vs* ρ_ℓ can be obtained by combining the mass conservation principle and the elastic constitutive relation for the liquid phase. The mass conservation is

$$\frac{\mathrm{d}}{\mathrm{d}t}\int\limits_{\Delta V} \rho_\ell \mathrm{d}v = 0$$

$$\dot{\rho}_\ell + \dot{\bar{u}}^\ell_{i,i}\rho_\ell = 0 \tag{6.48b}$$

By assuming that the volumetric behavior of the liquid is elastic, we have

$$\dot{\bar{u}}^\ell_{i,i} = -\frac{\dot{p}_\ell}{K_\ell} \tag{6.48c}$$

where K_ℓ is the bulk modulus of the liquid. Combining (6.48b) and (6.48c):

$$\frac{\dot{\rho}_\ell}{\rho_\ell} = \frac{\dot{p}_\ell}{K_\ell} \tag{6.48d}$$

Similarly, a constitutive equation for $\dot{\rho}_s$ can be derived. Similarly to (6.48b), we have

$$\dot{\rho}_s + \dot{\bar{u}}^s_{i,i}\rho_s = 0 \tag{6.48e}$$

As indicated in Table 6.1, the solid phase is subjected to an average stress of $\bar{\boldsymbol{\sigma}}^s = \boldsymbol{\sigma}' - p(\alpha - n)\boldsymbol{\delta}$ and an intrinsic stress of $\bar{\boldsymbol{\sigma}}^s/(1-n)$. Hence, we have

$$\frac{\dot{\rho}_s}{\rho_s} = -\dot{\bar{u}}^s_{i,i} = -\frac{\dot{\bar{\sigma}}^s_{kk}}{3(1-n)K_s} = -\frac{\dot{\sigma}'_{kk} - 3\dot{p}(\alpha - n)}{3(1-n)K_s} = -\frac{3K_T\dot{u}^s_{i,i} - 3\dot{p}(\alpha - n)}{3(1-n)K_s}$$
$$= -\frac{K_T\dot{u}^s_{i,i} - \dot{p}(\alpha - n)}{(1-n)K_s}$$

where K_s is the bulk modulus of the solid phase and K_T is the bulk modulus of the solid skeleton. From (6.11d), $K_T/K_s = 1 - \alpha$, which leads to

$$\frac{\dot{\rho}_s}{\rho_s} = \frac{1}{1-n}\left[\frac{\dot{p}}{K_s}(\alpha - n) - (1-\alpha)\dot{u}^s_{i,i}\right] \tag{6.48f}$$

In many cases (e.g., sand particles made of quartz), the solid grains may be assumed to be incompressible compared to the liquid phase, i.e., $K_s \approx \infty$. For example, Mei and Foda (1982) state that typically, solids (i.e., sand particles) are 25 times less compressible than water. With such an assumption, we have

$$\dot{\rho}_s = 0 \tag{6.48g}$$

As for $\mathbf{f}^d$ vs $(\mathbf{u}^\ell - \mathbf{u}^s)$, the generalized Darcy's law (6.16) is used:

$$f_i^d = S^2 n^2 \bar{k}_{ij}^{-1}(\dot{u}_j^\ell - \dot{u}_j^s) \tag{6.49}$$

The $p_c = (p_g - p_\ell)$ vs S relationship is assumed to be

$$\dot{S} = -r_{SWCC}\dot{p}_c = r_{SWCC}\dot{p}_\ell \tag{6.50a}$$

where $r_{SWCC} \geq 0$ is negative of the slope of the soil-water characteristic curve (Fig. 6.2), i.e.,

$$r_{SWCC} = -\frac{\partial S}{\partial p_c} \tag{6.50b}$$

6.3.3 Governing Equations

The balance equations, simplified based on the approximations and constitutive equations presented above, are summarized below. In (6.46a), using (6.48f) for $\dot{\rho}_s$ and diving the remaining terms by ρ_s, the mass conservation equation for solid becomes:

$$\frac{\dot{p}}{K_s}(\alpha - n) - (1 - \alpha)\dot{u}^s_{i,i} - \dot{n} + (1 - n)\dot{u}^s_{i,i} - r_s(1 - n) = 0$$

$$\frac{\dot{p}}{K_s}(\alpha - n) + (\alpha - n)\dot{u}^s_{i,i} - \dot{n} - r_s(1 - n) = 0 \tag{6.51a}$$

In (6.46b), expressing $\dot{\rho}_\ell$ in terms of $\dot{p}_\ell$ from (6.48d), $\dot{S}$ in terms of $\dot{p}_\ell$ from (6.50a), and dividing the equation by ρ_ℓ, the mass conservation for liquid becomes:

$$\left(\frac{S}{K_\ell} + r_{SWCC}\right)n\dot{p}_\ell + S\dot{n} + Sn\dot{u}^\ell_{i,i} + \beta_\ell r_s(1 - n)\frac{\rho_s}{\rho_\ell} - r_\ell Sn = 0 \tag{6.51b}$$

In (6.46e), noting that $p_{,i} = \chi p^\ell_{,i}$ (6.47a) and using (6.49), the linear momentum equation for the solid phase becomes

$$\sigma'_{ij,j} - (\alpha - n)\chi p^\ell_{,i} + (1 - n)\rho_s b^s_i - (1 - n)\rho_s \ddot{u}^s_i - (1 - n)\rho_s r_s \dot{u}^s_i + S^2 n^2 \bar{k}^{-1}_{ij}(\dot{u}^\ell_j - \dot{u}^s_j) = 0 \tag{6.51c}$$

The linear momentum equation for the liquid phase (6.46f) remains unchanged as:

$$-\chi n p^\ell_{,i} + Sn\rho_\ell b^\ell_i - Sn\rho_\ell \ddot{u}^\ell_i + (1 - n)\rho_s \beta_\ell r_s \dot{u}^\ell_i - Sn\rho_\ell r_\ell \dot{u}^\ell_i - S^2 n^2 \bar{k}^{-1}_{ij}(\dot{u}^\ell_j - \dot{u}^s_j) = 0 \tag{6.51d}$$

The linear momentum equation for the mixture (6.46i) becomes

$$\sigma'_{ij,j} - \alpha\chi p^\ell_{,i} + \rho\bar{\bar{b}}_i - \rho\ddot{\bar{\bar{u}}}_i + a_s\dot{u}_s + a_\ell\dot{u}_\ell = 0 \tag{6.51e}$$

where the coefficients a_s and a_ℓ are given in (6.45c) and $\bar{\bar{\mathbf{b}}}$ and $\ddot{\bar{\bar{\mathbf{u}}}}$ are mass-averaged quantities, given by (6.45d) (with the terms involving the gas phase dropped).

Noting that $\dot{p} = \chi\dot{p}^\ell$ (6.47a), we combine (6.51a) and (6.51b) to eliminate $\dot{n}$ as

$$S(\alpha - n)\dot{u}^s_{i,i} + Sn\dot{u}^\ell_{i,i} + \frac{n\dot{p}_\ell}{Q} + f_{s1} = 0 \tag{6.52a}$$

where

$$\frac{1}{Q} = \frac{S}{K_\ell} + \frac{S\chi(\alpha - n)}{nK_s} + r_{SWCC}$$

$$f_{s1} = -Sr_s(1 - n) + \beta_\ell r_s(1 - n)\frac{\rho_s}{\rho_\ell} - r_\ell Sn \tag{6.52b}$$

6.3.4 Summary of Balance Equations for the Approximate u—u—p Formulation

The linear momentum equation for the solid phase, fluid phase and mixture, respectively, are ((6.51c), (6.51d), and (6.51f))

$$\sigma'_{ij,j} - (\alpha - n)\chi p^{\ell}_{,i} + (1-n)\rho_s b^s_i - (1-n)\rho_s \ddot{u}^s_i - (1-n)\rho_s r_s \dot{u}^s_i + S^2 n^2 \bar{k}^{-1}_{ij}(\dot{u}^{\ell}_j - \dot{u}^s_j) = 0 \tag{6.53a}$$

$$-\chi n p^{\ell}_{,i} + Sn\rho_{\ell} b^{\ell}_i - Sn\rho_{\ell}\ddot{u}^{\ell}_i + (1-n)\rho_s \beta_{\ell} r_s \dot{u}^{\ell}_i - Sn\rho_{\ell} r_{\ell}\dot{u}^{\ell}_i - S^2 n^2 \bar{k}^{-1}_{ij}(\dot{u}^{\ell}_j - \dot{u}^s_j) = 0 \tag{6.53b}$$

$$\sigma'_{ij,j} - \alpha\chi p^{\ell}_{,i} + \rho\bar{\bar{b}}_i - \rho\ddot{\bar{\bar{u}}}_i + a_s \dot{u}_s + a_{\ell}\dot{u}_{\ell} = 0 \tag{6.53c}$$

The coefficients a_s and a_{ℓ} are given in (6.45c). $\bar{\bar{\mathbf{b}}}$ and $\ddot{\bar{\bar{\mathbf{u}}}}$ are mass-averaged quantities, given by (6.45d) (with the terms involving the gas phase dropped).

From the mass conservation equation for the solid (6.51a), we have

$$\dot{n} = (\alpha - n)\dot{u}^s_{i,i} - r_s(1-n) + \frac{\dot{p}}{K_s}(\alpha - n) \tag{6.53d}$$

The combined mass conservation equation is (6.52a)

$$S(\alpha - n)\dot{u}^s_{i,i} + Sn\dot{u}^{\ell}_{i,i} + \frac{n\dot{p}_{\ell}}{Q} + f_{s1} = 0 \tag{6.53e}$$

where Q and f_{s1} are given by (6.52b).

A summary of the unknown variables is presented in Table 6.4 and the types and number of available equations are summarized in Table 6.5. The relevant constitutive equations are given by (6.48a), (6.48f), (6.48d), (6.50a), and (6.49) as listed in Table 6.5.

In the finite element implementation of the equations, only the linear momentum equation for the solid (6.53a), and liquid phases (6.53b), and the mass conservation for the mixture (6.53e) need to be spatially discretized. The primary unknowns at the nodes (see the remark below) will then be $(\mathbf{u}^s, \mathbf{u}^{\ell}, p^{\ell})$. Among the remaining variables (i.e., secondary variables), S, n, ρ_s, and ρ_{ℓ} are obtained using (6.50a), (6.53d), (6.48f), and (6.48d), respectively. Once the skeleton strain $\boldsymbol{\varepsilon}^s$ is calculated from $\mathbf{u}^s$, the effective stress $\boldsymbol{\sigma}'$ is calculated by integrating (6.48a).

Table 6.4 Unknowns in the approximate $u - u - p$ formulation

Variable	$\boldsymbol{\sigma}'$	p_{ℓ}	S	n	ρ_s	ρ_{ℓ}	$\mathbf{u}^s$	$\mathbf{u}^{\ell}$	$\mathbf{f}^d$	Total
No. unknowns	6	1	1	1	1	1	3	3	3	20

Table 6.5 Equations typically available in the $u - u - p$ formulation

Principle	Relation	No. equations
Linear momentum: solid	Equation (6.53a)	3
Linear momentum: liquid	Equation (6.53b)	3
Mass conservation: solid	Equation (6.53d)	1
Mass conservation: combined	Equation (6.53e)	1
Constitutive equation	$\boldsymbol{\sigma}'$ *vs* $\boldsymbol{\varepsilon}^s$ (6.48a)	6
Constitutive equation	p_s *vs* ρ_s (6.48f)	1
Constitutive equation	p_ℓ *vs* ρ_ℓ (6.48d)	1
Constitutive equation	$p_c = (p_g - p_\ell)$ *vs* S (6.50a)	1
Darcy's law for fluid drag	$\mathbf{f}^d$ *vs* $(\mathbf{u}^\ell - \mathbf{u}^s)$ (6.49)	3
Total		20

The equations associated with the case of a *saturated porous media* are obtained as special cases from the equations summarized in this section with $\chi = S = 1$. The equations may be further specialized for modeling *saturated sand* subjected to a dynamic loading such as the *earthquake loading* (Problem 6.8).

Remark. To avoid hour-glass modes, it is common to use in the finite element implementation of the equations different orders of approximation within elements for the displacement and liquid pressure variables. For example, in using the eight-noded quadrilateral element (2D8), the displacements $(\mathbf{u}^s, u^\ell)$ are expressed in terms of the unknowns at all eight nodes, whereas the liquid pressure is expressed only in terms of the unknowns at the four corner nodes (Anandarajah 1993a).

6.4 Balance Principles for Unsaturated Porous (Three-Phase) Materials: The Approximate *u–p* Formulation

6.4.1 Approximations

The finite element implementation of the $u - u - p$ formulation described by (6.53a)–(6.53e) would require spatial discretization of the primary unknowns $\mathbf{u}^s$, $\mathbf{u}^\ell$, and p_ℓ (see, for example, Anandarajah 1993a). For instance, in a three-dimensional analysis, the number of unknowns in certain nodes will be 7. In large scale analyses, the required computational effort can be overwhelming.

There are many processes (e.g., deformation and flow processes under earthquake loading, wave loading, etc.) where the acceleration of the fluid phase relative to the solid phase can be neglected (Zienkiewicz 1982; Zienkiewicz and Shiomi 1984). Such an approximation will help reduce the primary unknowns from $(\mathbf{u}^s, \mathbf{u}^\ell, p^\ell)$ to $(\mathbf{u}^s, p^\ell)$. The resulting formulation is known as the $u - p$ formulation.

This is achieved as follows: introducing the relative displacement between the liquid and solid phases $\mathbf{w}$, we (a) express the liquid variables set $(\mathbf{u}^\ell, \dot{\mathbf{u}}^\ell, \ddot{\mathbf{u}}^\ell)$ in terms of $(\mathbf{w}, \dot{\mathbf{w}}, \ddot{\mathbf{w}})$, (b) neglect the terms involving $\mathbf{w}$ and (c) eliminate the terms involving $\dot{\mathbf{w}}$. We also make a few other assumptions.

From (6.34),

$$\mathbf{w} = \mathbf{u}^\ell - \mathbf{u}^s; \quad \dot{\mathbf{w}} = \dot{\mathbf{u}}^\ell - \dot{\mathbf{u}}^s; \quad \ddot{\mathbf{w}} = \ddot{\mathbf{u}}^\ell - \ddot{\mathbf{u}}^s \tag{6.54}$$

We also assume that $\mathbf{b}^s = \mathbf{b}^\ell = \mathbf{b}$; it follows from (6.45d) that $\bar{\bar{\mathbf{b}}} = \mathbf{b}$.

The linear momentum equations for the solid phase, fluid phase, and mixture, respectively, become ((6.53a)–(6.53c)):

$$\sigma'_{ij,j} - (\alpha - n)\chi p^\ell_{,i} + (1-n)\rho_s b^s_i - (1-n)\rho_s \ddot{u}^s_i - (1-n)\rho_s r_s \dot{u}^s_i + S^2 n^2 \bar{k}^{-1}_{ij} \dot{w}_j = 0 \tag{6.55a}$$

$$-\chi n p^\ell_{,i} + S n \rho_\ell b^\ell_i - S n \rho_\ell \ddot{u}^s_i - S n \rho_\ell \ddot{w}_i + f_{s2} \dot{u}^s_i + f_{s2} \dot{w}_i - S^2 n^2 \bar{k}^{-1}_{ij} \dot{w}_j = 0 \tag{6.55b}$$

$$\sigma'_{ij,j} - \alpha \chi p^\ell_{,i} + \rho b_i - \rho \frac{\ddot{u}^s_i \bar{\rho}_s + \ddot{u}^s_i \bar{\rho}_\ell + \ddot{w} \bar{\rho}_\ell}{\rho} + a_s \dot{u}_s + a_\ell \dot{u}_s + a_\ell \dot{w} = 0 \tag{6.55c}$$

where

$$f_{s2} = (1-n)\rho_s \beta_\ell r_s - S n \rho_\ell r_\ell \tag{6.55d}$$

and a_s and a_ℓ are given in (6.45c).

The combined mass conservation equation (6.53e) is

$$S(\alpha - n)\dot{u}^s_{i,i} + S n \dot{u}^s_{i,i} + S n \dot{w}_{i,i} + \frac{n \dot{p}_\ell}{Q} + f_{s1} = 0 \tag{6.55e}$$

where Q and f_{s1} are given by (6.52b). We neglect the term containing $\ddot{w}$ in (6.55b). Noting that $\mathbf{b}^s = \mathbf{b}^\ell = \mathbf{b}$, we evaluate $\dot{w}_i$ as

$$\dot{w}_i = k^*_{ij} \left[-\chi n p^\ell_{,j} + S n \rho_\ell b_j - S n \rho_\ell \ddot{u}^s_j + f_{s2} \dot{u}^s_j \right] \tag{6.55f}$$

where

$$k^{*^{-1}}_{ij} = S^2 n^2 \bar{k}^{-1}_{ij} - f_{s2} \delta_{ij} \tag{6.55g}$$

Neglecting the term containing $\ddot{w}$ and substituting for $\dot{w}_i$ from (6.55f), (6.55c) becomes

$$\sigma'_{ij,j} - \chi \alpha^*_{ij} p^\ell_{,j} + \rho^*_{ij} b_j - \rho^*_{ij} \ddot{u}^s_j + a^*_{ij} \dot{u}^s_j = 0 \tag{6.55h}$$

where

$$\rho^*_{ij} = \rho\delta_{ij} + a_\ell k^*_{ij} Sn\rho_\ell; \quad \alpha^*_{ij} = \alpha\delta_{ij} + a_\ell k^*_{ij} n; \\ a^*_{ij} = a_s\delta_{ij} + a_\ell\delta_{ij} + a_\ell f_{s2} k^*_{ij} \tag{6.55i}$$

Evaluating $\dot{w}_{i,i}$ from (6.55f) and substituting into (6.55e):

$$S\alpha\dot{u}^s_{i,i} + Sn\left\{k^*_{ij}\left[-\chi np^\ell_{,j} + Sn\rho_\ell b_j - Sn\rho_\ell \ddot{u}^s_j + f_{s2}\dot{u}^s_j\right]\right\}_{,i} + \frac{n\dot{p}_\ell}{Q} + f_{s1} = 0 \tag{6.55j}$$

6.4.2 Summary of Balance Equations for the Approximate u–p Formulation

The linear momentum equations for the solid phase, fluid phase, and mixture, respectively, are ((6.55a), (6.55b), and (6.55h))

$$\sigma'_{ij,j} - (\alpha - n)\chi p^\ell_{,i} + (1-n)\rho_s b_i - (1-n)\rho_s \ddot{u}^s_i - (1-n)\rho_s r_s \dot{u}^s_i + S^2 n^2 \bar{k}^{-1}_{ij} \dot{w}_j = 0 \tag{6.56a}$$

$$-\chi np^\ell_{,i} + Sn\rho_\ell b_i - Sn\rho_\ell \ddot{u}^s_i + f_{s2}\dot{u}^s_i + f_{s2}\dot{w}_i - S^2 n^2 \bar{k}^{-1}_{ij} \dot{w}_j = 0 \tag{6.56b}$$

$$\sigma'_{ij,j} - \chi\alpha^*_{ij} p^\ell_{,j} + \rho^*_{ij} b_j - \rho^*_{ij} \ddot{u}^s_j + a^*_{ij} \dot{u}^s_j = 0 \tag{6.56c}$$

$$b_i = b^\ell_i = b^s_i$$

where ρ^*_{ij}, α^*_{ij}, and a^*_{ij} are given by (6.55i), and f_{s2} is given by (6.55d).

From the mass equation for the solid (6.53d):

$$\dot{n} = (\alpha - n)\dot{u}^s_{i,i} - r_s(1-n) + \frac{\dot{p}}{K_s}(\alpha - n) \tag{6.56d}$$

The combined mass conservation equation is (6.55j)

$$S\alpha\dot{u}^s_{i,i} + Sn\left\{k^*_{ij}\left[-\chi np^\ell_{,j} + Sn\rho_\ell b_j - Sn\rho_\ell \ddot{u}^s_j + f_{s2}\dot{u}^s_j\right]\right\}_{,i} + \frac{n\dot{p}_\ell}{Q} + f_{s1} = 0 \tag{6.56e}$$

where Q and f_{s1} are given by (6.52b).

The velocity of liquid relative to the solid phase (6.55f) is

$$\dot{w}_i = k^*_{ij}\left[-\chi np^\ell_{,j} + Sn\rho_\ell b_j - Sn\rho_\ell \ddot{u}^s_j + f_{s2}\dot{u}^s_j\right] \tag{6.56f}$$

Table 6.6 Unknowns in the approximate $u-p$ formulation

Variable	σ'	p_ℓ	S	n	ρ_s	ρ_ℓ	$\mathbf{u}^s$	$\mathbf{w}$	$\mathbf{f}^d$	Total
No. unknowns	6	1	1	1	1	1	3	3	3	20

Table 6.7 Equations typically available in the $u-p$ formulation

Principle	Relation	No. equations
Linear momentum: solid	Equation (6.56a)	3
Linear momentum: liquid	Equation (6.56b)	3
Mass conservation: solid	Equation (6.56d)	1
Mass conservation: combined	Equation (6.56e)	1
Constitutive equation	$\boldsymbol{\sigma}'$ *vs* $\boldsymbol{\varepsilon}^s$ (6.48a)	6
Constitutive equation	p_s *vs* ρ_s (6.48f)	1
Constitutive equation	p_ℓ *vs* ρ_ℓ (6.48d)	1
Constitutive equation	$p_c = (p_g - p_\ell)$ *vs* S (6.50a)	1
Darcy's law for fluid drag	$\mathbf{f}^d$ *vs* $(\mathbf{u}^\ell - \mathbf{u}^s)$ (6.49)	3
Total		20

where k^*_{ij} is given by (6.55g).

The unknown variables of the problem in this formulation and the equations available for the solution are summarized in Tables 6.6 and 6.7.

In the finite element implementation of the equations, only the linear momentum equation for the mixture (6.56c) and the mass conservation for the mixture (6.56e) need to be spatially discretized. The primary unknowns at the nodes will then be $(\mathbf{u}^s, p^\ell)$. Among the remaining variables (i.e., secondary variables), S, n, ρ_s, ρ_ℓ, and $\dot{\mathbf{w}}$ are obtained using (6.50a), (6.56d), (6.48f), (6.48d), and (6.56f), respectively. Once the skeleton strain $\boldsymbol{\varepsilon}^s$ is calculated from $\mathbf{u}^s$, the effective stress $\boldsymbol{\sigma}'$ is calculated by integrating (6.48a).

The case of a saturated porous media is obtained as a special case of the above case of the unsaturated media with $\chi = S = 1$. The equations can be further simplified to address the fully coupled problem involving saturated sand. The $u-p$ formulation of this problem is addressed in Problem 6.9.

The formulations presented by different investigators differ from each other slightly. Here we point out the differences between the equations presented in this section and those of Schrefler and Simoni (1995) and Zienkiewicz et al. (1990a, b, 1999). Setting $r_\ell = r_s = 0$, one obtains $f_{s1} = f_{s2} = 0$, $a_s = a_\ell = 0$, $k^*_{ij} = k_{ij}/(S^2 n^2)$, $\alpha^*_{ij} = \alpha\delta_{ij}$, $a^*_{ij} = o_{ij}$, and $\rho^*_{ij} = \rho\delta_{ij}$, where o_{ij} is the null tensor. The linear momentum equation for the mixture (6.56c) then is

$$\sigma'_{ij,j} - \chi\alpha p^\ell_{,i} + \rho b_i - \rho\ddot{u}^s_i = 0 \tag{6.57a}$$

Which is identical to that in Zienkiewicz et al. (1999) and in Schrefler and Simoni (1995).

The mass conservation equation for the mixture (6.56e) is

$$\alpha\dot{u}^s_{i,i} + \frac{1}{S}\left\{k_{ij}\left[-\frac{\chi}{S}p^\ell_{,j} + \rho_\ell b_j - \rho_\ell\ddot{u}^s_j\right]\right\}_{,i} + \left[\frac{n}{K_\ell} + \frac{\chi(\alpha - n)}{K_s}\right]\dot{p}_\ell = 0 \tag{6.57b}$$

The corresponding equation in Zienkiewicz et al. (1999) is

$$\alpha \ddot{u}^s_{i,i} + \left\{k_{ij}\left[-p^\ell_{,j} + S\rho_\ell b^\ell_j - S\rho_\ell \ddot{u}^s_j\right]\right\}_{,i} + \left[\frac{nS}{K_\ell} + \frac{(\alpha - n)\chi}{K_s}\right]\dot{p} = 0 \tag{6.57c}$$

which is different from (6.57b); observe where S and χ appear in the two equations. The equation in Schrefler and Simoni (1995) is

$$\alpha S \ddot{u}^s_{i,i} + \left\{k_{ij}\left[-p^\ell_{,j} + \rho_\ell b^\ell_j - \rho_\ell \ddot{u}^s_j\right]\right\}_{,i} + \left[\frac{Sn}{K_\ell} + \frac{S^2(\alpha - n)}{K_s}\right]\dot{p}_\ell = 0 \tag{6.57d}$$

Equations (6.57b) and (6.57d) are identical when $\chi = S$. This is due to the assumption of $\chi = S$ in the effective stress equation (6.37) used by Schrefler and Simoni (1995). Furthermore, Schrefler and Simoni (1995) consider thermal effects, which are neglected here. They neglect the phase changes from solid to fluid, which are considered here. Apart from these differences, the formulation presented in this book is similar to that of Schrefler and Simoni (1995). For the case of the saturated particulate material ($S = 1$), it is easily verified that (6.57b), (6.57c), and (6.57d) coincide with each other.

6.5 Terzaghi's One-Dimensional Consolidation Theory

In this section, we will see how the widely used Terzaghi's one-dimensional consolidation theory is obtained as a special case of the general porous media theory presented in this chapter. We will start with the equations associated with the $u - p$ formulation (Sect. 6.4.2) and make appropriate assumptions to arrive at the one-dimensional consolidation equations.

The material is assumed to be fully saturated; hence $S = \chi = 1$ and $r_{SWCC} = 0$. It follows from (6.47a) that $p = p_\ell$. It is further assumed that for the material of interest, $\alpha = 1$ (e.g., soils). As in Sect. 6.4.2, setting $r_\ell = r_s = 0$ (no phase changes), one obtains $f_{s1} = f_{s2} = 0$, $a_s = a_\ell = 0$, $k^*_{ij} = k_{ij}/n^2 = k\delta_{ij}/n^2$ (isotropic material), $\alpha^*_{ij} = \alpha\delta_{ij} = \delta_{ij}$, $a^*_{ij} = o_{ij}$, and $\rho^*_{ij} = \rho\delta_{ij}$, where o_{ij} is a null tensor. The solid and liquid phases are assumed to be incompressible; then $K_s = K_\ell = \infty \Rightarrow 1/Q = 0$. It is also assumed that the problem is static (i.e., $\ddot{\mathbf{u}}^s = \ddot{\mathbf{w}} = 0$) and the body forces are absent ($\mathbf{b}^s = \mathbf{b}^\ell = 0$). Then the linear momentum equation for the liquid phase (6.56b) and the mixture (6.56c), respectively, are

$$-np_{,i} - n^2\bar{k}^{-1}_{ij}\dot{w}_j = 0 \tag{6.58a}$$

$$\sigma'_{ij,j} - p_{,i} = 0 \tag{6.58b}$$

The mass conservation (6.56e) is

$$\dot{u}^s_{i,i} - \left\{\bar{k}p_{,i}\right\}_{,i} = 0 \tag{6.58c}$$

The problem is assumed to be one-dimensional in direction 3, i.e.,

$$\dot{w}_1 = \dot{w}_2 = 0$$

$$p_{,1} = p_{,2} = 0$$

Noting that $\varepsilon_v = u^s_{i,i}$ where ε_v is the volumetric strain, and $\bar{k} = k/\gamma_\ell$, (6.58c) becomes

$$\frac{\partial \varepsilon_v}{\partial t} = \frac{k}{\gamma_\ell}\frac{\partial^2 p}{\partial x_3^2} \tag{6.58d}$$

We assume that the only stress component that changes during loading is σ_{33}. Equation (6.58b) in direction 3 is

$$\frac{\partial \sigma'_{33}}{\partial x_3} = \frac{\partial p}{\partial x_3} \Rightarrow \frac{\partial \sigma'_{33}}{\partial t} = \frac{\partial p}{\partial t} \tag{6.58e}$$

Let us assume that $\Delta e = a_v \Delta\sigma_{33}$, where a_v is coefficient of compressibility. Let us further assume that a_v remains a constant (i.e., independent of the changes in stress and time). Noting that $\Delta e = (1 + e_0)\Delta\varepsilon_v$, where e is the void ratio and e_0 is its initial value, we have

$$\Delta\varepsilon_v = \frac{a_v}{1 + e_0}\Delta\sigma_{33}$$

$$\frac{\partial \varepsilon_v}{\partial t} = \frac{a_v}{1 + e_0}\frac{\partial \sigma_{33}}{\partial t} = m_v\frac{\partial \sigma_{33}}{\partial t}; \quad m_v = \frac{a_v}{1 + e_0} \tag{6.58f}$$

where m_v is the coefficient of volumetric compressibility. Now we combine (6.58e), (6.58f), and (6.58d):

$$\begin{aligned} \frac{\partial p}{\partial t} &= \frac{\partial \sigma'_{33}}{\partial t} = \frac{1}{m_v}\frac{\partial \varepsilon_v}{\partial t} = \frac{k}{m_v\gamma_\ell}\frac{\partial^2 p}{\partial x_3^2} = C_v\frac{\partial^2 p}{\partial x_3^2} \\ \frac{\partial p}{\partial t} &= C_v\frac{\partial^2 p}{\partial x_3^2} \end{aligned} \tag{6.58g}$$

$$C_v = \frac{k}{m_v\gamma_\ell} \tag{6.58h}$$

where C_v is the coefficient of consolidation. Equation (6.58g) is the consolidation equation from Terzaghi's one-dimensional consolidation theory.

Problems

Problem 6.1 In the approximate analysis presented in Sect. 6.1.2 (6.6)–(6.12), various assumptions have been made. List these assumptions. Then present a critical discussion about each of these assumptions, stating with reasons whether you agree or disagree. If you disagree, state how you would improve.

Problem 6.2 Considering a saturated soil with immobile solid phase and the equilibrium of the skeleton shown in Fig. 6.3b, derive (6.15d).

Problem 6.3 Considering an unsaturated soil with immobile solid phase, show that

(a) The true velocity is related to the average velocity by $\dot{w}^{ave} = Sn\dot{u}_w$
(b) The drag force per unit volume is given by $f^d = n^2S^2\bar{k}^{-1}\dot{w}$

Problem 6.4 Prove the relationship given in (6.20a).

Problem 6.5 Prove the relationship given in (6.25).

Problem 6.6 Prove the relationship given in (6.26a).

Partial answer:
We will use the following result for the box product (or triple scalar product) λ (refer to any book on continuum mechanics for proof):

$$\lambda = \mathbf{a}.\mathbf{b} \times \mathbf{c} = \varepsilon_{ijk}a_ib_jc_k = \begin{vmatrix} a_1 & a_2 & a_3 \\ b_1 & b_2 & b_3 \\ c_1 & c_2 & c_3 \end{vmatrix}$$

where ε_{ijk} is known as the alternating tensor or the permutation symbol, whose coefficients are as follows: $\varepsilon_{ijk} = \mathbf{1}$ if i, j, k are even permutation of 1, 2, 3; $\varepsilon_{ijk} = -1$ if i, j, k are odd permutation of 1, 2, 3; $\varepsilon_{ijk} = 0$ otherwise.

If **a**, **b**, and **c** are vectors representing the sides of a parallelepiped, then λ represents its volume.

Let

$$F_{ij} = x_{i,J} = \frac{\partial x_i}{\partial X_j}.$$

Note that we used lower and upper case subscripts to denote material and spatial coordinates. Then

$$J = |\mathbf{F}| = \begin{vmatrix} F_{11} & F_{12} & F_{13} \\ F_{21} & F_{22} & F_{23} \\ F_{31} & F_{32} & F_{33} \end{vmatrix} = \varepsilon_{IJK}x_{1,I}x_{2,J}x_{3,K}$$

Now take the material derivative of J. Noting that a determinant is zero when two of its rows are identical, arrive at the answer.

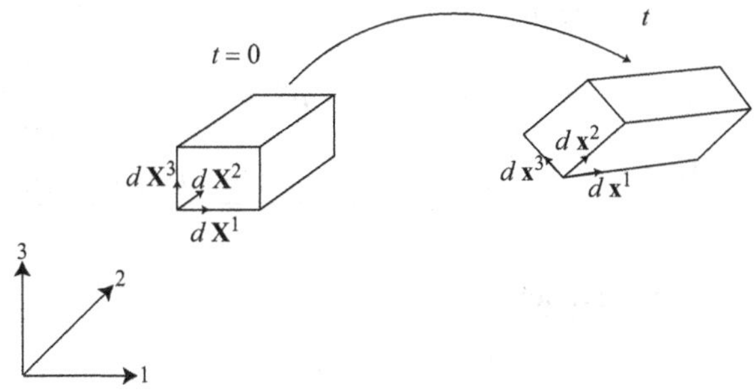

Fig. 6.6 Distortion of a parallelepiped

Problem 6.7 Prove the relationship given in (6.26b).

Partial Answer:
Consider a parallelepiped in the original configuration with its sides parallel to the coordinates axes with lengths (dX_1, dX_2, dX_2) as shown in Fig. 6.6. Denoting the base vectors (i.e., unit vectors) along the axes as $(\mathbf{e}_1, \mathbf{e}_2, \mathbf{e}_3)$, the initial volume of the parallelepiped is

$$dV = (dX_1\mathbf{e}_1) \times (dX_2\mathbf{e}_2).(dX_3\mathbf{e}_3) = dX_1 dX_2 dX_3$$

Now since the motion considered is assumed to be continuous, the particles in the initial parallelepiped remain in it while it moves with its shape becoming distorted as shown in Fig. 6.1. Its volume in the current configuration is

$$dv = (d\mathbf{x}^1) \times (d\mathbf{x}^2).(d\mathbf{x}^3) = \varepsilon_{ijk} dx_i^1 dx_j^2 dx_k^3$$

Take the material derivative of dv and simplify using (6.26a). Then noting that $d\mathbf{X}^1 = (dX_1, 0, 0)$ etc., arrive at the required result.

Problem 6.8 A problem of considerable interest in earthquake analysis is fully coupled modeling of saturated sand (for predicting possible sand liquefaction). The relevant equations for the $u - u - p$ formulation may be obtained from those in Sect. 6.3 (summarized in Sect. 6.3.4) as a special case under the following assumptions: (a) The material is fully saturated $(\chi = S = 1)$, (b) the phase changes are negligible $(r_\ell = r_s = 0)$, and (c) the change in porosity is negligible $\dot{n} = 0$. Show that the relevant equations are as follows:

Linear momentum for the solid phase:

$$\sigma'_{ij,j} - (\alpha - n)p_{,i} + (1 - n)\rho_s b_i^s - (1 - n)\rho_s \ddot{u}_i^s + n^2 \bar{k}_{ij}^{-1}(\dot{u}_j^\ell - \dot{u}_j^s) = 0$$

Linear momentum for the liquid phase:

$$-np_{,i} + n\rho_\ell b_i^\ell - n\rho_\ell \ddot{u}_i^\ell - n^2 \bar{k}_{ij}^{-1}(\dot{u}_j^\ell - \dot{u}_j^s) = 0$$

Linear momentum for the mixture:

$$\sigma'_{ij,j} - \alpha p_{,i} + \rho \bar{\bar{b}}_i - \rho \ddot{\bar{\bar{u}}}_i = 0$$

$$\bar{\bar{\mathbf{b}}} = \frac{\mathbf{b}^s(1-n)\rho_s + \mathbf{b}^\ell n\rho_\ell}{\rho}; \quad \ddot{\bar{\bar{u}}}_i = \frac{\ddot{u}_i^s(1-n)\rho_s + \ddot{u}_i^\ell n\rho_\ell}{\rho}$$

Combined mass conservation:

$$(\alpha - n)\dot{u}^s_{i,i} + n\dot{u}^\ell_{i,i} + \frac{n\dot{p}}{Q} = 0$$

$$\frac{\mathbf{1}}{Q} = \frac{1}{K_\ell} + \frac{(\alpha - n)}{nK_s}$$

Problem 6.9 Considering the problem involving saturated sand addressed in Problem 6.8, simplify the equations in Sect. 6.4 (summarized in Sect. 6.4.2) and show that the equations associated with the $u-p$ formulation of the fully coupled problem are:

Linear momentum for the solid phase:

$$\sigma'_{ij,j} - (\alpha - n)p_{,i} + (1-n)\rho_s b_i - (1-n)\rho_s \ddot{u}_i^s + n^2 \bar{k}_{ij}^{-1}\dot{w}_j = 0$$

Linear momentum for the liquid phase:

$$-np_{,i} + n\rho_\ell b_i - n\rho_\ell \ddot{u}_i^s - n^2 \bar{k}_{ij}^{-1}\dot{w}_j = 0$$

Linear momentum for the mixture:

$$\sigma'_{ij,j} - \alpha p_{,i} + \rho b_i - \rho \ddot{u}_i^s = 0$$

The combined mass conservation:

$$\alpha \dot{u}^s_{i,i} + \left\{\bar{k}_{ij}\left[-p_{,j} + \rho_\ell b_j^\ell - \rho_\ell \ddot{u}_j^s\right]\right\}_{,i} + \frac{n\dot{p}}{Q} = 0$$

$$\frac{1}{Q} = \frac{1}{K_\ell} + \frac{(\alpha - n)}{nK_s}$$

$$\rho = (1-n)\rho_s + n\rho_\ell$$

$$b_i = b_i^\ell = b_i^s$$

Chapter 7
Finite Element Analysis of Porous Media

The equations governing the mechanics of porous media have been presented in Chap. 6. The general formulations (i.e., the full formulation presented in Sect. 6.2.4) were specialized to obtain two specific formulations: (a) The *u–u–p* formulation (Sect. 6.3.4) and (b) the *u–p* formulation (Sect. 6.4.2). In the present chapter, the weak forms and finite element equations are developed for the *u–p* formulation. The finite element equations for the full and *u–u–p* formulations may be obtained by following a procedure similar to that in this chapter (see Problem 7.1 for the *u–u–p* formulation).

7.1 The Boundary Value Problem Definition

To define the boundary value problem properly, we repeat the relevant equations (from Sect. 6.4.2) in this section. The linear momentum equation for the mixture (6.56c) and the combined mass conservation equation (6.56e), respectively, are

$$\sigma'_{ij,j} - \chi\alpha^*_{ij}p^\ell_{,j} + \rho^*_{ij}b_j - \rho^*_{ij}\ddot{u}^s_j + a^*_{ij}\dot{u}^s_j = 0 \tag{7.1a}$$

$$S\alpha\dot{u}^s_{i,i} + Sn\left\{k^*_{ij}\left[-\chi n p^\ell_{,j} + Sn\rho_\ell b_j - Sn\rho_\ell\ddot{u}^s_j + f_{s2}\dot{u}^s_j\right]\right\}_{,i} + \frac{n\dot{p}_\ell}{Q} + f_{s1} = 0 \tag{7.1b}$$

where

$$\begin{aligned}&\rho^*_{ij} = \rho\delta_{ij} + a_\ell k^*_{ij}Sn\rho_\ell; \quad \alpha^*_{ij} = \alpha\delta_{ij} + a_\ell k^*_{ij}n; \quad a^*_{ij} = a_s\delta_{ij} + a_\ell\delta_{ij} + a_\ell f_{s2}k^*_{ij};\\ &k^{*^{-1}}_{ij} = S^2n^2\bar{k}^{-1}_{ij} - f_{s2}\delta_{ij}\end{aligned} \tag{7.1c}$$

$$f_{s1} = -Sr_s(1-n) + \beta_\ell r_s(1-n)\frac{\rho_s}{\rho_\ell} - r_\ell Sn \tag{7.1d}$$

A. Anandarajah, *Computational Methods in Elasticity and Plasticity: Solids and Porous Media*, DOI 10.1007/978-1-4419-6379-6_7,

$$f_{s2} = (1-n)\rho_s\beta_\ell r_s - Sn\rho_\ell r_\ell \tag{7.1e}$$

$$\frac{1}{Q} = \frac{S}{K_\ell} + \frac{S\chi(\alpha-n)}{nK_s} + r_{SWCC} \tag{7.1f}$$

$$k_{ij}^{*^{-1}} = S^2 n^2 \bar{k}_{ij}^{-1} - f_{s2}\delta_{ij} \tag{7.1g}$$

$$\begin{aligned} a_s &= -r_s\rho_s(1-n) \\ a_\ell &= (1-n)\rho_s\beta_\ell r_s - Sn\rho_\ell r_\ell \\ a_g &= \beta_g r_s(1-n)\rho_s + r_\ell Sn\rho_\ell \end{aligned} \tag{7.1h}$$

$$b_i = b_i^\ell = b_i^s$$

$$p = p^\ell$$

The definitions of α, χ, ρ and k_{ij} are found in (6.11d), (6.16d), (6.35d), and (6.37b). The constitutive equations are summarized in Sect. 6.3.2. Equations (7.1a) and (7.1b) are spatially discretized by the finite element method to evaluate the spatial values of the primary variables $(\mathbf{u}^s, p^\ell)$. Among the secondary variables, n (6.56d) and $\dot{\mathbf{w}}$ (6.56f) are obtained using

$$\dot{n} = (\alpha-n)\dot{u}_{i,i}^s - r_s(1-n) + \frac{\dot{p}^\ell}{K_s}(\alpha-n) \tag{7.1i}$$

$$\dot{w}_i = k_{ij}^*\left[-\chi np_{,j}^\ell + Sn\rho_\ell b_j^\ell - Sn\rho_\ell \ddot{u}_j^s + f_{s2}\ddot{u}_j^s\right] \tag{7.1j}$$

$\boldsymbol{\varepsilon}^s$ is obtained from $\mathbf{u}^s$ using the strain-displacement relations (Chap. 3). The other secondary variables S, ρ_s, ρ_ℓ, and $\boldsymbol{\sigma}'$ are obtained by integrating the constitutive equations (Table 6.8).

The definition of the boundary value problem requires specifications of boundary and initial conditions. As there are two types of primary variables, there are two sets of boundary conditions to be fulfilled. The boundary conditions based on $\mathbf{u}^s$ are

$$\mathbf{t} = \hat{\mathbf{t}} \quad \text{on} \quad \Gamma_t \tag{7.2a}$$

$$\mathbf{u}^s = \hat{\mathbf{u}}^s \quad \text{on} \quad \Gamma_u \tag{7.2b}$$

where $\Gamma = \Gamma_t \cup \Gamma_u$, $\hat{\mathbf{t}}$, is the prescribed traction on Γ_t and $\hat{\mathbf{u}}^s$ is the prescribed skeleton displacement on Γ_u.

The boundary conditions based on p^ℓ are

$$q_n^* = \hat{q}_n^* \quad \text{on} \quad \Gamma_q \tag{7.2c}$$

$$p^\ell = \hat{p}^\ell \quad \text{on} \quad \Gamma_p \tag{7.2d}$$

where $q_n^* = \mathbf{q}^{*\mathrm{T}}\mathbf{n}$; $\mathbf{q}^{*\mathrm{T}} = \mathbf{k}^*\mathbf{g}$; $\mathbf{g}^\mathrm{T} = \left\{p^\ell_{,1}, p^\ell_{,2}, p^\ell_{,3}\right\}$ and $\Gamma = \Gamma_q \cup \Gamma_p$. $\hat{q}_n^*$ is the prescribed liquid flux through Γ_q and $\hat{p}^\ell$ is the prescribed liquid pressure on Γ_p. We will identify which ones are natural and which ones are essential boundary conditions later, after determining the weak forms. The description of weak forms is presented in Chap. 5.

In the following sections, we present the weak forms and finite element formulations for the problem; for additional information, the reader may consult Selvadurai (1996), Lewis and Schrefler (1998), Schrefler and Simoni (1995), and Zienkiewicz et al. (1990b).

7.2 The Weak Forms

We will first find the weak form of (7.1a) by following the procedure presented in Chap. 5. Let us multiply (7.1a) by a test function $\mathbf{v}$, integrate over the volume of the domain and set it to zero as

$$\int_\Omega v_i\left\{\sigma'_{ij,j} - \chi\alpha^*_{ij}p^\ell_{,j} + b^m_i\right\}\mathrm{d}v = 0 \tag{7.3a}$$

where

$$b^m_i = \rho^*_{ij}b_j - \rho^*_{ij}\ddot{u}^s_j + a^*_{ij}\dot{u}^s_j \tag{7.3b}$$

Continuing on, it can be shown that

$$\int_\Omega v_i\sigma'_{ij,j}\,\mathrm{d}v = -\int_\Omega \bar{\boldsymbol{\varepsilon}}^\mathrm{T}_v\boldsymbol{\sigma}'_v\,\mathrm{d}v + \int_\Gamma \mathbf{v}^\mathrm{T}\mathbf{t}'\,\mathrm{d}s \tag{7.3c}$$

$$\int_\Omega v_i\alpha^*_{ij}p^\ell_{,j}\,\mathrm{d}v = -\int_\Omega \bar{\boldsymbol{\varepsilon}}^\mathrm{T}_v\boldsymbol{\alpha}_v p^\ell\,\mathrm{d}v + \int_\Gamma \mathbf{v}^\mathrm{T}\mathbf{p}_s\,\mathrm{d}s \tag{7.3d}$$

$$\int_\Omega v_i b^m_i\,\mathrm{d}v = \int_\Omega \mathbf{v}^\mathrm{T}\mathbf{b}^m\,\mathrm{d}v = \int_\Omega \mathbf{v}^\mathrm{T}[\boldsymbol{\rho}^*\mathbf{b} - \boldsymbol{\rho}^*\ddot{\mathbf{u}}^s + \mathbf{a}^*\dot{\mathbf{u}}^s]\mathrm{d}v \tag{7.3e}$$

where

$$t'_{ij} = \sigma'_{ij}n_j$$

$$\bar{\varepsilon}_{ij} = \frac{\partial(v_i)}{\partial x_j};\quad \bar{\boldsymbol{\varepsilon}}^\mathrm{T}_v = \{\bar{\varepsilon}_{11}, \bar{\varepsilon}_{22}, \bar{\varepsilon}_{33}, 2\bar{\varepsilon}_{12}, 2\bar{\varepsilon}_{13}, 2\bar{\varepsilon}_{23}\}$$

$$\boldsymbol{\sigma}'^{\mathrm{T}}_v = \{\sigma'_{11}, \sigma'_{22}, \sigma'_{33}, \sigma'_{12}, \sigma'_{13}, \sigma'_{23}\}$$

$$\boldsymbol{\alpha}^{\mathrm{T}}_v = \{\alpha^*_{11}, \alpha^*_{22}, \alpha^*_{33}, \alpha^*_{12}, \alpha^*_{13}, \alpha^*_{23}\}$$

$$\mathbf{p}^{\mathrm{T}}_s = \{p^\ell n_1, p^\ell n_2, p^\ell n_3\}$$

Note that $\boldsymbol{\alpha}_v$ is the vector form of the tensor $\boldsymbol{\alpha}$. Similar notations are used for $\boldsymbol{\sigma}'$ and $\boldsymbol{\varepsilon}$.

Equation (7.3a) then becomes (after multiplying through by a negative sign):

$$\begin{aligned}&\int_\Omega \bar{\boldsymbol{\varepsilon}}^{\mathrm{T}}_v \boldsymbol{\sigma}'_v \,\mathrm{d}v - \chi \int_\Omega \bar{\boldsymbol{\varepsilon}}^{\mathrm{T}}_v \boldsymbol{\alpha}_v p^\ell \,\mathrm{d}v - \int_\Omega \mathbf{v}^{\mathrm{T}}[\boldsymbol{\rho}^*\mathbf{b} - \boldsymbol{\rho}^*\ddot{\mathbf{u}}^s + \mathbf{a}^*\dot{\mathbf{u}}^s]\mathrm{d}v \\ &\quad - \int_\Gamma \mathbf{v}^T \mathbf{t}' \,ds + \chi \int_\Gamma \mathbf{v}^T \mathbf{p}_s \,ds = 0\end{aligned} \tag{7.4a}$$

Combining the two boundary terms and noting that $\boldsymbol{\sigma} = \boldsymbol{\sigma}' - \chi p^\ell \boldsymbol{\delta}$ (6.12b) and (6.47a) and $\mathbf{t} = \boldsymbol{\sigma}\mathbf{n}$, we have

$$-\int_\Gamma \mathbf{v}^{\mathrm{T}}\mathbf{t}' \,\mathrm{d}s + \chi \int_\Gamma \mathbf{v}^{\mathrm{T}}\mathbf{p}_s \,\mathrm{d}s = -\int_\Gamma v_i \left[\sigma'_{ij} - \chi p^\ell \delta_{ij}\right] n_j \,\mathrm{d}s = -\int_\Gamma \mathbf{v}^{\mathrm{T}}\mathbf{t}\,\mathrm{d}s$$

Equation (7.4a) may be written as

$$\int_\Omega \bar{\boldsymbol{\varepsilon}}^{\mathrm{T}}_v \boldsymbol{\sigma}'_v \,\mathrm{d}v - \chi \int_\Omega \bar{\boldsymbol{\varepsilon}}^{\mathrm{T}}_v \boldsymbol{\alpha}_v p^\ell \,\mathrm{d}v - \int_\Omega \mathbf{v}^{\mathrm{T}}[\boldsymbol{\rho}^*\mathbf{b} - \boldsymbol{\rho}^*\ddot{\mathbf{u}}^s + \mathbf{a}^*\dot{\mathbf{u}}^s]\mathrm{d}v - \int_\Gamma \mathbf{v}^{\mathrm{T}}\mathbf{t}\,\mathrm{d}s = 0 \tag{7.4b}$$

This is the weak form of (7.1a).

Now we find the weak form of (7.1b). Following the same procedure as above and using a scalar test function ϕ

$$\int_\Omega \phi \left[S\alpha \dot{u}^s_{i,i} - Sn^2\chi\left\{k^*_{ij} p^\ell_{,j}\right\}_{,i} + Sn\left\{k^*_{ij} b^n_j\right\}_{,i} + \frac{n\dot{p}_\ell}{Q} + f_{s1}\right]\mathrm{d}v \tag{7.4c}$$

where $b^n_j = Sn\rho_\ell b^\ell_j - Sn\rho_\ell \ddot{u}^s_j + f_{s2}\dot{u}^s_j$

Expanding the first term, we have

$$\int_\Omega S\alpha\phi \dot{u}^s_{i,i} \,\mathrm{d}v = \int_\Omega S\alpha\phi \dot{\varepsilon}^s_{kk} \,\mathrm{d}v = \int_\Omega S\alpha\phi \mathbf{m}^{\mathrm{T}} \dot{\boldsymbol{\varepsilon}}^s_v \,\mathrm{d}v \tag{7.5a}$$

where

$$\mathbf{m} = \{1, 1, 1, 0, 0, 0\} \tag{7.5b}$$

$$\boldsymbol{\varepsilon}_v^{sT} = \left\{\varepsilon_{11}^s, \varepsilon_{22}^s, \varepsilon_{33}^s, \varepsilon_{12}^s, \varepsilon_{13}^s, \varepsilon_{23}^s\right\} \tag{7.5c}$$

Let us denote

$$a_i = k_{ij}^* p_{,j}^\ell \quad \text{or} \quad \mathbf{a} = \mathbf{k}^* \mathbf{g} \tag{7.6a}$$

where

$$\mathbf{g}^{\mathrm{T}} = \left\{p_{,1}^\ell, p_{,2}^\ell, p_{,3}^\ell\right\} \tag{7.6b}$$

Switching variables by integration by parts, the second term in (7.4c) becomes:

$$\begin{aligned}
\int_\Omega Sn^2\chi\phi\left\{k_{ij}^* p_{,j}^\ell\right\}_{,i} \mathrm{d}v &= \int_\Omega Sn^2\chi\phi a_{i,i}\, \mathrm{d}v \\
&= -\int_\Omega Sn^2\chi\left[a_1\phi_{,1} + a_2\phi_{,2} + a_3\phi_{,3}\right]\mathrm{d}v \\
&\quad + \int_\Gamma Sn^2\chi\phi[a_1 n_1 + a_2 n_2 + a_3 n_3]\mathrm{d}s \\
&= -\int_\Omega Sn^2\chi\boldsymbol{\psi}\mathbf{k}^*\mathbf{g}\, \mathrm{d}v + \int_\Gamma Sn^2\chi\phi q_n^*\, \mathrm{d}s
\end{aligned} \tag{7.6c}$$

where

$$\begin{aligned}
\boldsymbol{\psi}^{\mathrm{T}} &= \{\phi_{,1}, \phi_{,2}, \phi_{,3}\} \\
q_n^* &= \mathbf{q}^{*\mathrm{T}}\mathbf{n}; \quad \mathbf{q}^{*\mathrm{T}} = \mathbf{k}^*\mathbf{g}
\end{aligned} \tag{7.6d}$$

The third term in (7.4c) becomes

$$\begin{aligned}
\int_\Omega Sn\phi\left\{k_{ij}^* b_j^n\right\}_{,i} \mathrm{d}v &= \int_\Omega Sn\phi\left\{k_{ij}^*\left(Sn\rho_\ell b_j^\ell - Sn\rho_\ell \ddot{u}_j^s + f_{s2}\dot{u}_j^s\right)\right\}_{,i} \mathrm{d}v \\
&= 0 - \int_\Omega S^2 n^2 \rho_\ell \phi k_{ij}^* \ddot{u}_{j,i}^s\, \mathrm{d}v + \int_\Omega Snf_{s2}\phi k_{ij}^* \dot{u}_{j,i}^s\, \mathrm{d}v \\
&= -\int_\Omega S^2 n^2 \rho_\ell \phi \mathbf{k}_v^{*\mathrm{T}} \ddot{\boldsymbol{\varepsilon}}^s\, \mathrm{d}v + \int_\Omega Snf_{s2}\phi \mathbf{k}_v^{*\mathrm{T}} \dot{\boldsymbol{\varepsilon}}^s\, \mathrm{d}v
\end{aligned} \tag{7.7a}$$

where

$$\mathbf{k}_v^{*\mathrm{T}} = \{k_{11}^*, k_{22}^*, k_{33}^*, k_{12}^*, k_{13}^*, k_{23}^*\} \tag{7.7b}$$

The fourth and fifth terms in (7.4c) stay unchanged. Assembling the parts, (7.4c) becomes

$$\int_\Omega S\alpha\phi^{\mathrm{T}}\mathbf{m}^{\mathrm{T}}\dot{\boldsymbol{\varepsilon}}^s\,\mathrm{d}v + \int_\Omega Sn^2\chi\boldsymbol{\psi}^{\mathrm{T}}\mathbf{k}^*\mathbf{g}\,\mathrm{d}v - \int_\Gamma Sn^2\chi\phi q_n^*\,\mathrm{d}s - \int_\Omega S^2n^2\rho_\ell\phi\mathbf{k}_v^{*\mathrm{T}}\ddot{\boldsymbol{\varepsilon}}^s\,\mathrm{d}v$$
$$+ \int_\Omega Snf_{s2}\phi\mathbf{k}_v^{*\mathrm{T}}\dot{\boldsymbol{\varepsilon}}^s\,\mathrm{d}v + \int_\Omega \frac{n}{Q}\phi\dot{p}_\ell\,\mathrm{d}v + \int_\Omega \phi f_{s1}\,\mathrm{d}v = 0 \tag{7.8}$$

This is the weak form of (7.1b).

7.3 The Natural and Forced Boundary Conditions

With respect to the primary variable $\mathbf{u}^s$, from the boundary term in (7.4b), it can be verified (using the criteria described in Chap. 5) that (7.2a) is the natural boundary condition and (7.2b) is the forced boundary condition. Similarly, with respect to the primary variable p_ℓ, it may be seen from the boundary term in (7.8) that (7.2c) is the natural boundary condition and (7.2d) is the forced boundary condition.

7.4 The Finite Element Discretization and Element Matrices

With $\mathbf{v}=\delta\mathbf{u}$ and $\phi=\delta p$ (leaving the superscript 's' and subscript 'ℓ' out for simplicity of notations), the weak form of the linear momentum equation of the mixture (7.4b), and the mass conservation of the mixture (7.8), respectively, are

$$\int_\Omega \delta\boldsymbol{\varepsilon}_v^{\mathrm{T}}\boldsymbol{\sigma}'_v\,\mathrm{d}v - \chi\int_\Omega \delta\boldsymbol{\varepsilon}_v^{\mathrm{T}}\boldsymbol{\alpha}_v p\,\mathrm{d}v - \int_\Omega \delta\mathbf{u}^{\mathrm{T}}[\boldsymbol{\rho}^*\mathbf{b} - \boldsymbol{\rho}^*\ddot{\mathbf{u}} + \mathbf{a}^*\dot{\mathbf{u}}]\mathrm{d}v - \int_\Gamma \delta\mathbf{u}^{\mathrm{T}}\mathbf{t}\,\mathrm{d}s = 0 \tag{7.9a}$$

$$\int_\Omega (S\alpha)\delta p^{\mathrm{T}}\mathbf{m}^{\mathrm{T}}\dot{\boldsymbol{\varepsilon}}_v\,\mathrm{d}v + \int_\Omega (Sn^2\chi)\delta\mathbf{g}^{\mathrm{T}}\mathbf{k}^*\mathbf{g}\,\mathrm{d}v - \int_\Gamma (Sn^2\chi)\delta p q_n^*\,\mathrm{d}s$$
$$- \int_\Omega [S^2n^2\rho_\ell]\delta p\mathbf{k}_v^{*\mathrm{T}}\ddot{\boldsymbol{\varepsilon}}_v\,\mathrm{d}v + \int_\Omega (Snf_{s2})\delta p\mathbf{k}_v^{*\mathrm{T}}\dot{\boldsymbol{\varepsilon}}_v\,\mathrm{d}v \tag{7.9b}$$
$$+ \int_\Omega \left(\frac{n}{Q}\right)\delta p\dot{p}\,\mathrm{d}v + \int_\Omega \delta p f_{s1}\,\mathrm{d}v = 0$$

where

$$\varepsilon_{ij} = \frac{\partial u_i}{\partial x_j}; \quad \boldsymbol{\varepsilon}_v^{\mathrm{T}} = \{\varepsilon_{11}, \varepsilon_{22}, \varepsilon_{33}, 2\varepsilon_{12}, 2\varepsilon_{13}, 2\varepsilon_{23}\} \tag{7.9c}$$

$$\mathbf{g}^{\mathrm{T}} = \{p_{,1}, p_{,2}, p_{,3}\} \tag{7.9d}$$

$$\boldsymbol{\sigma}'^{\mathrm{T}}_v = \{\sigma'_{11}, \sigma'_{22}, \sigma'_{33}, \sigma'_{12}, \sigma'_{13}, \sigma'_{23}\} \tag{7.9e}$$

$$\boldsymbol{\alpha}_v^{\mathrm{T}} = \{\alpha^*_{11}, \alpha^*_{22}, \alpha^*_{33}, \alpha^*_{12}, \alpha^*_{13}, \alpha^*_{23}\} \tag{7.9f}$$

$$\mathbf{k}_v^{*\mathrm{T}} = \{k^*_{11}, k^*_{22}, k^*_{33}, k^*_{12}, k^*_{13}, k^*_{23}\} \tag{7.9g}$$

The approximations to **u** and p must be C_0 continuous, and satisfy the forced boundary conditions. The element formulations can be achieved in one of two ways: (a) use the same order of spatial approximations for **u** and p, but use reduced integration to avoid hour-glass modes or (b) use a higher order approximation for **u** than for p to avoid hour-glass modes. With this in mind, we develop the element equations here assuming different approximations for **u** and p as

$$\mathbf{u} = \mathbf{N}_u \hat{\mathbf{u}}; \quad p = \mathbf{N}_p \hat{\mathbf{p}} \tag{7.10}$$

where $\mathbf{N}_u$ and $\mathbf{N}_p$ are the element shape functions for **u** and p, respectively. $\hat{\mathbf{u}}$ and $\hat{\mathbf{p}}$ are the nodal unknowns corresponding to **u** and P, respectively. The gradients of **u** (7.9c) and p (7.9d) are expressed as

$$\boldsymbol{\varepsilon} = \mathbf{B}_u \hat{\mathbf{u}}; \quad \mathbf{g} = \mathbf{B}_p \hat{\mathbf{p}} \tag{7.11}$$

Using the approximations given by (7.10) and (7.11), and the linearized version of the stress-strain relation (6.48a), (7.9a), and (7.9b) at the element level become

$$\begin{aligned}
&\delta\hat{\mathbf{u}}^{\mathrm{T}}\left[\int_{\Omega_e} \mathbf{B}_u^{\mathrm{T}}\mathbf{D}\mathbf{B}_u \,\mathrm{d}v\right]\hat{\mathbf{u}} - \delta\hat{\mathbf{u}}^{\mathrm{T}}\left[\int_{\Omega_e} \chi\mathbf{B}_u^{\mathrm{T}}\boldsymbol{\alpha}_v\mathbf{N}_p \,\mathrm{d}v\right]\hat{\mathbf{p}} \\
&- \delta\hat{\mathbf{u}}^{\mathrm{T}}\left[\int_{\Omega_e} \mathbf{N}_u^{\mathrm{T}}\boldsymbol{\rho}^*\mathbf{b} \,\mathrm{d}v\right] + \delta\hat{\mathbf{u}}^{\mathrm{T}}\left[\int_{\Omega_e} \mathbf{N}_u^{\mathrm{T}}\boldsymbol{\rho}^*\mathbf{N}_u \,\mathrm{d}v\right]\ddot{\hat{\mathbf{u}}} \\
&- \delta\hat{\mathbf{u}}^{\mathrm{T}}\left[\int_{\Omega_e} \mathbf{N}_u^{\mathrm{T}}\mathbf{a}^*\mathbf{N}_u \,\mathrm{d}v\right]\dot{\hat{\mathbf{u}}} - \delta\hat{\mathbf{u}}^{\mathrm{T}}\left[\int_{\Gamma_e} \mathbf{N}_u^{\mathrm{T}}\hat{\mathbf{t}} \,\mathrm{d}s\right] = 0
\end{aligned} \tag{7.12}$$

$$
\begin{aligned}
&\delta\hat{\mathbf{p}}^{\mathrm{T}}\left[\int_{\Omega_e}\left(S\alpha\right)\mathbf{N}_p^{\mathrm{T}}\mathbf{m}^{\mathrm{T}}\mathbf{B}_u\,\mathrm{d}v\right]\dot{\hat{\mathbf{u}}} \\
&+\delta\hat{\mathbf{p}}^{\mathrm{T}}\left[\int_{\Omega_e}\left(Sn^2\chi\right)\mathbf{B}_p^{\mathrm{T}}\mathbf{k}^*\mathbf{B}_p\,\mathrm{d}v\right]\hat{\mathbf{p}}-\delta\hat{\mathbf{p}}^{\mathrm{T}}\left[\int_{\Gamma_e}\left(Sn^2\chi\right)\mathbf{N}_p^{\mathrm{T}}q_n^*\,\mathrm{d}s\right] \\
&-\delta\hat{\mathbf{p}}^{\mathrm{T}}\left[\int_{\Omega_e}\left(S^2n^2\rho_\ell\right)\mathbf{N}_p^{\mathrm{T}}\mathbf{k}_v^{*\mathrm{T}}\mathbf{B}_u\,\mathrm{d}v\right]\ddot{\hat{\mathbf{u}}}+\delta\hat{\mathbf{p}}^{\mathrm{T}}\left[\int_{\Omega_e}\left(Snf_{s2}\right)\mathbf{N}_p^{\mathrm{T}}\mathbf{k}_v^{*T}\mathbf{B}_u\,\mathrm{d}v\right]\dot{\hat{\mathbf{u}}} \\
&+\delta\hat{\mathbf{p}}^{\mathrm{T}}\left[\int_{\Omega_e}\left(\frac{n}{Q}\right)\mathbf{N}_p^{\mathrm{T}}\mathbf{N}_p\,\mathrm{d}v\right]\dot{\hat{\mathbf{p}}}+\delta\hat{\mathbf{p}}^{\mathrm{T}}\left[\int_{\Omega_e}\mathbf{N}_p^{\mathrm{T}}f_{s1}\,\mathrm{d}v\right]=0 \qquad (7.13)
\end{aligned}
$$

These weak forms lead to the following element matrix equation:

$$
\begin{bmatrix}\mathbf{M}_{uu} & 0\\ \mathbf{M}_{pu} & 0\end{bmatrix}\begin{bmatrix}\ddot{\hat{\mathbf{u}}}\\ \ddot{\hat{\mathbf{p}}}\end{bmatrix}+\begin{bmatrix}\mathbf{C}_{uu} & 0\\ \mathbf{C}_{pu} & \mathbf{C}_{pp}\end{bmatrix}\begin{bmatrix}\dot{\hat{\mathbf{u}}}\\ \dot{\hat{\mathbf{p}}}\end{bmatrix}+\begin{bmatrix}\mathbf{K}_{uu} & \mathbf{K}_{up}\\ 0 & \mathbf{K}_{pp}\end{bmatrix}\begin{bmatrix}\hat{\mathbf{u}}\\ \hat{\mathbf{p}}\end{bmatrix}=\begin{bmatrix}\mathbf{f}_u\\ \mathbf{f}_p\end{bmatrix} \qquad (7.14a)
$$

where

$$
\begin{aligned}
&\mathbf{M}_{uu}=\int_{\Omega_e}\mathbf{N}_u^{\mathrm{T}}\boldsymbol{\rho}^*\mathbf{N}_u\,\mathrm{d}v;\quad \mathbf{C}_{uu}=-\int_{\Omega_e}\mathbf{N}_u^{\mathrm{T}}\mathbf{a}^*\mathbf{N}_u\,\mathrm{d}v;\quad \mathbf{K}_{uu}=\int_{\Omega_e}\mathbf{B}_u^{\mathrm{T}}\mathbf{D}\mathbf{B}_u\,\mathrm{d}v;\\
&\mathbf{K}_{up}=-\int_{\Omega_e}\chi\mathbf{B}_u^{\mathrm{T}}\boldsymbol{\alpha}_v\mathbf{N}_p\,\mathrm{d}v;\quad \mathbf{f}_u=\int_{\Omega_e}\mathbf{N}_u^{\mathrm{T}}\boldsymbol{\rho}^*\mathbf{b}\,\mathrm{d}v+\int_{\Gamma_e}\mathbf{N}_u^{\mathrm{T}}\hat{\mathbf{t}}\,\mathrm{d}s \qquad (7.14b)
\end{aligned}
$$

$$
\begin{aligned}
&\mathbf{M}_{pu}=-\int_{\Omega_e}(S^2n^2\rho_\ell)\mathbf{N}_p^{\mathrm{T}}\mathbf{k}_v^{*\mathrm{T}}\mathbf{B}_u\mathrm{d}v;\ \mathbf{C}_{pu}=\int_{\Omega_e}(S\alpha)\mathbf{N}_p^{\mathrm{T}}\mathbf{m}^{\mathrm{T}}\mathbf{B}_u\mathrm{d}v+\int_{\Omega_e}(Snf_{s2})\mathbf{N}_p^{\mathrm{T}}\mathbf{k}_v^{*\mathrm{T}}\mathbf{B}_u\mathrm{d}v;\\
&\mathbf{C}_{pp}=\int_{\Omega_e}\left(\frac{n}{Q}\right)\mathbf{N}_p^{\mathrm{T}}\mathbf{N}_p\mathrm{d}v;\ \mathbf{K}_{pp}=\int_{\Omega_e}(Sn^2\chi)\mathbf{B}_p^{\mathrm{T}}\mathbf{k}^*\mathbf{B}_p\mathrm{d}v;\\
&\mathbf{f}_p=\int_{\Gamma_e}(Sn^2\chi)\mathbf{N}_p^{\mathrm{T}}q_n^*\mathrm{d}s-\int_{\Omega_e}\mathbf{N}_p^{\mathrm{T}}f_{s1}\,\mathrm{d}v \qquad (7.14c)
\end{aligned}
$$

7.5 Special Cases

7.5.1 Unsaturated Porous Media Without Phase Changes

Setting $r_\ell = r_s = 0$ (no phase changes), we obtain $f_{s1} = f_{s2} = 0$ (7.1d) and (7.1e), $a_s = a_\ell = 0$ (7.1h), $k^*_{ij} = \bar{k}_{ij}/(S^2n^2)$ (7.1g), $\alpha^*_{ij} = \alpha\delta_{ij}$ (7.1c), $a^*_{ij} = o_{ij}$ (7.1c), and $\rho^*_{ij} = \rho\delta_{ij}$

(7.1c), where o_{ij} is a null tensor. The linear momentum equation for the mixture (7.1a) and the combined mass conservation equation (7.1b), respectively, are

$$\sigma'_{ij,j} - \chi\alpha p^{\ell}_{,i} + \rho b_i - \rho \ddot{u}^s_i = 0 \tag{7.15a}$$

$$S\alpha\dot{u}^s_{i,i} + Sn\left\{\bar{k}_{ij}\left[-\chi n p^{\ell}_{,j} + Sn\rho_\ell b_j - Sn\rho_\ell \ddot{u}^s_j\right]\right\}_{,i} + \frac{n\dot{p}_\ell}{Q} = 0 \tag{7.15b}$$

where

$$\frac{1}{Q} = \frac{S}{K_\ell} + \frac{S\chi(\alpha - n)}{nK_s} + r_{SWCC} \tag{7.15c}$$

The definitions of α, χ, ρ, and k_{ij} are found in (6.11d), (6.16d), (6.35d), and (6.37b), respectively. The constitutive equations are summarized in Sect. 6.3.2. Equations (7.15a) and (7.15b) are spatially discretized by the finite element method to evaluate the spatial values of the primary variables $(\mathbf{u}^s, p^\ell)$. Among the secondary variables, n (6.56d) and $\mathbf{w}$ (6.56f) are obtained using

$$\dot{n} = (\alpha - n)\dot{u}^s_{i,i} + \frac{\dot{p}}{K_s}(\alpha - n) \tag{7.15d}$$

$$\dot{w}_i = \frac{\bar{k}_{ij}}{S^2 n^2}\left[-\chi n p^{\ell}_{,j} + Sn\rho_\ell b^\ell_j - Sn\rho_\ell \ddot{u}^s_j\right] \tag{7.15e}$$

$\boldsymbol{\varepsilon}^s$ is obtained from $\mathbf{u}^s$ using the strain-displacement relations (Chap. 3). The other secondary variables S, ρ_s, ρ_ℓ, and $\boldsymbol{\sigma}'$ are obtained by integrating the constitutive equations. The details of the boundary conditions remain the same as in (7.2a)–(7.2d). The weak forms of (7.15a) and (7.15b) are

$$\int_\Omega \delta\boldsymbol{\varepsilon}_v^{\mathrm{T}}\boldsymbol{\sigma}'_v\,\mathrm{d}v - \chi\alpha\int_\Omega \delta\boldsymbol{\varepsilon}_v^{\mathrm{T}}\mathbf{m}p\,\mathrm{d}v - \int_\Omega \delta\mathbf{u}^{\mathrm{T}}[\boldsymbol{\rho}\mathbf{b} - \boldsymbol{\rho}\ddot{\mathbf{u}}]\mathrm{d}v - \int_\Gamma \delta\mathbf{u}^{\mathrm{T}}\mathbf{t}\,\mathrm{d}s = 0 \tag{7.16a}$$

$$\begin{aligned}&\int_\Omega (S\alpha)\delta p^{\mathrm{T}}\mathbf{m}^{\mathrm{T}}\dot{\boldsymbol{\varepsilon}}_v\,\mathrm{d}v + \int_\Omega (Sn^2\chi)\delta\mathbf{g}^{\mathrm{T}}\bar{\mathbf{k}}\mathbf{g}\,\mathrm{d}v - \int_\Gamma (Sn^2\chi)\delta p q_n \\ &- \int_\Omega [S^2n^2\rho_\ell]\delta p\bar{\mathbf{k}}_v^{\mathrm{T}}\ddot{\boldsymbol{\varepsilon}}_v\,\mathrm{d}v + \int_\Omega \left(\frac{n}{Q}\right)\delta p\dot{p}\,\mathrm{d}v\end{aligned} \tag{7.16b}$$

where

$$\varepsilon_{ij} = \frac{\partial u_i}{\partial x_j}; \quad \boldsymbol{\varepsilon}_v^{\mathrm{T}} = \{\varepsilon_{11}, \varepsilon_{22}, \varepsilon_{33}, 2\varepsilon_{12}, 2\varepsilon_{13}, 2\varepsilon_{23}\} \tag{7.16c}$$

$$\mathbf{g}^{\mathrm{T}} = \{p_{,1}, p_{,2}, p_{,3}\} \tag{7.16d}$$

$$\boldsymbol{\sigma}'^{\mathrm{T}}_{v} = \{\sigma'_{11}, \sigma'_{22}, \sigma'_{33}, \sigma'_{12}, \sigma'_{13}, \sigma'_{23}\} \tag{7.16e}$$

$$\bar{\mathbf{k}}^{\mathrm{T}}_{v} = \{\bar{k}_{11}, \bar{k}_{22}, \bar{k}_{33}, \bar{k}_{12}, \bar{k}_{13}, \bar{k}_{23}\} \tag{7.16f}$$

$$q_n = \mathbf{q}^{\mathrm{T}}\mathbf{n}; \quad \mathbf{q}^{\mathrm{T}} = \bar{\mathbf{k}}\mathbf{g} \tag{7.16g}$$

$$\mathbf{m}^{\mathrm{T}} = \{1, 1, 1, 0, 0, 0\} \tag{7.16h}$$

These weak forms lead to the following element matrix equation:

$$\begin{bmatrix} \mathbf{M}_{uu} & 0 \\ \mathbf{M}_{pu} & 0 \end{bmatrix} \begin{bmatrix} \ddot{\hat{\mathbf{u}}} \\ \ddot{\hat{\mathbf{p}}} \end{bmatrix} + \begin{bmatrix} 0 & 0 \\ \mathbf{C}_{pu} & \mathbf{C}_{pp} \end{bmatrix} \begin{bmatrix} \dot{\hat{\mathbf{u}}} \\ \dot{\hat{\mathbf{p}}} \end{bmatrix} + \begin{bmatrix} \mathbf{K}_{uu} & \mathbf{K}_{up} \\ 0 & \mathbf{K}_{pp} \end{bmatrix} \begin{bmatrix} \hat{\mathbf{u}} \\ \hat{\mathbf{p}} \end{bmatrix} = \begin{bmatrix} \mathbf{f}_u \\ \mathbf{f}_p \end{bmatrix} \tag{7.17a}$$

where

$$\begin{aligned} \mathbf{M}_{uu} &= \int_{\Omega_e} \mathbf{N}_u^{\mathrm{T}} \boldsymbol{\rho} \mathbf{N}_u \,\mathrm{d}v; \quad \mathbf{K}_{uu} = \int_{\Omega_e} \mathbf{B}_u^{\mathrm{T}} \mathbf{D} \mathbf{B}_u \,\mathrm{d}v; \\ \mathbf{K}_{up} &= \int_{\Omega_e} \chi\alpha \mathbf{B}_u^{\mathrm{T}} \mathbf{m} \mathbf{N}_p \,\mathrm{d}v; \quad \mathbf{f}_u = \int_{\Omega_e} \mathbf{N}_u^{\mathrm{T}} \boldsymbol{\rho} \mathbf{b} \,\mathrm{d}v + \int_{\Gamma_e} \mathbf{N}_u^{\mathrm{T}} \hat{\mathbf{t}} \,\mathrm{d}s \end{aligned} \tag{7.17b}$$

$$\begin{aligned} \mathbf{M}_{pu} &= -\int_{\Omega_e} (S^2 n^2 \rho_\ell) \mathbf{N}_p^{\mathrm{T}} \bar{\mathbf{k}}_v^{\mathrm{T}} \mathbf{B}_u \,\mathrm{d}v; \quad \mathbf{C}_{pu} = \int_{\Omega_e} (S\alpha) \mathbf{N}_p^{\mathrm{T}} \mathbf{m}^{\mathrm{T}} \mathbf{B}_u \,\mathrm{d}v \\ \mathbf{C}_{pp} &= \int_{\Omega_e} \left(\frac{n}{Q}\right) \mathbf{N}_p^{\mathrm{T}} \mathbf{N}_p \,\mathrm{d}v; \quad \mathbf{K}_{pp} = \int_{\Omega_e} (S n^2 \chi) \mathbf{B}_p^{\mathrm{T}} \bar{\mathbf{k}} \mathbf{B}_p \,\mathrm{d}v; \\ \mathbf{f}_p &= \int_{\Gamma_e} (S n^2 \chi) \mathbf{N}_p^{\mathrm{T}} q_n \,\mathrm{d}s \end{aligned} \tag{7.17c}$$

Note that, compared to the general equation given by (7.14a), $\mathbf{C}_{uu}$ has vanished and the force term $\mathbf{f}_p$ does not contain the term involving f_{s1}.

7.5.2 Saturated Porous Media Without Phase Changes

The equations for this case are obtained in a straightforward manner from those presented in Sect. 7.5.1 (7.17a–7.17c) with $S = \chi = 1$.

7.5.3 Drained Deformation of Saturated Porous Media Without Phase Changes

This is the case where the liquid pressure does not change with time. The matrix equations for this case are obtained from those presented in Sect. 7.5.1 by dropping all terms involving the time derivatives of $\hat{\mathbf{p}}$ (i.e., setting $\mathbf{C}_{pp} = 0$ in (7.17a)). Hence, (7.17a) becomes

$$\begin{bmatrix} \mathbf{M}_{uu} & 0 \\ 0 & 0 \end{bmatrix} \begin{bmatrix} \ddot{\hat{\mathbf{u}}} \\ \ddot{\hat{\mathbf{p}}} \end{bmatrix} + \begin{bmatrix} 0 & 0 \\ \mathbf{C}_{pu} & 0 \end{bmatrix} \begin{bmatrix} \dot{\hat{\mathbf{u}}} \\ \dot{\hat{\mathbf{p}}} \end{bmatrix} + \begin{bmatrix} \mathbf{K}_{uu} & \mathbf{K}_{up} \\ 0 & \mathbf{K}_{pp} \end{bmatrix} \begin{bmatrix} \hat{\mathbf{u}} \\ \hat{\mathbf{p}} \end{bmatrix} = \begin{bmatrix} \mathbf{f}_u \\ \mathbf{f}_p \end{bmatrix} \tag{7.18a}$$

where the expressions for the matrices are given by (7.17b) and (7.17c) with $S = \chi = 1$. It may be noted that in the second equation in (7.18a), the terms $\mathbf{C}_{pu}\dot{\hat{\mathbf{u}}}$ and $\mathbf{K}_{pp}\hat{\mathbf{p}}$ come from the $S\alpha\dot{u}^s_{i,i}$ and $S\alpha\dot{w}_{i,i}$ terms, respectively, in (7.1b) (along with (7.1j)). Typically, when the problem can be approximated as a "drained" problem, the permeability tends to be very high (e.g., coarse sand) and that $\dot{w}_{i,i} >> \dot{u}^s_{i,i}$. In such a case, the $\mathbf{C}_{pu}\dot{\hat{\mathbf{u}}}$ term can be neglected. Also, assuming that the effective stresses have been calculated at the beginning of the analysis (accounting for the initial pore water pressure), and noting that the pore water pressure does not change during the analysis, the $\mathbf{K}_{up}\hat{\mathbf{p}}$ can be dropped from the first equation of (7.18a). In this case, (7.18a) breaks down into two uncoupled equations (i.e., the solution of $\hat{\mathbf{u}}$ is independent of the value of $\hat{\mathbf{p}}$, and vice versa):

$$\begin{aligned} \mathbf{M}_{uu}\ddot{\hat{\mathbf{u}}} + \mathbf{K}_{uu}\hat{\mathbf{u}} &= \mathbf{f}_u \\ \mathbf{K}_{pp}\hat{\mathbf{p}} &= \mathbf{f}_p \end{aligned} \tag{7.18b}$$

where the first equation of (7.18b) is the equilibrium equation for the skeleton deformation, and the second equation of (7.18b) is the steady-state seepage equation. If the latter is not needed, it can be excluded from (7.18b).

7.5.4 Undrained Deformation of Saturated Porous Media Without Phase Changes

In this case, the coefficients of the permeability tensor $\bar{k}_{ij}$ are so small that the relative motion of the liquid phase with respect to the solid phase can be neglected. The middle term in (7.15b) can then be dropped. In addition, we neglect the term involving the boundary flux q_n (7.16b) and (7.17c). Equation (7.17a) then becomes

$$\begin{bmatrix} \mathbf{M}_{uu} & 0 \\ 0 & 0 \end{bmatrix} \begin{bmatrix} \ddot{\hat{\mathbf{u}}} \\ \ddot{\hat{\mathbf{p}}} \end{bmatrix} + \begin{bmatrix} 0 & 0 \\ \mathbf{C}_{pu} & \mathbf{C}_{pp} \end{bmatrix} \begin{bmatrix} \dot{\hat{\mathbf{u}}} \\ \dot{\hat{\mathbf{p}}} \end{bmatrix} + \begin{bmatrix} \mathbf{K}_{uu} & \mathbf{K}_{up} \\ 0 & 0 \end{bmatrix} \begin{bmatrix} \hat{\mathbf{u}} \\ \hat{\mathbf{p}} \end{bmatrix} = \begin{bmatrix} \mathbf{f}_u \\ 0 \end{bmatrix} \tag{7.19a}$$

where it may be seen that the terms involving $\mathbf{M}_{pu}$, $\mathbf{K}_{pp}$, and $\mathbf{f}_p$ have been dropped. The second equation in (7.19a) becomes

$$\mathbf{C}_{pu}\dot{\hat{\mathbf{u}}} + \mathbf{C}_{pp}\dot{\hat{\mathbf{p}}} = 0 \tag{7.19b}$$

Noting that $\mathbf{C}_{pu}$ and $\mathbf{C}_{pp}$ are independent of time, (7.19b) can be integrated to yield

$$\mathbf{C}_{pu}\hat{\mathbf{u}} + \mathbf{C}_{pp}\hat{\mathbf{p}} = 0 \tag{7.19c}$$

The unknowns $\hat{\mathbf{p}}$ could then be solved from (7.19c) (provided $\mathbf{C}_{pp}$ is invertible), and substituted into the first of (7.19a) to solve for the unknowns $\hat{\mathbf{u}}$. For $\mathbf{C}_{pp}$ (7.17c) to be invertible, Q (7.15c) must be finite (i.e., some compressibility must exist). If the value of Q tends to be relatively large, numerical problems are likely to be encountered. In this case, (7.19a) may then be rearranged as

$$\begin{bmatrix} \mathbf{M}_{uu} & 0 \\ 0 & 0 \end{bmatrix} \begin{bmatrix} \ddot{\hat{\mathbf{u}}} \\ \ddot{\hat{\mathbf{p}}} \end{bmatrix} + \begin{bmatrix} 0 & 0 \\ 0 & 0 \end{bmatrix} \begin{bmatrix} \dot{\hat{\mathbf{u}}} \\ \dot{\hat{\mathbf{p}}} \end{bmatrix} + \begin{bmatrix} \mathbf{K}_{uu} & \mathbf{K}_{up} \\ \mathbf{C}_{pu} & \mathbf{C}_{pp} \end{bmatrix} \begin{bmatrix} \hat{\mathbf{u}} \\ \hat{\mathbf{p}} \end{bmatrix} = \begin{bmatrix} \mathbf{f}_u \\ 0 \end{bmatrix} \tag{7.19d}$$

Different orders of approximation can be used for $\hat{\mathbf{u}}$ and $\hat{\mathbf{p}}$ such that $n_u > n_p$, where n_u and n_p are number of unknowns in the variables $\hat{\mathbf{u}}$ and $\hat{\mathbf{p}}$, respectively (Babuska 1973; Brezzi 1974; Zienkiewicz and Taylor 1989).

For instance, consider the following two element formulations:

Element 1:

- four-noded quadrilateral element,
- $\hat{\mathbf{u}}$ and $\hat{\mathbf{p}}$ both are approximated by a four-point (corner point) shape function
- all element matrices integrated at four Gauss points

Element 2:

- eight-noded quadrilateral element,
- $\hat{\mathbf{u}}$ is approximated by an eight-point (corner and mid-side point) shape function,
- $\hat{\mathbf{p}}$ is approximated by a four-point (corner point) shape function
- Element matrices $\mathbf{M}_{uu}$, $\mathbf{K}_{uu}$, $\mathbf{f}_u$ are integrated using nine Gauss points; all other matrices and vectors in (7.17b) and (7.17c) are integrated using four Gauss points (reduced integration)

As illustrated in Example 7.1, the use of Element 2 generally results in $n_u > n_p$, avoiding the hour-glass effect. On the other hand, the use of Element 1 results in $n_u \approx n_p$, leading to the hour-glass effect.

Remarks

1. The *u–u–p* formulation presented in Chap. 6 can also be used directly in finite element solutions. The finite element equations for this case has been presented by Zienkiewicz and Shiomi (1984). In Problem 7.1, you are asked to derive the equations step by step.

2. In Zienkiewicz and Shiomi (1984), various special cases of the generalized governing equations (including the *u–u–p* and *u–p* formulations) have been presented for a saturated material. For example, when the fluid is compressible, the liquid pressure variable can be eliminated at the element level. This may be done in two different ways: (1) Eliminate the pressure variable at the element level, simplify the differential equations and then develop the finite element equations or (2) take the finite element equations associated with the *u–u–p* formulation, eliminate the pressure variable and develop a modified matrix equations that involve only $\mathbf{u}^s$ and $\mathbf{u}^\ell$ as independent variables. While these two approaches are equivalent (Malkus and Hughes 1978), the former (called the "irreducible solution," see Chap. 5 for definition) requires the use of special methods for integrating the element stiffness matrices (e.g., reduced integration technique), the latter allows more efficient interpolation functions to be used (e.g., in a 2D analysis, use eight-point approximation for displacements and four-point approximation for the pore pressure, as discussed above in this section).

Example 7.1: One-Dimensional Consolidation of Saturated Soil

Question: A horizontal deposit of homogeneous, linear elastic saturated soil deposit (Fig. 7.1) is subjected to a surface surcharge of t_y kPa. The time-variation of t_y is

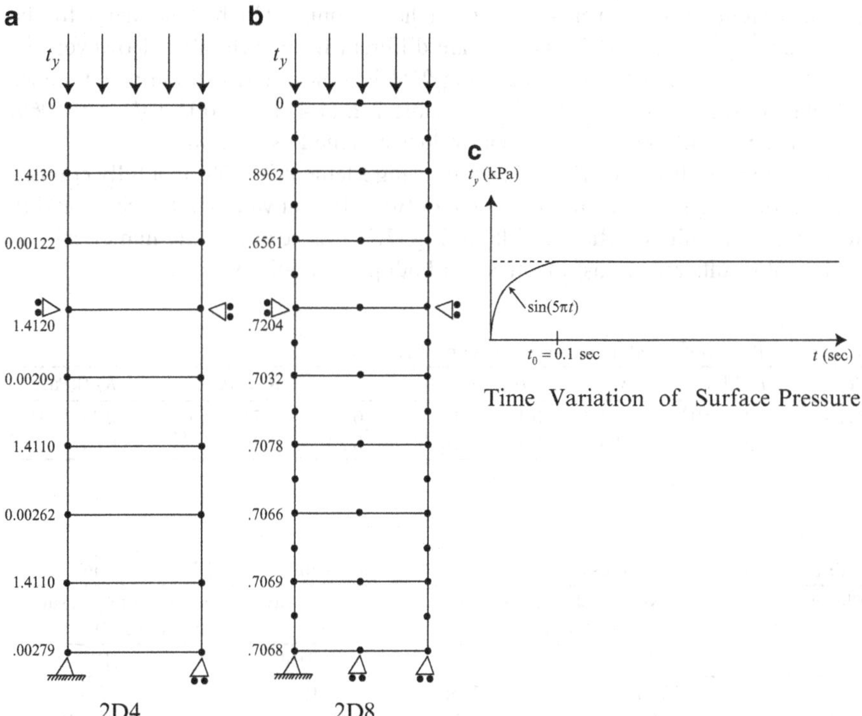

Fig. 7.1 One-dimensional consolidation analysis using 2D4 and 2D8 elements

shown in Fig. 7.1c, where the initial rise is a sine curve. The relevant soil properties are listed in Table 7.1. Determine the time-variation of the nodal pore water pressure values using (a) Element 1 (four-noded element) and (b) Element 2 (eight-noded element) and using the meshes shown in Fig. 7.1a and b, respectively. Repeat the analysis for two values of the solid and fluid bulk moduli as listed in Table 7.1. Use a time step of 0.003333 s. Compare and discuss the results.

Answer: The mesh and boundary conditions are shown in Fig. 7.1a and b. The pore pressure (p) values obtained using Property Set 1 (compressible) and Property Set 2 (incompressible) are listed in Table 7.2 for the first three corner nodes from the base for $t = 0.05$ s. The hour-glass effect may be clearly noticed when vary large value is used for the bulk modulus of the solid and fluid (Set 2). The results at each node are marked in Fig. 7.1a and b for $t = 0.05$ s.

The time-variations of the pore pressure p at the first two nodes from the base, obtained from the analyses employing a large value for the bulk moduli (incompressible) of the solid and fluid phases, are presented in Fig. 7.2. The results obtained using the 2D4 elements are presented in Fig. 7.2a and those obtained using the 2D8 elements are presented in Fig. 7.2b. (Note that the figures on the left present the results for the first 3 s and those on the right present the results for the entire analysis duration of 40 s.)

For the analysis using Element 2 (2D8), it is seen that the pore pressures at the two nodes remain equal to each other from the beginning. On the other hand, for the analysis using Element 1 (2D4), they are different from each other. However, the average for the elements obtained using 2D4 is almost equal to the nodal values calculated using 2D8, e.g., referring to Table 7.2, (1.411 + 0.00279)/2 = 0.7069, which is almost the same as the nodal values obtained using 2D8.

To further confirm the effectiveness of using Element 2 (2D8) in a fully coupled analysis of this type, the results obtained for two different values of the permeability are compared with analytical results in Fig. 7.3; it is seen that the numerical and analytical results are almost identical for both permeability values.

Table 7.1 Properties used in elastic fully coupled analysis

Set	E (kPa)	ν	$\bar{k}$ (m^4/(kN-s))	n	K_s (kPa)	K_ℓ (kPa)
Set 1	30×10^3	0.3	0.102×10^{-2}	0.3	2.74×10^7	0.11×10^7
Set 2	30×10^3	0.3	0.102×10^{-6}	0.3	2.74×10^{13}	0.11×10^{13}

Table 7.2 Values of pore pressure at $t = 0.05$ s calculated using 2D4 and 2D8 elements

Elevation (m)	Nodal values of pore pressure (property set 1)		Nodal values of pore pressure (property set 2)	
	2D4	2D8	2D4	2D8
0	0.7048	0.7066	0.00279	0.7068
3.75	0.7081	0.7066	1.411	0.7069
7.50	0.7070	0.7077	0.002616	0.7066

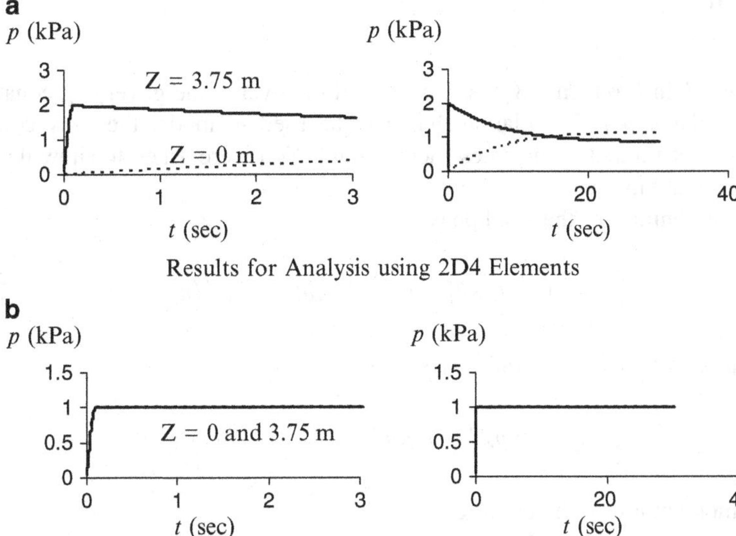

Fig. 7.2 Results for incompressible analysis using 2D4 and 2D8 elements

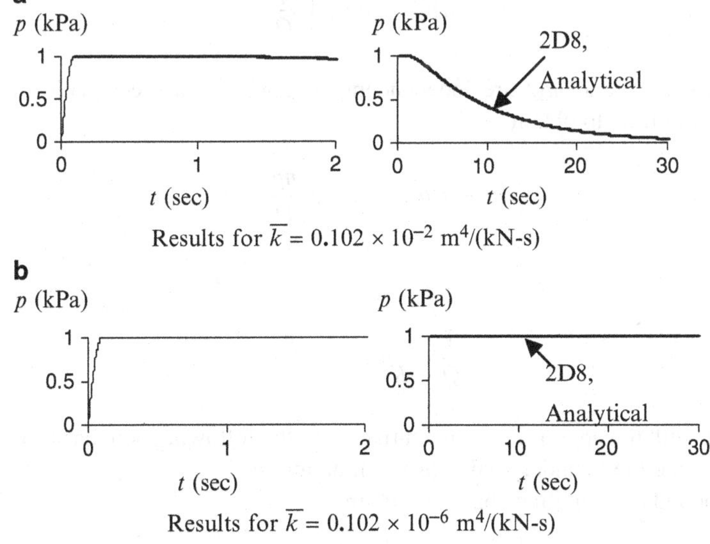

Fig. 7.3 Comparison of finite element results with analytical results

Problems

Problem 7.1 In Problem 6.8, you were asked to develop the governing equations based on the *u*–*u*–*p* formulation that may be used to model the fully coupled behavior of saturated particulates such as sand. You were asked to show that the relevant equations are

Linear momentum for the solid phase:

$$\sigma'_{ij,j} - (\alpha - n)p_{,i} + (1-n)\rho_s b_i^s - (1-n)\rho_s \ddot{u}_i^s + n^2 \bar{k}_{ij}^{-1}(\dot{u}_j^\ell - \dot{u}_j^s) = 0 \tag{7.20}$$

Linear momentum for the liquid phase:

$$-np_{,i} + n\rho_\ell b_i^\ell - n\rho_\ell \ddot{u}_i^\ell - n^2 \bar{k}_{ij}^{-1}(\dot{u}_j^\ell - \dot{u}_j^s) = 0 \tag{7.21}$$

Linear momentum for the mixture:

$$\sigma'_{ij,j} - \alpha p_{,i} + \rho \bar{\bar{b}}_i - \rho \ddot{\bar{\bar{u}}}_i = 0 \tag{7.22}$$

$$\bar{\bar{\mathbf{b}}} = \frac{\mathbf{b}^s(1-n)\rho_s + \mathbf{b}^\ell n\rho_\ell}{\rho}; \quad \ddot{\bar{\bar{u}}}_i = \frac{\ddot{u}_i^s(1-n)\rho_s + \ddot{u}_i^\ell n\rho_\ell}{\rho}; \quad \rho = (1-n)\rho_s + n\rho_\ell$$

Combined mass conservation:

$$(\alpha - n)\dot{u}_{i,i}^s + n\dot{u}_{i,i}^\ell + \frac{n\dot{p}}{Q} = 0$$

Since $(\alpha-n)$, n, and n/Q are independent of time, the above equation may be integrated in time to obtain

$$(\alpha - n)u_{i,i}^s + nu_{i,i}^\ell + \frac{np}{Q} = 0 \tag{7.23}$$

where

$$\frac{1}{Q} = \frac{1}{K_\ell} + \frac{(\alpha - n)}{nK_s} \tag{7.24}$$

Along with the appropriate constitutive laws, the following boundary conditions will complete the boundary value problem definition:

The boundary conditions based on $\mathbf{u}^s$ are

$$\mathbf{t} = \hat{\mathbf{t}} \quad \text{on} \quad \Gamma = \Gamma_t^s \tag{7.25a}$$

$$\mathbf{u}^s = \hat{\mathbf{u}}^s \quad \text{on} \quad \Gamma = \Gamma_u^s \tag{7.25b}$$

where $\Gamma = \Gamma_t^s \cup \Gamma_u^s$, $\hat{\mathbf{t}}$ is the prescribed traction on Γ_t^s and $\hat{\mathbf{u}}^s$ is the prescribed skeleton displacement on Γ_u^s.

The boundary conditions based on $\mathbf{u}^\ell$ are

$$p = \hat{p} \quad \text{on} \quad \Gamma = \Gamma_t^\ell \tag{7.26a}$$

$$\mathbf{u}^\ell = \hat{\mathbf{u}}^\ell \quad \text{on} \quad \Gamma = \Gamma_u^\ell \tag{7.26b}$$

where $\Gamma = \Gamma_t^\ell \cup \Gamma_u^\ell$, $\hat{p}$ is the prescribed liquid pressure on Γ_t^ℓ and $\hat{\mathbf{u}}^\ell$ is the prescribed liquid displacement on Γ_u^ℓ. (The exact form of (7.26) will be established below).

Part(a): Determine the weak form of the linear momentum equation for the solid phase.

Answer: Using a test function $\mathbf{v}$, the weak form becomes

$$\int_\Omega \bar{\boldsymbol{\varepsilon}}_v^{\mathrm{T}} \boldsymbol{\sigma}'_v \, \mathrm{d}v - (\alpha - n) \int_\Omega \bar{\boldsymbol{\varepsilon}}_v^{\mathrm{T}} \mathbf{m} p \, \mathrm{d}v - \int_\Omega \mathbf{v}^{\mathrm{T}} \mathbf{b}^m \, \mathrm{d}v - \int_\Gamma \mathbf{v}^{\mathrm{T}} \mathbf{t} \, \mathrm{d}s = 0 \tag{7.27}$$

where

$$t'_{ij} = \sigma'_{ij} n_j$$

$$\mathbf{t}' - \mathbf{p}^s = \boldsymbol{\sigma}'\mathbf{n} - p\mathbf{n} = \boldsymbol{\sigma}\mathbf{n} = \mathbf{t}$$

$$\bar{\boldsymbol{\varepsilon}}_{ij} = \frac{\partial(v_i)}{\partial x_j}; \quad \bar{\boldsymbol{\varepsilon}}_v^{\mathrm{T}} = \{\bar{\varepsilon}_{11}, \bar{\varepsilon}_{22}, \bar{\varepsilon}_{33}, 2\bar{\varepsilon}_{12}, 2\bar{\varepsilon}_{13}, 2\bar{\varepsilon}_{23}\}$$

$$\boldsymbol{\sigma}'^{\mathrm{T}}_v = \{\sigma'_{11}, \sigma'_{22}, \sigma'_{33}, \sigma'_{12}, \sigma'_{13}, \sigma'_{23}\}$$

$$\mathbf{m}^{\mathrm{T}} = \{1, 1, 1, 0, 0, 0\}$$

$$\mathbf{p}_s^{\mathrm{T}} = \{pn_1, pn_2, pn_3\}$$

$$b_i^m = (1-n)\rho_s b_i^s - (1-n)\rho_s \ddot{u}_i^s + n^2 \bar{k}_{ij}^{-1} (\dot{u}_j^\ell - \dot{u}_j^s)$$

Part(b): Determine the weak form of the linear momentum equation for the liquid phase.

Answer: Using a test function **w**, the weak form becomes

$$-n\int_{\Omega}\bar{\bar{\boldsymbol{\varepsilon}}}_v^{\mathrm{T}}\mathbf{m}p\mathrm{d}v-\int_{\Omega}\mathbf{w}^{\mathrm{T}}\bar{\mathbf{b}}^{m}\,\mathrm{d}v+\int_{\Gamma}w_n p\mathrm{d}s=0 \tag{7.28}$$

where $w_n=\mathbf{w}^{\mathrm{T}}\mathbf{n}$

$$\bar{\bar{\varepsilon}}_{ij}=\frac{\partial(w_i)}{\partial x_j};\quad \bar{\bar{\boldsymbol{\varepsilon}}}_v^{\mathrm{T}}=\{\bar{\bar{\varepsilon}}_{11},\bar{\bar{\varepsilon}}_{22},\bar{\bar{\varepsilon}}_{33},2\bar{\bar{\varepsilon}}_{12},2\bar{\bar{\varepsilon}}_{13},2\bar{\bar{\varepsilon}}_{23}\}$$

$$\bar{b}_i^{m}=n\rho_\ell b_i^\ell-n\rho_\ell\ddot{u}_i^\ell-n^2\bar{k}_{ij}^{-1}(\dot{u}_j^\ell-\dot{u}_j^s)$$

Part(c): Determine the weak form of the combined mass conservation equation.

Answer: Using a test function ϕ, the weak form becomes

$$\int_{\Omega}\phi\left\{(\alpha-n)\mathbf{m}^{\mathrm{T}}\boldsymbol{\varepsilon}_v^s+n\mathbf{m}^{\mathrm{T}}\boldsymbol{\varepsilon}_v^\ell+\frac{np}{Q}\right\}\mathrm{d}v=0 \tag{7.29}$$

where

$$(\boldsymbol{\varepsilon}_v^s)^{\mathrm{T}}=\{\varepsilon_{11}^s,\varepsilon_{22}^s,\varepsilon_{33}^s,2\varepsilon_{12}^s,2\varepsilon_{13}^s,2\varepsilon_{23}^s\}$$

$$(\boldsymbol{\varepsilon}_v^\ell)^{\mathrm{T}}=\{\varepsilon_{11}^\ell,\varepsilon_{22}^\ell,\varepsilon_{33}^\ell,2\varepsilon_{12}^\ell,2\varepsilon_{13}^\ell,2\varepsilon_{23}^\ell\}$$

Part(d): Identify the natural and essential boundary conditions.

Answer: Using the criteria discussed in Chap. 5:

- The natural boundary conditions are $\mathbf{t}=\hat{\mathbf{t}}$ on $\boldsymbol{\Gamma}_t^s$ (from (7.27), using $\mathbf{v}=\mathbf{u}^s$) and $p=\hat{p}$ on $\boldsymbol{\Gamma}_t^\ell$ (from (7.28), using $\mathbf{w}=\mathbf{u}^\ell$)
- The essential boundary conditions are $\mathbf{u}^s=\hat{\mathbf{u}}^s$ on Γ_u^s (from (7.27), using $\mathbf{v}=\mathbf{u}^s$) and $u_n^\ell=\hat{u}_n^\ell$ on $\boldsymbol{\Gamma}_u^\ell$ (from (7.28), using $\mathbf{w}=\mathbf{u}^\ell$)

Part(e): Based on the weak forms developed above, develop the finite element equations.

Answer: Using the following approximations

$$\mathbf{u}^s=\mathbf{N}_u^s\hat{\mathbf{u}}^s;\quad \mathbf{u}^\ell=\mathbf{N}_u^\ell\hat{\mathbf{u}}^\ell;\quad p=\mathbf{N}_p\hat{\mathbf{p}}$$

$$\boldsymbol{\varepsilon}^s=\mathbf{B}_u^s\hat{\mathbf{u}}^s;\quad \boldsymbol{\varepsilon}^\ell=\mathbf{B}_u^\ell\hat{\mathbf{u}}^\ell;\quad \mathbf{g}=\mathbf{B}_p\hat{\mathbf{p}}$$

and the linearized constitutive equation $\boldsymbol{\sigma}'_v = \mathbf{D}\boldsymbol{\varepsilon}^s_v$, the three weak forms presented above lead to the following matrix equation:

$$\begin{bmatrix} \mathbf{M}_{ss} & 0 & 0 \\ 0 & 0 & 0 \\ 0 & 0 & \mathbf{M}_{\ell\ell} \end{bmatrix} \begin{bmatrix} \ddot{\hat{\mathbf{u}}}^s \\ \ddot{\hat{\mathbf{p}}} \\ \ddot{\hat{\mathbf{u}}}^\ell \end{bmatrix} + \begin{bmatrix} \mathbf{C}_{ss} & 0 & \mathbf{C}_{s\ell} \\ 0 & 0 & 0 \\ \mathbf{C}_{\ell s} & 0 & \mathbf{C}_{\ell\ell} \end{bmatrix} \begin{bmatrix} \dot{\hat{\mathbf{u}}}^s \\ \dot{\hat{\mathbf{p}}} \\ \dot{\hat{\mathbf{u}}}^\ell \end{bmatrix} + \begin{bmatrix} \mathbf{K}_{ss} & \mathbf{K}_{sp} & 0 \\ \mathbf{K}_{ps} & \mathbf{K}_{pp} & \mathbf{K}_{p\ell} \\ 0 & \mathbf{K}_{\ell p} & 0 \end{bmatrix} \begin{bmatrix} \hat{\mathbf{u}}^s \\ \hat{\mathbf{p}} \\ \hat{\mathbf{u}}^\ell \end{bmatrix}$$

$$= \begin{bmatrix} \mathbf{f}_s \\ \mathbf{f}_p \\ \mathbf{f}_\ell \end{bmatrix}$$

The matrices involved in the first row of the above matrix equation are

$$\mathbf{k}^e_{ss} = \int_{\Omega_e} \mathbf{B}_s^{\mathrm{T}}\mathbf{D}\mathbf{B}_s \,\mathrm{d}v; \quad \mathbf{k}^e_{sp} = -(\alpha - n)\int_{\Omega_e} \mathbf{B}_s^{\mathrm{T}}\mathbf{m}\mathbf{N}_p \,\mathrm{d}v; \quad \mathbf{f}^e_{sb} = (1-n)\int_{\Omega_e} \mathbf{N}_s^{\mathrm{T}}\rho_s\mathbf{b}^s \,\mathrm{d}v$$

$$\mathbf{M}^e_{ss} = (1-n)\int_{\Omega_e} \mathbf{N}_s^{\mathrm{T}}\rho_s\mathbf{N}_s \,\mathrm{d}v; \quad \mathbf{C}^e_{s\ell} = -n^2\int_{\Omega_e} \mathbf{N}_s^{\mathrm{T}}\mathbf{k}^{-1}\mathbf{N}_\ell \,\mathrm{d}v; \quad \mathbf{C}^e_{ss} = n^2\int_{\Omega_e} \mathbf{N}_s^{\mathrm{T}}\mathbf{k}^{-1}\mathbf{N}_s \,\mathrm{d}v$$

$$\mathbf{f}^e_{st} = \int_{\Gamma^s_t} \mathbf{N}_s^{\mathrm{T}}\hat{\mathbf{t}} \,\mathrm{d}s; \quad \mathbf{f}^e_s = \mathbf{f}^e_{sb} + \mathbf{f}^e_{st}$$

The matrices involved in the second row of the above matrix equation are

$$\mathbf{k}^e_{ps} = -(\alpha - n)\int_{\Omega_e} \mathbf{N}_p^{\mathrm{T}}\mathbf{m}^{\mathrm{T}}\mathbf{B}_s \,\mathrm{d}v; \quad \mathbf{k}^e_{p\ell} = -n\int_{\Omega_e} \mathbf{N}_p^{\mathrm{T}}\mathbf{m}^{\mathrm{T}}\mathbf{B}_\ell \,\mathrm{d}v; \quad \mathbf{k}^e_{pp} = -n\int_{\Omega_e} \mathbf{N}_p^{\mathrm{T}}\frac{1}{Q}\mathbf{N}_p \,\mathrm{d}v$$

The matrices involved in the third row of the above matrix equation are

$$\mathbf{k}^e_{\ell p} = -n\int_{\Omega_e} \mathbf{B}_\ell^{\mathrm{T}}\mathbf{m}\mathbf{N}_p \,\mathrm{d}v; \quad \mathbf{f}^e_{\ell b} = n\int_{\Omega_e} \mathbf{N}_\ell^{\mathrm{T}}\rho_\ell\mathbf{b}^\ell \,\mathrm{d}v; \quad \mathbf{M}^e_{\ell\ell} = n\int_{\Omega_e} \mathbf{N}_\ell^{\mathrm{T}}\rho_\ell\mathbf{N}_\ell \,\mathrm{d}v$$

$$\mathbf{C}^e_{\ell\ell} = n^2\int_{\Omega_e} \mathbf{N}_\ell^{\mathrm{T}}\bar{\mathbf{k}}^{-1}\mathbf{N}_\ell \,\mathrm{d}v; \quad \mathbf{C}^e_{\ell s} = -n^2\int_{\Omega_e} \mathbf{N}_\ell^{\mathrm{T}}\bar{\mathbf{k}}^{-1}\mathbf{N}_s \,\mathrm{d}v; \quad \mathbf{f}^e_{\ell t} = -\int_{\Gamma^\ell_t} \mathbf{N}_\ell^{\mathrm{T}}\mathbf{n}\mathbf{N}_p \,\mathrm{d}s;$$

$$\mathbf{f}^e_\ell = \mathbf{f}^e_{\ell b} + \mathbf{f}^e_{\ell t}$$

Fig. 7.4 2D Plane strain embankment

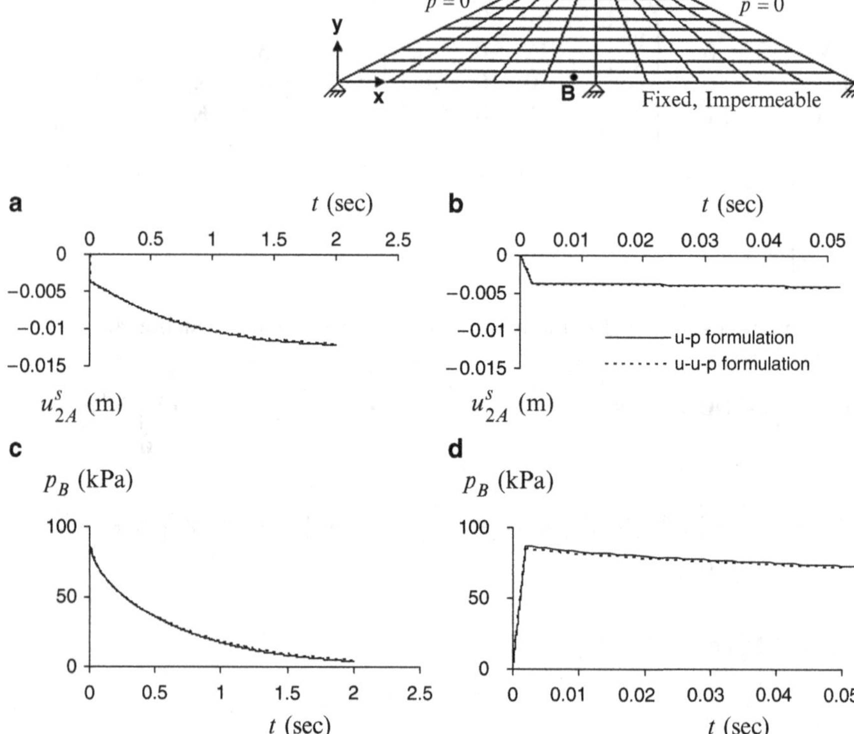

Fig. 7.5 Comparison of results obtained using $u-p$ and $u-u-p$ formulations

Part(f): Comment on the special structure of the above matrix equation

Problem 7.2 The 2D plane strain embankment shown in Fig. 7.4 is initially stress free. The embankment is to be loaded under a vertical gravity load with $b_y = -10$ kN/m^3. The properties given in Table 7.1 are to be used. The embankment has an impermeable boundary at the base. This boundary is also fixed ($u^s_x = u^s_y = 0$). The other three boundaries are free of traction ($\hat{\mathbf{t}} = \mathbf{0}$). The value of the liquid pressure is zero on these three boundaries. The gravity load is applied in 0.2 s. Analyze the problem using both the *u–p* and *u–u–p* formulations and compare the results. Use a time step of 0.002 s. State the boundary conditions clearly for each of the analysis types.

Partial answer: The results are compared in Fig. 7.5 in terms of the vertical displacement at point A and the pore liquid pressure at point B. Note that the figures on the right are the same as those on the left except for the time scale. It may be seen that, as expected, the differences of the results obtained using the two formulations are hardly noticeable.

Chapter 8
Methods of Nonlinear Analysis

Materials such as metals, soils, and rocks (e.g., lime stones) are inherently nonlinear and plastic. Except in a limited class of problems, the behavior of structures made of these materials cannot be predicted without the consideration of their nonlinear plastic stress–strain behavior. Contrary to linear elastic problems, nonlinear problems require iterative methods for obtaining the solution, both at the global (structure) and local (Gauss point) levels. There are several methods of carrying out the iterations. In this chapter, we will describe (1) a class of methods called the Newton's methods which form the basis for commonly used global and local iterative algorithms, and (2) Euler methods of solving initial value problems, which form the basis for the commonly used local iterative algorithms.

8.1 Nonlinear Finite Element Analysis

In Chap. 5, it was shown that, when the stress–strain relation is linearized as $\boldsymbol{\sigma} = \mathbf{D}\boldsymbol{\varepsilon}$, the finite element approximation of solid mechanics problems leads to a set of linear simultaneous equations for the nodal unknowns as

$$\mathbf{K}\hat{\mathbf{u}} = \mathbf{P} \tag{8.1}$$

$\mathbf{K}$ is the global stiffness matrix and $\hat{\mathbf{u}}$ is the global unknown nodal displacement vector. The product $\mathbf{K}\hat{\mathbf{u}}$ is the internal or "spring" force and $\mathbf{P}$ is the externally applied force. Equation (8.1) states that the internal spring force is equal to the externally applied force. This is a statement of equilibrium. Let us denote the spring force by

$$\mathbf{F}_s = \mathbf{K}\hat{\mathbf{u}} \tag{8.2}$$

Equation (8.2) takes a different form when the stress–strain relation is nonlinear. For linear materials, $\mathbf{K}$ in (8.2) is the gradient of $\mathbf{F}_s$ with respect to $\hat{\mathbf{u}}$, and is

A. Anandarajah, *Computational Methods in Elasticity and Plasticity: Solids and Porous Media*, DOI 10.1007/978-1-4419-6379-6_8,

independent of $\hat{\mathbf{u}}$. For nonlinear materials, the gradient is not a constant. The gradient is known as the global tangent stiffness matrix $\mathbf{K}^t$, which at the $(n+1)$th load step is obtained from element tangent stiffness matrices as

$$\mathbf{K}_{n+1}^t = \frac{\partial \mathbf{F}_s^{n+1}}{\partial \mathbf{u}} = \sum \frac{\partial \mathbf{f}_s^{n+1}}{\partial \mathbf{u}} = \sum \frac{\partial}{\partial \mathbf{u}} \left\{ \int \mathbf{B}^{\mathrm{T}} \boldsymbol{\sigma}_{n+1} \, \mathrm{d}v \right\} = \sum \int \mathbf{B}^{\mathrm{T}} \frac{\partial \boldsymbol{\sigma}_{n+1}}{\partial \boldsymbol{\varepsilon}_{n+1}} \frac{\partial \boldsymbol{\varepsilon}_{n+1}}{\partial \mathbf{u}_{n+1}} \mathrm{d}v$$
$$= \sum \int \mathbf{B}^{\mathrm{T}} \mathbf{D}_{n+1}^t \mathbf{B} \, \mathrm{d}v = \sum \mathbf{k}_{n+1}^t \tag{8.3a}$$

where

$$\mathbf{D}_{n+1}^t = \frac{\partial \boldsymbol{\sigma}_{n+1}}{\partial \boldsymbol{\varepsilon}_{n+1}} \tag{8.3b}$$

$\mathbf{D}_{n+1}^t$ is known as the material tangent stiffness matrix (or operator). The global spring force $\mathbf{F}_s$ is obtained from element spring forces $\mathbf{f}^s$, which at the $(n+1)$th load step is given by

$$\mathbf{f}_s^{n+1} = \int_{\Omega_i} \mathbf{B}^T \boldsymbol{\sigma}_{n+1} \, dv \tag{8.3c}$$

It follows from (8.2) and (8.3) that when the relation between stress and strain is nonlinear, as shown in Fig. 8.1a, the element relation $\mathbf{f}_s$ versus $\mathbf{u}$ and the global relation $\mathbf{F}_s$ versus $\mathbf{u}$ are also nonlinear. Let us say that $\mathbf{F}_s$ versus $\mathbf{u}$ schematically looks like the one shown in Fig. 8.1b.

Suppose that a static, monotonic load of $\mathbf{P}$ is applied to the system, and a corresponding displacement $\hat{\mathbf{u}}$ is required. Let us say that the stress and strain produced by the load in a typical element (specifically, at a typical Gauss point) are $\hat{\boldsymbol{\sigma}}$ and $\hat{\boldsymbol{\varepsilon}}$ respectively. The relationship between the two is governed by the curve shown in Fig. 8.1a. Since the strains and stresses at Gauss points and the displacements at nodes are unknowns, the exact locations of point A on the curves in Fig. 8.1a and b are unknowns, and need to be established iteratively.

For plastic materials, the load is divided up into a number of small increments, and within each increments, iterations are performed. The corresponding analysis is

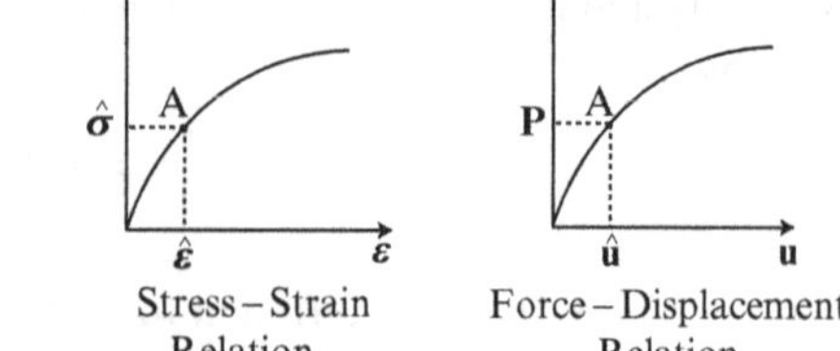

Fig. 8.1 Nonlinear stress–strain and force–displacement relations

called the incremental-iterative analysis. In such analyses, given the solution at a given stage, the solution corresponding to a given load increment is sought. In other words, in a nonlinear elastic analysis, we can apply the full load in one increment, whereas in an elasto-plastic analysis, we apply the load in several small increments. The iterations are applied to the full load in the former, whereas they are applied for each increment separately in the latter. Apart from this key difference, the details of the iterations are about the same. In the present chapter, we focus on the iterative method. The method presented in the present chapter is directly applicable to a nonlinear elastic analysis. In subsequent chapters, we will see how the iterative method can be applied to an elasto-plastic analysis.

Formally, the problem can be stated as follows: given the functional form $\hat{\boldsymbol{\sigma}}(\hat{\boldsymbol{\varepsilon}})$, determine the nodal displacement $\hat{\mathbf{u}}$ for a given (constant) load $\mathbf{P}$ such that

$$\mathbf{F}_s(\hat{\mathbf{u}}) = \mathbf{P} \tag{8.4}$$

The problem at hand is complex in that we have to go from the nonlinearity $\hat{\boldsymbol{\sigma}}$ versus $\hat{\boldsymbol{\varepsilon}}$ at the local level to the nonlinearity $\mathbf{F}_s$ versus $\hat{\mathbf{u}}$ at the global level. To understand some of the numerical iterative techniques available for us to use, let us first look at a simpler, scalar problem of finding $x = \hat{x}$ such that

$$f(\hat{x}) = \bar{f} \tag{8.5}$$

where $f(x)$ is a nonlinear function of x as shown in Fig. 8.2a, and $\bar{f}$ is a constant. Among a suite of techniques that are available in the numerical analysis literature, the two methods that are most suitable for finite element analysis are the

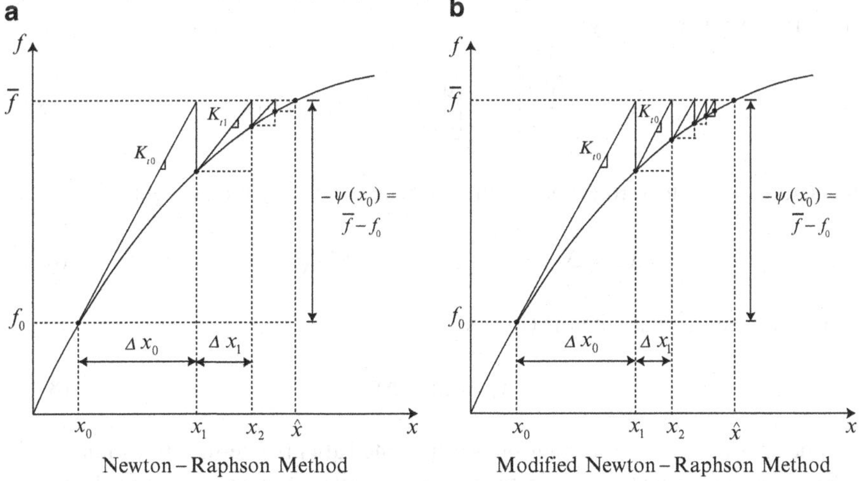

Fig. 8.2 Illustration of Newton–Raphson iterative methods

Newton–Raphson (full) method (also known as the Newton's method) and the modified Newton–Raphson method. These methods are briefly described in this section. For additional details, the reader is referred to standard texts on finite element, numerical or optimization methods (e.g., Zienkiewicz 1977; Griffiths and Smith 1991; Fletcher 1987).

8.2 The Newton–Raphson (Full) Method for Scalar Equations

The Newton–Raphson method begins with an estimate for $\hat{x}$ (say, x_0), and iteratively refines this estimate until (8.5) is satisfied within a specified tolerance. The algorithm for the iterative process is derived by the Taylor series expansion of the error (or residual). Let us define the error as:

$$\psi(x) = f(x) - \bar{f} \tag{8.6}$$

Denoting the next best estimate by x_1, let us express the value of $\psi(x_1)$ in Taylor series by expansion about x_0 as

$$\psi(x_1) = \psi(x_0) + \left[\frac{\mathrm{d}\psi}{\mathrm{d}x}\right]_{x_0} \Delta x + \frac{1}{2}\left[\frac{\mathrm{d}^2\psi}{\mathrm{d}x^2}\right]_{x_0} \Delta x^2 + \cdots \tag{8.7a}$$

where

$$\Delta x = x_1 - x_0 \tag{8.7b}$$

Truncating the series after the term with the first derivative of ψ, and denoting the derivative at $x = x_0$ by:

$$K_{t0} = \left[\frac{\mathrm{d}\psi}{\mathrm{d}x}\right]_{x_0} = \left[\frac{\mathrm{d}f}{\mathrm{d}x}\right]_{x_0} \quad \text{(from (8.6))}, \tag{8.8}$$

and setting $\psi(x_1)$ to zero (which is the ultimate desired result), we get

$$\Delta x = \Delta x_0 = -K_{t0}^{-1}\psi(x_0) = K_{t0}^{-1}(\bar{f} - f_0) \tag{8.9a}$$

The refined estimate then is (from (8.7b))

$$x_1 = x_0 + \Delta x_0 \tag{8.9b}$$

Since the Taylor series expansion was truncated after two terms, the estimate x_1 is still an approximation to $\hat{x}$, but in general x_1 is expected to be a closer to $\hat{x}$ than x_0 is.

The geometrical interpretation of the iterative process is shown in Fig. 8.2a. The procedure may be repeated until

$$|\psi(x_n)| < TOL \tag{8.10}$$

where TOL is the acceptable error. Since the slope of the curve is used to guide the iteration, it is seen that the process should quickly converge to the solution of the problem $\hat{x}$. In an example, presented below, we will show that the rate of convergence is approximately quadratic. Defining an error at ith iteration as

$$e_i = \|x_i - x_{\text{exact}}\| \tag{8.11a}$$

where $\| \ \|$ is a suitable norm (e.g., 2 norm), quadratic convergence implies

$$\frac{e_{i+1}}{e_i^2} \approx \text{constant.} \tag{8.11b}$$

For $f(x) \in C^2$ (continuous up to the second derivative), x_i and $\hat{x}$ are sufficiently close to each other and $|f'(\varsigma)| > k_1$ and $|f''(\varsigma)| < k_2$ for some k_1, k_2, and ς that is between x_i and $\hat{x}$, this property can be proved analytically as shown below in the remark (Luenberger 1984).

Remark

Theorem: *Let $f(x) \in C^2$ (continuous up to the second derivative), $f(x_*) = 0$ (ie., x_* is the solution of $f(x) = 0$), $f'(x_*) \neq 0$, and x_k and x_* are sufficiently close to each other. Then the sequence $\{x_k\}_{k=0}^{k=\infty}$ generated by Newton's method converges to x_* at a second-order rate.*

Proof The Taylor series for $f(x)$ is

$$f_x = f_k + f_k'(x - x_k) + R_x; \quad R_x \approx \frac{1}{2} f_k''(x - x_k)^2$$

where $f_k = f(x_k)$, etc. Neglecting the remainder R_x and setting $f_x = f_{k+1} = 0$, we get Newton's iterative equation (for $|f_k'| > 0$) as

$$x_{k+1} = x_k - \frac{f_k}{f_k'}$$

Noting that $f_* = 0$, the above equation may be rewritten as

$$x_{k+1} - x_* = x_k - x_* - \frac{f_k - f_*}{f_k'}$$

$$x_{k+1} - x_* = \frac{1}{f_k'}\left[f_* - f_k - f_k'(x_* - x_k)\right] = \frac{R_*}{f_k'} \approx \frac{f_k''}{2f_k'}(x_* - x_k)^2$$

By continuity of f_k'' (which follows from the requirement $f(x) \in C^2$), we have $|f_k''| \leq k_2$ for some finite k_2 (bounded). By this, along with the property $|f_k'| \geq k_1 > 0$, the above equation becomes

$$|x_{k+1} - x_*| = \frac{k_2}{2k_1}|x_k - x_*|^2 = c|x_k - x_*|^2; \quad c = \frac{k_2}{2k_1}$$

Now if $c|x_k - x_*| < 1$, which requires that x_k and x_* are sufficiently close, k_2 is sufficiently small and k_1 is sufficiently large, then

$$|x_{k+1} - x_*| < |x_k - x_*|$$

and hence the sequence $\{x_k\}_{k=0}^{k=\infty}$ converges to x_*. It follows from $|x_{k+1} - x_*| = c|x_k - x_*|^2$ that the rate of convergence is second order.

8.3 The Modified Newton–Raphson Method for Scalar Equations

In the finite element context, K_{t0} appearing in (8.8) and (8.9a) is the tangent stiffness matrix (8.3a). Thus, the tangent stiffness matrix must be evaluated and the full system of equations solved (8.9a) at every iteration. Both of these tasks (the former more than the latter in most plasticity problems) are highly computationally intensive. In the modified Newton–Raphson method, the tangent K_{t0} is evaluated once at the beginning of the iteration, and kept unchanged throughout the iteration. The process is geometrically depicted in Fig. 8.2b. Example 8.1 will demonstrate that the rate of convergence in the present case is much slower than that in the case of the full Newton–Raphson method. The rate of convergence in the present case is approximately linear, i.e.,

$$\frac{e_{i+1}}{e_i} \approx \text{constant.} \tag{8.12}$$

whereas it is quadratic in the full Newton's method described in the preceding section.

Example 8.1

Question: A scalar function is given by

$$f = \frac{x}{1+x}$$

Given that $f = 0.1$ at $x = 0.1111$, determine the value of x when $f = \bar{f} = 0.3$.

Table 8.1 Newton–Raphson convergence results

Iteration number	x_i	K_{ti}	$\bar{f}$	f_i	Δx_i	e_i	e_{i+1}/e_i	e_{i+1}/e_i^2
1	0.1111	0.8100	0.3	0.1000	0.2469	0.3175	–	–
2	0.3580	0.5422	0.3	0.2636	0.0671	0.0705	0.2222	0.7000
3	0.4251	0.4924	0.3	0.2983	0.0035	0.0035	0.0494	0.6997
4	0.4286	0.4900	0.3	0.3000	0.0000	0.0000	0.0020	0.5888

Table 8.2 Modified Newton–Raphson convergence results

Iteration number	x_i	K_{ti}	$\bar{f}$	f_i	Δx_i	e_i	e_{i+1}/e_i	e_{i+1}/e_i^2
1	0.1111	0.8100	0.3	0.1000	0.2469	0.3175	–	–
2	0.3580	0.8100	0.3	0.2636	0.0449	0.0705	0.2222	0.7000
3	0.4029	0.8100	0.3	0.2872	0.0158	0.0257	0.3636	5.150
4	0.4187	0.8100	0.3	0.2951	0.0060	0.0098	0.3840	14.96
5	0.4247	0.8100	0.3	0.2981	0.0023	0.0038	0.3908	39.67
6	0.4271	0.8100	0.3	0.2993	0.0009	0.0015	0.3932	102.0
7	0.4280	0.8100	0.3	0.2997	0.0004	0.0006	0.3938	260.0
8	0.4283	0.8100	0.3	0.2999	0.0001	0.0002	0.3934	659.9
9	0.4285	0.8100	0.3	0.3000	0.0001	0.0001	0.3913	1,668

Answer: The exact answer to this problem is

$$0.3 = \frac{\hat{x}}{1+\hat{x}} \Rightarrow \hat{x} = 0.42857$$

The gradient (tangent) of the function is given by

$$K_t = \frac{df}{dx} = \frac{1}{(1+x)^2}$$

The problem is solved both by the Newton–Raphson and modified Newton–Raphson methods, and the results are presented in Tables 8.1 and 8.2, respectively. Examination of the last two columns supports the earlier claims that the rate of convergence is quadratic for the full Newton–Raphson method and linear for the modified Newton–Raphson method.

8.4 The Newton–Raphson Method for Nonlinear System of Equations

The methods presented in the preceding sections can be easily extended to a nonlinear system of equations simply by appropriately treating the scalars as either vectors or matrices. We will illustrate the process for a system of two nonlinear equations

$$3x_1 + 2x_2 - 4x_1x_2 - 0.3 = 0 \tag{8.13a}$$

$$-2x_1 + 5x_2 - x_2^2 - 0.3 = 0 \tag{8.13b}$$

Let us arrange them into a form that can be related to the internal spring force and the externally applied load that we encounter in a nonlinear finite element analysis (8.4):

$$F_s^1 = 3x_1 + 2x_2 - 4x_1x_2 = 0.3 = P^1 \tag{8.14a}$$

$$F_s^2 = -2x_1 + 5x_2 - x_2^2 = 0.3 = P^2 \tag{8.14b}$$

Defining vectors $\mathbf{F}_s = \{F_s^1, F_s^2\}$ and $\mathbf{P} = \{P^1, P^2\}$, consider $\mathbf{F}_s$ as the internal spring force defined in (8.4) and $\mathbf{P}$ as the externally applied force (8.4). Then we see that the spring force not only depends on both x_1 and x_2 (i.e., coupled as in finite element equations), but also varies nonlinearly with x_1 and x_2. Nevertheless, the system can be solved by the Newton–Raphson method. Analogous to (8.6), we defining a residual vector as

$$\boldsymbol{\psi}(\mathbf{x}) = \mathbf{F}_s(\mathbf{x}) - \bar{\mathbf{F}} \tag{8.15}$$

where $\mathbf{P} = \bar{\mathbf{F}}$ is the externally applied force vector. Let $\mathbf{x}_0$ is the initial estimate of $\mathbf{x}$. Equations (8.7)–(8.9) are rewritten in vector–matrix forms as:

$$\boldsymbol{\psi}(\mathbf{x}_1) = \boldsymbol{\psi}(\mathbf{x}_0) + \left[\frac{\mathrm{d}\boldsymbol{\psi}}{\mathrm{d}\mathbf{x}}\right]_{\mathbf{x}_0} \Delta\mathbf{x} + \frac{1}{2}\Delta\mathbf{x}^{\mathrm{T}} \left[\frac{\mathrm{d}^2\boldsymbol{\psi}}{\mathrm{d}\mathbf{x}^2}\right]_{\mathbf{x}_0} \Delta\mathbf{x} + \cdots \tag{8.16a}$$

$$\Delta\mathbf{x} = \mathbf{x}_1 - \mathbf{x}_0 \tag{8.16b}$$

The tangent (or tangent stiffness in a finite element analysis) is now a matrix. Let us define this at any $\mathbf{x}$ as:

$$\mathbf{K}_{t0} = \frac{\mathrm{d}\boldsymbol{\psi}}{\mathrm{d}\mathbf{x}} = \frac{\mathrm{d}\mathbf{F}_s}{\mathrm{d}\mathbf{x}} = \begin{bmatrix} \frac{\mathrm{d}F_s^1}{\mathrm{d}x_1} & \frac{\mathrm{d}F_s^1}{\mathrm{d}x_2} \\ \frac{\mathrm{d}F_s^2}{\mathrm{d}x_1} & \frac{\mathrm{d}F_s^2}{\mathrm{d}x_2} \end{bmatrix} \quad \text{(from (8.15))}, \tag{8.17}$$

The remaining equations are

$$\Delta\mathbf{x} = \Delta\mathbf{x}_0 = -\mathbf{K}_{t0}^{-1}\boldsymbol{\psi}(\mathbf{x}_0) = \mathbf{K}_{t0}^{-1}(\bar{\mathbf{F}} - \mathbf{F}_0) \tag{8.18a}$$

$$\mathbf{x}_1 = \mathbf{x}_0 + \Delta\mathbf{x}_0 \tag{8.18b}$$

$$|\boldsymbol{\psi}(\mathbf{x}_n)| < TOL \tag{8.18c}$$

where $|\boldsymbol{\psi}(\mathbf{x}_n)|$ is a suitable norm (e.g., 2 norm) of the residual vector $\boldsymbol{\psi}(\mathbf{x}_n)$.

Example 8.2

Question: Solve the nonlinear equations given in (8.14) by the Newton–Raphson and modified Newton–Raphson methods.

Answer: The tangent stiffness matrix is

$$\mathrm{K}_t = \begin{bmatrix} \dfrac{\mathrm{d}F_s^1}{\mathrm{d}x_1} & \dfrac{\mathrm{d}F_s^1}{\mathrm{d}x_2} \\ \dfrac{\mathrm{d}F_s^2}{\mathrm{d}x_1} & \dfrac{\mathrm{d}F_s^2}{\mathrm{d}x_2} \end{bmatrix} = \begin{bmatrix} 3-4x_2 & -2-4x_1 \\ -2 & 5-2x_2 \end{bmatrix}$$

The results obtained by repeating (8.18) are presented in (a) Table 8.3 for Newton–Raphson method with $\mathbf{x}_0 = \{0, 0\}$, (b) Table 8.4 for modified Newton–

Table 8.3 Newton–Raphson convergence results: Example 8.2 with $\mathbf{x}_0 = \{0, 0\}$

Iteration number	x_i^1 / x_i^2	K_t^{11} / K_t^{21}	K_t^{12} / K_t^{22}	$\bar{F}^1$ / $\bar{F}^2$	F_i^1 / F_i^2	Δx_i^1 / Δx_i^2	e_i	e_{i+1}/e_i	e_{i+1}/e_i^2
1	0.0000	3.0000	−2.0000	0.3	0.0000	0.1909	0.3515		
	0.0000	−2.0000	5.0000	0.3	0.0000	0.1364			
2	0.1909	2.4545	−2.7636	0.3	0.1959	0.0895	0.1182	0.3363	0.9569
	0.1364	−2.0000	4.7273	0.3	0.2814	0.0418			
3	0.2804	2.2874	−8.1215	0.3	0.2850	0.0171	0.0195	0.1646	10.393
	0.1782	−2.0000	4.7273	0.3	0.2983	0.0077			
4	0.2975	2.2564	−8.1900	0.3	0.2995	0.0007	0.0007	0.0351	1.803
	0.1859	−2.0000	4.6282	0.3	0.2999	0.0003			
5	0.2981	2.2552	−8.1926	0.3	0.3000	0.0000	0.0000	0.0602	88.17
	0.1862	−2.0000	4.6276	0.3	0.3000	0.0000			

Table 8.4 Modified Newton–Raphson convergence results: Example 8.2 with $\mathbf{x}_0 = \{0, 0\}$

Iteration number	x_i^1 / x_i^2	K_t^{11} / K_t^{21}	K_t^{12} / K_t^{22}	$\bar{F}^1$ / $\bar{F}^2$	F_i^1 / F_i^2	Δx_i^1 / Δx_i^2	e_i	e_{i+1}/e_i	e_{i+1}/e_i^2
1	0.0000	3.0000	−2.0000	0.3	0.0000	0.1909	0.3515		
	0.0000	−2.0000	5.0000	0.3	0.0000	0.1364			
2	0.1909	8.0000	−2.0000	0.3	0.1959	0.0507	0.1182	0.3363	0.9569
	0.1364	−2.0000	5.0000	0.3	0.2814	0.0240			
3	0.2416	8.0000	−2.0000	0.3	0.2491	0.0244	0.0621	0.5254	4.444
	0.1604	−2.0000	5.0000	0.3	0.2929	0.0112			
4	0.2660	8.0000	−2.0000	0.3	0.2724	0.0132	0.0352	0.5676	9.138
	0.1716	−2.0000	5.0000	0.3	0.2963	0.0060			
16	0.2981	8.0000	−2.0000	0.3	0.2999	0.0000	0.0001	0.5462	>10
	0.1862	−2.0000	5.0000	0.3	0.3000	0.0000			

Table 8.5 Newton–Raphson convergence results: Example 8.2 with $\mathbf{x}_0 = \{1.0, 1.0\}$

Iteration	x_i^1	K_t^{11}	K_t^{12}	$\bar{F}^1$	F_i^1	Δx_i^1	e_i	e_{i+1}/e_i	e_{i+1}/e_i^2
number	x_i^2	K_t^{21}	K_t^{22}	$\bar{F}^2$	F_i^2	Δx_i^2			
1	1.000	1.0000	−6.0000	0.3	−8.000	0.0200	0.6610		
	1.000	−2.0000	8.0000	0.3	2.000	−0.5533			
2	1.020	1.2133	−6.0800	0.3	0.3443	−0.2340	0.2787	0.4216	0.6610
	0.4467	−2.0000	4.1067	0.3	−0.0062	−0.0394			
3	0.7860	1.3710	−5.1438	0.3	0.2631	−0.0357	0.0427	0.5493	0.2787
	0.4072	−2.0000	4.1855	0.3	0.2984	−0.0167			
4	0.7502	1.4378	−5.0010	0.3	0.2976	−0.0029	0.0032	1.779	0.0427
	0.3905	−2.0000	4.2189	0.3	0.2997	−0.0013			
5	0.7473	1.4431	−4.9893	0.3	0.3000	0.0000	0.0000	4.447	0.0032
	0.3892	−2.0000	4.2215	0.3	0.3000	0.0000			

Table 8.6 Modified Newton–Raphson convergence results: Example 8.2 with $\mathbf{x}_0 = \{1.0, 1.0\}$

Iteration	x_i^1	K_t^{11}	K_t^{12}	$\bar{F}^1$	F_i^1	Δx_i^1	e_i	e_{i+1}/e_i	e_{i+1}/e_i^2
number	x_i^2	K_t^{21}	K_t^{22}	$\bar{F}^2$	F_i^2	Δx_i^2			
1	1.000	−1.000	−6.0000	0.3	−8.00	0.0200	0.6610		
	1.000	−2.0000	8.0000	0.3	2.000	−0.5533			
2	1.020	−1.0000	−6.0000	0.3	0.3443	−0.1136	0.2787	0.4216	0.6378
	0.4467	−2.0000	8.0000	0.3	−0.0062	0.0263			
3	0.9064	−1.0000	−6.0000	0.3	0.0584	−0.0370	0.1798	0.6451	2.315
	0.4730	−2.0000	8.0000	0.3	0.3284	−0.0341			
4	0.8694	−1.0000	−6.0000	0.3	0.2043	−0.0340	0.1318	0.7333	4.079
	0.4389	−2.0000	8.0000	0.3	0.2629	−0.0103			
32	0.7474	−1.0000	−6.0000	0.3	0.2999	0.0000	0.0002	0.8226	>10
	0.3893	−2.0000	8.0000	0.3	0.3000	0.0000			

Raphson method with $\mathbf{x}_0 = \{0, 0\}$, (c) Table 8.5 for Newton–Raphson method with $\mathbf{x}_0 = \{1.0, 1.0\}$, and (d) Table 8.6 for modified Newton–Raphson method with $\mathbf{x}_0 = \{1.0, 1.0\}$. The error e is defined as the 2 norm:

$$e_i = \|\mathbf{x}_{\text{exact}} - \mathbf{x}_i\|_2 = \sqrt{(x^1_{\text{exact}} - x^1_i)^2 + (x^2_{\text{exact}} - x^2_i)^2}$$

It may be observed that the analysis with $\mathbf{x}_0 = \{0, 0\}$ converges to $\hat{\mathbf{x}} = \mathbf{x}_n = \{0.2981, 0.1862\}$ and that the analysis with $\mathbf{x}_0 = \{1.0, 1.0\}$ converges to a different solution $\hat{\mathbf{x}} = \mathbf{x}_n = \{0.7473, 0.3892\}$. The reader may verify that both of these solutions satisfy the equations. Based on the error ratios, the rate of convergence is unclear for Newton's method (Tables 8.3 and 8.5). It is clearly linear for the modified Newton's method (Tables 8.4 and 8.6). The fast convergence of the Newton's method (five iterations) relative to that of the modified Newton's method (>16 iterations) suggests that the rate of convergence of the Newton's method is better than linear.

Remarks

- Nonlinear functions have more than one stationary points. The starting point must be close to a given stationary point in order to locate that stationary point.
- In certain cases, the functions involved may not be defined when the values of the variables are outside certain ranges. For example, $f = (1 - x)^{1/2}$ is not defined for $x > 1$. During the search process, depending on the starting point, the variable may assume a value that is outside its valid range (Problem 8.1b).
- Depending on the nature and nonlinearity of the constitutive functions involved in an elasto-plastic model, the difficulties mentioned above may be encountered during the integration of material models.

8.5 Application of the Newton–Raphson Method to Nonlinear Finite Element Analysis

The procedure presented in this section is applicable to both the nonlinear elastic and elasto-plastic analyses. Three differences between these analysis types are: (1) The external load is applied in one increment in elastic analyses whereas it is applied in multiple increments in elasto-plastic analysis, (2) the elasto-plastic analyses involve additional variables $\boldsymbol{\zeta}$, known as the hardening (or plastic internal) variables, which are absent in elastic analyses, and (3) iterations are needed for establishing the stresses (and hardening variables) for a strain increment in elasto-plastic analyses, whereas no such iterations are needed in elastic analyses. In applying the algorithm presented in this section to elastic analyses, (1) you may apply the load in one increment, (2) ignore the part about hardening variables, and (3) calculate the stresses directly from the stress–strain relation without resorting to iterations.

The problem at hand is stated as follows: given $(\mathbf{F}_s^n, \boldsymbol{\sigma}_n, \boldsymbol{\zeta}_n, \boldsymbol{\varepsilon}_n, \mathbf{u}_n)$, determine $(\mathbf{F}_s^{n+1}, \boldsymbol{\sigma}_{n+1}, \boldsymbol{\zeta}_n, \boldsymbol{\varepsilon}_{n+1}, \mathbf{u}_{n+1})$ that corresponds to a new load of $\bar{\mathbf{F}}_{n+1}$. The solution process is very similar to that involved in the solution of a system of nonlinear equations presented in the preceding section. The global tangent stiffness $\mathbf{K}_{n+1}^t$ ((8.3a), Fig. 8.3b) is determined as

$$\mathbf{K}_{n+1}^t = \sum \mathbf{k}_{n+1}^t \quad \text{where } \mathbf{k}_{n+1}^t = \int \mathbf{B}^{\mathrm{T}} \mathbf{D}_{\mathbf{n+1}}^{\mathbf{t}} \mathbf{B} \, \mathrm{d}v \tag{8.19}$$

where $\mathbf{D}_{n+1}^t$ is the tangent operator shown schematically in Fig. 8.3a (8.3b) and $\mathbf{k}_{n+1}^t$ is the element tangent stiffness matrix formed using $\mathbf{D}_{n+1}^t$. The element spring force $\mathbf{f}_s^{n+1}$ is computed as

$$\mathbf{f}_s^{n+1} = \int \mathbf{B}^T \boldsymbol{\sigma}_{n+1} \, \mathrm{d}v \tag{8.20}$$

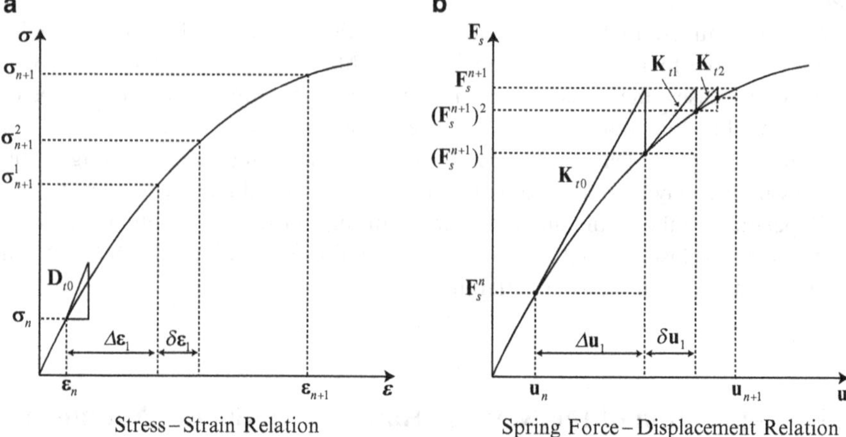

Fig. 8.3 Application of Newton–Raphson iterations in finite element analysis

The key steps are: (1) establish $(\mathbf{F}_s^n, \boldsymbol{\sigma}_n, \boldsymbol{\zeta}_n, \boldsymbol{\varepsilon}_n, \mathbf{u}_n)$, (2) determine the residual force $\boldsymbol{\psi}(\mathbf{u})$ (8.15), (3) determine the current global tangent stiffness matrix $\mathbf{K}^t$ (8.19) and solve for the iterative displacements ($\delta\mathbf{u}_0, \delta\mathbf{u}_1$, etc.) and the corresponding iterative strains, (4) update and find the incremental displacements ($\Delta\mathbf{u}_0, \Delta\mathbf{u}_1$, etc.) and the corresponding incremental strains, (5) use the constitutive model to find the current stresses ($\boldsymbol{\sigma}_{n+1}$) and hardening variables ($\boldsymbol{\zeta}_{n+1}$) corresponding to the incremental strains, (6) determine the current element spring force (8.20) and assemble to find the global spring force $\mathbf{F}_s^{n+1}$, (7) determine the residual force $\boldsymbol{\psi}_{n+1}(\mathbf{u})$, (8) check for convergence, and (9) stop if converged; continue otherwise.

The steps described above are what constitute the *global iterations*. In step 5, it is required to evaluate $\boldsymbol{\sigma}_{n+1}$ and $\boldsymbol{\zeta}_{n+1}$ corresponding to the incremental strain based on the specific constitutive model employed for the material at the specific Gauss point. In elasto-plasticity, this step involves iterations, which are referred to as the *local iterations*.

The exact global algorithm depends on the complexity and sophistication of the specific algorithm employed. The essential steps involved in a typical, simple global analysis are summarized in Algorithm 8.1. The local iterations are the primary subject of some of the subsequent chapters.

Algorithm 8.1. Global Iteration Strategy for Nonlinear Finite Element Analysis

Definition of new parameters: $N_G^{\max}$ is maximum number of global iterations permitted before execution is terminated, and N_G is the corresponding counter.

1. Establish $(\mathbf{F}_s^n, \boldsymbol{\sigma}_n, \boldsymbol{\zeta}_n, \boldsymbol{\varepsilon}_n, \mathbf{u}_n)$, $N_G^{\max}$, *TOL*, and material properties. Set $(\mathbf{F}_s^{n+1}, \boldsymbol{\sigma}_{n+1}, \boldsymbol{\zeta}_{n+1}, \boldsymbol{\varepsilon}_{n+1}, \mathbf{u}_{n+1}) = (\mathbf{F}_s^n, \boldsymbol{\sigma}_n, \boldsymbol{\zeta}_n, \boldsymbol{\varepsilon}_n, \mathbf{u}_n)$, $\Delta\boldsymbol{\varepsilon} = 0$, and $\Delta\mathbf{u} = 0$
2. Determine $\boldsymbol{\psi}_{n+1}(\mathbf{u}_{n+1}) = \mathbf{F}_s^{n+1}(\mathbf{u}_{n+1}) - \overline{\mathbf{F}}$. Set $N_G = 1$

(continued)

3. *Compute* $\mathbf{K}^t_{n+1}(\boldsymbol{\sigma}_{n+1}, \boldsymbol{\zeta}_{n+1}, \boldsymbol{\varepsilon}_{n+1})$, modify $\mathbf{K}^t_{n+1}$ and $\boldsymbol{\psi}(\mathbf{u}_{n+1})$ for specified displacements, solve $\mathbf{K}^t_{n+1}\delta\mathbf{u} = \boldsymbol{\psi}_{n+1}$ and find $\delta\mathbf{u}$ and $\delta\boldsymbol{\varepsilon}$
4. Update incremental quantities $\Delta\boldsymbol{\varepsilon} \leftarrow \Delta\boldsymbol{\varepsilon} + \delta\boldsymbol{\varepsilon}$, $\Delta\mathbf{u} \leftarrow \Delta\mathbf{u} + \delta\mathbf{u}$
5. Find $\Delta\boldsymbol{\sigma}$ and $\Delta\boldsymbol{\zeta}$ corresponding to $\Delta\boldsymbol{\varepsilon}$ using the constitutive law. Update stresses $\boldsymbol{\sigma}_{n+1} = \boldsymbol{\sigma}_n + \Delta\boldsymbol{\sigma}$ and hardening variables $\boldsymbol{\zeta}_{n+1} = \boldsymbol{\zeta}_n + \Delta\boldsymbol{\zeta}$
6. Update spring force $\mathbf{F}^{n+1}_s(\boldsymbol{\sigma}_{n+1})$
7. Determine residual spring force $\boldsymbol{\psi}_{n+1}(\mathbf{u}_{n+1}) = \mathbf{F}^{n+1}_s(\mathbf{u}_{n+1}) - \overline{\mathbf{F}}$ and the error $|\boldsymbol{\psi}|$
8. Check for convergence: if $|\boldsymbol{\psi}| \leq TOL$ go to step 11
9. if $(N_G > N_G^{\max})$ STOP
10. $N_G \leftarrow N_G + 1$, go to step 3
11. Update and establish required quantities $\boldsymbol{\varepsilon}_{n+1} = \boldsymbol{\varepsilon}_n + \Delta\boldsymbol{\varepsilon}$ and $\mathbf{u}_{n+1} \leftarrow \mathbf{u}_n + \Delta\mathbf{u}$. Continue with the next load increment if needed.

8.6 Application of the Newton–Raphson Method to Calibration of Constitutive Models

Algorithm 8.1 pertains to the global finite element analysis. In step (5), given $\Delta\boldsymbol{\varepsilon}$, stresses and plastic internal variables are to be calculated and updated as $\boldsymbol{\sigma}_{n+1} = \boldsymbol{\sigma}_n + \Delta\boldsymbol{\sigma}$ and $\boldsymbol{\zeta}_{n+1} = \boldsymbol{\zeta}_n + \Delta\boldsymbol{\zeta}$, respectively. While this task is trivial for elastic constitutive models, a suitable iterative procedure is needed for elasto-plastic constitutive models. A specific algorithm that is suitable for calibration of constitutive models is presented below. The algorithm, however, is applicable to both elastic and elasto-plastic models.

The Newton's strategy used here is the same as that in Algorithm 8.1, but concerns a case where the stresses, strains, and hardening variables are uniform within the specimen.

Then for a specimen with unit dimensions (see Fig. 8.5), the forces and displacements become equal to the stresses and strains, respectively. The problem is basically modeled using one finite element. The element stiffness matrix $\mathbf{k}$ and the tangent operator $\mathbf{D}$ are the same. The algorithm is presented for an increment; it can be easily extended for a complete loading involving several load increments.

The problem at hand is the following: given $(\boldsymbol{\sigma}_n, \boldsymbol{\varepsilon}_n, \boldsymbol{\zeta}_n, \Delta\boldsymbol{\varepsilon}, \text{and}\, \Delta\boldsymbol{\sigma})$ find $(\boldsymbol{\sigma}_{n+1}, \boldsymbol{\varepsilon}_{n+1}, \boldsymbol{\zeta}_{n+1})$, where $\Delta\boldsymbol{\varepsilon}$ is the specified strain increment vector, $\Delta\boldsymbol{\sigma}$ is the specified stress increment vector, $\boldsymbol{\zeta}_n$ is the hardening variables (in general, tensor of some order). Note that when a component of $\Delta\boldsymbol{\sigma}$ is specified, the corresponding component of $\Delta\boldsymbol{\varepsilon}$ is to be computed, and vice versa. Referring to Fig. 8.4, during Newton's iteration, an equation of the form

$$\mathbf{D}\delta\boldsymbol{\varepsilon} = \delta\boldsymbol{\sigma}$$

needs to be solved at the global level. When a component of $\delta\boldsymbol{\varepsilon}$ is specified, $\mathbf{D}$ and $\delta\boldsymbol{\sigma}$ are modified as follows:

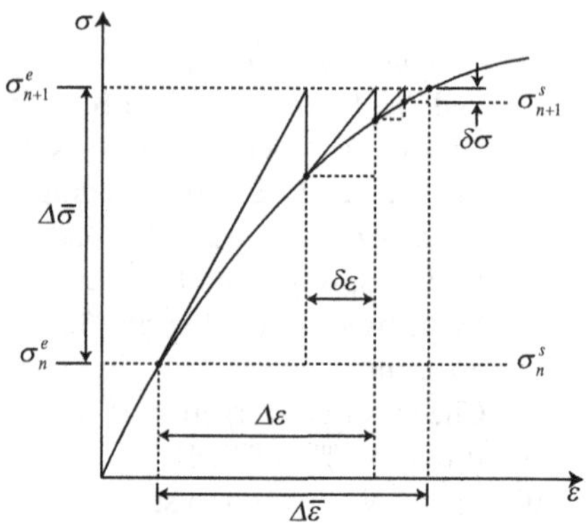

Fig. 8.4 Schematic of global iteration for model calibration

$$\begin{bmatrix} D_{11} & . & . & . & D_{16}. \\ . & . & . & . & . \\ . & . & x_0 & . & . \\ . & . & . & . & . \\ D_{16} & . & . & . & D_{66} \end{bmatrix} \begin{pmatrix} \delta\varepsilon_{11} \\ . \\ \delta\varepsilon_k \\ \\ . \end{pmatrix} = \begin{pmatrix} \delta\sigma_{11} \\ . \\ x_0\delta\hat{\varepsilon}_k \\ . \\ . \end{pmatrix}$$

where x_0 is a large number. Here $\delta\varepsilon_k = \Delta\varepsilon_k$ during the first iteration, and $\delta\varepsilon_k = 0$ during the subsequent iterations.

Referring to Fig. 8.4, $\boldsymbol{\sigma}^e$ and $\boldsymbol{\sigma}^s$ are, respectively, the external and internal (spring) "forces," and $\delta\boldsymbol{\sigma}$ is the difference between the two at any stage of the iteration. The iteration is considered converged when the following error norms on residual stresses and energy are less than specified tolerances

$$e_\sigma = \frac{\|\delta\boldsymbol{\sigma}\|}{\left\|\boldsymbol{\sigma}^e_{n+1}\right\|} \leq TOL_1 \text{ and } e_e = \delta\boldsymbol{\sigma}^{\mathrm{T}}\Delta\boldsymbol{\varepsilon} \leq TOL_2$$

Note that other suitable error measures may also be employed. The procedure is summarized in Algorithm 8.2.

Algorithm 8.2. Global Iteration Strategy for Model Calibration

Definition of new parameters: $N_G^{\max}$ is maximum number of global iterations permitted before execution is terminated, and N_G is the corresponding counter.

1. Establish, $(\boldsymbol{\sigma}_n, \boldsymbol{\varepsilon}_n, \boldsymbol{\zeta}_n, \Delta\bar{\boldsymbol{\sigma}}, \Delta\bar{\boldsymbol{\varepsilon}}), N_G^{\max}, TOL_1, TOL_2$, and material properties.
2. Initialize: $\boldsymbol{\sigma}^s_n = \boldsymbol{\sigma}_n,\ \boldsymbol{\sigma}^e_n = \boldsymbol{\sigma}_n,\ \Delta\boldsymbol{\sigma} = 0,\ \Delta\boldsymbol{\varepsilon} = 0,\ \boldsymbol{\sigma}^e_{n+1} = \boldsymbol{\sigma}^e_n + \Delta\hat{\boldsymbol{\sigma}}$, $\boldsymbol{\sigma}^s_{n+1} = \boldsymbol{\sigma}^s_n$, and $\delta\boldsymbol{\sigma} = \boldsymbol{\sigma}^e_{n+1} - \boldsymbol{\sigma}^s_{n+1}, \boldsymbol{\sigma}_{n+1} = \boldsymbol{\sigma}_n, \boldsymbol{\zeta}_{n+1} = \boldsymbol{\zeta}_n$, and $N_G = 1$.
3. *Compute* $\mathbf{D}^t_{n+1}(\boldsymbol{\sigma}_{n+1}, \boldsymbol{\zeta}_{n+1})$ (elastic for $N_G = 1$ and elasto-plastic for $N_G > 1$), modify $\mathbf{D}^t_{n+1}$ and $\delta\boldsymbol{\sigma}$ for fixed strain, solve $\mathbf{D}^t_{n+1}\delta\boldsymbol{\varepsilon} = \delta\boldsymbol{\sigma}$, and find $\delta\boldsymbol{\varepsilon}$.
4. Update incremental strain $\Delta\boldsymbol{\varepsilon} \leftarrow \Delta\boldsymbol{\varepsilon} + \delta\boldsymbol{\varepsilon}$.

(continued)

5. Find $\Delta\boldsymbol{\sigma}$ and $\Delta\boldsymbol{\zeta}$ corresponding to $\Delta\boldsymbol{\varepsilon}$.
6. Update spring force and stress: $\boldsymbol{\sigma}^s_{n+1} \leftarrow \boldsymbol{\sigma}^s_{n+1} + \Delta\boldsymbol{\sigma}$, $\boldsymbol{\sigma}_{n+1} \leftarrow \boldsymbol{\sigma}_{n+1} + \Delta\boldsymbol{\sigma}$, and $\boldsymbol{\zeta}_{n+1} \leftarrow \boldsymbol{\zeta}_n + \Delta\boldsymbol{\zeta}$.
7. Determine residual and errors: $\delta\boldsymbol{\sigma} = \boldsymbol{\sigma}^e_{n+1} - \boldsymbol{\sigma}^s_{n+1}$, $e_\sigma = \|\delta\boldsymbol{\sigma}\|/\|\boldsymbol{\sigma}^e_{n+1}\|$, $e_e = \delta\boldsymbol{\sigma}^{\mathrm{T}}\Delta\boldsymbol{\varepsilon}$.
8. For $N_G > 1$, check for convergence: if ($e_\sigma \leq TOL_1$ and $e_e \leq TOL_2$) go to step 11.
9. if ($N_G > N_G^{\max}$) STOP.
10. $N_G \leftarrow N_G + 1$, go to step 3.
11. Update and establish the required quantities: $\boldsymbol{\varepsilon}_{n+1} = \boldsymbol{\varepsilon}_n + \Delta\boldsymbol{\varepsilon}$. $\boldsymbol{\sigma}_{n+1}$ and $\boldsymbol{\zeta}_{n+1}$ are already available.

8.7 Solution of Initial Value Problems by Euler Methods

Most constitutive relationships are expressed in the form of rate relations, where the rate of stress and hardening variables are expressed as a function of the rate of strains. As we have seen in the preceding section, the finite element analysis requires incremental stress for a given incremental strain. This requires the rate constitutive relations to be integrated.

In mathematical terms, the integration problem is an initial value problem. To define some of the associated terminologies, let us consider the simple initial value problem
Given:

$$\dot{y}(t) = g[y(t)]; \quad y(0) = y_n; \quad t \in [0, T] \tag{8.21a}$$

Find:

$$y_{n+1}(t + \Delta t) = y_n + \Delta y \tag{8.21b}$$

where t is time and g is a smooth function. Let us assume that the function in (8.21a) is too complex to integrate exactly, and hence a numerical method is to be used. The exact value of y_{n+1} may be algorithmically approximated as

$$y_{n+1} = y_n + \Delta t g(y_{n+\theta}) \tag{8.22a}$$

$$y_{n+\theta} = \theta y_{n+1} + (1 - \theta) y_n; \quad \theta \in [0, 1] \tag{8.22b}$$

where θ is a scalar parameter. The method in (8.22) is referred to as the *generalized midpoint rule*. Depending on the value of θ, various well-known methods of numerical integration methods are generated, including:

$\theta = 0$: forward Euler (explicit) method
$\theta = (1/2)$: midpoint rule
$\theta = 1$: backward Euler (implicit) method

The important properties of any algorithm are: *consistency, stability, and accuracy*. The reader is referred to standard texts (e.g., Gear 1971; Hairer et al. 1987) for detailed discussions on these topics from a mathematics standpoint. Analyses pertaining to elasto-plastic applications may be found in Ortiz and Popov (1985) and Simo and Hughes (1998).

An algorithm having first-order accuracy is considered to be consistent. Considering the initial value problem stated in (8.21), an algorithm is (a) first-order accurate if the algorithmic approximation y_{n+1} (8.22) agrees with the exact value $y(t_{n+1})$ to within second-order terms in step size Δt, and (b) second-order accurate if the algorithmic y_{n+1} agrees with the exact value $y(t_{n+1})$ to within third-order terms in step size Δt. Second-order accuracy is not a requirement, but a desirable one from the point of view of efficiency. Ortiz and Popov (1985) show from a theoretical analysis that, when a generalized midpoint rule as in (8.22) is used, *a second-order accuracy is achieved for* $\theta = 1/2$.

The simple, uniaxial problem presented in Example 8.3 (Fig. 8.5) helps gain some insight into the computational process, and help demonstrate the differences between what are known as *the continuum and consistent operators*. Ortiz and Popov (1985) also theoretically show that *for unconditional stability of the numerical algorithms* $1/2 \leq \theta \leq 1$. However, on the basis of numerical experimentations, they find that for small time steps, a high accuracy is achieved for $\theta = 1/2$ as the theory indicates, but for large time steps, an optimal accuracy is obtained for $\theta > 1/2$.

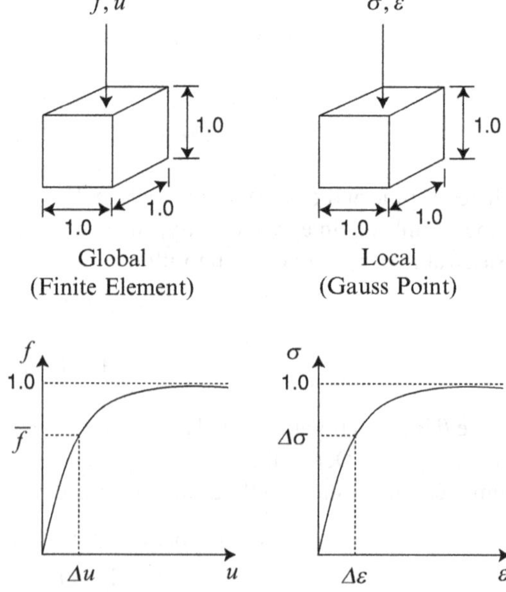

Fig. 8.5 Schematic of uniaxial load-deformation and stress–strain relations

Example 8.3

Problem Consider a uniaxial problem shown in Fig. 8.5. The stress–strain relation of the material is given by

$$\sigma = 1 - e^{-\alpha\varepsilon}; \quad \alpha > 0 \tag{8.23}$$

where α is a constant which controls the steepness of the stress–strain curve. Differentiating (8.23):

$$\dot{\sigma} = \alpha e^{-\alpha\varepsilon}\dot{\varepsilon} = \alpha r\dot{\varepsilon} \tag{8.24a}$$

where

$$r = e^{-\alpha\varepsilon} = 1 - \sigma \tag{8.24b}$$

Let us redefine this as a global/local iteration problem as described in Sect. 8.6 (Algorithm 8.2).

Global Problem Referring to Fig. 8.5, our global problem consists of a uniformly loaded specimen (one-element finite element model), subjected to a uniaxial external force f. The uniaxial displacement is u. Let the uniaxial stress and strain be σ and ε, respectively. The stress–strain behavior of the material is represented by the rate relation (8.24a). The spring force is denoted by f_s. Let $f_{s,n} = f_n = 0$ and $u_n = 0$. Noting that the body is a unit cube, it follows $f_s = \sigma$ and $u = \varepsilon$. Also the global tangent stiffness matrix and the material tangent operator are equal to each other, i.e., $K^t = D^t$.

Local Problem Given the strain increment $\Delta\varepsilon$ and the initial stresses, determine the corresponding stress increment according to the constitutive law given by (8.24a). Let $\sigma_n = 0$ and $\varepsilon_n = 0$.

Question: Determine the displacement u_{n+1} for a prescribed external load of $f_{n+1} = \bar{f}$.

Answer: At local points, let us use the generalized midpoint rule to approximate r as

$$\sigma_{n+1} = \sigma_n + \alpha r_{n+\theta}\Delta\varepsilon \tag{8.25a}$$

$$r_{n+\theta} = 1 - \sigma_{n+\theta}, \quad \text{where} \quad \sigma_{n+\theta} = \theta\sigma_{n+1} + (1-\theta)\sigma_n \tag{8.25b}$$

Combining (8.25a) and (8.25b):

$$\sigma_{n+1} = \sigma_n + \alpha[1 - \theta\sigma_{n+1} - (1-\theta)\sigma_n]\Delta\varepsilon \tag{8.26}$$

Recognizing $\sigma_n = 0$:

$$\sigma_{n+1} = [\alpha - \alpha\theta\sigma_{n+1}]\Delta\varepsilon$$

$$\sigma_{n+1} = R^{-1}(\alpha\Delta\varepsilon), \quad \text{where} \quad R = 1 + \alpha\theta\Delta\varepsilon \tag{8.27}$$

Differentiating (8.25a), and noting that quantities at increment number n are all constants, the rate of stress during the increment is

$$\dot{\sigma} = \alpha\dot{r}_{n+\theta}\Delta\varepsilon + \alpha r_{n+\theta}\dot{\varepsilon} \tag{8.28a}$$

$$\dot{\sigma} = -\alpha\theta\dot{\sigma}\Delta\varepsilon + \alpha r_{n+\theta}\dot{\varepsilon} \tag{8.28b}$$

$$\dot{\sigma} = R^{-1}[\alpha r_{n+\theta}]\dot{\varepsilon} \tag{8.28c}$$

or

$$\dot{\sigma} = D\dot{\varepsilon} \tag{8.29}$$

where

$$D = R^{-1}[\alpha r_{n+\theta}] \tag{8.30a}$$

D is the so-called consistent tangent operator.

From (8.24a)

$$\dot{\sigma} = \bar{D}\dot{\varepsilon}$$

where

$$\bar{D} = \alpha[1 - \sigma_{n+1}] \tag{8.30b}$$

$\bar{D}$ is known as the continuum tangent operator.

In summary, at the local level, the stress at point $n + 1$ is calculated for a given strain increment from (8.27)

$$\sigma_{n+1} = R^{-1}(\alpha\Delta\varepsilon), \quad \text{where} \quad R = 1 + \alpha\theta\Delta\varepsilon \tag{8.30c}$$

The global iterations are performed using either the consistent operator D given by (8.30a)

$$D^t = D = R^{-1}[\alpha r_{n+\theta}]; \quad r_{n+\theta} = 1 - \theta\sigma_{n+1} - (1 - \theta)\sigma_n \tag{8.30d}$$

or the continuum operator $\bar{D}$ given by (8.30b).

$$D^t = \bar{D} = \alpha[1 - \sigma_{n+1}] \tag{8.30e}$$

Forward Euler Method (Explicit) Let us first consider the forward Euler (explicit) method for which $\theta = 0$. Then $R = 1$ from (8.30c), $r_{n+\theta} = 1$ from (8.30d) and hence $D = \alpha$ from (8.30d). Noting that the stress is zero at the start, $\bar{D} = \alpha$ from (8.30e). During the first global iteration,

$$\delta u = \frac{\bar{f}}{D^t} = \frac{\bar{f}}{\alpha} = \delta\varepsilon \Rightarrow \Delta\varepsilon = \delta\varepsilon = \frac{\bar{f}}{\alpha}$$

Then at the local level, $\sigma_{n+1} = R^{-1}(\alpha\Delta\varepsilon) = \alpha\Delta\varepsilon$. At the global level, the spring force is calculated as $f_{s,n+1} = \sigma_{n+1} = \alpha\Delta\varepsilon = \bar{f}$. Hence, the process converges in one iteration. Inverting (8.23), the exact solution to strain is obtained as

$$\Delta\varepsilon_{\text{exact}} = -\frac{1}{\alpha}\ln(1 - \bar{f})$$

Hence,

$$\frac{\Delta\varepsilon_{\text{num}}}{\Delta\varepsilon_{\text{exact}}} = -\frac{\bar{f}}{\ln(1 - \bar{f})}$$

It is seen that the accuracy depends on the magnitude of $\bar{f}$ (i.e., the load step size). For example, for $\bar{f} = 0.2$, $\Delta\varepsilon_{\text{num}}/\Delta\varepsilon_{\text{exact}} = 0.896$ and for $\bar{f} = 0.5$, $\Delta\varepsilon_{\text{num}}/\Delta\varepsilon_{\text{exact}} = 0.721$.

Backward Euler Method (Implicit) The analysis is now repeated with the backward Euler (implicit) method for which $\theta = 1$. The analyses are performed with $\alpha = 1$. The results showing the convergence behavior are presented in Table 8.7 for the case using the continuum operator $\bar{D}$, and in Table 8.8 for the case using the consistent operator D. In both cases, converged value of u is 1.0.

It may be observed that method using consistent tangent operator achieves quadratic convergence (i.e., $e^{k+1}/(e^k)^2$ remains a constant) whereas that with continuum operator only achieves linear convergence (i.e., e^{k+1}/e^k remains a constant). Noting that the exact answer for u (obtained from (8.23) for $\sigma = 0.5$) is 0.6931, the accuracy is seen to be poor.

Table 8.9 presents the final results for a few other cases, each with different values for $\bar{f}$ and θ, using consistent or continuum operators. The following observations can be made: (1) when the midpoint rule is used ($\theta = 0.5$), the convergence behavior using continuum and consistent operators are similar, and the accuracy of converged results is better than the accuracy of the results using $\theta = 1.0$ (agreeing with the analysis of Ortiz and Popov 1985), and (2) for the case with $\theta = 1.0$, increase in step size significantly increases the number of required iterations for the case using continuum operator, but only slightly increases the number of required iterations for the case using consistent operator.

A fundamental difference between the mathematical initial value problem stated in (8.21) and the elasto-plastic integration problem of interest in this book is the fact

Table 8.7 Newton–Raphson convergence results for the case $\theta = 1$ and using continuum tangent operator (Example 8.3)

Iteration number k	u^k_{n+1}	$K^{t,k}_{n+1} = \bar{D}^k_{n+1}$	$\bar{f}$	$f^k_{s,n+1} = \sigma^k_{n+1}$	δu^k	$e^k = 1 - u^k_{n+1}$	$e^{k+1}/(e^k)^2$	e^{k+1}/e^k
1	0.0000	1.0000	0.5000	0.0000	0.5000	1.0000	0.50	0.5000
2	0.5000	0.6667	0.5000	0.3333	0.2500	0.5000	1.00	0.5000
3	0.7500	0.5714	0.5000	0.4286	0.1250	0.2500	2.00	0.5000
4	0.8750	0.5333	0.5000	0.4667	0.0625	0.1250	4.00	0.5000
5	0.9375	0.5161	0.5000	0.4839	0.0313	0.0625	8.00	0.5000
6	0.9688	0.5079	0.5000	0.4921	0.0156	0.0313	16.00	0.5000
7	0.9844	0.5039	0.5000	0.4961	0.0078	0.0156	31.99	0.5000
8	0.9922	0.5020	0.5000	0.4980	0.0039	0.0078	64.00	0.5000
9	0.9961	0.5010	0.5000	0.4990	0.0020	0.0039	127.99	0.5000
10	0.9980	0.5005	0.5000	0.4995	0.0010	0.0020	256.00	0.5000
11	0.9990	0.5002	0.5000	0.4998	0.0005	0.0010	512.00	0.5000
12	0.9995	0.5001	0.5000	0.4999	0.0002	0.0005	1,024.25	0.5001
13	0.9998	0.5001	0.5000	0.4999	0.0001	0.0002	2,046.00	0.4996
14	0.9999	0.5000	0.5000	0.5000	0.0001	0.0001	4,100.00	0.5002
15	0.9999	0.5000	0.5000	0.5000	0.0000	0.0001	8,192.00	0.5000

Table 8.8 Newton–Raphson convergence results for the case with $\theta = 1$ and using consistent tangent operator (Example 8.3)

Iteration number k	u^k_{n+1}	$K^{t,k}_{n+1} = D^k_{n+1}$	$\bar{f}$	$f^k_{s,n+1} = \sigma^k_{n+1}$	δu^k	$e^k = 1 - u^k_{n+1}$	$e^{k+1}/(e^k)^2$	e^{k+1}/e^k
1	0.0000	1.0000	0.5000	0.0000	0.5000	1.0000	0.5000	0.5000
2	0.5000	0.4444	0.5000	0.3333	0.3750	0.5000	0.5000	0.2500
3	0.8750	0.2844	0.5000	0.4667	0.1172	0.1250	0.5000	0.0625
4	0.9922	0.2520	0.5000	0.4980	0.0078	0.0078	0.5000	0.0039
5	1.0000	0.2500	0.5000	0.5000	0.0000	0.0000	0.0000	0.0000

Table 8.9 Newton–Raphson convergence results for different values of $\bar{f}$ and θ (Example 8.3)

Case number	$\bar{f}$	θ	Using continuum operator		Using consistent operator	
			No. iterations	$(u_{n+1}(\text{numerical}))/(u_{n+1}(\text{exact}))$	No. iterations	$(u_{n+1}(\text{numerical}))/(u_{n+1}(\text{exact}))$
1	0.5	1.0	15	1.443	5	1.443
2	0.5	0.5	3	0.962	5	0.962
3	0.95	1.0	239	6.340	9	6.340
4	0.95	0.5	3	0.604	5	0.604

that in the latter, the yield criterion must also be satisfied. In other words, as the stresses and hardening variables change, the stress point must continue to lie on the yield surface (called the consistency condition). Because the consistency is enforced, the accuracy is generally higher than what was achieved in this example.

Definition. *The* consistent tangent *operator is the stiffness tensor relating the rate of stress to the rate of strain derived from the algorithmic relation between the incremental stress and strain. The* continuum tangent, *on the other hand, is the stiffness tensor involved in the original relation between the rate of stress and the rate of strain.*

Remarks Consider the term R in (8.27). In a general elasto-plastic analysis where the equations are integrated implicitly, R will be a 6×6 matrix (or bigger depending on the specific algorithm), hence requiring inversion of a 6×6 matrix at every Gauss point. Also, note the appearance of the term $\dot{r}$ in (8.28a). The analogous term in the elasto-plastic analysis is the flow direction vector $\mathbf{r}$. Hence, the algorithm requires derivative of this with respect to stresses and hardening variables. When the hardening variables are scalars (as in the case of isotropic hardening models), the derivative of $\mathbf{r}$ with respect to the hardening variables can be easily obtained, but the derivatives with respect to the stresses can be difficult to obtain in some cases. The degree of difficulty increases when one or more of the hardening variables are a tensor of order 2 or higher. Nevertheless, these difficulties, which are inherent in the implicit algorithms employing consistent tangent operator, can generally compensate for the excellent computational efficiency of such algorithms.

Problems

Problem 8.1 Determine the value of x when $f = \bar{f}$ for the following two functions by both the Newton–Raphson method and modified Newton–Raphson method, and investigate the rate of convergence in each case.

(a) $f = 2x - x^2; \quad \bar{f} = 0.8$
(b) $f = 1 - (1 - x^2)^{1/2}; \quad \bar{f} = 0.2$

Discuss, if any, the difficulties of employing the Newton–Raphson method in each case.

Problem 8.2 Determine the solution to the following nonlinear set of equations using the Newton–Raphson and the modified Newton–Raphson methods.

(a) $5x_1 + 8x_2 - 3x_1^2 - 2x_1x_2 = 5$

$2x_1 + 6x_2 - 5x_2^2 - 3x_1x_2 = -3$

Use $\mathbf{x}_{in} = \{2, 2\}$

Ans: $\hat{\mathbf{x}} = \mathbf{x}_n = \{1.7549, 1.2169\}$

(b) $3x_1 - x_2 = -2$

$2x_1^3 - x_2 = 0$

Use $\mathbf{x}_{in} = \{1, 2\}$

Ans: $\hat{\mathbf{x}} = \mathbf{x}_n = \{2, 8\}$; Another solution : $\hat{\mathbf{x}} = \mathbf{x}_n = \{-0.5, -0.5\}$

(c) $x_1^2 - 2x_2 = 8$

$x_1^2 + x_2^2 = 16$

Use $\mathbf{x}_{in} = \{1, 1\}$

Problem 8.3 The stress–strain relationship of a nonlinear elastic material is given by the following rate equation:

$$\dot{\sigma}_{ijk\ell} = C_{ijk\ell}\dot{\varepsilon}_{k\ell}$$

$$C_{ijk\ell} = \left(K - \frac{2}{3}G\right)\delta_{ij}\delta_{k\ell} + G(\delta_{ik}\delta_{j\ell} + \delta_{i\ell}\delta_{jk})$$

$$K = \frac{E}{3(1-2\nu)}; \quad G = \frac{E}{2(1+\nu)}$$

$$E = \frac{E_0}{1 + aJ_\varepsilon}; \quad J_\varepsilon = \left(\frac{1}{2}e_{k\ell}e_{k\ell}\right)^{1/2}; \quad e_{k\ell} = \varepsilon_{k\ell} - \frac{1}{3}\varepsilon_{pp}\delta_{k\ell}$$

where $E_0 > 0$ and $a > 0$ are model constants. Note that E decreases from E_0 at $J_\varepsilon = 0$ to 0 when $J_\varepsilon \to \infty$. Poison's ratio is independent of the strain.

(a) Write down the backward Euler equation for calculating $\boldsymbol{\sigma}_{n+1}$.
(b) Derive the expression for the consistent tangent operator based on the algorithmic equation for the stress you established in step (a).
(c) Following Algorithm 8.2 and the computer coding guide provided in Appendix 3, develop a computer program to perform simulations of the behavior of a uniformly loaded specimen (see Example 8.3).
(d) With $E_0 = 200$ GPa, $a = 1{,}000$, and $\nu = 0.3$, calculate the stress–strain relation of a specimen subjected to a uniaxial stress loading ($\sigma_1 \neq 0$ and $\sigma_2 = \sigma_3 = 0$) until the uniaxial stress–strain curve becomes almost flat. Using the continuum and consistent operators in the analyses and varying the step size, investigate the convergence behavior and the result accuracy.
(e) Repeat step (d) under uniaxial strain ($\varepsilon_1 \neq 0$ and $\varepsilon_2 = \varepsilon_3 = 0$) loading. Compare the results with those obtained in step (d) and discuss the differences.

Chapter 9
Theory of Rate-Independent Elasto-Plasticity

The term *inelasticity* is used to describe any constitutive behavior other than elastic. Typically, the inelastic behavior includes viscoelastic, viscoplastic, and elasto-plastic behaviors. The uniaxial stress–strain relation shown in Fig. 9.1 is characteristic of many materials (e.g., aluminum, soil, etc.). For some materials (e.g., metals), the elastic part of the stress–strain relation is linear (and hence the elastic modulus is a constant). The nonlinearity results largely due to plasticity. However, the elastic modulus itself does change with loading for materials such as soil and rubber. A portion of the work done on a material by external forces is recoverable and the remaining portion is irrecoverable. The former is associated with inelastic strains and the latter with elastic strains. The behavior of inelastic materials is history- (or path-) dependent, whereas that of elastic materials is history independent.

In this chapter, we will first present the definition of different behaviors. We will devote the remainder of the chapter to the description of the rate-independent elasto-plastic behavior, including mathematical frameworks that can be used to model such a behavior.

9.1 Classes of Constitutive Behaviors

We will define different classes of constitutive behaviors and the associated fundamental concepts with respect to a uniaxial loading case and subsequently generalize to a multi-axial loading case. We will use simple rheological elements, shown in Fig. 9.2, to reinforce the precise meaning of different classes of behaviors.

The elastic element (Fig. 9.2a) represents a fully recoverable behavior. When the stress is increased from σ_A to σ_B (loading) and back to σ_A (unloading), the stress–strain relation during loading and unloading coincide with each other. Let us consider an infinitesimal loading (i.e., $\sigma_A - \sigma_B = \Delta\sigma$, where $\Delta\sigma$ is a small

A. Anandarajah, *Computational Methods in Elasticity and Plasticity: Solids and Porous Media*, DOI 10.1007/978-1-4419-6379-6_9,

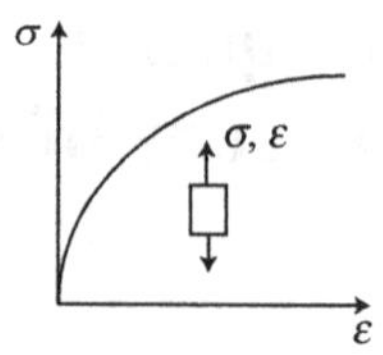

Fig. 9.1 Nonlinear uniaxial stress–strain relation

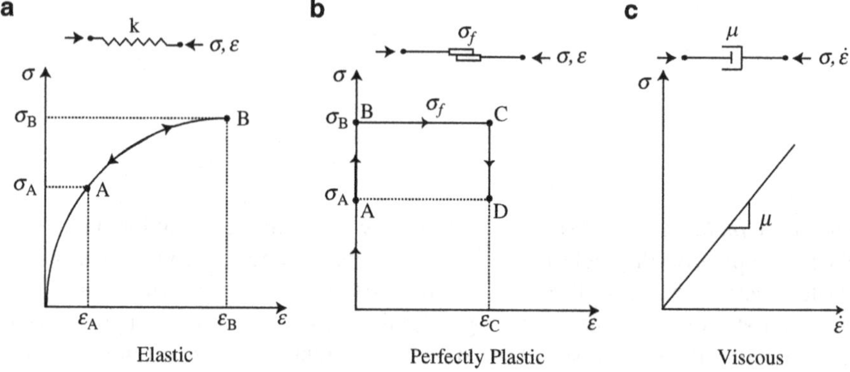

Fig. 9.2 Elementary rheological models

quantity), and calculate the net work done (per unit volume) on the specimen during the infinitesimal stress cycle:

$$\Delta W = \int_A^B \sigma \mathrm{d}\varepsilon + \int_B^A \sigma \mathrm{d}\varepsilon = 0 \tag{9.1}$$

The work put into the material during loading is fully recoverable. Elastic deformation is instantaneous; i.e., it takes place immediately upon application of the load.

A plastic element (Fig. 9.2b) is one where the strain remains zero until the stress exceeds a certain value (the failure stress σ_f) and increases suddenly (or instantaneously) thereafter. The behavior is also called the *rigid plastic* behavior. When the specimen is subjected to an infinitesimal stress cycle (A–B–C–D) with the strain from B to C controlled, the net work done on the specimen is

$$\Delta W = \int_A^B \sigma \mathrm{d}\varepsilon + \int_B^C \sigma \mathrm{d}\varepsilon + \int_C^D \sigma \mathrm{d}\varepsilon = 0 + \Delta\sigma\Delta\varepsilon + 0 = \Delta\sigma\Delta\varepsilon > 0 \tag{9.2}$$

There is thus a positive work loss during the stress cycle.

A viscous element (Fig. 9.2c) is one where the stress is a function of the rate of strain. For instance, in the linear viscous model portrayed in Fig. 9.2c (which is the most commonly used viscous model), the stress is linearly proportional to the rate of strain.

9.1.1 Viscoelasticity

A material exhibiting elements of both the elastic and viscous behaviors is known as the viscoelastic material. The stress–strain behavior of such materials may be modeled by combining the elastic and viscous elements in several ways as shown, for example, in Fig. 9.3. The series model shown in Fig. 9.3a is known as the Maxwell's model, and the parallel model shown in Fig. 9.3b is known as the Kelvin–Voigt model. To capture specific stress–strain behavior manifested by a given material, the elements shown in Fig. 9.3a and b may be combined in a suitable manner as in Fig. 9.3c.

The viscoelastic models may be used to represent (a) the rate-dependent behavior where the stress–strain relation is a function of the rate of strain (Fig. 9.4a), (b) the creep behavior where the strain increases under a constant applied load (Fig. 9.4b) and (c) the stress relaxation behavior where the stress decreases under a constant applied strain (Fig. 9.4c).

9.1.2 Rate-Independent Elasto-Plasticity

This class refers to behavior that contains elements of both elastic and plastic behaviors (Hill, 1950). Both elastic and plastic deformations occur instantaneously. The uniaxial behavior may be modeled by combining the elastic and plastic elements shown in Fig. 9.2. For instance, the series elastic/plastic model,

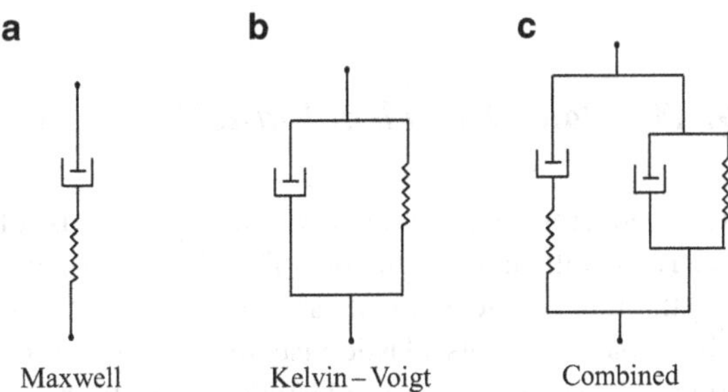

Fig. 9.3 Rheological models

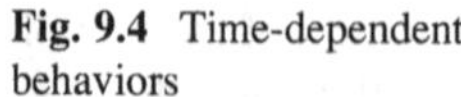

Fig. 9.4 Time-dependent behaviors

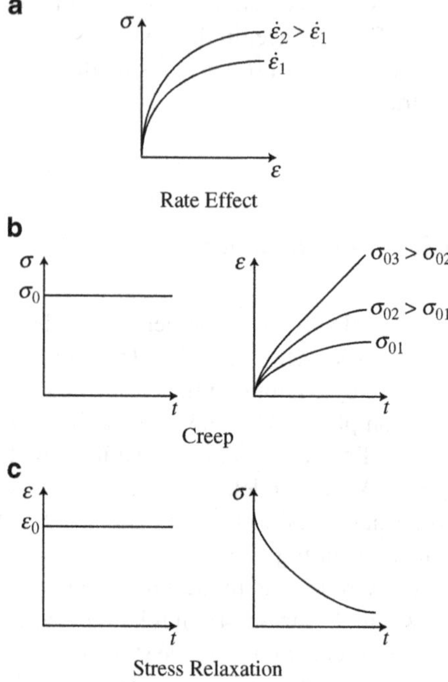

Fig. 9.5 Elastic-perfectly-plastic relation

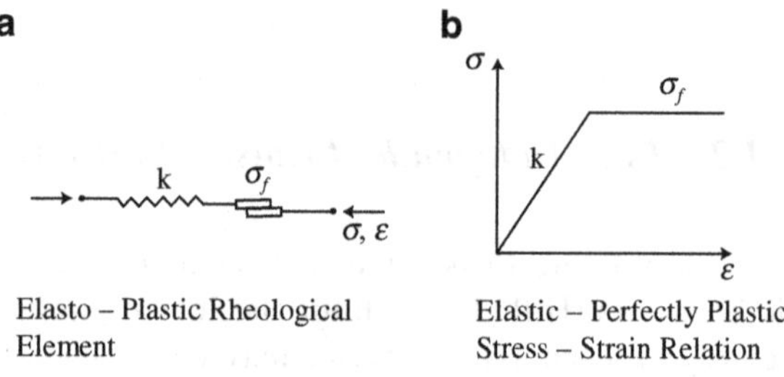

shown in Fig. 9.5a, corresponds to the elastic, perfectly-plastic behavior shown in Fig. 9.5b.

9.1.3 Rate-Dependent Elasto-Plasticity

The behavior that contains all three elements of elastic, viscous, and plastic behaviors falls in this category. The uniaxial behavior may be modeled by combining the three basic elements shown in Fig. 9.2. For instance, the model shown in Fig. 9.6a represents elastic/viscoplastic behavior, whereas the model shown in Fig. 9.6b represents elasto-plastic/viscoplastic behavior.

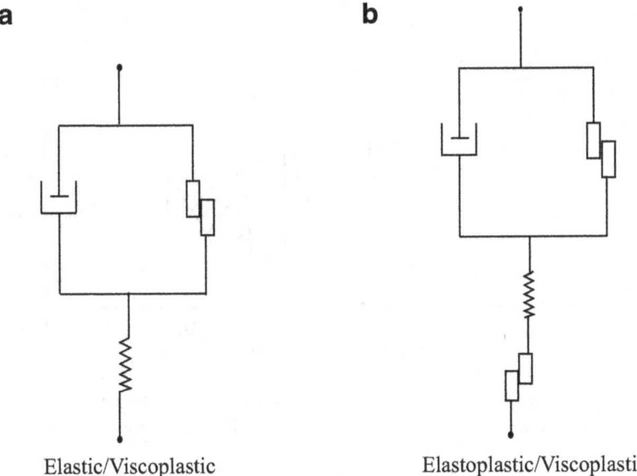

Fig. 9.6 Combinations of elastic, plastic and viscous elements

9.2 Theory of Rate-Independent Plasticity

9.2.1 Fundamentals

9.2.1.1 Uniaxial Behavior

A typical loading–unloading uniaxial stress–strain relation of an elasto-plastic material is schematically shown in Fig. 9.7a. Let us assume that upon loading from the zero stress level (point O), the material behaves elastically up to a stress σ_{y0} (point A) and then elasto-plastically thereafter.

When the material is unloaded from point A, it follows the path AO. When unloaded fully to point O and reloaded, the material follows the path OA as if the previous loading has never taken place. Hence the material does not exhibit memory in this range of loading.

When the material is unloaded from point B, it follows the path BD and, upon reloading, it follows the path DB. The behavior from B to D to B therefore is purely elastic. However, during reloading, the material remembered that the earlier unloading had taken place from point B. Hence the *material exhibits a memory* in the elasto-plastic range.

Upon reaching point B, the stress–strain relation follows the earlier backbone curve (in an approximate sense). Similar behaviors are observed for the loading/ unloading along paths CE, GI, and HJ.

The backbone curve continues to rise up to point F, falls from F to H and flattens beyond H. The slope gradually increases up to point F, becomes zero at F, becomes negative from F to H, and stays zero after H.

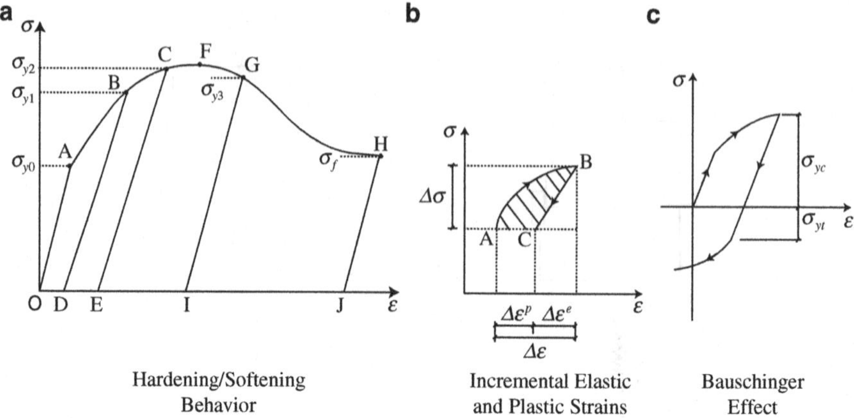

Fig. 9.7 Some fundamentals of elasto-plastic behavior

A number of key observations that are characteristics of an elasto-plastic behavior is made. During reloading from points D, E, I, and J, the behavior of the material is purely elastic along paths OA, DB, EC, IG, and JH. When the behavior changes from elastic to elasto-plastic, the material is said to *yield*. Thus, the material first yields at point A, and subsequently at points such as B, C, G, and H. The stress at which yielding occurs is referred to as the *yield stress*. The yield stress increases during loading from O to F ($\sigma_{y2} > \sigma_{y1} > \sigma_{y0}$) and decreases beyond point F ($\sigma_{y3} <$ the yield stress at point F). When the yield stress increases, the material is said to experience *hardening* (or *strain hardening*) and when it decreases, it is said to experience *softening* (or *strain softening*). To model the hardening/softening behavior, a mathematical rule, known as the *hardening rule*, is required. When the yield stress remains the same during plastic behavior, the behavior is called *perfectly-plastic*.

Consider an infinitesimal loading from point A to point B, as shown in Fig. 9.7b. Let the infinitesimal strain be $\Delta\varepsilon$, of which $\Delta\varepsilon^e$ is recovered upon unloading to point C and $\Delta\varepsilon^p$ is unrecoverable. $\Delta\varepsilon^e$ is called the elastic strain and $\Delta\varepsilon^p$ is called the plastic strain, where

$$\Delta\varepsilon = \Delta\varepsilon^e + \Delta\varepsilon^p \tag{9.3}$$

Note that the additivity presented in (9.3) in general holds only when the strains are small. During the stress cycle shown in Fig. 9.7b, the net work done is

$$\Delta W = \int_A^B \sigma \mathrm{d}\varepsilon + \int_B^C \sigma \mathrm{d}\varepsilon \tag{9.4a}$$

$$\int_A^B \sigma \mathrm{d}\varepsilon \approx \frac{1}{2}\Delta\sigma(\Delta\varepsilon^e + \Delta\varepsilon^p); \quad \int_B^C \sigma \mathrm{d}\varepsilon = -\frac{1}{2}\Delta\sigma\Delta\varepsilon^e \Rightarrow \Delta W = \Delta W^p \approx \frac{1}{2}\Delta\sigma\Delta\varepsilon^p \geq 0 \tag{9.4b}$$

This is the property that signifies plastic behavior; i.e., the network done during a stress cycle (infinitesimal or finite) is nonnegative. Geometrically, ΔW^p is the shaded area in Fig. 9.7b. The fact that the plastic work is nonnegative is the motivation for the stability postulates (to be discussed later in Sect. 9.3).

It should be clear by now that the plastic behavior is *history dependent and path dependent*. For instance, if the specimen you set out to test had been loaded by someone else along the path O–A–B–D–B–C–E (Fig. 9.7a), the stress–strain relation that you will measure is E–C–F–G–H, and not O–A–B–C–F–G–H. This is because the specimen has been previously loaded, and it remembers some of the previous loading history. The material has *memory*. The specimen remembers the previous maximum yield stress (σ_{y2} in this case), but it does not remember that it was subjected to loading/unloading along BD. This characteristic makes mathematical modeling somewhat simpler.

At and beyond point H, the material undergoes unlimited deformation with no additional stress. The material is said to experience *plastic flow*. Unlike fluid flow, the plastic flow is inviscid and the material behavior is rate-independent.

Let us consider the behavior of metals for a moment. As we have seen in Fig. 9.7a, when a specimen is loaded (say, in compression) beyond the initial yield stress (point A), its yield stress increases. It turns out that when the loading is then reversed (in tension) as shown in Fig. 9.7c, the material yields earlier in the opposite direction ($\sigma_{yc} > \sigma_{yt}$). This effect is referred to as the Bauschinger effect.

As an elasto-plastic material is history dependent, the stress–strain relation can only be sensibly defined at a given point (*the rate relation*) or at most in an infinitesimal sense. The latter is known as the *incremental stress–strain relation*.

Note that even for a material that is rate-independent, it is convenient to describe the relation between stresses and strains in terms of their time rates. The relation between the rate of stress and rate of strain simply means the slope of the stress–strain curve at the current point.

Let us consider the incremental stress–strain relation during a stress cycle. Let us define the overall (E), elastic,(E_e) and plastic (E_p) moduli as

$$\Delta\sigma = E\Delta\varepsilon; \quad \Delta\sigma = E_e\Delta\varepsilon^e; \quad \text{and} \quad \Delta\sigma = E_p\Delta\varepsilon^p \tag{9.5}$$

Then from (9.3):

$$\frac{\Delta\sigma}{E} = \frac{\Delta\sigma}{E_e} + \frac{\Delta\sigma}{E_p} \tag{9.6a}$$

or

$$\frac{1}{E} = \frac{1}{E_e} + \frac{1}{E_p} \Rightarrow E = \frac{E_e E_p}{E_e + E_p} \tag{9.6b}$$

Thus, the overall modulus can be represented in terms of elastic and plastic moduli.

9.2.1.2 Multi-axial Behavior

Many of the fundamentals described in the preceding section for the uniaxial loading case can be extended to the multi-axial loading case. The yield stress is a scalar quantity in uniaxial loading. In multi-axial loading it becomes a surface, known as the *yield surface*. For instance, consider a biaxial loading of a metal specimen (Fig. 9.8). Let us assume that the material is isotropic (i.e., the behavior is directionally independent). Let us say that a uniaxial load test yields the stress–strain relation shown in Fig. 9.7a. Now let us construct yield and failure surfaces for this material in the biaxial $\sigma_1 - \sigma_2$ space shown in Fig. 9.8.

When a specimen is subjected to loading where both σ_1 and σ_2 are varied in some known fashion, the plot of σ_1 versus σ_2 makes a trajectory in the $\sigma_1 - \sigma_2$ space, known as the *stress path* in the $\sigma_1 - \sigma_2$ space. When the specimen is loaded with a stress path along the positive σ_1-axis, the specimen will first yield when $\sigma_1 = \sigma_{y0}$ since this is a uniaxial loading. Similarly, when the specimen is loaded along a stress path that coincides with the positive σ_2-axis, the specimen will yield when $\sigma_2 = \sigma_{y0}$ because (a) this is a uniaxial loading as well and (b) the material is isotropic (thus, the yield stress is the same in different directions). During a virgin loading, the specimen will yield at $\sigma_1 = -\sigma_{y0}$ for loading along the negative σ_1-axis and at $\sigma_2 = -\sigma_{y0}$ for loading along the negative σ_2-axis.

Now the question is what happens when the specimen is loaded along the stress path, say, $\sigma_1 = \sigma_2$. It has been experimentally observed that for most materials, the yield points corresponding to loading along different stress paths are connected, a smooth surface is obtained. The surface is known as the *yield surface*. Hence when the material is loaded along the path $\sigma_1 = \sigma_2$, the material will yield when the stress point reaches the yield surface.

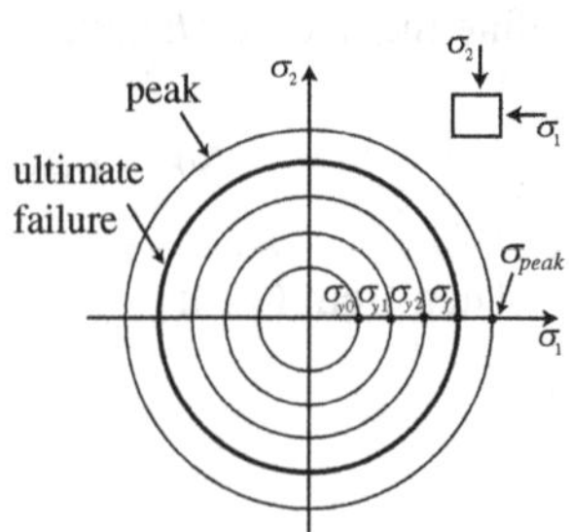

Fig. 9.8 Evolution of yield surface during multi-axial loading

For the ideal material under discussion here, let us assume that the yield surface is a circle with a radius σ_{y0}. (We will gradually make allowance for real behaviors that deviate from the assumptions made here.) This is the *initial yield surface* corresponding to the *initial yield stress* σ_{y0} associated with the uniaxial loading case.

It was noted in the uniaxial case that when the stress is increased beyond the initial yield stress, the yield stress increases (until the peak). This translates into growth of the size of *subsequent yield surfaces (*also known as *loading surfaces).* But the question is whether the growth is symmetrical or nonsymmetrical about the origin. It turns out that the answer depends on several factors such as the material type, past history, loading direction, etc. Let us first define some ideal cases of behaviors, as shown in Fig. 9.9. The real behavior usually is a combination of these ideal behaviors.

In Fig. 9.9a, as the stress point changes from 1 to 2 to 3, the yield surface expands in the stress space uniformly in all directions with its center (of similarity) and shape remaining the same. In general, the subsequent yield surfaces are similar about a point (similarity center) lying on or inside the surface. The geometrical shape remains the same; for example, if the initial surface is an ellipse with an aspect ratio a, the subsequent surfaces are also an ellipse with an aspect ratio a. Hence the change in the size of the yield surface can be mathematically modeled using just one scalar hardening variable. The type of *hardening* is known as the *isotropic hardening*.

If the yield surface translates in the stress space with no change in its size and shape, as shown in Fig. 9.9b, the hardening is known as the *kinematic hardening* (Prager 1955). In this case, a tensorial hardening variable is needed to mathematically model the variation of the surface center in the stress hyperspace. A combined isotropic/kinematic hardening is shown in Fig. 9.9c, where both the size and location of the yield surface change in the stress space.

For some materials, such as soil, a type of hardening where the yield surface rotates (known as the *rotational hardening*) has been found to represent the material behavior better than the classical isotropic or kinematic hardening types (Sekiguchi and Ohta 1977; Anandarajah and Dafalias 1986; Anandarajah 2008a).

A rule that prescribes the motion/change of the yield surface during loading is needed. This is known as the *hardening rule* (or the *hardening law*). It may be pointed out that this task is accomplished in a uniaxial case simply by prescribing the backbone stress–strain curve shown in Fig. 9.7a.

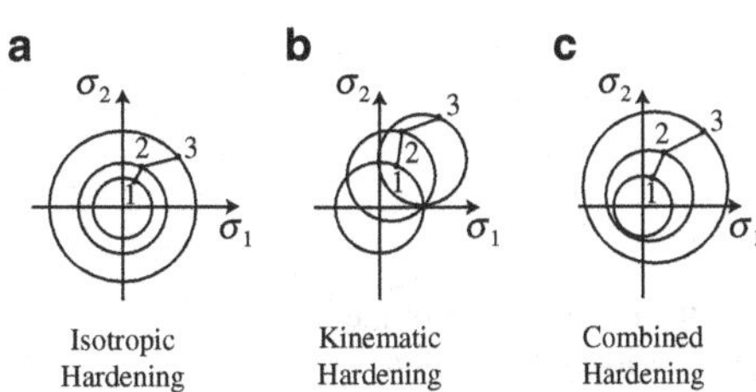

Fig. 9.9 Standard hardening behaviors

It is now easy to import the other concepts from the uniaxial to the multi-axial case. For example, the strain-softening is associated with the contraction of the yield surface in the case of isotropic hardening. It is somewhat more complex for kinematic hardening; the yield surface locally moves outward from the current point for hardening materials and inward for softening materials. We will give a somewhat more general definition of hardening/softening (due to Drucker 1950, 1951) later in Sect. 9.3.1.

The (ultimate) failure stress becomes a failure surface, such as the surface shown in Fig. 9.8b. Note that when the material does not exhibit softening, the failure surface is the outermost yield surface. Otherwise, the surface corresponding to "peak" is the outermost surface. It can be shown that, for consistent mathematical modeling, admissible yield and failure surfaces (for both strain-hardening and strain-softening materials) must be convex. We will elaborate more on this later in Sect. 9.3.4.

Another difficulty encountered in the multi-axial case has to do with the need for a *loading/unloading criterion*. This was trivial in the uniaxial case: for hardening material, loading is associated with an increase in stress and unloading with a decrease in stress. (This is somewhat complicated for a strain softening material).

Let us first consider the behavior of a hardening material. Suppose that we subject the material to an incremental stress loading. In the multi-axial case, the incremental stress is a tensor, having both the magnitude and direction. The behavior depends not only on the magnitude but also on the direction of the incremental stress. When the stress probing takes place entirely inside the yield surface (e.g., probe from points A to B in Fig. 9.10), the behavior is decidedly elastic. Consider the probes 1, 2, 3, and 4 from point C. The probes 1 and 2 are associated with plastic behavior since these probes are directed outward from the yield surface. The probe 4 is associated with elastic behavior since this probe is directed inward from the yield surface. The probe 3, known as *neutral loading*, is also associated with elastic behavior. Now considering a softening material, probe 4 is not necessarily elastic!

For a hardening material, (a) when the end point of the stress probe vector is outside the surface, the corresponding increment is associated with plastic behavior and (b) when it is inside, the corresponding increment is associated with elastic

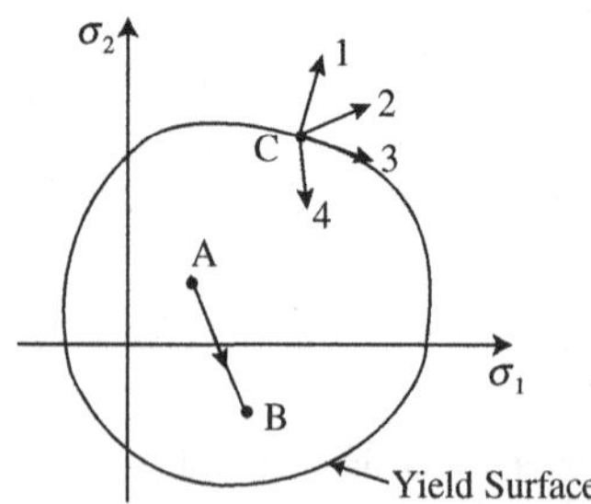

Fig. 9.10 Illustration of the need for loading criterion

behavior. Hence it seems that this property could be used to define loading/unloading events. There are two difficulties with this approach: (a) The stress increment is not known a priori in general (Note that in a finite element analysis, strain increment is specified and the corresponding stress increment is required!), and (b) the approach does not work for a softening material. Hence a more effective method is needed for defining loading/unloading.

9.2.2 Microscopic Interpretation of Plasticity and Hardening Behavior

The microscopic mechanisms that cause plastic deformation vary from material to material. In *metals*, the plastic deformation occurs due to movement of *dislocations* (Anderson et al. 1990). Movements of dislocations are resisted by obstacles, including impurities, grain boundaries, and other moving dislocations. The internal stress must exceed a certain threshold value to overcome the repulsion exerted by the obstacles to the movement of dislocations; the macroscopic equivalent of this threshold stress is the *yield stress*.

As the material deforms plastically, dislocations multiply, and interact with each other, making the motion of dislocations, and hence the plastic deformation, more difficult. This increases the *yield stress*. This is the underlying mechanisms of *work hardening or strain hardening*. The temperature has a significant influence on this process. When the temperature is raised to 0.3–0.4 times the melting temperature, the dislocation *rearrangement, and annihilation* begin to occur. The number of dislocations is decreased and the yield stress is lowered. At even higher temperatures, *recrystallization* takes place and the yield stress drops back to the initial value. The material is said to have been *hot-worked*. Due to its low melting point, lead is hot-worked at the room temperature. The yield stress remains very low and lead remains very soft. On the other hand, temperature must be raised to more than 900°C for mild steel to be hot-worked. *Cold-working* is the process of work hardening at low enough temperatures where recrystallization does not occur. For a more comprehensive discussion, the reader is referred to Anderson et al. (1990).

As pointed out earlier, metals exhibit Bauschinger effect (Fig. 9.7). This is attributed to internal stresses that develop on obstacles. A specimen subjected to an external stress is in equilibrium under the so-called *back stress* and internal stress on obstacles σ_0. When the direction of external stress is reversed, the dislocations cease to move until the internal stress on the obstacles is revered to $-2\sigma_0$ (Lubliner 1990). At this point, the dislocations can move backwards. The material exhibits *kinematic hardening*.

When metals are subjected to very large inelastic deformation (as, for example, in rolling or wire drawing), preferred orientation of crystallographic planes (or texture) is formed. In single crystals, the crystallographic planes orient themselves with respect to the maximum strains. The texture development in polycrystals is similar. However, since the individual grains cannot rotate freely, a complex deformation pattern involving lattice bending and fragmentation ensue (Dieter 1986).

In the above discussion, it has been tacitly assumed that metals are hardening materials. However, depending on the loading regimes and metal types, other forms of behaviors including perfect plasticity and strain softening are also observed. The perfect plasticity is generally considered to result from slip on a particular crystallographic plane. The strain-softening is associated with localization of slip on a microstructural feature. Once the internal stress becomes high enough to shear the microstructural feature, the yield stress experiences a corresponding decrease (Soboyejo 2003).

The physics of the plasticity of metals has so many other finer details. The description presented above is adequate for the purpose of phenomenological modeling that is of interest in this book. For further details, the reader may consult textbooks such as Soboyejo (2003) and Anderson et al. (1990).

To sum, *the plastic behavior in metals is related to changes taking place at the microscopic level such as the changes in the number of dislocations, build up of internal stresses at the microstructural features and texture development.*

In materials such as *soil*, the plastic deformation occurs due to interparticle contact *slippage* and particle-to-particle *rolling*. The internal slip and rolling are resisted by interparticle frictional forces (both in cohesionless and cohesive soil) and physico-chemical forces (only in cohesive soil). As in metals, applied stress must exceed a certain value – *the yield stress* – to overcome the resisting forces. The resisting forces depend on (a) certain *fixed properties* such as the interparticle friction and cohesion, (b) the *intensity of the current stresses* and (c) various characteristics of the *internal fabric* such as the interparticle spacing and anisotropy (based on contact normal distribution, particle orientation distribution, etc).

In the hardening/softening regime, the plastic deformation causes permanent changes in the internal fabric, which either increases the yield stress, resulting in *hardening*, or decreases the yield stress, resulting in *softening*. Hence, as in metals, *the plastic behavior in soil is related to changes taking place at the microscopic level such as the changes in fabric anisotropy and interparticle spacing* (which is represented macroscopically by the *void ratio*). The internal fabric controls the magnitude and direction of the plastic strain rate, which in turn changes the internal fabric (in the hardening/softening regime).

It is apparent from the above discussion that the permanent microscopic changes are a result of the inelastic strains. In the absence of inelastic strains, there should be no permanent changes at the microscopic level. This leads to the conclusion that there should be no change in quantities such as the yield stress and back stress when there are no plastic strains. The inverse, however, is not always true. For example, during perfect plasticity, there is unlimited plastic (deviatioric) strain, but the size and location of the yield surface remain unchanged. With this background, we introduce the concept of plastic internal variables (PIVs) below.

9.2.3 Plastic Internal Variables

In Chap. 4, an elastic behavior was defined as one where the stress $\boldsymbol{\sigma}$ depends only on the strain $\boldsymbol{\varepsilon}$ and not on the past thermodynamic history (Eringen 1967). In thermodynamics, with the temperature effects neglected, the material state is represented solely in terms of $\boldsymbol{\varepsilon}$; hence

$$\boldsymbol{\sigma} = \boldsymbol{\sigma}(\boldsymbol{\varepsilon}) \tag{9.7a}$$

In elasto-plastic materials (or in inelastic materials in general), the behavior depends also on the past history. Hence additional variables known as the *internal variables* or *hidden variables* ($\boldsymbol{\zeta}$) are introduced into the constitutive functionals to quantify the effect of the past history. The material state may now be quantified in terms of $(\boldsymbol{\varepsilon}, \boldsymbol{\zeta})$; hence (Coleman 1964):

$$\boldsymbol{\sigma} = \boldsymbol{\sigma}(\boldsymbol{\varepsilon}, \boldsymbol{\zeta}) \tag{9.7b}$$

In materials for which the microscopic mechanisms responsible for the plastic behavior are known (see Sect. 9.2.2), it is in principle possible to use physically meaningful variables (e.g., dislocation density for metals, fabric for soils, etc.) for $\boldsymbol{\zeta}$. However, even when such mechanisms are known, it is not always possible to develop quantitative relations between the variation of such physical variables and plastic strains. So, in many cases of constitutive models, it is common to equate $\boldsymbol{\zeta}$ with a *hardening variable* such as the size of the yield surface.

In some constitutive models, however, certain macroscopic variables are introduced to conceptually represent certain microscopic features. For example, in certain models on granular materials, a second order tensor is used to introduce a rotational hardening into the models. The second order tensor represents in a semi-quantitative manner the anisotropy based on internal fabric, defined in terms of the particle-to-particle contact normal distribution (Anandarajah 2008a, b).

Whatever the case might be, we introduce $\boldsymbol{\zeta}$ such that the rate of change of $\boldsymbol{\zeta}$ is zero when the rate of change of the plastic strain is zero. Let us formally introduce a set of *internal variables* (more specifically, *plastic internal variables* or PIVs for short) $\boldsymbol{\zeta} = \{\zeta_1, \zeta_2, ..\}$ such that

$$\dot{\boldsymbol{\zeta}} = 0 \quad \text{when} \quad \dot{\boldsymbol{\varepsilon}}^p = 0 \tag{9.7c}$$

As pointed out earlier, the inverse of (9.7c) is not true in general; for example, when a material experiences perfect plasticity, it undergoes large plastic strains with the size of the yield surface remaining a constant. Note that each $\zeta_1, \zeta_2, ..$ individually is tensor of some order, and $\boldsymbol{\zeta}$ is a generalized compact notation for a collection of such variables. For instance, ζ_1 may be a scalar parameter (e.g., the size of the yield surface), ζ_2 may be a second order tensor (e.g., back stress), etc.

Examples of commonly used, plastic strain-based PIVs are:

1. Accumulated deviatoric strain (in modeling strain hardening/softening)

$$\xi = \int dJ_{\boldsymbol{\varepsilon}} \tag{9.8a}$$

2. Accumulated plastic work (in modeling work hardening/softening)

$$W^p = \int \sigma_{ij} d\varepsilon_{ij}^p \tag{9.8b}$$

9.2.4 General Structure of Rate-Independent Plasticity

The emphasis of the presentations in this section is on the structure of a general elasto-plastic framework. A specific theory that can simulate the stress–strain behavior of a material at hand may be developed within this framework. Also the discussion is limited to the classical framework, which involves a single yield surface and a single plastic potential surface.

It may, however, be pointed out that the classical framework is inadequate for modeling some aspects of the stress–strain behavior of some materials. For example, to simulate the cyclic behavior of granular materials, multiple surfaces may need to be employed. Before developing such a complex model or understanding such an existing complex model, it is important to understand the classical framework. The material presented in this section will serve this role.

9.2.4.1 Material State

On the basis of the discussions presented in the preceding section, we formally define the material state in terms of two types of variables: (1) the stresses $\boldsymbol{\sigma}$ and (2) the PIVs $\boldsymbol{\zeta}$. All functions used in the constitutive equations will in general be a function of $\boldsymbol{\sigma}$ and $\boldsymbol{\zeta}$. We begin here with the definition of the yield surface (or yield function):

$$\phi(\boldsymbol{\sigma}, \boldsymbol{\zeta}) = 0 \tag{9.9a}$$

The yield surface representing the current material state is obtained for

$$\boldsymbol{\zeta} = \bar{\boldsymbol{\zeta}} \tag{9.9b}$$

where $\bar{\zeta}$ is the value of PIV at the current material state. The yield surface is schematically shown in Fig. 9.11. The domain inside the yield surface is the *purely elastic domain*, where the behavior is purely elastic. The domain outside the yield surface cannot be reached by a material state, and is therefore *inaccessible*. The yield function is defined such that

$$\phi(\boldsymbol{\sigma}, \zeta = \bar{\zeta}) = 0 \text{ on the surface} \tag{9.10a}$$

$$> 0 \text{ outside the surface} \tag{9.10b}$$

$$< 0 \text{ inside the surface} \tag{9.10c}$$

9.2.4.2 Loading/Unloading Criterion

Recalling the discussion we had earlier regarding loading and unloading, let us define a scalar quantity L as the scalar product between the rate of stress $\dot{\boldsymbol{\sigma}}$ and the unit normal $\mathbf{n}$ to the yield surface at the current stress point (Fig. 9.12)

$$L = \dot{\sigma}_{k\ell} n_{k\ell} \tag{9.11a}$$

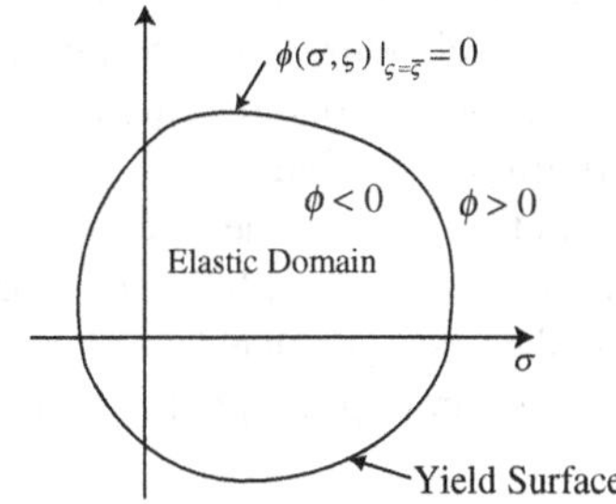

Fig. 9.11 Functional form of yield surface

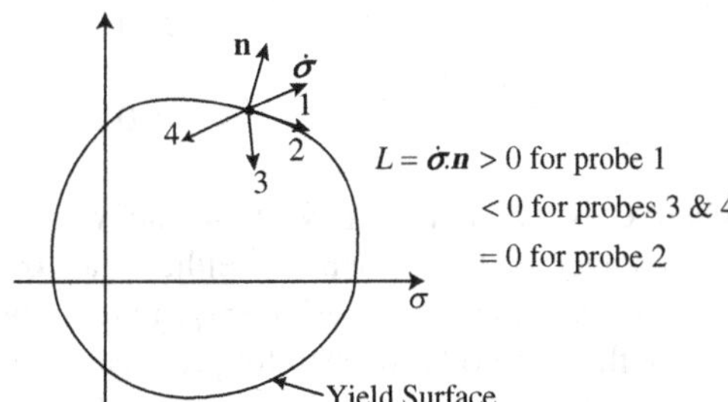

Fig. 9.12 Illustration of loading index

where

$$n_{k\ell} = \frac{1}{g}\frac{\partial\phi}{\partial\sigma_{k\ell}} \quad \text{and}\, g = \left|\frac{\partial\phi}{\partial\boldsymbol{\sigma}}\right| \tag{9.11b}$$

Remark One could use the vector $\partial\phi/\partial\boldsymbol{\sigma}$, instead of its normalized version $\mathbf{n}$. In this case, the expression for plastic modulus will be normalized appropriately to account for this as we will see later. When the unit normal $\mathbf{n}$ is used, the formulation is usually referred to as the *unit normal formulation*.

Now referring back to Fig. 9.12, it is seen that $L > 0$ for stress probe 1, $L = 0$ for stress probe 2, and $L < 0$ for stress probes 3 and 4. Thus, the scalar L may be used to define the loading/unloading events as

$$L > 0 : \text{Loading(elasto} - \text{plastic)} \tag{9.12a}$$

$$= 0 : \text{Neutral loading(elastic)} \tag{9.12b}$$

$$< 0 : \text{Unloading(elastic)} \tag{9.12c}$$

L is referred to as the *loading index*.

Remark In finite element applications of constitutive models, given a strain increment $\Delta\boldsymbol{\varepsilon}$, the corresponding stress increment $\Delta\boldsymbol{\sigma}$ needs to be calculated. It is first necessary to determine if the strain increment will cause elasto-plastic behavior or elastic behavior. Here the rate quantities must be converted to incremental quantities, and the loading/unloading criterion given by (9.12) is not useful. A different criterion is needed.

Let us denote the stress at the forward integration point $n+1$ by $\boldsymbol{\sigma}_{n+1}$ (i.e., $\boldsymbol{\sigma}_{n+1} = \boldsymbol{\sigma}_n + \Delta\boldsymbol{\sigma}$). It is seen from Fig. 9.12 that $\boldsymbol{\sigma}_{n+1}$ would lie outside the current yield surface for an elasto-plastic stress probe. Based on this, the loading/unloading events can in principle be defined as:

$$\phi(\boldsymbol{\sigma}_{n+1}, \boldsymbol{\zeta} = \bar{\boldsymbol{\zeta}}_n) = 0\,\text{neutral loading} \tag{9.12d}$$

$$> 0\,\text{loading} \tag{9.12e}$$

$$< 0\,\text{unloading (or elastic loading)} \tag{9.12f}$$

where $\bar{\boldsymbol{\zeta}}_n$ is the value of PIV at the backward integration (i.e., current) point n.

However, (9.12) is not useful either since $\boldsymbol{\sigma}_{n+1}$ is not known a priori. In addition, for a strain-softening material, $\boldsymbol{\sigma}_{n+1}$ may lie inside the initial yield surface. Hence yet another method is needed for practical computations. We will see in subsequent chapters that the so-called elastic predictor (or elastic trial stress), defined as

$\boldsymbol{\sigma}_{n+1}^{tri} = \boldsymbol{\sigma}_n + \mathbf{C}^e : \Delta\boldsymbol{\varepsilon}$, can be used for this purpose. $\boldsymbol{\sigma}_{n+1}^{tri}$ lies outside the surface for elasto-plastic loading and inside the surface for elastic loading.

9.2.4.3 Structure of Constitutive Equations

On the basis of observations made thus far, the rate of plastic strain may be represented as

$$\dot{\boldsymbol{\varepsilon}}_{ij}^{p} = f_{ij}(\boldsymbol{\sigma}, \zeta, \dot{\boldsymbol{\sigma}}, \dot{\zeta}) \text{ with} \tag{9.13a}$$

$$\dot{\boldsymbol{\zeta}} = 0 \quad \text{for } \dot{\boldsymbol{\varepsilon}}_{ij}^{p} = 0 \tag{9.13b}$$

$$\dot{\boldsymbol{\varepsilon}}_{ij}^{p} = 0 \quad \text{for } L \leq 0 \tag{9.13c}$$

In addition, we will add the following requirements:

- The stress–strain relation is rate-independent.
- The stress–strain relation is continuous across the yield surface (the *continuity condition*).

For the stress–strain relation to be rate-independent, the function f_{ij} in (9.13a) must be homogeneous of order 1 in $\dot{\boldsymbol{\sigma}}$ and $\dot{\boldsymbol{\zeta}}$ (see the definition below).

Definition A function $f(x)$ is homogeneous of order 1 if $f(nx) = nf(x)$. Examples of such functions are:

1. $f = \sqrt{x_i x_i}$
2. $f = \sqrt{A_{ij} x_i x_j}$
3. $f(\boldsymbol{\sigma}) = \dfrac{I^2}{J} = \dfrac{(\sigma_{k\ell}\delta_{k\ell})^2}{\left\{\frac{1}{2}\left(\sigma_{ij} - \frac{1}{3}(\sigma_{k\ell}\delta_{k\ell})\delta_{ij}\right)\left(\sigma_{ij} - \frac{1}{3}(\sigma_{k\ell}\delta_{k\ell})\delta_{ij}\right)\right\}^{1/2}}$

An example of a function that is homogeneous of order 1 in $\dot{\boldsymbol{\sigma}}$ is the loading index L (9.11a). The need for this requirement for ensuring the rate-independence is illustrated in Example 9.1.

Example 9.1

Question: Show that the uniaxial stress–strain relation

$$\dot{\boldsymbol{\varepsilon}} = a\dot{\boldsymbol{\sigma}} \tag{9.14a}$$

is rate-independent whereas

$$\dot{\boldsymbol{\varepsilon}} = a\dot{\boldsymbol{\sigma}}^2 \tag{9.14b}$$

is rate-dependent.

Answer: Let us consider the application of an infinitesimal stress $\delta\boldsymbol{\sigma}$ over a time period δt and calculate the corresponding strain by both of the above equations. Note that as $\delta t \to 0$,

$$\dot{\sigma} = \frac{\delta\sigma}{\delta t} = \frac{\mathrm{d}\sigma}{\mathrm{d}t}$$

and

$$\dot{\varepsilon} = \frac{\delta\varepsilon}{\delta t} = \frac{\mathrm{d}\varepsilon}{\mathrm{d}t}.$$

Then, from (9.14a), we have

$$\delta\varepsilon = a\delta\sigma$$

where it is seen that $\delta\boldsymbol{\varepsilon}$ depends only on $\delta\sigma$ and not on the rate of stress. On the other hand, from (9.14b), we have

$$\frac{\delta\varepsilon}{\delta t} = a\left(\frac{\delta\sigma}{\delta t}\right)^2 \Rightarrow \delta\varepsilon = a\frac{\delta\sigma}{\delta t}\delta\sigma = a\dot{\sigma}\delta\sigma$$

It is seen that $\delta\varepsilon$ depends not only on $\delta\sigma$ but also on the rate of stress (i.e., rate of loading!). Hence in the second case (9.14b), for a given stress increment, the calculated strain increment depends on the rate of loading. The function on the right side of (9.14a) is homogeneous of order one, whereas that on the right side of (9.14b) is not.

One possible form of (9.13a) that is homogeneous of order one in $\dot{\boldsymbol{\sigma}}$ and $\dot{\boldsymbol{\zeta}}$ is

$$\dot{\varepsilon}_{ij}^{p} = \langle L\rangle f_{ij}(\boldsymbol{\sigma}, \boldsymbol{\zeta}) \tag{9.15}$$

where $\dot{\boldsymbol{\sigma}}$ and $\dot{\boldsymbol{\zeta}}$ have been left out of f_{ij} and appear only within L. The loading index L (9.11a) is homogeneous of order one in $\dot{\boldsymbol{\sigma}}$ and independent of $\dot{\boldsymbol{\zeta}}$. $\langle . \rangle$ is the Heavyside function, defined as: $\langle L\rangle = L$ for $L > 0$ and $\langle L\rangle = 0$ for $L \leq 0$. Hence (9.15) also satisfies (9.13c).

L varies smoothly from its maximum value when $\mathbf{n}$ and $\dot{\boldsymbol{\sigma}}$ are collinear to zero when $\mathbf{n}$ and $\dot{\boldsymbol{\sigma}}$ are orthogonal to each other (probe 2, Fig. 9.12). In between, L varies smoothly according to the cosine function. The plastic strain rate also decreases gradually from its maximum value to zero with no discontinuity along the way and across the yield surface. Hence the use of L in the specific form as in (9.15) helps meet another requirement – the *continuity condition*. Such continuity is required for successful implementation of the constitutive model in numerical analyses.

Now we need a suitable functional form for $\dot{\boldsymbol{\zeta}}$ such that (9.13b) is satisfied. One possible way is to express $\dot{\boldsymbol{\varepsilon}}^p$ and $\dot{\boldsymbol{\zeta}}$ as:

$$\dot{\varepsilon}_{ij}^p = \langle L \rangle f_{ij}(\boldsymbol{\sigma}, \boldsymbol{\zeta}) \text{ and} \tag{9.16a}$$

$$\dot{\boldsymbol{\zeta}} = \langle L \rangle \mathbf{s} \tag{9.16b}$$

where it may be noted that when $L \leq 0$, both $\dot{\boldsymbol{\varepsilon}}^p$ and $\dot{\boldsymbol{\zeta}}$ are both zero. Specific form of $\mathbf{s}$ in (9.16b) will emerge out of the specific hardening rule to be employed for the material at hand.

It follows from (9.16a) that the function f_{ij} represents the direction and magnitude of the plastic strain rate tensor. Assuming that the direction can be derived from a potential function ψ, (9.16a) may be written as

$$\dot{\varepsilon}_{ij}^p = G(\boldsymbol{\sigma}, \boldsymbol{\zeta}) \langle L \rangle \frac{\partial \psi(\boldsymbol{\sigma}, \boldsymbol{\zeta})}{\partial \sigma_{ij}} \tag{9.17a}$$

The function

$$\psi(\boldsymbol{\sigma}, \boldsymbol{\zeta}) = 0 \tag{9.17b}$$

is known as the *plastic potential.*

Define a unit normal as:

$$\mathbf{r} = \frac{1}{\bar{g}} \frac{\partial \psi}{\partial \boldsymbol{\sigma}} \quad \text{where } \bar{g} = \left| \frac{\partial \psi}{\partial \boldsymbol{\sigma}} \right|, \tag{9.17c}$$

One specific form for (9.17a) is

$$\dot{\varepsilon}_{ij}^p = \frac{1}{K_p} \langle L \rangle r_{ij} \tag{9.18}$$

where K_p is referred to as the *plastic modulus.*

In another commonly used form for (9.17a), the plastic modulus is included within the loading index. With this definition, the plastic constitutive equations take the form:

$$\dot{\varepsilon}_{ij}^p = \langle L \rangle r_{ij} \tag{9.19a}$$

$$\dot{\boldsymbol{\zeta}} = \langle L \rangle \mathbf{s} \tag{9.19b}$$

where

$$L = \frac{1}{K_p} \dot{\sigma}_{k\ell} n_{k\ell} \tag{9.19c}$$

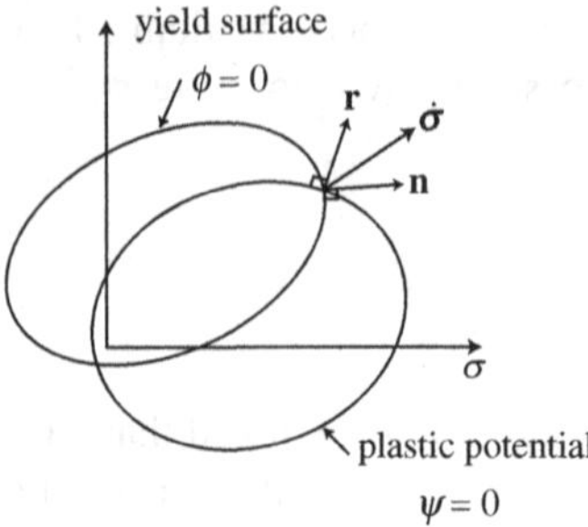

Fig. 9.13 Illustration of yield and plastic potentials

Using the definition that for a hardening material, the yield surface locally moves outward, we have $\dot{\sigma}_{k\ell}n_{k\ell} > 0$. Since $L > 0$ for plastic loading, it then follows that for a hardening material K_p must be positive. Similarly, for a softening material, the yield surface locally moves inward, which leads to $\dot{\sigma}_{k\ell}n_{k\ell} < 0$. K_p must be negative for a softening material.

The geometrical meanings of the yield and plastic potential surfaces are shown in Fig. 9.13. L defined in the manner of (9.19c) is also referred to as the *plastic consistency parameter*. (A widely-used notation for L is $\dot{\lambda}$).

9.2.4.4 Associated and Non-associated Flow Rules

When $\psi = \phi$, we have $\mathbf{n} = \mathbf{r}$, and the flow rule is known as the *associated* flow rule. When $\psi \neq \phi$, we have $\mathbf{n} \neq \mathbf{r}$, and the flow rule is known as the *nonassociated* flow rule. We will show later that the continuum tangent operator, and hence the global finite element stiffness matrix, will be symmetric in the case of associated-flow models and nonsymmetric in the case of nonassociated flow models. Also we will see later that the satisfaction of Drucker's stability postulate (Drucker 1950, 1951) requires normality ($\mathbf{n} = \mathbf{r}$) and hence the use of the associated flow rule. However, materials such as granular materials (e.g., sand) and rocks do exhibit nonassociated flow rule, and hence the nonassociated flow rules are commonly employed in the constitutive models used for these materials.

9.2.4.5 Consistency Condition

When $L > 0$, the plastic loading takes place and the stress point moves away (outward for hardening materials or inward for softening materials) from the yield surface. Regardless of which of the above events occurs, the stress point must continue to lie on the yield surface. Thus, the yield surface must change in size, shape and/or location in such a way that the current point is still on the yield surface. This requirement is known as the *consistency condition*, and is mathematically stated as:

$$\dot{\phi}(\boldsymbol{\sigma}, \boldsymbol{\zeta}) = 0 \tag{9.20a}$$

With the aid of the chain rule:

$$\frac{\partial \phi}{\partial \boldsymbol{\sigma}} \dot{\boldsymbol{\sigma}} + \frac{\partial \phi}{\partial \boldsymbol{\zeta}} \dot{\boldsymbol{\zeta}} = 0 \tag{9.20b}$$

From (9.19b), (9.19c), and (9.11b), we have

$$g K_p L + \frac{\partial \phi}{\partial \boldsymbol{\zeta}} \mathbf{s} L = 0 \tag{9.20c}$$

This leads to

$$K_p = -\frac{1}{g} \frac{\partial \phi}{\partial \boldsymbol{\zeta}} \mathbf{s} \tag{9.20d}$$

Thus, the plastic modulus is not an independent quantity. Once the yield surface and the hardening rules are selected, the plastic modulus is calculated from (9.20d). Alternatively, one could choose an expression for the plastic modulus, and develop a hardening rule from this. As discussed earlier,

$$\begin{aligned} K_p &\geq 0 \text{ for a hardening material} \\ &= 0 \text{ for a perfectly-plastic material} \\ &\leq 0 \text{ for a softening material} \end{aligned} \tag{9.20e}$$

When the formulation is not a unit normal formulation

$$\dot{\varepsilon}_{ij}^{p} = \langle L \rangle r_{ij}; \quad \dot{\boldsymbol{\zeta}} = \langle L \rangle \mathbf{s} \tag{9.20f}$$

$$L = \frac{1}{K_p} \dot{\sigma}_{k\ell} n_{k\ell} \tag{9.20g}$$

$$r_{ij} = \frac{\partial \psi}{\partial \sigma_{ij}}; \quad n_{ij} = \frac{\partial \phi}{\partial \sigma_{ij}} \tag{9.20h}$$

The consistency condition leads to

$$K_p = -\frac{\partial \phi}{\partial \boldsymbol{\zeta}} \mathbf{s} \tag{9.20i}$$

9.2.5 *Steps in Developing Constitutive Models Under the Classical Framework*

From the concepts and equations presented above, it is seen that development of a plasticity model would generally involve the following steps:

1. Specify the yield and potential functions that are suitable to use as the material state changes during loading.
2. Identify the hardening behaviors (e.g., isotropic, kinematic, combined, etc.) and hardening variables (e.g., yield surface size, dislocation density, plastic void ratio, etc.) that may be suitable for the material at hand, and either
   a. develop suitable mathematical equations (i.e., (9.16b)) for the evolution of each of the hardening variables during loading and derive the plastic modulus from the consistency equation, or
   b. specify an expression for the plastic modulus and develop an expression for the hardening function from the consistency equation.
3. Supplement the plasticity equations with elasticity (linear elastic, hyperelastic, etc.) equations.

9.2.6 *Elasto-Plastic Continuum Tangent Operator*

It was noted in Chap. 8 that nonlinear finite element analysis using the Newton–Raphson technique requires a material tangent operator (or tensor) and the value of incremental stress for a given value of incremental strain. We will first derive what is known as the continuum tangent operator.

We start with the elastic constitutive law

$$\dot{\sigma}_{ij} = E_{ijk\ell}\dot{\varepsilon}^{e}_{k\ell} \tag{9.21}$$

The total strain is decomposed into elastic and plastic parts as (under the small strain assumption)

$$\dot{\varepsilon}_{ij} = \dot{\varepsilon}^{e}_{ij} + \dot{\varepsilon}^{p}_{ij} \tag{9.22}$$

Equation (9.21) may now be written as

$$\dot{\sigma}_{ij} = E_{ijk\ell}\left(\dot{\varepsilon}_{k\ell} - \dot{\varepsilon}^{p}_{k\ell}\right) \tag{9.23a}$$

With the aid of (9.19a)

$$\dot{\sigma}_{ij} = E_{ijk\ell}\left(\dot{\varepsilon}_{k\ell} - Lr_{k\ell}\right) \tag{9.23b}$$

With the understanding that the term involving L in (9.23b) exists only when $L > 0$, the Heavyside function is dropped. Combining (9.19c) and (9.23b)

$$L = \frac{1}{K_p} n_{k\ell} \left[E_{k\ell pq} (\dot{\varepsilon}_{pq} - L r_{pq}) \right]$$

$$L \left[K_p + n_{k\ell} E_{k\ell pq} r_{pq} \right] = n_{k\ell} E_{k\ell pq} \dot{\varepsilon}_{pq}$$

$$L = \frac{n_{k\ell} E_{k\ell pq} \dot{\varepsilon}_{pq}}{K_p + n_{k\ell} E_{k\ell pq} r_{pq}} \tag{9.24}$$

Combining (9.23b) and (9.24)

$$\dot{\sigma}_{ij} = E_{ijk\ell} \left[\dot{\varepsilon}_{k\ell} - \frac{n_{rs} E_{rspq} \dot{\varepsilon}_{pq}}{K_p + n_{k\ell} E_{k\ell pq} r_{pq}} r_{k\ell} \right]$$

$$\dot{\sigma}_{ij} = \left[E_{ijk\ell} \dot{\varepsilon}_{k\ell} - \frac{E_{ijk\ell} n_{rs} E_{rspq} \dot{\varepsilon}_{pq} r_{k\ell}}{K_p + n_{k\ell} E_{k\ell pq} r_{pq}} \right] = \left[E_{ijk\ell} \dot{\varepsilon}_{k\ell} - \frac{(E_{ijab} r_{ab})(n_{rs} E_{rsk\ell}) \dot{\varepsilon}_{k\ell}}{K_p + n_{rs} E_{rspq} r_{pq}} \right]$$

$$\dot{\sigma}_{ij} = \left[E_{ijk\ell} - \frac{(E_{ijab} r_{ab})(n_{rs} E_{rsk\ell})}{K_p + n_{rs} E_{rspq} r_{pq}} \right] \dot{\varepsilon}_{k\ell} \tag{9.25}$$

or

$$\dot{\sigma}_{ij} = D_{ijk\ell} \dot{\varepsilon}_{k\ell} \tag{9.26a}$$

where the fourth-order tensor $D_{ijk\ell}$ is the continuum elasto-plastic tangent operator, and is given by

$$D_{ijk\ell} = E_{ijk\ell} - \frac{(E_{ijab} r_{ab})(n_{rs} E_{rsk\ell})}{K_p + n_{rs} E_{rspq} r_{pq}} \tag{9.26b}$$

Note that in the above operations, the dummy subscripts have been changed appropriately to avoid confusion and to retain the correct meaning.

It may be shown (Problem 9.8) that when the flow rule is associated, $D_{ijk\ell}$ is symmetric in the sense $D_{ijk\ell} = D_{k\ell ij}$. When the flow rule is nonassociated, $D_{ijk\ell}$ is non-symmetric. When $D_{ijk\ell}$ possesses this type of major symmetry, the 6×6 matrix (needed for the finite element analysis) derived from $D_{ijk\ell}$ is also symmetric (Chap. 4).

Remark For most materials, the classical elasto-plastic constitutive framework described above is inadequate for describing their stress–strain relationships under all loading conditions, especially under cyclic loading conditions. To address this problem, several more generalized and flexible frameworks have been developed during the past few decades; for example, the multi-surface framework of Mroz (1967) and the bounding surface framework (Dafalias and Popov 1974, 1975; Krieg 1975; Dafalias 1986; Dafalias and Herrmann 1986; Anandarajah and Dafalias 1986).

9.3 Stability Postulates, Convexity and Normality, and Uniqueness

A mathematical framework for developing a rate-independent elasto-plastic constitutive model has been presented in Sect. 9.2. Researchers have developed additional guidelines for making specific choices of the components of constitutive models including the yield surface, flow rule, and hardening rule. When these guidelines (or restrictions) are followed, the resulting models can be shown to possess certain desirable characteristics from numerical modeling point of view. For example, they can be proven to yield unique solutions to boundary value problems.

Some of the restrictions established in the early years are too restrictive in developing models for some materials (e.g., soil and concrete). For example, sand obeys non-associated flow rule, which is excluded from the "acceptable" behaviors by some of the earlier restrictions. Nevertheless, the restrictions serve as a baseline from which to develop suitable modifications to fit the behavior of the specific material at hand. For this reason, we present a brief treatment of the classical restrictions in the following sections.

The restrictions referred to above are known as the *stability postulates*. The widely-accepted postulates are those of Drucker (1950, 1951), now known as *Drucker's stability postulates*. The postulates lead to certain consequences; for example, they lead to the requirement that the yield surface must be convex, and the flow rule must be associative (normality). We discuss these issues in the following sections. A postulate that is applicable to strain-softening behavior was proposed by Il'iushin (1961) in strain space. The treatment of Il'iushin's postulate is beyond the scope of this book.

9.3.1 Definition of Hardening/Softening Materials

Recall that we characterized a material as hardening, softening or perfectly-plastic depending on whether the yield surface locally expands (or moves outward), shrinks (moves inward) or remains unchanged in size, respectively. Drucker (1950, 1951) characterizes a material as either hardening, softening or perfectly-plastic based on the inner product $\dot{\boldsymbol{\sigma}}.\dot{\boldsymbol{\varepsilon}}^p$, where $\dot{\boldsymbol{\sigma}}$ is the rate of stress and $\dot{\boldsymbol{\varepsilon}}^p$ is the rate of plastic strain:

$$\text{Hardening material}: \dot{\boldsymbol{\sigma}}.\dot{\boldsymbol{\varepsilon}}^p \geq 0 \tag{9.27a}$$

$$\text{Softening material}: \dot{\boldsymbol{\sigma}}.\dot{\boldsymbol{\varepsilon}}^p \leq 0 \tag{9.27b}$$

$$\text{Perfectly-plastic material}: \dot{\boldsymbol{\sigma}}.\dot{\boldsymbol{\varepsilon}}^p = 0 \tag{9.27c}$$

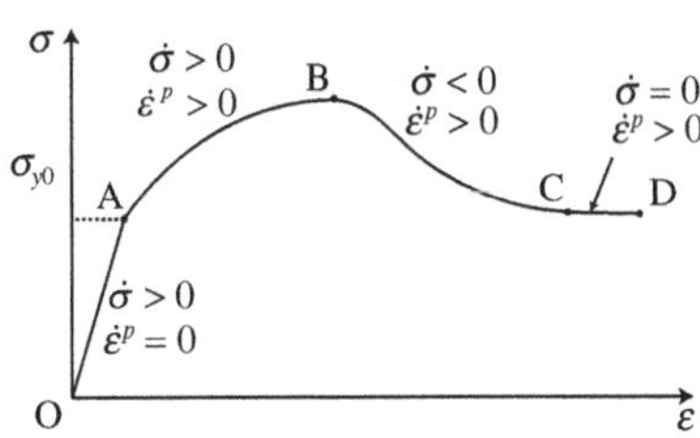

Fig. 9.14 Uniaxial stress–strain relation showing elastic, hardening, softening, and perfectly-plastic regions

Consider a uniaxial stress–strain relationship shown in Fig. 9.14, which consists of an elastic range (O–A), hardening range (A–B), softening range (B–C), and a perfectly plastic range (C–D). The plastic strain is zero in the elastic range (i.e., $\dot{\boldsymbol{\varepsilon}}^p = 0$) and increases in all other ranges (i.e., $\dot{\boldsymbol{\varepsilon}}^p > 0$). The stress increases in the elastic and hardening ranges (i.e., $\dot{\boldsymbol{\sigma}} > 0$), decreases in the softening range (i.e., $\dot{\sigma} < 0$), and remains unchanged in the perfectly plastic range (i.e., $\dot{\sigma} = 0$). It is now easy to verify that (9.27) holds in various regions. Drucker's definition allows the angle between $\dot{\boldsymbol{\sigma}}$ and $\dot{\boldsymbol{\varepsilon}}^p$ to be used to define the material behavior instead of the descriptions concerning the movement of the yield surface such as "outward movement" and "inward movement". (It must be pointed out that for materials such as granular materials, (9.27b) holds true even in the hardening regime).

9.3.2 *Definition of Stable/Unstable Behaviors and Drucker's Stability Postulates*

Drucker (1950, 1951) equates the hardening behavior with *stable behavior* and softening behavior with *unstable behavior*. Multiplying (9.27a) with $\mathrm{d}t^2$ where $\mathrm{d}t > 0$, we have for a stable material

$$\mathrm{d}t^2(\dot{\boldsymbol{\sigma}}.\dot{\boldsymbol{\varepsilon}}^p) \geq 0 \quad \Rightarrow \quad \mathrm{d}\boldsymbol{\sigma}.\mathrm{d}\boldsymbol{\varepsilon}^p \geq 0 \tag{9.28a}$$

The product with the elastic strain is (for a positive definite elastic compliance tensor $\mathbf{E}^e$)

$$\mathrm{d}\boldsymbol{\sigma}.\mathrm{d}\boldsymbol{\varepsilon}^e = \mathrm{d}\boldsymbol{\sigma} : \mathbf{E}^e : \mathrm{d}\boldsymbol{\sigma} > 0 \tag{9.28b}$$

Adding (9.28a) and (9.28b), we have

$$\mathrm{d}\boldsymbol{\sigma}.\mathrm{d}\boldsymbol{\varepsilon} = \mathrm{d}\boldsymbol{\sigma}.(\mathrm{d}\boldsymbol{\varepsilon}^e + \mathrm{d}\boldsymbol{\varepsilon}^p) > 0 \tag{9.28c}$$

Now considering an infinitesimal stress cycle as shown in Fig. 9.15, the infinitesimal work done during loading from point A to B is approximately $\mathrm{d}\boldsymbol{\sigma}.(\mathrm{d}\boldsymbol{\varepsilon}^e + \mathrm{d}\boldsymbol{\varepsilon}^p)/2$

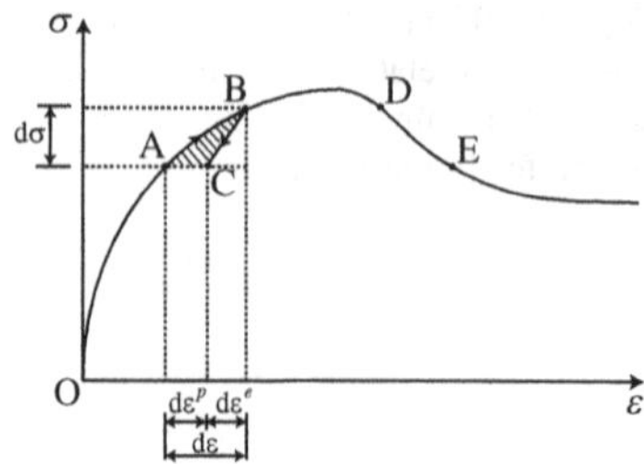

Fig. 9.15 Infinitesimal plastic work (shaded area)

and that during unloading from B to C is $-\,\mathrm{d}\boldsymbol{\sigma}.\mathrm{d}\boldsymbol{\varepsilon}^e/2$. Then the work done over an infinitesimal stress cycle is $\mathrm{d}\boldsymbol{\sigma}.\mathrm{d}\boldsymbol{\varepsilon}^p/2$.

On the basis of the forgoing, Drucker defines a stable material as follows. Let a body be in equilibrium under the applied external loads and the internal stress $\boldsymbol{\sigma}$. Now let the external load be changed slowly by an infinitesimal quantity. Let the corresponding change in stress, total strain and plastic strain be $\mathrm{d}\boldsymbol{\sigma}$, $\mathrm{d}\boldsymbol{\varepsilon}$ and $\mathrm{d}\boldsymbol{\varepsilon}^p$, respectively. Then *for a stable material*

$$\mathrm{d}\boldsymbol{\sigma}.\mathrm{d}\boldsymbol{\varepsilon} > 0\,(\text{positive})\text{ during an infinitesimal loading, and} \tag{9.29a}$$

$$\mathrm{d}\boldsymbol{\sigma}.\mathrm{d}\boldsymbol{\varepsilon}^p \geq 0\,(\text{non}-\text{negative})\text{ during an infinitesimal stress cycle, with the equality applied only when the changes are purely elastic} \tag{9.29b}$$

These requirements ((9.29a) and (9.29b)) are now known as the *Drucker's stability postulates.*

Referring to Fig. 9.15, it is easily seen that the shaded area is $\mathrm{d}\boldsymbol{\sigma}.\mathrm{d}\boldsymbol{\varepsilon}^p/2$, which is the infinitesimal plastic work. Hence, Drucker's postulate states that the infinitesimal plastic work must be non-negative for a stable material. Also it is easily verified by considering two points D and E on the strain softening part of the stress–strain curve that $\mathrm{d}\boldsymbol{\sigma}.\mathrm{d}\boldsymbol{\varepsilon} < 0$. Hence the first postulate (9.29a) excludes strain-softening from stable behaviors.

9.3.3 General Stability Postulates

In this section, we present the stability postulates formally and examine some of their consequences. We use Drucker's first postulate in its original form. The second postulate was presented by Martin (1975) in a different form; we will present this form here. This form is particularly suitable for proving normality and convexity. We will, however, show that Martin's form of the second postulate leads to Drucker's second postulate.

9.3.3.1 First Stability Postulate (Drucker's First Postulate)

Here we restate the Drucker's first postulate. When a material at equilibrium is subjected to slowly varying infinitesimal external forces, the induced infinitesimal stresses and strains must satisfy the following condition for a stable material

$$\mathrm{d}\boldsymbol{\sigma}.\mathrm{d}\boldsymbol{\varepsilon} > 0 \tag{9.30a}$$

9.3.3.2 Second Stability Postulate (Complementary Work Postulate)

Consider a stress cycle in stress space (Fig. 9.16) that starts from point a and ends at point A, where a ≡ A. The changes in stresses are not necessarily infinitesimal. For a stable material, the complementary work done around the stress cycle must be non positive, i.e.,

$$\oint \boldsymbol{\varepsilon}\mathrm{d}\boldsymbol{\sigma} \leq 0 \tag{9.30b}$$

9.3.3.3 Implications of the Complementary Work Postulate

In Fig. 9.17, we show a few cases of stress cycles in the uniaxial stress–strain space. It is easy to verify that the complementary work postulate is satisfied for the stress–strain relation shown in Fig. 9.17a. The shaded area is the net complementary work during the cycle, which is negative. The stress–strain curve shown represents known real materials and the postulate accepts the behavior as intended.

In Fig. 9.17b, we show a stress–strain behavior that has not been observed for real materials. The complementary work postulate excludes this from the acceptable behaviors since the net-complementary work done is positive.

The stress cycle shown in Fig. 9.17c takes place along an elastic loading–unloading path (assuming that the elastic modulus remains the same during loading and unloading). The net-complementary work done is zero. The behavior is acceptable by the complementary work postulate.

Shown in Fig. 9.17d is the typical behavior observed for most materials, where the yielding occurs before the original yield point (i.e., the yield point established at the beginning of the previous unloading) is reached. For the stress cycle shown, the

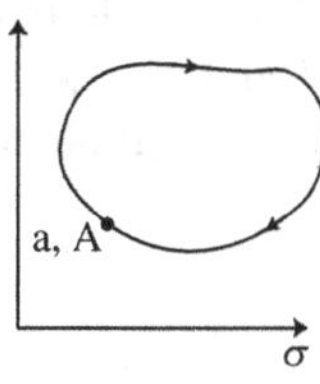

Fig. 9.16 Schematic of a stress cycle in stress space

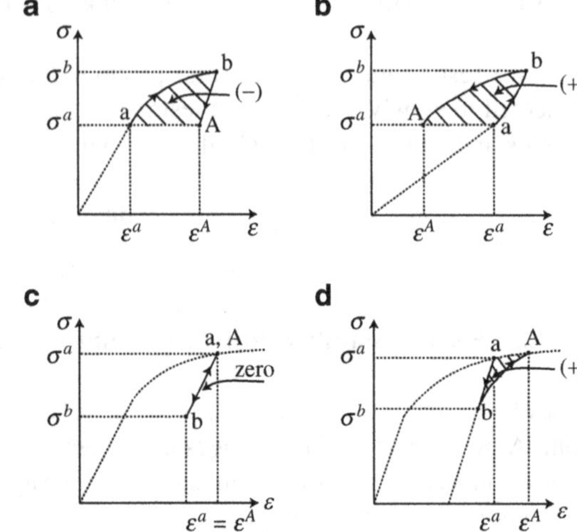

Fig. 9.17 Net complementary energies during stress cycles

net-complementary work is positive and hence the behavior is excluded from the acceptable behaviors.

In the four cases considered in Fig. 9.17, taking points a and b to be close to each other, $\mathrm{d}\sigma = \sigma^b - \sigma^a$. It may be easily verified that the first postulate accepts all of the behaviors shown in Fig. 9.17; $\mathrm{d}\sigma\mathrm{d}\varepsilon > 0$ both from a to b and from b to A for all cases shown. Hence the first postulate, while it excludes the strain-softening behavior, includes some of the other unacceptable behaviors; e.g., Fig. 9.17b. Hence the complementary work postulate is necessary for excluding these unacceptable behaviors.

Remark The Drucker's second postulate (9.29b) excludes the behavior shown in Fig. 9.17b. We will show below that the complementary work postulate (9.30b) implies Drucker's second postulate. However, Drucker's second postulate accepts the behavior shown in Fig. 9.17d, whereas the complementary work postulate (9.30b) rejects it. In this sense, the complementary work postulate given by (9.30b) is more restrictive.

9.3.3.4 A Consequence of the Second Postulate

We will now show that the complementary work postulate (9.30b) leads to Drucker's second postulate (9.29b). Let us consider a case where the stress changes from $\boldsymbol{\sigma}^a$ to $\boldsymbol{\sigma}^b$ along any arbitrary path. Let the strains at the two end points a and b be $\boldsymbol{\varepsilon}^a$ and $\boldsymbol{\varepsilon}^b$, respectively. The stress and strain quantities are related by the following identity (Problem 9.15):

$$\int_{\boldsymbol{\sigma}^a}^{\boldsymbol{\sigma}^b} \boldsymbol{\varepsilon}\mathrm{d}\boldsymbol{\sigma} + \int_{\boldsymbol{\varepsilon}^a}^{\boldsymbol{\varepsilon}^b} \boldsymbol{\sigma}\mathrm{d}\boldsymbol{\varepsilon} = \boldsymbol{\sigma}^b\boldsymbol{\varepsilon}^b - \boldsymbol{\sigma}^a\boldsymbol{\varepsilon}^a \tag{9.31}$$

Now let us consider a cyclic stress path such as the one shown in Fig. 9.16. For such a path, $\boldsymbol{\sigma}^a = \boldsymbol{\sigma}^b$, but $\boldsymbol{\varepsilon}^a \neq \boldsymbol{\varepsilon}^b = \boldsymbol{\varepsilon}^A$. Applying the identity given by (9.31)

$$\oint \boldsymbol{\varepsilon} \mathrm{d}\boldsymbol{\sigma} + \int_{\boldsymbol{\varepsilon}^a}^{\boldsymbol{\varepsilon}^A} \boldsymbol{\sigma} \mathrm{d}\boldsymbol{\varepsilon} = \boldsymbol{\sigma}^a(\boldsymbol{\varepsilon}^A - \boldsymbol{\varepsilon}^a) = \boldsymbol{\sigma}^a \int_{\boldsymbol{\varepsilon}^a}^{\boldsymbol{\varepsilon}^A} \mathrm{d}\boldsymbol{\varepsilon}$$

$$\oint \boldsymbol{\varepsilon} \mathrm{d}\boldsymbol{\sigma} + \int_{\boldsymbol{\varepsilon}^a}^{\boldsymbol{\varepsilon}^A} (\boldsymbol{\sigma} - \boldsymbol{\sigma}^a) \mathrm{d}\boldsymbol{\varepsilon} = 0$$

From the complementary work postulate, we have $\oint \boldsymbol{\varepsilon} \mathrm{d}\boldsymbol{\sigma} \leq 0$ and hence

$$\int_{\boldsymbol{\varepsilon}^a}^{\boldsymbol{\varepsilon}^A} (\boldsymbol{\sigma} - \boldsymbol{\sigma}^a) \mathrm{d}\boldsymbol{\varepsilon} \geq 0 \tag{9.32}$$

That is, the net work done along a *stress cycle along any arbitrary path is non negative.*

Now specializing (9.32) to an infinitesimal stress cycle shown in Fig. 9.17a,

$$\int_{\boldsymbol{\varepsilon}^a}^{\boldsymbol{\varepsilon}^A} (\boldsymbol{\sigma} - \boldsymbol{\sigma}^a) \mathrm{d}\boldsymbol{\varepsilon} = \int_{\boldsymbol{\varepsilon}^a}^{\boldsymbol{\varepsilon}^b} (\boldsymbol{\sigma} - \boldsymbol{\sigma}^a) \mathrm{d}\boldsymbol{\varepsilon} + \int_{\boldsymbol{\varepsilon}^b}^{\boldsymbol{\varepsilon}^A} (\boldsymbol{\sigma} - \boldsymbol{\sigma}^a) \mathrm{d}\boldsymbol{\varepsilon} \geq 0$$

$$\frac{1}{2} \mathrm{d}\boldsymbol{\sigma}(\mathrm{d}\boldsymbol{\varepsilon}^e + \mathrm{d}\boldsymbol{\varepsilon}^p) - \frac{1}{2} \mathrm{d}\boldsymbol{\sigma} \mathrm{d}\boldsymbol{\varepsilon}^e \geq 0$$

$$\mathrm{d}\boldsymbol{\sigma} \mathrm{d}\boldsymbol{\varepsilon}^p \geq 0 \tag{9.33}$$

This is Drucker's second postulate.

9.3.4 *Convexity and Normality*

9.3.4.1 Proof of Convexity and Normality

The existence of an yield surface in the stress space in the classical sense (yield surface containing a purely elastic region) is assumed. Consider the cyclic stress path shown in Fig. 9.18a, which consists of an elastic loading from 1 to 2, infinitesimal stress probing from 2 to 3 and elastic unloading from 3 to 1. The yield surface changes such that the current stress point continues to lie on the yield

Fig. 9.18 Convexity and normality

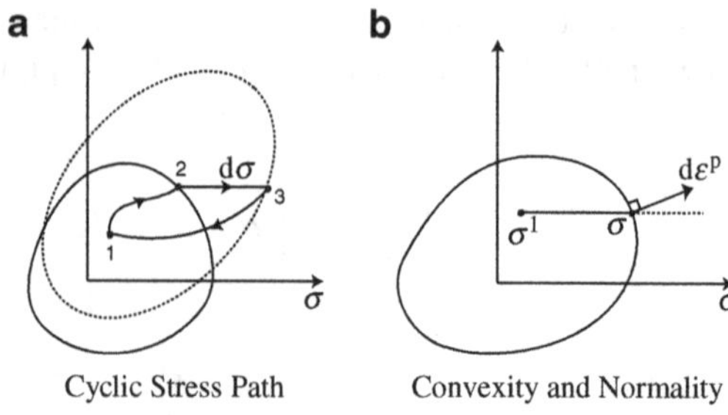

Table 9.1 Stresses and strains at key points during a stress cycle

Points	Elastic strain	Plastic strain	Stress
1	$\boldsymbol{\varepsilon}^{e1}$	$\boldsymbol{\varepsilon}^{p1}$	$\boldsymbol{\sigma}^1$
2	$\boldsymbol{\varepsilon}^{e2}$	$\boldsymbol{\varepsilon}^{p1}$	$\boldsymbol{\sigma}^2$
3	$\boldsymbol{\varepsilon}^{e2} + \mathrm{d}\boldsymbol{\varepsilon}^e$	$\boldsymbol{\varepsilon}^{p1} + \mathrm{d}\boldsymbol{\varepsilon}^p$	$\boldsymbol{\sigma}^2 + \mathrm{d}\boldsymbol{\sigma}$
1 on return path	$\boldsymbol{\varepsilon}^{e1}$	$\boldsymbol{\varepsilon}^{p1} + \mathrm{d}\boldsymbol{\varepsilon}^p$	$\boldsymbol{\sigma}^1$

surface during the stress probe. Let the stress probe be $\mathrm{d}\boldsymbol{\sigma}$, and the corresponding elastic and plastic strains be $\mathrm{d}\boldsymbol{\varepsilon}^e$ and $\mathrm{d}\boldsymbol{\varepsilon}^p$, respectively. The stress, total strain, elastic strain, and plastic strain at the three points and point 1 on the return path, are listed in Table 9.1. The material is assumed to have initial elastic and plastic strains, $\boldsymbol{\varepsilon}^{e1}$ and $\boldsymbol{\varepsilon}^{p1}$, respectively, at point 1.

Let us now evaluate the complementary work done around the cycle and simplify it as follows (Martin 1975):

$$\oint \boldsymbol{\varepsilon}\mathrm{d}\boldsymbol{\sigma} = \int_1^2 \boldsymbol{\varepsilon}\mathrm{d}\boldsymbol{\sigma} + \int_2^3 \boldsymbol{\varepsilon}\mathrm{d}\boldsymbol{\sigma} + \int_3^1 \boldsymbol{\varepsilon}\mathrm{d}\boldsymbol{\sigma}$$
$$= \int_1^2 (\boldsymbol{\varepsilon}^e + \boldsymbol{\varepsilon}^{p1})\mathrm{d}\boldsymbol{\sigma} + \int_2^3 (\boldsymbol{\varepsilon}^e + \boldsymbol{\varepsilon}^p)\mathrm{d}\boldsymbol{\sigma} + \int_3^1 (\boldsymbol{\varepsilon}^e + \boldsymbol{\varepsilon}^{p1} + \mathrm{d}\boldsymbol{\varepsilon}^p)\mathrm{d}\boldsymbol{\sigma}$$
$$= \left[\int_1^2 \boldsymbol{\varepsilon}^e\mathrm{d}\boldsymbol{\sigma} + \int_2^3 \boldsymbol{\varepsilon}^e\mathrm{d}\boldsymbol{\sigma} + \int_3^1 \boldsymbol{\varepsilon}^e\mathrm{d}\boldsymbol{\sigma}\right] + \oint \boldsymbol{\varepsilon}^{p1}\mathrm{d}\boldsymbol{\sigma} + \int_2^3 (\boldsymbol{\varepsilon}^p - \boldsymbol{\varepsilon}^{p1})\mathrm{d}\boldsymbol{\sigma} + \int_3^1 \mathrm{d}\boldsymbol{\varepsilon}^p\mathrm{d}\boldsymbol{\sigma}$$

where $\boldsymbol{\varepsilon}^e$ and $\boldsymbol{\varepsilon}^p$ are variable elastic and plastic strains. With the assumption of unvarying elastic properties, the first term (grouped within the bracket) is zero. Noting that $\boldsymbol{\varepsilon}^{p1}$ in the second term is a constant; it can be factored out of the integral to show that the second term is zero as well. The remaining two terms simplify as

$$\int_2^3 (\boldsymbol{\varepsilon}^p - \boldsymbol{\varepsilon}^{p1})\mathrm{d}\boldsymbol{\sigma} \approx O(\mathrm{d}\boldsymbol{\sigma}\mathrm{d}\boldsymbol{\varepsilon}^p)$$

$$\int_3^1 d\boldsymbol{\varepsilon}^p d\boldsymbol{\sigma} = d\boldsymbol{\varepsilon}^p \int_3^1 d\boldsymbol{\sigma} = (\boldsymbol{\sigma}^1 - \boldsymbol{\sigma} - d\boldsymbol{\sigma})d\boldsymbol{\varepsilon}^p = (\boldsymbol{\sigma}^1 - \boldsymbol{\sigma})d\boldsymbol{\varepsilon}^p - O(d\boldsymbol{\sigma}d\boldsymbol{\varepsilon}^p)$$

where the notation $O(d\boldsymbol{\sigma}d\boldsymbol{\varepsilon}^p)$ implies order of $d\boldsymbol{\sigma}d\boldsymbol{\varepsilon}^p$. When point 1 is arbitrarily chosen within the yield surface, the term $O(d\boldsymbol{\sigma}d\boldsymbol{\varepsilon}^p)$ is second order compared to $(\boldsymbol{\sigma} - \boldsymbol{\sigma}^1)d\boldsymbol{\varepsilon}^p$ and hence can be neglected. We then get

$$\oint \boldsymbol{\varepsilon} d\boldsymbol{\sigma} = (\boldsymbol{\sigma}^1 - \boldsymbol{\sigma})d\boldsymbol{\varepsilon}^p$$

By the complementary energy postulate (9.30b):

$$\oint \boldsymbol{\varepsilon} d\boldsymbol{\sigma} = (\boldsymbol{\sigma}^1 - \boldsymbol{\sigma})d\boldsymbol{\varepsilon}^p \leq 0$$

$$(\boldsymbol{\sigma} - \boldsymbol{\sigma}^1)d\boldsymbol{\varepsilon}^p \geq 0 \tag{9.34}$$

The condition given by (9.34) can be satisfied only when

- The infinitesimal plastic strain vector is normal to the yield surface (normality).
- The yield surface is convex (convexity).

Figure 9.18b shows a typical convex yield surface and the infinitesimal plastic strain vector that is normal to the yield surface. It is easy to verify that the angle between the $(\boldsymbol{\sigma} - \boldsymbol{\sigma}^1)$ vector and $d\boldsymbol{\varepsilon}^p$ vector will be less than 90° regardless of the position of point 1 within the yield surface.

If either of the conditions listed above (normality and convexity) is violated, the condition in (9.34) will be violated. Figure 9.19a shows how the failure to obey normality can violate (9.34) even when the yield surface is convex. Figure 9.19b shows how a nonconvex yield surface can violate (9.34) even when normality is obeyed.

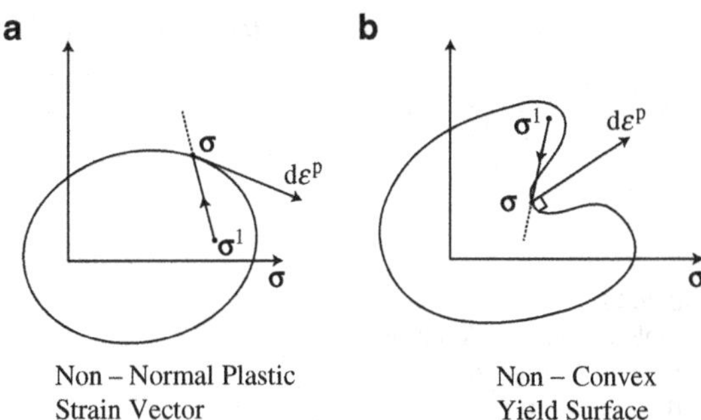

Fig. 9.19 Violation of normality or convexity

9.3.4.2 Applicability of Convexity and Normality to Softening and Perfectly-Plastic Materials

Equation (9.34) has been derived using only the complementary energy postulate (the second postulate, not the first postulate). The second postulate does not exclude the strain softening behavior; only the first postulate does. Consequently, (9.34) is satisfied by strain-softening materials as well, as illustrated in Fig. 9.20. Note that (a) the shaded area equals $\oint \varepsilon \mathrm{d}\sigma$ which is negative, satisfying the second postulate, and (b) $(\sigma - \sigma^1)\mathrm{d}\varepsilon^p \geq 0$. It is easy to see that even in the perfectly-plastic region, $(\sigma - \sigma^1)\mathrm{d}\varepsilon^p \geq 0$. (By following the same approach as above in this section, it can be systematically shown that $(\sigma - \sigma^1)\mathrm{d}\varepsilon^p \geq 0$ for a perfectly-plastic material; see Problem 9.12).

Hence the condition given by (9.34) is satisfied by hardening, softening, and perfectly-plastic materials. Thus, according to the first and second stability postulates ((9.30a) and (9.30b)), the *convexity and normality must be obeyed by all material models regardless of whether they are designed to represent hardening, softening or perfectly-plastic materials.*

The condition given by (9.34) is known as the *maximum plastic dissipation postulate*, and was independently proposed, before Drucker (1950, 1951), by several earlier researchers (von Mises 1928; Taylor 1947; Hill 1948).

9.3.4.3 Applicability of Convexity and Normality to Yield Surfaces with Corners

When an yield surface has a corner (a singular point) as in Fig. 9.21a, maximum plastic dissipation postulate (9.34) can still be satisfied provided that the infinitesimal plastic strain vector at the corner lies in the hypercone shown in Fig. 9.21a. However, the precise direction of $\mathrm{d}\boldsymbol{\varepsilon}^p$ must be uniquely defined. One way to define a unique direction for $\mathrm{d}\boldsymbol{\varepsilon}^p$ is to employ a number of separate smooth yield surfaces to define the corner (Koiter 1953) as shown in Fig. 9.21b, where two separate yield surfaces are used. The net plastic strain is taken to be the sum of the plastic strains from each of the yield surfaces as

$$\mathrm{d}\boldsymbol{\varepsilon}^p = \mathrm{d}\boldsymbol{\varepsilon}^{p1} + \mathrm{d}\boldsymbol{\varepsilon}^{p2}$$

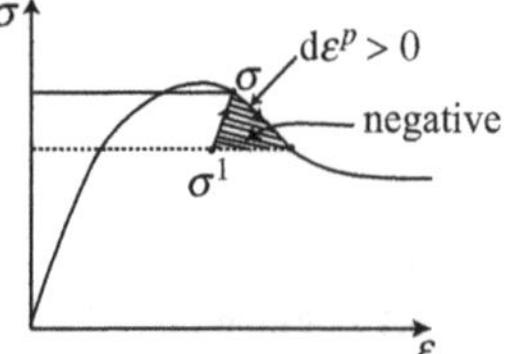

Fig. 9.20 Net complementary work during stress cycle for strain-softening material

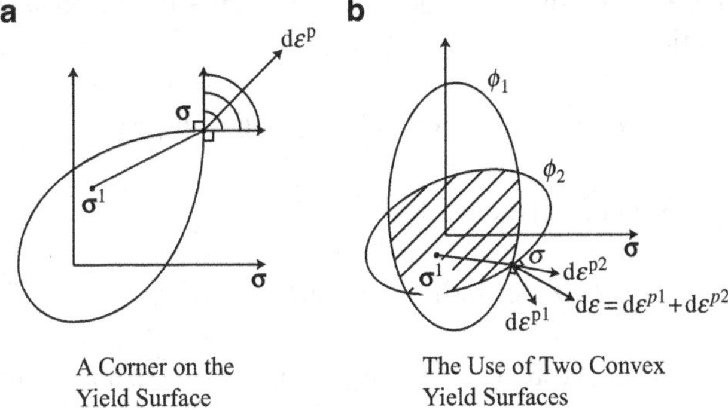

Fig. 9.21 Yield surfaces with corners

Fig. 9.22 An elasto-plastic body with elastic, hardening, and flow regions

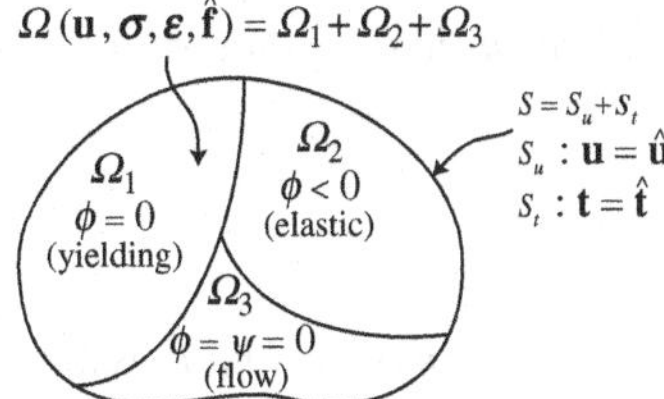

where $\mathrm{d}\boldsymbol{\varepsilon}^{p1}$ and $\mathrm{d}\boldsymbol{\varepsilon}^{p2}$ are infinitesimal plastic strains derived from the yield surfaces ϕ_1 and ϕ_2, respectively. $\mathrm{d}\boldsymbol{\varepsilon}^{p1}$ is normal to ϕ_1 and $\mathrm{d}\boldsymbol{\varepsilon}^{p2}$ is normal to ϕ_2. The elastic domain must be restricted to the domain that is inside of both ϕ_1 and ϕ_2 as shown in the figure. It is easy to see that maximum plastic dissipation postulate (9.34) will then be fulfilled.

9.3.5 Uniqueness of Boundary Value Solutions: Hardening Materials

Consider a body, shown in Fig. 9.22, which is in equilibrium under the applied external forces and the developed internal stresses. The boundary S is divided into two parts as $S = S_u + S_t$. S_u is the part where the displacement is prescribed as $\mathbf{u} = \hat{\mathbf{u}}$. S_t is the part where the traction is prescribed as $\mathbf{t} = \hat{\mathbf{t}}$. In addition, the externally applied body force $\hat{\mathbf{f}}$ acts within the domain Ω. The unknown variables are (1) the displacement $\mathbf{u}$, the stress $\boldsymbol{\sigma}$ and the strain $\boldsymbol{\varepsilon}$ within Ω, (2) $\mathbf{t}$ on S_u and (3) $\mathbf{u}$ on S_t; i.e.,

$$\text{Unknowns: } \{\boldsymbol{\sigma}, \boldsymbol{\varepsilon}, \mathbf{u} \text{ in } \Omega, \mathbf{u} \text{ on } S_t, \mathbf{t} \text{ on } S_u\} \tag{9.35}$$

Let us impose infinitesimal changes to the variables on the boundary (i.e., the boundary conditions) as

$$\hat{\mathbf{t}} \Rightarrow \hat{\mathbf{t}} + \mathrm{d}\hat{\mathbf{t}}; \quad \hat{\mathbf{u}} \Rightarrow \hat{\mathbf{u}} + \mathrm{d}\hat{\mathbf{u}}; \quad \hat{\mathbf{f}} \Rightarrow \hat{\mathbf{f}} + \mathrm{d}\hat{\mathbf{f}} \tag{9.36}$$

Let the changes in the unknowns be

$$\{\mathrm{d}\boldsymbol{\sigma}, \mathrm{d}\boldsymbol{\varepsilon}, \mathrm{d}\mathbf{u} \text{ in } \Omega, \ \mathrm{d}\mathbf{u} \text{ on } S_t, \ \mathrm{d}\mathbf{t} \text{ on } S_u\} \tag{9.37}$$

The question to be investigated is: Can the changes in unknown variables given in (9.37) be determined uniquely?

We will use the *virtual work principle* to carry out the investigation. The details of the virtual work principle are described in Appendix 1. The virtual work principle is written in terms of:

- *The statically admissible field (SAF)*:

$$\left\{\mathrm{d}\hat{\mathbf{f}}, \mathrm{d}\mathbf{t}, \mathrm{d}\boldsymbol{\sigma}\right\} \tag{9.38a}$$

- *The kinematically admissible field (KAF)*:

$$\{\mathrm{d}\mathbf{u}, \mathrm{d}\boldsymbol{\varepsilon}\} \tag{9.38b}$$

When a SAF satisfies the boundary condition $\mathrm{d}\mathbf{t} = \mathrm{d}\hat{\mathbf{t}}$ on S_t, it is referred to as a pertinent statically admissible field (PSAF). Similarly, when a KAF satisfies the boundary condition $\mathrm{d}\mathbf{u} = \mathrm{d}\hat{\mathbf{u}}$ on S_u, it is referred to as the pertinent kinematically admissible field (PKAF).

To prove uniqueness, let's first assume that the solution is nonunique. Then there must be at least two distinct solutions. Let these two solutions be:

$$\left\{\mathrm{d}\boldsymbol{\sigma}^1, \mathrm{d}\boldsymbol{\varepsilon}^1, \mathrm{d}\mathbf{u}^1 \text{ in } \Omega, \mathrm{d}\mathbf{u}^1 \text{ on } S_t, \mathrm{d}\mathbf{t}^1 \text{ on } S_u\right\}$$

$$\left\{\mathrm{d}\boldsymbol{\sigma}^2, \mathrm{d}\boldsymbol{\varepsilon}^2, \mathrm{d}\mathbf{u}^2 \text{ in } \Omega, \mathrm{d}\mathbf{u}^2 \text{ on } S_t, \mathrm{d}\mathbf{t}^2 \text{ on } S_u\right\}$$

Since these are supposed to be the real solutions, the variables form a PSAF and PKAF as:

$$\text{Solution1: PSAF:}\{\mathrm{d}\hat{\mathbf{f}}, \mathrm{d}\mathbf{t}^1, \mathrm{d}\boldsymbol{\sigma}^1\} \quad \text{PKAF:}\{\mathrm{d}\mathbf{u}^1, \mathrm{d}\boldsymbol{\varepsilon}^1\} \tag{9.39a}$$

$$\text{Solution2: PSAF:}\{\mathrm{d}\hat{\mathbf{f}}, \mathrm{d}\mathbf{t}^2, \mathrm{d}\boldsymbol{\sigma}^2\} \quad \text{PKAF:}\{\mathrm{d}\mathbf{u}^2, \mathrm{d}\boldsymbol{\varepsilon}^2\} \tag{9.39b}$$

Using the four sets written above, one can write four virtual work equations as

$$\int_S \mathrm{d}\mathbf{t}^1 \mathrm{d}\mathbf{u}^1 \mathrm{dS} + \int_\Omega \mathrm{d}\hat{\mathbf{f}} \mathrm{d}\mathbf{u}^1 \mathrm{d}\Omega = \int_\Omega \mathrm{d}\boldsymbol{\sigma}^1 \mathrm{d}\boldsymbol{\varepsilon}^1 \mathrm{d}\Omega \tag{9.40a}$$

$$\int_S \mathrm{d}\mathbf{t}^1 \mathrm{d}\mathbf{u}^2 \mathrm{dS} + \int_\Omega \mathrm{d}\hat{\mathbf{f}} \mathrm{d}\mathbf{u}^2 \mathrm{d}\Omega = \int_\Omega \mathrm{d}\boldsymbol{\sigma}^1 \mathrm{d}\boldsymbol{\varepsilon}^2 \mathrm{d}\Omega \tag{9.40b}$$

$$\int_S \mathrm{d}\mathbf{t}^2 \mathrm{d}\mathbf{u}^1 \mathrm{dS} + \int_\Omega \mathrm{d}\hat{\mathbf{f}} \mathrm{d}\mathbf{u}^1 \mathrm{d}\Omega = \int_\Omega \mathrm{d}\boldsymbol{\sigma}^2 \mathrm{d}\boldsymbol{\varepsilon}^1 \mathrm{d}\Omega \tag{9.40c}$$

$$\int_S \mathrm{d}\mathbf{t}^2 \mathrm{d}\mathbf{u}^2 \mathrm{dS} + \int_\Omega \mathrm{d}\hat{\mathbf{f}} \mathrm{d}\mathbf{u}^2 \mathrm{d}\Omega = \int_\Omega \mathrm{d}\boldsymbol{\sigma}^2 \mathrm{d}\boldsymbol{\varepsilon}^2 \mathrm{d}\Omega \tag{9.40d}$$

By summing up the equations as: (9.40a) – (9.40b) – (9.40c) + (9.40d), we have

$$\int_S \left(\mathrm{d}\mathbf{t}^1 - \mathrm{d}\mathbf{t}^2\right)\left(\mathrm{d}\mathbf{u}^1 - \mathrm{d}\mathbf{u}^2\right)\mathrm{d}S = \int_\Omega \left(\mathrm{d}\boldsymbol{\sigma}^1 - \mathrm{d}\boldsymbol{\sigma}^2\right)\left(\mathrm{d}\boldsymbol{\varepsilon}^1 - \mathrm{d}\boldsymbol{\varepsilon}^2\right)\mathrm{d}\Omega \tag{9.41}$$

Remark Noting that the virtual work equation is linear in $\mathrm{d}\boldsymbol{\sigma}$, $\mathrm{d}\boldsymbol{\varepsilon}$, $\mathrm{d}\mathbf{u}$, $\mathrm{d}\mathbf{t}$ and $\mathrm{d}\mathbf{f}$, (9.41) can be written down directly by defining PSAF and PKAF as the difference between corresponding variables.

The term on the left hand side of (9.41) is zero since $\mathrm{d}\mathbf{t}^1 = \mathrm{d}\mathbf{t}^2 = \mathrm{d}\hat{\mathbf{t}}$ on S_t and $\mathrm{d}\mathbf{u}^1 = \mathrm{d}\mathbf{u}^2 = \mathrm{d}\hat{\mathbf{u}}$ on S_u. Defining

$$I = \left(\mathrm{d}\boldsymbol{\sigma}^1 - \mathrm{d}\boldsymbol{\sigma}^2\right)\left(\mathrm{d}\boldsymbol{\varepsilon}^1 - \mathrm{d}\boldsymbol{\varepsilon}^2\right) \tag{9.42a}$$

Equation (9.41) then becomes

$$\int_\Omega I \mathrm{d}\Omega = 0 \tag{9.42b}$$

Splitting the total strain increments into elastic and plastic parts:

$$\begin{aligned} I &= I_1 + I_2; \\ I_1 &= \left(\mathrm{d}\boldsymbol{\sigma}^1 - \mathrm{d}\boldsymbol{\sigma}^2\right)\left(\mathrm{d}\boldsymbol{\varepsilon}^{e1} - \mathrm{d}\boldsymbol{\varepsilon}^{e2}\right); \\ I_2 &= \left(\mathrm{d}\boldsymbol{\sigma}^1 - \mathrm{d}\boldsymbol{\sigma}^2\right)\left(\mathrm{d}\boldsymbol{\varepsilon}^{p1} - \mathrm{d}\boldsymbol{\varepsilon}^{p2}\right) \end{aligned} \tag{9.43}$$

Using the inverse elastic constitutive law

$$\mathrm{d}\boldsymbol{\varepsilon}^e = \mathbf{E}^e : \mathrm{d}\boldsymbol{\sigma} \tag{9.44a}$$

and recalling that the elastic compliance tensor $\mathbf{E}^e$ is positive definite, we have

$$I_1 = \left(\mathrm{d}\boldsymbol{\sigma}^1 - \mathrm{d}\boldsymbol{\sigma}^2\right) : \mathbf{E}^e : \left(\mathrm{d}\boldsymbol{\sigma}^1 - \mathrm{d}\boldsymbol{\sigma}^2\right) \geq 0 \tag{9.44b}$$

As per the plastic behavior, five different possibilities for the solutions for $\mathrm{d}\boldsymbol{\sigma}^1$ and $\mathrm{d}\boldsymbol{\sigma}^2$ are shown in Fig. 9.23. We assume that the flow rule is *associated* and that the plastic constitutive relation is expressed in the form given by (9.19). Writing (9.19) in infinitesimal form, we have

$$\mathrm{d}\boldsymbol{\varepsilon}^p = \frac{1}{K_p}[\mathrm{d}\boldsymbol{\sigma} : \mathbf{n}]\mathbf{n} \tag{9.45}$$

We now investigate the sign of I_2 for each of the five cases shown in Fig. 9.23.

- *Case 1*: Here the two stress probes are tangent to the surface. The behavior is elastic in both cases and hence

$$\mathrm{d}\boldsymbol{\varepsilon}^{p1} = \mathrm{d}\boldsymbol{\varepsilon}^{p2} = 0 \Rightarrow I_2 = 0$$

- *Case 2*: Here both probes produce plastic strains. Assuming $K_p > 0$ (which is the case for a hardening material before reaching failure or peak, (9.20e))

$$I_2 = \frac{1}{K_p}\left(\mathrm{d}\boldsymbol{\sigma}^1 - \mathrm{d}\boldsymbol{\sigma}^2\right)\left(\mathrm{d}\boldsymbol{\sigma}^1 : \mathbf{n} - \mathrm{d}\boldsymbol{\sigma}^2 : \mathbf{n}\right)\mathbf{n} = \frac{1}{K_p}\left(\mathrm{d}\boldsymbol{\sigma}^1 : \mathbf{n} - \mathrm{d}\boldsymbol{\sigma}^2 : \mathbf{n}\right)^2 > 0$$

- *Case 3*: Here the behavior associated with both stress probes are elastic and hence

$$\mathrm{d}\boldsymbol{\varepsilon}^{p1} = \mathrm{d}\boldsymbol{\varepsilon}^{p2} = 0 \Rightarrow I_2 = 0$$

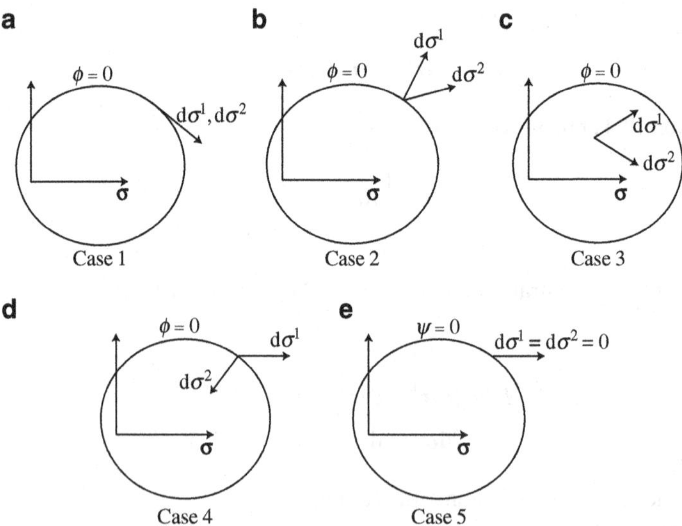

Fig. 9.23 Hardening material: different possible cases of the two solutions

- *Case 4*: Here $\mathrm{d}\boldsymbol{\varepsilon}^{p1} \neq 0$ and $\mathrm{d}\boldsymbol{\varepsilon}^{p2} = 0$. One has

$$I_2 = \frac{1}{K_p}\left(\mathrm{d}\boldsymbol{\sigma}^1 - \mathrm{d}\boldsymbol{\sigma}^2\right)\left[\mathrm{d}\boldsymbol{\sigma}^1 : \mathbf{n}\right]\mathbf{n} = \frac{1}{K_p}\left[(\mathrm{d}\boldsymbol{\sigma}^1 : \mathbf{n})^2 - (\mathrm{d}\boldsymbol{\sigma}^1 : \mathbf{n})(\mathrm{d}\boldsymbol{\sigma}^2 : \mathbf{n})\right]$$

  Noting that $(\mathrm{d}\boldsymbol{\sigma}^1 : \mathbf{n})^2 > 0$, $(\mathrm{d}\boldsymbol{\sigma}^1 : \mathbf{n}) > 0$ and $(\mathrm{d}\boldsymbol{\sigma}^2 : \mathbf{n}) < 0$, and that $K_p > 0$, it follows $I_2 > 0$. The same conclusion is reached when the stress probes are switched.
- *Case 5*: Here the current stress point is on the failure surface ψ where flow occurs. We will prove in the next section that $\mathrm{d}\boldsymbol{\sigma}^1 = \mathrm{d}\boldsymbol{\sigma}^2 = 0$, which leads to $I_2 = 0$.

Hence considering all possible cases, we have

$$I_2 \geq 0 \tag{9.46}$$

Combining (9.44b) and (9.46), it follows

$$I = I_1 + I_2 \geq 0 \tag{9.47}$$

However, from (9.42b), the integral of I must be zero. The only possibility then is

$$I = 0$$

Since $I_1 \geq 0$ and $I_2 \geq 0$, it follows that I_1 and I_2 must be individually zero. From $I_1 = 0$, we have

$$I_1 = \left(\mathrm{d}\boldsymbol{\sigma}^1 - \mathrm{d}\boldsymbol{\sigma}^2\right) : \mathbf{E}^e : \left(\mathrm{d}\boldsymbol{\sigma}^1 - \mathrm{d}\boldsymbol{\sigma}^2\right) = 0 \quad \Rightarrow \quad \mathrm{d}\boldsymbol{\sigma}^1 = \mathrm{d}\boldsymbol{\sigma}^2 \tag{9.48a}$$

From the elastic constitutive relation (9.44a), it follows

$$\mathrm{d}\boldsymbol{\varepsilon}^{e1} = \mathrm{d}\boldsymbol{\varepsilon}^{e2} \tag{9.48b}$$

From the plastic constitutive relation (9.45), we have

$$\mathrm{d}\boldsymbol{\varepsilon}^{p1} = \mathrm{d}\boldsymbol{\varepsilon}^{p2} \tag{9.48c}$$

Summing up the elastic and plastic strains, we get

$$\mathrm{d}\boldsymbol{\varepsilon}^1 = \mathrm{d}\boldsymbol{\varepsilon}^2 \tag{9.48d}$$

From the strain-displacement compatibility relations, it follows

$$\mathrm{d}\mathbf{u}^1 = \mathrm{d}\mathbf{u}^2 \tag{9.48e}$$

From the traction–stress relation $\mathbf{t} = \boldsymbol{\sigma} : \mathbf{n}$, we have

$$\mathrm{d}\mathbf{t}^1 = \mathrm{d}\mathbf{t}^2 \tag{9.48f}$$

We have thus established that the solution to the boundary value problem is unique when

- The plastic constitutive equation is given by (9.45), uses associated flow rule, and convex yield surface, and $K_p > 0$, and
- The elastic compliance tensor is positive definite.

9.3.6 Uniqueness of Boundary Value Solutions During Flow

In this section, we consider the case where the domain (or the structure) experiences flow in the sense that the displacement increases with no change in the external forces and prescribed displacements. More precisely, the material flow occurs in the sense that the displacement $\mathbf{du}$ increases with no change in the boundary conditions $\left\{\hat{\mathbf{u}} \text{ on } S_u, \hat{\mathbf{t}} \text{ on } S_t, \hat{\mathbf{f}} \text{ in } \Omega\right\}$.

We follow the approach presented by Martin (1975). We will first prove that when flow occurs, $\mathrm{d}\boldsymbol{\sigma} = 0$ at all points in the domain; i.e., the stresses remain unchanged. We will then show that the stresses can be uniquely determined at some points in the structure.

9.3.6.1 Proof that $\mathrm{d}\boldsymbol{\sigma} = 0$ During Flow of the Structure

Assume that $\mathrm{d}\boldsymbol{\sigma} \neq 0$ in general. The structure must then be at equilibrium under the external forces and the internal stress $\boldsymbol{\sigma} + \mathrm{d}\boldsymbol{\sigma}$, where $\mathrm{d}\boldsymbol{\sigma}$ is any arbitrary change. Since $\mathrm{d}\boldsymbol{\sigma}$ is arbitrary, it can also be zero, i.e., $\mathrm{d}\boldsymbol{\sigma} = 0$. On this basis, let us define two separate PSAF, one with $\boldsymbol{\sigma}$ and the other with $\boldsymbol{\sigma} + \mathrm{d}\boldsymbol{\sigma}$, where $\mathrm{d}\boldsymbol{\sigma} \neq 0$. Also define a PKAF with $\{\mathbf{du}, \mathrm{d}\boldsymbol{\varepsilon}\}$. That is,

PSAF-1: $\left\{\hat{\mathbf{f}}, \mathbf{t}, \boldsymbol{\sigma}\right\}$
PSAF-2: $\left\{\hat{\mathbf{f}}, \mathbf{t}, \boldsymbol{\sigma} + \mathrm{d}\boldsymbol{\sigma}\right\}$
PKAF: $\{\mathrm{d}\mathbf{u}, \mathrm{d}\boldsymbol{\varepsilon}\}$

With these static and kinematic fields, let us write down two virtual work equations as

$$\int_S \mathbf{t}\mathbf{d}\mathbf{u}\mathbf{d}\mathbf{S} + \int_\Omega \hat{\mathbf{f}}\mathbf{d}\mathbf{u}\mathbf{d}\Omega = \int_\Omega \boldsymbol{\sigma}\mathrm{d}\boldsymbol{\varepsilon}\mathrm{d}\Omega$$

$$\int_S \mathbf{t}\mathbf{d}\mathbf{u}\mathbf{d}\mathbf{S} + \int_\Omega \hat{\mathbf{f}}\mathbf{d}\mathbf{u}\mathbf{d}\Omega = \int_\Omega (\boldsymbol{\sigma}+\mathrm{d}\boldsymbol{\sigma})\mathrm{d}\boldsymbol{\varepsilon}\mathrm{d}\Omega$$

Subtracting the two equations, we get

$$\int_{\Omega} \mathbf{d}\boldsymbol{\sigma}\mathbf{d}\boldsymbol{\varepsilon} d\Omega = 0 \tag{9.49a}$$

Defining $I = \mathbf{d}\boldsymbol{\sigma}\mathbf{d}\boldsymbol{\varepsilon}$, we have

$$\int_{\Omega} I d\Omega = 0 \tag{9.49b}$$

Let us expand I as

$$I = \mathbf{d}\boldsymbol{\sigma}\mathbf{d}\boldsymbol{\varepsilon} = \mathbf{d}\boldsymbol{\sigma}\mathbf{d}\boldsymbol{\varepsilon}^e + \mathbf{d}\boldsymbol{\sigma}\mathbf{d}\boldsymbol{\varepsilon}^p = \mathbf{d}\boldsymbol{\sigma} : \mathbf{E}^e : \mathbf{d}\boldsymbol{\sigma} + \mathbf{d}\boldsymbol{\sigma}\mathbf{d}\boldsymbol{\varepsilon}^p$$

$$I = I_1 + I_2; \quad I_1 = \mathbf{d}\boldsymbol{\sigma} : \mathbf{E}^e : \mathbf{d}\boldsymbol{\sigma}; I_2 = \mathbf{d}\boldsymbol{\sigma}\mathbf{d}\boldsymbol{\varepsilon}^p$$

For a nonzero $\mathbf{d}\boldsymbol{\sigma}$ and a positive definite elastic compliance tensor $\mathbf{E}^e$, we have $I_1 \geq 0$. Since *normality must hold for both hardening and perfect plasticity* (Problem 9.12), it follows that $I_2 \geq 0$. Hence

$$I \geq 0 \tag{9.50}$$

Combining (9.49b) and (9.50),

$$I = \mathbf{d}\boldsymbol{\sigma}\mathbf{d}\boldsymbol{\varepsilon} = 0 \tag{9.51}$$

Equation (9.51) implies that individually $I_1 \geq 0$ and $I_2 \geq 0$. The former is possible only if

$$\mathbf{d}\boldsymbol{\sigma} = 0 \tag{9.52}$$

We have thus proved that the changes in stresses must be zero when the externally applied forces and prescribed displacements remain unchanged, and the case where the structure experiences plastic flow is no exception.

9.3.6.2 Uniqueness of Boundary Value Solutions When Flow Occurs

We assume that the boundary conditions remain unchanged and flow occurs in some parts of the body. We have already proved above (9.52) that the change in stress is zero at every point in the body during flow, i.e., $\mathbf{d}\boldsymbol{\sigma} = 0$. The change in elastic strains $\mathbf{d}\boldsymbol{\varepsilon}^e$ is therefore zero. The infinitesimal strain consists only of the plastic part. That is at every point in the body, we have

$$\mathbf{d}\boldsymbol{\varepsilon} = \mathbf{d}\boldsymbol{\varepsilon}^p$$

However, in the hardening regime, we have $\mathrm{d}\boldsymbol{\varepsilon}^p = 0$ since $\mathrm{d}\boldsymbol{\sigma} = 0$ at every point. Hence if a material point is at the elastic or hardening regime, the strain must be zero at that point. The strain occurs only at points that are on the failure surface.

The question of interest concerns whether the stresses can be determined uniquely at the material points. We assume that there are two distinct solutions. Let these solutions be

$$\left\{\boldsymbol{\sigma}^1, \mathrm{d}\boldsymbol{\varepsilon}^{p1}, \mathrm{d}\mathbf{u}^1 \text{ in } \Omega, \mathrm{d}\mathbf{u}^1 \text{ on } S_t, \mathbf{t}^1 \text{ on } S_u\right\}$$

$$\left\{\boldsymbol{\sigma}^2, \mathrm{d}\boldsymbol{\varepsilon}^{p2}, \mathrm{d}\mathbf{u}^2 \text{ in } \Omega, \mathrm{d}\mathbf{u}^2 \text{ on } S_t, \mathbf{t}^2 \text{ on } S_u\right\}$$

Since these are supposed to be the real solutions, the variables form a PSAF and PKAF as:

$$\text{Solution 1: PSAF}: \left\{\hat{\mathbf{f}}, \mathbf{t}^1, \boldsymbol{\sigma}^1\right\} \quad \text{PKAF}: \left\{\mathrm{d}\mathbf{u}^1, \mathrm{d}\boldsymbol{\varepsilon}^{p1}\right\} \tag{9.53a}$$

$$\text{Solution 2: PSAF}: \left\{\hat{\mathbf{f}}, \mathbf{t}^2, \boldsymbol{\sigma}^2\right\} \quad \text{PKAF}: \left\{\mathrm{d}\mathbf{u}^2, \mathrm{d}\boldsymbol{\varepsilon}^{p2}\right\} \tag{9.53b}$$

As done in the preceding section, we write four virtual work equations ((9.40a)–(9.40d)), algebraically sum them, and simplify them using the boundary conditions to obtain

$$\int_\Omega \left(\boldsymbol{\sigma}^1 - \boldsymbol{\sigma}^2\right)\left(\mathrm{d}\boldsymbol{\varepsilon}^{p1} - \mathrm{d}\boldsymbol{\varepsilon}^{p2}\right)\mathrm{d}\Omega = 0 \tag{9.54a}$$

Letting $I = (\boldsymbol{\sigma}^1 - \boldsymbol{\sigma}^2)(\mathrm{d}\boldsymbol{\varepsilon}^{p1} - \mathrm{d}\boldsymbol{\varepsilon}^{p2})$, (9.54a) becomes

$$\int_\Omega I\mathrm{d}\Omega = 0 \tag{9.54b}$$

Consider the six possibilities shown in Fig. 9.24. In all cases, the failure surface is convex and the flow rule obeys normality. The failure surfaces in Cases 1, 2, and 3 do not have any flat parts, whereas the ones in Cases 4–6 do.

- *Case 1*: Both points 1 and 2 are on the surface.

$$\left(\boldsymbol{\sigma}^1 - \boldsymbol{\sigma}^2\right)\mathrm{d}\boldsymbol{\varepsilon}^{p1} \geq 0; \left(\boldsymbol{\sigma}^1 - \boldsymbol{\sigma}^2\right)\mathrm{d}\boldsymbol{\varepsilon}^{p2} \leq 0 \Rightarrow I \geq 0 \tag{9.55a}$$

Since $\mathrm{d}\boldsymbol{\varepsilon}^{p1}$ and $\mathrm{d}\boldsymbol{\varepsilon}^{p2}$ are arbitrary, let's assume that $\mathrm{d}\boldsymbol{\varepsilon}^{p1} \neq \mathrm{d}\boldsymbol{\varepsilon}^{p2} \neq 0$. Then (9.54b) and (9.55a) implies $I = 0 \Rightarrow \boldsymbol{\sigma}^1 = \boldsymbol{\sigma}^2$. The stresses can be uniquely determined.

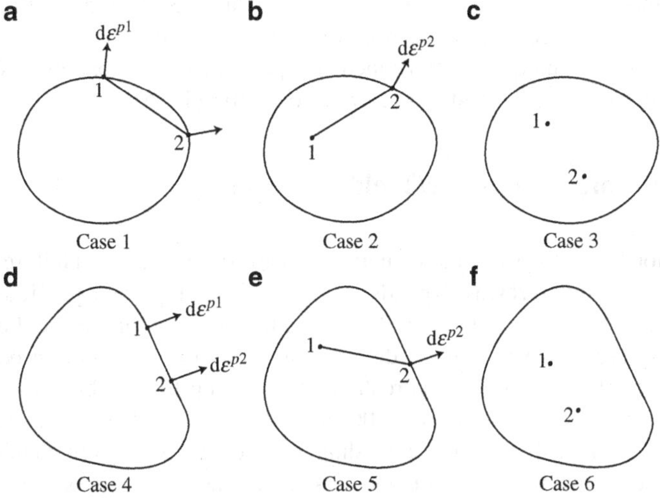

Fig. 9.24 Structure at flow: different possible cases of the two solutions

- *Case 2*: Point 2 is on the surface and point 1 is inside; hence $d\boldsymbol{\varepsilon}^{p1} = 0$ and $d\boldsymbol{\varepsilon}^{p2} \neq 0$.

$$I = -\left(\boldsymbol{\sigma}^1 - \boldsymbol{\sigma}^2\right)d\boldsymbol{\varepsilon}^{p2} \geq 0 \tag{9.55b}$$

  which along with (9.54b) implies $I = 0 \Rightarrow \boldsymbol{\sigma}^1 = \boldsymbol{\sigma}^2$. It is easy to verify that the outcome remains unchanged when points 1 and 2 are switched. The stresses can thus be uniquely determined.
- *Case 3*: Both points are inside the surface. $d\boldsymbol{\varepsilon}^{p1} = d\boldsymbol{\varepsilon}^{p2} = 0$ and (9.54b) implies that $\boldsymbol{\sigma}^1$ and $\boldsymbol{\sigma}^2$ can assume any arbitrary values. The stresses cannot be uniquely determined.
- *Case 4*: Both points 1 and 2 are on the flat part of the surface.

$$\left(\boldsymbol{\sigma}^1 - \boldsymbol{\sigma}^2\right)d\boldsymbol{\varepsilon}^{p1} = 0; \left(\boldsymbol{\sigma}^1 - \boldsymbol{\sigma}^2\right)d\boldsymbol{\varepsilon}^{p2} = 0 \Rightarrow I = 0 \tag{9.55c}$$

  That is, for any arbitrary values of $\boldsymbol{\sigma}^1$ and $\boldsymbol{\sigma}^2$, we have $I = 0$. The stresses cannot be uniquely determined.
- *Case 5*: This case is similar to Case 2; the stresses can be uniquely determined.
- *Case 6:* This case is similar to Case 3; the stresses cannot be uniquely determined.

In conclusion, when the yield surface is convex and the constitutive model employs an associated flow rule, there will be *some regions where the stress can be uniquely determined* (Cases 1, 2, and 5) and there will be *other regions where the stress cannot be uniquely determined* (Cases 3, 4, and 6). When the stresses cannot be uniquely determined, the strains and displacements cannot be determined uniquely either. In these cases (Cases 3, 4, and 6), the stresses, strains and displacements depend on the previous history. In Cases 1, 2, and 5, the stresses are uniquely determined independent of the loading history. For further discussion, the reader is referred to Martin (1975).

9.4 Examples of Failure/Yield Criteria

It is common to use the same mathematical shapes for the yield and failure surfaces. The shape of yield surfaces depends on the material type and previous loading history. Among the two, the material type has the dominant influence. For metals, Bridgeman (1952) experimentally demonstrated that the stress–strain behavior is unaffected by the presence of hydrostatic pressure, up to very high values of the pressure (as high as 367,500 psi or 24,000 atm). On the other hand, Coulomb (1773) presented his famous friction law according to which the shear strength of frictional materials (soils, rocks, etc.) on a plane depends on the normal stress on that plane. Since the shear strength is integral part of the stress–strain behavior of frictional materials, the stress–strain behavior of these materials also depends on the hydrostatic pressure. The shape and location of the yield surface in the stress space changes with loading (Sect. 9.2.1). Appropriate hardening rules must be established in constitutive models to account for this. In this section, we present a number of initial shapes that have been proposed in the past for yield surface of materials.

9.4.1 Von Mises Criterion

While it is generally called von Mises (1913) yield criterion, there is evidence that it has been suggested by Maxwell in 1856, and published later by Huber (1904). For this reason, this criterion is sometimes referred to as the Maxwell–Huber–von Mises criterion.

The Mises yield criterion assumes that yielding occurs when the value of the deviatoric stress invariant exceeds a certain value that is independent of the hydrostatic pressure. Mathematically, the yield surface can be written as

$$\phi(\mathbf{s}, k) = s_{ij}s_{ij} - \frac{2}{3}k^2 = 0 \tag{9.56}$$

or

$$J - \frac{1}{\sqrt{3}}k = 0 \tag{9.57}$$

where k is defined as a positive quantity. J is the deviatoric stress invariant (Chap. 2). k will serve as the PIV in the constitutive model that employs this yield criterion.

For equibiaxial (or triaxial) loading ($\sigma_1 \neq \sigma_2 = \sigma_3$), (9.57) becomes

$$\frac{1}{\sqrt{3}}|q| - \frac{1}{\sqrt{3}}k = 0 \quad (9.58a)$$

or

$$|q| = k \quad (9.58b)$$

where $q = \sigma_1 - \sigma_3$ (Chap. 2).

For the case of uniaxial tension or compression loading where $\sigma_3 = 0$,

$$|\sigma_1| = |\sigma_y| = k \quad (9.59a)$$

where σ_y is the yield stress in uniaxial tension or compression. The parameter k is, therefore, the yield stress in uniaxial tension or compression (hence the use of the factor 2/3 in (9.56)).

For equibiaxial (or triaxial) loading, noting that $q = \sigma_1 - \sigma_3 = 2\tau_{\max}$, where $\tau_{\max}$ is the maximum shear stress, we have

$$2|\tau_{\max}| = k \quad (9.59b)$$

Let τ_y be the yield stress in simple shear test. Then it can be verified that

$$\tau_y - \frac{1}{\sqrt{3}}k = 0 \Rightarrow k = \sqrt{3}\tau_y \quad (9.59c)$$

Also recall from Chap. 2 that the octahedral shear stress is given by

$$\tau_{\text{oct}} = \frac{1}{3}\left[(\sigma_{11} - \sigma_{22})^2 + (\sigma_{22} - \sigma_{33})^2 + (\sigma_{33} - \sigma_{11})^2 + 6(\sigma_{12}^2 + \sigma_{23}^2 + \sigma_{13}^2)\right]^{1/2} \quad (9.60)$$

Noting that $\tau_{\text{oct}} = \sqrt{(2/3)}J$, the Mises criterion can also be written in terms of the components of the stress tensor in the following form:

$$\phi(\sigma_{ij}, k) = \frac{3}{2}\tau_{\text{oct}}^2 - \frac{1}{3}k^2 = 0$$

$$\phi(\sigma_{ij}, k) = \frac{1}{6}\left[(\sigma_{11} - \sigma_{22})^2 + (\sigma_{22} - \sigma_{33})^2 + (\sigma_{33} - \sigma_{11})^2 + 6(\sigma_{12}^2 + \sigma_{23}^2 + \sigma_{13}^2)\right] - \frac{1}{3}k^2 = 0 \quad (9.61)$$

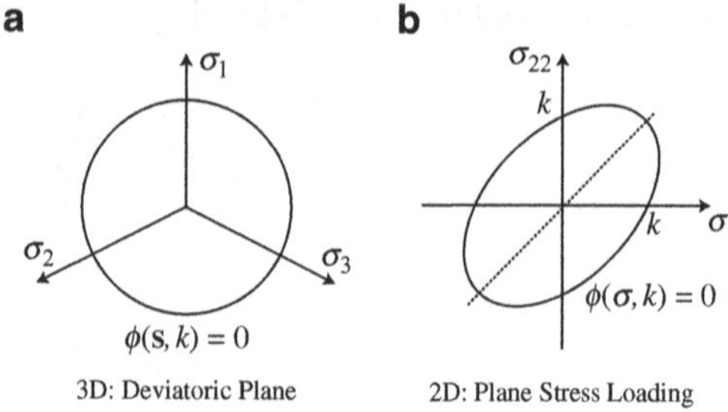

Fig. 9.25 Shape of von Mises yield surface in two- and three-dimensional loading cases

It follows from (9.57) that the shape of the yield surface is a cylinder in the three-dimensional stress space with its axis coinciding with the space diagonal. The projection of the cylinder is a circle in the octahedral plane, as shown in Fig. 9.25a.

Consider a plane stress loading in the principal directions: $\sigma_{12} = \sigma_{23} = \sigma_{13} = \sigma_{33} = 0$. Equation (9.61) simplifies to

$$2\sigma_{11}^2 + 2\sigma_{22}^2 - 2\sigma_{11}\sigma_{22} - 2k^2 = 0$$

or

$$\left(\frac{\sigma_{11}}{k}\right)^2 - \left(\frac{\sigma_{11}}{k}\right)\left(\frac{\sigma_{22}}{k}\right) + \left(\frac{\sigma_{22}}{k}\right)^2 = 1 \tag{9.62}$$

which is equation of an ellipse in the $\sigma_{11} - \sigma_{22}$ space with an inclined axes as shown in Fig. 9.25b.

To model the kinematic hardening behavior, Prager (1955) modified (9.56) as follows:

$$\phi(\sigma_{ij}, \alpha_{ij}, k) = \left(s_{ij} - \alpha_{ij}\right)\left(s_{ij} - \alpha_{ij}\right) - \frac{2}{3}k^2 = 0 \tag{9.63}$$

where α_{ij} is called the *back stress*. When α_{ij} is a null tensor, (9.63) degenerates to (9.56), with the center of the circle shown in Fig. 9.25a located at the origin in the octahedral plane; otherwise, the center is displaced from the origin by α_{ij}, as shown in Fig. 9.26. The back stress α_{ij} becomes the kinematic hardening variable (one of the PIVs). A suitable hardening rule, as discussed in Sect. 9.5, is required to model its evolution during loading.

9.4.2 *Tresca Criterion*

Following the work of Coulomb (1773) in soil mechanics, Tresca (1864) proposed an yield criterion for metals, which is now known as Tresca yield criterion. Tresca yield surface is the oldest yield criterion proposed for metals.

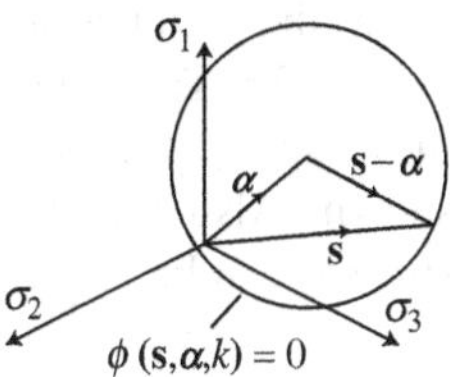

Fig. 9.26 Von Mises yield surface with back stress

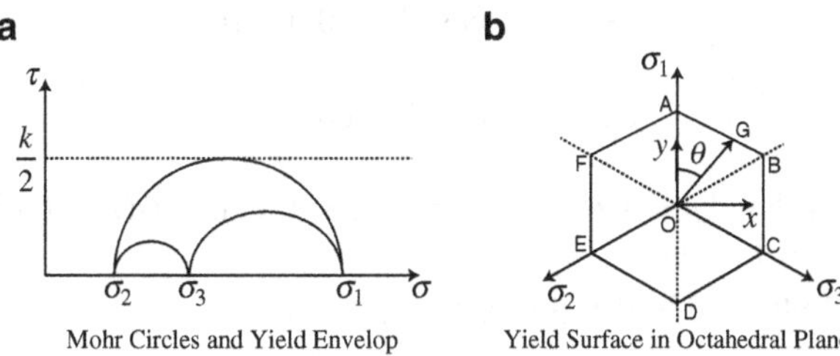

Fig. 9.27 Tresca yield surface in three-dimensional principal stress space

In the Tresca yield criterion, the material is considered to experience yielding when the maximum shear stress reaches a limit as

$$\tau_{\max} = \frac{k}{2} \tag{9.64}$$

where $k/2$ is the yield stress based on shear stress.

In the principal stress space, the Tresca yield criterion can be written as

$$\tau_{\max} = \frac{1}{2}\max[|\sigma_1 - \sigma_2|, |\sigma_2 - \sigma_3|, |\sigma_1 - \sigma_3|] = \frac{k}{2} \tag{9.65}$$

Consider the ordering of the principal stresses based on their magnitudes as

$$\sigma_2 \leq \sigma_3 \leq \sigma_1 \tag{9.66}$$

σ_1, σ_2, and σ_3 are, respectively, the major, minor, and intermediate principal stresses. The maximum shear stress (Fig. 9.27a) then is

$$\tau_{\max} = (\sigma_1 - \sigma_2)/2 \tag{9.67a}$$

The yield criterion (9.65) becomes

$$\sigma_1 - \sigma_2 = k \tag{9.67b}$$

As we will show below, the yield surface in the principal stress space is a regular hexagonal cylinder with its axis coinciding with the space diagonal. The projection of this cylinder has six linear sides as shown in Fig. 9.27b.

In the octahedral plane, the sector representing (9.66) is AOB shown in Fig. 9.27b. The side of the yield surface in this sector is given by (9.67b). To see the shape of this side, let us introduce a rectangular $x - y$ coordinate system as shown. Also recall the polar coordinates introduced in Chap. 2. The principal stresses can be expressed in terms of θ (Fig. 9.27b), r_0 and p as

$$\sigma_1 = r_0 \cos\theta + p \tag{9.68a}$$

$$\sigma_2 = -r_0 \cos(\theta - 60°) + p \tag{9.68b}$$

$$\sigma_3 = -r_0 \cos(\theta + 60°) + p \tag{9.68c}$$

where $r_0 = 2J/\sqrt{3}$ and $p = (\sigma_1 + \sigma_2 + \sigma_3)/3$. Noting that $OG = J\sqrt{2}$ (Chap. 2), it follows that $OG = r_0\sqrt{3/2}$.

Expressing x and y in terms of r_0 and θ:

$$x = OG \sin\theta = r_0\sqrt{3/2}\sin\theta$$

$$y = OG \cos\theta = r_0\sqrt{3/2}\cos\theta$$

Then

$$\sigma_1 = \sqrt{\frac{2}{3}}y + p$$

$$\sigma_2 = -\frac{r_0}{2}\cos\theta - \frac{r_0\sqrt{3}}{2}\sin\theta + p = -\frac{1}{2}\sqrt{\frac{2}{3}}y - \frac{\sqrt{3}}{2}\sqrt{\frac{2}{3}}x + p = -\frac{1}{\sqrt{6}}y - \frac{\sqrt{3}}{\sqrt{6}}x + p$$

Equation (9.67b) becomes

$$\frac{3}{\sqrt{6}}y + \frac{\sqrt{3}}{\sqrt{6}}x = k \Rightarrow y = -\frac{1}{\sqrt{3}}x + \sqrt{\frac{2}{3}}k \tag{9.69}$$

It is seen that the yield surface is a line in this sector (i.e., sector AOB). When $x = 0$, $y = k\sqrt{2/3}$; hence

$$OA = \sqrt{\frac{2}{3}}k$$

On line OB, denoting $OB = \ell$, $x = \ell\cos 30° = \ell\sqrt{3}/2$; $y = \ell \sin 30° = \ell/2$. Substituting these into (9.69)

$$\frac{\ell}{2} = -\frac{1}{\sqrt{3}}\frac{\ell\sqrt{3}}{2} + \sqrt{\frac{2}{3}}k \quad \Rightarrow \quad \ell = \sqrt{\frac{2}{3}}k \Rightarrow OB = \sqrt{\frac{2}{3}}k \tag{9.70}$$

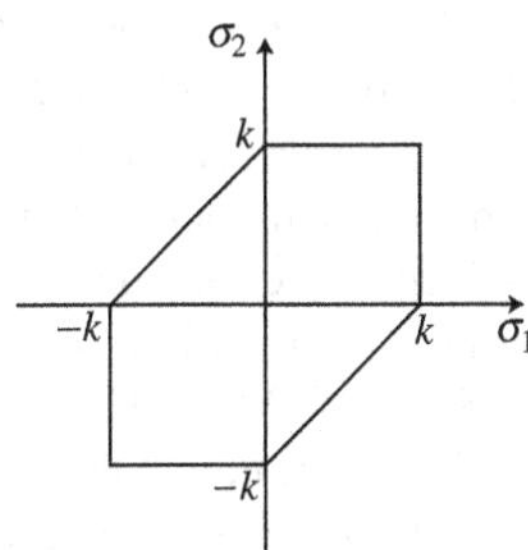

Fig. 9.28 Tresca yield surface for plane stress case

Hence AB is parallel to the σ_3-axis. The shapes and positions of the other sides can be similarly established. The net result is that the shape of the Tresca yield surface is a *regular hexagon* in the octahedral plane as shown in Fig. 9.27b.

Also note that ℓ (9.70) is independent of p. Hence the yield surface in the stress space is a *regular hexagonal cylinder* whose axis is the space diagonal.

Consider a plane stress loading in the principal stress space with $\sigma_3 = 0$. The yield criterion in (9.65) simplifies to

$$\sigma_1 - \sigma_2 = \pm k \tag{9.71a}$$

$$\sigma_2 = \pm k \tag{9.71b}$$

$$\sigma_1 = \pm k \tag{9.71c}$$

The yield surface bounded by these lines in the $\sigma_1 - \sigma_2$ space is an irregular hexagon as shown in Fig. 9.28.

If a specimen is loaded in uniaxial tension with $\sigma_2 = \sigma_3 = 0$, (9.71a) and (9.71c) become

$$\sigma_1 = k = \sigma_y \tag{9.72}$$

Hence the constant k appearing in the Tresca yield criterion (9.65) is also equal to the uniaxial yield (normal) stress (as the constant k appearing in the von Mises criterion, (9.56)). Recall from (9.64), k is twice the yield shear stress; a simple shear test may be performed to find this.

9.4.3 *Mohr–Coulomb Criterion*

The yield and failure surfaces are pressure-dependent for geologic materials such as soil, rock and concrete. The original Coulomb's friction law (Coulomb 1773) is a failure criterion for frictional materials such as sand. The Mohr–Coulomb criterion is an extension of Coulomb's law to frictional–cohesive materials (clays, rocks, etc.).

The Mohr–Coulomb criterion is actually a failure criterion. When used in an elastic, perfectly-plastic constitutive model, it serves both as the yield surface and the failure surface. However, hardening/softening behaviors may be incorporated into the constitutive model, and in such a case, the surface, with the parameters appropriately defined, serves as both the yield (i.e. loading) and failure surfaces. The terms "yield surface" and "failure surface" are used interchangeably in this section.

Since the Mohr–Coulomb criterion is used for pressure-dependent materials, and that it is common in the treatment of such materials to take the *compressive stresses to be positive*, we will adopt this sign convention in this section.

Consider the ordering of the principal stresses based on their magnitudes as

$$\sigma_2 \leq \sigma_3 \leq \sigma_1 \tag{9.73}$$

the Mohr circles and the failure envelop in the $\sigma - \tau$ space are shown in Fig. 9.29a. (Recall that the Tresca yield surface is parallel to the σ-axis in the $\sigma - \tau$ space, Fig. 9.27a.)

According to the Mohr–Coulomb criterion, considering a failure plane, the relationship between the shear strength τ_p and normal stress σ_n on that plane is

$$\tau_p = \sigma_n \tan\phi + c \tag{9.74}$$

where ϕ and c are the friction angle and cohesion of the material, respectively. Referring to Fig. 9.29b, writing the radius OA in terms of the stresses, we have

$$\frac{\sigma_1 - \sigma_2}{2} = \left(\frac{\sigma_1 + \sigma_2}{2}\right)\sin\phi + c\cos\phi \tag{9.75}$$

Rearranging this, it can be easily shown that

$$\sigma_2 = k_A\sigma_1 - 2c\sqrt{k_A} \tag{9.76a}$$

where

$$k_A = \frac{1 - \sin\phi}{1 + \sin\phi} \tag{9.76b}$$

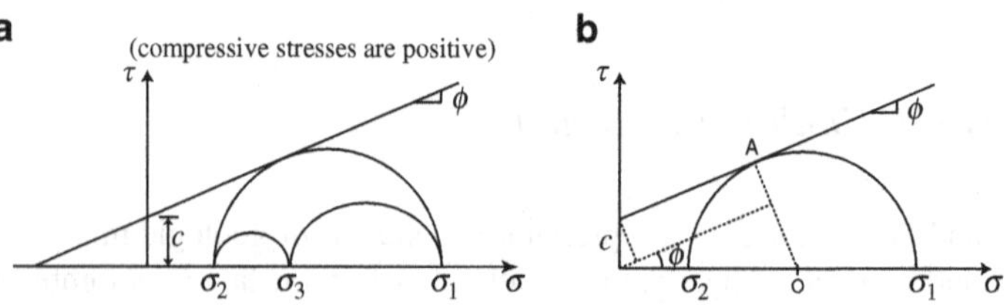

Fig. 9.29 Mohr circles and the Mohr–Coulomb failure criterion

By following the approach as in the section on Tresca yield surface (Sect. 9.4.2), let us introduce a rectangular $x-y$ coordinate system as shown in Fig. 9.27b. σ_1 and σ_2 are expressed in terms of x and y as

$$\sigma_1 = \sqrt{\frac{2}{3}}y + p$$

$$\sigma_2 = -\frac{1}{\sqrt{6}}y - \frac{\sqrt{3}}{\sqrt{6}}x + p$$

where p is the mean normal pressure. Substituting these into (9.76a),

$$-\frac{1}{\sqrt{6}}y - \frac{\sqrt{3}}{\sqrt{6}}x + p = k_A\sqrt{\frac{2}{3}}y + k_A p - 2c\sqrt{k_A}$$

$$y = -\frac{\sqrt{3}}{(1+2k_A)}x + \frac{\sqrt{6}(1-k_A)p + 2\sqrt{6}c\sqrt{k_A}}{1+2k_A} \tag{9.77}$$

At a given value of p, the yield surface in this sector (i.e., sector AOB), therefore, is a straight line as shown in Fig. 9.30. At $x = 0$,

$$y = \frac{\sqrt{6}(1-k_A)p + 2\sqrt{6}c\sqrt{k_A}}{1+2k_A}$$

Hence, referring to Fig. 9.30,

$$OA = \frac{\sqrt{6}(1-k_A)p + 2\sqrt{6}c\sqrt{k_A}}{1+2k_A} \tag{9.78}$$

On line OG, denoting $OG = \ell$, $x = \ell\cos 30° = \ell\sqrt{3}/2$; $y = \ell\sin 30° = \ell/2$. Substituting these into (9.77)

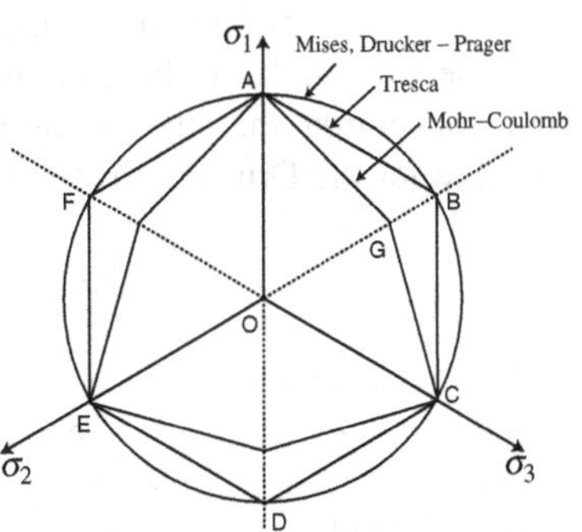

Fig. 9.30 Comparison of yield surfaces

$$\frac{\ell}{2} = -\frac{\sqrt{3}}{(1+2k_A)}\frac{\ell\sqrt{3}}{2} + \frac{\sqrt{6}(1-k_A)p + 2\sqrt{6}c\sqrt{k_A}}{1+2k_A}$$

$$\ell\frac{2+k_A}{1+2k_A} = \frac{\sqrt{6}(1-k_A)p + 2\sqrt{6}c\sqrt{k_A}}{1+2k_A}$$

$$OG = \ell = \frac{\sqrt{6}(1-k_A)p + 2\sqrt{6}c\sqrt{k_A}}{2+k_A} \tag{9.79}$$

It is seen that $OA \neq OG$ since the denominators in (9.78) and (9.79) are unequal. To quantify the difference, let us define a length ratio as

$$n = \frac{OG}{OA} = \frac{1+2k_A}{2+k_A} = \frac{1+2\frac{1-\sin\phi}{1+\sin\phi}}{2+\frac{1-\sin\phi}{1+\sin\phi}} = \frac{3-\sin\phi}{3+\sin\phi} \tag{9.80}$$

It is thus seen that the size of the yield surface in the negative σ_2-axis is smaller than that in the positive σ_1-axis. For example, for $\phi = 30°$, $n = 0.7142$. It may be shown that the shapes in the other sectors are as shown in Fig. 9.30.

Noting that OA and OG are functions of p, it is easy to see that the yield surface is an irregular hexagonal cone in the stress space whose axis is the space diagonal. When $c = 0$, its apex is at the origin, since $OA = OG = 0$ ((9.78) and (9.79)).

It is also easy to verify that when $\phi = 0$, the above equations become identical to those for the Tresca yield surface: $\sin\phi = 0$, $k_A = 1$ and $OA = OG = 2\sqrt{2/3}c$. Noting from Figs. 9.27a and 9.29a that $2c = k$, $OA = OG = \sqrt{2/3}k$, which is the same as that given by (9.70).

9.4.4 Drucker–Prager Criterion

The Drucker–Prager failure/yield criterion (Drucker and Prager 1952) is a generalization of the Mohr–Coulomb criterion. The criterion is useful for representing the yielding behavior of pressure sensitive materials such as rock, soil and concrete. As in the section on the Mohr–Coulomb failure criterion (Sect. 9.4.3), let us adopt the sign convention that the compressive stresses are positive. The mathematical equation for the Drucker–Prager failure criterion is

$$\phi(\sigma_{ij}, \alpha, k) = J - \alpha I - k = 0 \tag{9.81}$$

This is understood to be a straight line in the $I - J$ space with the slope α and intercept k, and a circle in the octahedral plane for a constant value of I. The surface is a cone in the stress space whose axis is the space diagonal. The shape is shown in the octahedral plane and an arbitrary meridional plane in Fig. 9.31a and b,

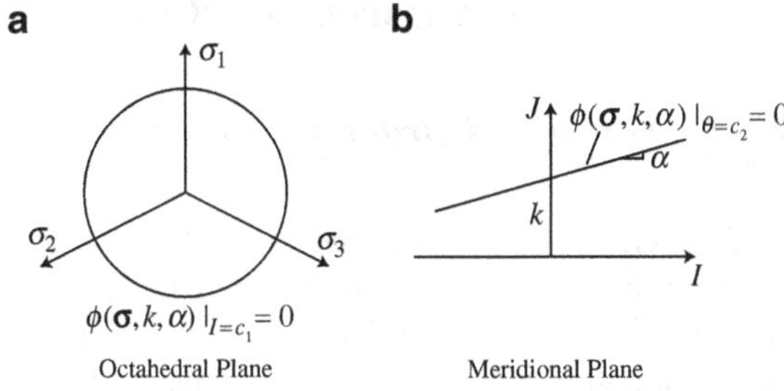

Fig. 9.31 Shape of Drucker–Prager yield surface

respectively. α and k are functions of the friction and cohesion of the material, respectively. These parameters can be related to the actual friction angle and cohesion of the material as described below.

The Mohr–Coulomb failure criterion is given by (9.74). Considering a equibiaxial loading with $\sigma_1 > \sigma_2 = \sigma_3$, the relationship between σ_1 and σ_2, is given by (9.76a). Let us now write the invariants in terms of the stresses for this equibiaxial case as

$$I = \sigma_1 + 2\sigma_2 = \sigma_1 + 2\sigma_1 k_A - 4c\sqrt{k_A} \tag{9.82a}$$

$$J = \frac{1}{\sqrt{3}}(\sigma_1 - \sigma_2) = \frac{1}{\sqrt{3}}\left[\sigma_1 - k_A\sigma_1 + 2c\sqrt{k_A}\right] \tag{9.82b}$$

Let us find σ_1 from (9.82a) as

$$\sigma_1 = \frac{I + 4c\sqrt{k_A}}{1 + 2k_A}$$

and substitute into (9.82b) to obtain

$$J = \frac{1}{\sqrt{3}}\left[\frac{I + 4c\sqrt{k_A}}{1 + 2k_A}\right](1 - k_A) + \frac{2c\sqrt{k_A}}{\sqrt{3}} \tag{9.83}$$

Rearranging this, it follows

$$\alpha = \frac{2\sin\phi}{\sqrt{3}(3 - \sin\phi)} \quad \text{and} \quad k = \frac{6c\cos\phi}{\sqrt{3}(3 - \sin\phi)} \tag{9.84}$$

Compared to the von Mises yield criterion, the Drucker–Prager criterion embodies just one additional feature: the (linear) pressure dependency. While the Mohr–Coulomb yield surface has corners in the octahedral plane (Fig. 9.30), the Drucker–Prager surface is smooth, rendering computer implementation much easier.

9.5 Examples of Hardening Rules

9.5.1 *Isotropic Hardening Rules*

As discussed in Sect. 9.2.1 (see Fig. 9.9), the hardening behavior can be classified into isotropic, kinematic, and combined hardening behaviors. Isotropic hardening is one where the surface expands or contracts uniformly in all directions in the stress space without undergoing a change in its shape and center. The von Mises criteria (9.56) can be written as

$$\phi(\sigma_{ij}, k) = \bar{\phi}(\sigma_{ij}) - \frac{2}{3}k^2 = 0 \tag{9.85}$$

where the yield stress k dictates the size of the surface. In the past, for metals, the following two plastic strain measures have been used to model the evolution of k with loading ((9.8a) and (9.8b)):

$$\xi^p = \int \left[\frac{1}{2}\mathrm{d}e^p_{ij}\mathrm{d}e^p_{ij}\right]^{1/2} \tag{9.86a}$$

and

$$W^p = \int \sigma_{ij}\mathrm{d}\varepsilon^p_{ij} \tag{9.86b}$$

where ξ^p is equivalent or effective deviatoric plastic strain (cumulative length of the trajectory in the deviatoric plastic strain space) and W^p is the cumulative plastic work. Note that different paths may give the same values of ξ^p or W^p. Hence these parameters do not uniquely relate to the stress or strain path, but are uniquely related to the plastic deformation experienced by the material.

The size parameter k is expressed as a function of either ξ^por W^p; i.e.,

$$k = k(\xi^p) \tag{9.87a}$$

or

$$k = k(W^p) \tag{9.87b}$$

Equation (9.87a) is known as the *strain hardening law* and (9.87b) is known as the *work hardening law*.

By differentiating (9.87a) and (9.87b):

$$\dot{k} = \frac{dk}{d\xi^p}\dot{\xi}^p = g(\xi^p)\dot{\xi}^p \tag{9.88a}$$

$$\dot{k} = \frac{dk}{dW^p}\dot{W}^p = h(W^p)\dot{W}^p \tag{9.88b}$$

where g and h are the slopes of the $k-\xi^p$ and $k-W^p$ relations respectively, as shown in Fig. 9.32. Figure 9.32a shows nonlinear hardening behaviors and Fig. 9.32b shows linear hardening behaviors. k_0 is the initial yield stress.

We will consider the strain hardening law for further discussion here. One may adopt one of the many relations proposed in the past, or propose a new one based on the observed behavior of the specific material at hand. Consider, for instance, a hyperbolic relation as follows:

$$k = k_0 + \frac{m(k_f - k_0)\xi^p}{(k_f - k_0) + m\xi^p} \tag{9.89}$$

Taking the rate of (9.89), we have

$$\dot{k} = g\dot{\xi}^p \tag{9.90a}$$

$$g = \frac{m(k_f - k_0)^2}{\left[(k_f - k_0) + m\xi^p\right]^2} \tag{9.90b}$$

It is seen that at $\xi^p = 0$, $g = m$. As $\xi^p \to \infty$, it follows that $k \to \infty$ and $g \to 0$. This function will be used in a model in subsequent chapters on integration of constitutive equations.

Finally, note that other yield functions such as the Drucker–Prager yield functions may be similarly treated. For instance, in the Drucker–Prager model, when the parameters k and α change, the surface remains a circle centered around the origin in the octahedral plane (hence undergoing an isotropic hardening in this

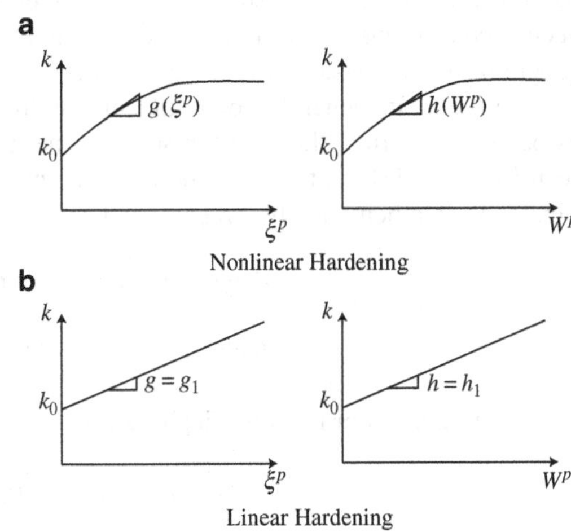

Fig. 9.32 Schematic of linear and nonlinear isotropic hardening functions

plane). Such a hardening behavior is not possible to be identified in the meridional plane.

Also note that in (9.86a) and (9.86b), variations are possible. For instance, one could use total plastic strain increments, instead of deviatoric strain increments, in the integral in (9.86a). Both measures are identical for models employing yield surfaces that are independent of the hydrostatic pressure, because the volumetric plastic strain is predicted to be zero by such models. They are, however, different for models employing pressure-dependent yield surfaces such as the Drucker–Prager yield surface, where the volumetric plastic strain is predicted to be non zero.

9.5.2 Kinematic Hardening Rules

While in isotropic hardening the size of the yield surface changes with its center and shape remaining unchanged, in pure kinematic hardening the center changes with its size and shape remaining unchanged. Generally, kinematic hardening is modeled by introducing a tensorial PIV, known as the back stress, and defining a suitable evolution rule for it. In pressure-dependent material like soil, this back stress may have both the volumetric and deviatoric components. In pressure-independent materials like metals, since the yield surface is independent of the hydrostatic stress, the back stress is a pure deviatoric tensor.

A common form of modified von Mises yield criterion with back stress is given by (9.63), where α_{ij} is the back stress that varies with plastic strain and k is a scalar constant that remains unchanged during plastic behavior (in pure kinematic hardening). Among the classical kinematic hardening rules, Prager's (1955) rule and Ziegler's (Shield and Ziegler 1958; Ziegler 1959) rule have become widely used. While Prager assumes that the direction of the rate of change of α_{ij} coincides with the direction of the plastic strain rate $\dot{\boldsymbol{\varepsilon}}^p$, Ziegler assumes that it coincides with the vector connecting α_{ij} to σ_{ij}. These directions are schematically shown in a generalized stress space in Fig. 9.33a, where it is seen that the two directions are not the same in general. For a model employing the von Mises surface and associated flow rule, they are the same in the general three-dimensional case, as seen in Fig. 9.33b, but not for the two-dimensional plane stress case, as seen in Fig. 9.33c. Mathematically, Prager's and Ziegler's rules can be written as

$$\text{Prager's rule: } \dot{\alpha}_{ij} = c_1(\boldsymbol{\sigma}, \boldsymbol{\zeta})\dot{\varepsilon}^p_{ij} \tag{9.91a}$$

$$\text{Ziegler's rule : } \dot{\alpha}_{ij} = \dot{\mu}(\sigma_{ij} - \alpha_{ij}) \tag{9.91b}$$

One suitable form for $\dot{\mu}$ in Ziegler's rule is

$$\dot{\mu} = c_2(\boldsymbol{\sigma}, \boldsymbol{\zeta})\dot{\xi}^p \tag{9.91c}$$

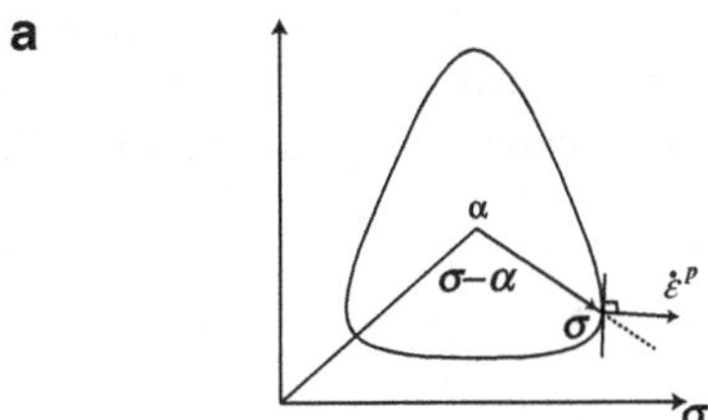

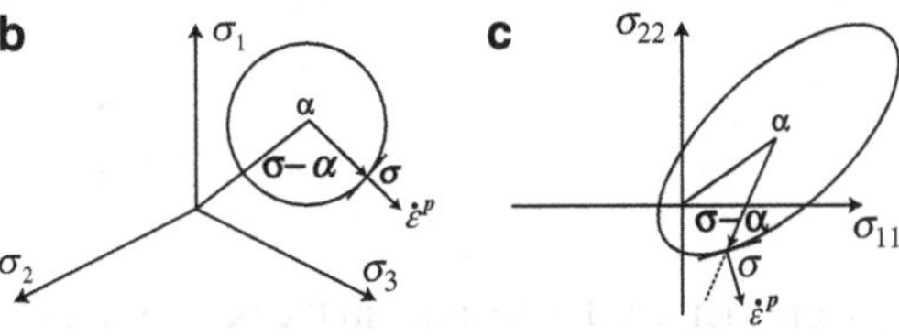

Fig. 9.33 Schematic of directions used in Prager's and Ziegler's kinematic hardening rules

where $\dot{\xi}^p$ is the deviatoric invariant of the rate of plastic strain. When c_1 in Prager's rule or c_2 in Ziegler's rule are independent of the state variables $(\boldsymbol{\sigma}, \boldsymbol{\zeta})$, the hardening rule is linear; otherwise it is nonlinear.

Another widely-used kinematic hardening rule is the nonlinear rule originally proposed by Armstrong and Frederick (1966) and further developed by Chaboche (1977, 1986). Let us consider the von Mises yield function given by (9.63) along with the associated flow rule. In the Armstrong–Frederick kinematic hardening rule, the evolution of the back stress is modeled using the following equation:

$$\dot{\alpha}_{ij} = \frac{2}{3} c \dot{\varepsilon}^p_{ij} - \frac{2}{\sqrt{3}} \gamma \alpha_{ij} \dot{\xi}^p \tag{9.92a}$$

where ξ^p is the cumulative deviatoric plastic strain (9.86a). c and γ are model parameters. Hence the direction of $\dot{\alpha}_{ij}$ is obtained as a combination of the directions of the yield surface normal and α_{ij}. Expanding (9.92a) with the aid of the plastic constitutive equation (9.20f), it can be shown (Problem 9.18) that

$$\dot{\alpha}_{ij} = \dot{\lambda} \bar{s}_{ij}; \quad \bar{s}_{ij} = \frac{4}{3} c b_{ij} - \frac{2}{\sqrt{3}} \gamma \alpha_{ij} \bar{r}^d \tag{9.92b}$$

where

$$b_{ij} = s_{ij} - \alpha_{ij}$$

and

$$\bar{r}^d = \left(\frac{1}{2} r_{ij} r_{ij} \right)^{1/2}$$

(Note the use of a bar over the hardening function $\bar{s}_{ij}$ to avoid confusing this with the deviatoric stress tensor s_{ij}.) Using the consistency condition, an expression for the plastic modulus can be obtained (Problem 9.18)

$$K_p = \frac{8}{9}k(2ck - 3\gamma ka); \quad a = (s_{k\ell} - \alpha_{k\ell})\alpha_{k\ell} \tag{9.92c}$$

Failure occurs when $K_p = 0$, which also corresponds to (9.20i) $\bar{\mathbf{s}} = 0$ and hence to $\dot{\boldsymbol{\alpha}} = 0$. Setting $K_p = 0$, we have

$$a = \frac{2}{3}\frac{c}{\gamma}$$

Consider a radial loading in the octahedral plane. In this case, the tensors $\mathbf{b}$ and $\boldsymbol{\alpha}$ have the same direction. Hence, we have

$$a = (\mathbf{s} - \boldsymbol{\alpha})\boldsymbol{\alpha} = \frac{|\mathbf{s} - \boldsymbol{\alpha}|}{|\boldsymbol{\alpha}|}\boldsymbol{\alpha}.\boldsymbol{\alpha} = |\mathbf{s} - \boldsymbol{\alpha}||\boldsymbol{\alpha}| = \sqrt{\frac{2}{3}}k|\boldsymbol{\alpha}|$$

Combining the above two equations, we have an expression for the failure value of the magnitude of $\boldsymbol{\alpha}$ as

$$|\boldsymbol{\alpha}_f| = \sqrt{\frac{2}{3}}\frac{c}{\gamma} \tag{9.92d}$$

The failure value of the deviatoric stress invariant J_f is then obtained as

$$|\mathbf{s}_f - \boldsymbol{\alpha}_f| = \sqrt{\frac{2}{3}}k$$

$$J_f = \left(\frac{1}{2}\mathbf{s}_f.\mathbf{s}_f\right)^{1/2} = \sqrt{\frac{1}{3}}\left(k + \frac{c}{\gamma}\right) \tag{9.92e}$$

For the case of uniaxial tension (or compression) loading, we have

$$\sigma_f = J_f\sqrt{3} = k + \frac{c}{\gamma} \tag{9.92f}$$

Hence the model parameters c and γ not only control the hardening behavior, but also the failure stress.

9.6 Examples of Classical Constitutive Equations

9.6.1 Levy–Mises Equations

In 1870, Saint–Venant proposed a two-dimensional plasticity theory where he assumed that the work hardening is negligible and that the principal axes of strain increment coincide with the principal axes of stresses. The elastic strains were neglected, and hence total strains were equal to the plastic strains. The theory was later generalized to the three-dimensional loading case independently by Levy (1870, 1871) and von Mises (1913). The theory is now called the Levy–Mises theory. Noting that the principal directions of stresses are the same as the principal directions of deviatoric stresses, the Levy–Mises stress–strain rate relation can be written as

$$\dot{\varepsilon}_{ij} = \dot{\lambda} s_{ij} \tag{9.93}$$

It follows from (9.93) that

$$\dot{\varepsilon}_{kk} = \dot{\lambda} s_{kk} = 0$$

The parameter $\dot{\lambda}$ can be determined in two alternate ways as follows. First, as suggested by Mises, let's use the von Mises yield criterion. Writing (9.93) as

$$\dot{\varepsilon}_{ij}\dot{\varepsilon}_{ij} = \dot{\lambda}^2 s_{ij} s_{ij}$$

and using (9.56), we have

$$\dot{\varepsilon}_{ij}\dot{\varepsilon}_{ij} = \frac{2}{3}\dot{\lambda}^2 k^2$$

or

$$\dot{\lambda} = \sqrt{3}\frac{\dot{J}_{\varepsilon}}{k} \tag{9.94}$$

where

$$\dot{J}_{\varepsilon} = \left(\frac{1}{2}\dot{\varepsilon}_{ij}\dot{\varepsilon}_{ij}\right)^{1/2} \tag{9.95}$$

is a scalar equivalent strain rate. Alternatively, (9.93) can be written as

$$\dot{\varepsilon}_{ij} s_{ij} = \dot{\lambda} s_{ij} s_{ij}$$

which, with the aid of (9.56), becomes

$$\dot{\varepsilon}_{ij}s_{ij} = \frac{2}{3}\dot{\lambda}k^2$$

or

$$\dot{\lambda} = \frac{3}{2}\frac{\dot{\varepsilon}_{ij}s_{ij}}{k^2} \tag{9.96}$$

Note that the Levy–Mises equations model rigid plastic behavior (because elastic strains and work hardening are neglected). They cannot be compared with equations associated with hardening plasticity. For example, when one compares (9.93) with (9.19), the flow direction r_{ij} in (9.19) may be identified with s_{ij} in (9.94), but $<L>$ in (9.19) and $\dot{\lambda}$ in (9.93) are not the same. When work hardening is neglected, K_p is zero. In addition, when work hardening is absent, the stress increment becomes zero, and hence the dot product in (9.19c) becomes zero. Hence, the loading index in (9.19c) becomes undefined. In other words, in the case of perfect plasticity, the loading index is not defined. On the other hand, $\dot{\lambda}$ in (9.94) and (9.96) are expressed in terms of strain rate, and hence if the strain rate is specified, $\dot{\lambda}$ is defined.

When strain increments are specified, (9.94) or (9.96) can be used to determine $\dot{\lambda}$, which can then be substituted in (9.93) to find s_{ij}. On the other hand, if s_{ij} is given, neither (9.93) nor any of the others can be used to find $\dot{\varepsilon}_{ij}$. This is consistent with our understanding of the behavior of perfectly-plastic materials: at yield, the strain can take any value.

9.6.2 *Prandtl–Reuss Equations*

The Levy–Mises equations ignore elastic strains. Prandtl (1924) and Reuss (1930) proposed modified equations to include elastic strains in the theory. The equations are:

$$\dot{e}^e_{ij} = \frac{1}{2G}\dot{s}_{ij}; \quad \dot{\varepsilon}^e_{kk} = \frac{1}{K}\dot{p} \tag{9.97a}$$

$$\dot{e}^p_{ij} = \dot{\lambda}s_{ij}; \quad \dot{\varepsilon}^p_{kk} = 0 \tag{9.97b}$$

$$\dot{e}_{ij} = \frac{1}{2G}\dot{s}_{ij} + \dot{\lambda}s_{ij} \tag{9.97c}$$

and

$$\dot{\varepsilon}_{kk} = \frac{1}{K}\dot{p} \tag{9.97d}$$

where $\dot{e}_{ij}$, $\dot{\varepsilon}_{kk}$, G, K are, respectively, the deviatoric strain rate, volumetric strain rate, shear modulus, and bulk modulus. Recall that the plastic volumetric strain rate is zero in Levy–Mises theory. As in the preceding section, $\dot{\lambda}$ may be determined in two alternate ways for the case of perfect plasticity. With a suitable hardening law, the Prantl–Reuss theory can be used for work hardening plasticity as well.

Problems

Problem 9.1 List four distinct types of constitutive behaviors. Give key features that signify these behaviors.

Problem 9.2 Describe the following: (a) Yield stress, (b) Hardening, (c) Softening, and (d) Bauschinger effect.

Problem 9.3 Describe the following types of hardening behaviors: (a) Isotropic hardening, (b) kinematic hardening, and (c) combined hardening.

Problem 9.4 List one microscopic mechanism that is responsible for plastic behavior in (a) metals and (b) soils.

Problem 9.5 Explain the rationale for assuming that the rate of change of PIVs is zero when the rate of plastic strain is zero.

Problem 9.6 Explain why the rate of plastic strain must be homogeneous of order one function of the rate of stress for the stress–strain behavior to be rate-independent.

Determine if the following functionals are homogeneous of order one in their arguments.

(a) $f(\boldsymbol{\sigma}) = \frac{1}{3}\sin^{-1}\left[\frac{3\sqrt{3}}{2}\left(\frac{S}{J}\right)^3\right]$ where

$$s_{ij} = \sigma_{ij} - \frac{1}{3} I \delta_{ij}$$

$$J = \left(\frac{1}{2} s_{ij} s_{ij}\right)^{1/2}$$

$$S = \left(\frac{1}{3} s_{ik} s_{k\ell} s_{\ell i}\right)^{1/3}$$

(b) $f(\dot{\boldsymbol{\sigma}}) = \dfrac{a_{ij}\dot{\sigma}_{ik}\dot{\sigma}_{kj}}{(\dot{\sigma}_{k\ell}\dot{\sigma}_{k\ell})^{1/2}}$

Problem 9.7 Explain what the consistency condition is.

In a certain constitutive model, the plastic constitutive equation is given by (9.19). The model uses associated flow rule. The key equations are

$$\phi(I, J, \xi) = I^2 + \left(\frac{J}{N}\right)^2 - 3\xi I = 0$$

$$\dot{\xi} = c\xi\dot{\varepsilon}^P_{kk}$$

Show that the plastic modulus is given by

$$K_p = \frac{9c\xi I}{\left[3(2I - 3\xi)^2 + \frac{2J^2}{N^4}\right]}(2I - 3\xi)$$

Problem 9.8 The continuum tangent operator is given by (9.26b). Defining

$$\bar{\boldsymbol{\sigma}} = \left\{ \sigma_{11} \quad \sigma_{22} \quad \sigma_{33} \quad \sigma_{12} \quad \sigma_{23} \quad \sigma_{13} \right\}$$

$$\bar{\boldsymbol{\varepsilon}} = \left\{ \varepsilon_{11} \quad \varepsilon_{22} \quad \varepsilon_{33} \quad \gamma_{12} \quad \gamma_{23} \quad \gamma_{13} \right\}$$

$$\underset{6\times1}{\bar{\boldsymbol{\sigma}}} = \underset{6\times6}{\bar{\mathbf{C}}}\ \underset{6\times1}{\bar{\boldsymbol{\varepsilon}}}$$

Show that $\bar{\mathbf{C}}$ is symmetric when the flow rule is associated and non-symmetric when the flow rule is nonassociated.

Problem 9.9 Considering that a material obeys an isotropic hardening/softening rule and using the definition that the yield surface expands, remains unchanged or contracts depending on whether the material experiences hardening, flow (perfect plasticity) or softening, show that

$K_p \geq 0$ for a hardening material
$= 0$ for a perfectly-plastic material
≤ 0 for a softening material

Problem 9.10 Many real materials do not obey Drucker's postulates – explain.

Which of the Drucker's postulates listed in (9.29a) and (9.29b) exclude strain-softening behaviors from the acceptable behaviors?

Do Drucker's postulates ((9.29a) and (9.29b)) exclude the behaviors shown in Fig. 9.17b and 9.19d? Explain.

Does the complementary energy postulate (9.30b) exclude the behaviors shown in Fig. 9.17b and 9.19d? Explain.

Problem 9.11 Show that a constitutive law that is based on (9.19) and uses an associated flow rule will satisfy Drucker's first and second postulates ((9.29a) and (9.29b)) provided $K_p > 0$ and $\mathbf{C}^e$ is positive definite.

Problem 9.12 By following the approach in Sect. 9.3.4, show that convexity and normality hold on the failure surface.

Problem 9.13 Considering the von Mises yield criterion (9.56), show that

$$k = \sqrt{3}|\tau|$$

where τ is the yield shear stress in pure shear loading. Note that in a pure shear loading

$$\boldsymbol{\sigma}^* = \begin{bmatrix} \sigma_1 \\ -\sigma_1 \\ 0 \end{bmatrix}$$

Problem 9.14 Consider a tension–torsion test of a thin-walled tube, where an element is subjected to

$$\bar{\boldsymbol{\sigma}} = \{\sigma_{11} \quad \sigma_{22} \quad \sigma_{33} \quad \sigma_{12} \quad \sigma_{23} \quad \sigma_{13}\} = \{\sigma \quad 0 \quad 0 \quad \tau \quad 0 \quad 0\}$$

Derive equations for the yield surface in the $\sigma - \tau$ space according to Tresca and von Mises yield criteria. Plot the yield surface in the $\sigma - \tau$ space for $\sigma = 0$ to k, where k is the yield stress in uniaxial tension.

Problem 9.15 Considering any arbitrary path (not necessarily monotonic) from point a to point b, prove that the following identity holds:

$$\int_{\boldsymbol{\sigma}^a}^{\boldsymbol{\sigma}^b} \boldsymbol{\varepsilon} \mathrm{d}\boldsymbol{\sigma} + \int_{\boldsymbol{\varepsilon}^a}^{\boldsymbol{\varepsilon}^b} \boldsymbol{\sigma} \mathrm{d}\boldsymbol{\varepsilon} = \boldsymbol{\sigma}^b \boldsymbol{\varepsilon}^b - \boldsymbol{\sigma}^a \boldsymbol{\varepsilon}^a$$

Hint: Show this first for a uniaxial, monotonic path. Extend the results for a multi-axial monotonic path. Then break up a general path from a to b into piece-wise segments of monotonic paths and prove the results for any path by summation.

Problem 9.16 A circular shaft of diameter 0.1 m is subjected to a torque $T = 50\, kN\, m$, bending moment $M = 20\, kN\, m$ and an axial tensile force $P = 100\, kN$. Determine if the shaft is safe by using (a) the von Mises failure criterion (9.61) and (b) the Tresca failure criterion (9.65). The uniaxial failure stress is $k = 700\, MPa$.

Problem 9.17 A cubical soil specimen is subjected to the principal stresses $\sigma_1 = 200\, kPa$, $\sigma_2 = 60\, kPa$, and $\sigma_3 = 100\, kPa$. The cohesion and friction angle of the soil are $c = 50\, kPa$ and $\phi = 30°$. Determine whether the stress point is inside or outside the (a) Mohr–Coulomb failure surface and (b) the Drucker–Prager failure surface (9.81) (assuming circular in the octahedral plane).

Problem 9.18 Consider the von Mises yield function given by (9.63) and the associated flow rule. Consider the plastic constitutive laws given by (9.19a), (9.19b) and (9.19c). Expanding the Armstrong–Frederick kinematic hardening rule (9.92a) with the aid of the plastic constitutive equation (9.20f), show that

$$\dot{\alpha}_{ij} = \dot{\lambda} \bar{s}_{ij}; \quad \bar{s}_{ij} = \frac{4}{3} c b_{ij} - \frac{2}{\sqrt{3}} \gamma \alpha_{ij} \bar{r}^d$$

Then using the consistency condition (9.20a), show that the plastic modulus is given by

$$K_p = \frac{8}{9} k(2ck - 3\gamma ka); \quad a = (s_{k\ell} - \alpha_{k\ell})\alpha_{k\ell}$$

Chapter 10
Methods of Integrating Elasto-Plastic Constitutive Equations

The presentation in this chapter is limited to rate-independent elasto-plastic constitutive models. As we have seen in Chap. 8, one of the important aspects of an elasto-plastic finite element analysis is the evaluation of the *incremental stress* vector $\Delta\boldsymbol{\sigma}$ for a given *incremental strain* vector $\Delta\boldsymbol{\varepsilon}$. In addition, the Newton–Raphson based global iterative schemes require a material *tangent modulus matrix* (or *tangent stiffness operator*) $\mathbf{D}^t$. While this task is fairly trivial in an elastic analysis (either linear or nonlinear), it is nontrivial in a typical elasto-plastic analysis. In this chapter, we present three methods of integrating rate-independent elasto-plastic constitutive laws (*the two-step Euler method, cutting-plane method, and closest point projection method or CPPM*). The presentation of the theories and equations is kept as general as possible so that the underlying concepts can be applied to any specific rate-independent elasto-plastic model (e.g., models employing isotropic, kinematic, and rotational hardening, models employing single and multiple yield surfaces, pressure-independent and pressure-dependent models, etc.). The specifics of the equations will vary based on the details of each model.

10.1 Problem Statement

The general equations associated with a rate-independent elasto-plastic constitutive model, developed within the classical framework, are presented in Chap. 9. It was pointed out in Chap. 8 that the solution of nonlinear finite element equations by the Newton–Raphson method (or by one of its variants) requires at each Gauss point the tangent modular matrix $\mathbf{D}^t$ and the incremental stress vector $\Delta\boldsymbol{\sigma}$ defined as

$$\mathbf{D}^t = \frac{\partial\boldsymbol{\sigma}}{\partial\boldsymbol{\varepsilon}} \text{ and } \Delta\boldsymbol{\sigma} = \int_t^{t+\Delta t} \dot{\boldsymbol{\sigma}}\mathrm{d}t \tag{10.1}$$

A. Anandarajah, *Computational Methods in Elasticity and Plasticity: Solids and Porous Media*, DOI 10.1007/978-1-4419-6379-6_10,

where $\dot{\boldsymbol{\sigma}}$ is the stress rate vector, $\dot{\boldsymbol{\varepsilon}}$ is the strain rate vector, and t is time. By combining the elastic and plastic constitutive equations, we have

$$\dot{\boldsymbol{\varepsilon}} = \dot{\boldsymbol{\varepsilon}}^e + \dot{\boldsymbol{\varepsilon}}^p; \quad \dot{\boldsymbol{\varepsilon}}^p = \dot{\lambda}\mathbf{r} \tag{10.2a}$$

$$\dot{\boldsymbol{\sigma}} = \mathbf{C}\dot{\boldsymbol{\varepsilon}}^e = \mathbf{C}[\dot{\boldsymbol{\varepsilon}} - \dot{\boldsymbol{\varepsilon}}^p] = \mathbf{C}\dot{\boldsymbol{\varepsilon}} - \dot{\lambda}\mathbf{C}\mathbf{r}$$

where

$$\dot{\lambda} = \dot{\lambda}(\boldsymbol{\sigma},\boldsymbol{\zeta}) = \frac{1}{K_p}\dot{\boldsymbol{\sigma}}^{\mathrm{T}}\mathbf{n}; \quad \mathbf{n}(\boldsymbol{\sigma},\boldsymbol{\zeta}) = \frac{\partial\phi(\boldsymbol{\sigma},\boldsymbol{\zeta})}{\partial\boldsymbol{\sigma}}; \tag{10.2b}$$

$$\dot{\boldsymbol{\zeta}} = \dot{\lambda}\mathbf{s} = \mathbf{g}(\boldsymbol{\sigma},\boldsymbol{\zeta}); \quad \mathbf{s} = \mathbf{s}(\boldsymbol{\sigma},\boldsymbol{\zeta}); \quad \mathbf{r}(\boldsymbol{\sigma},\boldsymbol{\zeta}) = \frac{\partial\psi(\boldsymbol{\sigma},\boldsymbol{\zeta})}{\partial\boldsymbol{\sigma}}; \tag{10.2c}$$

$$\phi(\boldsymbol{\sigma},\boldsymbol{\zeta}) = 0; \quad \psi(\boldsymbol{\sigma},\boldsymbol{\zeta}) = 0 \tag{10.2d}$$

$\dot{\boldsymbol{\varepsilon}}^p$ is the plastic strain rate vector, $\mathbf{C}$ is the elastic tangent modular matrix, $\dot{\lambda}$ is the consistency parameter, $\mathbf{r}$ is the flow direction (vector normal to the plastic potential ψ), $\mathbf{n}$ is the yield direction (vector normal to the yield surface, defined by the function ϕ), $\boldsymbol{\zeta}$ is the plastic internal variable (PIV) or the hardening variable (which in general consists of scalar and/or tensorial PIVs such as the back-stress), $\mathbf{s}$ is the direction of the rate of PIV and

$$K_p = -(\partial_{\boldsymbol{\zeta}}\phi)^{\mathrm{T}}\mathbf{s} \tag{10.2e}$$

is the plastic modulus. Also, recall the combined elasto-plastic stress–strain rate equation (Chap. 9)

$$\dot{\boldsymbol{\sigma}} = \bar{\mathbf{D}}\dot{\boldsymbol{\varepsilon}} \tag{10.2f}$$

where $\bar{\mathbf{D}}$ is the elasto-plastic continuum tangent operator.

The problem at hand is stated as follows: Given $(\boldsymbol{\sigma}_n, \boldsymbol{\zeta}_n, \Delta\boldsymbol{\varepsilon})$, find $(\mathbf{D}^t_{n+1}, \boldsymbol{\sigma}_{n+1}, \boldsymbol{\zeta}_{n+1})$, where n denotes the increment number (Fig. 10.1). (It is understood from Algorithms 8.1 and 8.2 that $\mathbf{D}^t_{n+1}$ and $\boldsymbol{\sigma}_{n+1}$ are required at different stages during a given global iteration.) The process is, therefore, a strain-driven process, i.e., given $\Delta\boldsymbol{\varepsilon}$, we need to calculate $\Delta\boldsymbol{\sigma}$ and $\Delta\boldsymbol{\zeta}$.

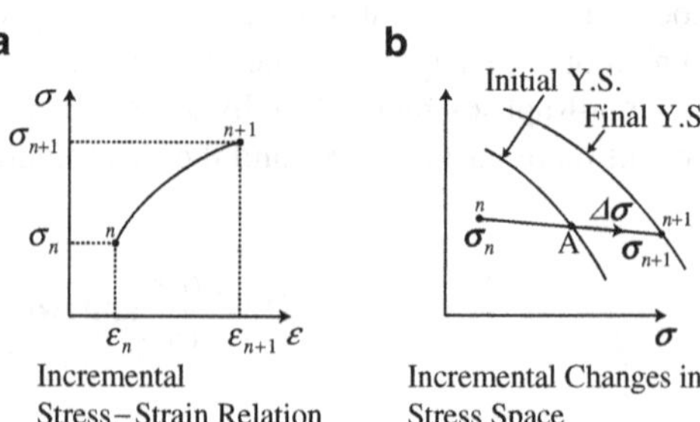

Fig. 10.1 Geometrical representation of stress and strain increments

The problem falls in the category of initial value problems. The reader may wish to review the details of the Euler integration methods presented in Chap. 8.

10.2 Use of Continuum Versus Consistent Tangent Operators

It was shown in Chap. 8 that when the backward Euler (implicit) method is used to integrate an initial value problem (such as the one we have), the use of the *consistent tangent operator* significantly increases the rate of convergence relative to the use of the *continuum tangent operator*.

First consider the problem of finding the tangent stiffness matrix. It was shown in Chap. 9 that $\dot{\boldsymbol{\sigma}} = \bar{\mathbf{D}}\dot{\boldsymbol{\varepsilon}}$ where

$$\bar{\mathbf{D}} = \left[\mathbf{C} - \frac{(\mathbf{Cr})(\mathbf{Cn})^{\mathrm{T}}}{\mathbf{n}^{\mathrm{T}}\mathbf{Cr} + K_p}\right] \tag{10.3a}$$

and

$$\dot{\lambda} = \frac{\mathbf{n}^{\mathrm{T}}\mathbf{C}\dot{\boldsymbol{\varepsilon}}}{\mathbf{n}^{\mathrm{T}}\mathbf{Cr} + K_p} \tag{10.3b}$$

$\bar{\mathbf{D}}$ is the continuum elasto-plastic tangent operator. Note that to avoid confusion in notations, we use $\mathbf{D}$ for the consistent operator and $\bar{\mathbf{D}}$ for the continuum operator.

The functions involved in (10.2a) are in general too complex to perform the integration in (10.1) analytically in finding $\Delta\boldsymbol{\sigma}$. Thus, a suitable numerical method is needed. Noting that $\boldsymbol{\sigma}_{n+1} = \boldsymbol{\sigma}_n + \Delta\boldsymbol{\sigma}$ and $\boldsymbol{\varepsilon}_{n+1} = \boldsymbol{\varepsilon}_n + \Delta\boldsymbol{\varepsilon}$, the tangent operator at any point during the increment can be evaluated if $\boldsymbol{\sigma}_{n+1}$ versus $\boldsymbol{\varepsilon}_{n+1}$ is known. An analytical $\boldsymbol{\sigma}_{n+1}$ versus $\boldsymbol{\varepsilon}_{n+1}$ relation, however, is not available. Based on the numerical algorithm employed, an algorithmic $\boldsymbol{\sigma}_{n+1}$ versus $\boldsymbol{\varepsilon}_{n+1}$ relation (which is approximate) is available. This relation can be differentiated to obtain an approximation to the tangent operator as

$$\mathbf{D}^t = \frac{\partial[\boldsymbol{\sigma}_n + \Delta\boldsymbol{\sigma}]}{\partial[\boldsymbol{\varepsilon}_n + \Delta\boldsymbol{\varepsilon}]} \tag{10.4}$$

where $\mathbf{D}^t$ is referred to as the *elasto-plastic consistent tangent modular matrix* or simply the *elasto-plastic consistent tangent operator*. The operator is called consistent because it is consistent with the algorithm used to compute the stresses. Although $\mathbf{D}^t$ exists, it is not always easy to compute because the computation of $\mathbf{D}^t$ requires gradients of quantities such as $\mathbf{r}$, $\mathbf{n}$, etc. with respect to $\boldsymbol{\sigma}$ and $\boldsymbol{\zeta}$. These gradients may be difficult to derive for complex models.

10.3 Brief History

Most of the early studies were concerned with the implementation of metal plasticity models. While exact integration is possible in some cases of perfect plasticity (e.g., Marques 1984; Loret and Prevost 1986; Krieg and Krieg 1977), most hardening plasticity models require a numerical method. One step forward Euler schemes (Krieg and Krieg 1977; Shreyer et al. 1979), one step backward schemes (Shreyer et al. 1979; Nguyen 1977) and sub-stepping methods (Nayak and Zienkiewicz 1972; Bushnell 1977; Mondkar and Powell 1977; Nyssen 1981; Sloan 1987; Herrmann, et al. 1987) have been employed in the past. Most methods of calculating the incremental stress gradually cause the final stress to "drift" off the yield surface. This requires a method to bring the stress back to the yield surface. The classical method for achieving this is the *radial return algorithm* of Wilkins (1964). Further work on radial return algorithm has been conducted by Krieg and Krieg (1977), Shreyer et al. (1979), and Yoder and Whirley (1984).

A more general class of very powerful algorithms, of which the radial return algorithm is a subclass, is known as the elastic predictor–plastic corrector algorithms. Here the calculations are performed essentially as a two-step process. In the first step, the inelastic response of the material is *frozen*, the full length of the strain increment is applied as an elastic strain, and the corresponding stress increment, known as the elastic predictor, is calculated. In the second step, the elastic trial stress is relaxed back onto the yield surface by applying an adequate amount of inelastic strain. This process is known as the *return mapping*. The procedure appears to have been first suggested by Wilkins (1964) and later developed and applied to many problems by Simo and Taylor (1985), Simo and Taylor (1986), Simo et al. (1988), Simo and Ortiz (1985), Ortiz and Papov (1985), Simo and Govindjee (1988), and Ortiz and Simo (1986). The return mapping is tantamount to finding the *closest point projection* of the trial stress (i.e., the elastic predictor) onto the (convex) yield surface. For this reason, the algorithm is also referred to as the *closest point projection method* (CPPM). Since the final stress is forced to lie on the yield surface, the algorithm enforces consistency at the end of the time step. The method is a backward Euler (implicit) method.

CPPM is applied in conjunction with Newton's iterative technique. It was recognized by researchers that the use of the continuum tangent operator destroys the quadratic convergence associated with Newton's method. Simo and Taylor (1985) demonstrated that with the use of the consistent tangent operator, the quadratic convergence can be preserved. The consistent tangent operator was also derived by Runesson and Samuelsson (1985). It appears that the optimal method available at present for implementing elasto-plastic constitutive models is to use CPPM to integrate the rate equations at the Gauss points in finding the stress increment for a given strain increment, and to use consistent tangent operator in the global Newton's iterations. CPPM is an implicit method, and requires gradients of $\mathbf{r}$ and $\mathbf{n}$ with respect to $\boldsymbol{\sigma}$ and $\boldsymbol{\zeta}$. Hence, as the numerical results presented in subsequent chapters would demonstrate, the method is computationally more intensive than other comparable methods (e.g., sub-stepping, cutting-plane, etc.) for small step sizes, but is less intensive for large step sizes. The main advantage is that the method for the most part is stable and provides optimal accuracy.

In some cases, however, the method fails to converge as we will see in some of the subsequent chapters. The factors that lead to convergence difficulties are: (1) large step size, (2) high nonlinearity of hardening rules, (3) high curvature of the yield and potential surfaces, (4) existence of regions in the stress or hardening variable hyperspace where elastic or plastic properties are undefined, and (5) discontinuities of constitutive functions. In cases where the above factors are involved, other supplementary methods may be needed to overcome the convergence difficulties. For example, line search strategies may be employed in some cases (Armero and Pérez-Foguet 2002; Pérez-Foguet and Armero 2002).

Kojić and Bathe (1987) developed another implicit method known as the "effective stress function" method, which is similar to the CPPM. The method was later generalized to what is known as the "governing parameter method." The method and its applications are summarized in the recent book Kojić and Bathe (2005). De Borst and Heeres (2002) extended CPPM to generalized plasticity models (i.e., those without an explicitly defined yield surfaces).

Another class of algorithms known as the cutting-plane algorithm was proposed by Ortiz and Simo (1986). The algorithm has the advantage that it does not require the evaluation of the gradients of $\mathbf{r}$ and $\mathbf{n}$ with respect to $\boldsymbol{\sigma}$ and $\boldsymbol{\zeta}$. It has the disadvantage that a consistent tangent operator cannot be easily found. Thus, the computational saving in not finding the gradients of $\mathbf{r}$ and $\mathbf{n}$ during the stress calculations is counteracted by the loss of efficiency due to the use of the continuum operator during the global Newton's iteration. However, when the model is not simple enough to permit the easy derivation of the gradients of $\mathbf{r}$ and $\mathbf{n}$, the cutting-plane algorithm may be a good choice.

A good source for the description of various integration methods is the book by Crisfield (1991). More elaborate coverage of CPPM may be found in Simo and Hughes (1998). Another recent book on the subject is the one by Dunne and Petrinic (2005).

In applying the algorithms developed for metals to materials such as geologic materials (soil, rock, powders, etc.), the differences between these materials must be recognized and dealt with appropriately. The essential differences stem from the fact that (1) in geologic materials, elastic and plastic properties (e.g., yield surface) are pressure-dependent, (2) geologic materials exhibit dilatancy behavior (volume change during shear), and (3) geologic materials generally exhibit highly nonlinear hardening behavior. Powerful algorithms such as CPPM developed for metals will still work in their original forms for geologic materials for the most part, provided that the variable elastic properties are appropriately accounted for.

In metal plasticity, the elastic properties normally remain constant during loading, i.e., they are neither a function of plastic strains nor a function of stresses. As a result, when the plastic corrector is found iteratively, the elastic trial stress remains unchanged. Most of the algorithms developed for metal plasticity are based on this fact. However, in geologic materials, the elastic moduli are a function of stress and are modeled as such in most constitutive models, e.g., the bulk modulus is a linear function of the mean normal pressure in the Cam-clay model of soils (Schofield and Wroth 1968). Thus, the elastic predictor does not remain a constant, which complicates the direct application of CPPM to geologic materials. However, if the elastic properties are approximated (say, using the generalized midpoint rule) then the general structure of CPPM may be retained. In implicit representation, the elastic predictor varies during the iteration, whereas in explicit representation, the elastic predictor remains a constant during the iteration. Alternatively, the elastic parts may be integrated exactly and the CPPM equations modified accordingly (Borja 1991). We will demonstrate this in subsequent chapters on geologic materials. Examples of the application of CPPM to geologic materials may be found in Borja and Lee (1990), Hashash and Whittle (1992), Jeremic and Sture (1997), Manzari and Nour (1997), Manzari and Prachathananukit (2001), and Wang et al. (2004).

In the following sections, after a brief discussion on the "elastic predictor–plastic corrector" concept, the following three methods are described:

1. Two-step Euler method with sub-stepping
2. Cutting-plane method
3. Closest Point Projection Method (CPPM)

10.4 Elastic Predictor–Plastic Corrector Methods

10.4.1 Definitions

Regardless of the specific method of integration employed, the elastic predictor is used to check the loading/unloading event. Integrating (10.2a), we have

$$\int_{t}^{t+\Delta t} \dot{\boldsymbol{\sigma}} = \int_{t}^{t+\Delta t} \boldsymbol{C}\dot{\boldsymbol{\varepsilon}} - \int_{t}^{t+\Delta t} \dot{\lambda}\boldsymbol{C}\boldsymbol{r} \tag{10.5a}$$

$$\Delta\boldsymbol{\sigma} = \Delta\boldsymbol{\sigma}^{\text{ep}} + \Delta\boldsymbol{\sigma}^{\text{pc}} \tag{10.5b}$$

where

$$\Delta\boldsymbol{\sigma}^{\text{ep}} = \int_{t}^{t+\Delta t} \mathrm{C}\dot{\boldsymbol{\varepsilon}} \quad \text{and} \quad \Delta\boldsymbol{\sigma}^{\text{pc}} = -\int_{t}^{t+\Delta t} \dot{\lambda}\mathrm{Cr} \tag{10.5c}$$

Hence,

$$\boldsymbol{\sigma}_{n+1} = \boldsymbol{\sigma}_n + \Delta\boldsymbol{\sigma}^{\text{ep}} + \Delta\boldsymbol{\sigma}^{\text{pc}} = \boldsymbol{\sigma}^{\text{tr}}_{n+1} + \Delta\boldsymbol{\sigma}^{\text{pc}} \tag{10.5d}$$

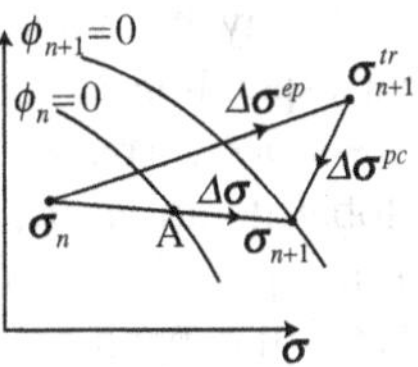

Fig. 10.2 Geometrical illustrations of elastic predictor and plastic corrector

where

$$\boldsymbol{\sigma}^{\mathrm{tr}}_{n+1} = \boldsymbol{\sigma}_n + \Delta\boldsymbol{\sigma}^{\mathrm{ep}} \tag{10.5e}$$

It follows from (10.5a) that $\Delta\boldsymbol{\sigma}$ can be split up into two parts, $\Delta\boldsymbol{\sigma}^{\mathrm{ep}}$ and $\Delta\boldsymbol{\sigma}^{\mathrm{pc}}$. This leads to (10.5d) and (10.5e), where $\boldsymbol{\sigma}_{n+1}$ is expressed as a sum of $\boldsymbol{\sigma}^{\mathrm{tr}}_{n+1}$ and $\Delta\boldsymbol{\sigma}^{\mathrm{pc}}$ (Fig. 10.2).

Now consider a scenario where the inelastic response of the material is assumed to be "frozen" and the entire length of strain increment is assumed to produce a purely elastic response. Then, $\dot{\lambda} = 0$ and $\Delta\boldsymbol{\sigma}^{\mathrm{pc}} = 0$, and thus $\boldsymbol{\sigma}_{n+1} = \boldsymbol{\sigma}^{\mathrm{tr}}_{n+1}$. For this reason, $\boldsymbol{\sigma}^{\mathrm{tr}}_{n+1}$ is called the *elastic predictor* and $\Delta\boldsymbol{\sigma}^{\mathrm{pc}}$ is called the *plastic corrector*. $\Delta\boldsymbol{\sigma}^{\mathrm{ep}}$ is the portion of the stress increment that goes with the elastic predictor. It must be noted that $\Delta\boldsymbol{\sigma}^{\mathrm{ep}}$ is not the stress increment associated with the elastic strain increment, i.e., $\Delta\boldsymbol{\sigma}^{\mathrm{ep}} \neq \int_t^{t+\Delta t} \mathbf{C}\dot{\boldsymbol{\varepsilon}}^e$, except when the response is purely elastic. The geometrical interpretations of the elastic predictor and plastic corrector are shown in Fig. 10.2.

Regardless of the method of integration used, there is a need to determine if a given strain increment will cause plastic deformation. It can be shown (see Appendix 5) that $\boldsymbol{\sigma}^{\mathrm{tr}}_{n+1}$ may be used to determine if a step is elastic or elasto-plastic as follows:

$$\phi^{\mathrm{tr}}_{n+1}(\boldsymbol{\sigma}^{\mathrm{tr}}_{n+1}, \zeta_n) > 0 \Longrightarrow \text{Elasto-Plastic} \tag{10.6a}$$

$$\phi^{\mathrm{tr}}_{n+1}(\boldsymbol{\sigma}^{\mathrm{tr}}_{n+1}, \zeta_n) \leq 0 \Longrightarrow \text{Elastic} \tag{10.6b}$$

$\boldsymbol{\sigma}^{\mathrm{tr}}_{n+1}$ is also known as the "trial stress," hence the use of the superscript "tr."

In the methods presented below, it is assumed that the strain increment has already been determined to cause elasto-plastic behavior.

10.4.2 Global Iterative Strategy: Incremental Versus Iterative Strains

The global iterative strategy applicable for an elasto-plastic finite element analysis is presented in Algorithm 8.1 (Chap. 8). A special version of this algorithm that is suitable for model calibration is presented in Algorithm 8.2 (Chap. 8).

It may be noted that in Algorithms 8.1 and 8.2, the incremental stress $\Delta\boldsymbol{\sigma}$ corresponding to the incremental strain $\Delta\boldsymbol{\varepsilon}$ is computed, rather than the iterative stress $\delta\boldsymbol{\sigma}$ corresponding to the iterative strain $\delta\boldsymbol{\varepsilon}$. In other words, during every global iteration, the integration of stresses always starts from point n rather than from a point between n and $n+1$. If this is not done, the final value of the calculated $\Delta\boldsymbol{\sigma}$ will depend on the stress path followed during the iteration, and hence on the specific algorithm employed. For further discussion on this, the reader is referred to Crisfield (1991), Key et al. (1980), and Marques (1984).

10.5 Two-Step Euler Integration Method with Sub-stepping

10.5.1 General Description

The two-step Euler method is explicit. The calculations are performed in two steps for the purpose finding an appropriate sub-step size with the objective of controlling the error during each sub-step. By integrating (10.2f) and (10.2c), we have

$$\Delta\boldsymbol{\sigma} = \int_t^{t+\Delta t} \dot{\boldsymbol{\sigma}}\, \mathrm{d}t = \int_t^{t+\Delta t} \bar{\mathbf{D}}\dot{\boldsymbol{\varepsilon}}\, \mathrm{d}t \tag{10.7a}$$

$$\Delta\boldsymbol{\zeta} = \int_t^{t+\Delta t} \dot{\boldsymbol{\zeta}} = \int_t^{t+\Delta t} \mathbf{g}(\boldsymbol{\sigma},\boldsymbol{\zeta})\mathrm{d}t \tag{10.7b}$$

Applying the forward Euler (explicit) method to (10.7a), we have

$$\Delta\boldsymbol{\sigma}^1 = \bar{\mathbf{D}}_n\Delta\boldsymbol{\varepsilon};\quad \boldsymbol{\sigma}^1_{n+1} = \boldsymbol{\sigma}_n + \Delta\boldsymbol{\sigma}^1 = \boldsymbol{\sigma}_n + \bar{\mathbf{D}}_n\Delta\boldsymbol{\varepsilon} \tag{10.8}$$

$\boldsymbol{\sigma}^1_{n+1}$ is taken as an estimate for $\boldsymbol{\sigma}_{n+1}$. In some cases, the hardening rule can be integrated exactly. Here we will use the forward Euler method to integrate (10.7b) as well. Since an estimate of stress $\boldsymbol{\sigma}^1_{n+1}$ is now available, we integrate (10.7b) using $\boldsymbol{\sigma}^1_{n+1}$ as

$$\boldsymbol{\zeta}^1_{n+1} = \boldsymbol{\zeta}_n + \mathbf{g}(\boldsymbol{\sigma}^1_{n+1},\boldsymbol{\zeta}_n)\Delta t \tag{10.9}$$

As shown schematically in Fig. 10.3, the error introduced on $\boldsymbol{\sigma}$ by this method of integration is $\delta\boldsymbol{\sigma}$. The error $\delta\boldsymbol{\sigma}$ decreases as the step size decreases, and thus sub-stepping should improve the accuracy. A better method of integrating (10.7a) is to use a trapezoidal rule as

$$\Delta\boldsymbol{\sigma} = \frac{1}{2}[\bar{\boldsymbol{D}}_n + \bar{\boldsymbol{D}}_{n+1}]\Delta\boldsymbol{\varepsilon}$$

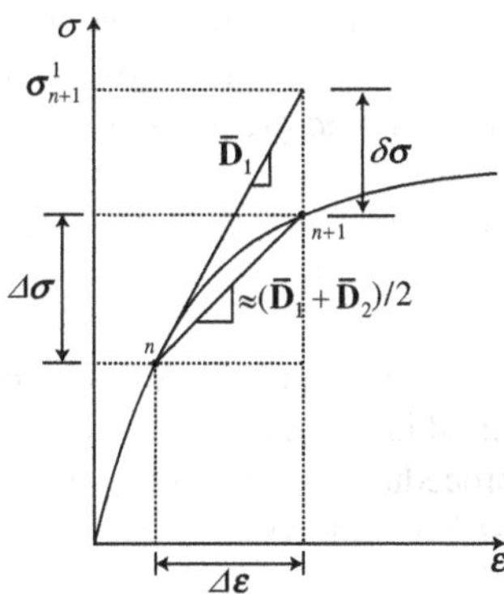

Fig. 10.3 Geometrical representation of quantities associated with the two-step Euler method

This requires $\boldsymbol{\sigma}_{n+1}$ which is not available yet. However, the estimate $\boldsymbol{\sigma}_{n+1}^{1}$ may be used, and a better estimate for $\boldsymbol{\sigma}_{n+1}$ may be obtained as follows:

$$\Delta\boldsymbol{\sigma}^2 = \frac{1}{2}\left[\bar{\mathbf{D}}_n(\boldsymbol{\sigma}_n, \boldsymbol{\zeta}_n) + \bar{\mathbf{D}}_{n+1}(\boldsymbol{\sigma}_{n+1}^1, \boldsymbol{\zeta}_{n+1}^1)\right]\Delta\boldsymbol{\varepsilon}; \tag{10.10a}$$

$$\boldsymbol{\sigma}_{n+1}^2 = \boldsymbol{\sigma}_n + \Delta\boldsymbol{\sigma}^2 = \boldsymbol{\sigma}_n + \frac{1}{2}\left[\bar{\mathbf{D}}_n(\boldsymbol{\sigma}_n, \boldsymbol{\zeta}_n) + \bar{\mathbf{D}}_{n+1}(\boldsymbol{\sigma}_{n+1}^1, \boldsymbol{\zeta}_{n+1}^1)\right]\Delta\boldsymbol{\varepsilon} \tag{10.10b}$$

$\boldsymbol{\sigma}_{n+1}^2$ is then used to obtain a corresponding estimate $\boldsymbol{\zeta}_{n+1}^2$ according to

$$\boldsymbol{\zeta}_{n+1}^2 = \boldsymbol{\zeta}_n + \mathbf{g}(\boldsymbol{\sigma}_{n+1}^2, \boldsymbol{\zeta}_{n+1}^1)\Delta t \tag{10.11}$$

(Note that the superscript 2 denotes "second estimate" and not "the squaring of the term") An estimate for errors may now be made as

$$\delta\boldsymbol{\sigma} \approx \boldsymbol{\sigma}_{n+1}^2 - \boldsymbol{\sigma}_{n+1}^1 \tag{10.12a}$$

$$\delta\boldsymbol{\zeta} \approx \boldsymbol{\zeta}_{n+1}^2 - \boldsymbol{\zeta}_{n+1}^1 \tag{10.12b}$$

The values of $\delta\boldsymbol{\sigma}$ and $\delta\boldsymbol{\zeta}$ may be used to estimate the number of sub-steps to be used to achieve a certain accuracy. Nyssen (1981) and Sloan (1987) used this approach and realized good success.

In a related secant method, the procedure presented above is repeated within the same increment to reduce the errors given in (10.12a) and (10.12b) to within acceptable tolerances. For example, Herrmann et al. (1987) developed such a technique with sub-stepping for the implementation of the isotropic bounding surface model. In the studies presented in Anandarajah and Dafalias (1986) and Anandarajah et al. (1995), the secant technique was used to integrate an anisotropic bounding surface model, which involved a tensorial PIV; again good success was realized.

Sloan (1987) proposed a more systematic method of carrying out the two-step procedure presented above. A scalar measure of the error, E, is calculated from (10.12a) and (10.12b) as

$$E = \max\left\{\frac{||\delta\boldsymbol{\sigma}||}{||\boldsymbol{\sigma}^2_{n+1}||};\frac{||\delta\boldsymbol{\zeta}||}{||\boldsymbol{\zeta}^2_{n+1}||}\right\} \tag{10.13}$$

The strain increment that comes from the parent finite element program is applied in small increments. The sizes of these small increments are determined by an iterative procedure based on the value of E. The current strain increment $\Delta\boldsymbol{\varepsilon}^c$ is modified to $\Delta\boldsymbol{\varepsilon}^m$ as follows:

$$\Delta\boldsymbol{\varepsilon}^m = \beta\Delta\boldsymbol{\varepsilon}^c \tag{10.14a}$$

where

$$\beta = 0.8\left[\frac{TOL}{E}\right]^{1/2} \tag{10.14b}$$

Sloan recommends limiting the value of β at any stage during the iterative process as $\beta_{\min} \le \beta \le \beta_{\max}$, where $\beta_{\max} = 2$ and $\beta_{\min} = 0.1$. However, it was found in the applications presented in subsequent chapters that when β is near 1.0, the progress of the iterative process was very slow. The following range is found to work better: $0.1 \le \beta \le 0.5$. Here, the current step is at least divided in half before another trial. The most suitable value may depend on the constitutive model.

Algorithm 10.1. The two-step Euler method of integrating constitutive models to evaluate incremental stress

Definition of (new) parameters: $N_L^{\max}$ and $N_G^{\max}$ are the maximum number of iterations permitted for the inner and outer loops in the algorithm before execution is terminated, and N_L and N_G are the corresponding counters. T is a scale parameter; when $T = 1$, full length of the strain vector has been applied. ΔT is the step size.

1. Establish $\Delta\boldsymbol{\varepsilon}, \boldsymbol{\zeta}_n, \boldsymbol{\sigma}_n, TOL, \beta_{\min}, \beta_{\max}, N_L^{\max}, N_G^{\max}$ and constitutive model parameters
2. Compute $\bar{\mathbf{D}}_n(\boldsymbol{\sigma}_n, \boldsymbol{\zeta}_n)$
3. Initialize parameters $T = 0$, $\Delta T = 1.0$, $N_L = 0$, and $N_G = 0$
4. Compute $\boldsymbol{\sigma}^1_{n+1}, \boldsymbol{\zeta}^1_{n+1}, \bar{\mathbf{D}}_{n+1}(\boldsymbol{\sigma}^1_{n+1}, \boldsymbol{\zeta}^1_{n+1}), \boldsymbol{\sigma}^2_{n+1}, \boldsymbol{\zeta}^2_{n+1}, \delta\boldsymbol{\sigma}, \delta\boldsymbol{\zeta}$, and E
5. If ($E < TOL$) go to step 8
6. Set $N_L \leftarrow N_L + 1$. If ($N_L > N_L^{\max}$) STOP
7. Compute β, $\beta \leftarrow \max(\beta, \beta_{\min})$, $\beta \leftarrow \min(\beta, \beta_{\max})$, $\Delta\boldsymbol{\varepsilon} \leftarrow \beta\Delta\boldsymbol{\varepsilon}$, $\Delta T \leftarrow \beta\Delta T$ and go to step 4
8. Set $N_G \leftarrow N_G + 1$. If ($N_G > N_G^{\max}$) STOP
9. Set $T \leftarrow T + \Delta T$. If ($T \ge 1.0$) go to 12

(continued)

10. If $(T + \Delta T > 1)$ then $x = (1 - T)/\Delta T$, $\Delta T \leftarrow x\Delta T$, and $\Delta\boldsymbol{\varepsilon} \leftarrow x\Delta\boldsymbol{\varepsilon}$
11. Set $\boldsymbol{\sigma}_n \leftarrow \boldsymbol{\sigma}^2_{n+1}$ and $\boldsymbol{\zeta}_n \leftarrow \boldsymbol{\zeta}^2_{n+1}$, and go to step 3
12. Set $\boldsymbol{\sigma}_{n+1} \leftarrow \boldsymbol{\sigma}^2_{n+1}$ and $\boldsymbol{\zeta}_{n+1} \leftarrow \boldsymbol{\zeta}^2_{n+1}$, and return to the main program

Note that the following two aspects are left out of this algorithm, and the reader is encouraged to modify the algorithm to incorporate these: (1) the step involving the evaluation of the elastic predictor and the loading/unloading check and (2) the step involving the determination of the point of intersection of the yield surface and the elastic predictor increment $\Delta\boldsymbol{\sigma}^{ep}$.

The iterative procedure then proceeds as follows: Start with the full length of $\Delta\boldsymbol{\varepsilon}$, estimate $\delta\boldsymbol{\sigma}$, and determine the value of E. If $E > TOL$, then calculate a value for β, and determine $\Delta\boldsymbol{\varepsilon}^m$. Repeat the process until $E \leq TOL$. Update the stresses and PIVs, and find the remaining length of strain to be applied. Start with the current value of the step size, and iteratively find the optimal size. Repeat the process until the full length of strain is applied. Algorithm 10.1 summarizes the steps.

10.5.2 Intersection of Stress Increment Vector with Yield Surface

The two-step Euler method presents one difficulty, not encountered in backward Euler procedures presented later. The elasto-plastic calculations must begin from a point on the surface. When $\boldsymbol{\sigma}_n$ lies inside the yield surface, first a determination is made as to whether the elastic predictor is inside or outside the surface (as is to be done as part of the loading/unloading check regardless of the integration method to be employed). If it is inside, the entire length of strain increment only invokes elastic behavior, and the elastic predictor is the required quantity $\boldsymbol{\sigma}_{n+1}$. The PIV remains unchanged. However, when the elastic predictor is outside the yield surface, the intersection of the elastic stress increment $\Delta\boldsymbol{\sigma}^{\mathrm{ep}}$ with the yield surface must first be established. The strain increment is divided up into two parts: the initial part that produces purely elastic stress, and the remaining part that produces elasto-plastic stress. This is not necessary when the $\boldsymbol{\sigma}_n$ lies on the yield surface (which may be known from the previous incremental analysis) and the elastic predictor is outside the yield surface. When the step size is very large, it is possible for the initial stress to lie on the yield surface, and the elastic predictor to cross the yield surface, and end up on the outside of the surface. The possibility for this occurrence also needs to be checked and acted on appropriately.

Referring to Fig. 10.4, the stress $\boldsymbol{\sigma}^*$ at any point between $\boldsymbol{\sigma}_n$ and $\boldsymbol{\sigma}^{\mathrm{tr}}_{n+1}$ may be written as

$$\boldsymbol{\sigma}^* = \boldsymbol{\sigma}_n + \alpha_1\left[\boldsymbol{\sigma}^{\mathrm{tr}}_{n+1} - \boldsymbol{\sigma}_n\right]; \quad \alpha_1 \in [0, 1] \tag{10.15}$$

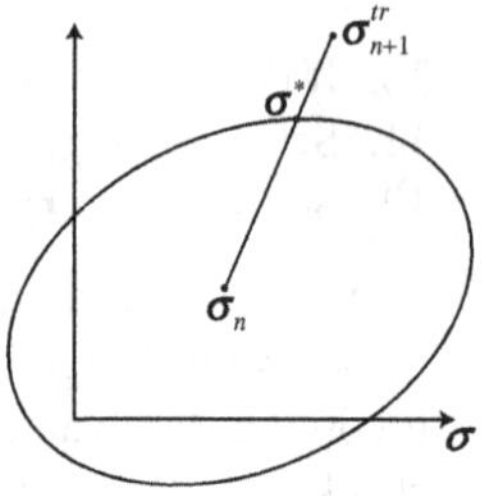

Fig. 10.4 Schematic of the point of intersection

The goal is to find α_1 such that

$$\phi(\boldsymbol{\zeta}_n, \boldsymbol{\sigma}^*) = 0 \tag{10.16}$$

While in some simple cases, it may be possible to analytically find α_1, in most cases, a numerical technique needs to be employed. Newton's method (Chap. 8) is one convenient method.

The use of Newton's method requires gradient of ϕ with respect to α_1, which may be computed using the chain rule as

$$\frac{\partial \phi}{\partial \alpha_1} = \frac{\partial \phi}{\partial \boldsymbol{\sigma}^*}\frac{\partial \boldsymbol{\sigma}^*}{\partial \alpha_1} = \mathbf{n}^{*\mathrm{T}}\left[\boldsymbol{\sigma}^{\mathrm{tr}}_{n+1} - \boldsymbol{\sigma}_n\right] \tag{10.17}$$

It is left for the reader to modify Algorithm 10.1 to achieve this.

10.6 The Cutting-Plane Method

10.6.1 Introduction

The cutting-plane algorithm was proposed by Ortiz and Simo (1986). The algorithm is based on the concept of what is known as the "operator splitting," where the constitutive evolution equations are split into two parts–elastic and plastic. The elastic part is first integrated. This is basically the same as what is normally done in estimating the elastic predictor in the elastic predictor–plastic corrector algorithms. The PIVs remain unchanged during this first step. The stresses at the end of the elastic step are used as initial variables for the plastic step. The plastic set of equations are now solved. The plastic equations suggest that the stress trajectories to be followed during the relaxation phase follow the steepest descent path corresponding to the yield surface. While this return relaxation path may be computed analytically for some, simple models (e.g., the isotropic hardening von Mises models), it needs to be computed numerically for most models.

One method was suggested by Ortiz and Simo (1986), where the yield surface is linearized around the current values of the stresses and PIVs. This is tantamount to making a series of straight intersection or "cut" with the plane $\phi = 0$ onto which the current values of the stresses and PIVs are projected to find the next values of the stresses and PIVs–hence the name the *cutting-plane algorithm*. This provides a method for computing the consistency parameter, and hence the stresses and PIVs, iteratively.

The cutting-plane method is simple, and does not require computation of the gradients of $\mathbf{r}$, $\mathbf{n}$, and $\mathbf{s}$. Hence, the method is simpler to implement, and computationally efficient on a per iteration basis. However, the algorithm does not lend itself to finding the consistent tangent operator. The adverse effects of the use of continuum operator counteract the benefits gained by not computing the gradients of $\mathbf{r}, \mathbf{n}$, and $\mathbf{s}$ during the stress computations. When the model is not simple enough to permit the derivation of the gradients of $\mathbf{r}, \mathbf{n}$, and $\mathbf{s}$ easily, the cutting-plane algorithm may be a good choice.

One word of caution in applying the cutting-plane algorithm to pressure-dependent materials is that the elastic models employed in such materials are normally nonconservative. Thus, during a stress cycle, energy may be generated. Implicit in the cutting-plane algorithm is the assumption that the sum of the elastic work done during the elastic predictor phase and the plastic corrector phase is equal to the elastic work done during the increment. Thus, unless measures are taken to rectify this, the step size must be small enough to achieve accurate results.

Furthermore, experience suggests that in some models (e.g., two-surface models with highly nonlinear hardening and/or pressure-dependent elastic properties), convergence problems may be encountered unless the step size is sufficiently small.

10.6.2 General Description

The constitutive equations presented in (10.2a) and (10.2b) are split up into elastic and plastic parts as

$$
\begin{array}{rcccl}
\dot{\boldsymbol{\varepsilon}} = \dot{\boldsymbol{\varepsilon}}^e + \dot{\boldsymbol{\varepsilon}}^p & & \dot{\boldsymbol{\varepsilon}} = \dot{\hat{\boldsymbol{\varepsilon}}} & & \dot{\boldsymbol{\varepsilon}} = \dot{\boldsymbol{\varepsilon}}^e + \dot{\boldsymbol{\varepsilon}}^p = \mathbf{0} \\
\dot{\boldsymbol{\sigma}} = \mathbf{C}[\dot{\boldsymbol{\varepsilon}} - \dot{\boldsymbol{\varepsilon}}^p] & & \dot{\boldsymbol{\sigma}} = \mathbf{C}\dot{\boldsymbol{\varepsilon}} & & \dot{\boldsymbol{\sigma}} = -\mathbf{C}\dot{\boldsymbol{\varepsilon}}^p \\
\dot{\boldsymbol{\varepsilon}}^p = \dot{\lambda}\mathbf{r} & \Rightarrow & \dot{\boldsymbol{\varepsilon}}^p = \mathbf{0} & + & \dot{\boldsymbol{\varepsilon}}^p = \dot{\lambda}\mathbf{r} \\
\dot{\boldsymbol{\zeta}} = \dot{\lambda}\mathbf{s} & & \dot{\boldsymbol{\zeta}} = \mathbf{0} & & \dot{\boldsymbol{\zeta}} = \dot{\lambda}\mathbf{s} \\
\textit{Total} & & \textit{Elastic} & & \textit{Plastic}
\end{array}
\qquad (10.18)
$$

where $\dot{\hat{\boldsymbol{\varepsilon}}}$ is the specified strain rate. The elastic part may be viewed as that governing the response of the material when the *inelastic response is frozen*. Of course, this will be the actual response when the behavior of the material associated with the strain increment is purely elastic. Otherwise, the integration of these equations provides the elastic predictor, which is then used as initial conditions for the integrations of the plastic equations.

Now, consider the integration of the plastic equations. Combining the second and third equations of the plastic set in (10.18), we have

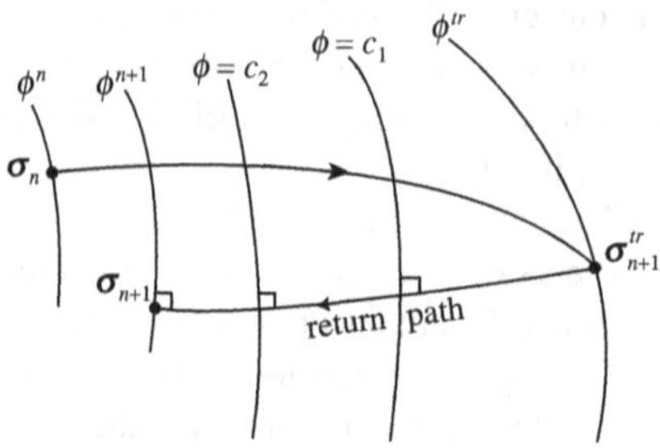

Fig. 10.5 The cutting-plane method: return path with associated flow rule and $\mathbf{C} = \mathbf{I}$

$$\dot{\boldsymbol{\sigma}} = -\dot{\lambda}\boldsymbol{Cr} \tag{10.19a}$$

Rewriting the evolution equation for the PIV

$$\dot{\boldsymbol{\zeta}} = \dot{\lambda}\boldsymbol{s} \tag{10.19b}$$

Equations (10.19a) and (10.19b) are the relaxation equations for the stresses and PIVs, respectively. These equations must be integrated to find the respective quantities. When $\mathbf{C} = \mathbf{I}$ (i.e., identity matrix) and the flow is associated ($\mathbf{n} = \mathbf{r}$, it is seen from (10.19a) that the return path is normal to the yield surface as shown in Fig. 10.5. The stress trajectories follow the steepest descent paths corresponding to the yield function. When $\mathbf{C} \neq \mathbf{I}$, as is the usual case, the steepest descent direction is defined with respect to $\mathbf{C}$. For a model with von Mises yield surface and isotropic hardening, the return path is trivially obtained, since it is simply the radial line connecting $\boldsymbol{\sigma}^{\text{tr}}_{n+1}$ and yield surface center. For most models, the return path needs to be determined numerically. The yield surface is in the form

$$\phi(\boldsymbol{\sigma}, \boldsymbol{\zeta}) = 0 \tag{10.20}$$

The yield surface is linearized around the current values of the stresses $\boldsymbol{\sigma}^i_{n+1}$ and PIVs $\boldsymbol{\zeta}^i_{n+1}$ as

$$\phi^{i+1} = \phi^i + \frac{\partial \phi^i}{\partial \boldsymbol{\sigma}^i_{n+1}} \delta \boldsymbol{\sigma}^i + \frac{\partial \phi^i}{\partial \boldsymbol{\zeta}^i_{n+1}} \delta \boldsymbol{\zeta}^i \approx 0 \tag{10.21}$$

Recalling that $\mathbf{n}^i = (\partial \phi^i / \partial \boldsymbol{\sigma}^i_{n+1})$, (10.21) becomes

$$\phi^{i+1} = \phi^i + \mathbf{n}^{i\mathrm{T}} \delta \boldsymbol{\sigma}^i + \frac{\partial \phi^i}{\partial \boldsymbol{\zeta}^i_{n+1}} \delta \boldsymbol{\zeta}^i \approx 0 \tag{10.22}$$

where $\phi^{i+1} = \phi(\boldsymbol{\sigma}^{i+1}_{n+1}, \boldsymbol{\zeta}^{i+1}_{n+1})$ and $\phi^i = \phi(\boldsymbol{\sigma}^i_{n+1}, \boldsymbol{\zeta}^i_{n+1})$. Discretizing (10.19a) and (10.19b) by the forward Euler method (explicit), we have

$$\boldsymbol{\sigma}^{i+1}_{n+1} = \boldsymbol{\sigma}^i_{n+1} + \delta \boldsymbol{\sigma}^i, \quad \text{where} \quad \delta \boldsymbol{\sigma}^i = -\delta \lambda^i \mathrm{C}^i \mathbf{r}^i \tag{10.23a}$$

and

$$\boldsymbol{\zeta}_{n+1}^{i+1} = \boldsymbol{\zeta}_{n+1}^{i} + \delta\boldsymbol{\zeta}^{i}, \quad \text{where} \quad \delta\boldsymbol{\zeta}^{i} = \delta\lambda^{i}\mathbf{s}^{i} \tag{10.23b}$$

Substituting (10.23a) and (10.23b) into (10.22), and recognizing that

$$K_p^i = -\left[\frac{\partial\phi^i}{\partial\boldsymbol{\zeta}_{n+1}^i}\right]^{\mathrm{T}} \mathbf{s}^i$$

we get

$$\delta\lambda^i = \frac{\phi^i}{\left[\mathbf{n}^{\mathrm{T}}\mathbf{C}\mathbf{r} + K_p\right]^i} \tag{10.24}$$

The procedure is repeated until convergence, and the stresses and PIVs are accumulated as

$$\boldsymbol{\sigma}_{n+1}^{i+1} = \boldsymbol{\sigma}_{n+1}^{\mathrm{tr}} + \sum \delta\lambda^i\mathbf{C}^i\mathbf{r}^i \tag{10.25a}$$

$$\boldsymbol{\zeta}_{n+1}^{i+1} = \boldsymbol{\zeta}_{n} + \sum \delta\lambda^i\mathbf{s}^i \tag{10.25b}$$

The following convergence criteria may be used to terminate the iterations:

$$e_\sigma = \left\|\boldsymbol{\sigma}_{n+1}^{i+1} - \boldsymbol{\sigma}_{n+1}^{i}\right\| / \left\|\boldsymbol{\sigma}_{n+1}^{i+1}\right\| \le TOL_\sigma \tag{10.26a}$$

$$e_\varsigma = \left\|\boldsymbol{\zeta}_{n+1}^{i+1} - \boldsymbol{\zeta}_{n+1}^{i}\right\| / \left\|\boldsymbol{\zeta}_{n+1}^{i+1}\right\| \le TOL_\varsigma \tag{10.26b}$$

$$\phi^{i+1} \le TOL_\phi \tag{10.26c}$$

The algorithm is interpreted as follows. Consider a simple case of perfect plasticity ($\delta\boldsymbol{\zeta}^i = 0$) with $\mathbf{C} = \mathbf{I}$ (where $\mathbf{I}$ is the identity matrix) and associated flow rule ($\mathbf{n} = \mathbf{r}$). Consider a straight cut on ϕ^i such that its normal coincides with $\mathbf{r}^i$ as shown in Fig. 10.6. Then from (10.23a), $\delta\boldsymbol{\sigma}^i = -\delta\lambda^i\mathbf{r}^i$. Thus, a move is made from $\boldsymbol{\sigma}_{n+1}^i$ to $\boldsymbol{\sigma}_{n+1}^{i+1}$ along $\mathbf{r}^i$, which, according to (10.22), minimizes the value of the yield function (when $\delta\lambda^i$ is calculated from (10.24)). Generalizing this notion, it may be said that a series of "cuts" with ϕ^i are made. $\boldsymbol{\sigma}_{n+1}^i$ and $\boldsymbol{\zeta}_{n+1}^i$ are projected onto these cutting planes in order to find optimal values for $\boldsymbol{\sigma}_{n+1}^{i+1}$ and $\boldsymbol{\zeta}_{n+1}^{i+1}$; hence the name "cutting-plane algorithm."

The algorithm is explicit in that the quantities at point $i + 1$ are found using quantities at point i. The algorithm is easy to implement. Most importantly, the algorithm does not require gradients of $\mathbf{r}, \mathbf{n}$, and $\mathbf{s}$ to be found (as is the case with implicit algorithms to be described in Sect. 10.7), which makes it very attractive for models with complex yield surfaces and hardening rules. The convergence is expected to be quadratic.

The evaluation of the consistent tangent operator requires differentiating (10.25a) and (10.25b). Due to the summation nature of the equation, closed-form

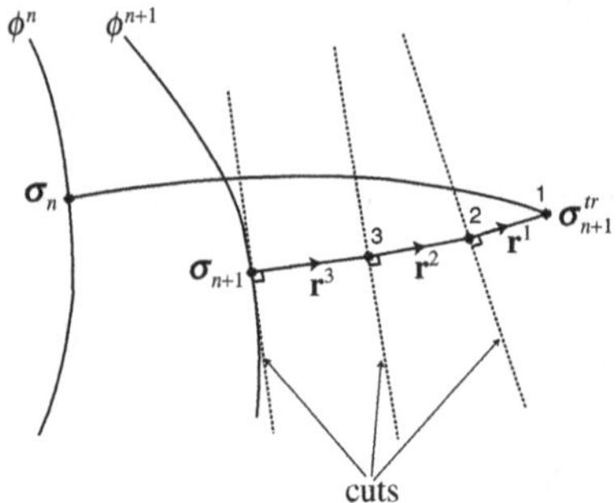

Fig. 10.6 Return path according to the cutting-plane algorithm

equation for consistent tangent operator is difficult to find–this is the major shortcoming of the method, as the use of continuum tangent operator can slow down the global finite element iterations.

Since the elastic part of the equations is integrated separately, the algorithm allows analytical integration of the elastic part (whenever possible). However, an assumption that is implicit in the method is that the elastic relations be conservative. This is in general not true for pressure-dependent materials due to the coupling of elastic moduli with mean pressure p. Thus, accuracy may deteriorate with increased level of elastic nonlinearity and strain increment size.

It may be noted that, unlike in the case of the two-step Euler method described in the preceding section, the cutting-plane algorithm does not require the point of intersection of the elastic predictor increment $\Delta\boldsymbol{\sigma}^{\text{ep}}$ and yield surface to be found.

The theory presented in this section is quite general and can be applied to any specific constitutive model. The implementation procedure is summarized in Algorithm 10.2.

Algorithm 10.2. The cutting-plane method of integrating constitutive models to evaluate incremental stress

Definition of (new) parameters: $N_L^{\max}$ is the maximum number of iterations permitted before execution is terminated, and N_L is the corresponding counter. TOL is a small, machine-dependent number (it could be set equal to TOL_ϕ).

1. Establish $\Delta\boldsymbol{\varepsilon}$, $\boldsymbol{\zeta}_n$, $\boldsymbol{\sigma}_n$, TOL_σ, TOL_ς, TOL_ϕ, TOL, $N_L^{\max}$, and the constitutive model parameters
2. Initialize parameters: $N_L = 0$ and $\boldsymbol{\zeta}_{n+1} = \boldsymbol{\zeta}_n$
3. Compute elastic predictor $\boldsymbol{\sigma}^{\text{tr}}_{n+1}$ and $\phi^{\text{tr}}(\boldsymbol{\sigma}^{\text{tr}}_{n+1}, \boldsymbol{\zeta}_n)$, and set $\boldsymbol{\sigma}_{n+1} = \boldsymbol{\sigma}^{\text{tr}}_{n+1}$ and $\phi^i = \phi^{\text{tr}}$
4. Perform loading/unloading check: if ($\phi^{\text{tr}} < TOL$) go to step 9
5. Compute $\delta\lambda^i$, $\delta\boldsymbol{\sigma}^i$ and $\delta\boldsymbol{\zeta}^i$ and find updated quantities $\boldsymbol{\sigma}_{n+1}$ and $\boldsymbol{\zeta}_{n+1}$
6. Find errors e_σ, e_ς and $\phi^i(\boldsymbol{\sigma}_{n+1}, \boldsymbol{\zeta}_{n+1})$ and perform convergence checks. If converged, go to step 9
7. Set $N_L \leftarrow N_L + 1$. If ($N_L > N_L^{\max}$) STOP
8. Go to step 5
9. Return to the main program

10.7 Closest Point Projection Method (CPPM)

10.7.1 Introduction

CPPM is an elastic predictor–plastic corrector method that seeks to satisfy consistency at the end of the increment. Hence, CPPM is a *backward Euler, implicit procedure*. The method is robust and allows analytical derivation of the consistent tangent operator. Thus, for large step sizes, the method is both stable and efficient relative to the two-step Euler method or the cutting-plane method described in the preceding sections. One drawback of the method is that it requires gradients of $\mathbf{r}, \mathbf{n}$, and $\mathbf{s}$ with respect to $\boldsymbol{\sigma}$ and $\boldsymbol{\zeta}$, which may be difficult to evaluate for certain complex models. Even when it is possible to do so, the procedure may be computationally less efficient than the other methods for smaller step sizes.

10.7.2 Description of CPPM: Constant Elastic Properties

First, the method is described by considering a material with constant elastic properties such as metals. In this case, the elastic predictor (Figs. 10.2 and 10.7) remains a constant during the iteration. Referring to Fig. 10.2, the goal is to find $\boldsymbol{\sigma}_{n+1}$ and $\boldsymbol{\zeta}_{n+1}$ by the following backward Euler (implicit) equations

$$\boldsymbol{\sigma}_{n+1} = \boldsymbol{\sigma}_n + \Delta\boldsymbol{\sigma}^{\text{ep}} + \Delta\boldsymbol{\sigma}^{\text{pc}} = \boldsymbol{\sigma}^{\text{tr}}_{n+1} + \Delta\boldsymbol{\sigma}^{\text{pc}} \tag{10.27a}$$

$$\Delta\boldsymbol{\sigma}^{\text{ep}} = \int_t^{t+\Delta t} \mathbf{C}\dot{\boldsymbol{\varepsilon}} = \mathbf{C}\Delta\boldsymbol{\varepsilon} \tag{10.27b}$$

$$\Delta\boldsymbol{\sigma}^{\text{pc}} = -\int_t^{t+\Delta t} \dot{\lambda}\mathbf{C}\mathbf{r} = -\Delta\lambda\mathbf{C}\mathbf{r}^{n+1} \tag{10.27c}$$

$$\boldsymbol{\zeta}_{n+1} = \boldsymbol{\zeta}_n + \Delta\boldsymbol{\zeta} \tag{10.28a}$$

$$\Delta\boldsymbol{\zeta} = \Delta\lambda\mathbf{s}^{n+1} \tag{10.28b}$$

such that

$$\phi^{n+1}(\boldsymbol{\sigma}_{n+1}, \boldsymbol{\zeta}_{n+1}) = 0 \tag{10.28c}$$

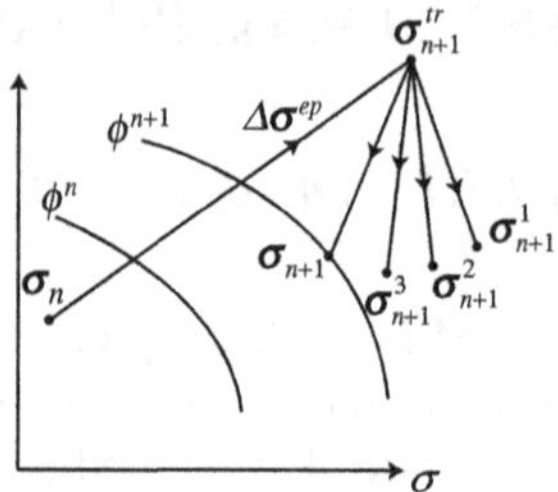

Fig. 10.7 Schematic of procedure associated with CPPM: The case of constant elastic properties

To sum, the equations for calculating $\boldsymbol{\sigma}_{n+1}$ and $\boldsymbol{\zeta}_{n+1}$ are

$$\boldsymbol{\sigma}_{n+1} = \boldsymbol{\sigma}_n + \mathbf{C}\Delta\boldsymbol{\varepsilon} - \Delta\lambda\mathbf{C}\mathbf{r}^{n+1}(\boldsymbol{\sigma}_{n+1}, \boldsymbol{\zeta}_{n+1}) \tag{10.29a}$$

$$\boldsymbol{\zeta}_{n+1} = \boldsymbol{\zeta}_n + \Delta\lambda\mathbf{s}^{n+1}(\boldsymbol{\sigma}_{n+1}, \boldsymbol{\zeta}_{n+1}) \tag{10.29b}$$

Since $\boldsymbol{\sigma}_{n+1}$ and $\boldsymbol{\zeta}_{n+1}$ are functions of themselves, the procedure is iterative.

Remark.
Alternatively, (10.29a) (and (10.29b) if $\boldsymbol{\zeta}$ is also a tensor) can be split up into a spherical and deviatoric parts as follows. Recall:

$$C_{ijk\ell} = \left(K - \frac{2}{3}G\right)\delta_{ij}\delta_{k\ell} + G\left(\delta_{ik}\delta_{j\ell} + \delta_{i\ell}\delta_{jk}\right)$$

Consider the inner product:

$$\begin{aligned} C_{ijk\ell}r_{k\ell} &= C_{ijk\ell}\left[r^d_{k\ell} + \frac{1}{3}r_{pp}\delta_{k\ell}\right] \\ &= \left[\left(K - \frac{2}{3}G\right)\delta_{ij}\delta_{k\ell} + G\left(\delta_{ik}\delta_{j\ell} + \delta_{i\ell}\delta_{jk}\right)\right]\left[r^d_{k\ell} + \frac{1}{3}r_{pp}\delta_{k\ell}\right] \\ &= \left(K - \frac{2}{3}G\right)r_{pp}\delta_{ij} + 2Gr^d_{ij} + \frac{2}{3}Gr_{pp}\delta_{ij} \\ &= Kr_{pp}\delta_{ij} + 2Gr^d_{ij} \end{aligned}$$

Hence, the second order tensor $C_{ijk\ell}r_{k\ell}$ is split up into spherical and deviatoric parts. Similarly, it follows:

$$C_{ijk\ell}\Delta\boldsymbol{\varepsilon}_{k\ell} = K\Delta\boldsymbol{\varepsilon}_{pp}\delta_{ij} + 2G\Delta e_{ij}; \quad \Delta e_{ij} = \Delta\varepsilon_{ij} - \frac{1}{3}\Delta\varepsilon_{pp}\delta_{ij}$$

where Δe_{ij} is the deviatoric part of the strain increment $\Delta\varepsilon_{ij}$. Based on these, (10.29a) can be split up into spherical and deviatoric parts as

$$p_{n+1} = p_n + K\Delta\varepsilon_{pp} - K\Delta\lambda r^{n+1}_{pp}$$

$$\bar{\mathbf{s}}_{n+1} = \bar{\mathbf{s}}_n + 2G\Delta\mathbf{e} - 2G\Delta\lambda\mathbf{r}^d_{n+1}$$

where

$$\bar{\mathbf{s}}_{n+1} = \boldsymbol{\sigma}_{n+1} - p_{n+1}\boldsymbol{\delta}$$

is the deviatoric part of $\boldsymbol{\sigma}_{n+1}$. In pressure-independent models, $r_{pp}^{n+1} = 0$, and hence the variation of p is purely elastic. Hence, only the deviatoric equation needs to be included in the iterative process.

Definition 10.1. *When the iteration is based on the stress tensor $\boldsymbol{\sigma}$ as in (10.29a), we will refer to the formulation as the* σ-space *formulation. When the iteration is based on the spherical–deviatoric stress decomposition (as in the remark above), we will refer to the formulation as the S-space formulation.*

For simple models (e.g., the von Mises model with a simple isotropic hardening rule as in a subsequent chapter), it is possible to combine the above equations into one simple scalar (typically nonlinear) equation in $\Delta\lambda$. The equation can then be solved by a suitable procedure, and used in (10.29a) and (10.29b) to compute the stress and PIV at point $n+1$. For example, this is possible for the von Mises model with a simple isotropic hardening rule. We need to use the S-space formulation to achieve this (see the remark above). We will demonstrate this for a simple von Mises model in a subsequent chapter.

For pressure-dependent materials also, one can use either the σ-space formulation or the S-space formulation. As we will see in subsequent chapters, the latter may lead to simpler formulations (involving a simple scalar equation in $\Delta\lambda$ as above) in some cases (e.g., when the yield surface is a circle in the octahedral plane, the hardening law is isotropic and the flow rule is associated).

In the remaining portion of this chapter, we will discuss the details of CPPM in the full σ-space. The method proceeds as follows. An initial estimate of $\Delta\lambda$ is made by a suitable procedure. For example, the explicit method used in the cutting-plane algorithm (10.24) can be used for this purpose, according to which

$$\Delta\lambda_1 = \frac{\phi^{\text{tr}}}{\left[\mathbf{n}^{\mathrm{T}}\mathbf{C}\mathbf{r} + K_p\right]^{\text{tr}}} \tag{10.30}$$

As indicated in the above equation, the quantities evaluated at the elastic predictor are used to determine a value for $\Delta\lambda_1$. Then using $\mathbf{r}^{\text{tr}}$ and $\mathbf{n}^{\text{tr}}$, the first estimate for $\boldsymbol{\sigma}_{n+1}$ and $\boldsymbol{\zeta}_{n+1}$, say $\boldsymbol{\sigma}_{n+1}^{1}$ (Fig. 10.7) and $\boldsymbol{\zeta}_{n+1}^{1}$, are found by (10.29a) and (10.29b). These quantities will not satisfy (10.28c) because $\mathbf{r}^{\text{tr}} \neq \mathbf{r}^{n+1}$ and $\mathbf{n}^{\text{tr}} \neq \mathbf{n}^{n+1}$. The following residuals are defined:

$$\mathbf{R}_\sigma = \boldsymbol{\sigma}_{n+1} - \left[\boldsymbol{\sigma}_n + \mathbf{C}\Delta\boldsymbol{\varepsilon} - \Delta\lambda\mathbf{C}\mathbf{r}^{n+1}\right] = \mathbf{R}_\sigma(\boldsymbol{\sigma}_{n+1}, \boldsymbol{\zeta}_{n+1}, \Delta\lambda) \tag{10.31a}$$

$$\mathbf{R}_\varsigma = \boldsymbol{\zeta}_{n+1} - \left[\boldsymbol{\zeta}_n + \Delta\lambda\mathbf{s}^{n+1}\right] = \mathbf{R}_\varsigma(\boldsymbol{\sigma}_{n+1}, \boldsymbol{\zeta}_{n+1}, \Delta\lambda) \tag{10.31b}$$

$$R_\phi = \phi^{n+1}(\boldsymbol{\sigma}_{n+1}, \boldsymbol{\zeta}_{n+1}) = \mathbf{R}_\phi(\boldsymbol{\sigma}_{n+1}, \boldsymbol{\zeta}_{n+1}) \tag{10.31c}$$

Remark. In (10.31b), the backward Euler representation is used for $\boldsymbol{\zeta}_{n+1}$. However, in some cases, it is possible to directly integrate the hardening law to obtain an analytical equation for $\boldsymbol{\zeta}_{n+1}$ in terms of $\boldsymbol{\sigma}_{n+1}$ and $\Delta\lambda$ (an example will be presented in a subsequent chapter on the von Mises model). Such representations will not only reduce the number of local iterations necessary to achieve certain accuracy, but also will result in a better overall accuracy upon convergence.

During the first iteration, the residuals are calculated using $\boldsymbol{\sigma}^1_{n+1}$, $\boldsymbol{\zeta}^1_{n+1}$, and $\Delta\lambda_1$ for $\boldsymbol{\sigma}_{n+1}$ and $\boldsymbol{\zeta}_{n+1}$ ((10.29a) and (10.29b)). This entails the calculation of $\mathbf{r}^{n+1}$ and $\mathbf{n}^{n+1}$ in (10.31a)–(10.31c) at $\boldsymbol{\sigma}^1_{n+1}$ and $\boldsymbol{\zeta}^1_{n+1}$.

At this point, we specialize the equations for a case where $\boldsymbol{\zeta}$ and $\mathbf{R}^1_\varsigma$ are scalars, say, ς and R^1_ς; extension to tensorial $\boldsymbol{\zeta}$ and $\mathbf{R}^1_\varsigma$ is straightforward and will be demonstrated through specific examples in subsequent chapters. The specialized equations to be derived are still very general in that they are applicable to any standard plasticity model with one scalar PIV.

The residuals (10.31a–c) are now linearized around $\boldsymbol{\sigma}^1_{n+1}$ and $\boldsymbol{\zeta}^1_{n+1}$ as follows (using tensor notations for clarity)

$$R^1_{\sigma_{ij}} + \left[\frac{\partial R_{\sigma ij}}{\partial \sigma_{k\ell}}\right]_1 \delta\sigma_{k\ell} + \left[\frac{\partial R_{\sigma ij}}{\partial \varsigma}\right]_1 \delta\varsigma + \left[\frac{\partial R_{\sigma ij}}{\partial \Delta\lambda}\right]_1 \delta\lambda = 0 \tag{10.32a}$$

$$R^1_\varsigma + \left[\frac{\partial R_\varsigma}{\partial \sigma_{k\ell}}\right]_1 \delta\sigma_{k\ell} + \left[\frac{\partial R_\varsigma}{\partial \varsigma}\right]_1 \delta\varsigma + \left[\frac{\partial R_\varsigma}{\partial \Delta\lambda}\right]_1 \delta\lambda = 0 \tag{10.32b}$$

$$R^1_\phi + \left[\frac{\partial R_\phi}{\partial \sigma_{k\ell}}\right]_1 \delta\sigma_{k\ell} + \left[\frac{\partial R_\phi}{\partial \varsigma}\right]_1 \delta\varsigma = 0 \tag{10.32c}$$

Where the quantities $\left[\frac{\partial R_{\varsigma_{ij}}}{\partial \sigma_{k\ell}}\right]_1$ etc., are evaluated at $(\boldsymbol{\sigma}^1_{n+1}, \boldsymbol{\zeta}^1_{n+1})$.

Let us introduce simpler notations as

$$A_{ijk\ell} = \left[\frac{\partial R_{\sigma_{ij}}}{\partial \sigma_{k\ell}}\right]_1; \quad B_{ij} = \left[\frac{\partial R_{\sigma_{ij}}}{\partial \varsigma}\right]_1; \quad F_{ij} = \left[\frac{\partial R_{\sigma_{ij}}}{\partial \Delta\lambda}\right]_1 \tag{10.33a}$$

$$H_{k\ell} = \left[\frac{\partial R_\varsigma}{\partial \sigma_{k\ell}}\right]_1; \quad \omega = \left[\frac{\partial R_\varsigma}{\partial \varsigma}\right]_1; \quad \beta = \left[\frac{\partial R_\varsigma}{\partial \Delta\lambda}\right]_1 \tag{10.33b}$$

$$E_{k\ell} = \left[\frac{\partial R_\phi}{\partial \sigma_{k\ell}}\right]_1; \quad \gamma = \left[\frac{\partial R_\phi}{\partial \varsigma}\right]_1 \tag{10.33c}$$

Equations (10.32a–c) are written as

$$R^1_{\sigma_{ij}} + A_{ijk\ell}\,\delta\sigma_{k\ell} + B_{ij}\,\delta\varsigma + F_{ij}\,\delta\lambda = 0 \tag{10.34a}$$

$$R^1_{\varsigma} + H_{k\ell}\,\delta\sigma_{k\ell} + \omega\,\delta\varsigma + \beta\,\delta\lambda = 0 \tag{10.34b}$$

$$R^1_{\phi} + E_{k\ell}\,\delta\sigma_{k\ell} + \gamma\,\delta\varsigma = 0 \tag{10.34c}$$

Let us repeat (10.31a–c) with the first one written in tensor notation as

$$R_{\sigma_{ij}} = \sigma^{n+1}_{ij} - \sigma^{n}_{ij} - C_{ijpq}\Delta\varepsilon_{pq} + \Delta\lambda C_{ijpq} r^{n+1}_{pq} \tag{10.35a}$$

$$R_{\varsigma} = \varsigma_{n+1} - \varsigma_n - \Delta\lambda s^{n+1} \tag{10.35b}$$

$$R_{\phi} = \phi^{n+1}(\sigma_{n+1}, \zeta_{n+1}) \tag{10.35c}$$

The coefficients of (10.34a–c) are derived as follows. It is understood that numerical values of the residuals are to be evaluated at $(\boldsymbol{\sigma}^1_{n+1}, \boldsymbol{\zeta}^1_{n+1})$. The subscript 1 is left out for simplicity of notations. We will treat the elasticity tensor **C** as a variable for generality. Except for the quantities at n, all others are variables. Differentiating (10.35a–c), we derive equations for the coefficients of (10.34a–c) as

$$A_{ijk\ell} = \frac{\partial R_{\boldsymbol{\sigma}_{ij}}}{\partial \sigma_{k\ell}} = \delta_{ik}\delta_{j\ell} - \frac{\partial C_{ijpq}}{\partial \sigma_{k\ell}}\left[\Delta\varepsilon_{pq} - \Delta\lambda r^{n+1}_{pq}\right] + \Delta\lambda C_{ijpq}\frac{\partial r^{n+1}_{pq}}{\partial \sigma_{k\ell}} \tag{10.36a}$$

$$B_{ij} = \frac{\partial R_{\sigma_{ij}}}{\partial \varsigma} = \Delta\lambda C_{ijpq}\frac{\partial r^{n+1}_{pq}}{\partial \varsigma} \tag{10.36b}$$

$$F_{ij} = \frac{\partial R_{\sigma_{ij}}}{\partial \Delta\lambda} = C_{ijpq} r^{n+1}_{pq} \tag{10.36c}$$

$$H_{k\ell} = \frac{\partial R_{\varsigma}}{\partial \sigma_{k\ell}} = -\Delta\lambda\frac{\partial s^{n+1}}{\partial \sigma_{k\ell}} \tag{10.36d}$$

$$\omega = \frac{\partial R_{\varsigma}}{\partial \varsigma} = 1 - \Delta\lambda\frac{\partial s^{n+1}}{\partial \varsigma} \tag{10.36e}$$

$$\beta = \frac{\partial R_{\varsigma}}{\partial \Delta\lambda} = -s^{n+1} - \Delta\lambda\frac{\partial s^{n+1}}{\partial \Delta\lambda} \tag{10.36f}$$

$$E_{k\ell} = \frac{\partial R_{\phi}}{\partial \sigma_{k\ell}} = \frac{\partial \phi^{n+1}}{\partial \sigma_{k\ell}} = n^{n+1}_{k\ell} \tag{10.36g}$$

$$\gamma = \frac{\partial R_{\phi}}{\partial \varsigma} = \frac{\partial \phi^{n+1}}{\partial \varsigma} \tag{10.36h}$$

It is seen that the quantities needed for evaluating the coefficients of (10.34a–c) are

- $\Delta\boldsymbol{\varepsilon}$
- Values of $\mathbf{C}, \mathbf{r}, \mathbf{n}, \partial\phi/\partial\boldsymbol{\zeta}, \Delta\lambda$ and $\mathbf{s}$ at the forward point $n+1$ at the current iteration
- Gradients $\partial\mathbf{C}/\partial\boldsymbol{\sigma}, \partial\mathbf{r}/\partial\boldsymbol{\sigma}, \partial\mathbf{r}/\partial\boldsymbol{\zeta}, \partial\mathbf{s}/\partial\boldsymbol{\sigma}, \partial\mathbf{s}/\partial\boldsymbol{\zeta}$, and $\partial\mathbf{s}/\partial\Delta\lambda$

Equations (10.34a–c) are solved simultaneously to find the three unknowns $\delta\sigma_{ij}$, $\delta\varsigma$, and $\delta\lambda$. The solution strategies are discussed in a subsequent section. The new estimates for $\boldsymbol{\sigma}_{n+1}$, ς_{n+1}, and $\Delta\lambda$ are found as

$$\boldsymbol{\sigma}^2_{n+1} = \boldsymbol{\sigma}^1_{n+1} + \delta\boldsymbol{\sigma} \tag{10.37a}$$

$$\varsigma^2_{n+1} = \varsigma^2_{n+1} + \delta\varsigma \tag{10.37b}$$

$$\Delta\lambda_2 = \Delta\lambda_1 + \delta\lambda \tag{10.37c}$$

Note that superscript 2 on these variables does not denote "square," but rather the "second estimate."

The residuals $\mathbf{R}^1_\sigma$, R^1_ς, and R^1_ϕ may still not be sufficiently small. The procedure is repeated using the new estimates $\boldsymbol{\sigma}^2_{n+1}$ (Fig. 10.7), ς^2_{n+1}, and $\Delta\lambda_2$. During this second iteration, the residuals are calculated using $\boldsymbol{\sigma}^2_{n+1}, \boldsymbol{\zeta}^2_{n+1}$, and $\Delta\lambda_2$ for $\boldsymbol{\sigma}_{n+1}$ and $\boldsymbol{\zeta}_{n+1}$ in (10.31a–c). $\mathbf{r}^{n+1}$ and $\mathbf{n}^{n+1}$ are calculated using $\boldsymbol{\sigma}^2_{n+1}$ and $\boldsymbol{\zeta}^2_{n+1}$. The plastic corrector $\Delta\boldsymbol{\sigma}^{\text{pc}}$ gradually converges to the correct value within a tolerance as shown schematically in Fig. 10.7. The following convergence criteria may be used:

$$e_\sigma = \left\|\mathbf{R}^{i+1}_\sigma\right\| / \left\|\boldsymbol{\sigma}^i_{n+1}\right\| \le TOL_\sigma \tag{10.38a}$$

$$e_\varsigma = \left\|R^{i+1}_\varsigma\right\| / \left\|\boldsymbol{\zeta}^i_{n+1}\right\| \le TOL_\varsigma \tag{10.38b}$$

$$R^{i+1}_\phi = \phi^{i+1} \le TOL_\phi \tag{10.38c}$$

Once convergence is reached, the stresses and PIVs computed using the update formulas (10.37a) and (10.37b) are almost the same as those computed using (10.29a) and (10.29b). That is,

$$\boldsymbol{\sigma}^k_{n+1} \approx \boldsymbol{\sigma}_n + \mathbf{C}\Delta\boldsymbol{\varepsilon} - \Delta\lambda_k \mathbf{C}\mathbf{r}^{n+1}(\boldsymbol{\sigma}^k_{n+1}, \varsigma^k_{n+1}) \tag{10.39a}$$

$$\varsigma^k_{n+1} \approx \varsigma_n + \Delta\lambda_k \mathbf{s}^{n+1}(\boldsymbol{\sigma}^k_{n+1}, \varsigma^k_{n+1}) \tag{10.39b}$$

where k is the iteration number at which the convergence occurred. Once convergence is achieved, the left and right hand sides of the above equations are equal to each other. Hence, dropping the iteration number, we have

$$\boldsymbol{\sigma}_{n+1} = \boldsymbol{\sigma}_n + \mathbf{C}\Delta\boldsymbol{\varepsilon} - \Delta\lambda\mathbf{C}\mathbf{r}^{n+1}(\boldsymbol{\sigma}_{n+1}, \varsigma_{n+1}) \tag{10.40a}$$

$$\varsigma_{n+1} = \varsigma_n + \Delta\lambda s^{n+1}(\boldsymbol{\sigma}_{n+1}, \varsigma_{n+1}) \tag{10.40b}$$

These represent the "algorithmic" equations for the calculation of $\boldsymbol{\sigma}_{n+1}$ and $\boldsymbol{\zeta}_{n+1}$. In addition to these, there is the consistency equation

$$R_\phi = \phi^{n+1}(\boldsymbol{\sigma}_{n+1}, \boldsymbol{\zeta}_{n+1}) = \mathbf{R}_\phi(\boldsymbol{\sigma}_{n+1}, \boldsymbol{\zeta}_{n+1}) = 0 \tag{10.41}$$

By taking the rate of (10.40a), (10.40b), and (10.41), and combining the equations to eliminate $\dot{\varsigma}$ and $\dot{\lambda}$, an equation relating $\dot{\boldsymbol{\sigma}}$ to $\dot{\boldsymbol{\varepsilon}}$ can be derived in the form

$$\dot{\boldsymbol{\sigma}} = \mathbf{D}^t\dot{\boldsymbol{\varepsilon}} \tag{10.42}$$

where $\mathbf{D}^t$ is the consistent tangent operator. The expression for $\mathbf{D}^t$ is derived in a subsequent section.

10.7.3 Description of CPPM: Variable Elastic Properties

When the elastic properties vary during the iteration (as in geologic materials where the properties are pressure-dependent), the elastic predictor does not remain the same during the iterations. This is because the elastic modular matrix cannot be taken out of the integral

$$\Delta\boldsymbol{\sigma}^{\text{ep}} = \int_t^{t+\Delta t} \mathbf{C}\dot{\boldsymbol{\varepsilon}} \neq \mathbf{C}\int_t^{t+\Delta t} \dot{\boldsymbol{\varepsilon}} = \mathbf{C}\Delta\varepsilon \tag{10.43}$$

In addition, the elastic modular matrix cannot be taken out of the integral in carrying out the integration in (10.27c) either, i.e.,

$$\Delta\boldsymbol{\sigma}^{\text{pc}} = -\int_t^{t+\Delta t} \dot{\lambda}\mathbf{C}\mathbf{r} \neq -\mathbf{C}\int_t^{t+\Delta t} \dot{\lambda}\mathbf{r} \approx -\Delta\lambda\mathbf{C}\mathbf{r}^{n+1} \tag{10.44}$$

There are two ways to handle variable elastic properties:

1. Approximate the elastic properties using, say, the generalized midpoint rule or the generalized trapezoidal rule, and integrate (10.43) and (10.44), or
2. Integrate (10.43) analytically and replace the total strain increment $\Delta\boldsymbol{\varepsilon}$ with elastic strain increment by subtracting the plastic part $\Delta\lambda\mathbf{r}$

Let us consider the first method with midpoint rule for the elastic properties. Equations (10.43) and (10.44) are integrated as

$$\Delta\boldsymbol{\sigma}^{\mathrm{ep}} = \int_{t}^{t+\Delta t} \boldsymbol{C}\dot{\boldsymbol{\varepsilon}} \approx \boldsymbol{C}^{n+\theta}\Delta\varepsilon \tag{10.45a}$$

$$\Delta\boldsymbol{\sigma}^{\mathrm{pc}} = -\int_{t}^{t+\Delta t} \dot{\lambda}\mathbf{C}\mathbf{r} \approx -\Delta\lambda\mathbf{C}^{n+\theta}\mathbf{r}^{n+1} \tag{10.45b}$$

where

$$\boldsymbol{C}^{n+\theta} = \boldsymbol{C}(K_{n+\theta}, G_{n+\theta}) \tag{10.46a}$$

$$K_{n+\theta} = K(\boldsymbol{\sigma}_{n+\theta}) \tag{10.46b}$$

$$G_{n+\theta} = G(\boldsymbol{\sigma}_{n+\theta}) \tag{10.46c}$$

$$\boldsymbol{\sigma}_{n+\theta} = (1-\theta)\boldsymbol{\sigma}_n + \theta\boldsymbol{\sigma}_{n+1}\,; \quad \theta \in [0,1] \tag{10.46d}$$

The method employs explicit elastic properties when $\theta = 0$ and implicit elastic properties when $\theta > 0$. The iterative behavior is shown in Fig. 10.8a, where it is noted that both the elastic predictor and plastic corrector change until convergence. (Note that iterations are required even in finding the elastic predictor.)

Experience indicates that, due to the fact that the mean normal pressure p could become negative during the iteration, the explicit method (which uses only p_n, not p_{n+1} performs better than the implicit method).

Now let us consider the second method where the elastic relation is integrated analytically. Depending on the elastic model involved, a suitable method must be devised to integrate the equations. The plastic corrector is obtained by simply replacing the strain increments in the elastic predictors by elastic strain increments. These are obtained by subtracting the plastic strain increments from the total strain increments. The elastic predictor is used (1) for determining whether the strain increment causes plastic behavior, and (2) as the first estimate of $\boldsymbol{\sigma}_{n+1}$ needed during the iterative procedure. Other than this, the notion of the elastic predictor does not strictly apply here, since the incremental equations are coupled and the integration is done directly from point n to $n+1$, as shown schematically in Fig. 10.8b. Further details are given in subsequent chapters on geologic materials.

From an algorithmic stand point, the major difference between the two methods of handling the pressure-dependent elastic properties is the following: In the first method, the form of the residual equations (10.31a–c) is independent of the specific nature of the elastic relations. The influence of the specific relationship between the elastic properties and pressure enters through the derivative of $\mathbf{C}$ with respect to $\boldsymbol{\sigma}$. In the second method, the form of the residual equations changes with the specific relationship between the elastic properties and pressure. Hence, when the elastic relation is changed, the residual equations need to be re-derived.

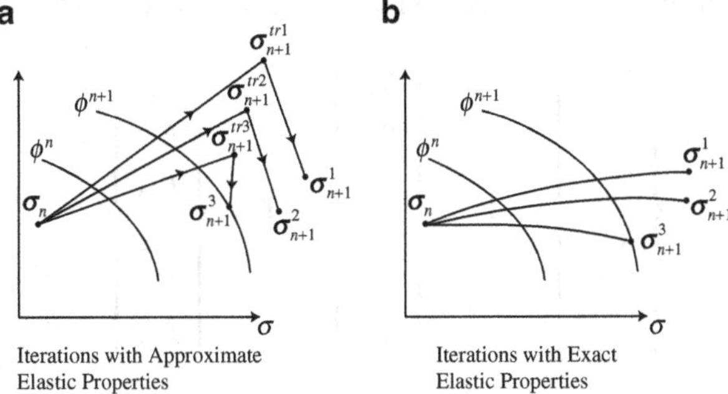

Fig. 10.8 Schematic of procedure associated with CPPM: The case of variable elastic properties

10.7.4 Solution Strategies

10.7.4.1 Simplification due to Symmetries

The number of numerical operations can be greatly reduced if existing symmetries are recognized and taken advantage of. Recognizing that the stress and strain tensors are symmetric, we will use Voigt's notation and place their elements in vectors as

$$\bar{\boldsymbol{\sigma}} = \{ \sigma_{11} \quad \sigma_{22} \quad \sigma_{33} \quad \sigma_{12} \quad \sigma_{23} \quad \sigma_{13} \} \tag{10.47a}$$

$$\bar{\boldsymbol{\varepsilon}} = \{ \varepsilon_{11} \quad \varepsilon_{22} \quad \varepsilon_{33} \quad \gamma_{12} \quad \gamma_{23} \quad \gamma_{13} \} \tag{10.47b}$$

where $\gamma_{12} = 2\varepsilon_{12}$, $\gamma_{23} = 2\varepsilon_{23}$, and $\gamma_{13} = 2\varepsilon_{13}$. The use of engineering strains renders the method of forming the stress vector to be different from the method of forming the strain vector from their tensorial counterparts. With this difference in mind, in the following, we will refer to second order tensors as either a "stress-like" quantity or a "strain-like" quantity. We use similar representations for the incremental stresses and strains.

The equations in (10.34a–c) can be written in reduced order matrix– vector form as follows. Define:

$$\mathbf{A} = \begin{bmatrix} A_{11} & A_{12} & A_{13} & A_{14} & A_{15} & A_{16} \\ A_{21} & A_{22} & A_{23} & A_{24} & A_{25} & A_{26} \\ A_{31} & A_{32} & A_{33} & A_{34} & A_{35} & A_{36} \\ A_{41} & A_{42} & A_{43} & A_{44} & A_{45} & A_{46} \\ A_{51} & A_{52} & A_{53} & A_{54} & A_{55} & A_{56} \\ A_{61} & A_{62} & A_{63} & A_{64} & A_{65} & A_{66} \end{bmatrix} \tag{10.48a}$$

For details on the generation of the coefficients of **A** from $A_{ijk\ell}$, refer to Appendix 6.

$$\mathbf{R}_\sigma = \begin{Bmatrix} R^1_{\sigma 11} \\ R^1_{\sigma 22} \\ R^1_{\sigma 33} \\ R^1_{\sigma 12} \\ R^1_{\sigma 23} \\ R^1_{\sigma 13} \end{Bmatrix}; \quad \mathbf{B} = \begin{Bmatrix} B_{11} \\ B_{22} \\ B_{33} \\ B_{12} \\ B_{23} \\ B_{13} \end{Bmatrix}; \quad \mathbf{F} = \begin{Bmatrix} F_{11} \\ F_{22} \\ F_{33} \\ F_{12} \\ F_{23} \\ F_{13} \end{Bmatrix} \tag{10.48b}$$

$$\mathbf{H}^{\mathrm{T}} = \left\{ H_{11} \quad H_{22} \quad H_{33} \quad 2H_{12} \quad 2H_{23} \quad 2H_{13} \right\} \tag{10.48c}$$

$$\mathbf{E}^{\mathrm{T}} = \left\{ E_{11} \quad E_{22} \quad E_{33} \quad 2E_{12} \quad 2E_{23} \quad 2E_{13} \right\} \tag{10.48d}$$

Then (10.34a–c) may be written in a matrix-vector form as

$$\mathbf{A}\,\delta\boldsymbol{\sigma} + \mathbf{B}\,\delta\varsigma + \mathbf{F}\,\delta\lambda = -\mathbf{R}^1_{\boldsymbol{\sigma}} \tag{10.49a}$$

$$\mathbf{H}^{\mathrm{T}}\,\delta\boldsymbol{\sigma} + \omega\,\delta\varsigma + \beta\,\delta\lambda = -R^1_{\varsigma} \tag{10.49b}$$

$$\mathbf{E}^{\mathrm{T}}\,\delta\boldsymbol{\sigma} + \gamma\,\delta\varsigma = -R^1_{\phi} \tag{10.49c}$$

10.7.4.2 Solution Method 1

The three matrix–vector equations in (10.49a–c) can be arranged in a single matrix–vector equation as

$$\begin{bmatrix} \underset{6\times 6}{\mathbf{A}} & \underset{6\times 1}{\mathbf{B}} & \underset{6\times 1}{\mathbf{F}} \\ \underset{1\times 6}{\mathbf{H}^{\mathrm{T}}} & \underset{1\times 1}{\omega} & \underset{1\times 1}{\beta} \\ \underset{1\times 6}{\mathbf{E}^{\mathrm{T}}} & \underset{1\times 1}{\gamma} & \underset{1\times 1}{0} \end{bmatrix} \begin{bmatrix} \underset{6\times 1}{\delta\boldsymbol{\sigma}} \\ \underset{1\times 1}{\delta\varsigma} \\ \underset{1\times 1}{\delta\lambda} \end{bmatrix} = \begin{bmatrix} \underset{6\times 1}{-\mathbf{R}^1_{\sigma}} \\ \underset{1\times 1}{-R^1_{\varsigma}} \\ \underset{1\times 1}{-R^1_{\phi}} \end{bmatrix} \tag{10.50}$$

The above equation system can be easily solved.

10.7.4.3 Solution Method 2

Equations (10.49a–c) can be directly manipulated to obtain the solution as follows: Let us first find an expression for $\delta\varsigma$. From (10.49b), we have

$$\delta\varsigma = -\frac{1}{\omega}\left\{R^1_{\varsigma} + \mathbf{H}^{\mathrm{T}}\,\delta\boldsymbol{\sigma} + \beta\,\delta\lambda\right\} \tag{10.51a}$$

Substitute this in (10.49a), it follows:

$$\boldsymbol{0} = \mathbf{R}_\sigma^1 + \mathbf{A}\,\delta\boldsymbol{\sigma} - \mathbf{B}\frac{1}{\omega}\left\{R_\varsigma^1 + \mathbf{H}^\mathrm{T}\,\delta\boldsymbol{\sigma} + \beta\,\delta\lambda\right\} + \mathbf{F}\,\delta\lambda$$

$$\delta\boldsymbol{\sigma} = -\left[\mathbf{A} - \frac{1}{\omega}\mathbf{B}\mathbf{H}^\mathrm{T}\right]^{-1}\left\{\mathbf{R}_\sigma^1 - \frac{R_\varsigma^1}{\omega}\mathbf{B} - \frac{\beta}{\omega}\mathbf{B}\,\delta\lambda + \mathbf{F}\,\delta\lambda\right\}$$

Let

$$\mathbf{R} = \left[\mathbf{A} - \frac{1}{\omega}\mathbf{B}\mathbf{H}^\mathrm{T}\right]^{-1} \tag{10.51b}$$

Then

$$\delta\boldsymbol{\sigma} = \mathbf{R}\left\{-\mathbf{R}_\sigma^1 + \frac{R_\varsigma^1}{\omega}\mathbf{B} + \frac{\beta}{\omega}\mathbf{B}\,\delta\lambda - \mathbf{F}\,\delta\lambda\right\} \tag{10.51c}$$

From (10.49c) and (10.51a), we have

$$R_\phi^1 + \mathbf{E}^\mathrm{T}\,\delta\boldsymbol{\sigma} - \gamma\frac{1}{\omega}\left\{R_\varsigma^1 + \mathbf{H}^\mathrm{T}\,\delta\boldsymbol{\sigma} + \beta\,\delta\lambda\right\} = 0$$

Substituting for $\delta\boldsymbol{\sigma}$ from (10.51c), we get

$$R_\phi^1 + \left[\mathbf{E}^\mathrm{T} - \frac{\gamma}{\omega}\mathbf{H}^\mathrm{T}\right]\mathbf{R}\left\{-\mathbf{R}_\sigma^1 + \frac{R_\varsigma^1}{\omega}\mathbf{B} + \frac{\beta}{\omega}\mathbf{B}\,\delta\lambda - \mathbf{F}\,\delta\lambda\right\} - \gamma\frac{1}{\omega}\left\{R_\varsigma^1 + \beta\,\delta\lambda\right\} = 0 \tag{10.51d}$$

Noting that $\mathbf{E} = \mathbf{n}$ (10.36a), define a modified vector $\bar{\mathbf{n}}$ as

$$\bar{\mathbf{n}} = \mathbf{E} - \frac{\gamma}{\omega}\mathbf{H} = \mathbf{n} - \frac{\gamma}{\omega}\mathbf{H} \tag{10.51e}$$

Also introduce the following notation

$$\bar{\mathbf{r}}^* = \mathbf{F} - \frac{\beta}{\omega}\mathbf{B} \tag{10.51f}$$

Equation (10.51d) is rewritten as

$$R_{\phi}^{1} + \bar{\mathbf{n}}^{\mathrm{T}}\mathbf{R}\left\{-\mathbf{R}_{\sigma}^{1} + \frac{R_{\varsigma}^{1}}{\omega}\mathbf{B} - \bar{\mathbf{r}}^{*}\,\delta\lambda\right\} - \gamma\frac{1}{\omega}\left\{R_{\varsigma}^{1} + \beta\,\delta\lambda\right\} = 0$$

$$\delta\lambda = \frac{R_{\phi}^{1} - \bar{\mathbf{n}}^{\mathrm{T}}\mathbf{R}\mathbf{R}_{\sigma}^{1} + \frac{R_{\varsigma}^{1}}{\omega}\bar{\mathbf{n}}^{\mathrm{T}}\mathbf{R}\mathbf{B} - \frac{\gamma}{\omega}R_{\varsigma}^{1}}{\frac{\gamma\beta}{\omega} + \bar{\mathbf{n}}^{\mathrm{T}}\mathbf{R}\bar{\mathbf{r}}^{*}} \tag{10.52}$$

Let us introduce the following quantities (the reason for which will be explained later)

$$\bar{\mathbf{r}} = \mathbf{C}^{-1}\bar{\mathbf{r}}^{*} = \mathbf{C}^{-1}\left[\mathbf{F} - \frac{\beta}{\omega}\mathbf{B}\right] = \mathbf{C}^{-1}\mathbf{F} - \frac{\beta}{\omega}\mathbf{C}^{-1}\mathbf{B} = \bar{\mathbf{F}} - \frac{\beta}{\omega}\bar{\mathbf{B}}, \tag{10.53a}$$

where

$$\bar{\mathbf{F}} = \mathbf{C}^{-1}\mathbf{F} \quad \text{and} \quad \bar{\mathbf{B}} = \mathbf{C}^{-1}\mathbf{B} \tag{10.53b}$$

Also, define $x_0 = \dfrac{1}{\omega}$ and a modified elastic operator (10.53c)

$$\bar{\mathbf{C}} = \mathbf{R}\mathbf{C} \tag{10.53d}$$

Now, modify (10.52) in terms of these variables as follows. First insert $\mathbf{I} = \mathbf{C}\mathbf{C}^{-1}$ as

$$\delta\lambda = \frac{R_{\phi}^{1} - \bar{\mathbf{n}}^{\mathrm{T}}\mathbf{R}\mathbf{R}_{\sigma}^{1} + x_0 R_{\varsigma}^{1}\bar{\mathbf{n}}^{\mathrm{T}}\mathbf{R}\mathbf{C}\mathbf{C}^{-1}\mathbf{B} - x_0\gamma R_{\varsigma}^{1}}{x_0\gamma\beta + \bar{\mathbf{n}}^{\mathrm{T}}\mathbf{R}\mathbf{C}\mathbf{C}^{-1}\bar{\mathbf{r}}^{*}} \tag{10.54}$$

Simplify this in terms of the new quantities (i.e., quantities with an over bar)

$$\delta\lambda = \frac{R_{\phi}^{1} - \bar{\mathbf{n}}^{\mathrm{T}}\mathbf{R}\mathbf{R}_{\sigma}^{1} + x_0 R_{\varsigma}^{1}\bar{\mathbf{n}}^{\mathrm{T}}\bar{\mathbf{C}}\bar{\mathbf{B}} - x_0\gamma R_{\varsigma}^{1}}{x_0\gamma\beta + \bar{\mathbf{n}}^{\mathrm{T}}\bar{\mathbf{C}}\bar{\mathbf{r}}} \tag{10.55}$$

Now, using (10.51f), $\delta\boldsymbol{\sigma}$ is obtained from (10.51c) as

$$\begin{aligned}
\delta\boldsymbol{\sigma} &= \mathbf{R}\left\{-\mathbf{R}_{\sigma}^{1} + \frac{R_{\varsigma}^{1}}{\omega}\mathbf{B} - \bar{\mathbf{r}}^{*}\,\delta\lambda\right\} \\
&= -\mathbf{R}\mathbf{R}_{\sigma}^{1} + \frac{R_{\varsigma}^{1}}{\omega}\mathbf{R}\mathbf{B} - \mathbf{R}\bar{\mathbf{r}}^{*}\,\delta\lambda \\
&= -\mathbf{R}\mathbf{R}_{\sigma}^{1} + \frac{R_{\varsigma}^{1}}{\omega}\mathbf{R}\mathbf{C}\mathbf{C}^{-1}\mathbf{B} - \mathbf{R}\mathbf{C}\mathbf{C}^{-1}\bar{\mathbf{r}}^{*}\,\delta\lambda \\
&= -\mathbf{R}\mathbf{R}_{\sigma}^{1} + x_0 R_{\varsigma}^{1}\bar{\mathbf{C}}\bar{\mathbf{B}} - \bar{\mathbf{C}}\bar{\mathbf{r}}\,\delta\lambda
\end{aligned} \tag{10.56}$$

From (10.49c)

$$\delta\varsigma = -\frac{1}{\gamma}\left[R_{\phi}^{1} + \mathbf{E}^{\mathrm{T}}\,\delta\boldsymbol{\sigma}\right]$$

$$= -\frac{1}{\gamma}\left[R_{\phi}^{1} + \mathbf{n}^{\mathrm{T}}\,\delta\boldsymbol{\sigma}\right] \tag{10.57}$$

10.7.4.4 Summary of Equations for Solution Method 2

$$\delta\lambda = \frac{R_{\phi}^{1} - \bar{\mathbf{n}}^{\mathrm{T}}\mathbf{R}\mathbf{R}_{\sigma}^{1} + x_0 R_{\varsigma}^{1}\bar{\mathbf{n}}^{\mathrm{T}}\bar{\mathbf{C}}\bar{\mathbf{B}} - x_0\gamma R_{\varsigma}^{1}}{x_0\gamma\beta + \bar{\mathbf{n}}^{\mathrm{T}}\bar{\mathbf{C}}\bar{\mathbf{r}}} \tag{10.58a}$$

$$\delta\boldsymbol{\sigma} = -\mathbf{R}\mathbf{R}_{\sigma}^{1} + x_0 R_{\varsigma}^{1}\bar{\mathbf{C}}\bar{\mathbf{B}} - \bar{\mathbf{C}}\bar{\mathbf{r}}\,\delta\lambda \tag{10.58b}$$

$$\delta\varsigma = -\frac{1}{\gamma}\left[R_{\phi}^{1} + \mathbf{n}^{\mathrm{T}}\,\delta\boldsymbol{\sigma}\right] \tag{10.58c}$$

where

$$\bar{\mathbf{r}} = \bar{\mathbf{F}} - \frac{\beta}{\omega}\bar{\mathbf{B}} \tag{10.59a}$$

$$\bar{\mathbf{n}} = \mathbf{n} - \frac{\gamma}{\omega}\mathbf{H} \tag{10.59b}$$

$$\bar{\mathbf{F}} = \mathbf{C}^{-1}\mathbf{F} \text{ and } \bar{\mathbf{B}} = \mathbf{C}^{-1}\mathbf{B} \tag{10.59c}$$

$$x_0 = \frac{1}{\omega} \tag{10.59d}$$

$$\bar{\mathbf{C}} = \mathbf{R}\mathbf{C} \tag{10.59e}$$

$$\mathbf{R} = \left[\mathbf{A} - \frac{1}{\omega}\mathbf{B}\mathbf{H}^{\mathrm{T}}\right]^{-1} \tag{10.59f}$$

All of the quantities in (10.58)–(10.59) can be calculated from **A, B, F, H, E** (= **n**), $\omega, \beta,$ and γ (10.36), the elastic modular matrix **C** and the residuals $\mathbf{R}_{\sigma}^{1}$, R_{ς}^{1}, and R_{ϕ}^{1}. Note that the method requires the inversion of a 6×6 matrix (10.59f). It can be verified that as $\Delta\lambda \to 0$, we have $\bar{\mathbf{C}} \to \mathbf{C}$ (from (10.36a), (10.36d), (10.59f), and (10.59e)) and $\bar{\mathbf{n}} \to \mathbf{n}$ (from (10.36d) and (10.59b)).

The reason for the particular form of $\bar{\mathbf{r}}$, $\bar{\mathbf{F}}$, and $\bar{\mathbf{B}}$ will now be explained. From (10.59c) and (10.36c), we have

$$\bar{\mathbf{F}} = \mathbf{C}^{-1}\mathbf{C}\mathbf{r} = \mathbf{r} \tag{10.60a}$$

From (10.59c) and (10.36b), we get

$$\bar{\mathbf{B}} = \Delta\lambda\mathbf{C}^{-1}\mathbf{C}\frac{\partial \mathbf{r}^{n+1}}{\partial \varsigma} = \Delta\lambda\frac{\partial \mathbf{r}^{n+1}}{\partial \varsigma} \tag{10.60b}$$

Then from (10.53a), it follows:

$$\bar{\mathbf{r}} = \mathbf{r} - \Delta\lambda\frac{\beta}{\omega}\frac{\partial \mathbf{r}^{n+1}}{\partial \varsigma} \tag{10.60c}$$

Thus, $\bar{\mathbf{r}}$ is a modified flow direction and as $\Delta\lambda \to 0$, $\bar{\mathbf{r}} \to \mathbf{r}$, just as $\bar{\mathbf{n}} \to \mathbf{n}$.

The steps involved in integrating a constitutive law with constant elastic properties by CPPM are summarized in Algorithm 10.3.

Algorithm 10.3. CPPM for Integrating constitutive models to evaluate incremental stress

Definition of (new) parameters: $N_L^{\max}$ is the maximum number of iterations permitted before execution is terminated, and N_L is the corresponding counter. *TOL* is a small, machine-dependent number (it could be set equal to TOL_ϕ).

1. Establish $\Delta\boldsymbol{\varepsilon}$, $\boldsymbol{\zeta}_n$, $\boldsymbol{\sigma}_n$, TOL_σ, TOL_ς, TOL_ϕ, TOL, $N_L^{\max}$, and the constitutive model parameters
2. Initialize parameters: $N_L = 0$ and $\boldsymbol{\zeta}_{n+1} = \boldsymbol{\zeta}_n$
3. Compute elastic predictor $\boldsymbol{\sigma}_{n+1}^{\text{tr}}$, C, and $\phi^{\text{tr}}(\boldsymbol{\sigma}_{n+1}^{\text{tr}}, \boldsymbol{\zeta}_n)$. Perform loading/unloading check: if ($\phi^{\text{tr}} < TOL$), set $\boldsymbol{\sigma}_{n+1} = \boldsymbol{\sigma}_{n+1}^{\text{tr}}$ and go to step 11
4. Set $\boldsymbol{\sigma}_{n+1} = \boldsymbol{\sigma}_{n+1}^{\text{tr}}$ and calculate $\mathbf{r}_{n+1}$, $\mathbf{n}_{n+1}$, and K_p^{n+1}
5. Set $R_\phi^1 = \phi^{\text{tr}}$. Compute initial value for $\Delta\lambda$ (10.30). Compute $\boldsymbol{\sigma}_{n+1}$ (10.40a) and $\boldsymbol{\zeta}_{n+1}$ (10.40b). Set $\boldsymbol{\sigma}_{n+1} = \boldsymbol{\sigma}_{n+1}^1$ and $\boldsymbol{\zeta}_{n+1} = \boldsymbol{\zeta}_{n+1}^1$
6. Begin iterations. Compute $R_\phi^1 = \phi(\boldsymbol{\sigma}_{n+1}, \boldsymbol{\zeta}_{n+1})$, $\mathbf{r}_{n+1}$, and $\mathbf{n}_{n+1}$
7. Compute $\boldsymbol{\sigma}_{n+1}^2$ (10.40a) and $\boldsymbol{\zeta}_{n+1}^2$ (10.40b). Compute $\mathrm{R}_\sigma^1 = \boldsymbol{\sigma}_{n+1}^1 - \boldsymbol{\sigma}_{n+1}^2$ and $R_\varsigma^1 = \boldsymbol{\zeta}_{n+1}^1 - \boldsymbol{\zeta}_{n+1}^2$
8. Find errors e_σ, e_ς, and e_ϕ (10.38) and perform convergence check. If converged, go to step 11
9. Set $N_L \leftarrow N_L + 1$. If ($N_L > N_L^{\max}$) STOP
10. Compute $\delta\lambda$, $\delta\boldsymbol{\sigma}$, and $\delta\boldsymbol{\zeta}$ ((10.50) or (10.58)). Update $\boldsymbol{\sigma}_{n+1}^1 \leftarrow \boldsymbol{\sigma}_{n+1}^1 + \delta\boldsymbol{\sigma}$, $\boldsymbol{\zeta}_{n+1}^1 = \boldsymbol{\zeta}_{n+1}^1 + \delta\boldsymbol{\zeta}$, and $\Delta\lambda \leftarrow \Delta\lambda + \delta\lambda$. Set $\boldsymbol{\sigma}_{n+1} = \boldsymbol{\sigma}_{n+1}^1$ and $\boldsymbol{\zeta}_{n+1} = \boldsymbol{\zeta}_{n+1}^1$. Go to step 6
11. Return to the main program

10.7.5 Elasto-Plastic Consistent Tangent Operator: Constant Elastic Properties

The algorithmic equations in tensor notation are ((10.40a), (10.40b), and (10.41))

$$\sigma_{ij}^{n+1} = \sigma_{ij}^{n} + C_{ijk\ell}\Delta\varepsilon_{k\ell} - \Delta\lambda C_{ijk\ell} r_{k\ell}^{n+1} \tag{10.61a}$$

$$\varsigma_{n+1} = \varsigma_n + \Delta\lambda s^{n+1} \tag{10.61b}$$

$$\phi^{n+1} = 0 \tag{10.61c}$$

Taking the rate of these equations and treating $\Delta\boldsymbol{\varepsilon}_{ij}$ as a variable, we have

$$\begin{aligned}\dot{\sigma}_{ij} = {} & \frac{\partial C_{ijk\ell}}{\partial \sigma_{pq}}\dot{\sigma}_{pq}\Delta\varepsilon_{k\ell} + C_{ijk\ell}\dot{\varepsilon}_{k\ell} - \Delta\lambda\frac{\partial C_{ijk\ell}}{\partial \sigma_{pq}}\dot{\sigma}_{pq} r_{k\ell}^{n+1} - \Delta\lambda C_{ijk\ell}\frac{\partial r_{k\ell}^{n+1}}{\partial \sigma_{pq}}\dot{\sigma}_{pq} \\ & - \Delta\lambda C_{ijk\ell}\frac{\partial r_{k\ell}^{n+1}}{\partial \varsigma}\dot{\varsigma} - C_{ijk\ell} r_{k\ell}^{n+1}\dot{\lambda}\end{aligned} \tag{10.62a}$$

$$\dot{\varsigma} = \Delta\lambda\frac{\partial s^{n+1}}{\partial \sigma_{pq}}\dot{\sigma}_{pq} + \Delta\lambda\frac{\partial s^{n+1}}{\partial \varsigma}\dot{\varsigma} + \left[s^{n+1} + \Delta\lambda\frac{\partial s^{n+1}}{\partial \Delta\lambda}\right]\dot{\lambda} \tag{10.62b}$$

$$0 = \frac{\partial \phi^{n+1}}{\partial \sigma_{pq}}\dot{\sigma}_{pq} + \frac{\partial \phi^{n+1}}{\partial \varsigma}\dot{\varsigma} \tag{10.62c}$$

Rearrange the equations as

$$\mathbf{R}_{\sigma}^{*1} + \mathbf{A}^{*}\dot{\boldsymbol{\sigma}} + \mathbf{B}^{*}\dot{\varsigma} + \mathbf{F}^{*}\dot{\lambda} = 0 \tag{10.63a}$$

$$R_{\varsigma}^{*1} + \mathbf{H}^{*\mathrm{T}}\dot{\boldsymbol{\sigma}} + \omega^{*}\dot{\varsigma} + \beta^{*}\dot{\lambda} = 0 \tag{10.63b}$$

$$R_{\phi}^{*1} + \mathbf{E}^{*\mathrm{T}}\dot{\boldsymbol{\sigma}} + \gamma^{*}\dot{\varsigma} = 0 \tag{10.63c}$$

By comparing (10.63) with (10.49), and the terms in (10.62) with those in (10.36), it is easy to verify that:

$$\mathrm{A}^* = \mathrm{A}, \mathrm{B}^* = \mathrm{B}, \mathrm{F}^* = \mathrm{F}, \mathrm{H}^* = \mathrm{H}, \mathrm{E}^* = \mathrm{E},\ \omega^* = \omega, \beta^* = \beta \text{ and } \gamma^* = \gamma$$

The residuals are given by

$$R_{\sigma}^{*1} = -C_{ijk\ell}\dot{\varepsilon}_{k\ell}; \quad R_{\varsigma}^{*1} = 0; \quad R_{\phi}^{*1} = 0 \tag{10.64}$$

From (10.58a) and (10.58b), it follows:

$$\dot{\lambda} = \frac{\bar{\mathbf{n}}^{\mathrm{T}}\mathbf{R}\mathbf{C}\dot{\boldsymbol{\varepsilon}}}{x_0\gamma\beta + \bar{\mathbf{n}}^{\mathrm{T}}\bar{\mathbf{C}}\bar{\mathbf{r}}} = \frac{\bar{\mathbf{n}}^{\mathrm{T}}\bar{\mathbf{C}}\dot{\boldsymbol{\varepsilon}}}{x_0\gamma\beta + \bar{\mathbf{n}}^{\mathrm{T}}\bar{\mathbf{C}}\bar{\mathbf{r}}} \tag{10.65a}$$

$$\dot{\boldsymbol{\sigma}} = \mathbf{R}\mathbf{C}\dot{\boldsymbol{\varepsilon}} - \bar{\mathbf{C}}\bar{\mathbf{r}}\dot{\lambda} \tag{10.65b}$$

Combining the two, we have

$$\dot{\boldsymbol{\sigma}} = \mathbf{R}\mathbf{C}\dot{\boldsymbol{\varepsilon}} - \bar{\mathbf{C}}\bar{\mathbf{r}}\frac{\bar{\mathbf{n}}^{\mathrm{T}}\bar{\mathbf{C}}\dot{\boldsymbol{\varepsilon}}}{x_0\gamma\beta + \bar{\mathbf{n}}^{\mathrm{T}}\bar{\mathbf{C}}\bar{\mathbf{r}}} = \left[\bar{\mathbf{C}} - \frac{(\bar{\mathbf{C}}\bar{\mathbf{r}})(\bar{\mathbf{n}}^{\mathrm{T}}\bar{\mathbf{C}})}{x_0\gamma\beta + \bar{\mathbf{n}}^{\mathrm{T}}\bar{\mathbf{C}}\bar{\mathbf{r}}}\right]\dot{\boldsymbol{\varepsilon}} = \mathbf{D}\dot{\boldsymbol{\varepsilon}} \tag{10.66a}$$

where

$$\mathbf{D} = \left[\bar{\mathbf{C}} - \frac{(\bar{\mathbf{C}}\bar{\mathbf{r}})(\bar{\mathbf{n}}^{\mathrm{T}}\bar{\mathbf{C}})}{x_0\gamma\beta + \bar{\mathbf{n}}^{\mathrm{T}}\bar{\mathbf{C}}\bar{\mathbf{r}}}\right] \tag{10.66b}$$

This is the expression for the consistent tangent operator.

A number of observations can be made. Noting from (10.36f) and (10.36h) that

$$\beta = -s^{n+1} - \Delta\lambda\frac{\partial s^{n+1}}{\partial\Delta\lambda}$$

and

$$\gamma = \frac{\partial\phi^{n+1}}{\partial\varsigma}$$

it follows from (10.2e) that $x_0\gamma\beta$ is a modified plastic modulus, say, K_p^*. Thus, the expression for the consistent tangent operator is in the same form as that for the continuum tangent operator (10.3a), with modified elastic tangent operator $\bar{\mathbf{C}}$, flow direction $\bar{\mathbf{r}}$, yield direction $\bar{\mathbf{n}}$, and plastic modulus K_p^*. It can be verified that as $\Delta\lambda \to 0$, $\bar{\mathbf{C}} \to \mathbf{C}$, $\bar{\mathbf{r}} \to \mathbf{r}$, $\bar{\mathbf{n}} \to \mathbf{n}$, and $K_p^* \to K_p$. The continuum operator is recovered from the consistent operator.

Even when the model employs an associated flow rule (i.e., $\mathbf{r} = \mathbf{n}$), it follows that $\bar{\mathbf{r}} \neq \bar{\mathbf{n}}$ and hence in general $\bar{\mathbf{C}} \neq \bar{\mathbf{C}}^{\mathrm{T}}$. Thus, $\mathbf{D}$ is nonsymmetric in general. In some simple cases, however, $\mathbf{D}$ could still be symmetric (e.g., isotropic hardening von Mises model described in a subsequent chapter).

10.7.6 Convergence Difficulties

CPPM is based on Newton's iterative scheme and suffers from the drawback inherent in Newton's methods. For instance, the method converges only when the initial guess is near the solution. How near must the initial guess be varies with the inherent degree of nonlinearity of the hardening rules, yield functions, etc. There are several other reasons why the method may lead to divergence (variables gradually increasing to infinity) or non-convergence (oscillations).

In general, CPPM encounters convergence difficulties in the following situations:

- At the local level, iterations are performed in establishing the plastic corrector from the trial point (Fig. 10.2). Hence, non-convergence is possible when the trial point is too far from the solution. For example, consider a model that simulates a complex combination of hardening and softening in different regimes. The trial stress may end up in the "softening" regime when the final solution is actually supposed to be in the "hardening" regime. This may lead to divergence or non-convergence.
- In some cases, the change in a hardening variable during the local iteration may simply be too large resulting in divergence or non-convergence. For example, in the Cam-clay model, the hardening variable representing the yield surface size (I_0) may change from a positive initial value to a negative value during the iteration when I_0 must never be negative. This may lead to divergence or non-convergence.
- Convergence difficulties may be encountered when the solution lies near corners on the yield and plastic potential surfaces. For example, in the Drucker-Prager and sliding–rolling models, the above surfaces are conical and the apex of the cones is one such corner. In the sliding–rolling model, the yield surface is a composite consisting of a cone and a plane normal to the axis; the intersection between the two again introduces corners. Special care is needed to deal with corners.
- Convergence difficulties may be encountered when the yield and plastic potential surfaces are noncircular in the octahedral plane. For accurate matching between experimental and theoretical stress–strain relations, noncircular surfaces are often required for geologic materials. Convergence difficulties have been noted by others in the past near points of high curvature of the yield and potential surfaces.
- In pressure-dependent models (Drucker–Prager, Cam-clay, and sliding–rolling models), the elastic and plastic moduli are functions of the mean normal pressure. The elastic moduli become undefined when the magnitude of the mean normal pressure is less than a limit (either zero or a negative value depending on the functional form between the moduli and the mean normal pressure). The plastic modulus changes from positive to negative at a limit pressure. Consequently, convergence difficulties are encountered around the limiting pressure.

The convergence difficulties mentioned above have been noted and treated in some of the recent papers and in this book in the subsequent chapters. The convergence problems around corners and regions of high curvature of the yield surface have been discussed by De Borst (1986), Crisfield (1987), Bicanic and

Pearce (1996), and Perez-Foguet and Armero (2002). Various techniques for dealing with such problems have been proposed and tested, e.g., use improved predictors rather than the elastic trial stresses as starting points for the plastic corrector iterations (De Souza Neto et al. 1994; Bicanic and Pearce 1996); use line search in the iterative process (Crisfield 1987; Dutko et al. 1993; Bicanic and Pearce 1996; Armero and Perez-Foguet 2002; Perez-Foguet and Armero 2002); use sub-stepping (Pérez-Foguet et al. 2001), etc.

These convergence difficulties will be discussed at appropriate places in the following chapters in the context of specific models where the corresponding difficulties are generally encountered.

10.8 Accuracy and Efficiency of Different Methods

Three different methods of implementing a standard plasticity model have been presented in the preceding sections. We will now briefly comment on the accuracy and efficiency of these methods.

10.8.1 Accuracy

Except for the two-step Euler method with sub incrementing, the remaining two methods (the cutting plane and CPPM) are variants of the elastic predictor–plastic corrector method. In the two-step Euler method, the stress rate is integrated as

$$\int_{t}^{t+\Delta t} \dot{\boldsymbol{\sigma}} = \int_{t}^{t+\Delta t} \bar{\mathbf{D}}\dot{\boldsymbol{\varepsilon}} \approx \frac{1}{2}[\bar{\mathbf{D}}_n + \bar{\mathbf{D}}_{n+1}]\Delta\boldsymbol{\varepsilon}$$

Hence, the trapezoidal rule is used in an approximate sense (it is approximate because $\bar{\mathbf{D}}_{n+1}$ is computed using an estimate of $\boldsymbol{\sigma}_{n+1}$). However, the results should be accurate when the step size decreases, which is achieved based on a certain criteria. However, the consistency is not enforced. Hence, in models with more complicated hardening rules and yield surfaces, the current stress state may gradually drift away from the yield surface.

In the remaining two methods (the cutting plane and CPPM), instead of controlling the step size, the consistency is enforced at the end of the increment. In the cutting-plane method, this is achieved in an explicit manner (i.e., at the beginning of each iteration, quantities computed at the beginning of the step are used), whereas in CPPM, this is achieved in an implicit manner. The satisfaction of consistency should improve accuracy.

In CPPM, however, the variation of the other variables during the increment is not considered. The plastic corrector is computed as

$$\Delta\boldsymbol{\sigma}^{\mathrm{pc}} = -\int_{t}^{t+\Delta t} \dot{\lambda}\mathbf{C}\mathbf{r} \approx \Delta\lambda\mathbf{C}\mathbf{r}^{n+1}$$

The variation of $\mathbf{r}$ during the increment is neglected. In addition, when the elastic properties are pressure-dependent, the variation of $\mathbf{C}$ during the increment is neglected (when elastic properties are approximated as in (10.45a)). The accuracy, therefore, depends on the step size.

10.8.2 Efficiency

The algorithmic details are too dissimilar and complicated to make a step-by-step comparison of the computations involved. One method of comparing the efficiencies is to compare the computer CPU time consumption. Here too, the outcome depends on the details of the coding. In general, for smaller step sizes, the two-step method is more efficient than the cutting-plane method and CPPM. However, for larger step sizes, the trend is opposite – the cutting-plane method and CPPM are much more efficient. The reason is that when the step size is large, the two-step Euler method divides the increment into very small steps and hence takes a large number of overall iterations, whereas CPPM and the cutting-plane method always use one step and converges in a few iterations. We will present results supporting these assertions in subsequent chapters.

Problems

Problem 10.1 Describe the differences between implicit and explicit integration methods. Is the two-step Euler method described in this chapter an implicit or explicit method? Explain your answer.

Problem 10.2 List two merits and two shortcomings of the two-step Euler method of integrating constitutive laws.

Problem 10.3 Explain the basis for the name "cutting-plane algorithm." Is the cutting-plane method described in this chapter an implicit or explicit method? Explain your answer.

Problem 10.4 List two merits and two shortcomings of the cutting-plane method of integrating constitutive laws.

Problem 10.5 List two merits and two shortcomings of CPPM of integrating constitutive laws.

Problem 10.6 List two key differences between the cutting-plane algorithm and CPPM.

Problem 10.7 Even when the constitutive model uses associated flow rule, the consistent tangent matrix involved in CPPM can be nonsymmetric. Explain.

Chapter 11
The von Mises Model and Its Integration

Among the many rate-independent, elasto-plastic constitutive laws available today for describing the stress–strain behavior of materials, the models employing the von Mises yield surface, and the associated flow rule are some of the widely used and simplest. While isotropic hardening models are generally adequate for monotonic loading applications, more sophisticated models (e.g., nonlinear kinematic hardening models, multi-surface models, bounding surface models, etc.) are generally needed for cyclic loading applications.

In the present chapter, we first describe (Sects. 11.1–11.4) an isotropic hardening von Mises model and its integration using the closest point projection method (CPPM) presented in Chap. 10. Both linear and nonlinear hardening rules (Chap. 9) are considered. The models are integrated using both the S-space and σ-space formulations of CPPM. The overall computational efficiency of the S-space formulation is generally higher than that of the σ-space formulation. For the models described in this chapter, the S-space formulation is adequate. However, when the model is very complex, the σ-space formulation may be the only one that can be used. With this in mind, the σ-space formulation is also described in this chapter. We then present a kinematic hardening von Mises model and describe its integration (Sects. 11.5–11.7) using the S-space formulation of CPPM. We use the Armstrong–Frederick nonlinear kinematic hardening rule (Chap. 9).

11.1 An Isotropic Hardening von Mises Model

The elastic stress–strain relation is given by

$$\dot{\boldsymbol{\sigma}} = \mathbf{C}\dot{\boldsymbol{\varepsilon}}^e \tag{11.1a}$$

where

$$C_{ijk\ell} = \left(K - \frac{2}{3}G\right)\delta_{ij}\delta_{k\ell} + G\left(\delta_{ik}\delta_{j\ell} + \delta_{i\ell}\delta_{jk}\right) \tag{11.1b}$$

A. Anandarajah, *Computational Methods in Elasticity and Plasticity: Solids and Porous Media*, DOI 10.1007/978-1-4419-6379-6_11,

G is shear modulus, K is bulk modulus, $\boldsymbol{\delta}$ is Kronecker delta, $\mathbf{C}$ is continuum elastic tangent operator, $\dot{\boldsymbol{\sigma}}$ is rate of stress, and $\dot{\boldsymbol{\varepsilon}}^e$ is rate of elastic strain.

Two stress invariants I and J are defined as (Chap. 2)

$$I = \sigma_{k\ell}\delta_{k\ell} \tag{11.2a}$$

$$s_{ij} = \sigma_{ij} - \frac{1}{3}I\delta_{ij} \tag{11.2b}$$

$$J = \left(\frac{1}{2}s_{k\ell}s_{k\ell}\right)^{1/2} \tag{11.2c}$$

The yield surface is a regular cylinder in the stress space with its axis coinciding with the space diagonal. Its projection in the octahedral plane is a circle, given by

$$\phi(\mathbf{s}, k) = s_{ij}s_{ij} - \frac{2}{3}k^2 = 0 \tag{11.3}$$

where the scalar k is the (only) hardening variable (PIV) in the model, geometrically representing the size of the yield surface. Hence the PIV of the model is

$$\varsigma \equiv k \tag{11.4}$$

Assuming an associated flow rule, the equations involved in the plastic constitutive relation are

$$\dot{\boldsymbol{\varepsilon}}^p = \dot{\lambda}\mathbf{r}; \quad \mathbf{r} = \frac{\partial\psi}{\partial\boldsymbol{\sigma}}; \quad \dot{\lambda} = \frac{1}{K_p}\mathbf{n}\dot{\boldsymbol{\sigma}}; \quad \mathbf{n} = \frac{\partial\phi}{\partial\boldsymbol{\sigma}}; \quad \phi = \psi; \quad \mathbf{r} = \mathbf{n} \tag{11.5}$$

where $\dot{\boldsymbol{\varepsilon}}^p$ is the plastic strain rate and $\dot{\lambda}$ is the consistency parameter (or the loading index). The algorithms will be derived without assuming $\phi = \psi$. This allows the algorithms presented in this chapter to be easily extended to constitutive models based on nonassociated flow rules.

We will consider two different hardening rules: (1) A linear isotropic hardening and (2) a nonlinear isotropic hardening (based on hyperbolic relation presented in Chap. 9). The rate of change of k is given by

$$\dot{k} = g\dot{\xi}^p \tag{11.6a}$$

where ξ^p is the cumulative deviatoric plastic strain invariant, defined below.

For the *linear hardening* model, we assume that

$$g = g_1 \tag{11.6b}$$

where g_1 is a constant.

For the *nonlinear hardening* model, the following hyperbolic relation is used (Chap. 9):

$$k = k_0 + \frac{m(k_f - k_0)\xi^p}{(k_f - k_0) + m\xi^p} \tag{11.6c}$$

where k_0 (initial yield stress of a virgin specimen), m (hardening parameter) and k_f (failure stress) are model parameters.

Taking the rate of (11.6c) and comparing the resulting equation with (11.6a), it follows that

$$g = g_2 = \frac{m(k_f - k_0)^2}{\left[(k_f - k_0) + m\xi^p\right]^2} = \frac{a_1}{(a_2 + b\xi^p)^2};$$

$$b = m; \quad a_2 = k_f - k_0; \quad a_1 = m(k_f - k_0)^2 = ba_2^2 \tag{11.6d}$$

where

$$\dot{\xi}^p = \left[\frac{1}{2}\dot{e}^p_{ij}\dot{e}^p_{ij}\right]^{1/2}; \quad \dot{e}_{ij} = \dot{\varepsilon}_{ij} - \frac{1}{3}\varepsilon_{kk}\delta_{ij} \tag{11.6e}$$

From (11.5), (11.6e) becomes

$$\dot{\xi}^p = \left[\frac{1}{2}\dot{e}^p_{ij}\dot{e}^p_{ij}\right] = \left[\frac{1}{2}r^d_{ij}r^d_{ij}\right]^{1/2}\dot{\lambda} = \bar{r}^d\dot{\lambda}; \quad \bar{r}^d = \left[\frac{1}{2}r^d_{ij}r^d_{ij}\right]^{1/2} \tag{11.7a}$$

where r^d_{ij} is the deviatoric part of r_{ij}

$$r^d_{ij} = r_{ij} - \frac{1}{3}r_{pp}\delta_{ij} \tag{11.7b}$$

Combining (11.6a) and (11.7a), for both linear and nonlinear hardening models, we have

$$\dot{k} = g\bar{r}^d\dot{\lambda} = \bar{g}\dot{\lambda} \tag{11.7c}$$

Thus, the hardening function **s** in the equations presented in Chap. 9 and 10 is a scalar, given by

$$s = \bar{g} = g\bar{r}^d \tag{11.7d}$$

Now, let us take the gradients of the yield function with respect to the state variables. First, note the following. Differentiating (11.2b), it follows that

$$\frac{\partial s_{k\ell}}{\partial \sigma_{ij}} = \delta_{ki}\delta_{\ell j} - \frac{1}{3}\delta_{ij}\delta_{k\ell} \tag{11.8a}$$

Taking the derivative of ϕ with respect to s_{ij}, we have

$$\frac{\partial \phi}{\partial s_{k\ell}} = 2\delta_{ik}\delta_{j\ell}s_{ij} = 2s_{k\ell}$$

Now, using the above two equations, it follows that

$$\begin{aligned} r_{ij} &= \frac{\partial \phi}{\partial \sigma_{ij}} = \frac{\partial \phi}{\partial s_{k\ell}}\frac{\partial s_{k\ell}}{\partial \sigma_{ij}} \quad \text{(using the chain rule)} \\ &= 2s_{k\ell}(\delta_{ki}\delta_{\ell j} - \frac{1}{3}\delta_{ij}\delta_{k\ell}) \\ &= 2s_{ij} \quad (\text{since } s_{kk} = 0) \end{aligned} \tag{11.8b}$$

It is, therefore, seen that r_{ij} is purely deviatoric (i.e., its volumetric part is zero); hence it follows that

$$r_{ij}^{d} = r_{ij} \tag{11.8c}$$

It follows from (11.8b) that, in the octahedral plane, r_{ij} and s_{ij} are collinear. Combining (11.7a), (11.8b) and (11.8c), we have

$$\bar{r}^{d} = \left[\frac{1}{2}r_{ij}^{d}r_{ij}^{d}\right]^{1/2} = \left[\frac{1}{2}r_{ij}r_{ij}\right]^{1/2} = 2\left[\frac{1}{2}s_{ij}s_{ij}\right]^{1/2} = 2J \tag{11.9}$$

The derivative of ϕ with respect to k is

$$\frac{\partial \phi}{\partial k} = -\frac{4}{3}k \tag{11.10}$$

From the consistency equation (refer to Chap. 9 for more details) and making use of (11.7d) and (11.10), we have

$$K_p = -(\partial_\zeta \phi)^{\mathrm{T}}\mathbf{s} \tag{11.11a}$$

$$K_p = \frac{4}{3}kg\bar{r}^{d} \tag{11.11b}$$

11.2 Application of CPPM to Isotropic Hardening von Mises Model: The σ-Space Formulation

11.2.1 Euler Equations

Since the hardening rules chosen in this chapter are simple, we do have an exact expression for k as a function of ξ^p. This allows the Euler approximation to be made at the level of ξ^p instead of k. This is however, not possible when the hardening rule is complex. Here we will derive the equations for both cases and examine the efficiency of each at the end.

The relevant backward Euler equations ((10.29a), (10.29b), and (10.28c), Chap. 10) are written as

$$\boldsymbol{\sigma}_{n+1} = \boldsymbol{\sigma}_n + \mathbf{C}\Delta\boldsymbol{\varepsilon} - \Delta\lambda\mathbf{C}\mathbf{r}^{n+1}(\boldsymbol{\sigma}_{n+1}, k_{n+1}) \tag{11.12a}$$

$$k_{n+1} = \bar{k}(\boldsymbol{\sigma}_{n+1}, k_{n+1}, \Delta\lambda) \tag{11.12b}$$

$$\phi^{n+1}(\boldsymbol{\sigma}_{n+1}, k_{n+1}) = 0 \tag{11.12c}$$

Approximation 1: Euler Approximation at the Level of k ("Euler").
Here, the backward Euler approximation is made at the level of k. From (11.7c), we have

$$\bar{k} = k_n + \Delta\lambda\bar{g}^{n+1}(\boldsymbol{\sigma}_{n+1}, k_{n+1}) \tag{11.13a}$$

Approximation 2: Euler Approximation at the Level of ξ^p ("Exact").
The exact expressions for k as a function of ξ^p are listed below:

Linear hardening (by integrating (11.61) with (11.6b) used for g):

$$\bar{k} = k_{ex} = k_0 + g_1\xi^p \tag{11.14a}$$

Nonlinear hardening (from (11.6c)):

$$\bar{k} = k_{ex} = k_0 + \frac{m(k_f - k_0)\xi^p}{(k_f - k_0) + m\xi^p} \tag{11.14b}$$

Discretizing (11.7a) by the backward Euler approximation, we have

$$\xi^p_{n+1} = \xi^p_n + \Delta\xi; \quad \Delta\xi = \bar{r}^d_{n+1}\Delta\lambda \quad \Rightarrow \quad \xi^p_{n+1} = \xi^p_n + \bar{r}^d_{n+1}\Delta\lambda \tag{11.14c}$$

Using (11.14c), the exact equation for the case of linear hardening (11.14a) is

$$\bar{k}(\boldsymbol{\sigma}, k, \Delta\lambda) = k_{ex}(\boldsymbol{\sigma}, k, \Delta\lambda) = k_0 + g_1\left[\xi^p_n + \bar{r}^d_{n+1}\Delta\lambda\right] \tag{11.15a}$$

and for the nonlinear hardening (11.14b) is

$$\bar{k}(\boldsymbol{\sigma}, k, \Delta\lambda) = k_{ex}(\boldsymbol{\sigma}, k, \Delta\lambda) = k_0 + \frac{m(k_f - k_0)\left[\xi_n^p + \bar{r}_{n+1}^d \Delta\lambda\right]}{(k_f - k_0) + m\left[\xi_n^p + \bar{r}_{n+1}^d \Delta\lambda\right]} \tag{11.15b}$$

11.2.2 The Residual Equations and Their Linearizations

The residual equations are given by (10.31a–c)

$$\mathbf{R}_\sigma = \boldsymbol{\sigma}_{n+1} - \left[\boldsymbol{\sigma}_n + \mathbf{C}\Delta\boldsymbol{\varepsilon} - \Delta\lambda\mathbf{C}\mathbf{r}^{n+1}\right] = \mathbf{R}_\sigma(\boldsymbol{\sigma}_{n+1}, k_{n+1}, \Delta\lambda) \tag{10.16a}$$

$$\mathbf{R}_\varsigma = k_{n+1} - \bar{k} = \mathbf{R}_\varsigma(\sigma_{n+1}, k_{n+1}, \Delta\lambda) \tag{10.16b}$$

$$R_\phi = \phi^{n+1}(\boldsymbol{\sigma}_{n+1}, k_{n+1}) = \mathbf{R}_\phi(\boldsymbol{\sigma}_{n+1}, k_{n+1}) \tag{10.16c}$$

The linearizations of the above equations are (10.34a–c)

$$R_{\sigma ij}^1 + A_{ijk\ell}\delta\sigma_{k\ell} + B_{ij}\delta k + F_{ij}\delta\lambda = 0 \tag{11.17a}$$

$$R_\varsigma^1 + H_{k\ell}\delta\sigma_{k\ell} + \omega\delta k + \beta\delta\lambda = 0 \tag{11.17b}$$

$$R_\phi^1 + E_{k\ell}\delta\sigma_{k\ell} + \gamma\delta k = 0 \tag{11.17c}$$

The coefficients of the above equations are given as (10.36a–h):

$$A_{ijk\ell} = \frac{\partial R_{\sigma ij}}{\partial \sigma_{k\ell}} = \delta_{ik}\delta_{j\ell} - \frac{\partial C_{ijpq}}{\partial \sigma_{k\ell}}\left[\Delta\varepsilon_{pq} - \Delta\lambda r_{pq}^{n+1}\right] + \Delta\lambda C_{ijpq}\frac{\partial r_{pq}^{n+1}}{\partial \sigma_{k\ell}} \tag{11.18a}$$

$$B_{ij} = \frac{\partial R_{\sigma ij}}{\partial k} = \Delta\lambda C_{ijpq}\frac{\partial r_{pq}^{n+1}}{\partial k} \tag{11.18b}$$

$$F_{ij} = \frac{\partial R_{\sigma ij}}{\partial \Delta\lambda} = C_{ijpq} r_{pq}^{n+1} \tag{11.18c}$$

$$H_{k\ell} = \frac{\partial R_\varsigma}{\partial \sigma_{k\ell}} = -\frac{\partial \bar{k}}{\partial \sigma_{k\ell}} \tag{11.18d}$$

$$\omega = \frac{\partial R_\varsigma}{\partial k} = 1 - \frac{\partial \bar{k}}{\partial k} \tag{11.18e}$$

$$\beta = \frac{\partial R_\varsigma}{\partial \Delta\lambda} = -\frac{\partial \bar{k}}{\partial \Delta\lambda} \tag{11.18f}$$

$$E_{k\ell} = \frac{\partial R_\phi}{\partial \sigma_{k\ell}} = \frac{\partial \phi^{n+1}}{\partial \sigma_{k\ell}} = n_{k\ell}^{n+1} \tag{11.18g}$$

$$\gamma = \frac{\partial R_\phi}{\partial k} = \frac{\partial \phi^{n+1}}{\partial k} \tag{11.18h}$$

11.2.3 The Coefficients for the Isotropic Hardening von Mises Model

For the von Mises model of interest here, the elastic properties are pressure independent. Hence, it follows that

$$C_{ijpq} = \text{constant} \quad \text{and} \quad \frac{\partial C_{ijpq}}{\partial \sigma_{k\ell}} = 0$$

The other quantities needed for the evaluation of the coefficients in (11.18a–h) may be evaluated from the surface properties r_{pq}^{n+1}, $\frac{\partial r_{pq}^{n+1}}{\partial \sigma_{k\ell}}$, and $\frac{\partial r_{pq}^{n+1}}{\partial k}$, and hardening properties $\bar{k}$, $\frac{\partial \bar{k}}{\partial \sigma_{k\ell}}$, $\frac{\partial \bar{k}}{\partial k}$, and $\frac{\partial \bar{k}}{\partial \Delta\lambda}$. It may be noted that the quantities $\sigma_{k\ell}$ and k in these derivatives are to be equated with $\sigma_{k\ell}^{n+1}$ and k_{n+1}; i.e.,

$$\frac{\partial \bar{k}}{\partial \sigma_{k\ell}} = \frac{\partial \bar{k}}{\partial \sigma_{k\ell}^{n+1}},$$

etc. (for convenience, the subscript or superscript "$n+1$" has been dropped). These quantities are derived here from the equations presented in Sect. 11.1.

$$r_{pq}^{n+1} = 2s_{pq}^{n+1} \qquad \text{(from 11.8b)}$$

$$\frac{\partial r_{pq}^{n+1}}{\partial \sigma_{k\ell}} = 2\left(\delta_{pk}\delta_{q\ell} - \frac{1}{3}\delta_{pq}\delta_{k\ell}\right) \qquad \text{(from 11.8b and using 11.8a)}$$

$$\frac{\partial r_{pp}^{n+1}}{\partial k} = 0$$

For Approximation 1 ("Euler") of the hardening variable, we have

$$\bar{k} = k_n + \Delta\lambda \bar{g}^{n+1}(\boldsymbol{\sigma}_{n+1}, k_{n+1}) \qquad \text{(from 11.13a)}$$

$$\bar{g}^{n+1} = g^{n+1}\bar{r}^{d}_{n+1} \qquad \text{(From 11.7d)}$$

$$\frac{\partial \bar{g}^{n+1}}{\partial \sigma_{k\ell}} = \bar{r}^{d}_{n+1}\frac{\partial g^{n+1}}{\partial \sigma_{k\ell}} + g^{n+1}\frac{\partial \bar{r}^{d}_{n+1}}{\partial \sigma_{k\ell}}$$

$$\frac{\partial \bar{g}^{n+1}}{\partial k} = \bar{r}^{d}_{n+1}\frac{\partial g^{n+1}}{\partial k} + g^{n+1}\frac{\partial \bar{r}^{d}_{n+1}}{\partial k}$$

$$\frac{\partial \bar{g}^{n+1}}{\partial \Delta\lambda} = \bar{r}^{d}_{n+1}\frac{\partial g^{n+1}}{\partial \Delta\lambda} + g^{n+1}\frac{\partial \bar{r}^{d}_{n+1}}{\partial \Delta\lambda}$$

$$\bar{r}^{d}_{n+1} = \left[\frac{1}{2}r^{d}_{ij}r^{d}_{ij}\right]^{1/2} = \left[\frac{1}{2}r_{ij}r_{ij}\right]^{1/2} \qquad \text{(From 11.7a and 11.8c)}$$

$$\frac{\partial \bar{r}^{d}_{n+1}}{\partial \sigma_{k\ell}} = \frac{1}{2\bar{r}^{d}_{n+1}}\frac{2}{2}\frac{\partial r_{ij}}{\partial \sigma_{k\ell}}r_{ij} = \frac{1}{2\bar{r}^{d}_{n+1}}2r_{ij}(\delta_{ik}\delta_{j\ell} - \frac{1}{3}\delta_{ij}\delta_{k\ell}) = \frac{1}{\bar{r}^{d}_{n+1}}r_{k\ell}$$

$$\frac{\partial \bar{r}^{d}_{n+1}}{\partial k} = 0$$

$$\frac{\partial \bar{r}^{d}_{n+1}}{\partial \Delta\lambda} = 0$$

The gradient of g depends on the specific hardening rule as follows:
Linear hardening

$$g^{n+1} = g_1 = \text{constant}; \quad \frac{\partial g^{n+1}}{\partial \sigma_{k\ell}} = 0; \quad \frac{\partial g^{n+1}}{\partial k} = 0; \quad \frac{\partial g^{n+1}}{\partial \Delta\lambda} = 0$$

Based on the above, we have

$$\frac{\partial \bar{g}^{n+1}}{\partial \sigma_{k\ell}} = \frac{g_1}{\bar{r}^{d}_{n+1}}r_{k\ell}; \quad \frac{\partial \bar{g}^{n+1}}{\partial k} = 0; \quad \frac{\partial \bar{g}^{n+1}}{\partial \Delta\lambda} = 0$$

Nonlinear hardening

$$\xi^p = \xi^p_n + \Delta\xi^p = \xi^p_n + \bar{r}^{d}_{n+1}\Delta\lambda \qquad \text{(From 11.14c)}$$

$$g^{n+1} = g_2^{n+1} = \frac{m(k_f - k_0)^2}{\left[(k_f - k_0) + m\xi^p\right]^2} = \frac{a_1}{\left(a_2 + b\xi^p_{n+1}\right)^2} = \frac{a_1}{\left(a_2 + b\xi^p_n + b\bar{r}^{d}_{n+1}\Delta\lambda\right)^2}$$

(From 11.6d)

$$\frac{\partial g^{n+1}}{\partial \sigma_{k\ell}} = -\frac{2a_1 b\Delta\lambda}{\left(a_2 + b\xi_n^p + b\bar{r}_{n+1}^d \Delta\lambda\right)^3}\frac{\partial \bar{r}_{n+1}^d}{\partial \sigma_{k\ell}} = -\frac{2a_1 b\Delta\lambda}{\bar{r}_{n+1}^d\left(a_2 + b\xi_{n+1}^p\right)^3} r_{k\ell}$$

$$\frac{\partial g^{n+1}}{\partial k} = 0$$

$$\frac{\partial g^{n+1}}{\partial \Delta\lambda} = -\frac{2a_1 b\bar{r}_{n+1}^d}{\left(a_2 + b\xi_n^p + b\bar{r}_{n+1}^d \Delta\lambda\right)^3} = -\frac{2a_1 b\bar{r}_{n+1}^d}{\left(a_2 + b\xi_{n+1}^p\right)^3}$$

$$\frac{\partial \bar{g}^{n+1}}{\partial \sigma_{k\ell}} = \left[\frac{g^{n+1}}{\bar{r}_{n+1}^d} - \frac{2a_1 b\Delta\lambda}{\left(a_2 + b\xi_{n+1}^p\right)^3}\right] r_{k\ell}; \quad \frac{\partial \bar{g}^{n+1}}{\partial k} = 0; \quad \frac{\partial \bar{g}^{n+1}}{\partial \Delta\lambda} = -\frac{2a_1 b\left(\bar{r}_{n+1}^d\right)^2}{\left(a_2 + b\xi_{n+1}^p\right)^3}$$

For Approximation 2 ("Exact") of the hardening variable, we have:

Linear Hardening

$$\bar{k} = g^{n+1}\left[\xi_n^p + \bar{r}_{n+1}^d \Delta\lambda\right] \qquad \text{(From 11.15a)}$$

$$\frac{\partial \bar{k}}{\partial \sigma_{k\ell}} = \frac{\partial g^{n+1}}{\partial \sigma_{k\ell}}\left[\xi_n^p + \bar{r}_{n+1}^d \Delta\lambda\right] + \Delta\lambda g^{n+1}\frac{\partial \bar{r}_{n+1}^d}{\partial \sigma_{k\ell}}$$

$$\frac{\partial \bar{k}}{\partial k} = \frac{\partial g^{n+1}}{\partial k}\left[\xi_n^p + \bar{r}_{n+1}^d \Delta\lambda\right] + \Delta\lambda g^{n+1}\frac{\partial \bar{r}_{n+1}^d}{\partial k}$$

$$\frac{\partial \bar{k}}{\partial \Delta\lambda} = \frac{\partial g^{n+1}}{\partial \Delta\lambda}\left[\xi_n^p + \bar{r}_{n+1}^d \Delta\lambda\right] + g^{n+1}\bar{r}_{n+1}^d$$

Nonlinear hardening

$$\bar{k} = k_0 + \frac{m(k_f - k_0)\xi_{n+1}^p}{a_2 + b\xi_{n+1}^p} = k_0 + \frac{m(k_f - k_0)\left[\xi_n^p + \bar{r}_{n+1}^d \Delta\lambda\right]}{(k_f - k_0) + m\left[\xi_n^p + \bar{r}_{n+1}^d \Delta\lambda\right]}$$

By taking the derivative of this equation and simplifying with the aid of the relations in (11.6d), we have

$$\frac{\partial \bar{k}}{\partial \sigma_{k\ell}} = \frac{m(k_f - k_0)}{\left(a_2 + b\xi_{n+1}^p\right)^2}\left[\Delta\lambda\left(a_2 + b\xi_{n+1}^p\right) - m\Delta\lambda\xi_{n+1}^p\right]\frac{\partial \bar{r}_{n+1}^d}{\partial \sigma_{k\ell}} = \frac{a_1\Delta\lambda}{\left(a_2 + b\xi_{n+1}^p\right)^2}\frac{\partial \bar{r}_{n+1}^d}{\partial \sigma_{k\ell}}$$

$$\frac{\partial \bar{k}}{\partial k} = \frac{m(k_f - k_0)}{\left(a_2 + b\xi_{n+1}^p\right)^2}\left[\Delta\lambda\left(a_2 + b\xi_{n+1}^p\right) - m\Delta\lambda\xi_{n+1}^p\right]\frac{\partial \bar{r}_{n+1}^d}{\partial k} = \frac{a_1\Delta\lambda}{\left(a_2 + b\xi_{n+1}^p\right)^2}\frac{\partial \bar{r}_{n+1}^d}{\partial k}$$

$$\frac{\partial \bar{k}}{\partial k} = \frac{m(k_f - k_0)}{\left(a_2 + b\xi^p_{n+1}\right)^2}\left[\left(a_2 + b\xi^p_{n+1}\right) - m\xi^p_{n+1}\right]\bar{r}^d_{n+1} = \frac{a_1}{\left(a_2 + b\xi^p_{n+1}\right)^2}\bar{r}^d_{n+1}$$

Substituting the above expressions in (11.18a–h), it follows that:

$$\begin{aligned}
A_{ijk\ell} &= \delta_{ik}\delta_{j\ell} + \Delta\lambda C_{ijpq}\frac{\partial r^{n+1}_{pq}}{\partial \sigma_{k\ell}} \\
&= \delta_{ik}\delta_{j\ell} + 2\Delta\lambda C_{ijpq}\left(\delta_{pk}\delta_{q\ell} - \frac{1}{3}\delta_{pq}\delta_{k\ell}\right) \\
&= \delta_{ik}\delta_{j\ell} + 2\Delta\lambda\left[\left(K - \frac{2}{3}G\right)\delta_{ij}\delta_{pq} + G\left(\delta_{ip}\delta_{jq} + \delta_{iq}\delta_{jp}\right)\right]\left(\delta_{pk}\delta_{q\ell} - \frac{1}{3}\delta_{pq}\delta_{k\ell}\right) \\
&= \delta_{ik}\delta_{j\ell} + 2\Delta\lambda\left[\left(K - \frac{2}{3}G\right)\left(\delta_{ij}\delta_{k\ell} - \delta_{ij}\delta_{k\ell}\right) + G\left(\delta_{ik}\delta_{j\ell} + \delta_{i\ell}\delta_{jk}\right) - \frac{2}{3}G\delta_{ij}\delta_{k\ell}\right] \\
&= \delta_{ik}\delta_{j\ell} + 2\Delta\lambda G\left(\delta_{ik}\delta_{j\ell} + \delta_{i\ell}\delta_{jk}\right) - \frac{4}{3}G\delta_{ij}\delta_{k\ell}\Delta\lambda
\end{aligned}$$

$$B_{ij} = \Delta\lambda C_{ijpq}\frac{\partial r^{n+1}_{pq}}{\partial k} = 0$$

$$\begin{aligned}
F_{ij} &= \frac{\partial R_{\sigma ij}}{\partial \Delta\lambda} = C_{ijpq}r^{n+1}_{pq} \\
&= \left[\left(K - \frac{2}{3}G\right)\delta_{ij}\delta_{pq} + G\left(\delta_{ip}\delta_{jq} + \delta_{iq}\delta_{jp}\right)\right]2s_{pq} \\
&= 4Gs_{ij}
\end{aligned}$$

For Approximation 1 ("Euler") of the hardening variable, we have:

Linear Hardening:

$$H_{k\ell} = -\frac{\partial \bar{k}}{\partial \sigma_{k\ell}} = -\Delta\lambda\frac{\partial \bar{g}^{n+1}}{\partial \sigma_{k\ell}} = -\Delta\lambda\frac{g_1}{\bar{r}^d_{n+1}}r_{k\ell}$$

$$\omega = 1 - \frac{\partial \bar{k}}{\partial k} = 1 - \Delta\lambda\frac{\partial \bar{g}^{n+1}}{\partial k} = 1$$

$$\beta = -\frac{\partial \bar{k}}{\partial \Delta\lambda} = -\bar{g}^{n+1} - \Delta\lambda\frac{\partial \bar{g}^{n+1}}{\partial \Delta\lambda} = -\bar{g}^{n+1}$$

Nonlinear Hardening:

$$H_{k\ell} = -\frac{\partial \bar{k}}{\partial \sigma_{k\ell}} = -\Delta\lambda\frac{\partial \bar{g}^{n+1}}{\partial \sigma_{k\ell}} = -\Delta\lambda\left[\frac{g^{n+1}}{\bar{r}^d_{n+1}} - \frac{2a_1 b\Delta\lambda}{\left(a_2 + b\xi^p_{n+1}\right)^3}\right]r_{k\ell}$$

$$\omega = 1 - \frac{\partial \bar{k}}{\partial k} = 1 - \Delta\lambda \frac{\partial \bar{g}^{n+1}}{\partial k} = 1$$

$$\beta = -\frac{\partial \bar{k}}{\partial \Delta\lambda} = -\bar{g}^{n+1} - \Delta\lambda \frac{\partial \bar{g}^{n+1}}{\partial \Delta\lambda} = -\bar{g}^{n+1} + \Delta\lambda \frac{2a_1 b \left(\bar{r}^d_{n+1}\right)^2}{\left(a_2 + b\xi^p_{n+1}\right)^3}$$

For Approximation 2 ("Exact") of the hardening variable, we have:

Linear Hardening:

$$H_{k\ell} = -\frac{\partial \bar{k}}{\partial \sigma_{k\ell}} = -\frac{\partial g^{n+1}}{\partial \sigma_{k\ell}} \left[\xi^p_n + \bar{r}^d_{n+1}\Delta\lambda\right] - \Delta\lambda g^{n+1} \frac{\partial \bar{r}^d_{n+1}}{\partial \sigma_{k\ell}}$$

$$\omega = 1 - \frac{\partial \bar{k}}{\partial k} = 1 - \frac{\partial g^{n+1}}{\partial k} \left[\xi^p_n + \bar{r}^d_{n+1}\Delta\lambda\right] - \Delta\lambda g^{n+1} \frac{\partial \bar{r}^d_{n+1}}{\partial k}$$

$$\beta = -\frac{\partial \bar{k}}{\partial \Delta\lambda} = -\frac{\partial g^{n+1}}{\partial \Delta\lambda} \left[\xi^p_n + \bar{r}^d_{n+1}\Delta\lambda\right] - g^{n+1} \bar{r}^d_{n+1}$$

Nonlinear Hardening:

$$H_{k\ell} = -\frac{\partial \bar{k}}{\partial \sigma_{k\ell}} = -\frac{a_1 \Delta\lambda}{\left(a_2 + b\xi^p_{n+1}\right)^2} \frac{\partial \bar{r}^d_{n+1}}{\partial \sigma_{k\ell}} = -\frac{a_1 \Delta\lambda}{\left(a_2 + b\xi^p_{n+1}\right)^2} \frac{1}{\bar{r}^d_{n+1}} r_{k\ell}$$

$$\omega = 1 - \frac{\partial \bar{k}}{\partial k} = 1 - \frac{a_1 \Delta\lambda}{\left(a_2 + b\xi^p_{n+1}\right)^2} \frac{\partial \bar{r}^d_{n+1}}{\partial k} = 1$$

$$\beta = -\frac{\partial \bar{k}}{\partial \Delta\lambda} = -\frac{a_1}{\left(a_2 + b\xi^p_{n+1}\right)^2} \bar{r}^d_{n+1}$$

The remaining parameters are given by:

$$E_{k\ell} = \frac{\partial R_\phi}{\partial \sigma_{k\ell}} = \frac{\partial \phi^{n+1}}{\partial \sigma_{k\ell}} = r_{k\ell} = 2s_{k\ell}$$

$$\gamma = \frac{\partial \phi^{n+1}}{\partial k} = -\frac{4}{3}k$$

11.2.4 The Implementation Details

The symmetry of stresses and strains allow the tensorial equations in (11.17a– c) to be implemented in reduced order matrix-vector forms; the details are presented in

Appendix 6. The three matrix-vector equations in (11.17a–c) can be arranged in a single matrix-vector equation system as

$$\begin{bmatrix} \underset{6\times6}{\mathbf{A}} & \underset{6\times1}{\mathbf{B}} & \underset{6\times1}{\mathbf{F}} \\ \underset{1\times6}{\mathbf{H}^{\mathrm{T}}} & \underset{1\times1}{\omega} & \underset{1\times1}{\beta} \\ \underset{1\times6}{\mathbf{E}^{\mathrm{T}}} & \underset{1\times1}{\gamma} & \underset{1\times1}{0} \end{bmatrix} \begin{bmatrix} \underset{6\times1}{\delta\boldsymbol{\sigma}} \\ \underset{1\times1}{\delta k} \\ \underset{1\times1}{\delta\lambda} \end{bmatrix} = \begin{bmatrix} \underset{6\times1}{-\mathbf{R}_{\sigma}^{1}} \\ \underset{1\times1}{-R_{\varsigma}^{1}} \\ \underset{1\times1}{-R_{\phi}^{1}} \end{bmatrix} \tag{11.19}$$

Equation (11.19) can either be directly solved as a system or manipulated to obtain the following equations (10.58a)–(10.59e):

$$\delta\lambda = \frac{R_{\phi}^{1} - \bar{\mathbf{n}}^{\mathrm{T}}\mathbf{R}\mathbf{R}_{\sigma}^{1} + x_0 R_{\varsigma}^{1}\bar{\mathbf{n}}^{\mathrm{T}}\bar{\mathbf{C}}\bar{\mathbf{B}} - x_0\gamma R_{\varsigma}^{1}}{x_0\gamma\beta + \bar{\mathbf{n}}^{\mathrm{T}}\bar{\mathbf{C}}\bar{\mathbf{r}}} \tag{11.20a}$$

$$\delta\sigma = -\mathbf{R}\mathbf{R}_{\sigma}^{1} + x_0 R_{\varsigma}^{1}\bar{\mathbf{C}}\bar{\mathbf{B}} - \bar{\mathbf{C}}\bar{\mathbf{r}}\delta\lambda \tag{11.20b}$$

$$\delta k = -\frac{1}{\gamma}\left[R_{\phi}^{1} + \mathbf{n}^{\mathrm{T}}\delta\boldsymbol{\sigma}\right] \tag{11.20c}$$

where

$$\bar{\mathbf{r}} = \bar{\mathbf{F}} - \frac{\beta}{\omega}\bar{\mathbf{B}} \tag{11.21a}$$

$$\bar{\mathbf{n}} = \mathbf{n} - \frac{\gamma}{\omega}\mathbf{H} \tag{11.21b}$$

$$\overline{\mathbf{F}} = \mathbf{C}^{-1}\mathbf{F} \quad \text{and} \quad \overline{\mathbf{B}} = \mathbf{C}^{-1}\mathbf{B}. \tag{11.21c}$$

$$x_0 = \frac{1}{\omega} \tag{11.21d}$$

$$\overline{\mathbf{C}} = \mathbf{R}\mathbf{C} \tag{11.21e}$$

$$\mathbf{R} = \left[\mathbf{A} - \frac{1}{\omega}\mathbf{B}\mathbf{H}^{\mathrm{T}}\right]^{-1} \tag{11.21f}$$

The method requires an initial value for $\Delta\lambda$. This may be calculated from (10.30) as

$$\Delta\lambda_1 = \frac{\phi^{tr}}{\left[\mathbf{n}^{\mathrm{T}}\mathbf{C}\mathbf{r} + K_p\right]^{tr}} \tag{11.22}$$

The new estimates for $\boldsymbol{\sigma}_{n+1}$, k_{n+1}, and $\Delta\lambda$ are found as (10.37a–c):

$$\boldsymbol{\sigma}_{n+1}^{2} = \boldsymbol{\sigma}_{n+1}^{1} + \delta\boldsymbol{\sigma} \tag{11.23a}$$

$$k_{n+1}^{2} = k_{n+1}^{1} + \delta k \tag{11.23b}$$

$$\Delta\lambda_2 = \Delta\lambda_1 + \delta\lambda \tag{11.23c}$$

Note that superscript 2 on these variables does not denote "squared term," but rather the "second estimate."

The following convergence criteria may be used (10.38a)–(10.38c):

$$e_\sigma = \left\|\mathbf{R}_\sigma^{i+1}\right\| / \left\|\boldsymbol{\sigma}_{n+1}^{i}\right\| \leq TOL_\sigma \tag{11.24a}$$

$$e_\varsigma = \left\|R_\varsigma^{i+1}\right\| / \left\|k_{n+1}^{i}\right\| \leq TOL_\varsigma \tag{11.24b}$$

$$R_\phi^{i+1} = \phi^{i+1} \leq TOL_\phi \tag{11.24c}$$

The procedure is summarized in Algorithm 11.1.

The consistent tangent operator is given by (10.66b):

$$\mathbf{D} = \overline{\mathbf{C}} - \frac{(\overline{\mathbf{C}}\bar{\mathbf{r}})(\bar{\mathbf{n}}^{\mathrm{T}}\overline{\mathbf{C}})}{x_0\gamma\beta + \bar{\mathbf{n}}^{\mathrm{T}}\overline{\mathbf{C}}\bar{\mathbf{r}}} \tag{11.25}$$

It was shown in Chap. 10 that when $\Delta\lambda = 0$, the consistent and continuum operators become identical to each other. Also, when the "Euler" approximation is used to integrate the hardening rule, the consistent operator $\mathbf{D}$ is found to be symmetric even when $\Delta\lambda \neq 0$ (Problem 11.1). (Recall that it was pointed out in Chap. 10 that even when an associated flow rule is employed, the consistent operator is in general non symmetric.)

Algorithm 11.1. CPPM for integrating the constitutive relations from the isotropic hardening von Mises model

Definition of (new) parameters: $N_L^{\max}$ is the maximum number of iterations permitted before execution is terminated, and N_L is the corresponding counter. TOL is a small, machine-dependent number (it could be set equal to TOL_ϕ). Iflag = 1 if converged, 0 otherwise.

1. Establish $\Delta\boldsymbol{\varepsilon}, k_n, \boldsymbol{\sigma}_n, TOL_\sigma, TOL_\varsigma, TOL_\phi, TOL, N_L^{\max}$, and the constitutive model parameters
2. Initialize parameters: $N_L = 0$ and $k_{n+1} = k_n$

(continued)

3. Compute elastic predictor $\boldsymbol{\sigma}_{n+1}^{tr}$ and $\phi^{tr}(\boldsymbol{\sigma}_{n+1}^{tr}, k_n)$. Perform loading/unloading check: if ($\phi^{tr} < TOL$), set $\boldsymbol{\sigma}_{n+1} = \boldsymbol{\sigma}_{n+1}^{tr}$ and go to step 12
4. Set $\boldsymbol{\sigma}_{n+1} = \boldsymbol{\sigma}_{n+1}^{tr}$ and calculate $\mathbf{r}_{n+1}$, $\bar{r}_{n+1}^{d}$, and K_p^{n+1}
5. Set $R_\phi^1 = \phi^{tr}$. Compute trial value of $\Delta\lambda$. Compute $\xi_{n+1}^{p} = \xi_n^p + \Delta\lambda \bar{r}_{n+1}^{d}$ and $\bar{r}_{n+1}^{d}$. Compute $\boldsymbol{\sigma}_{n+1}^{1} = \boldsymbol{\sigma}_n + \mathbf{C}\Delta\boldsymbol{\varepsilon} - \Delta\lambda\mathbf{C}\mathbf{r}_{n+1}$ and k_{n+1}^1 by either Approximation 1 or 2. Set $\boldsymbol{\sigma}_{n+1} = \boldsymbol{\sigma}_{n+1}^{1}$ and $k_{n+1} = k_{n+1}^{1}$
6. Begin iterations. Compute $R_\phi^1 = \phi(\boldsymbol{\sigma}_{n+1}, k_{n+1})$, $\mathbf{r}_{n+1}$ and $\bar{r}_{n+1}^{d}$
7. Compute $\boldsymbol{\sigma}_{n+1}^{2} = \boldsymbol{\sigma}_n + \mathbf{C}\Delta\boldsymbol{\varepsilon} - \Delta\lambda\mathbf{C}\mathbf{r}_{n+1}$. Compute $\xi_{n+1}^{p} = \xi_n^p + \Delta\lambda \bar{r}_{n+1}^{d}$ and $\bar{r}_{n+1}^{d}$. Compute k_{n+1}^{2} either by Approximation 1 or 2
8. Compute $\boldsymbol{R}_\sigma^1 = \boldsymbol{\sigma}_{n+1}^{1} - \boldsymbol{\sigma}_{n+1}^{2}$ and $R_\varsigma^1 = k_{n+1}^1 - k_{n+1}^2$
9. Find errors e_σ, $e_\varsigma$and e_ϕ, and perform convergence check. If converged, set Iflag = 1 and go to step 12
10. Set $N_L \leftarrow N_L + 1$. If $\left(N_L > N_L^{\max}\right)$, set Iflag = 0 and go to step 12
11. Compute $\delta\lambda$, $\delta\boldsymbol{\sigma}$, and δk. Update $\boldsymbol{\sigma}_{n+1}^{1} \leftarrow \boldsymbol{\sigma}_{n+1}^{1} + \delta\boldsymbol{\sigma}$, $k_{n+1}^1 = k_{n+1}^1 + \delta k$ and $\Delta\lambda \leftarrow \Delta\lambda + \delta\lambda$. Set $\boldsymbol{\sigma}_{n+1} = \boldsymbol{\sigma}_{n+1}^{1}$ and $k_{n+1} = k_{n+1}^{1}$. Go to step 6
12. Return to the main program

11.3 Application of CPPM to Isotropic Hardening von Mises Model: The *S*-Space Formulation

As shown in Sect. 10.7.2 (Chap. 10), the Euler equation for stress (11.12a) can be decomposed into spherical and deviatoric parts as

$$p_{n+1} = p_n + K\Delta\varepsilon_{pp} - K\Delta\lambda r_{pp}^{n+1} \tag{11.26a}$$

$$\mathbf{s}_{n+1} = \mathbf{s}_n + 2G\Delta\mathbf{e} - 2G\Delta\lambda\mathbf{r}_{n+1}^{d} \tag{11.26b}$$

p_{n+1} is the spherical part of $\boldsymbol{\sigma}_{n+1}$, $\mathbf{s}_{n+1} = \boldsymbol{\sigma}_{n+1} - p_{n+1}\boldsymbol{\delta}$ is the deviatoric part of $\boldsymbol{\sigma}_{n+1}$ and $\Delta\mathbf{e} = \Delta\boldsymbol{\varepsilon} - \Delta\boldsymbol{\varepsilon}_{pp}\boldsymbol{\delta}/3$ is the incremental deviatoric strain. In pressure-independent models such as the von Mises model considered in this chapter, $r_{pp}^{n+1} = 0$. Hence the variation of p is purely elastic, and is determined as

$$\Delta p_{n+1} = K\Delta\varepsilon_{kk} = K\Delta\varepsilon_v \tag{11.27}$$

where $\Delta\varepsilon_v$ is the incremental volumetric strain. Hence only the deviatoric equation needs to be included in the iterative process. As we will see below, the solution of iterative equations simplifies to the solution of a simple scalar equation on $\Delta\lambda$, rendering the process computationally very effective.

In the preceding sections, we considered two approximations for k_{n+1} as follows:

1. Approximation 1 ("Euler") for the Hardening Variable: The backward Euler approximation is used for k as $k_{n+1} = \bar{k} = k_n + \Delta\lambda \bar{g}^{n+1}(\boldsymbol{\sigma}_{n+1}, k_{n+1})$ (11.13a).
2. Approximation 2 ("Exact") for the Hardening Variable: The exact relation between k_{n+1} and ξ^p_{n+1} is used (11.14b), but the backward Euler approximation is used for ξ^p as $\xi^p_{n+1} = \xi^p_n + \bar{r}^d_{n+1}\Delta\lambda$ (11.14c).

Since the equations associated with these cases are largely different from each other, we will consider them separately below.

11.3.1 *Approximation 1 for the Hardening Variable: Euler Approximation at the Level of* k *("Euler")*

11.3.1.1 Calculation of Stresses and PIVs

Working in the deviatoric space as shown in Fig. 11.1a and making use of (11.8b), the elasto-plastic stress is given by (using symbolic notations)

$$\mathbf{s}_{n+1} = \mathbf{s}_n + 2G(\Delta\mathbf{e} - \Delta\mathbf{e}^p) = \mathbf{s}^{tr} - 2G\Delta\lambda\mathbf{r}_{n+1} = \mathbf{s}^{tr} - 4G\Delta\lambda\mathbf{s}_{n+1}, \quad \text{where } \mathbf{s}^{tr} = \mathbf{s}_n + 2G\Delta\mathbf{e} \tag{11.28a}$$

or

$$\mathbf{s}_{n+1} = \frac{1}{1 + 4G\Delta\lambda}\mathbf{s}^{tr} = \frac{\mathbf{s}^{tr}}{x_0} \tag{11.28b}$$

where

$$x_0 = 1 + 4G\Delta\lambda \tag{11.28c}$$

Noting that x_0 is a scalar, (11.28b) implies that $\mathbf{s}$ and $\mathbf{s}^{tr}$ are collinear, and that $\mathbf{s}$ is obtained simply by scaling $\mathbf{s}^{tr}$. This is the *classical radial return* (Wilkins 1964) as shown in Fig. 11.1b.

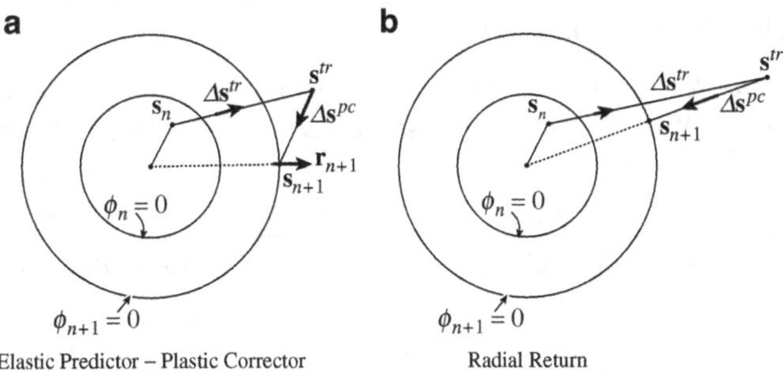

Fig. 11.1 The elastic predictor-plastic corrector in the deviatoric stress space

The yield stress at the forward point is obtained as (from (11.13a) and making use of (11.9) and (11.7c))

$$k_{n+1} = k_n + \Delta\lambda \bar{r}^d_{n+1} g_{n+1} = k_n + 2\Delta\lambda J_{n+1} g_{n+1} \tag{11.29a}$$

From (11.28b), it follows that

$$J_{n+1} = \frac{J^{tr}}{x_0}; \quad J_{n+1} = \left(\frac{1}{2}\mathbf{s}_{n+1} : \mathbf{s}_{n+1}\right)^{1/2}; \quad J^{tr} = \left(\frac{1}{2}\mathbf{s}^{tr} : \mathbf{s}^{tr}\right)^{1/2} \tag{11.29b}$$

Combining (11.29a) and (11.29b), we have

$$k_{n+1} = k_n + \frac{2\Delta\lambda J^{tr} g_{n+1}}{x_0} \tag{11.29c}$$

Let us now proceed to develop equations for evaluating $\Delta\lambda$ for linear and nonlinear hardening models separately below.

Linear Hardening: In the case of linear hardening, $g_{n+1} = g_1 =$ constant (11.6b). Let us combine (11.28b) and (11.29c) with the equation of the yield surface given by (11.3) as

$$\mathbf{s}_{n+1}.\mathbf{s}_{n+1} - \frac{2}{3}k^2 = 0$$

$$\frac{\mathbf{s}^{tr}.\mathbf{s}^{tr}}{x_0^2} - \frac{2}{3}\left[k_n + \frac{2\Delta\lambda J^{tr} g_{n+1}}{x_0}\right]^2 = 0$$

$$\frac{\sqrt{2}J^{tr}}{x_0} - \frac{\sqrt{2}}{\sqrt{3}}\left[k_n + 2\Delta\lambda J^{tr} g_{n+1}/x_0\right] = 0$$

$$\sqrt{3}J^{tr} - k_n(1 + 4G\Delta\lambda) - 2\Delta\lambda J^{tr} g_1 = 0$$

$$\Delta\lambda = \frac{\sqrt{3}J^{tr} - k_n}{2J^{tr} g_1 + 4Gk_n} \tag{11.30}$$

Thus $\Delta\lambda$ is analytically obtained. The stress and yield stress can now be obtained from (11.28b) and (11.29c), respectively.

Non-Linear Hardening: In this case, the hardening function is nonlinear and is given by (11.6d). With the aid of (11.14c), (11.9), and (11.29b), we have

$$g_{n+1} = g_2 = \frac{m(k_f - k_0)^2}{\left[(k_f - k_0) + m\xi^p_{n+1}\right]^2} = \frac{a_1}{(a_2 + b\xi^p_{n+1})^2} = \frac{a_1}{(a_2 + b\xi^p_n + b\bar{r}^d_{n+1}\Delta\lambda)^2}$$

$$= \frac{a_1}{(a_2 + b\xi^p_n + 2bJ_{n+1}\Delta\lambda)^2} = \frac{a_1}{(a_2 + b\xi^p_n + 2b\frac{J^{tr}}{x_0}\Delta\lambda)^2} = \frac{a_1 x_0^2}{(a_2 x_0 + b\xi^p_n x_0 + 2bJ^{tr}\Delta\lambda)^2}$$

$$g_{n+1} = \frac{a_1 x_0^2}{\left(a_2 x_0 + b\xi_n^p x_0 + 2bJ^{tr}\Delta\lambda\right)^2} \tag{11.31}$$

Equation (11.29c) now becomes

$$k_{n+1} = k_n + \frac{2\Delta\lambda J^{tr}}{x_0} \frac{a_1 x_0^2}{\left(a_2 x_0 + b\xi_n^p x_0 + 2bJ^{tr}\Delta\lambda\right)^2} \tag{11.32}$$

Combine (11.28b) and (11.32) with the equation of the yield surface given by (11.3), we arrive at

$$J_{n+1} - \frac{k_{n+1}}{\sqrt{3}} = 0$$

$$\frac{J^{tr}}{x_0} - \frac{1}{\sqrt{3}} k_n - \frac{2\Delta\lambda J^{tr}}{\sqrt{3} x_0} \frac{a_1 x_0^2}{\left(a_2 x_0 + b\xi_n^p x_0 + 2bJ^{tr}\Delta\lambda\right)^2} = 0$$

$$J^{tr}\sqrt{3} - k_n(1 + 4G\Delta\lambda) - 2\Delta\lambda J^{tr} a_1 \frac{(1 + 16G^2\Delta\lambda^2 + 8G\Delta\lambda)}{\left(a_2 + 4Ga_2\Delta\lambda + b\xi_n^p + 4Gb\xi_n^p\Delta\lambda + 2b\Delta\lambda J^{tr}\right)^2} = 0$$

Let us define:

$$\begin{aligned} &\alpha_1 = \sqrt{3}J^{tr} - k_n; \quad \alpha_2 = -4Gk_n; \quad \beta_1 = a_2 + b\xi_n^p; \quad \text{and} \\ &\beta_2 = 4Ga_2 + 4Gb\xi_n^p + 2bJ^{tr} \end{aligned} \tag{11.33}$$

Then the above equation becomes:

$$(\alpha_1 + \alpha_2\Delta\lambda)(\beta_1 + \beta_2\Delta\lambda) = 2\Delta\lambda J^{tr} a_1 (1 + 16G^2\Delta\lambda^2 + 8G\Delta\lambda)$$

which upon expansion becomes the following cubic equation in $\Delta\lambda$:

$$B_1\Delta\lambda^3 + B_2\Delta\lambda^2 + B_3\Delta\lambda + B_4 = 0 \tag{11.34}$$

where

$$B_1 = \alpha_2\beta_2^2 - 32G^2 J^{tr} a_1 \tag{11.35a}$$

$$B_2 = 2\beta_1\beta_2\alpha_2 + \alpha_1\beta_2^2 - 16GJ^{tr} a_1 \tag{11.35b}$$

$$B_3 = \alpha_2\beta_1^2 + 2\beta_1\beta_2\alpha_1 - 2J^{tr} a_1 \tag{11.35c}$$

$$B_4 = \alpha_1\beta_1^2 \tag{11.35d}$$

Equation (11.34) is to be solved to obtain a value of $\Delta\lambda$ (which must be real and positive). The stress and yield stress can now be obtained from (11.28b) and (11.32), respectively.

11.3.1.2 Calculation of the Consistent Tangent Operator

From (11.28a), (11.28b), and (11.28c), we have

$$s_{ij} = \frac{s_{ij}^{tr}}{x_0} = \frac{s_{ij}^{n} + 2G\Delta e_{ij}}{x_0} = \frac{s_{ij}^{n} + 2G\Delta e_{ij}}{1 + 4G\Delta\lambda} \tag{11.36}$$

Differentiating (11.36), it follows that

$$\dot{s}_{ij} = \frac{2G\dot{e}_{ij}}{x_0} - \frac{4G}{x_0^2}(s_{ij}^{n} + 2G\Delta e_{ij})\dot{\lambda} \tag{11.37}$$

To further expand (11.37), we need an expression for $\dot{\lambda}$, which depends on the type of hardening as follows:

Linear Hardening.
From (11.30), we have

$$\Delta\lambda = \frac{\sqrt{3}J^{tr} - k_n}{2J^{tr}g_1 + 4Gk_n}$$

$$\dot{\lambda} = \frac{1}{(2J^{tr}g_1 + 4Gk_n)^2}\left[(2J^{tr}g_1 + 4Gk_n)\sqrt{3}\dot{J}^{tr} - \left(\sqrt{3}J^{tr} - k_n\right)2g_1\dot{J}^{tr}\right]$$

$$\dot{\lambda} = \frac{k_n(g_1 + 2\sqrt{3}G)}{(2J^{tr}g_1 + 4Gk_n)^2}\dot{J}^{tr} \tag{11.38}$$

$$J^{tr} = \left[\frac{1}{2}s_{k\ell}^{tr}s_{k\ell}^{tr}\right]^{1/2}; \quad s_{k\ell}^{tr} = s_{k\ell}^{n} + 2G\Delta e_{k\ell}$$

$$\dot{J}^{tr} = \frac{1}{2J^{tr}}\frac{2}{2}s_{k\ell}^{tr}\dot{s}_{k\ell}^{tr} = \frac{1}{2J^{tr}}s_{k\ell}^{tr}(2G\dot{e}_{k\ell}) = \frac{Gs_{k\ell}^{tr}\dot{e}_{k\ell}}{J^{tr}} \tag{11.39}$$

Combining (11.38) and (11.39):

$$\dot{\lambda} = \frac{k_n(g_1 + 2\sqrt{3}G)}{(2J^{tr}g_1 + 4Gk_n)^2}\frac{Gs_{k\ell}^{tr}\dot{e}_{k\ell}}{J^{tr}} = \bar{f}_1 s_{k\ell}^{tr}\dot{e}_{k\ell} \tag{11.40a}$$

where

$$\bar{f}_1 = \frac{k_n G(g_1 + 2\sqrt{3}G)}{J^{tr}(2J^{tr}g_1 + 4Gk_n)^2} \tag{11.40b}$$

Nonlinear Hardening:
From (11.33), it follows that

$$\alpha_1 = \sqrt{3}J^{tr} - k_n; \quad \alpha_2 = -4Gk_n; \quad \beta_1 = a_2 + b\xi_n^{p}; \quad \text{and} \quad \beta_2 = 4Ga_2 + 4Gb\xi_n^{p} + 2bJ^{tr}$$
$$\dot{\alpha}_1 = \sqrt{3}\dot{J}^{tr}$$
$$\dot{\alpha}_2 = 0$$
$$\dot{\beta}_1 = 0$$
$$\dot{\beta}_2 = 2b\dot{J}^{tr}$$

From (11.35a–d), we have

$$\dot{B}_1 = \dot{\alpha}_2\beta_2^2 + 2\alpha_2\beta_2\dot{\beta}_2 - 32G^2a_1\dot{J}^{tr} = \overline{B}_1\dot{J}^{tr} \tag{11.41a}$$

where

$$\overline{B}_1 = 4\alpha_2\beta_2 b - 32G^2a_1 \tag{11.41b}$$

$$\dot{B}_2 = 2\beta_1\dot{\beta}_2\alpha_2 + \dot{\alpha}_1\beta_2^2 + 2\alpha_1\beta_2\dot{\beta}_2 - 16Ga_1\dot{J}^{tr} = \overline{B}_2\dot{J}^{tr} \tag{11.41c}$$

where

$$\overline{B}_2 = 4\beta_1\alpha_2 b + \sqrt{3}\beta_2^2 + 4\alpha_1\beta_2 b - 16Ga_1 \tag{11.41d}$$

$$\dot{B}_3 = 2\beta_1\dot{\beta}_2\alpha_1 + 2\beta_1\beta_2\dot{\alpha}_1 - 2a\dot{J}^{tr} = \overline{B}_3\dot{J}^{tr} \tag{11.41e}$$

where

$$\overline{B}_3 = 4\beta_1\alpha_1 b + 2\sqrt{3}\beta_1\beta_2 - 2a_1 \tag{11.41f}$$

$$\dot{B}_4 = \dot{\alpha}_1\beta_1^2 = \overline{B}_4\dot{J}^{tr} \tag{11.41g}$$

where

$$\overline{B}_4 = \sqrt{3}\beta_1^2 \tag{11.41h}$$

Differentiating (11.34) and using (11.39), it follows that

$$(3B_1\Delta\lambda^2 + 2B_2\Delta\lambda + B_3)\dot{\lambda} + (\dot{B}_1\Delta\lambda^3 + \dot{B}_2\Delta\lambda^2 + \dot{B}_3\Delta\lambda + \dot{B}_4) = 0$$

$$\dot{\lambda} = -\frac{\overline{B}_1\Delta\lambda^3 + \overline{B}_2\Delta\lambda^2 + \overline{B}_3\Delta\lambda + \overline{B}_4}{3B_1\Delta\lambda^2 + 2B_2\Delta\lambda + B_3}\dot{J}^{tr} = -\frac{\overline{B}_1\Delta\lambda^3 + \overline{B}_2\Delta\lambda^2 + \overline{B}_3\Delta\lambda + \overline{B}_4}{3B_1\Delta\lambda^2 + 2B_2\Delta\lambda + B_3}\frac{Gs_{k\ell}^{tr}\dot{e}_{k\ell}}{J^{tr}}$$

or

$$\dot{\lambda} = \bar{f}_1 s_{k\ell}^{tr}\dot{e}_{k\ell} \tag{11.42a}$$

where

$$\bar{f}_1 = -\frac{\overline{B}_1\Delta\lambda^3 + \overline{B}_2\Delta\lambda^2 + \overline{B}_3\Delta\lambda + \overline{B}_4}{3B_1\Delta\lambda^2 + 2B_2\Delta\lambda + B_3}\frac{G}{J^{tr}} \tag{11.42b}$$

Consistent Tangent Operator
Combining (11.37) and (11.40a) for linear hardening or (11.42a) for nonlinear hardening, we get

$$\dot{s}_{ij} = \frac{2G\dot{e}_{ij}}{x_0} - \frac{4G}{x_0^2}(s_{ij}^n + 2G\Delta e_{ij})\bar{f}_1 s_{k\ell}^{tr}\dot{e}_{k\ell}$$

Then the stress–strain rate relation is obtained as (making use of (11.27)) follows:

$$\begin{aligned}
\dot{\sigma}_{ij} &= \dot{p}\delta_{ij} + \dot{s}_{ij} \\
&= K\dot{\varepsilon}_v\delta_{ij} + \frac{2G\dot{e}_{ij}}{x_0} - \frac{4G}{x_0^2}(s_{ij}^n + 2G\Delta e_{ij})\bar{f}_1 s_{k\ell}^{tr}\dot{e}_{k\ell} \\
&= K\dot{\varepsilon}_v\delta_{ij} + \frac{2G}{x_0}(\dot{\varepsilon}_{ij} - \frac{1}{3}\dot{\varepsilon}_v\delta_{ij}) - \frac{4G}{x_0^2}(s_{ij}^n + 2G\Delta e_{ij})\bar{f}_1 s_{k\ell}^{tr}(\dot{\varepsilon}_{k\ell} - \frac{1}{3}\dot{\varepsilon}_v\delta_{k\ell}) \\
&= \left\{K - \frac{2G}{3x_0}\right\}\dot{\varepsilon}_v\delta_{ij} + \frac{2G}{x_0}\dot{\varepsilon}_{ij} - \frac{4G}{x_0^2}(s_{ij}^n + 2G\Delta e_{ij})\bar{f}_1 s_{k\ell}^{tr}\dot{\varepsilon}_{k\ell} \quad (\text{since } s_{k\ell}^{tr}\delta_{k\ell} = 0) \\
&= \left\{K - \frac{2G}{3x_0}\right\}\dot{\varepsilon}_{k\ell}\delta_{k\ell}\delta_{ij} + \frac{G}{x_0}(\delta_{ik}\delta_{j\ell} + \delta_{i\ell}\delta_{jk})\dot{\varepsilon}_{k\ell} - \frac{4G\bar{f}_1}{x_0^2}s_{ij}^{tr}s_{k\ell}^{tr}\dot{\varepsilon}_{k\ell}
\end{aligned}$$

Hence we have,

$$\dot{\sigma}_{ij} = D_{ijk\ell}\dot{\varepsilon}_{k\ell} \tag{11.43a}$$

where

$$D_{ijk\ell} = \left\{K - \frac{2G}{3x_0}\right\}\delta_{k\ell}\delta_{ij} + \frac{G}{x_0}(\delta_{ik}\delta_{j\ell} + \delta_{i\ell}\delta_{jk}) - \frac{4G\bar{f}_1}{x_0^2}s_{ij}^{tr}s_{k\ell}^{tr} \tag{11.43b}$$

x_0, s_{ij}^{tr}, and $\bar{f}_1$ are given by (11.28c), (11.28a), and (11.40b) (for linear hardening) or (11.42b) (for nonlinear hardening), respectively. $D_{ijk\ell}$ in (11.43b) is the required consistent tangent operator.

11.3.2 Approximation 2 for the Hardening Variable: Euler Approximation at the Level of ξ^p ("Exact")

11.3.2.1 Calculation of Stresses and PIVs

The expressions for quantities $\mathbf{s}_{n+1}$, $\mathbf{s}^{tr}$ and x_0 remain the same as those in (11.28a) (11.28b), and (11.28c). Since both approximations ("Exact" and "Euler") coincide for the case of linear hardening, we will only consider the case of *nonlinear hardening* here.

From (11.12b) and (11.15b), we have

$$k_{n+1} = \bar{k}(\boldsymbol{\sigma}, k, \Delta\lambda) = k_{ex}(\boldsymbol{\sigma}, k, \Delta\lambda) = k_0 + \frac{m(k_f - k_0)\left[\xi_n^p + \bar{r}_{n+1}^d \Delta\lambda\right]}{(k_f - k_0) + m\left[\xi_n^p + \bar{r}_{n+1}^d \Delta\lambda\right]} \tag{11.44a}$$

$$k_{n+1} = k_0 + \frac{c_1\left[\xi_n^p + \bar{r}_{n+1}^d \Delta\lambda\right]}{a_2 + b\left[\xi_n^p + \bar{r}_{n+1}^d \Delta\lambda\right]}; \quad c_1 = m(k_f - k_0); \quad a_2 = k_f - k_0; \quad b = m \tag{11.44b}$$

Equation (11.3) can be written as

$$J_{n+1}\sqrt{3} = k_{n+1} \tag{11.45}$$

Combining (11.45), (11.29b), and (11.44b)

$$\frac{J^{tr}}{x_0}\sqrt{3} = k_0 + \frac{c_1\left[\xi_n^p + \bar{r}_{n+1}^d \Delta\lambda\right]}{a_2 + b\left[\xi_n^p + \bar{r}_{n+1}^d \Delta\lambda\right]} \tag{11.46a}$$

$$\left(J^{tr}\sqrt{3} - k_0 x_0\right)\left(a_2 + b\xi_n^p + b\bar{r}_{n+1}^d \Delta\lambda\right) = x_0\left(c_1\xi_n^p + c_1\bar{r}_{n+1}^d \Delta\lambda\right) \tag{11.46b}$$

From (11.9 and 11.29b), it follows that

$$\bar{r}_{n+1}^d = 2J^{tr}/x_0 \tag{11.47}$$

Using this relation, (11.46b) becomes

$$\left(J^{tr}\sqrt{3} - k_0 x_0\right)\left(a_2 x_0 + b x_0 \xi_n^p + 2J^{tr} b\Delta\lambda\right) = x_0\left(c_1\xi_n^p x_0 + 2\Delta\lambda J^{tr} c_1\right) \tag{11.48}$$

Combining (11.48) with (11.28c), we get

$$\begin{aligned}&\left(J^{tr}\sqrt{3} - k_0 - 4G\Delta\lambda k_0\right)\left(a_2 + b\xi_n^p + 4G\Delta\lambda a_2 + 4G\Delta\lambda b\xi_n^p + 2J^{tr} b\Delta\lambda\right)\\ &\quad = \left(c_1\xi_n^p + 4Gc_1\Delta\lambda\xi_n^p + 2\Delta\lambda J^{tr} c_1\right)(1 + 4G\Delta\lambda)\end{aligned} \tag{11.49}$$

Upon expanding and grouping, (11.49) results in the following quadratic equation

$$B_1\Delta\lambda^2 + B_2\Delta\lambda + B_3 = 0 \tag{11.50}$$

where

$$B_1 = -8Gk_0\left(2Ga_2 + 2Gb\xi_n^p + J^{tr}b\right) - 8Gc_1\left(2G\xi_n^p + J^{tr}\right) \tag{11.51a}$$

$$B_2 = -4Gk_0\left(a_2 + b\xi_n^p\right) + 2\left(J^{tr}\sqrt{3} - k_0\right)\left(2Ga_2 + 2Gb\xi_n^p + J^{tr}b\right) - 8c_1G\xi_n^p - 2J^{tr}c_1 \tag{11.51b}$$

$$B_3 = \left(J^{tr}\sqrt{3} - k_0\right)\left(a_2 + b\xi_n^p\right) - c_1\xi_n^p \tag{11.51c}$$

The quadratic equation given by (11.50) can be solved for $\Delta\lambda$ (which must be real and positive). The stress and yield stress can now be obtained from (11.28b) and (11.44b), respectively.

11.3.2.2 Calculation of the Consistent Tangent Operator

The expression for the rate of deviatoric stress is the same as the one given by (11.37). The rate $\dot{\lambda}$ is needed. This can be found by differentiating (11.50). To solve for $\dot{\lambda}$, the rates $\dot{B}_1$, $\dot{B}_2$, and $\dot{B}_3$ are required. Let us first find these by differentiating (11.51a)–(11.51c) as follows:

$$\dot{B}_1 = -8G(k_0b + c_1)\dot{J}^{tr} = \overline{B}_1\dot{J}^{tr} \tag{11.52a}$$

where

$$\overline{B}_1 = -8G(k_0b + c_1) \tag{11.52b}$$

$$\dot{B}_2 = 2\sqrt{3}\left(2Ga_2 + 2Gb\xi_n^p + J^{tr}b\right)\dot{J}^{tr} + 2b\left(J^{tr}\sqrt{3} - k_0\right)\dot{J}^{tr} - 2c_1\dot{J}^{tr} = \overline{B}_2\dot{J}^{tr} \tag{11.52c}$$

where

$$\overline{B}_2 = 4\sqrt{3}Ga_2 + 4\sqrt{3}Gb\xi_n^p + 4\sqrt{3}bJ^{tr} - 2bk_0 - 2c_1 \tag{11.52d}$$

$$\dot{B}_3 = \sqrt{3}(a_2 + b\xi_n^p)\dot{J}^{tr} = \overline{B}_3\dot{J}^{tr} \tag{11.52e}$$

where

$$\overline{B}_3 = \sqrt{3}\left(a_2 + b\xi_n^p\right) \tag{11.52f}$$

Differentiating (11.50) and using (11.52a–f), it follows that

$$2B_1(\Delta\lambda)\dot{\lambda} + \Delta\lambda^2\dot{B}_1 + B_2\dot{\lambda} + \dot{B}_2\Delta\lambda + \dot{B}_3 = 0$$

$$(2B_1\Delta\lambda + B_2)\dot{\lambda} + \left(\Delta\lambda^2\overline{B}_1 + \Delta\lambda\overline{B}_2 + \overline{B}_3\right)\dot{J}^{tr} = 0$$

Using (11.39), we get

$$\dot{\lambda} = -\frac{\overline{B}_1\Delta\lambda^2 + \overline{B}_2\Delta\lambda + \overline{B}_3}{2B_1\Delta\lambda + B_2}\dot{J}^{tr} = -\frac{\overline{B}_1\Delta\lambda^2 + \overline{B}_2\Delta\lambda + \overline{B}_3}{2B_1\Delta\lambda + B_2}\frac{Gs^{tr}_{k\ell}\dot{e}_{k\ell}}{J^{tr}}$$

or

$$\dot{\lambda} = \bar{f}_1 s^{tr}_{k\ell}\dot{e}_{k\ell} \tag{11.53a}$$

where

$$\bar{f}_1 = -\frac{\overline{B}_1\Delta\lambda^2 + \overline{B}_2\Delta\lambda + \overline{B}_3}{2B_1\Delta\lambda + B_2}\frac{G}{J^{tr}} \tag{11.53b}$$

The consistent tangent operator is now calculated from (11.43b) using (11.53b) for $\bar{f}_1$. The parameters x_0 and s^{tr}_{ij} are given by (11.28c) and (11.28a), respectively.

11.4 Sample Numerical Results from the Isotropic Hardening von Mises Model

In this section, we first present some sample calculations for finding the incremental stress and PIV for a given incremental strain. We then present simulations of the behavior of a one-element problem (stress–strain behavior) and a multi-element problem (load-deformation behavior).

11.4.1 Calculation of Incremental Quantities

The nonlinear isotropic hardening law presented in this chapter (11.6c) is considered. The values used for the model parameters, relevant quantities at the initial point n and incremental strain are listed below. The unit is GPa for stresses, yield stresses, and moduli. We will work in the principal stress space.

$$k_0 = 0.25, \quad k_f = 0.4, \quad m = 20, \quad E = 200, \quad \nu = 0.3$$

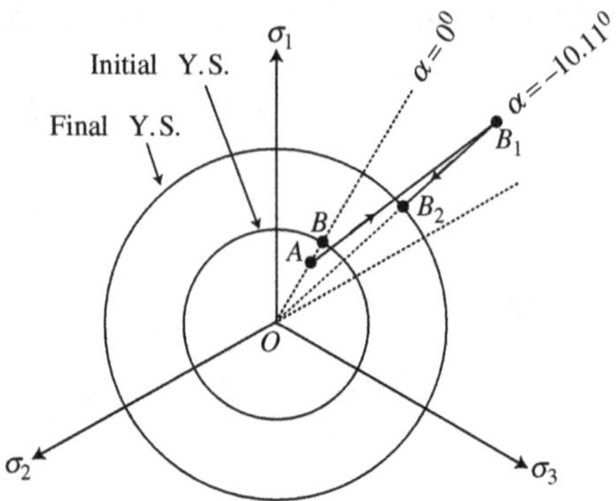

Fig. 11.2 The yield surfaces and stress points in the octahedral plane

$$\boldsymbol{\sigma}_n = \{\, 0.1 \quad 0.05 \quad 0.075 \,\}$$

$$k_n = 0.25$$

$$\Delta\boldsymbol{\varepsilon} = \{\, 0.03 \quad -0.028 \quad 0.01 \,\}$$

We will perform the calculations first using the S-space formulation with both the "Euler" and "Exact" methods of integrating the hardening rules. The calculations are then carried out using the σ-space formulation. For all of these methods, the initial sets of calculations are identical. They are performed using the steps in Sect. 11.3.1. The initial and final yield surfaces and the initial, trial, and final stresses are marked in an octahedral plane in Fig. 11.2 (the explanations are provided in the following sections).

$$K = \frac{E}{3(1-2\nu)} = 166.67; \quad G = \frac{E}{2(1+\nu)} = 76.92$$

$$I_n = 0.225, \quad p_n = \frac{I_n}{3} = 0.075, \quad J_n = 0.025, \quad \mathbf{s}_n = \{\, 0.025 \quad -0.025 \quad 0 \,\},$$
$$\phi^n = -0.04042$$

To mark the point representing the initial stress, we calculate the Lode angle α (Chap. 2). The value of α is 0°. The initial stress then plots at Point A, where $OA = |\mathbf{s}_n| = \sqrt{2}J_n = 0.03536$. Noting that $OB = k_n\sqrt{2/3} = 0.2041$, it is seen that the initial point is inside the surface. Note that α is calculated only for the purpose of marking the initial point; it is not required by the algorithm.

Since $\phi^n < 0$, the initial point is inside the surface (as we have already noted above). We now calculate the elastic trial stress to determine if the strain increment leads to an elastic or elasto-plastic behavior.

$$\Delta\varepsilon_{pp} = 0.012$$

$$\Delta\mathbf{e} = \Delta\boldsymbol{\varepsilon} - \frac{1}{3}\Delta\varepsilon_{pp}\boldsymbol{\delta} = \{\, 0.026 \quad -0.032 \quad 0.006 \,\}$$

$$\mathbf{s}^{tr} = \mathbf{s}^{n} + 2G\Delta\mathbf{e} = \{\, 4.0249 \quad -4.9481 \quad 0.92310 \,\}$$

$$p^{tr} = p_n + K\Delta\varepsilon_{pp} = 2.0750$$

$$\boldsymbol{\sigma}^{tr} = \mathbf{s}^{tr} + p^{tr}\boldsymbol{\delta} = \{\, 6.0999 \quad -2.8731 \quad 2.9981 \,\}$$

$$J^{tr} = 4.5572$$

$$\phi^{tr} = 41.4945$$

$$(\xi^{p})^{tr} = \xi^{p}_{n} = 0$$

$$k^{tr} = k_n$$

Since $\phi^{tr} > 0$, the incremental behavior is elasto-plastic. The Lode angle α at the trial point is $-10.11°$. This places the elastic trial stress at Point B_1 as shown in Fig. 11.2, where $OB_1 = \sqrt{2}J^{tr} = 6.4448$.

11.4.1.1 The *S*-Space Formulation with Approximation 1 (Sect. 11.3.1) for Integration of Hardening Rule ("Euler")

Here the procedure presented in Sect. 11.3.1 is followed. We now calculate a value for $\Delta\lambda$ such that (a) the trial stress point is brought back to the surface, and (b) the expansion of the surface is properly accounted for. This involves the solution of the cubic equation (11.34) as follows.

$$B_1 = -4402590.08, \quad B_2 = 391075.38, \quad B_3 = 517.98 \quad \text{and} \quad B_4 = 0.1719,$$

$$\Delta\lambda = 0.090139$$

Now the final value for the stress and PIV are calculated as follows:

$$x_0 = 1 + 4G\Delta\lambda = 28.7349$$

$$\boldsymbol{\sigma}_{n+1} = \frac{\mathbf{s}^{tr}}{x_0} + p^{tr}\boldsymbol{\delta} = \{\, 2.2151 \quad 1.9028 \quad 2.1071 \,\}$$

$$J_{n+1} = 0.1586$$

$$\Delta\boldsymbol{\sigma} = \{\, 2.1151 \quad 1.8528 \quad 2.0321 \,\}$$

$$\xi^p_{n+1} = \xi^p_n + \Delta\xi = 0 + \bar{r}^d_{n+1}\Delta\lambda = 2J_{n+1}\Delta\lambda = 0.0286$$

$$a_1 = m(k_f - k_0)^2 = 0.45, \quad a_2 = 0.15, \quad b = 20 \qquad \text{(from 11.6d)}$$

$$g_{n+1} = \frac{a_1}{\left(a_2 + b\xi^p_{n+1}\right)^2} = 0.8637$$

$$k_{n+1} = k_n + g_{n+1}\bar{r}^d_{n+1}\Delta\lambda = k_n + 2g_{n+1}J_{n+1}\Delta\lambda = 0.2747$$

The Lode angle α at the final point $n+1$ is $-10.11°$. This places the point on the final surface at Point B_2 as shown in Fig. 11.2, where $OB_2 = \sqrt{2}J_{n+1} = k_{n+1}\sqrt{2/3} = 0.2243$. Note that Point B_2 is on the radial line OB_1.

The required solution is:

$$\Delta\boldsymbol{\sigma} = \{\, 2.1151 \quad 1.8528 \quad 2.0321 \,\}$$

$$\Delta\lambda = 0.090139$$

$$\xi^p_{n+1} = 0.0286$$

$$k_{n+1} = 0.2747$$

11.4.1.2 The *S*-Space Formulation with Approximation 2 (Sect. 11.3.2) for Integration of Hardening Rule ("Exact")

Here the procedure presented in Sect. 11.3.2 is followed. We now calculate a value for $\Delta\lambda$ which will bring the trial stress point back to the surface with the expansion of the surface considered. This entails the solution of the quadratic equation (11.50) as follows.

$$B_1 = -25985.75, \quad B_2 = 1707.17 \quad \text{and} \quad \mathbf{B}_3 = 1.1465,$$

The two solutions are: –0.0006648 and 0.06636

Choosing the positive value, we have $\Delta\lambda = 0.06636$

Now the final value for the stress and PIV are calculated as follows:

$$x_0 = 1 + 4G\Delta\lambda = 21.4188$$

$$\boldsymbol{\sigma}_{n+1} = \frac{\mathbf{s}^{tr}}{x_0} + p^{tr}\boldsymbol{\delta} = \{\, 2.2629 \quad 1.8439 \quad 2.1181 \,\}$$

$$J_{n+1} = 0.2128$$

$$\Delta\boldsymbol{\sigma} = \{2.1629 \quad 1.7939 \quad 2.0431\}$$

$$\xi^p_{n+1} = \xi^p_n + \Delta\xi = 0 + \bar{r}^d_{n+1}\Delta\lambda = 2J_{n+1}\Delta\lambda = 0.0282$$

$$c_1 = 3, \quad a_2 = 0.15, \quad b = 20 \qquad \text{(from 11.44b)}$$

$$k_{n+1} = k_0 + \frac{c_1\xi^p_{n+1}}{a_2 + b\xi^p_{n+1}} = 0.3685$$

The Lode angle α at the final point $n+1$ is $-10.11°$ which places the point on the final surface at Point B_2 as shown in Fig. 11.2, where $OB_2 = \sqrt{2}J_{n+1} = k_{n+1}\sqrt{2/3} = 0.3$. Note that Point B_2 is on the radial line OB_1.

The required solution is:

$$\Delta\boldsymbol{\sigma} = \{2.1629 \quad 1.7939 \quad 2.0431\}$$

$$\Delta\lambda = 0.06636$$

$$\xi^p_{n+1} = 0.0282$$

$$k_{n+1} = 0.3685$$

11.4.1.3 The σ-Space Formulation

Here the procedure presented in Sect. 11.2 is followed. First a trial value for $\Delta\lambda$ is calculated at the elastic trial stress point as follows (note: the trial point is the same as $n+1$ in the equations given below):

$$\mathbf{r}_{n+1} = 2\mathbf{s}_{n+1} = \{8.0499 \quad -9.8962 \quad 1.8462\}$$

$$\bar{r}^d_{n+1} = 2J_{n+1} = 9.1144$$

$$a_1 = m(k_f - k_0)^2 = 0.45, \quad a_2 = 0.15, \quad b = 20 \qquad \text{(from 11.6d)}$$

$$g_{n+1} = \frac{a_1}{(a_2 + b\xi^p_{n+1})^2} = 20$$

$$K_p = \frac{4}{3}k_{n+1}g_{n+1}\bar{r}^d_{n+1} = 60.76$$

$$\mathbf{n}^{\mathrm{T}}\mathbf{Cr} = 4Gr_J^2 = 16GJ_{n+1}^2 = 25560.71$$

$$\Delta\lambda = \frac{\phi^{n+1}}{\left[\mathbf{n}^{\mathrm{T}}\mathbf{Cr} + K_p\right]^{n+1}} = 0.0016195$$

The stress and PIV are recalculated using this value of $\Delta\lambda$ (Step 5, Algorithm 11.1) as

$$\boldsymbol{\sigma}_{n+1} = \boldsymbol{\sigma}_n + \mathrm{C}\Delta\boldsymbol{\varepsilon} - \Delta\lambda\mathrm{C}\mathbf{r}_{n+1} = \{4.0943 \quad -0.4074 \quad 2.5381\}$$

$$\xi_{n+1}^p = \xi_n^p + \Delta\lambda\bar{r}_{n+1}^d = 0.014761$$

$$a_1 = 0.45, \quad a_2 = 0.15, \quad b = 20\,(\text{from above})$$

$$g_{n+1} = \frac{a_1}{(a_2 + b\xi_{n+1}^p)^2} = 2.2702$$

$$k_{n+1} = k_n + g_{n+1}\bar{r}_{n+1}^d\Delta\lambda = 0.2835$$

At this point, the iterations (Steps 6–11 in Algorithm 11.1) begin and continue until convergence. The key results, during each iteration, are listed in Table 11.1. The final value of $\Delta\lambda$ (which is not listed in the table) is 0.0901.

The required solution is:

$$\Delta\boldsymbol{\sigma} = \{2.1151 \quad 1.8528 \quad 2.0321\}$$

$$\Delta\lambda = 0.090139$$

$$\xi_{n+1}^p = 0.0286$$

$$k_{n+1} = 0.2747$$

Table 11.1 Results during the σ-space iteration

Iter. no.	σ_{n+1}^1			k^1	σ_{n+1}^2			k^2	e_s	e_ς	e_ϕ
	σ_{11}	σ_{22}	σ_{33}		σ_{11}	σ_{22}	σ_{33}				
1	4.09	−0.41	2.54	0.28	5.09	−1.6	2.77	0.29	0.33	0.14×10^{-1}	0.65×10^{2}
2	3.09	0.83	2.31	0.29	4.34	−0.70	2.59	0.28	0.51	0.11×10^{-1}	0.40×10
3	2.59	1.44	2.19	0.28	3.87	−0.13	2.49	0.28	0.56	0.24×10^{-1}	0.40×10
4	2.35	1.73	2.14	0.27	3.54	0.28	2.41	0.28	0.52	0.23×10^{-1}	0.92
5	2.25	1.86	2.11	0.27	3.09	0.83	2.31	0.28	0.37	0.16×10^{-1}	0.17
6	2.22	1.90	2.11	0.27	2.47	1.60	2.16	0.28	0.11	0.50×10^{-2}	0.15×10^{-1}
7	2.22	1.90	2.11	0.27	2.22	1.90	2.11	0.27	0.32×10^{-2}	0.21×10^{-3}	0.15×10^{-3}
8	2.22	1.90	2.11	0.27	2.22	1.90	2.11	0.27	0.44×10^{-6}	0.10×10^{-6}	0.10×10^{-8}
9	2.22	1.90	2.11	0.27	2.22	1.90	2.11	0.27	0.25×10^{-13}	0.10×10^{-14}	0.24×10^{-16}

During each iteration, the coefficient matrix in (11.19) is calculated and (11.19) is solved for the iterative changes $\delta\boldsymbol{\sigma}$, δk, and $\delta\lambda$. For the sake of completeness, the equation and the solution are listed below for the first iteration.

$$\begin{bmatrix} 1.3322 & -0.1661 & -0.1661 & 0 & 621.32 \\ -0.1661 & 1.3322 & -0.1661 & 0 & -763.81 \\ -0.1661 & -0.1661 & 1.3322 & 0 & 142.49 \\ -0.00005 & 0.00006 & 0.00001 & 1 & -0.1470 \\ 4.0385 & -4.9647 & 0.9262 & -0.378 & 0 \end{bmatrix} \begin{bmatrix} \delta\sigma_{11} \\ \delta\sigma_{22} \\ \delta\sigma_{33} \\ \delta k \\ \delta\lambda \end{bmatrix} = \begin{bmatrix} -0.9995 \\ 1.2287 \\ -0.2292 \\ -0.00399 \\ 10.4006 \end{bmatrix}$$

$$\delta\boldsymbol{\sigma} = \{-1.0043 \quad 1.2346 \quad -0.2303\}$$

$$\delta k = 0.00446$$

$$\delta\lambda = 0.00403$$

11.4.1.4 Comparison of Results and Discussion

By comparing the results from the three methods presented above, the following observations can be made:

1. The solutions from the S-space formulation with “Euler” approximation for the hardening rule and the σ-space formulation with “Euler” approximation for the hardening rule are identical (within numerical tolerance). These two methods are fundamentally the same and the results are expected to be the same.
2. The solutions from the S-space formulation with “Euler” approximation for the hardening rule and the S-space formulation with “Exact” approximation for the hardening rule are different from each other; the latter is expected to be more accurate than the former (further results to support this will be presented in Sect. 11.4.2).
3. The σ-space formulation is seen to involve significantly more computations than the S-space formulation; hence, whenever possible, the S-space formulation is preferred over the σ-space formulation.

11.4.2 Simulation of Stress–Strain Relation and Analysis of a Boundary-Value Problem

In this section, we analyze two boundary-value problems. The results are used to examine some aspects of the model behavior (of the isotropic hardening model) and the integration algorithms presented in the chapter.

The problems chosen are the following:

1. A single-element finite element problem, simulating the behavior of a uniformly-loaded laboratory specimen (Fig. 11.3). The eight-noded brick element is used (Chap. 5).
2. A multi-element finite element (two-dimensional plane strain) problem, representing a typical boundary-value problem (e.g., the metal-forging problem), where the stresses and strains are spatially nonuniform (Fig. 11.4). The eight-noded two-dimensional element is used (Chap. 5).

The model parameters used in the analyses are listed in Table 11.2. Several runs are made; the main characteristics of these runs are summarized in Table 11.3 for the one-element problem and in Table 11.4 for the multi-element problem.

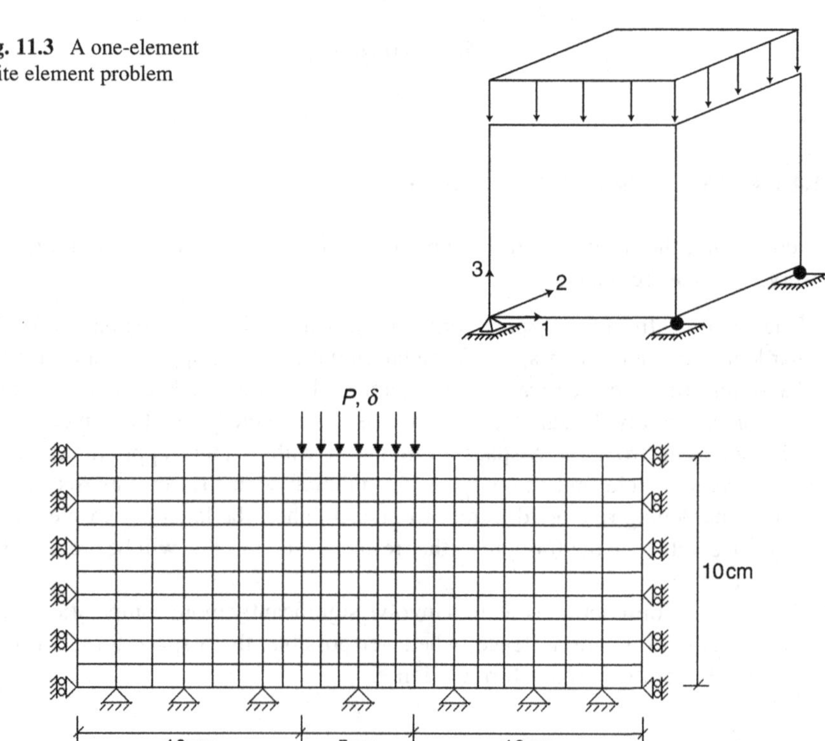

Fig. 11.3 A one-element finite element problem

Fig. 11.4 A multi-element finite element problem

Table 11.2 Isotropic hardening von Mises model parameters

Hardening type	E (GPa)	ν	m (GPa)	k_f (GPa)	$\frac{k_0}{k_f}$
Linear	200	0.3	20	n/a	n/a
Nonlinear	200	0.3	20	0.4	0.625

Table 11.3 Computer runs made on the single-element problem

No.	Hardening	σ-space/ S-space	Hardening variable approximation	No. load steps NINC	$\hat{\varepsilon}_3$ (%)	$\hat{\sigma}_3$ (GPa)	Tangent operator
S-1	Nonlinear	σ-space	"Euler" (App. 1)	100	20	–	Consistent
S-2	Nonlinear	S-space	"Euler" (App. 1)	100	20	–	Consistent
S-3	Nonlinear	S-space	"Exact" (App. 2)	100	20	–	Consistent
S-4	Nonlinear	S-space	"Exact" (App. 2)	10	20	–	Consistent
S-5	Nonlinear	S-space	"Euler" (App. 1)	10	20	–	Consistent
S-6	Linear	S-space	"Exact" (App. 1)	100	–	0.4	Consistent
S-7	Nonlinear	S-space	"Exact" (App. 1)	100	–	0.35	Consistent
S-8	None	S-space	–	100	1.5	0	Consistent
S-9	Nonlinear	S-space	"Exact" (App. 1)	100	± 1	–	Consistent
S-10	Nonlinear	S-space	"Exact" (App. 1)	100	–	± 0.35	Consistent

Table 11.4 Computer runs made on the multi-element problem

No.	Hardening	σ-space/ S-space	Hardening variable approximation	Number of load steps NINC	$\hat{\delta}$ (cm)	$\hat{\sigma}$ (GPa)	Tangent operator
M-1	Nonlinear	σ-space	"Euler" (App. 1)	200	20	–	Consistent
M-2	Nonlinear	S-space	"Euler" (App. 1)	200	20	–	Consistent
M-3	Nonlinear	S-space	"Exact" (App. 2)	200	20	–	Consistent
M-4	Nonlinear	S-space	"Exact" (App. 2)	100	20	–	Consistent
M-5	Nonlinear	S-space	"Exact" (App. 2)	50	20	–	Consistent
M-6	Nonlinear	S-space	"Exact" (App. 2)	200	20	–	Continuum
M-7	Nonlinear	S-space	"Exact" (App. 2)	100	–	1.4	Consistent

11.4.2.1 Comparison of Results Obtained Using the σ-Space and S-Space Formulations

The body is assumed to be stress free at the beginning. The one-element problem is subjected to a uniaxial loading. The principal stresses in directions 1 and 2 are kept zero. The body is loaded in direction 3 in strain-controlled mode to a strain of $\hat{\varepsilon}_3 = 20\%$ in 100 steps.

The (uniaxial) stress–strain relations calculated by the S-space and σ-space formulations are plotted together in Fig. 11.5a (Runs S-1 and S-2, Table 11.3). The "Euler" method is used to integrate the hardening rule. The global iterations are performed using the consistent tangent operator. As expected, the results are identical within numerical tolerance. Note that the initial elastic part does not appear in Fig. 11.5a because the elastic strain before yielding is too small. The key aspects such as the initial yielding (at $\sigma_3 = 0.25\,\text{GPa}$), hardening and failure (at $\sigma_3 = 0.4\,\text{GPa}$) may be noted.

The multi-element problem is subjected to a uniform displacement $\hat{\delta} = 20\,\text{cm}$ on the surface as shown in Fig. 11.4. The load at the mid node P is calculated and plotted with the corresponding displacement δ in Fig. 11.5b. The displacement is applied in 200 steps. The body is assumed to be stress free at the beginning. The analyses are carried out using both the S-space and σ-space formulations (Runs M-1

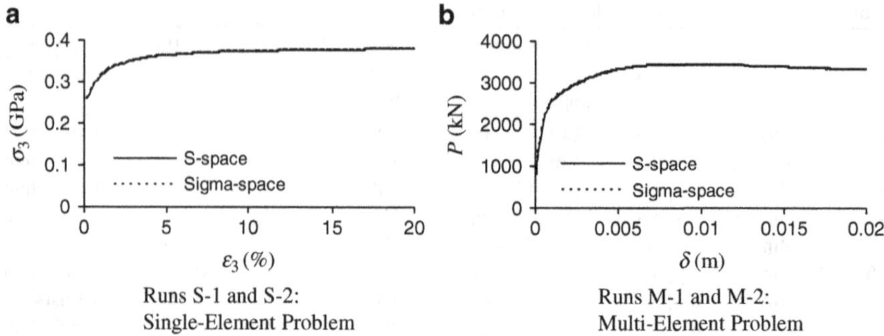

Fig. 11.5 Comparison of results obtained using the σ-space and S-space formulations for the one-element problem

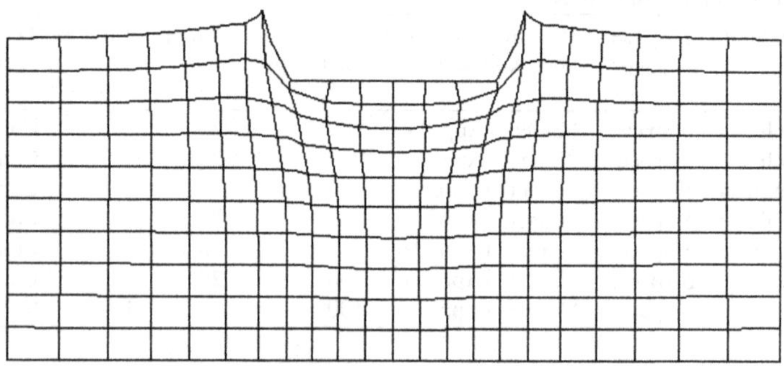

Fig. 11.6 Deformed configuration at failure (run M-2)

and M-2) to integrate the constitutive relations. The "Euler" method is used to integrate the hardening rule. The global iterations are performed using the consistent tangent operator. As expected, the results are identical within numerical tolerance. The deformed configuration at failure is shown in Fig. 11.6, where the occurrence of flow is apparent at the ends of the loaded area.

11.4.2.2 The Influence of the Method of Integrating the Hardening Rule and the Load Step Size

The results concerning the influence of the method of integrating the hardening rule in relation to the load step size are presented in Figs. 11.7 and 11.8. In Fig. 11.7, let us first consider the results obtained with displacement-controlled loading. (The results obtained with load-controlled loading are discussed in the next section.) The results presented in Fig. 11.7a are obtained using strain-controlled loading (Table 11.3).

Figure 11.7 shows that when the step size is fine enough (100 for the one-element problem and 200 for the multi-element problem, in the present case), the

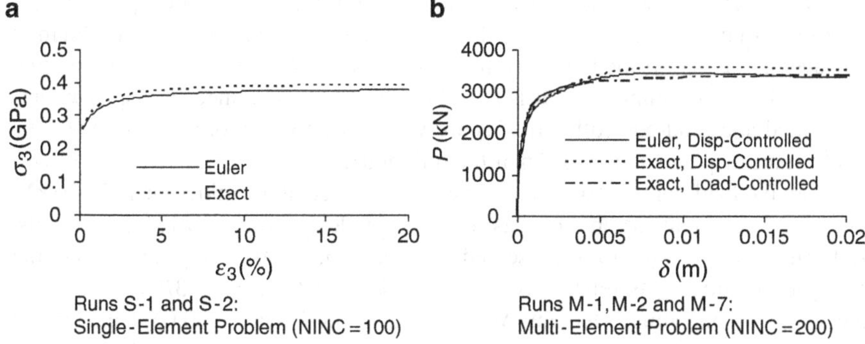

Fig. 11.7 The influence of the method of integrating hardening rule on the results (S-space formulation)

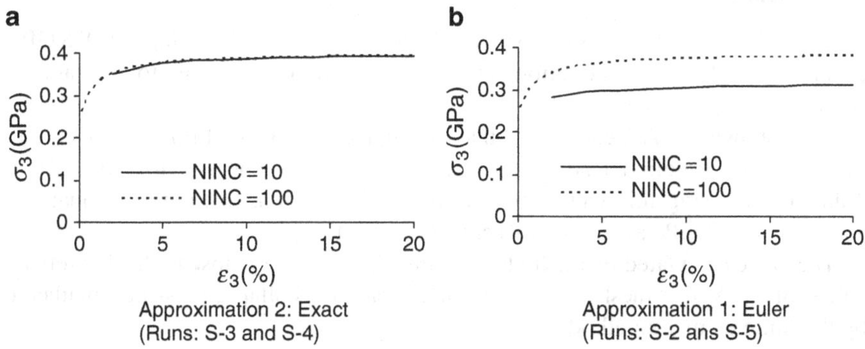

Fig. 11.8 The influence of load step size and method of integrating hardening rule on the stress–strain relation (one-element problem)

results obtained using the "Exact" and "Euler" methods for the integration of hardening rule are very close to each other. On the other hand the results shown in Fig. 11.8 indicate that as the step size increases, while the "Exact" method still gives very good results (Fig. 11.8a), the "Euler" method gives poor results (Fig. 11.8b). The multi-element problem could not be analyzed using large step sizes due to convergence difficulties.

11.4.2.3 Comparison Between the Finite Element and Analytical Collapse Pressures

For the multi-element problem shown in Fig. 11.4, analytical equations are available for the collapse pressure. These analytical equations have been obtained for a problem where the body is loaded with a uniformly distributed pressure (as opposed to uniform displacement). The analysis of the multi-element problem is repeated in load-controlled mode (run M-7 in Table 11.4) and the results are plotted in Fig. 11.7b for comparison. It is seen that the results obtained in displacement-controlled and load-controlled modes are very close to each other in terms of the load versus displacement relation at the mid node.

From the load-controlled analysis, the collapse load at the mid node is 3,420 kN. The finite element mesh consists of eight-noded two-dimensional elements. Referring to the equations presented in Chap. 5, it may be shown that the collapse pressure σ_0 is related to the mid node load P by $\sigma_0 = 3P/\ell$, where ℓ is the width of one finite element. With $P = 3,420\,\text{kN}$ and $\ell = 0.05/6\,\text{m}$, we get $\sigma_0 = 1.23 \times 10^6\, kPa = 1.23\,\text{GPa}$.

Several early researchers have developed equations for the collapse pressure; two examples are given below:

- Prandtl (1921): $\sigma_0 = 2.57\sigma_f$
- Terzaghi (1943): $\sigma_0 = 2.85\sigma_f$

where σ_f is the uniaxial failure stress. With $\sigma_f = k_f = 0.4\,\text{GPa}$, $\sigma_0 = 1.028\,\text{GPa}$ according to Prandtl's equation, and $\sigma_0 = 1.14\,\text{GPa}$ according to Terzaghi's equation.

It is shown in Appendix 9.1 that for plane strain condition (A9.13), the Tresca failure criterion leads to $\sigma_f = k_f/0.866 = 0.4/0.866 = 0.462$ GPa (from Table 11.2). Using this value, $\sigma_0 = 1.187$ GPa according to Prandtl's equation, and $\sigma_0 = 1.317$ GPa according to Terzaghi's equation.

The value calculated by the finite element method is very close to the theoretical values of σ_0. A finer mesh will somewhat decrease the collapse pressure calculated by the finite element method.

11.4.2.4 The Effect of Using Consistent or Continuum Operators on Computational Efficiency

The computational aspects of the convergence behavior are examined in Fig. 11.9. The results correspond to the multi-element problem. The results presented in Fig. 11.9a are obtained using the consistent operator for the global iteration. Figure 11.9a presents the variation of the number of global iterations with the load factor (a load factor of 1 corresponds to the full prescribed displacement of 20 cm). The results are presented for two runs: one with 100 load steps and the other with 50 load steps. In both cases, the number of global iterations increases with the load factor. The range of the global iterations per step is 4–9. The number of iterations is slightly higher in the early stages for the analysis with 50 load steps than that with 100 load steps. As seen in Fig. 11.9b, the use of the continuum operator increases the number of iterations significantly relative to the use of the consistent operator. (Note that 200 global steps were needed to carry out the analysis with

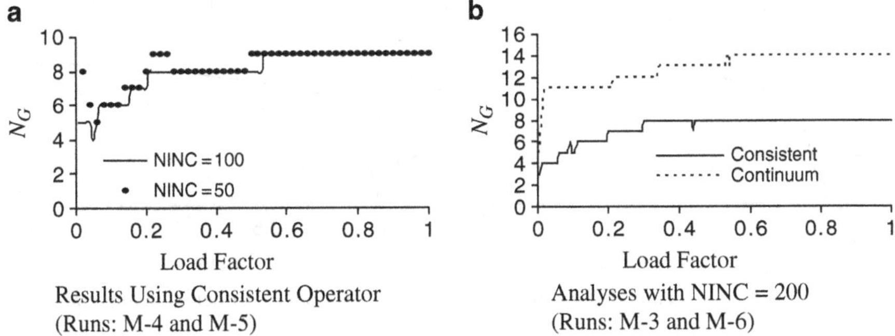

Fig. 11.9 Results on computational efficiency for the multi-element problem (nonlinear hardening, *S*-space formulation, exact integration for hardening variables)

the continuum operator – the reason why the comparison is done with 200 global steps for both.)

11.4.2.5 Typical Results with Linear and Nonlinear Hardening, and Perfect Plasticity

In this section, we compare the typical behaviors of isotropic hardening von Mises model with linear, nonlinear and no (i.e., perfect plasticity) hardening. To show the initial elastic behavior and the transition from elastic to elasto-plastic, only the initial part of the stress–strain curve is shown for each case in Fig. 11.10. The cases of linear, nonlinear and no hardening are shown in Figs. 11.10a, b, and c, respectively.

The differences in the behaviors are self evident. With linear hardening, the yield stress increases linearly whereas with nonlinear hardening, the yield stress increases nonlinearly. In the elastic perfectly-plastic case, the behavior remains elastic until the stress reaches a value of 0.4 GPa, at which point, the material experiences plastic flow. The linear hardening model cannot simulate failure whereas the nonlinear hardening model as well as the elastic perfectly-plastic model can.

11.4.2.6 The Behavior of the Model During Cyclic Loading

The one-element body is subjected to two cycles of loading. Figure 11.11 shows the behavior calculated under strain-controlled loading with strain amplitude of $\pm\,1\%$. Figure 11.11b shows the behavior calculated under a stress-controlled loading with stress amplitude of $\pm\,0.35$ GPa. The nonlinear hardening is employed. The body is assumed to be initially stress free.

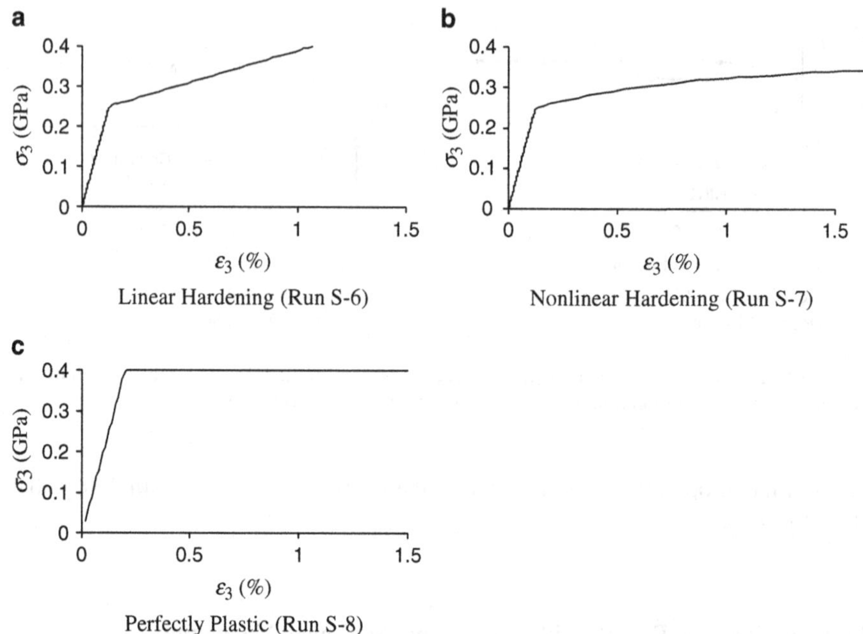

Fig. 11.10 Results obtained using linear hardening, nonlinear hardening and no hardening (perfect plasticity)

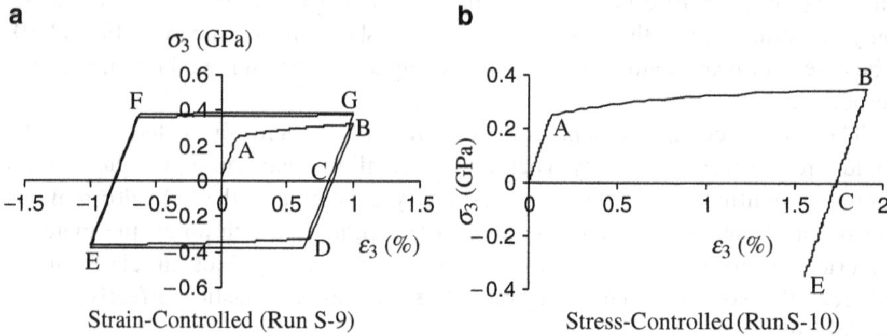

Fig. 11.11 Comparison of isotropic hardening model behaviors from strain-controlled and stress-controlled cyclic loading

During the strain-controlled loading, the behavior is elastic at the beginning (OA). The behavior then changes to elasto-plastic at the initial yield stress of 0.25 GPa (Point A). The yield stress increases to 0.35 GPa before the stress reversal occurs (Point B). The initial behavior during the stress reversal is elastic unloading (BD). The stress point reaches the yield surface on the opposite side (Point D) and the material yields again. The stress–strain curve is almost flat until the strain limit of -1% is reached (Point E). The above unloading–loading behavior repeats for every half cycle, giving rise to a hysteresis loop seen in the figure. The area within the loop created during a cycle (e.g., the first cycle BDEFG) represents the energy dissipation for the cycle.

Consider now the behavior during the stress-controlled loading (Fig. 11.11b). The behavior during the first quarter of the cycle (OAB) is similar to that during the strain-controlled loading. It comprises the initial elastic part (OA), followed by the elasto-plastic part (AB). During the second quarter of the cycle (BC), the behaviors are again similar. However, the behaviors differ thereafter (after C).

During the stress-controlled loading, the yield stress has increased to 0.35 GPa (Point B) during the first quarter of the cycle in tension. In isotropic hardening, the yield surface expands symmetrically about the center (see Fig. 11.2). Hence the yield stress has increased to 0.35 GPa in compression also. Hence during the second (BC) and third quarters (CE), the stress point never goes outside the yield surface and the behavior remains elastic. When the stress is reversed during the fourth (EC) and fifth quarters (CB), the behavior similarly remains elastic. It is clear that, as long as the stress amplitude of the cyclic loading remains a constant, the behavior continues to remain elastic, and the material no longer dissipates energy.

Hence the model is not suitable for application in cyclic loading situations where the stress amplitude remains the same or decreases during cycling. The observation applies to any isotropic hardening, classical model (one that contains a purely elastic domain within the yield surface).

This limitation of classical models triggered a surge in research a few decades back, resulting in a few different (but inter-related) methods of modeling the plastic behavior and energy dissipation during stress-controlled cyclic loading. Among these, the most notable ones are the bounding surface framework proposed by Dafalias and Papov (1975) and Krieg (1975), the multi-surface framework proposed by Mroz (1967) and the endochronic framework proposed by Valanis (1971). While these frameworks were originally developed with

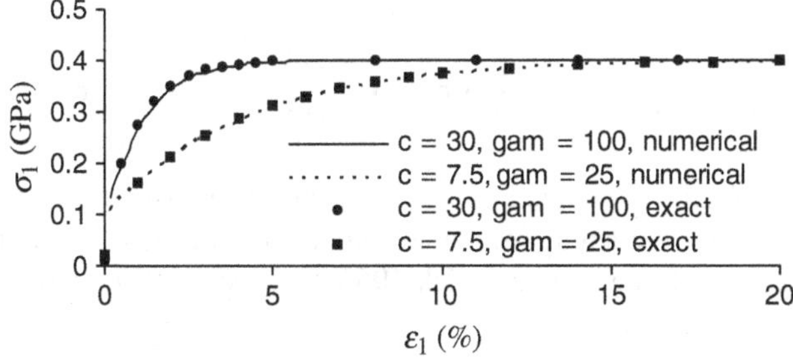

Fig. 11.12 Uniaxial behavior of the kinematic hardening von mises model under monotonic loading

metals in mind, they have later found extensive usage in soils: see Dafalias and Herrmann (1986) and Anandarajah and Dafalias (1986) for application of the bounding surface framework, Prevost (1978) for application of the multi-surface framework and Valanis and Reed (1982) for application of the endochronic framework. These frameworks themselves are sometimes inadequate, requiring a combination of additional features such as the rotational and kinematic hardening features for modeling the hysteresis and evolution of anisotropy properly; see Anandarajah (2008a) for an example of such a model developed for granular materials.

For some materials, kinematic hardening von Mises models may be adequate. In the following section, we present a systematic treatment of the von Mises model that incorporates the Armstrong–Frederick (Armstrong and Frederick 1966) nonlinear kinematic hardening rule.

11.5 A Kinematic Hardening von Mises Model

The elastic stress–strain relation is given by (11.1a) and (11.1b). The elastic properties K and G are assumed to be constants (i.e., independent of stresses and strains). The yield surface is assumed to be a circle with respect to a point in the octahedral plane, represented by the back stress $\boldsymbol{\alpha}$, and is given by

$$\phi(\mathbf{s}, \boldsymbol{\alpha}) = \left(s_{ij} - \alpha_{ij}\right)\left(s_{ij} - \alpha_{ij}\right) - \frac{2}{3}k^2 = 0 \tag{11.54}$$

k is assumed to remain a constant. Hence a pure kinematic hardening is employed. Hence the PIV of the model is a tensor and is equal to the back stress $\boldsymbol{\alpha}$ as

$$\boldsymbol{\zeta} \equiv \boldsymbol{\alpha} \tag{11.55}$$

With an associated flow rule, the plastic constitutive relation is given by the following equations:

$$\dot{\boldsymbol{\varepsilon}}^p = \dot{\lambda}\mathbf{r}; \quad \mathbf{r} = \frac{\partial \psi}{\partial \boldsymbol{\sigma}}; \quad \dot{\lambda} = \frac{1}{K_p}\mathbf{n}\dot{\boldsymbol{\sigma}}; \quad \mathbf{n} = \frac{\partial \phi}{\partial \boldsymbol{\sigma}}; \quad \phi = \psi; \quad \mathbf{r} = \mathbf{n} \tag{11.56}$$

where $\dot{\boldsymbol{\varepsilon}}^p$ is the plastic strain rate and $\dot{\lambda}$ is the consistency parameter (or the loading index). The kinematic hardening behavior is modeled using the Armstrong–Frederick hardening rule (Armstrong and Frederick 1966; Chaboche 1977, 1986)

$$\dot{\alpha}_{ij} = \frac{2}{3}c\dot{\varepsilon}_{ij}^p - \frac{2}{\sqrt{3}}\gamma\alpha_{ij}\dot{\xi}^p \tag{11.57}$$

where ξ^p is the cumulative deviatoric plastic strain invariant (11.6e). c and γ are model parameters. An analysis of the hardening rule has been presented in Chap. 9. Based on (11.56), (11.7a), and (11.8c), and noting that $r_{ij} = 2(s_{ij} - \alpha_{ij})$ (derived below), the hardening rule can be written as

$$\dot{\alpha}_{ij} = \dot{\lambda}\bar{s}_{ij}; \quad \bar{s}_{ij} = \frac{4}{3}cb_{ij} - \frac{2}{\sqrt{3}}\gamma\alpha_{ij}\bar{r}^d \tag{11.58a}$$

where

$$b_{ij} = s_{ij} - \alpha_{ij} \tag{11.58b}$$

and

$$\bar{r}^d = \left(\frac{1}{2}r_{ij}r_{ij}\right)^{1/2} \tag{11.58c}$$

(Note the use of a bar over the hardening function $\bar{s}_{ij}$ to avoid confusing this with the deviatoric stress tensor s_{ij}.) Using the consistency condition (11.11a), the following equation for the plastic modulus can be derived:

$$K_p = \frac{8}{9}k(2ck - 3\gamma ka); \quad a = (s_{k\ell} - \alpha_{k\ell})\alpha_{k\ell} \tag{11.59}$$

By differentiating the yield function, we have

$$r_{ij} = \frac{\partial\phi}{\partial\sigma_{ij}} = \frac{\partial\phi}{\partial s_{k\ell}}\frac{\partial s_{k\ell}}{\partial\sigma_{ij}} = 2(s_{k\ell} - \alpha_{k\ell})\left(\delta_{ki}\delta_{\ell j} - \frac{1}{3}\delta_{k\ell}\delta_{ij}\right) = 2\left(s_{ij} - \alpha_{ij}\right) = 2b_{ij} \tag{11.60a}$$

$$\frac{\partial\phi}{\partial\alpha_{ij}} = -2b_{ij} \tag{11.60b}$$

From (11.60a) and (11.54), we have

$$\bar{r}^d = \frac{2}{\sqrt{3}}k \tag{11.60c}$$

Note that r_{ij} is purely deviatoric. Noting that

$$C_{ijk\ell}r_{k\ell} = 4G(s_{ij} - \alpha_{ij}) = 4Gb_{ij} \tag{11.61a}$$

the continuum tangent operator (Chap. 9) is given by

$$\bar{D}_{ijk\ell} = C_{ijk\ell} - \frac{16G^2 b_{ij} b_{k\ell}}{K_p + 8G b_{pq} b_{pq}} \tag{11.61b}$$

11.6 Application of CPPM to Kinematic Hardening von Mises Model: The *S*-Space Formulation

11.6.1 Calculation of Stress and Back Stress

The derivations in this section follow those in Sect. 11.3. Noting that r_{ij} is purely deviatoric, it is understood that the volumetric plastic strain rate is zero. The S-space CPPM equations for the stresses then are (Sect. 10.7.2, Chap. 10)

$$p_{n+1} = p_n + K\Delta\varepsilon_{pp} \tag{11.62a}$$

$$s_{ij}^{n+1} = s_{ij}^n + 2G\Delta e_{ij} - 2G\Delta\lambda r_{ij}^{n+1} = s_{ij}^{tr} - 2G\Delta\lambda r_{ij}^{n+1} \tag{11.62b}$$

$$s_{ij}^{tr} = s_{ij}^n + 2G\Delta e_{ij} \tag{11.62c}$$

where p_{n+1} is the spherical part of $\boldsymbol{\sigma}_{n+1}$, $\mathbf{s}_{n+1} = \boldsymbol{\sigma}_{n+1} - p_{n+1}\boldsymbol{\delta}$ is the deviatoric part of $\boldsymbol{\sigma}_{n+1}$, $\Delta\mathbf{e} = \Delta\boldsymbol{\varepsilon} - \Delta\varepsilon_{pp}\boldsymbol{\delta}/3$ is the incremental deviatoric strain, and $\mathbf{s}^{tr}$ is the elastic trial deviatoric stress. From (11.57), the backward Euler equation for the back stress is

$$\alpha_{ij}^{n+1} = \alpha_{ij}^n + \frac{2}{3}c\Delta\lambda r_{ij}^{n+1} - \frac{2}{\sqrt{3}}\gamma\Delta\lambda\bar{r}^d\alpha_{ij}^{n+1} \tag{11.63}$$

As with the case of the isotropic hardening model (Sect. 11.3), a scalar equation for $\Delta\lambda$ can be derived in the present case as well. With the aid of (11.60a) and (11.60c), let us expand (11.63) and simplify as

$$\alpha_{ij}^{n+1} = \alpha_{ij}^n + \frac{4}{3}c\Delta\lambda\left(s_{ij}^{n+1} - \alpha_{ij}^{n+1}\right) - \frac{2}{\sqrt{3}}\gamma\Delta\lambda\left(\frac{2}{\sqrt{3}}k\right)\alpha_{ij}^{n+1}$$

$$\alpha_{ij}^{n+1} = \frac{1}{y_0}\left[\alpha_{ij}^n + \frac{4c\Delta\lambda}{3}s_{ij}^{n+1}\right] \tag{11.64a}$$

where

$$y_0 = 1 + \frac{4c\Delta\lambda}{3} + \frac{4k\gamma\Delta\lambda}{3} = 1 + \frac{4}{3}(c + k\gamma)\Delta\lambda \tag{11.64b}$$

By expanding (11.62b) and simplifying the resulting equation with the aid of (11.64b), it follows

$$\begin{aligned} s_{ij}^{n+1} &= s_{ij}^{tr} - 4G\Delta\lambda\left(s_{ij}^{n+1} - \alpha_{ij}^{n+1}\right) \\ &= s_{ij}^{tr} - 4G\Delta\lambda s_{ij}^{n+1} + \frac{4G\Delta\lambda}{y_0}\left[\alpha_{ij}^{n} + \frac{4c\Delta\lambda}{3}s_{ij}^{n+1}\right] \end{aligned}$$

$$s_{ij}^{n+1} = \frac{1}{x_0}\left[s_{ij}^{tr} + \frac{4G\Delta\lambda}{y_0}\alpha_{ij}^{n}\right] \tag{11.65a}$$

where

$$x_0 = 1 + 4G\Delta\lambda - \frac{16}{3}\frac{Gc\Delta\lambda^2}{y_0} \tag{11.65b}$$

Using (11.64a) and (11.65a), let us simplify b_{ij}^{n+1} as

$$\begin{aligned} b_{ij}^{n+1} &= s_{ij}^{n+1} - \alpha_{ij}^{n+1} \\ &= s_{ij}^{n+1} - \frac{1}{y_0}\left[\alpha_{ij}^{n} + \frac{4c\Delta\lambda}{3}s_{ij}^{n+1}\right] \\ &= \left[1 - \frac{4c\Delta\lambda}{3y_0}\right]s_{ij}^{n+1} - \frac{1}{y_0}\alpha_{ij}^{n} \\ &= \frac{1}{x_0}\left[1 - \frac{4c\Delta\lambda}{3y_0}\right]\left[s_{ij}^{tr} + \frac{4G\Delta\lambda}{y_0}\alpha_{ij}^{n}\right] - \frac{1}{y_0}\alpha_{ij}^{n} \\ &= a_1 s_{ij}^{tr} + a_2\alpha_{ij}^{n} \end{aligned} \tag{11.66a}$$

where

$$a_1 = \frac{1}{x_0}\left[1 - \frac{4c\Delta\lambda}{3y_0}\right] \tag{11.66b}$$

and

$$a_2 = \frac{4G\Delta\lambda}{x_0 y_0}\left[1 - \frac{4c\Delta\lambda}{3y_0}\right] - \frac{1}{y_0} \tag{11.66c}$$

Let us now substitute (11.66a) into (11.54) as

$$\phi = b_{ij}^{n+1} b_{ij}^{n+1} - \frac{2}{3} k^2 = 0$$

$$\phi = \left(a_1 s_{ij}^{tr} + a_2 \alpha_{ij}^n \right) \left(a_1 s_{ij}^{tr} + a_2 \alpha_{ij}^n \right) - \frac{2}{3} k^2 = 0$$

$$\phi = 2a_1^2 J_{tr}^2 + 2a_2^2 J_\alpha^2 + 2a_1 a_2 J_{\alpha s} = 0 \tag{11.67a}$$

where

$$J_{tr} = \left(\frac{1}{2} s_{ij}^{tr} s_{ij}^{tr} \right)^{1/2}; \quad J_\alpha = \left(\frac{1}{2} \alpha_{ij}^n \alpha_{ij}^n \right)^{1/2}; \quad J_{\alpha s} = s_{ij}^{tr} \alpha_{ij}^n \tag{11.67b}$$

Noting that J_{tr}, J_α, and $J_{\alpha s}$ depend only on the quantities at the backward point n, and a_1 and a_2 are functions of $\Delta\lambda$, it follows that (11.67a) is a nonlinear scalar equation on $\Delta\lambda$. The equation may be solved by a suitable method; here we will use Newton's iterative method. Newton's method requires the gradient of the function ϕ with respect to $\Delta\lambda$, which is evaluated below.

Differentiating (11.64b), (11.65b), (11.66b), (11.66c), and (11.67a), we have

$$\frac{\partial y_0}{\partial(\Delta\lambda)} = \frac{4}{3}(c + k\gamma) \tag{11.68a}$$

$$\frac{\partial x_0}{\partial(\Delta\lambda)} = 4G - \frac{32}{3} \frac{Gc\Delta\lambda}{y_0} + \frac{16}{3} \frac{Gc\Delta\lambda^2}{y_0^2} \frac{\partial y_0}{\partial(\Delta\lambda)} \tag{11.68b}$$

$$\frac{\partial a_1}{\partial(\Delta\lambda)} = \left[-\frac{1}{x_0^2} + \frac{4c\Delta\lambda}{3x_0^2 y_0} \right] \frac{\partial x_0}{\partial(\Delta\lambda)} + \frac{4c\Delta\lambda}{3x_0 y_0^2} \frac{\partial y_0}{\partial(\Delta\lambda)} - \frac{4c}{3x_0 y_0} \tag{11.68c}$$

$$\begin{aligned} \frac{\partial a_2}{\partial(\Delta\lambda)} = & \frac{4G}{x_0 y_0} - \frac{32Gc\Delta\lambda}{3x_0 y_0^2} + \left[-\frac{4G\Delta\lambda}{x_0^2 y_0} + \frac{16Gc\Delta\lambda^2}{3x_0^2 y_0^2} \right] \frac{\partial x_0}{\partial(\Delta\lambda)} \\ & + \left[-\frac{4G\Delta\lambda}{x_0 y_0^2} + \frac{32Gc\Delta\lambda^2}{3x_0 y_0^3} + \frac{1}{y_0^2} \right] \frac{\partial y_0}{\partial(\Delta\lambda)} \end{aligned} \tag{11.68d}$$

$$K_\phi = \frac{\partial\phi}{\partial(\Delta\lambda)} = \left[4a_1 J_{tr}^2 + 2a_2 J_{\alpha s} \right] \frac{\partial a_1}{\partial(\Delta\lambda)} + \left[4a_2 J_\alpha^2 + 2a_1 J_{\alpha s} \right] \frac{\partial a_2}{\partial(\Delta\lambda)} \tag{11.69}$$

Once $\Delta\lambda$ is calculated, (11.65a) and (11.64a) can be used to calculate S_{ij}^{n+1} and α_{ij}^{n+1}, respectively.

11.6.2 Calculation of Consistent Tangent Operator

Taking the rate of (11.67a), we have

$$\dot{\phi} = \frac{\partial \phi}{\partial(\Delta\lambda)}\dot{\lambda} + 4a_1^2 J_{tr}\dot{J}_{tr} + 2a_1a_2\dot{J}_{\alpha s} = 0 \tag{11.70a}$$

From (11.39), (11.62c), and (11.67b), we have

$$\dot{J}^{tr} = \frac{G s_{k\ell}^{tr}\dot{e}_{k\ell}}{J^{tr}}; \quad \dot{J}_{\alpha s} = \dot{s}_{ij}^{tr}\alpha_{ij}^n = 2G\dot{e}_{ij}\alpha_{ij}^n \tag{11.70b}$$

Combining (11.70a) and (11.70b), it follows

$$\dot{\lambda} = -\frac{1}{K_\phi}\left[4a_1^2 J_{tr}\frac{G s_{k\ell}^{tr}\dot{e}_{k\ell}}{J^{tr}} + 4a_1a_2G\dot{e}_{k\ell}\alpha_{k\ell}^n\right] = \bar{f}_{k\ell}\dot{e}_{k\ell} \tag{11.71a}$$

where

$$\bar{f}_{k\ell} = -\frac{1}{K_\phi}\left[4a_1^2 G s_{k\ell}^{tr} + 4a_1a_2G\alpha_{k\ell}^n\right] \tag{11.71b}$$

K_ϕ is given by (11.69). Taking the rate of (11.65a) and simplifying it using the above equations, we have

$$\dot{s}_{ij} = \frac{\dot{s}_{ij}^{tr}}{x_0} - \frac{s_{ij}^{tr}}{x_0^2}\frac{\partial x_0}{\partial(\Delta\lambda)}\dot{\lambda} + \left[\frac{4G}{x_0y_0} - \frac{4G\Delta\lambda}{x_0^2y_0}\frac{\partial x_0}{\partial(\Delta\lambda)} - \frac{4G\Delta\lambda}{x_0y_0^2}\frac{\partial y_0}{\partial(\Delta\lambda)}\right]\alpha_{ij}^n\dot{\lambda}$$

$$\dot{s}_{ij} = \frac{2G\dot{e}_{ij}}{x_0} - \left[\frac{s_{ij}^{tr}}{x_0^2}\frac{\partial x_0}{\partial(\Delta\lambda)} - x_1\alpha_{ij}^n\right]\bar{f}_{k\ell}\dot{e}_{k\ell}$$

$$\dot{s}_{ij} = \frac{2G\dot{e}_{ij}}{x_0} - \bar{g}_{ij}\bar{f}_{k\ell}\dot{e}_{k\ell} \tag{11.72}$$

where

$$x_1 = \frac{4G}{x_0y_0} - \frac{4G\Delta\lambda}{x_0^2y_0}\frac{\partial x_0}{\partial(\Delta\lambda)} - \frac{4G\Delta\lambda}{x_0y_0^2}\frac{\partial y_0}{\partial(\Delta\lambda)} \tag{11.73a}$$

and

$$\bar{g}_{ij} = \frac{s_{ij}^{tr}}{x_0^2}\frac{\partial x_0}{\partial(\Delta\lambda)} - x_1\alpha_{ij}^n \tag{11.73b}$$

Now we can synthesize (11.72) and (11.62a) to derive an equation for the consistent tangent operator as follows:

$$\begin{aligned}\dot{\sigma}_{ij} &= \dot{p}\delta_{ij} + \dot{s}_{ij} \\ &= K\dot{\varepsilon}_{pp}\delta_{ij} + \frac{2G\dot{e}_{ij}}{x_0} - \bar{g}_{ij}\bar{f}_{k\ell}\dot{e}_{k\ell} \\ &= K\delta_{ij}\dot{\varepsilon}_{pp} + \left[\frac{G}{x_0}\left(\delta_{ik}\delta_{j\ell} + \delta_{i\ell}\delta_{jk}\right) - \bar{g}_{ij}\bar{f}_{k\ell}\right]\left(\dot{\varepsilon}_{k\ell} - \frac{1}{2}\dot{\varepsilon}_{pp}\delta_{k\ell}\right) \\ &= \left[\left(K - \frac{2G}{3x_0}\right)\delta_{ij}\delta_{k\ell} + \frac{G}{x_0}\left(\delta_{ik}\delta_{j\ell} + \delta_{i\ell}\delta_{jk}\right) - \bar{g}_{ij}\bar{f}_{k\ell}\right]\dot{\varepsilon}_{k\ell}\end{aligned}$$

$$\dot{\sigma}_{ij} = D_{ijk\ell}\dot{\varepsilon}_{k\ell} \tag{11.74a}$$

where $D_{ijk\ell}$ is the elasto-plastic consistent tangent operator, given by

$$D_{ijk\ell} = \left(K - \frac{2G}{3x_0}\right)\delta_{ij}\delta_{k\ell} + \frac{G}{x_0}\left(\delta_{ik}\delta_{j\ell} + \delta_{i\ell}\delta_{jk}\right) - \bar{g}_{ij}\bar{f}_{k\ell} \tag{11.74b}$$

In simplifying the above equation, use has been made of the fact that $\bar{f}_{k\ell}$ is purely deviatoric and hence $\bar{f}_{k\ell}\delta_{k\ell} = 0$. It may be noted that with $x_0 = 1$, the first two terms in (11.74b) correspond to the elastic stiffness tensor $C_{ijk\ell}$ (11.1b). When $\Delta\lambda = 0$ and $\Delta\boldsymbol{\varepsilon} = 0$, it can be shown that the consistent operator given by (11.74b) degenerates into the continuum operator given by (11.61b) (Problem 11.5).

11.6.3 Exact Solution for the Case of Uniaxial Loading

Consider the behavior of a specimen subjected to a uniaxial stress loading. In this case, we have

$$\boldsymbol{\sigma} = \begin{bmatrix}\sigma_1\\0\\0\end{bmatrix};\quad p = \frac{\sigma_1}{3};\quad \mathbf{s} = \begin{bmatrix}2\sigma_1/3\\-\sigma_1/3\\-\sigma_1/3\end{bmatrix};\quad \boldsymbol{\alpha} = \begin{bmatrix}\alpha_1\\-\alpha_1/2\\-\alpha_1/2\end{bmatrix};\quad \boldsymbol{\varepsilon}^p = \mathbf{e}^p = \begin{bmatrix}\varepsilon_1^p\\-\varepsilon_1^p/2\\-\varepsilon_1^p/2\end{bmatrix} \tag{11.75a}$$

$$\dot{\xi}^p = \left[\frac{1}{2}\dot{e}_{ij}^p\dot{e}_{ij}^p\right]^{1/2} = \left[\frac{1}{2}\left\{\left(\dot{\varepsilon}_1^p\right)^2 + \frac{1}{4}\left(\dot{\varepsilon}_1^p\right)^2 + \frac{1}{4}\left(\dot{\varepsilon}_1^p\right)^2\right\}\right]^{1/2} = \frac{\sqrt{3}}{2}\dot{\varepsilon}_1^p \tag{11.75b}$$

From (11.57), during loading where $\dot{\varepsilon}_1^p > 0$ and $\alpha_1 > 0$, it follows that

$$\dot{\alpha}_1 = \frac{2}{3}c\dot{\varepsilon}_1^p - \frac{2}{\sqrt{3}}\gamma\alpha_1\left(\frac{\sqrt{3}}{2}\dot{\varepsilon}_1^p\right) = \frac{1}{3}(2c - 3\gamma\alpha_1)\dot{\varepsilon}_1^p$$

$$\int_{\alpha_{10}}^{\alpha_1} \frac{3\dot{\alpha}_1}{2c - 3\gamma\alpha_1} = \int_{\varepsilon_{10}^p}^{\varepsilon_1^p} \dot{\varepsilon}_1^p$$

$$\alpha_1 = \frac{2c}{3\gamma} + \left(\alpha_{10} - \frac{2c}{3\gamma}\right)\exp\{-\gamma(\varepsilon_1^p - \varepsilon_{10}^p)\}$$

$$\alpha_1 = \frac{2c}{3\gamma} + \left(\alpha_{10} - \frac{2c}{3\gamma}\right)\exp\left\{-\gamma\left(\varepsilon_1 - \frac{\sigma_1}{E} - \varepsilon_{10}^p\right)\right\} \tag{11.76}$$

Expanding the yield function given by (11.54) on the basis of (11.75a), we have

$$\left(\frac{2\sigma_1}{3} - \alpha_1\right)^2 + 2\left(\frac{\sigma_1}{3} - \frac{\alpha_1}{2}\right)^2 = \frac{2}{3}k^2$$

$$\sigma_1 = \frac{3}{2}\alpha_1 + k \tag{11.77}$$

During reversed loading, where $\dot{\varepsilon}_1^p < 0$ and $\alpha_1 > 0$, it can be shown that

$$\alpha_1 = -\frac{2c}{3\gamma} + \left(\alpha_{10} + \frac{2c}{3\gamma}\right)\exp\left\{+\gamma\left(\varepsilon_1 - \frac{\sigma_1}{E} - \varepsilon_{10}^p\right)\right\} \tag{11.78a}$$

$$\sigma_1 = \frac{3}{2}\alpha_1 - k \tag{11.78b}$$

11.7 Sample Numerical Results from Kinematic Hardening von Mises Model

In this section, we evaluate the model behavior and the robustness of the integration method in obtaining the stress–strain relationship. The model parameters chosen for the study are listed in Table 11.5. It was shown in Chap. 9 that, for the Armstrong–Frederick kinematic hardening rule, the uniaxial failure stress is given by

$$\sigma_f = k + \frac{c}{\gamma} \tag{11.79}$$

Table 11.5 Kinematic hardening von Mises model parameters

Hardening type	E (GPa)	ν	k (GPa)	c (GPa)	γ
Set 1	200	0.3	0.1	30	100
Set 2	200	0.3	0.1	3	10

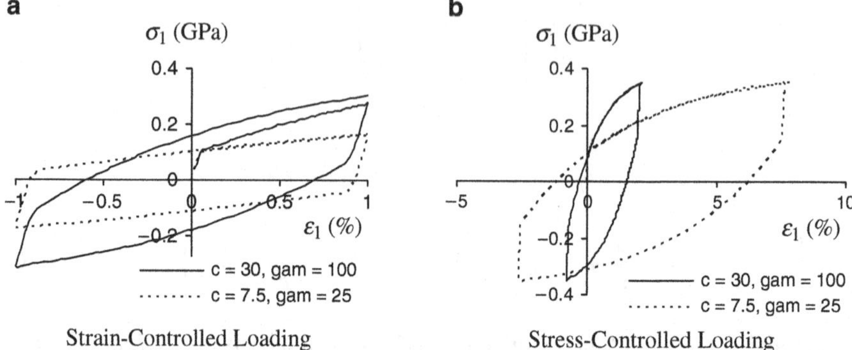

Fig. 11.13 Uniaxial behavior of the kinematic hardening von mises model under cyclic loading involving one cycle

Hence for both the model parameter Set 1 and Set 2, the failure stress is 0.4 GPa. However, the values of γ, which governs the stiffness of the stress–strain relation, are different.

The model behaviors shown in Fig. 11.12 are obtained by applying 20% uniaxial strain in 50 increments. The exact results are obtained using (11.76) and (11.77). It may be noted that (a) as expected, the uniaxial failure stress is calculated to be 0.4 GPa for both model parameter sets, (b) as the value of γ increases, the stress–strain relation becomes stiffer, and (c) the numerical results compare very well with the exact results.

Next the model behavior is simulated during a cyclic loading (involving one cycle). The model behavior is simulated under both the strain-controlled and stress-controlled loading, each using model parameter Set 1 and Set 2 listed in Table 11.5. The results are presented in Fig. 11.13.

It is illustrative to compare the behaviors with those of the isotropic hardening model (Fig. 11.11). Firstly, let us compare the behaviors under the strain-controlled loading. During load reversal, the range of elastic behavior is much larger in the isotropic hardening model (Fig. 11.11) than in the kinematic hardening model. Consequently the shapes of the hysteresis loops are different from each other. Secondly, it may be observed that the behaviors under the stress-controlled loading are very different from each other as well. The isotropic hardening model is incapable of simulating a hysteresis loop, whereas the kinematic hardening model predicts realistic hysteresis loops. This is perhaps the key reason for employing a kinematic hardening rule. Also note that, as expected, the behavior becomes stiffer as the value of γ increases.

Problems

Problem 11.1 Considering the isotropic hardening von Mises model described in this chapter (Sect. 11.1), show that the consistent operator (11.25) is symmetric (even when $\Delta\lambda \neq 0$) when the "Euler" method is used to integrate the hardening rule (11.13a).

Problem 11.2 Verify that the consistent tangent operator $\mathbf{D}$ given by (11.43b) for the S-space formulation is symmetric for both the "Euler" and "Exact" methods of integrating the hardening rule.

Problem 11.3 Considering the nonlinear isotropic hardening rule (11.6c) and using the following model parameters $k_0 = 0.25$, $k_f = 0.4$, $m = 20$, $E = 200$, $\nu = 0.3$, and the initial state variables $\boldsymbol{\sigma}_n = \{0.1 \quad 0.05 \quad 0.075\}$ and $k_n = 0.25$, determine the incremental stress and PIV for the following incremental strain: $\Delta\boldsymbol{\varepsilon} = \{0.03 \quad 0.01 \quad -0.028\}$
Use the S-space formulation (Sect. 11.3). Perform the integration using both

1. the "Euler" integration for the hardening rule (see Sects. 11.3.1 and 11.4.1, Approximation 1)
2. the "Exact" integration for the hardening rule (see Sects. 11.3.2 and 11.4.1, Approximation 2)

Sketch the initial and final yield surfaces on the octahedral plane. Mark the initial, trial and final stress points on the octahedral plane.

Problem 11.4 In Problem 8.3, you are asked to develop a computer program based on Algorithm 8.2 and the coding guide in Appendix 3 to calibrate a constitutive model. In that program, you called a subroutine called "MISES," where a nonlinear elastic model is implemented, (a) to calculate the incremental stress for a given incremental strain, and (b) to calculate the consistent or continuum operator needed for the global iteration.

Modify this program to implement the various versions of the von Mises model presented in the current chapter as follows:

1. Implement the S-space formulation of the isotropic hardening von Mises model with nonlinear hardening with the hardening rule integrated by the "Euler" method.
   (a) Using the properties in Table 11.2, reproduce the results in Fig. 11.5a to verify your code.
   (b) Investigate the influence of the load step size on the convergence behavior and accuracy of results. (You may make one run with a large number of steps and use the results as "exact" in the accuracy evaluations.)

   (c) Modify your code to include an option of using the continuum operator along with the consistent operator that you already have. Investigate the convergence behavior using both the continuum and consistent operators.
2. Implement the S-space formulation of the isotropic hardening von Mises model with nonlinear hardening with the hardening rule integrated by the "Exact" method.
   (a) Using the properties in Table 11.2, reproduce the results in Fig. 11.7a to verify your code.
   (b) Investigate the influence of the load step size on the convergence behavior and accuracy of results.
   (c) Investigate the convergence behavior using both the continuum and consistent operators.
3. Implement the σ-space formulation of the isotropic hardening von Mises model with nonlinear hardening with the hardening rule integrated by the "Euler" method.
   (a) Using the properties in Table 11.2, reproduce the results in Fig. 11.5a to verify your code.
   (b) Investigate the influence of the load step size on the convergence behavior and accuracy of results.
   (c) Investigate the convergence behavior using both the continuum and consistent operators.
4. Implement the S-space formulation of the kinematic hardening von Mises model with the Armstrong–Frederick rule described in this chapter.
   (a) Using the properties in Table 11.5, reproduce the results in Figs. 11.12 and 11.13 to verify your code.
   (b) Investigate the influence of the load step size on the convergence behavior and accuracy of results.
   (c) Modify your code to include an option of using the continuum operator along with the consistent operator that you already have. Investigate the convergence behavior using both the continuum and consistent operators.

Problem 11.5 Show that when $\Delta\lambda = 0$ and $\Delta\boldsymbol{\varepsilon} = 0$, the consistent tangent operator involved in the kinematic hardening model (11.74b) degenerates into the continuum operator given by (11.61b).

Chapter 12
The Modified Cam-Clay Model and its Integration

In Chap. 11, the details of two models based on the von Mises yield and failure surfaces have been presented. One of the models simulates the isotropic hardening behavior and the other simulates the kinematic hardening behavior. The latter has the potential for simulating the plastic behavior of materials during a cyclic loading. These von Mises models may be used to model the total stress-based constitutive behavior of soil. However, the effective stress-based constitutive behavior of soil involves two other aspects that the von Mises models presented in Chap. 11 are incapable of simulating. These aspects are (1) the dependence of elastic and plastic stress–strain relations on mean normal pressure ("pressure-dependency") and (2) the development of plastic volumetric strain during deviatoric stress loading (plastic dilatancy). The Cam-clay model (Schofield and Wroth 1968) is intended to fill this gap.

The original model (Schofield and Wroth 1968) has a singularity at the point where the yield surface crosses the mean normal pressure axis. The model was later modified to address this (Roscoe and Burland 1968). This later model, now known as *the modified Cam-clay model*, employs an ellipse for the yield surface, eliminating the singularity.

Only the modified Cam-clay model is considered in this chapter. The model has the following limitations: (1) It is satisfactory only for clays, (2) it is satisfactory only in the normally consolidated and lightly overconsolidated regions, (3) it ignores anisotropy, and (4) it is not suitable for problems involving cyclic loading. The model is thus good for "soft" clays under monotonic loading situations. Nevertheless, in describing the stress–strain behavior of cohesive soil, the model represents a significant advancement from the simple von Mises models described in Chap. 11. Also the theoretical details of the model are fairly simple. For these reasons, we present a detailed computational treatment of this model in this chapter.

We will integrate the model using all three of the integration methods presented in Chap. 10: the two-step Euler method, the cutting-plane method and the closest point projection method (CPPM).

The pressure-dependency of elastic properties in the model requires an appropriate treatment of the elastic relations in the implementation of CPPM. We will

A. Anandarajah, *Computational Methods in Elasticity and Plasticity: Solids and Porous Media*, DOI 10.1007/978-1-4419-6379-6_12,

present two methods, one based on the use of Euler approximations for the elastic properties and the other based on the exact integration of the elastic stress–strain relations.

We present three different implementations of CPPM: (a) CPPM in σ-space formulation with approximate elastic properties, (b) CPPM in σ -space formulation with exact elastic properties, and (c) CPPM in S-space formulation with approximate elastic properties.

For general details of the integration methods, the reader is referred to Chap. 10. A brief history of the different methods is also presented in Chap. 10.

12.1 Summary of the Modified Cam-Clay Model Equations

For details on the model behavior, the reader is referred to some of the earlier books on the model (e.g., Schofield and Wroth 1968; Atkinson and Bransby 1978). In most books on the model, the formulation is given in the triaxial stress space (i.e., the $p-q$ space). However, for application to finite element analysis, the formulation in the invariant space is required. In this regard, we present a formulation in the stress invariant space below. Only the final equations related to the constitutive relations are presented here.

Assuming isotropy, the elastic stress–strain relation is expressed as

$$\dot{\boldsymbol{\sigma}} = \mathbf{C}\dot{\boldsymbol{\varepsilon}}^e \tag{12.1a}$$

where

$$C_{ijk\ell} = \left(K - \frac{2}{3}G\right)\delta_{ij}\delta_{k\ell} + G\left(\delta_{ik}\delta_{j\ell} + \delta_{i\ell}\delta_{jk}\right) \tag{12.1b}$$

G and K are the shear and bulk moduli, respectively. $\boldsymbol{\delta}$ is the Kronecker delta. $\mathbf{C}$ is the elastic stiffness tensor. $\dot{\boldsymbol{\sigma}}$ and $\dot{\boldsymbol{\varepsilon}}^e$ are the rate of stress and rate of elastic strain, respectively. The bulk (K) and shear (G) moduli are related to the mean normal stress p as

$$K = \frac{(1+e_0)}{\kappa}p = \bar{K}_0 p \quad \text{and} \quad G = \frac{3(1-2\nu)}{2(1+\nu)}K = \bar{G}_0 p \tag{12.2a}$$

where

$$\bar{K}_0 = \frac{(1+e_0)}{\kappa} \quad \text{and} \quad \bar{G}_0 = \frac{3(1-2\nu)}{2(1+\nu)}\bar{K}_0 \tag{12.2b}$$

ν, e_0, and κ are Poisson's ratio, the initial void ratio, and the swelling index (slope of $\ln p$ versus e relation during isotropic unloading), respectively.

Three stress invariants I, J, and α are defined as in Chap. 2:

$$I = \sigma_{k\ell}\delta_{k\ell} \tag{12.3a}$$

$$s_{ij} = \sigma_{ij} - \frac{1}{3}I\delta_{ij}; \quad J = \left(\frac{1}{2}s_{k\ell}s_{k\ell}\right)^{1/2} \tag{12.3b}$$

$$S = \left(\frac{1}{3}s_{ij}s_{jk}s_{ki}\right)^{1/3} \tag{12.3c}$$

$$-\frac{\pi}{6} \leq \alpha = \frac{1}{3}\sin^{-1}\left[\frac{3\sqrt{3}}{2}\left(\frac{S}{J}\right)^3\right] \leq \frac{\pi}{6} \tag{12.3d}$$

The yield surface is an ellipse in the $I - J$ space, given by

$$\phi(I, J, \alpha, p_0) = I^2 + \left(\frac{J}{N}\right)^2 - 3p_0 I = 0 \tag{12.4}$$

where

$$N = \frac{M}{3\sqrt{3}}; M = g(n, \alpha)M_c; g(n, \alpha) = \frac{2n}{1 + n - (1 - n)\sin 3\alpha}; n = \frac{M_e}{M_c}.$$

M_c and M_e are the slopes of the critical state line in the $p - q$ space (Chap. 2) in triaxial compression and extension, respectively. p_0 is the preconsolidation (mean) pressure, which is the only (scalar) hardening variable in the model. Hence, ζ (the PIV defined in Chap. 9) is a scalar as

$$\varsigma \equiv p_0. \tag{12.5}$$

Geometrically, p_0 is the mean normal pressure at which the yield surface intersects the p-axis.

The plastic constitutive relations are given by:

$$\dot{\boldsymbol{\varepsilon}}^p = \dot{\lambda}\mathbf{r}; \quad \mathbf{r} = \frac{\partial \psi}{\partial \boldsymbol{\sigma}}; \quad \dot{\lambda} = \frac{1}{K_p}\mathbf{n}\dot{\boldsymbol{\sigma}}; \quad \mathbf{n} = \frac{\partial \phi}{\partial \boldsymbol{\sigma}} \tag{12.6}$$

where $\dot{\boldsymbol{\varepsilon}}^p$ is the plastic strain rate and $\dot{\lambda}$ is the consistency parameter (or the loading index). The Cam-clay model employs associated flow rule. This leads to $\phi = \psi$ and $\mathbf{r} = \mathbf{n}$. However, for generality, equations will be derived without this assumption so that the algorithms presented in this chapter may be easily extended to constitutive models based on non-associated flow rules.

The hardening rule is expressed as

$$\dot{p}_0 = \frac{1 + e_0}{\lambda^* - \kappa}p_0\dot{\varepsilon}_v^p = C_1 p_0\dot{\varepsilon}_v^p = h\dot{\varepsilon}_v^p = h\dot{\lambda}r_{kk} = s\dot{\lambda} \tag{12.7a}$$

where

$$C_1 = \frac{1+e_0}{\lambda^* - \kappa} \text{ is a constant, and } h = C_1 p_0 \text{ and } s = C_1 p_0 r_{kk} \tag{12.7b}$$

λ^* is the compression index (slope of $\ln p$ versus e relation during virgin isotropic loading) and $\dot{\varepsilon}_v^p$ is the rate of volumetric plastic strain. (Note that λ is the usual notation for the compression index, but to avoid confusion with the consistency parameter $\dot{\lambda}$, the symbol λ^* is used for the compression index.) Volumetric strain is obtained from the strain tensor as

$$\varepsilon_v = \varepsilon_{k\ell}\delta_{k\ell} \tag{12.8}$$

Applying the consistency condition and making use of (12.7a), we derive an equation for the plastic modulus as

$$\dot{\phi} = \frac{\partial \phi}{\partial \boldsymbol{\sigma}}\dot{\boldsymbol{\sigma}} + \frac{\partial \phi}{\partial p_0}\dot{p}_0 = 0 \quad \Rightarrow \quad K_p = -h\frac{\partial \phi}{\partial p_0} r_{kk} \tag{12.9a}$$

where r_{kk} is the volumetric part of $\mathbf{r}$ obtained as

$$r_{kk} = r_{kl}\delta_{kl} \tag{12.9b}$$

An expression for $\mathbf{r}$ is developed in Appendix 7. For convenience, the final result is repeated here:

$$r_{ij} = \frac{\partial \phi}{\partial I}\delta_{ij} + \frac{\partial \phi}{\partial J}\frac{s_{ij}}{2J} + \frac{\partial \phi}{\partial \alpha}\frac{\sqrt{3}}{2J\cos 3\alpha}\left[\frac{s_{ir}s_{rj}}{J^2} - \frac{2}{3}\delta_{ij} - \frac{3}{2}\left(\frac{S}{J}\right)^3\frac{s_{ij}}{J}\right] \tag{12.10}$$

The surface properties $\partial_I\phi$, $\partial_J\phi$, and $\partial_\alpha\phi$ are derived in Appendix 8.1. The final equations are again repeated here for convenience.

$$\frac{\partial \phi}{\partial I} = 2I - 3p_0 \tag{12.11a}$$

$$\frac{\partial \phi}{\partial J} = \frac{2J}{N^2} \tag{12.11b}$$

$$\frac{\partial \phi}{\partial \alpha} = J\bar{g}\frac{\partial \phi}{\partial J}; \quad \text{where } \bar{g} = -\frac{3(1-n)\cos 3\alpha}{1+n-(1-n)\sin 3\alpha} \tag{12.11c}$$

The results presented in this chapter have been obtained with the values listed in Table 12.1 for the model parameters and the void ratio of the soil.

Table 12.1 Cam-clay model parameters used in the study presented in this chapter

λ	κ	M_c	$n = M_e/M_c$	ν	e_0
0.14	0.03	1.2	1.0 or 0.8	0.3	1.1

12.2 The Two-Step Euler Integration Method with Substepping

12.2.1 General Description

The details of the two-step Euler method has been described in Sect. 10.5 (Chap. 10) and the reader is advised to read that section before continuing with the description presented in this section. The equations presented in Chap. 10 are specialized for the Cam-clay model below.

The elasto-plastic rate constitutive equations are of the form

$$\dot{\boldsymbol{\sigma}} = \bar{\mathbf{D}}\dot{\boldsymbol{\varepsilon}}$$

$$\dot{p}_0 = C_1 p_0 \dot{\varepsilon}_v^P$$

where $\bar{\mathbf{D}}$ is the continuum tangent operator. By integrating these equations, we have

$$\int_t^{t+\Delta t} \dot{\boldsymbol{\sigma}} = \int_t^{t+\Delta t} \bar{\mathbf{D}}\dot{\boldsymbol{\varepsilon}} \tag{12.12a}$$

$$\int_t^{t+\Delta t} \dot{p}_0 = \int_t^{t+\Delta t} C_1 p_0 \dot{\varepsilon}_v^p \tag{12.12b}$$

or

$$\int_t^{t+\Delta t} \dot{p}_0 = \int_t^{t+\Delta t} C_1 p_0 r_v \dot{\lambda} \tag{12.12c}$$

Applying the forward Euler (explicit) method to (12.12a)

$$\Delta\boldsymbol{\sigma}^1 = \bar{\mathbf{D}}_n \Delta\boldsymbol{\varepsilon}; \quad \boldsymbol{\sigma}_{n+1}^1 = \boldsymbol{\sigma}_n + \Delta\boldsymbol{\sigma}^1 = \boldsymbol{\sigma}_n + \bar{\mathbf{D}}_n \Delta\boldsymbol{\varepsilon} \tag{12.13}$$

where $\boldsymbol{\sigma}_{n+1}^1$ is a first estimate for $\boldsymbol{\sigma}_{n+1}$ (hence the use of the superscript "1"). If the hardening rule is too complex to integrate, it is integrated just as the stress rate equation is integrated. However, because of the simplicity of (12.12b), let us integrate the hardening rule as follows:

$$\int_{t}^{t+\Delta t} \frac{\dot{p}_0}{p_0} = C_1 \int_{t}^{t+\Delta t} \dot{\varepsilon}_v^P = C_1 \int_{t}^{t+\Delta t} \left[\dot{\varepsilon}_v - \dot{\varepsilon}_v^e\right] \tag{12.14a}$$

$$p_{0,1}^{n+1} = p_0^n \exp\left[C_1\left(\Delta\varepsilon_v - \Delta\varepsilon_v^e\right)\right] = p_0^n \exp\left[C_1\left(\Delta\varepsilon_v - \frac{\Delta p^1}{K_n}\right)\right] \tag{12.14b}$$

where Δp^1 is the incremental mean normal pressure, derived from $\Delta\boldsymbol{\sigma}^1$ given in (12.13), and K_n is the bulk modulus evaluated at point n. Thus, $p_{0,1}^{n+1}$ is an estimate of p_0^{n+1} obtained using the stress $\boldsymbol{\sigma}^1$.

The error introduced on $\boldsymbol{\sigma}$ in this method of integration is denoted by $\delta\boldsymbol{\sigma}$. The error $\delta\boldsymbol{\sigma}$ decreases as the step size decreases, and thus substepping should improve the accuracy.

A better method of integrating (12.12a) is to use a trapezoidal rule as

$$\Delta\boldsymbol{\sigma} = \frac{1}{2}[\bar{\mathbf{D}}_n + \bar{\mathbf{D}}_{n+1}]\Delta\boldsymbol{\varepsilon}$$

This requires $\boldsymbol{\sigma}_{n+1}$, which is not available yet. However, the estimate $\boldsymbol{\sigma}_{n+1}^1$ may be used, and a better estimate (a second estimate) for $\boldsymbol{\sigma}_{n+1}$ may be obtained as follows:

$$\Delta\boldsymbol{\sigma}^2 = \frac{1}{2}\left[\bar{\mathbf{D}}_n(\boldsymbol{\sigma}_n) + \bar{\mathbf{D}}_{n+1}(\boldsymbol{\sigma}_{n+1}^1)\right]\Delta\boldsymbol{\varepsilon}; \tag{12.15a}$$

$$\boldsymbol{\sigma}_{n+1}^2 = \boldsymbol{\sigma}_n + \Delta\boldsymbol{\sigma}^2 = \boldsymbol{\sigma}_n + \frac{1}{2}\left[\bar{\mathbf{D}}_n(\boldsymbol{\sigma}_n) + \bar{\mathbf{D}}_{n+1}(\boldsymbol{\sigma}_{n+1}^1)\right]\Delta\boldsymbol{\varepsilon} \tag{12.15b}$$

$\boldsymbol{\sigma}_{n+1}^2$ is then used to obtain a corresponding estimate for p_0^{n+1} according to

$$p_{0,2}^{n+1} = p_0^n \exp\left[C_1\left(\Delta\varepsilon_v - \frac{\Delta p^2}{\bar{K}}\right)\right], \quad \text{where} \quad \bar{K} = \frac{1}{2}\left[K_n(p_n) + K(p_{n+1}^2)\right] \tag{12.16}$$

(Note that the superscript 2 denotes "2nd estimate" and not "the square of a term.") An estimate for the error may now be made as

$$\delta\boldsymbol{\sigma} \approx \boldsymbol{\sigma}_{n+1}^2 - \boldsymbol{\sigma}_{n+1}^1 \tag{12.17}$$

$\delta\boldsymbol{\sigma}$ is used to estimate the number of substeps to be used to achieve a certain accuracy. Nyssen (1981) and Sloan (1987) used this approach and realized good success. The method was used in varied forms in many of the early implementations of more complex models. For example, Herrmann et al. (1987) used a similar substepping technique for the implementation of the isotropic bounding surface model. In the studies presented in Anandarajah and Dafalias (1986) and Anandarajah et al. (1995), the technique was used to integrate an anisotropic bounding surface model, which involves a tensorial PIV. Again, good success was realized.

Algorithm 12.1. Two-Step Euler Method for Integrating the Constitutive Relations from the Modified Cam-Clay Model

Definition of (new) parameters: $N_L^{\max}$ and $N_{G1}^{\max}$ are the maximum number of iterations permitted for the inner and outer loops in the algorithm before the execution is terminated, and N_L and N_{G1} are the corresponding counters. T is a scale parameter; when $T = 1$, full length of the strain vector has been applied. ΔT is the step size.

1. Establish $\Delta\boldsymbol{\varepsilon}$, p_0^n, $\boldsymbol{\sigma}_n$, TOL, $\beta_{\min}$, $\beta_{\max}$, $N_L^{\max}$, $N_{G1}^{\max}$ and the constitutive model parameters.
2. Compute $\bar{\mathbf{D}}_n(\boldsymbol{\sigma}_n, p_0^n)$.
3. Initialize parameters as $T = 0$, $\Delta T = 1.0$, $N_L = 0$, and $N_G = 0$.
4. Calculate $\boldsymbol{\sigma}_{n+1}^1$, $p_{0,1}^{n+1}$, $\bar{\mathbf{D}}_{n+1}(\boldsymbol{\sigma}_{n+1}^1, p_{0,1}^{n+1})$, $\boldsymbol{\sigma}_{n+1}^2$, $p_{0,2}^{n+1}$, $\delta\boldsymbol{\sigma}$ and E.
5. If $(E<TOL)$, go to step 8.
6. Set $N_L \leftarrow N_L + 1$. If $\left(N_L > N_L^{\max}\right)$ STOP.
7. Compute β, set $\beta \leftarrow \max(\beta, \beta_{\min})$, $\beta \leftarrow \min(\beta, \beta_{\max})$, $\Delta\boldsymbol{\varepsilon} \leftarrow \beta\Delta\boldsymbol{\varepsilon}$, and $\Delta T \leftarrow \beta\Delta T$, and go to step 4.
8. Set $N_{G1} \leftarrow N_{G1} + 1$. If $\left(N_{G1} > N_{G1}^{\max}\right)$, STOP.
9. Set $T \leftarrow T + \Delta T$. If $(T \geq 1.0)$, go to 12.
10. If $(T + \Delta T > 1)$ then set $x = (1 - T)/\Delta T$, $\Delta T \leftarrow x\Delta T$ and $\Delta\boldsymbol{\varepsilon} \leftarrow x\Delta\boldsymbol{\varepsilon}$.
11. Set $\boldsymbol{\sigma}_n \leftarrow \boldsymbol{\sigma}_{n+1}^2$ and $p_0^n \leftarrow p_{0,2}^{n+1}$, and go to step 3.
12. Set $\boldsymbol{\sigma}_{n+1} \leftarrow \boldsymbol{\sigma}_{n+1}^2$ and $p_0^{n+1} \leftarrow p_{0,2}^{n+1}$, and return to the main program.

Note that the following two steps are left out of this algorithm, and the reader is encouraged to modify the algorithm to incorporate them: (1) The step involving the evaluation of the elastic predictor and the loading/unloading check, and (2) the step involving the determination of the point of intersection of the yield surface and the elastic predictor increment $\Delta\boldsymbol{\sigma}^{ep}$.

The method of Sloan (1987) in evaluating the number of substeps is described here. A scalar measure of the error on $\boldsymbol{\sigma}$ is made as

$$E = \frac{\|\delta\boldsymbol{\sigma}\|}{\left\|\boldsymbol{\sigma}_{n+1}^2\right\|} \tag{12.18}$$

The strain increment that comes from the parent finite element program is applied in small increments, with the size of the increment determined by an iterative procedure. The current strain increment, $\Delta\boldsymbol{\varepsilon}^c$, is modified to $\Delta\boldsymbol{\varepsilon}^m$, as follows:

$$\Delta\boldsymbol{\varepsilon}^{\mathrm{m}} = \beta\Delta\boldsymbol{\varepsilon}^{\mathrm{c}} \tag{12.19a}$$

where

$$\beta = 0.8\left[\frac{TOL}{E}\right]^{1/2} \tag{12.19b}$$

Sloan recommends limiting the value of β at any stage during the iterative process as $\beta_{\min} \le \beta \le \beta_{\max}$, where $\beta_{\max} = 2$ and $\beta_{\min} = 0.1$. However, it was found in the applications presented here that when β is near 1.0, the progress of the iterative process was very slow. The following range is found to work better: $0.1 \le \beta \le 0.5$. Here, the current step is at least divided in half before another trial.

The iterative procedure then proceeds as follows: start with the full length of $\Delta\boldsymbol{\varepsilon}$, estimate $\delta\boldsymbol{\sigma}$, and determine the value of E. If $E > TOL$, then calculate a value for β and determine $\Delta\boldsymbol{\varepsilon}^{\mathrm{m}}$. Repeat the process until $E \le TOL$. Update the stresses and PIVs, and find the remaining length of strain to be applied. Start with the current value of the step size, and iteratively find the optimal size. Repeat the process until the full length of strain is applied. Algorithm 12.1 summarizes the steps.

When points n and $n+1$ are inside and outside the yield surface, respectively, the intersection between the elastic predictor and the yield surface needs to be evaluated; the reader is referred to Sect. 10.5.2 for details. Algorithm 12.1 needs to be modified to incorporate this.

12.2.2 Equations in Triaxial Space

The finite element analysis requires Algorithm 12.1 be implemented in the general invariant space. To check the coding and to gain some insight into the computational process, the equations are specialized for triaxial loading. The calculations involved can then be performed by hand.

The model behavior is independent of the Lode angle (12.3d). The following triaxial stress variables are adequate for representing the behavior:

$$p = \frac{1}{3}(\sigma_1 + 2\sigma_3) \quad \text{and} \quad q = \sigma_1 - \sigma_3$$

where p and q are the mean normal pressure and deviatoric stress, respectively. The corresponding triaxial strain variables are

$$\varepsilon_v = \varepsilon_{kk} = \varepsilon_1 + 2\varepsilon_3 \quad \text{and} \quad \varepsilon_d = \frac{2}{3}(\varepsilon_1 - \varepsilon_3)$$

where ε_v and ε_d are the volumetric and deviatoric strains, respectively. ε_v and ε_d are energy conjugates of p and q in the sense that

$$\sigma_1\varepsilon_1 + 2\sigma_3\varepsilon_3 = p\varepsilon_v + q\varepsilon_q$$

Assuming associated flow rule, the continuum tangent operator is given by

$$\bar{\mathbf{D}} = \left[\mathbf{C} - \frac{(\mathbf{Cr})(\mathbf{Cr})^T}{K_p + \mathbf{r}^T\mathbf{Cr}}\right] \tag{12.20a}$$

Letting

$$\mathbf{a} = \mathbf{Cr}, \quad K_p^* = \mathbf{r}^T\mathbf{Cr} \quad \text{and} \quad K_p^{**} = K_p + K_p^* \tag{12.20b}$$

equation (12.20a) becomes

$$\bar{\mathbf{D}} = \left[\mathbf{C} - \frac{\mathbf{aa}^T}{K_p^{**}}\right] \tag{12.20c}$$

In triaxial space, we can write

$$\boldsymbol{\sigma} = \{\sigma_1, \sigma_3, \sigma_3, 0, 0, 0\} \tag{12.21a}$$

$$\boldsymbol{\varepsilon} = \{\varepsilon_1, \varepsilon_3, \varepsilon_3, 0, 0, 0\} \tag{12.21b}$$

$$\mathbf{r} = \{r_{11}, r_{33}, r_{33}, 0, 0, 0\} \tag{12.21c}$$

From (12.10) and the equations given in Appendix 8.1, noting that $\partial_\alpha\phi = 0$ for triaxial loading, we have

$$\begin{aligned} r_{ij} &= \frac{\partial\phi}{\partial I}\delta_{ij} + \frac{\partial\phi}{\partial J}\frac{s_{ij}}{2J} \\ &= (6p - 3p_0)\delta_{ij} + \frac{27}{M_c^2}s_{ij} \end{aligned}$$

where $s_{11} = \frac{2}{3}q$ and $s_{33} = -\frac{1}{3}q$.

Thus,

$$r_{11} = 6p - 3p_0 + \frac{18}{M_c^2}q \quad \text{and} \quad r_{33} = 6p - 3p_0 - \frac{9}{M_c^2}q. \tag{12.21d}$$

An expression for $\mathbf{a}$ (12.20b) can be derived as follows:

$$a_{ij} = C_{ijk\ell}r_{k\ell} = \left[\bar{K}\delta_{ij}\delta_{k\ell} + G(\delta_{ik}\delta_{j\ell} + \delta_{i\ell}\delta_{jk}\right]r_{k\ell} = \bar{K}r_{kk} + 2Gr_{ij} \tag{12.22a}$$

where

$$\bar{K} = K - \frac{2}{3}G \tag{12.22b}$$

Denoting $\mathbf{a} = \{a_1, a_3, a_3, 0, 0, 0\}$, we have

$$a_1 = \bar{K}r_{kk} + 2Gr_{11} \quad \text{and} \quad a_3 = \bar{K}r_{kk} + 2Gr_{33} \tag{12.22c}$$

By combining (12.7b) and (12.9a), it follows that

$$K_p = -C_1 p_0 r_{kk} \frac{\partial \phi}{\partial p_0}$$

From the equations in Appendix 8.1, we have

$$K_p = -9C_1 p p_0 r_{kk} \tag{12.23}$$

Noting that

$$r_{ij}C_{ijk\ell}r_{k\ell} = r_{ij}\left[\bar{K}r_{kk}\delta_{ij} + 2Gr_{ij}\right] = \bar{K}r_{kk}^2 + 2G\left(r_{11}^2 + r_{33}^2\right) \tag{12.24a}$$

it follows that

$$K_p^* = \mathbf{r}^T\mathbf{C}\mathbf{r} = \bar{K}r_{kk}^2 + 2G\left(r_{11}^2 + r_{33}^2\right) \tag{12.24b}$$

Among the 36 coefficients in the 6×6 $\mathbf{C}$ and $\bar{\mathbf{D}}$ matrices, only the nine components in the 3×3 sub-matrix in the top left hand corner are relevant here because the shear components of $\boldsymbol{\sigma}$ and $\boldsymbol{\varepsilon}$ are zero.

From $C_{ijk\ell} = \bar{K}\delta_{ij}\delta_{k\ell} + G(\delta_{ik}\delta_{j\ell} + \delta_{i\ell}\delta_{jk})$, the relation needed for the computation of $\Delta\boldsymbol{\sigma}^{\text{ep}}$ (using (10.5c) and Euler approximation for the elastic properties) is

$$\begin{pmatrix} \Delta\sigma_1^{ep} \\ \Delta\sigma_3^{ep} \\ \Delta\sigma_3^{ep} \end{pmatrix} = \begin{bmatrix} \bar{K}+2G & \bar{K} & \bar{K} \\ \bar{K} & \bar{K}+2G & \bar{K} \\ \bar{K} & \bar{K} & \bar{K}+2G \end{bmatrix} \begin{pmatrix} \Delta\varepsilon_1 \\ \Delta\varepsilon_3 \\ \Delta\varepsilon_3 \end{pmatrix} = \begin{pmatrix} (\bar{K}+2G)\Delta\varepsilon_1 + 2\bar{K}\Delta\varepsilon_3 \\ \bar{K}\Delta\varepsilon_1 + 2(\bar{K}+G)\Delta\varepsilon_3 \\ \bar{K}\Delta\varepsilon_1 + 2(\bar{K}+G)\Delta\varepsilon_3 \end{pmatrix} \tag{12.25a}$$

The trial elastic stresses are calculated as

$$\sigma_1^{tr} = \sigma_n + \Delta\sigma_1^{ep} \tag{12.25b}$$

$$\sigma_3^{tr} = \sigma_n + \Delta\sigma_3^{ep} \tag{12.25c}$$

Considering only the relevant 3 × 3 sub-matrix, note that

$$\mathbf{aa}^T = \begin{bmatrix} a_1^2 & a_1a_3 & a_1a_3 \\ a_1a_3 & a_3^2 & a_3^2 \\ a_1a_3 & a_3^2 & a_3^2 \end{bmatrix} \tag{12.26a}$$

Denoting the relevant 3 × 3 elasto-plastic continuum operator by $\bar{\mathbf{D}}^*$ (12.20c), we have

$$\bar{\mathbf{D}}^* = \begin{bmatrix} \bar{K}+2G-a_1^2/K_p^{**} & \bar{K}-a_1a_3/K_p^{**} & \bar{K}-a_1a_3/K_p^{**} \\ \bar{K}-a_1a_3/K_p^{**} & \bar{K}+2G-a_3^2/K_p^{**} & \bar{K}-a_3^2/K_p^{**} \\ \bar{K}-a_1a_3/K_p^{**} & \bar{K}-a_3^2/K_p^{**} & \bar{K}+2G-a_3^2/K_p^{**} \end{bmatrix}$$
$$= \begin{bmatrix} D_1 & D_3 & D_3 \\ D_3 & D_2 & D_4 \\ D_3 & D_4 & D_2 \end{bmatrix} \tag{12.26b}$$

The elasto-plastic incremental stresses and the corresponding total stresses are given by

$$\Delta\boldsymbol{\sigma} = \bar{\mathbf{D}}\Delta\boldsymbol{\varepsilon} \tag{12.27a}$$

$$\Delta\sigma_1 = D_1\Delta\varepsilon_1 + 2D_3\Delta\varepsilon_3 \tag{12.27b}$$

$$\Delta\sigma_3 = D_3\Delta\varepsilon_1 + (D_2 + D_4)\Delta\varepsilon_3 \tag{12.27c}$$

$$\sigma_1^k = \sigma_1^n + \Delta\sigma_1 \tag{12.27d}$$

$$\sigma_3^k = \sigma_3^n + \Delta\sigma_3 \tag{12.27e}$$

The PIV is integrated as

$$\Delta p = (\Delta\sigma_1 + 3\Delta\sigma_3)/3 \tag{12.28a}$$

$$\Delta\varepsilon^e = \Delta p/K \tag{12.28b}$$

$$p_0^{n+1} = p_0^n \exp\left[C_1\left(\Delta\varepsilon_{kk} - \Delta\varepsilon_{kk}^{\mathrm{e}}\right)\right] \tag{12.28c}$$

Example 12.1 Two-Step Euler Method.

Question: Given $\sigma_1^n = 114$, $\sigma_3^n = 78$, $p_0^n = 100$, $\Delta\varepsilon_1 = 0.01$, and $\Delta\varepsilon_3 = -0.001088$, determine the step size ΔT at the end of the first inner iteration loop in Algorithm 12.1. Use $TOL = 1.0 \times 10^{-3}$, $\beta_{\min} = 0.1$, and $\beta_{\max} = 0.5$. Use the Euler approximation for the elastic properties with $\theta = 0.5$ (10.46). Use the values listed in

Table 12.1 (with $n = 1$) for the model parameters. Assume triaxial conditions with direction 1 taken as the axial direction.

Answer: Calculations are performed using the specialized triaxial equations presented in the preceding section. The initial point is schematically shown in Fig. 12.1. The step-by-step results are presented in Table 12.2, where the 2nd column lists the quantities calculated on the basis of $\sigma_1^n = 114$, $\sigma_3^n = 78$, and $p_0^n = 100$. The stresses at the end of this step are $\sigma_1 = 131.206$ and $\sigma_3 = 71.927$, which correspond to $\boldsymbol{\sigma}_{n+1}^1$ in (12.13). The PIV is $p_0 = 115.526$, which corresponds to $p_{0,1}^{n+1}$ in (12.14b). The calculations are then repeated in an identical manner, but by using $\sigma_1^n = 131.206$, $\sigma_3^n = 71.927$, and $p_0^n = 115.526$. The results are listed in column 3. The stresses at the end of this step correspond to $\boldsymbol{\sigma}_{n+1}^* = \boldsymbol{\sigma}_n + \bar{\mathbf{D}}_{n+1}(\boldsymbol{\sigma}_{n+1}^1)\Delta\boldsymbol{\varepsilon}$. The calculated stresses are $\sigma_1 = 129.382$ and $\sigma_3 = 90.004$. The better estimate of the stress then is (12.15b)

$$\boldsymbol{\sigma}_{n+1}^2 = \boldsymbol{\sigma}_n + \Delta\boldsymbol{\sigma}^2 = \boldsymbol{\sigma}_n + \frac{1}{2}\left[\bar{\mathbf{D}}_n(\boldsymbol{\sigma}_n) + \bar{\mathbf{D}}_{n+1}(\boldsymbol{\sigma}_{n+1}^1)\right]\Delta\boldsymbol{\varepsilon} \tag{12.29a}$$

$$= \frac{1}{2}(\boldsymbol{\sigma}_{n+1}^1 + \boldsymbol{\sigma}_{n+1}^*) \tag{12.29b}$$

The calculated values are $\sigma_1 = 130.294$ and $\sigma_3 = 80.965$. The error ((12.17) and (12.18)) may now be calculated as follows:

$$\|\delta\boldsymbol{\sigma}\| = \left\{\delta\sigma_1^2 + 2\delta\sigma_3^2\right\}^{1/2} = \left\{(130.294 - 131.206)^2 + 2(80.965 - 71.927)^2\right\}^{1/2}$$
$$= 12.81$$

$$\left\|\boldsymbol{\sigma}_{n+1}^2\right\| = \left\{\sigma_1^2 + 2\sigma_3^2\right\}^{1/2} = \left\{(130.294)^2 + 2(80.965)^2\right\}^{1/2} = 173.46$$

$$E = \frac{12.81}{173.46} = 0.0738$$

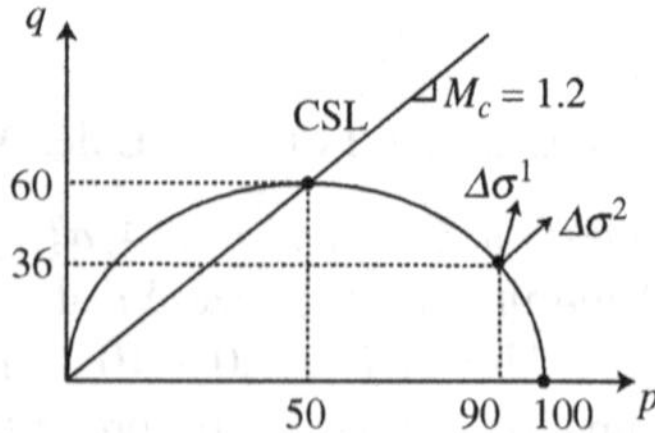

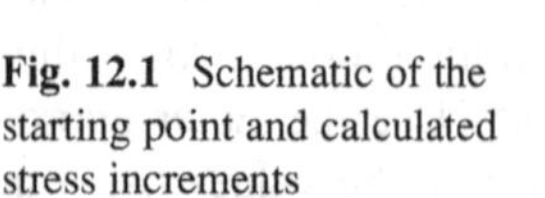
Fig. 12.1 Schematic of the starting point and calculated stress increments

Table 12.2 The two-step Euler method: calculation of stresses and PIVs (Example 12.1)

Quantity	Values	
	At point 1	At point 2
σ_1	114.0	131.206
σ_3	78.0	71.927
p	90.0	91.686
q	36.0	59.279
p_0	100.0	115.526
$\Delta\varepsilon_1$	0.010000	0.010000
$\Delta\varepsilon_3$	−0.001088	−0.001088
K	6,300.0	6,359.03
G	2,907.0	2,961.48
$\bar{K}$	4,362.0	4,443.75
$\Delta\sigma_1^{tr}$	92.268	–
$\Delta\sigma_3^{tr}$	27.803	–
σ_1^{tr}	206.268	–
σ_3^{tr}	105.803	–
p^{tr}	139.291	–
q^{tr}	100.465	–
ϕ^{tr}	1,2482.0 > 0	–
r_{11}	690.0	944.53
r_{33}	15.0	−116.95
r_{kk}	720.0	610.63
a_{11}	7.152×10^6	8.308×10^6
a_{33}	3.228×10^6	1.725×10^6
K_p	1.1139×10^9	1.112×10^9
K_p^*	5.0319×10^9	7.271×10^9
K_p^{**}	5.0319×10^9	8.383×10^9
D_1	1,852.411	2,133.284
D_2	8,480.703	10,011.900
D_3	605.545	2,734.584
D_4	2,666.703	4,088.945
$\Delta\sigma_1$	17.206	15.382
$\Delta\sigma_3$	−6.073	12.004
σ_1^{n+1}	131.206	129.382
σ_3^{n+1}	71.927	90.004
p_{n+1}	91.686	104.817
q_{n+1}	57.279	62.658
p_0^{n+1}	115.526	–

If a value of 1.0×10^{-3} is used for TOL, the step size is too large, and needs to be subdivided as follows (12.19b):

$$\beta = 0.8 \left[\frac{TOL}{E} \right]^{1/2} = 0.0931$$

Because of the restriction that $\beta > \beta_{\min} = 0.1$, the step size to be used is $\Delta T = 0.1$.

12.2.3 Sample Model Simulations

To examine the accuracy and robustness of the two-step Euler method of integration with substepping, sample simulations of the stress–strain behavior are performed. A one-element finite element model (with appropriate boundary conditions to simulate the behavior of a laboratory specimen) is used in the study.

The undrained behavior is simulated by modifying the effective stress–strain relation using a combined bulk modulus Γ. Let us express the rate of change of pore pressure as

$$\dot{u} = \Gamma \dot{\varepsilon}_{kk}$$

The effective stress–strain relation is given by

$$\dot{\sigma}_{ij} = D_{ijk\ell}\dot{\varepsilon}_{k\ell}$$

Combining the two, the following equation relating the total stress rate $\dot{\sigma}^t_{ij}$ to the strain rate $\dot{\varepsilon}_{k\ell}$ is obtained:

$$\dot{\sigma}^t_{ij} = \dot{\sigma}_{ij} + \dot{u}\delta_{ij} = D_{ijk\ell}\dot{\varepsilon}_{k\ell} + \Gamma\delta_{ij}\dot{\varepsilon}_{kk} = D_{ijk\ell}\dot{\varepsilon}_{k\ell} + \Gamma\delta_{ij}\delta_{k\ell}\dot{\varepsilon}_{k\ell} = \left[D_{ijk\ell} + \Gamma\delta_{ij}\delta_{k\ell}\right]\dot{\varepsilon}_{k\ell}$$

The simulations presented in this chapter have been obtained with $\Gamma = 1.0 \times 10^8$ kPa.

First, a series of analyses are performed to examine the model behavior during triaxial consolidated-drained and consolidated-undrained tests. The specimens are initially isotropically consolidated, unloaded to the desired initial confining pressure, and sheared under drained or undrained condition. The Cam-clay model parameters listed in Table 12.1 are used. All simulations are performed using $p_0 = 100$. The unit for stress-like and stiffness quantities is kPa, e.g., $p_0 = 100$. The results from a series of drained simulations are all plotted in Fig. 12.2 and a series of undrained simulations are plotted in Fig. 12.3.

The following values are employed for the control parameters in Algorithms 12.1 and 8.2: $TOL = 1.0 \times 10^{-3}$ (Algorithm 8.2), $N_G^{\max} = N_{G1}^{\max} = N_L^{\max} = 100$, $\Gamma = 1.0 \times 10^8$, $TOL = 1.0 \times 10^{-5}$ (Algorithm 12.1), $\beta_{\min} = 0.1$, $\beta_{\max} = 0.5$. The Euler approximation is used for the elastic properties with $\theta = 0.5$ (10.46).

Three triaxial compression tests are simulated, each with one of the following overconsolidation ratios: OCR $= 1$ (initial p of 100), 1.6667 (initial p of 60), and 5.0 (initial p of 20). All tests start from a point on the p- axis. The tests are conducted in strain-controlled loading mode with $\Delta\varepsilon_1 = \Delta\hat{\varepsilon}_1$. A sufficient level of strain is applied to cause complete failure. The results are compared with the "exact" solutions, which are numerical results obtained using the most accurate integration method considered later in this chapter (CPPM with exact integration for elastic properties and PIVs) with sufficiently small step sizes ($\Delta\varepsilon_1 = 0.05\%$). Step sizes smaller than $\Delta\varepsilon_1 = 0.05$ yield virtually identical results.

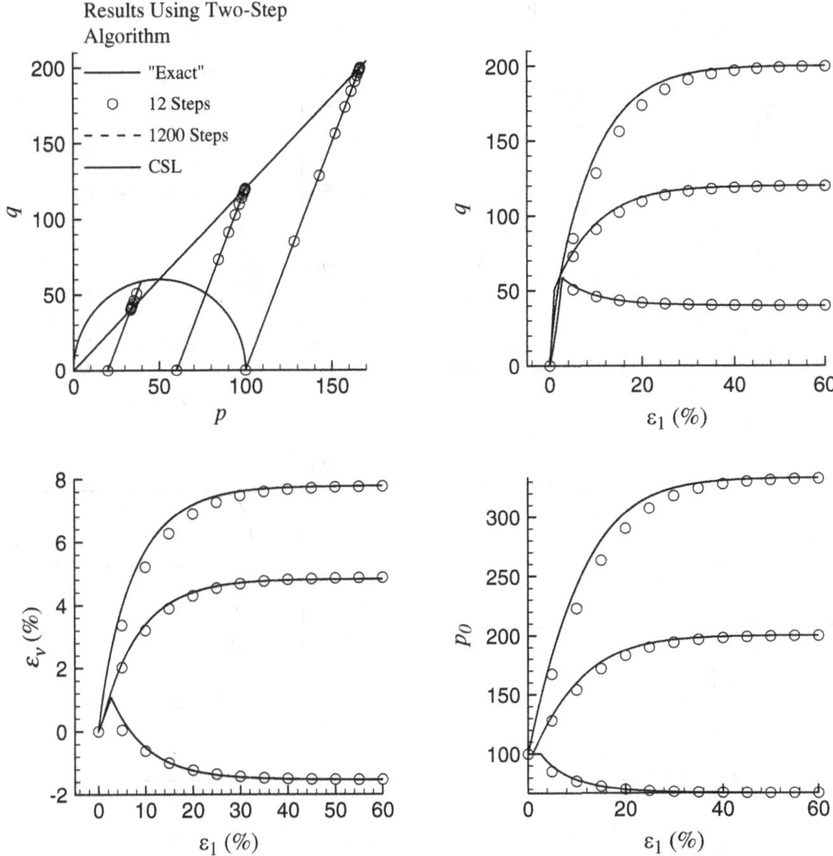

Fig. 12.2 Sample drained triaxial model simulations using the two-step Euler method

The results obtained with a step size $\Delta\varepsilon_1 = 5\%$ (which led to a total of 12 and 6 steps in drained and undrained, respectively) and a step size $\Delta\varepsilon_1 = 0.05\%$ (which led to a total of 1,200 and 600 steps in drained and undrained, respectively) are compared with the "exact" solutions. Observe that in all cases, the results with a step size $\Delta\varepsilon_1 = 0.05\%$ and the "exact" results are indistinguishable. The results with a step size $\Delta\varepsilon_1 = 5\%$ are further discussed below.

In all cases, convergence occurs. Many key aspects of the algorithm may be noted. Consider first the drained simulations. In the beginning of loading, the behavior is elasto-plastic for the test with OCR = 1 and elastic for tests with OCR = 1.67 and OCR = 5. Observe that for the test with OCR = 1.667, the elastic trial stress lies outside the yield surface. Thus, the point of intersection had to be found and the calculations performed in two steps.

The following observations may be made from the results of drained tests (Fig. 12.2): (1) For the tests with OCR = 1 and 1.67, the specimens experience

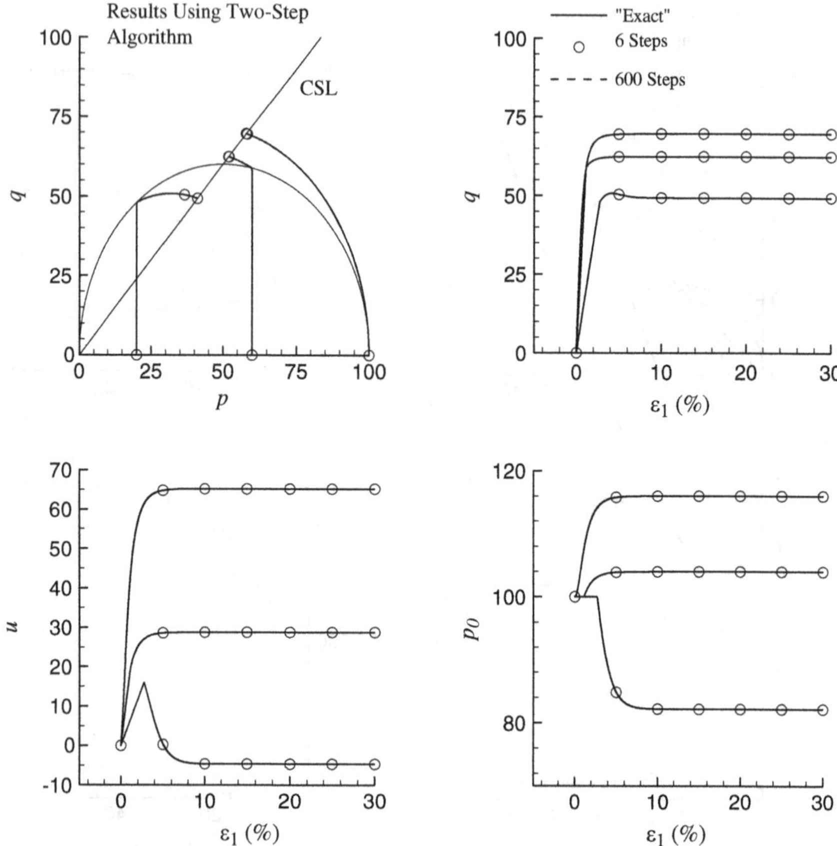

Fig. 12.3 Sample undrained triaxial model simulations using the two-step Euler method

volumetric compaction and the yield surface expands, and (2) for the test with OCR = 5, the specimen experiences (after a small elastic volumetric compaction) plastic volumetric dilation and the yield surface contracts. The integration algorithm is able to capture these behaviors well.

Considering the undrained behaviors, when the behavior is elastic, the effective stress path in the $p - q$ space is supposed to be vertical. It is seen from Fig. 12.3 that for OCR = 1.67 and 5, the effective stress path remains vertical until it reaches the yield surface. After that, the stress path leans to the left for the test with OCR = 1.67 and to the right for the test with OCR = 5. This is due to the fact that the specimen with OCR = 1.67 builds up positive plastic volumetric strain (compaction) whereas that with OCR = 5 builds up negative plastic volumetric strain (dilation). For the test with OCR = 1, the stress path leans to the left from the beginning. The positive plastic volumetric strain leads to expansion of the yield surface and vise-versa. The results presented in Fig. 12.3 confirm this. Also note the mild strain softening

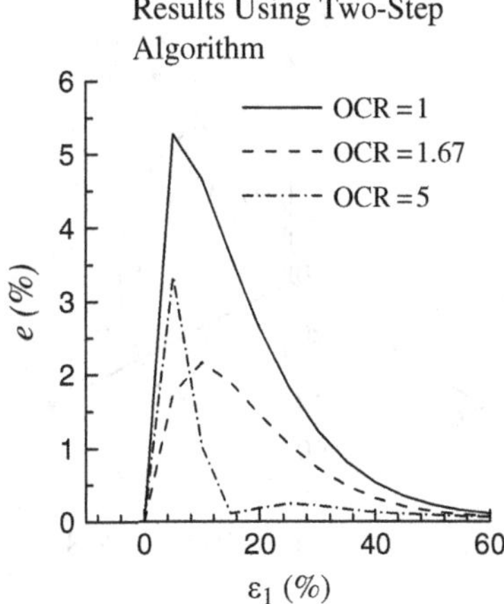

Fig. 12.4 Error on stresses calculated by the two-step Euler method

exhibited by the specimen with OCR = 5. These behaviors have been captured by the integration algorithm very well.

Considering the fact that the step size is very large, the agreement between the "exact" results and results with the step size $\Delta\varepsilon_1 = 5\%$ is very good. Some level of discrepancy is observable in the case of drained tests, but the agreement is extremely close in the case of undrained tests. Figure 12.4 presents the variation of the error e on computed stresses with the vertical strain for the drained tests; e is computed as

$$e = 100\left\|\boldsymbol{\sigma}_{exact} - \boldsymbol{\sigma}_{approximate}\right\| / \left\|\boldsymbol{\sigma}_{exact}\right\|\% \tag{12.30}$$

In all cases, the error is less than 5.5%. This is consistent with the visual observation that can be made from Fig. 12.2. The results must be evaluated with the following in mind: (1) The elastic properties are approximated using the Euler method with $\theta = 0.5$ (10.46) and (2) the hardening rule is integrated exactly (12.16).

The ability of the algorithm to handle three-dimensional stress paths is examined by considering a drained circular stress path. The results are presented in Fig. 12.5, along with the circular stress path and a cross section of the yield surface in the octahedral plane. Note that, because of symmetry about the σ_1-axis, only half the cross section is given. For discussions about the octahedral plane, the reader is referred to Chap. 2.

Considering a polar coordinate system (r_0, θ) in the octahedral plane, it may be shown that (Chap. 2)

$$\sigma_1 = r_0 \cos\theta + p \tag{12.31a}$$

$$\sigma_2 = -r_0 \cos(\theta - 60°) + p \tag{12.31b}$$

$$\sigma_3 = -r_0 \cos(\theta + 60°) + p \tag{12.31c}$$

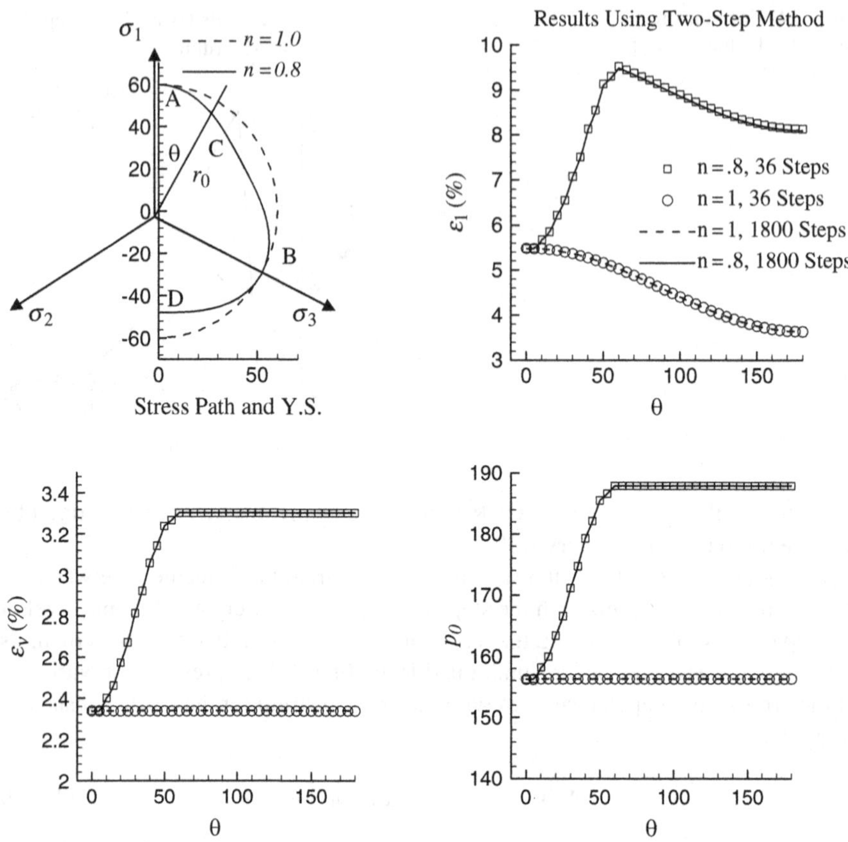

Fig. 12.5 Sample drained three-dimensional model simulations using two-step Euler method

where $(\sigma_1 + \sigma_2 + \sigma_3)/3 = p$ and $r_0 = 2J/\sqrt{3}$. For triaxial compression, $\theta = 0$, 120° or 240° and $|q| = \sqrt{3}J = 3r_0/2$. Using these relations, coordinates for a circular stress path are calculated at 18 points in the range $0 \leq \theta \leq 180°$, with $r_0 = 60$ and $p = 100$. The starting point on the circular path has the coordinates $\sigma_1 = 160$ and $\sigma_2 = \sigma_3 = 70$. This corresponds to $p = 100$ and $q = 90$. Starting from $p = 100$ and $q = 0$, the initial point on the circular path is reached in two increments in a stress-controlled loading mode. Then the loading is continued along the circular stress path; the results presented in Fig. 12.5 correspond only to this circular stress path. The strains at $\theta = 0$ are those produced during the initial loading from the p-axis to the starting point on the circular stress path.

In the octahedral plane, the shape of the yield surface is circular for $n = M_e/M_c = 1$ and noncircular for $n = M_e/M_c < 1$; its radius is smaller for $\alpha = -\pi/6$ (points C and D) than for $\alpha = \pi/6$ (points A and B). The dotted curve represents a circle in the octahedral plane. Now consider a loading where the stress path traverses a circle in the octahedral plane with the mean normal pressure p remaining a constant.

For $n = 1$, the stress probe vectors will be tangent to the surface. This corresponds to neutral loading, which is associated with a purely elastic behavior. The PIV remains unchanged. The strains are elastic and small. These aspects may be observed in Fig. 12.5. Also note that as p remains a constant, the loading does not produce any volumetric strain.

However, for $n < 1$, the stress probe vectors make an angle less than 90° with the flow direction $\mathbf{r}$ (except at some points such as point A). This produces plastic deformations, which are larger than that associated with a purely elastic behavior. A change in PIV is also produced. These aspects may be observed in Fig. 12.5 as well.

The yield surface expands for $0 \leq \theta \leq 60°$. From this point on, the stress probe vectors are directed inward from the yield surface, making an angle greater than 90° with the flow direction $\mathbf{r}$, producing only elastic strains. The PIV remains unchanged. These aspects of the model behavior may also be observed in Fig. 12.5.

Results using 36 steps are almost identical to those using 1,800 steps. The algorithm appears to perform well for this three-dimensional loading path. The efficiency of the method will be discussed later in the chapter, where the performances of different methods of integration are compared with each other.

12.3 The Cutting Plane Method

12.3.1 Introduction

The cutting plane algorithm is described in Sect. 10.6 (Chap. 10); the reader is advised to read that first. Only the final equations are presented here.

In this method, the stresses and PIV are computed by the iterative equations

$$\boldsymbol{\sigma}_{n+1}^{i+1} = \boldsymbol{\sigma}_{n+1}^{tr} + \sum \delta\lambda^i \mathbf{C}^i \mathbf{r}^i \tag{12.32a}$$

$$\boldsymbol{\zeta}_{n+1}^{i+1} = \boldsymbol{\zeta}_n + \sum \delta\lambda^i \mathbf{s}^i \tag{12.32b}$$

where

$$\delta\boldsymbol{\sigma}^i = -\delta\lambda^i \mathbf{C}^i \mathbf{r}^i \tag{12.32c}$$

$$\delta\boldsymbol{\zeta}^i = \delta\lambda^i \mathbf{s}^i \tag{12.32d}$$

i in the above equations is the iteration number. At each iteration, the value $\delta\lambda^i$ is computed from

$$\delta\lambda^i = \frac{\phi^i}{\left[\mathbf{n}^T\mathbf{C}\mathbf{r} + K_p\right]^i} \tag{12.33}$$

The following convergence criteria may be used to terminate the iterations:

$$e_\sigma = \left\|\boldsymbol{\sigma}_{n+1}^{i+1} - \boldsymbol{\sigma}_{n+1}^{i}\right\| / \left\|\boldsymbol{\sigma}_{n+1}^{i+1}\right\| \le TOL_\sigma \tag{12.34a}$$

$$e_\varsigma = \left\|\boldsymbol{\zeta}_{n+1}^{i+1} - \boldsymbol{\zeta}_{n+1}^{i}\right\| / \left\|\boldsymbol{\zeta}_{n+1}^{i+1}\right\| \le TOL_\varsigma \tag{12.34b}$$

$$\phi^{i+1} \le TOL_\phi \tag{12.34c}$$

In applying the above equations to the modified Cam-clay model, it is recognized that $\boldsymbol{\zeta} \equiv p_0$ and $\mathbf{n} = \mathbf{r}$. The general expression for $\mathbf{r}$ is presented in Appendix 7, with the Cam-clay model specific equations presented in Appendix 8.1.

The following key points may be recalled from Sect. 10.6: (1) The method is explicit at each iteration in that it requires only the forward-point (i.e., the $n + 1$ th) values of quantities at the current iteration i and not $i + 1$, (2) the method does not require the gradients of quantities such as $\mathbf{n}$, $\mathbf{r}$ and $\mathbf{s}$ (as does CPPM to be described next), (3) the consistent tangent operator cannot in general be evaluated, (4) the method does not require the intersection of elastic predictor and yield surface to be found (as does the two-step Euler method described in the preceding section), and (5) the convergence is expected to be quadratic at the local level (i.e., during the calculation of the incremental stresses and PIVs) as will be demonstrated in Example 12.2.

The implementation procedure is summarized in Algorithm 12.2. In Algorithm 12.2, the exact elastic predictor (see Appendix 8.4) is employed, whereas the explicit representation is used in the evaluation of the $\mathbf{C}$ matrix needed in the relaxation process. The convergence behavior is examined in Table 12.4 (Example 12.3).

Algorithm 12.2 The Cutting Plane Method for Integrating the Constitutive Relations from the Modified Cam-Clay Model

Definition of (new) parameters: $N_L^{\max}$ is the maximum number of iterations permitted before execution is terminated, and N_L is the corresponding counter. TOL is a small, machine-dependent number (it could be set equal to TOL_ϕ). If converged, Iflag = 1 and 0 otherwise.

1. Establish $\Delta\boldsymbol{\varepsilon}, p_0^n, \boldsymbol{\sigma}_n, TOL_\sigma, TOL_\varsigma, TOL_\phi, TOL, N_L^{\max}$, and the constitutive model parameters.
2. Initialize parameters as $N_L = 0$ and $p_0^{n+1} = p_0^n$.
3. Compute elastic predictor $\boldsymbol{\sigma}_{n+1}^{\mathrm{tr}}$ (Appendix 8.4) and $\phi^{\mathrm{tr}}(\boldsymbol{\sigma}_{n+1}^{tr}, p_0^n)$, and set $\boldsymbol{\sigma}_{n+1} = \boldsymbol{\sigma}_{n+1}^{tr}$ and $\phi^i = \phi^{tr}$.

(continued)

4. Perform loading/unloading check: if (ϕ^{tr}<TOL) go to step 9.
5. Compute $\delta\lambda^i$, $\delta\boldsymbol{\sigma}^i$, and δp_0^i, and find updated quantities $\boldsymbol{\sigma}_{n+1}$ and p_0^{n+1}.
6. Find errors e_σ, e_ς, and $\phi^i(\boldsymbol{\sigma}_{n+1}, p_0^{n+1})$ and perform convergence checks: $e_\sigma \leq TOL_\sigma$, $e_\varsigma \leq TOL_\varsigma$, and $\phi^{i+1} \leq TOL_\phi$. If converged, set iflag = 1 and go to step 9.
7. Set $N_L \leftarrow N_L + 1$. If $(N_L > N_L^{\max})$, set iflag = 0 and go to step 9.
8. Go to step 5.
9. Return to the main program.

Note: C^i required here is evaluated using K and G calculated using the current value of $\boldsymbol{\sigma}_{n+1}$.

12.3.2 Equations in Triaxial Space

The equations needed for performing the calculations in the triaxial space are among the suite of equations derived in Sect. 12.2.2. These equations are used in solving Example 12.2 (and are to be used in solving Problems 12.4 and 12.5).

Example 12.2 Local Iterations by the Cutting-Plane Method.

Question: Given $\sigma_1^n = 114, \sigma_3^n = 78, p_0^n = 100, \Delta\varepsilon_1 = 0.01$, and $\Delta\varepsilon_3 = -0.0008876$, determine the corresponding stresses and PIV by the cutting-plane algorithm. Carry out the iterations a few times and examine the convergence behavior. Use the values listed in Table 12.1 for the model parameters. Assume triaxial conditions with direction 1 taken as the axial direction.

Answer: First, the elastic predictor is calculated using the exact elastic predictor presented in Appendix 8.4 as follows:

$$p^n = 90,\ q^n = 36,\ \Delta\varepsilon_{kk} = 0.008225,\ x_1 = \exp[\bar{K}_0 \Delta\varepsilon_{kk}] = 1.7784,$$
$$p^{tr} = p^n x_1 = 160.06$$

$$G_{ave} = \frac{\bar{G}_0 p^n [x_1 - 1]}{\bar{K}_0 \Delta\varepsilon_{kk}} = 3930.49$$

$$\sigma_1^{tr} = \sigma_1^n + (p^{tr} - p^n) + 2G_{ave}(\Delta\varepsilon_1 - \Delta\varepsilon_{kk}/3) = 241.12$$

$$\sigma_3^{tr} = \sigma_3^n + (p^{tr} - p^n) + 2G_{ave}(\Delta\varepsilon_3 - \Delta\varepsilon_{kk}/3) = 119.53$$

$$q^{tr} = 121.59$$

$$\phi^{tr} = 1.789 \times 10^5$$

The iterative calculations to find the plastic corrector are then performed using the specialized triaxial equations presented in Sect. 12.2.2 and the specific cutting-plane equations given by (12.32a)–(12.34c). The step-by-step results are presented in Table 12.3. The errors ϕ, e_σ, and e_ς are reduced to 0.24×10^{-6}, 0.12×10^{-5}, and 0.30×10^{-6}, respectively in five iterations. The rate of convergence with respect to e_σ and e_ς are found to be quadratic because $(e_\sigma)_{k+1}/(e_\sigma)_k^2$ and $(e_\varsigma)_{k+1}/(e_\varsigma)_k^2$ remain almost constant during the iterations. The rate of convergence is not linear because $(e_\sigma)_{k+1}/(e_\sigma)_k$ and $(e_\varsigma)_{k+1}/(e_\varsigma)_k$ vary during the iterations. (Note that, instead of using the exact solution to define the errors such as e_σ in examining the rate of convergence as done in Chap. 8; the quantities calculated by (12.34) are directly used, where consecutive iterative solutions are used to define the errors.)

The cutting-plane algorithm can be easily incorporated with Algorithm 8.2 to perform model calibration. In triaxial space, the details are particularly simple and amenable to hand-calculations. It is, however, effectively done with a simple computer program. Adopting the notations presented in Sect. 12.2.2, the equation to be solved in step 3 of Algorithm 8.2 simplifies to

$$\begin{bmatrix} D_1 & D_3 & D_3 \\ D_3 & D_2 & D_4 \\ D_3 & D_4 & D_2 \end{bmatrix} \begin{pmatrix} \delta\varepsilon_1 \\ \delta\varepsilon_3 \\ \delta\varepsilon_3 \end{pmatrix} = \begin{pmatrix} \delta\sigma_1 \\ \delta\sigma_3 \\ \delta\sigma_3 \end{pmatrix}$$

which leads to

$$\delta\varepsilon_1 = \frac{1}{x_1}\left[\delta\sigma_3 - \frac{D_2 + D_4}{2D_3}\delta\sigma_1\right] \quad \text{and} \quad \delta\varepsilon_3 = \frac{1}{2x_1}\left[\delta\sigma_1 - \frac{D_1}{D_3}\delta\sigma_3\right]$$

where

$$x_1 = D_3 - \frac{D_1}{2D_3}(D_2 + D_4)$$

The application of the global/local procedure is treated in Example 12.3 and Problem 12.5.

Example 12.3 Global Iterations (Algorithm 8.2) Using the Cutting-Plane Method for the Local Iterations (Algorithm 12.2).

Question: A strain-controlled, triaxial drained compression loading (CD test) is to be conducted with $\sigma_1^n = 114$, $\sigma_3^n = 78$, and $p_0^n = 100$. The initial point is shown in Fig. 12.1. Determine the stresses and strains at the end of the incremental loading with $\Delta\hat{\varepsilon}_1 = 0.05$ and $\Delta\hat{\sigma}_3 = 0.0$. Use Algorithm 8.2 for global iterations and Algorithm 12.2 for local iterations (the cutting-plane method). Use the specific triaxial equations developed in Sect. 12.2.2 to perform the calculations. Examine the global convergence behavior of the algorithm. Use the model parameters listed in Table 12.1.

Table 12.3 The cutting-plane method: local iterations to find stresses and PIVs (Example 12.2)

Quantity	Iter. 1	Iter. 2	Iter. 3	Iter. 4	Iter. 5	Iter. 6	Iter. 7
p	160.06	113.48	97.49	94.75	94.66	94.66	94.66
q	121.59	72.12	53.49	50.09	49.98	49.98	49.98
r_{11}	2,180.20	1,258.47	917.24	855.78	853.70	853.69	853.69
r_{33}	−99.56	−93.69	−85.65	−83.42	−83.34	−83.34	−83.34
r_{kk}	1,981.07	1,071.09	745.94	688.93	687.01	687.01	687.01
K	11,204.17	7,943.31	6,824.06	6,632.42	6,626.00	6,626.00	6,626.00
G	5,169.92	3,665.27	3,148.82	3,060.39	3,057.43	3,057.42	3,057.42
$\overline{K}$	7,757.55	5,499.79	4,724.85	4,592.16	4,587.72	4,587.71	4,587.71
a_{11}	0.38E+08	0.15E+08	0.93E+07	0.84E+07	0.84E+07	0.84E+07	0.84E+07
a_{33}	0.14E+08	0.52E+07	0.30E+07	0.27E+07	0.26E+07	0.26E+07	0.26E+07
K_p	0.55E+10	0.23E+10	0.14E+10	0.13E+10	0.13E+10	0.13E+10	0.13E+10
K_p^*	0.80E+11	0.18E+11	0.80E+10	0.67E+10	0.67E+10	0.67E+10	0.67E+10
K_p^{**}	0.85E+11	0.20E+11	0.94E+10	0.80E+10	0.80E+10	0.80E+10	0.80E+10
$\delta\lambda$	0.21E−05	0.19E−05	0.54E−06	0.20E−07	0.23E−10	0.30E−16	0.38E−21
s	0.38E+07	0.22E+07	0.16E+07	0.15E+07	0.15E+07	0.15E+07	0.15E+07
$\delta\sigma_1$	−79.56	−28.41	−5.00	−0.17	0.00	0.00	0.00
$\delta\sigma_3$	−30.09	−9.78	−1.61	−0.05	0.00	0.00	0.00
δp_0	7.94	4.15	0.86	0.03	0.00	0.00	0.00
σ_1^{n+1}	161.55	133.14	128.14	127.97	127.97	127.97	127.97
σ_3^{n+1}	89.44	79.66	78.05	78.00	78.00	78.00	78.00
p_0	107.94	112.09	112.95	112.98	112.98	112.98	112.98
ϕ	0.38E+05	0.51E+04	0.16E+03	0.18E+00	0.24E−06	0.30E−11	0.30E−11
e_σ	0.44E+00	0.18E+00	0.32E−01	0.11E−02	0.12E−05	0.16E−11	0.20E−16
e_ς	0.74E−01	0.37E−01	0.76E−02	0.26E−03	0.30E−06	0.39E−12	0.49E−17
$(e_\sigma)_{k+1}/(e_\sigma)_k$	–	0.41E+00	0.18E+00	0.34E−01	0.11E−02	0.13E−05	0.13E−04
$(e_\varsigma)_{k+1}/(e_\varsigma)_k$	–	0.50E+00	0.21E+00	0.35E−01	0.11E−02	0.13E−05	0.13E−04
$(e_\sigma)_{k+1}/(e_\sigma)_k^2$	–	0.94E+00	0.99E+00	0.10E+01	0.10E+01	0.10E+01	0.77E+07
$(e_\varsigma)_{k+1}/(e_\varsigma)_k^2$	–	0.68E+01	0.55E+01	0.46E+01	0.43E+01	0.43E+01	0.32E+08

Answer: First, the variables are initialized as follows (Step 2 of Algorithm 8.2): $\sigma_1^n = \sigma_1^{n+1} = 114$, $\sigma_3^n = \sigma_3^{n+1} = 78$, and $p_0^n = p_0^{n+1} = 100$; $\sigma_{E1}^{n+1} = 114 + 0 = 114$, $\sigma_{E3}^{n+1} = 78 + 0 = 78$,

$$\sigma_{S1}^n = \sigma_{S1}^{n+1} = \sigma_1^n = 114,$$

$$\sigma_{S3}^n = \sigma_{S3}^{n+1} = \sigma_3^n = 78,$$

$$\delta\sigma_1 = \sigma_{E1}^{n+1} - \sigma_{S1}^{n+1} = 0 \quad \text{and} \quad \delta\sigma_3 = \sigma_{E3}^{n+1} - \sigma_{S3}^{n+1} = 0$$

Note that in the above equations, the subscripts "*E*" and "*S*" indicate "external force" and "spring force (internal)," respectively. The values of various quantities during the global iteration are listed in Table 12.4. Also recall that when the loading is of the mixed type as in the present case (the stress is controlled in direction 3 and the strain is controlled in direction 1), the **D** matrix and the $\delta\boldsymbol{\sigma}$ vector are modified as described in Sect. 8.6. In the present case, the coefficients D_1 and $\delta\sigma_1$ are modified; the modified coefficients are denoted as D_1^* and $\delta\sigma_1^*$ in Table 12.4.

It may be noted that the rate of convergence at the global level is linear with respect to e_σ (notice that $(e_\sigma)_{k+1}/(e_\sigma)_k$ remains almost constant and $(e_\sigma)_{k+1}/(e_\sigma)_k^2$ does not) and it took 22 iterations to reduce e_σ to below 1.0×10^{-3}. This is expected as the tangent operator is continuum and not consistent.

12.3.3 Sample Model Simulations

Sample model simulations are performed using a uniformly loaded single-element finite element model. The problems analyzed in this section are identical to those in Sect. 12.2.3, and the reader is referred to that section for the definition of the problems. The problems are analyzed here using the cutting-plane algorithm (Algorithm 12.2) with the following control parameters: $N_L^{\max} = 100$, $TOL = 1.0 \times 10^{-10}$, $TOL_\sigma = 0.01$, $TOL_\varsigma = 0.01$, $TOL_\phi = 1.0 \times 10^{-5} p_0$, and $\Gamma = 1.0 \times 10^8$ (combined bulk modulus used to simulate undrained behavior as described in Sect. 12.2.3).

The results are presented in Figs. 12.6 and 12.7 for drained and undrained simulations, respectively. The results obtained with a step size $\Delta\varepsilon_1 = 5\%$ (which led to a total of 12 and six steps in drained and undrained, respectively) and a step size $\Delta\varepsilon_1 = 0.05\%$ (which led to a total of 1,200 and 600 steps in drained and undrained, respectively) are compared with the "exact" solutions. As in the case of the two-step Euler method, in all cases, the results with a step size $\Delta\varepsilon_1 = 0.05\%$ and the "exact" results are indistinguishable. The results with a step size $\Delta\varepsilon_1 = 5\%$ are further discussed below.

Table 12.4 Global iterations by Algorithm 8.2 using the cutting-plane method for the local iterations (Example 12.3)

Quantity	Iter. 1	Iter. 2	Iter. 3	Iter. 4	Iter. 5	Iter. 22
D_1	10,164.00	3,865.90	3,796.82	3,737.38	3,689.37	3,526.61
D_2	10,164.00	9,592.86	10,170.47	10,594.99	10,910.11	11,883.46
D_3	4,368.00	4,800.17	4,828.20	4,828.31	4,816.90	4,717.38
D_4	4,368.00	4,081.43	4,353.22	4,547.14	4,687.36	5,096.18
D_1^*	0.100E+21	0.100E+21	0.100E+21	0.100E+21	0.100E+21	0.100E+21
$\delta\sigma_1^*$	0.500E+19	0.000E+00	0.000E+00	0.000E+00	0.000E+00	0.000E+00
$\delta\varepsilon_1$	0.500E−01	−0.131E−18	−0.953E−19	−0.701E−19	−0.522E−19	−0.551E−21
$\delta\varepsilon_3$	−0.150E−01	0.137E−02	0.987E−03	0.726E−03	0.542E−03	0.584E−05
$\Delta\varepsilon_1$	0.500E−01	0.500E−01	0.500E−01	0.500E−01	0.500E−01	0.500E−01
$\Delta\varepsilon_3$	−0.150E−01	−0.137E−01	−0.127E−01	−0.119E−01	−0.114E−01	−0.971E−02
$\Delta\sigma_1$	24.12	29.65	33.73	36.78	39.08	46.40
$\Delta\sigma_3$	−18.69	−14.33	−11.00	−8.45	−6.50	−0.08
Δp_0	35.97	39.51	42.11	44.05	45.52	50.23
σ_1^n+1	138.12	143.65	147.73	150.78	153.08	160.40
σ_3^n+1	59.31	63.67	67.00	69.55	71.50	77.92
p_0^n+1	135.97	139.51	142.11	144.05	145.52	150.23
σ_{S1}^n+1	138.12	143.65	147.73	150.78	153.08	160.40
σ_{S3}^n+1	59.31	63.67	67.00	69.55	71.50	77.92
$\delta\sigma_1$	−24.12	−29.65	−33.73	−36.78	−39.08	−46.40
$\delta\sigma_3$	18.69	14.33	11.00	8.45	6.50	0.08
e_σ	0.167E+00	0.128E+00	0.981E−01	0.753E−01	0.579E−01	0.681E−03
e_e	−0.562E+00	−0.392E+00	−0.279E+00	−0.202E+00	−0.148E+00	−0.148E–02
$(e_\sigma)_k+1/(e_\sigma)_k$	–	0.767E+00	0.768E+00	0.768E+00	0.769E+00	0.770E+00
$(e_\sigma)_k+1/(e_\sigma)_k^2$	–	0.460E+01	0.601E+01	0.783E+01	0.102E+02	0.871E+03

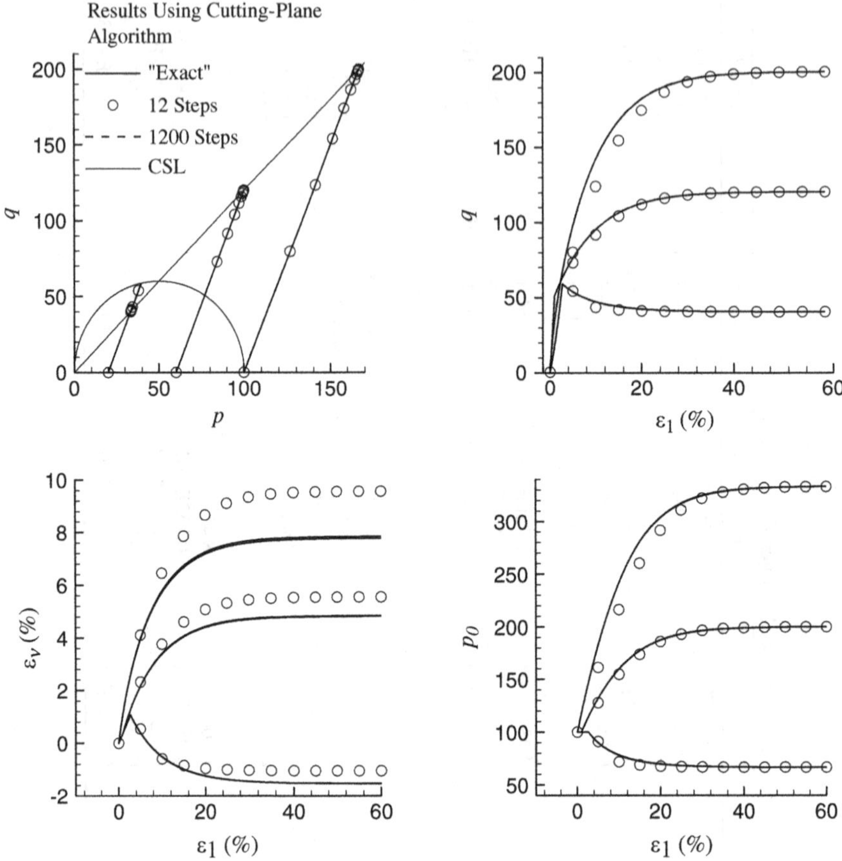

Fig. 12.6 Sample numerical simulations of triaxial drained behavior using the cutting-plane algorithm

As per the robustness of the algorithm, convergence was achieved in all cases considered. Except for accuracy, the performance of the method is very similar to the performance of the two-step Euler method discussed in Sect. 12.2.3; the reader is advised to read that section first. While the results seem to be almost as accurate as those obtained with the two-step Euler method in many respects, accuracy of volumetric strain calculated during the drained simulations is poorer with the step size $\Delta\varepsilon_1 = 5\%$ (Fig. 12.6).

During the simulation of the octahedral plane circular stress path (such as the one shown in Fig. 12.5), the method performed as good as the two-step Euler method. The results using 36 steps and 1,800 steps gave indistinguishable results. The results, however, slightly varied from those obtained using the two-step Euler method (Sect. 12.2). The results are not presented as they do not reveal new information.

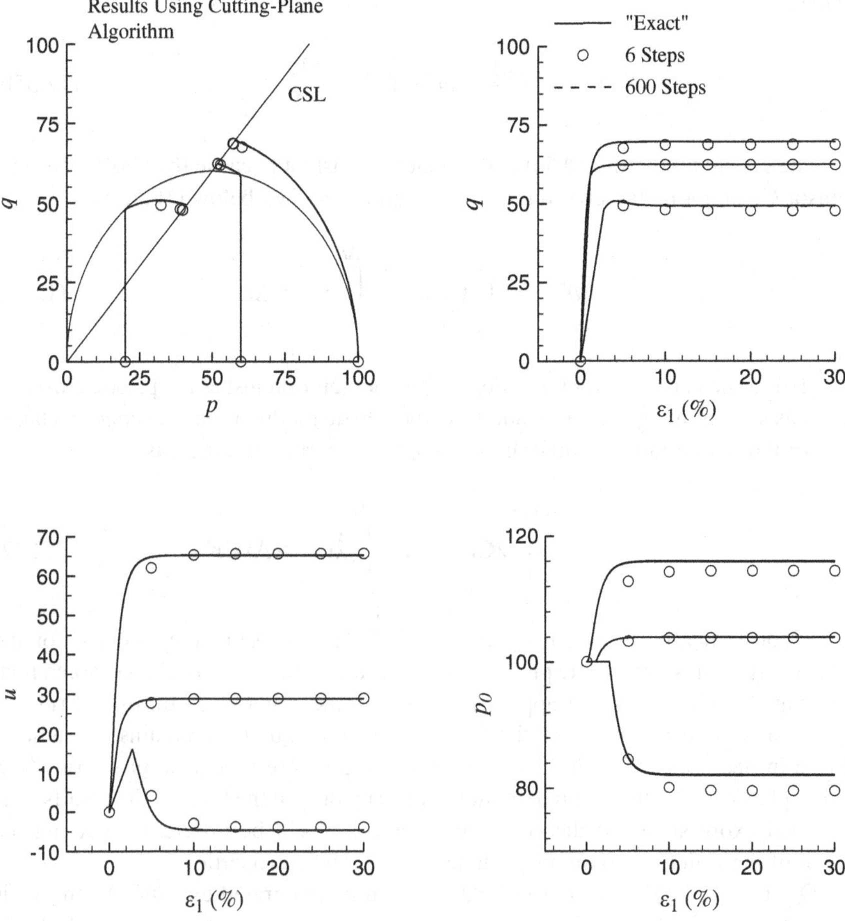

Fig. 12.7 Sample numerical simulations of triaxial undrained behavior using the cutting-plane algorithm

12.4 The Closest Point Projection (Implicit) Method with Approximate Elastic Properties: The σ-Space Formulation

The CPPM is described in Sect. 10.7 (Chap. 10) and the reader is advised to read that first. The method is applied to a simple isotropic hardening von Mises model in Chap. 11. In regard to the implementation of CPPM, the key difference between the von Mises model and the modified Cam-clay model is that in the latter, the elastic properties are pressure-dependent. Recall from (12.2a) and (12.2b) that

$$K = \bar{K}_0 p \quad \text{and} \quad G = \bar{G}_0 p \tag{12.35a}$$

where

$$\bar{K}_0 = \frac{(1+e_0)}{\kappa} \quad \text{and} \quad \bar{G}_0 = \frac{3(1-2\nu)}{2(1+\nu)}\bar{K}_0 \tag{12.35b}$$

The elastic predictor (10.5e) is no longer a constant because the elastic stiffness tensor **C** cannot be factored out of the integral as shown below (10.5c):

$$\Delta\boldsymbol{\sigma}^{ep} = \int_{t}^{t+\Delta t} \mathbf{C}\dot{\boldsymbol{\varepsilon}} \neq \mathbf{C}\int_{t}^{t+\Delta t} \dot{\boldsymbol{\varepsilon}} = \mathbf{C}\Delta\boldsymbol{\varepsilon} \tag{12.36}$$

Hence, the elastic predictor varies during the iteration just as the plastic corrector does as shown in Fig. 12.8a. In addition, the elastic modular matrix cannot be taken out of the integral in the equation for the plastic corrector either, as

$$\Delta\boldsymbol{\sigma}^{\mathrm{pc}} = -\int_{t}^{t+\Delta t} \dot{\lambda}\mathbf{C}\mathbf{r} \neq -\boldsymbol{C}\int_{t}^{t+\Delta t} \dot{\lambda}\mathbf{r} \approx -\Delta\lambda\mathbf{C}\mathbf{r}^{n+1} \tag{12.37}$$

Algorithmically, however, if an explicit Euler representation is made for the elastic properties, the elastic predictor remains a constant. It may also be noted that, as long as an Euler type of approximation (i.e., the generalized mid-point rule) is used for the elastic properties, the form of the main equations remains the same as those in Sect. 10.7. The influence of the mean pressure on elastic properties (and other plastic properties, for that matter) shows up only at the level of the details. The detailed expressions for the coefficients in 10.36 will be different reflecting the form of the Euler approximation made for the elastic properties.

On the other hand, the elastic relations can be integrated exactly. When this is done, the form of the main equations changes somewhat. The elastic and plastic parts are treated in an integral manner and the variation of the incremental stress during the iteration is schematically shown in Fig. 12.8b (Borja 1991).

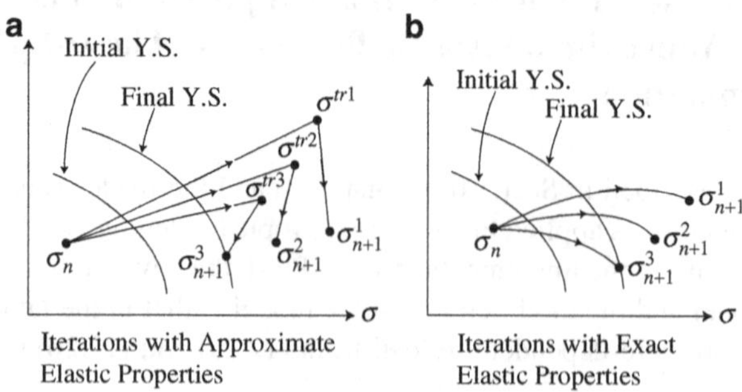

Fig. 12.8 Schematic of iterations involved in CPPM using approximate and exact elastic properties

To summarize, there are two ways to handle variable elastic properties:

1. Approximate the elastic properties using, say, the generalized midpoint rule, and integrate (12.36) and (12.37), or
2. Integrate (12.36) exactly using the actual relations between the elastic properties and pressure (Appendix 8.4), and modify (12.37) accordingly.

In the former, we will present the formulation with the generalized mid-point Euler representation for the elastic properties. By choosing the appropriate value for θ, the method either becomes an explicit, implicit, or generalized mid-point method. However, as will be discussed later, the explicit method generally works better.

The hardening rule given by (12.7a) can also be represented in two different ways: (1) Approximated by the Euler method or (2) integrated exactly. For consistency, we will use the Euler approximation with the Euler method of handling the elastic properties listed above (the 1st method), and integrate exactly with the exact method of handling the elastic properties listed above (the 2nd method).

Recall that we discussed two different formulations of CPPM in Chap. 10: (1) The σ-space formulation and (2) the S-space formulation (i.e., the deviatoric space formulation). When the yield surface is a circle in the octahedral plane, the modified Cam-clay model can be implemented by either formulation. When the yield surface is noncircular in the octahedral plane, the σ-space formulation is more suitable.

The formulation based on the Euler elastic properties is presented below, and that based on the exact elastic properties is presented in the next section (Sect. 12.5).

12.4.1 Description of CPPM with Approximate Elastic Properties

The σ-space formulation is considered. The generalized mid-point rule is used for the approximation of elastic properties. Equation (12.36) is integrated as

$$\Delta\boldsymbol{\sigma}^{\text{ep}} = \int_{t}^{t+\Delta t} \mathbf{C}\dot{\boldsymbol{\varepsilon}} \approx \mathbf{C}^{n+\theta}\Delta\varepsilon \tag{12.38}$$

where

$$\mathbf{C}^{n+\theta} = \mathbf{C}(K_{n+\theta}, G_{n+\theta}) \tag{12.39a}$$

$$K_{n+\theta} = K(\boldsymbol{\sigma}_{n+\theta}) \tag{12.39b}$$

$$G_{n+\theta} = G(\boldsymbol{\sigma}_{n+\theta}) \tag{12.39c}$$

and

$$\boldsymbol{\sigma}_{n+\theta} = (1-\theta)\boldsymbol{\sigma}_n + \theta\boldsymbol{\sigma}_{n+1}; \quad \theta \in [0, 1] \tag{12.39d}$$

The method employs explicit elastic properties when $\theta = 0$ and implicit elastic properties when $\theta > 0$. The iterative behavior is shown in Fig. 12.8a, where it is noted that both elastic predictor and the plastic corrector change until convergence. Note that, when $\theta > 0$, iterations are required even in finding the elastic predictor.

The Backward Euler Equations

The Euler approximation is used for both the elastic properties and hardening rule. To this end, the relevant backward Euler equations ((10.29a), (10.29b), and (10.28c)) are written as follows:

$$\boldsymbol{\sigma}_{n+1} = \boldsymbol{\sigma}_n + \mathbf{C}^{n+\theta}\Delta\boldsymbol{\varepsilon} - \Delta\lambda\mathbf{C}^{n+\theta}\mathbf{r}^{n+1}(\boldsymbol{\sigma}_{n+1}, p_0^{n+1}) \tag{12.40a}$$

$$p_0^{n+1} = p_0^n + \Delta\lambda s_{n+1} \tag{12.40b}$$

$$\phi^{n+1}(\boldsymbol{\sigma}_{n+1}, p_0^{n+1}) = 0 \tag{12.40c}$$

The Residual Equations and their Linearizations

The residual equations are given by ((10.31a)–(10.31c)):

$$\mathbf{R}_\sigma = \boldsymbol{\sigma}_{n+1} - \left[\boldsymbol{\sigma}_n + \mathbf{C}^{n+\theta}\Delta\boldsymbol{\varepsilon} - \Delta\lambda\mathbf{C}^{n+\theta}\mathbf{r}^{n+1}\right] = \mathbf{R}_\sigma(\boldsymbol{\sigma}_{n+1}, p_0^{n+1}, \Delta\lambda) \tag{12.41a}$$

$$\mathbf{R}_\varsigma = p_0^{n+1} - p_0^n + \Delta\lambda s_{n+1} = \mathbf{R}_\varsigma(\boldsymbol{\sigma}_{n+1}, p_0^{n+1}, \Delta\lambda) \tag{12.41b}$$

$$R_\phi = \phi^{n+1}(\boldsymbol{\sigma}_{n+1}, p_0^{n+1}) = \mathbf{R}_\phi(\boldsymbol{\sigma}_{n+1}, p_0^{n+1}) \tag{12.41c}$$

The linearizations of the above equations are ((10.34a)–(10.34c)):

$$R^1_{\sigma ij} + A_{ijk\ell}\delta\sigma_{k\ell} + B_{ij}\delta p_0 + F_{ij}\delta\lambda = 0 \tag{12.42a}$$

$$R^1_\varsigma + H_{k\ell}\delta\sigma_{k\ell} + \omega\delta p_0 + \beta\delta\lambda = 0 \tag{12.42b}$$

$$R^1_\phi + E_{k\ell}\delta\sigma_{k\ell} + \gamma\delta p_0 = 0 \tag{12.42c}$$

The coefficients of the above equations are given as ((10.36a)–(10.36h)):

$$A_{ijk\ell} = \frac{\partial R_{\sigma_{ij}}}{\partial\sigma_{k\ell}} = \delta_{ik}\delta_{j\ell} - \frac{\partial C^{n+\theta}_{ijpq}}{\partial\sigma_{k\ell}}\left[\Delta\varepsilon_{pq} - \Delta\lambda r^{n+1}_{pq}\right] + \Delta\lambda C^{n+\theta}_{ijpq}\frac{\partial r^{n+1}_{pq}}{\partial\sigma_{k\ell}} \tag{12.43a}$$

$$B_{ij} = \frac{\partial R_{\sigma_{ij}}}{\partial k} = \Delta\lambda C^{n+\theta}_{ijpq} \frac{\partial r^{n+1}_{pq}}{\partial k} \tag{12.43b}$$

$$F_{ij} = \frac{\partial R_{\sigma_{ij}}}{\partial \Delta\lambda} = C^{n+\theta}_{ijpq} r^{n+1}_{pq} \tag{12.43c}$$

$$H_{k\ell} = \frac{\partial R_{\varsigma}}{\partial \sigma_{k\ell}} = -\Delta\lambda \frac{\partial s^{n+1}}{\partial \sigma_{k\ell}} \tag{12.43d}$$

$$\omega = \frac{\partial R_{\varsigma}}{\partial p_0} = 1 - \Delta\lambda \frac{\partial s^{n+1}}{\partial p_0} \tag{12.43e}$$

$$\beta = \frac{\partial R_{\varsigma}}{\partial \Delta\lambda} = -s^{n+1} \tag{12.43f}$$

$$E_{k\ell} = \frac{\partial R_{\phi}}{\partial \sigma_{k\ell}} = \frac{\partial \phi^{n+1}}{\partial \sigma_{k\ell}} = n^{n+1}_{k\ell} \tag{12.43g}$$

$$\gamma = \frac{\partial R_{\phi}}{\partial p_0} = \frac{\partial \phi^{n+1}}{\partial p_0} \tag{12.43h}$$

The Coefficients for the Modified Cam-Clay Model

Term Involving the Elastic Modular Matrix.
The elastic stiffness tensor is written as

$$C^{n+\theta}_{ijpq} = \left(K_{n+\theta} - \frac{2}{3} G_{n+\theta}\right) \delta_{ij}\delta_{pq} + G_{n+\theta}\left(\delta_{ip}\delta_{jq} + \delta_{iq}\delta_{jp}\right) \tag{12.44a}$$

where

$$K_{n+\theta} = \bar{K}_0 p^{n+\theta}; \quad G_{n+\theta} = \bar{G}_0 p^{n+\theta}; \tag{12.44b}$$

$$\bar{K}_0 = \frac{(1 + e_0)}{\kappa} \quad \text{and} \quad \bar{G}_0 = \frac{3(1 - 2v)}{2(1 + v)} \bar{K}_0 \tag{12.44c}$$

From

$$\boldsymbol{\sigma}_{n+\theta} = (1 - \theta)\boldsymbol{\sigma}_n + \theta\boldsymbol{\sigma}_{n+1}; \quad \theta \in [0, 1] \tag{12.44d}$$

we have

$$p^{n+\theta} = (1-\theta)p^n + \theta p^{n+1}; \quad \theta \in [0,1] \tag{12.44e}$$

Equation (12.44a) is rearranged as

$$C_{ijpq}^{n+\theta} = K_{n+\theta}\left[\left(1-\frac{2}{3}\frac{\bar{G}_0}{\bar{K}_0}\right)\delta_{ij}\delta_{pq} + \frac{\bar{G}_0}{\bar{K}_0}\left(\delta_{ip}\delta_{jq}+\delta_{iq}\delta_{jp}\right)\right] \tag{12.45a}$$

In (12.45a), $K_{n+\theta}$ is the only term that depends on the stress. Differentiating (12.44b), it follows that

$$\frac{\partial K_{n+\theta}}{\partial \sigma_{k\ell}^{n+1}} = \bar{K}_0 \frac{\partial p^{n+\theta}}{\partial \sigma_{k\ell}^{n+1}} = \bar{K}_0 \theta \frac{\partial p^{n+1}}{\partial \sigma_{k\ell}^{n+1}} = \frac{1}{3}\bar{K}_0\theta\delta_{k\ell} = \frac{K_{n+\theta}\theta}{3p^{n+\theta}}\delta_{k\ell} \tag{12.45b}$$

Hence we have

$$\frac{\partial C_{ijpq}^{n+\theta}}{\partial \sigma_{k\ell}} = \frac{K_{n+\theta}\theta}{3p^{n+\theta}}\left[\left(1-\frac{2}{3}\frac{\bar{G}_0}{\bar{K}_0}\right)\delta_{ij}\delta_{pq} + \frac{\bar{G}_0}{\bar{K}_0}\left(\delta_{ip}\delta_{jq}+\delta_{iq}\delta_{jp}\right)\right]\delta_{k\ell} \tag{12.45c}$$

and

$$t_{ijk\ell} = \frac{\partial C_{ijpq}^{n+\theta}}{\partial \sigma_{k\ell}}\left[\Delta\varepsilon_{pq} - \Delta\lambda r_{pq}^{n+1}\right] = y_1\left[t_{ijk\ell}^1 - \Delta\lambda t_{ijk\ell}^2\right] \tag{12.45d}$$

where

$$t_{ijk\ell}^1 = \left[\left(1-\frac{2}{3}\frac{\bar{G}_0}{\bar{K}_0}\right)\delta_{ij}\Delta\varepsilon_{pp} + 2\frac{\bar{G}_0}{\bar{K}_0}\Delta\varepsilon_{ij}\right]\delta_{k\ell} \tag{12.45e}$$

$$t_{ijk\ell}^2 = \left[\left(1-\frac{2}{3}\frac{\bar{G}_0}{\bar{K}_0}\right)\delta_{ij}r_{pp} + 2\frac{\bar{G}_0}{\bar{K}_0}r_{ij}\right]\delta_{k\ell} \tag{12.45f}$$

and

$$y_1 = \frac{K_{n+\theta}\theta}{3p^{n+\theta}} \tag{12.45g}$$

The Remaining Terms. The remaining quantities may be evaluated from the surface properties r_{pq}^{n+1}, $\frac{\partial r_{pq}^{n+1}}{\partial \sigma_{k\ell}}$, and $\frac{\partial r_{pq}^{n+1}}{\partial p_0}$, and the hardening properties s^{n+1}, $\frac{\partial s^{n+1}}{\partial \sigma_{k\ell}}$, and $\frac{\partial s^{n+1}}{\partial p_0}$. Expression for r_{pq}^{n+1} is given in Appendices 7 and 8.1. Expressions for $\frac{\partial r_{pq}^{n+1}}{\partial \sigma_{k\ell}}$ and $\frac{\partial r_{pq}^{n+1}}{\partial p_0}$ are derived in Appendix 8.2. Equations for s^{n+1}, $\frac{\partial s^{n+1}}{\partial \sigma_{k\ell}}$, and $\frac{\partial s^{n+1}}{\partial p_0}$ are presented in Appendix 8.3. Once these quantities are calculated, the remaining quantities in (12.43a)–(12.43h) may be found noting the following:

$$
\begin{aligned}
C^{n+\theta}_{ijpq}\frac{\partial r^{n+1}_{pq}}{\partial \sigma_{k\ell}} &= \left[\left(K_{n+\theta}-\frac{2}{3}G_{n+\theta}\right)\delta_{ij}\delta_{pq}+G_{n+\theta}\left(\delta_{ip}\delta_{jq}+\delta_{iq}\delta_{jp}\right)\right]\frac{\partial r^{n+1}_{pq}}{\partial \sigma_{k\ell}} \\
&= \left(K_{n+\theta}-\frac{2}{3}G_{n+\theta}\right)\delta_{ij}\frac{\partial r^{n+1}_{pp}}{\partial \sigma_{k\ell}}+2G_{n+\theta}\frac{\partial r^{n+1}_{ij}}{\partial \sigma_{k\ell}}
\end{aligned}
\tag{12.46a}
$$

It is shown in Appendix 8.2 that

$$
\frac{\partial r^{n+1}_{pq}}{\partial p_0} = -3\delta_{pq}
$$

It then follows that

$$
\begin{aligned}
C^{n+\theta}_{ijpq}\frac{\partial r^{n+1}_{pq}}{\partial p_0} &= -3\left[\left(K_{n+\theta}-\frac{2}{3}G_{n+\theta}\right)\delta_{ij}\delta_{pq}+G_{n+\theta}\left(\delta_{ip}\delta_{jq}+\delta_{iq}\delta_{jp}\right)\right]\delta_{pq} \\
&= -9K_{n+\theta}\delta_{ij}
\end{aligned}
\tag{12.46b}
$$

Also note that

$$
C^{n+\theta}_{ijpq}r^{n+1}_{pq} = \left(K_{n+\theta}-\frac{2}{3}G_{n+\theta}\right)\delta_{ij}r_{pp}+2G_{n+\theta}r_{ij}
\tag{12.46c}
$$

The Implementation Details

The symmetry of stresses and strains allow the tensorial equations in (12.42a)–(12.42c) to be implemented in reduced order matrix-vector forms; the details are presented in Appendix 6. The three matrix-vector equations in (12.42a)–(12.42c) can be arranged in a single matrix-vector equation system as

$$
\begin{bmatrix}
\underset{6\times 6}{\mathbf{A}} & \underset{6\times 1}{\mathbf{B}} & \underset{6\times 1}{\mathbf{F}} \\
\underset{1\times 6}{\mathbf{H}^T} & \underset{1\times 1}{\omega} & \underset{1\times 1}{\beta} \\
\underset{1\times 6}{\mathbf{E}^T} & \underset{1\times 1}{\gamma} & \underset{1\times 1}{0}
\end{bmatrix}
\begin{bmatrix}
\underset{6\times 1}{\delta\sigma} \\
\underset{1\times 1}{\delta p_0} \\
\underset{1\times 1}{\delta\lambda}
\end{bmatrix}
=
\begin{bmatrix}
\underset{6\times 1}{-\mathbf{R}^1_\sigma} \\
\underset{1\times 1}{-R^1_\varsigma} \\
\underset{1\times 1}{-R^1_\phi}
\end{bmatrix}
\tag{12.47}
$$

Equation (12.27) can either be directly solved as a system or manipulated to obtain the following ((10.58a)–(10.59f)):

$$\delta\lambda = \frac{R_\phi^1 - \bar{\mathbf{n}}^T \mathbf{R} \mathbf{R}_\sigma^1 + x_0 R_\varsigma^1 \bar{\mathbf{n}}^T \bar{\mathbf{C}} \bar{\mathbf{B}} - x_0 \gamma R_\varsigma^1}{x_0 \gamma \beta + \bar{\mathbf{n}}^T \bar{\mathbf{C}} \bar{\mathbf{r}}} \tag{12.48a}$$

$$\delta\boldsymbol{\sigma} = -\mathbf{R}\mathbf{R}_\sigma^1 + x_0 R_\varsigma^1 \bar{\mathbf{C}} \bar{\mathbf{B}} - \bar{\mathbf{C}} \bar{\mathbf{r}} \delta\lambda \tag{12.48b}$$

$$\delta p_0 = -\frac{1}{\gamma}\left[R_\phi^1 + \mathbf{n}^T \delta\boldsymbol{\sigma}\right] \tag{12.48c}$$

where

$$\bar{\mathbf{r}} = \bar{\mathbf{F}} - \frac{\beta}{\omega} \bar{\mathbf{B}} \tag{12.48d}$$

$$\bar{\mathbf{n}} = \mathbf{n} - \frac{\gamma}{\omega} \mathbf{H} \tag{12.48e}$$

$$\bar{\mathbf{F}} = \mathbf{C}^{-1}\mathbf{F} \quad \text{and} \quad \bar{\mathbf{B}} = \mathbf{C}^{-1}\mathbf{B}. \tag{12.48f}$$

$$x_0 = \frac{1}{\omega} \tag{12.48g}$$

$$\bar{\mathbf{C}} = \mathbf{R}\mathbf{C} \tag{12.48h}$$

$$\mathbf{R} = \left[\mathbf{A} - \frac{1}{\omega}\mathbf{B}\mathbf{H}^{\mathrm{T}}\right]^{-1} \tag{12.48i}$$

The method requires an initial value for $\Delta\lambda$, which may be calculated from (10.30)

$$\Delta\lambda_1 = \frac{\phi^{tr}}{\left[\mathbf{n}^T \mathbf{C} \mathbf{r} + K_p\right]^{tr}} \tag{12.49}$$

The new estimates for $\boldsymbol{\sigma}_{n+1}$, p_0^{n+1}, and $\Delta\lambda$ are found as ((10.37a)–(10.37c)):

$$\boldsymbol{\sigma}_{n+1}^2 = \boldsymbol{\sigma}_{n+1}^1 + \delta\boldsymbol{\sigma} \tag{12.50a}$$

$$p_0^{n+1,2} = p_0^{n+1,1} + \delta p_0 \tag{12.50b}$$

$$\Delta\lambda_2 = \Delta\lambda_1 + \delta\lambda \tag{12.50c}$$

Note that superscript 2 on these variables does not denote "the square of the variable," but rather the "second estimate."

The following convergence criteria may be used ((10.38a)–(10.38c)):

$$e_\sigma = \|\mathbf{R}_\sigma^{i+1}\| / \|\boldsymbol{\sigma}_{n+1}^{i}\| \leq TOL_\sigma \tag{12.51a}$$

$$e_\varsigma = \|R_\varsigma^{i+1}\| / \|p_0^{n+1,i}\| \leq TOL_\varsigma \tag{12.51b}$$

$$R_\phi^{i+1} = \phi^{i+1} \leq TOL_\phi \tag{12.51c}$$

The procedure is summarized in Algorithm 12.3.
The consistent tangent operator is given by (10.66b):

$$\mathbf{D} = \bar{\mathbf{C}} - \frac{(\bar{\mathbf{C}}\bar{\mathbf{r}})(\bar{\mathbf{n}}^T\bar{\mathbf{C}})}{x_0\gamma\beta + \bar{\mathbf{n}}^T\bar{\mathbf{C}}\bar{\mathbf{r}}} \tag{12.52}$$

It was shown in Chap. 10 that when $\Delta\lambda = 0$, the consistent and continuum operators become identical to each other. Also recall that it was pointed out in Chap. 10 that even when an associated flow rule is employed, the consistent operator is in general not symmetric.

Algorithm 12.3 CPPM for Integrating the Constitutive Relations from the Modified Cam-Clay Model with Approximate Elastic Properties

Definition of (new) parameters: $N_L^{\max}$ is the maximum number of iterations permitted before the execution is terminated, and N_L is the corresponding counter. TOL is a small, machine-dependent number (it could be set equal to TOL_ϕ). Iflag = 1 if converged and 0 otherwise.

1. Establish $\Delta\boldsymbol{\varepsilon}$, p_0^n, $\boldsymbol{\sigma}_n$, TOL_σ, TOL_ς, TOL_ϕ, TOL, $N_L^{\max}$, and constitutive model parameters.
2. Initialize parameters as $N_L = 0$ and $p_0^{n+1} = p_0^n$.
3. Compute elastic predictor $\boldsymbol{\sigma}_{n+1}^{tr}$ and $\phi^{tr}(\boldsymbol{\sigma}_{n+1}^{tr}, p_0^n)$. Perform loading/unloading check: if $(\phi^{tr} < TOL)$, set $\boldsymbol{\sigma}_{n+1} = \boldsymbol{\sigma}_{n+1}^{tr}$ and go to step 12.
4. Set $\boldsymbol{\sigma}_{n+1} = \boldsymbol{\sigma}_{n+1}^{tr}$ and calculate $p_{n+\theta}$, $K_{n+\theta}$, $G_{n+\theta}$, s_{n+1}, and $\mathbf{r}_{n+1}$.
5. Set $R_\phi^1 = \phi^{tr}$. Compute $\Delta\lambda$, $\boldsymbol{\sigma}_{n+1}^1 = \boldsymbol{\sigma}_n + \mathbf{C}^{n+\theta}\Delta\boldsymbol{\varepsilon} - \Delta\lambda\mathbf{C}^{n+\theta}\mathbf{r}^{n+1}$ and $p_{0,1}^{n+1} = p_0^n + \Delta\lambda s^{n+1}$. Set $\sigma_{n+1} = \sigma_{n+1}^1$ and $p_0^{n+1} = p_{0,1}^{n+1}$.
6. Begin iterations. Compute $R_\phi^1 = \phi(\boldsymbol{\sigma}_{n+1}, p_0^{n+1})$, $p_{n+\theta}$, $K_{n+\theta}$, $G_{n+\theta}$, s_{n+1}, and $\mathbf{r}^{n+1}$.
7. Compute $\boldsymbol{\sigma}_{n+1}^2 = \boldsymbol{\sigma}_n + \mathbf{C}^{n+\theta}\Delta\boldsymbol{\varepsilon} - \Delta\lambda\mathbf{C}^{n+\theta}\mathbf{r}^{n+1}$ and $p_{0,2}^{n+1} = p_0^n + \Delta\lambda s^{n+1}$.
8. Compute $\mathbf{R}_{\boldsymbol{\sigma}}^1 = \boldsymbol{\sigma}_{n+1}^1 - \boldsymbol{\sigma}_{n+1}^2$ and $R_\varsigma^1 = p_{0,1}^{n+1} - p_{0,2}^{n+1}$.
9. Find errors e_σ, e_ς, and e_ϕ and perform convergence check: $e_\sigma \leq TOL_\sigma$, $e_\varsigma \leq TOL_\varsigma$, and $\phi^{i+1} \leq TOL_\phi$. If converged, set iflag = 1 and go to step 12.
10. Set $N_L \leftarrow N_L + 1$. If $(N_L > N_L^{\max})$, set iflag = 0 and go to step 12.

(continued)

11. Compute $\delta\lambda$, $\delta\boldsymbol{\sigma}$, and δp_0. Update $\boldsymbol{\sigma}_{n+1}^1 \leftarrow \boldsymbol{\sigma}_{n+1}^1 + \delta\boldsymbol{\sigma}$, $p_{0,1}^{n+1} \leftarrow p_{0,1}^{n+1} + \delta p_0$, and $\Delta\lambda \leftarrow \Delta\lambda + \delta\lambda$. Set $\boldsymbol{\sigma}_{n+1} = \boldsymbol{\sigma}_{n+1}^1$ and $p_0^{n+1} = p_{0,1}^{n+1}$. Go to step 6.
12. Return to the main program.

12.4.2 Equations in Triaxial Space

As shown in Sect. 12.2.2, only the first three components of the vectors such as $\boldsymbol{\sigma}$ and $\boldsymbol{\varepsilon}$ are relevant here because the shear components of $\boldsymbol{\sigma}$ and $\boldsymbol{\varepsilon}$ are zero. In addition, because of the assumption of the triaxial condition, the 2nd and 3rd components of $\boldsymbol{\sigma}$ and $\boldsymbol{\varepsilon}$ (and a few other vectors) are equal to each other. Among the 36 coefficients in the 6×6 matrices such as $\mathbf{C}$, only the nine components in the 3×3 sub-matrix in the top left hand corner are relevant.

Quantities Needed for the Calculation of Stresses

Let $f_1 = 1 - \frac{2\bar{G}_0}{3\bar{K}_0}$ and $f_2 = \frac{2\bar{G}_0}{\bar{K}_0}$, and recall that $y_1 = \frac{K_{n+\theta}\theta}{3p^{n+\theta}}$.

Then from (12.45d)–(12.45f), we have

$$t_{11} = y_1 \left[f_1 \left\{ \Delta\varepsilon_{kk} - \Delta\lambda r_{kk}^{n+1} \right\} + f_2 \left\{ \Delta\varepsilon_{11} - \Delta\lambda r_{11}^{n+1} \right\} \right]$$

$$t_{33} = y_1 \left[f_1 \left\{ \Delta\varepsilon_{kk} - \Delta\lambda r_{kk}^{n+1} \right\} + f_2 \left\{ \Delta\varepsilon_{33} - \Delta\lambda r_{33}^{n+1} \right\} \right]$$

From (A8.25a) and noting that $L_{ijpq}^3 = 0$ because $\partial_\alpha\phi = 0$, it follows that

$$\frac{\partial r_{ij}}{\partial \sigma_{pq}} = L_{ijpq}^1 + L_{ijpq}^2$$

$$= \frac{\partial[\partial_I\phi]}{\partial\sigma_{pq}}\delta_{ij} + \frac{\partial[\partial_J\phi]}{\partial\sigma_{pq}}\frac{s_{ij}}{2J} + \frac{\partial_J\phi}{2J}\frac{\partial s_{ij}}{\partial\sigma_{pq}} - \frac{\partial_J\phi}{2J^2}\frac{\partial J}{\partial\sigma_{pq}}s_{ij}$$

Let us further simplify this for triaxial conditions using the surface properties presented in Appendix 8.1, and (A8.14), and (A8.15) as follows:

$$\frac{\partial r_{ij}}{\partial \sigma_{pq}} = 2\delta_{ij}\delta_{pq} + \frac{2}{N^2}\frac{s_{ij}}{2J}\frac{s_{pq}}{2J} + \frac{2J}{N^2}\frac{1}{2J}\left[\delta_{ip}\delta_{jq} - \frac{1}{3}\delta_{ij}\delta_{pq}\right] - \frac{2J}{N^2}\frac{1}{2J^2}\frac{s_{pq}}{2J}s_{ij}$$

$$= 2\delta_{ij}\delta_{pq} + \frac{1}{N^2}\left[\delta_{ip}\delta_{jq} - \frac{1}{3}\delta_{ij}\delta_{pq}\right]$$

$$\frac{\partial r_{kk}}{\partial \sigma_{pq}} = 6\delta_{pq}$$

It then follows that

$$C^{n+\theta}_{ijk\ell}\frac{\partial r_{k\ell}}{\partial \sigma_{pq}} = 6\bar{K}_{n+\theta}\delta_{ij}\delta_{pq} + 4G_{n+\theta}\delta_{ij}\delta_{pq} + \frac{2G_{n+\theta}}{N^2}\left[\delta_{ip}\delta_{jq} - \frac{1}{3}\delta_{ij}\delta_{pq}\right]$$
$$= K_1\delta_{ij}\delta_{pq} + \frac{2G_{n+\theta}}{N^2}\delta_{ip}\delta_{jq}$$

where $\bar{K} = K_{n+\theta} - \frac{2}{3}G_{n+\theta}$ and $K_1 = 6\bar{K}_{n+\theta} + 4G_{n+\theta} - \frac{2G_{n+\theta}}{3N^2}$
From 12.43b and A8.29c, we have

$$B_{ij} = \frac{\partial R_{\sigma_{ij}}}{\partial p_0} = \Delta\lambda C^{n+\theta}_{ijpq}\frac{\partial r^{n+1}_{pq}}{\partial p_0} = -3\Delta\lambda C^{n+\theta}_{ijpq}\delta_{pq} = -9K_{n+\theta}\Delta\lambda\delta_{ij}$$

$$\bar{B}_{ij} = -3\Delta\lambda\delta_{ij}$$

From (12.43d) and (A8.31), we have

$$H_{k\ell} = -\Delta\lambda\frac{\partial s^{n+1}}{\partial \sigma_{k\ell}} = -C_1p_0^{n+1}\Delta\lambda\frac{\partial r^{n+1}_{pp}}{\partial \sigma_{k\ell}} = -6C_1p_0^{n+1}\Delta\lambda\delta_{k\ell}$$

Then it follows that

$$\frac{1}{\omega}\mathbf{BH}^T = x_0^*\begin{bmatrix}1 & 1 & 1\\ 1 & 1 & 1\\ 1 & 1 & 1\end{bmatrix}$$

where $x_0^* = 54C_1x_0p_0^{n+1}K_{n+\theta}\Delta\lambda^2$
Synthesizing these equations, it follows that

$$\mathbf{R} = \left[\mathbf{A} - \frac{1}{\omega}\mathbf{BH}^T\right]^{-1}$$

$$\mathbf{A} - \frac{1}{\omega}\mathbf{BH}^T$$
$$= \begin{bmatrix}1 & 0 & 0\\ 0 & 1 & 0\\ 0 & 0 & 1\end{bmatrix} - \begin{bmatrix}t_{11} & t_{11} & t_{11}\\ t_{33} & t_{33} & t_{33}\\ t_{33} & t_{33} & t_{33}\end{bmatrix}$$
$$+ \Delta\lambda\begin{bmatrix}K_1 + 2G_{n+\theta}/N^2 & K_1 & K_1\\ K_1 & K_1 + 2G_{n+\theta}/N^2 & K_1\\ K_1 & K_1 & K_1 + 2G_{n+\theta}/N^2\end{bmatrix}$$

$$-x_0^*\begin{bmatrix}1&1&1\\1&1&1\\1&1&1\end{bmatrix} = \text{(say)}\begin{bmatrix}T_1&T_3&T_3\\T_4&T_2&T_4\\T_4&T_4&T_2\end{bmatrix}$$

where

$$T_1 = 1 - t_{11} + \Delta\lambda\left(K_1 + \frac{2G_{n+\theta}}{N^2}\right) - x_0^*$$

$$T_2 = 1 - t_{33} + \Delta\lambda\left(K_1 + \frac{2G_{n+\theta}}{N^2}\right) - x_0^*$$

$$T_3 = -t_{11} + \Delta\lambda K_1 - x_0^*$$

$$T_4 = -t_{33} + \Delta\lambda K_1 - x_0^*$$

Let us denote

$$\mathbf{R} = \begin{bmatrix}T_1&T_3&T_3\\T_4&T_2&T_4\\T_4&T_4&T_2\end{bmatrix}^{-1} = \begin{bmatrix}S_1&S_3&S_3\\S_4&S_2&S_5\\S_4&S_5&S_2\end{bmatrix}$$

and define $\alpha_1 = \frac{T_2}{T_4}$, $\alpha_2 = \frac{T_1}{T_3}$, $Z_1 = 2 - \alpha_2 - \alpha_1\alpha_2$, $Z_2 = 2 - \alpha_2(1+\alpha_1)$, and $Z_3 = 2 + \alpha_1^2\alpha_2 - \alpha_2 - 2\alpha_1$

Then it can be shown that

$$S_1 = -\frac{(1+\alpha_1)}{Z_1T_3};\quad S_2 = \frac{(\alpha_1\alpha_2 - 1)}{Z_3T_4};\quad S_3 = \frac{(1-\alpha_1)}{Z_3T_4};\quad S_4 = \frac{1}{Z_2T_3} \quad \text{and}$$
$$S_5 = \frac{(1-\alpha_2)}{Z_3T_4}$$

Also, let us define

$$\bar{\mathbf{C}} = \mathbf{RC} = \begin{bmatrix}\bar{C}_1&\bar{C}_3&\bar{C}_3\\\bar{C}_4&\bar{C}_2&\bar{C}_5\\\bar{C}_4&\bar{C}_5&\bar{C}_2\end{bmatrix}$$

and $\bar{K}_1 = \bar{K}_{n+\theta} + 2G_{n+\theta}$. Then it can be shown that

$$\bar{C}_1 = S_1\bar{K}_1 + 2S_3\bar{K};\quad \bar{C}_2 = (S_4+S_5)\bar{K} + S_2\bar{K}_1;\quad \bar{C}_3 = (S_1+S_3)\bar{K} + S_3\bar{K}_1$$

$$\bar{C}_4 = (S_2+S_5)\bar{K} + S_4\bar{K}_1;\quad \bar{C}_5 = (S_2+S_4)\bar{K} + S_5\bar{K}_1$$

$$r_{11} = 6p^{n+1} - 3p_0^{n+1} + 18q^{n+1}/M^2$$

$$r_{33} = 6p^{n+1} - 3p_0^{n+1} - 9q^{n+1}/M^2$$

$$\bar{\mathbf{n}} = \mathbf{n} - \frac{\gamma}{\omega}\mathbf{H} = \mathbf{r} - 9p^{n+1}x_0(6\Delta\lambda C_1 p_0^{n+1}\boldsymbol{\delta})$$

$$\bar{n}_{11} = n_{11} - x_1; \quad \bar{n}_{33} = n_{33} - x_1; \quad \text{where} \quad x_1 = 54C_1 p^{n+1} p_0^{n+1} x_0 \Delta\lambda$$

$$\bar{\mathbf{r}} = \mathbf{r} - \Delta\lambda\frac{\beta}{\omega}\frac{\partial \mathbf{r}}{\partial p_0^{n+1}} = \mathbf{r} - \Delta\lambda x_0(-C_1 p_0^{n+1} r_{kk}^{n=1})(-3\boldsymbol{\delta})$$

$$\bar{r}_{11} = r_{11} - x_2; \quad \bar{r}_{33} = r_{33} - x_2; \quad \text{where} \quad x_2 = 3C_1 p_0^{n+1} r_{kk}^{n+1} x_0 \Delta\lambda$$

$$(\mathbf{RR}_{\sigma_{11}}) = S_1 R_{\sigma_{11}} + 2S_3 R_{\sigma_{33}}; \quad (\mathbf{RR}_{\sigma_{33}}) = S_4 R_{\sigma_{11}} + S_2 R_{\sigma_{33}} + S_5 R_{\sigma_{33}}$$

$$(\bar{\mathbf{C}}\bar{\mathbf{B}})_{11} = \bar{C}_1 \bar{B}_{11} + 2\bar{C}_3 \bar{B}_{33}; \quad (\bar{\mathbf{C}}\bar{\mathbf{B}})_{33} = \bar{C}_4 \bar{B}_{11} + \bar{C}_2 \bar{B}_{33} + \bar{C}_5 \bar{B}_{33}$$

$$(\bar{\mathbf{C}}\bar{\mathbf{r}})_{11} = C_{r11} = \bar{C}_1 \bar{r}_{11} + 2\bar{C}_3 \bar{r}_{33}; \quad (\bar{\mathbf{C}}\bar{\mathbf{r}})_{33} = C_{r33} = \bar{C}_4 \bar{r}_{11} + \bar{C}_2 \bar{r}_{33} + \bar{C}_5 \bar{r}_{33}$$

$$\bar{K}_p^* = \bar{\mathbf{n}}^{\mathrm{T}}\bar{\mathbf{C}}\bar{\mathbf{r}} = \bar{n}_{11}(\bar{C}_1 \bar{r}_{11} + 2\bar{C}_3 \bar{r}_{33}) + 2\bar{n}_{33}(\bar{C}_4 \bar{r}_{11} + \bar{C}_2 \bar{r}_{33} + \bar{C}_5 \bar{r}_{33})$$

$$\bar{K}_p = x_0 \gamma \beta$$

$$\gamma = -9p^{n+1}$$

$$\beta = -s^{n+1} = -C_1 p_0^{n+1} r_{kk}^{n+1}$$

$$\bar{K}_p^{**} = \bar{K}_p + K_p^*$$

$$\bar{\mathbf{n}}^T \mathbf{RR}_\sigma = \bar{n}_{11}(S_1 R_{\sigma_{11}} + 2S_3 R_{\sigma_{33}}) + 2\bar{n}_{33}(S_4 R_{\sigma_{11}} + S_2 R_{\sigma_{33}} + S_5 R_{\sigma_{33}})$$

$$\bar{\mathbf{n}}^T \bar{\mathbf{C}}\bar{\mathbf{B}} = \bar{n}_{11}(\bar{C}_1 \bar{B}_{11} + 2\bar{C}_3 \bar{B}_{33}) + 2\bar{n}_{33}(\bar{C}_4 \bar{B}_{11} + \bar{C}_2 \bar{B}_{33} + \bar{C}_5 \bar{B}_{33})$$

Quantities Needed for the Calculation of Consistent Elasto-Plastic Tangent Operator
Let us write (12.52) as,

$$\mathbf{D} = \bar{\mathbf{C}} - \frac{(\bar{\mathbf{C}}\bar{\mathbf{r}})(\bar{\mathbf{n}}^T \bar{\mathbf{C}})}{x_0 \gamma \beta + \bar{\mathbf{n}}^T \bar{\mathbf{C}}\bar{\mathbf{r}}} = \bar{\mathbf{C}} - \frac{(\bar{\mathbf{C}}\bar{\mathbf{r}})(\bar{\mathbf{C}}^T \bar{\mathbf{n}})^T}{\bar{K}_p^{**}}$$

where,

$$(\bar{\mathbf{C}}^T \bar{\mathbf{n}})_{11} = C_{n11} = \bar{C}_1 \bar{n}_{11} + 2\bar{C}_4 \bar{n}_{33}; \quad (\bar{\mathbf{C}}^T \bar{\mathbf{n}})_{33} = C_{n33} = \bar{C}_3 \bar{n}_{11} + (\bar{C}_2 + \bar{C}_5)\bar{n}_{33}$$

Hence we have,

$$(\bar{\mathbf{C}}^T\bar{\mathbf{r}})(\bar{\mathbf{C}}^T\bar{\mathbf{n}})^T = \begin{bmatrix} C_{r11}C_{n11} & C_{r11}C_{n33} & C_{r11}C_{n33} \\ C_{r33}C_{n11} & C_{r33}C_{n33} & C_{r33}C_{n33} \\ C_{r33}C_{n11} & C_{r33}C_{n33} & C_{r33}C_{n33} \end{bmatrix} = (\text{say}) \begin{bmatrix} E_1 & E_3 & E_3 \\ E_4 & E_2 & E_5 \\ E_4 & E_5 & E_2 \end{bmatrix}$$

$$\mathbf{D} = \begin{bmatrix} D_1 & D_3 & D_3 \\ D_4 & D_2 & D_5 \\ D_4 & D_5 & D_2 \end{bmatrix} = \begin{bmatrix} \bar{C}_1 & \bar{C}_3 & \bar{C}_3 \\ \bar{C}_4 & \bar{C}_2 & \bar{C}_5 \\ \bar{C}_4 & \bar{C}_5 & \bar{C}_2 \end{bmatrix} - \frac{1}{\bar{K}_p^{**}} \begin{bmatrix} E_1 & E_3 & E_3 \\ E_4 & E_2 & E_5 \\ E_4 & E_5 & E_2 \end{bmatrix}$$

$$D_1 = \bar{C}_1 - E_1/\bar{K}_p^{**}; \quad D_2 = \bar{C}_2 - E_2/\bar{K}_p^{**}; \quad D_3 = \bar{C}_3 - E_3/\bar{K}_p^{**};$$
$$D_4 = \bar{C}_4 - E_4/\bar{K}_p^{**}; \quad D_5 = \bar{C}_5 - E_5/\bar{K}_p^{**}$$

The use of these equations to calculate stresses are illustrated in Example 12.4, and to carry out global iterations in Example 12.5. These are also the subjects in Problems 12.6 and 12.7, respectively.

Example 12.4 Local Iterations by CPPM with Approximate Elastic Properties.

Question: Given $\sigma_1^n = 114$, $\sigma_3^n = 78$, $p_0^n = 100$, $\Delta\varepsilon_1 = 0.01$, and $\Delta\varepsilon_3 = -0.001418$, determine the corresponding stresses and the PIV by CPPM with approximate elastic properties using $\theta = 0.5$. Carry out the iterations a few times and examine the convergence behavior. Use the values listed in Table 12.1 for the model parameters. Assume triaxial conditions.

Answer: As pointed out earlier, when $\theta > 0$, iterations are required even in finding the elastic predictor. The elastic predictor is determined by a simple, direct iteration procedure. The results are $\sigma_1^{\text{tr}} = 233.32$, $\sigma_3^{\text{tr}} = 108.69$, and $p^{\text{tr}} = 150.24$. The elastic properties are then computed as $p_{n+\theta} = 120.11$, $K_{n+\theta} = 8{,}408.3$, and $G_{n+\theta} = 3{,}881.0$. Then the following parameters are computed: $f_1 = 0.6923$, $f_2 = 0.9231$, $N = 0.2309$, $r_{11} = 2159.26$, $r_{33} = -177.49$, $A_1 = 2.73 \times 10^7$, $A_3 = 0.912 \times 10^7$, $\bar{K}_p = 4.66 \times 10^9$, $\bar{K}_p^* = 55.63 \times 10^9$, $\bar{K}_p^{**} = 60.28 \times 10^9$, $R_\phi = 165{,}000.8$, $\Delta\lambda = R_\phi/\bar{K}_p^{**} = 2.74 \times 10^{-6}$, $\sigma_1^1 = 158.7$, $\sigma_3^1 = 83.7$, $s = 3.45 \times 10^6$, and $p_{0,1} = 109.43$.

With these initial estimates, the iterations begin and the results are presented in Table 12.5, with the main results listed in Table 12.5a and b, and the errors listed in Table 12.5c. The results converged in four iterations to within an error of 1.0×10^{-3}. When the error gets too small, the ratios presented in the tables do not behave very well. As the results converged very quickly, it is hard to conclude which ratio remains a constant. However, it appears that the rate of convergence is between quadratic and cubic.

Table 12.5a Local iterations (Algorithm 12.3) using CPPM with approximate elastic properties (Example 12.4)

Quantity	Iter. 1	Iter. 2	Iter. 3	Iter. 4	Iter. 5
p^{n+1}	108.72	97.23	94.89	94.97	94.97
q^{n+1}	74.99	53.26	51.08	50.90	50.91
R_{ϕ}	0.344E+05	0.445E+04	0.109E+03	0.200E+00	0.205E−04
$p^{n+\theta}$	99.36	93.61	92.45	92.48	92.48
$K_{n+\theta}$	6,955.07	6,553.01	6,471.17	6,473.93	6,473.89
$G_{n+\theta}$	3,210.26	3,024.68	29,812.91	2,988.18	2,988.16
$\bar{K}_{n+\theta}$	4,814.90	4,536.56	4,479.90	4,481.81	4,481.78
$\bar{K}_1$	11,235.42	10,585.92	10,453.71	10,458.17	10,458.11
K_1	1,602.15	1,509.53	1,490.68	1,491.32	1,491.31
r_{11}	1,261.32	911.89	866.25	864.37	864.38
r_{33}	−144.66	−86.72	−91.46	−90.09	−90.11
r_{kk}	972.01	738.44	683.33	684.19	684.17
A_1	0.128E+08	0.887E+07	0.824E+07	0.823E+07	0.823E+07
A_3	0.375E+07	0.283E+07	0.251E+07	0.253E+07	0.253E+07
s	0.203E+07	0.159E+07	0.149E+07	0.149E+07	0.149E+07
σ_1^2	177.73	147.67	130.45	128.91	128.91
σ_3^2	93.12	83.02	78.61	78.00	78.00
$p_{0,1}$	105.56	110.61	113.60	113.92	113.92
$R_{\sigma_{11}}$	−19.02	−14.94	−1.51	0.00	0.00
$R_{\sigma_{33}}$	−9.40	−3.54	−0.74	0.00	0.00
R_{ς}	3.87	1.80	0.26	0.00	0.00
γ	−0.978E+03	−0.875E+03	−0.854E+03	−0.855E+03	−0.855E+03
β	−0.203E+07	−0.159E+07	−0.149E+07	−0.149E+07	−0.149E+07
$\partial_{p_0} s$	−0.246E+03	−0.522E+04	−0.652E+04	−0.651E+04	−0.651E+04
x_0	0.999E+00	0.966E+00	0.944E+00	0.943E+00	0.943E+00
χ_0^*	0.588E–02	0.329E–01	0.601E–01	0.627E–01	0.627E–01
x_1	0.336E+02	0.729E+02	0.962E+02	0.983E+02	0.983E+02
x_2	0.167E+02	0.307E+02	0.385E+02	0.394E+02	0.394E+02
y_1	0.117E+02	0.117E+02	0.117E+02	0.117E+02	0.117E+02
t_{11}	0.107E+00	0.599E−01	0.297E−01	0.269E−01	0.269E−01
t_{33}	0.254E−01	0.893E−02	0.109E−02	−0.281E−05	−0.487E−06

Table 12.5b Local iterations using CPPM (Algorithm 12.3) with approximate elastic properties: continuation from Table 12.5a

Quantity	Iter. 1	Iter. 2	Iter. 3	Iter. 4	Iter. 5
T_1	0.122E+01	0.168E+01	0.195E+01	0.197E+01	0.197E+01
T_2	0.130E+01	0.173E+01	0.198E+01	0.200E+01	0.200E+01
T_3	−0.108E+00	−0.827E−01	−0.761E−01	−0.756E−01	−0.756E−01
T_4	−0.269E−01	−0.317E−01	−0.475E−01	−0.487E−01	−0.487E−01
S_1	0.822E+00	0.598E+00	0.514E+00	0.508E+00	0.508E+00
S_2	0.769E+00	0.580E+00	0.506E+00	0.501E+00	0.501E+00
S_3	0.698E−01	0.292E−01	0.203E−01	0.197E−01	0.197E−01
S_4	0.173E−01	0.112E−01	0.127E−01	0.127E−01	0.127E−01
S_5	0.173E−01	0.112E−01	0.127E−01	0.127E−01	0.127E−01
$\bar{C}_1$	0.991E+04	0.659E+04	0.556E+04	0.549E+04	0.549E+04
$\bar{C}_2$	0.881E+04	0.624E+04	0.541E+04	0.535E+04	0.535E+04
$\bar{C}_3$	0.508E+04	0.315E+04	0.261E+04	0.257E+04	0.257E+04
$\bar{C}_4$	0.398E+04	0.280E+04	0.246E+04	0.244E+04	0.244E+04

(continued)

Table 12.5b (continued)

Quantity	Iter. 1	Iter. 2	Iter. 3	Iter. 4	Iter. 5
$\bar{C}_5$	0.398E+04	0.280E+04	0.246E+04	0.244E+04	0.244E+04
$\bar{n}_{11}$	0.123E+04	0.839E+03	0.770E+03	0.766E+03	0.766E+03
$\bar{n}_{33}$	−0.178E+03	−0.160E+03	−0.188E+03	−0.188E+03	−0.188E+03
$\bar{r}_{11}$	0.124E+04	0.881E+03	0.828E+03	0.825E+03	0.825E+03
$\bar{r}_{33}$	−0.161E+03	−0.117E+03	−0.130E+03	−0.129E+03	−0.129E+03
$\bar{K}_p^*$	0.121E+11	0.380E+10	0.264E+10	0.258E+10	0.258E+10
$\bar{K}_p$	0.199E+10	0.134E+10	0.120E+10	0.120E+10	0.120E+10
$\bar{K}_p^{**}$	0.141E+11	0.514E+10	0.384E+10	0.378E+10	0.378E+10
B_{11}	−0.821E−05	−0.201E−04	−0.275E−04	−0.280E−04	−0.280E−04
B_{33}	−0.821E−05	−0.201E−04	−0.275E−04	−0.280E−04	−0.280E−04
$\bar{\mathbf{n}}^T\mathbf{RR}_\sigma$	−0.181E+05	−0.694E+04	−0.468E+03	−0.731E+00	−0.734E−06
$\bar{\mathbf{n}}^T\bar{\mathbf{C}}\bar{\mathbf{B}}$	−0.153E+03	−0.141E+03	−0.121E+03	−0.120E+03	−0.120E+03
$\delta\lambda$	0.395E−05	0.246E−05	0.197E−06	0.148E−09	0.694E−14
$\delta\sigma_1$	−25.97	−3.79	−0.04	0.00	0.00
$\delta\sigma_3$	−4.25	−1.61	0.14	0.00	0.00
δp_0	2.98	1.45	0.06	0.00	0.00
$\delta\lambda$	0.669E−05	0.915E−05	0.935E−05	0.935E−05	0.935E−05
σ_1^1	132.74	128.94	128.91	128.91	128.91
σ_3^1	79.48	77.86	78.00	78.00	78.00
$p_{0,1}$	112.41	113.86	113.92	113.92	113.92

Table 12.5c Local iterations using CPPM (Algorithm 12.3) with approximate elastic properties: errors

Quantity	Iter. 1	Iter. 2	Iter. 3	Iter. 4	Iter. 5
e_σ	0.117E+00	0.906E−01	0.108E−01	0.224E−04	0.175E−09
e_ς	0.354E−01	0.160E−01	0.224E−02	0.472E−05	0.635E−10
e_ϕ	0.344E+05	0.445E+04	0.109E+03	0.200E+00	0.205E−04
$(e_\sigma)_{k+1}/(e_\sigma)_k$	–	0.773E+00	0.120E+00	0.207E−02	0.782E−05
$(e_\varsigma)_{k+1}/(e_\varsigma)_k$	–	0.453E+00	0.140E+00	0.211E−02	0.135E−04
$(e_\sigma)_{k+1}/(e_\sigma)_k^2$	–	0.659E+01	0.132E+01	0.191E+00	0.349E+00
$(e_\varsigma)_{k+1}/(e_\varsigma)_k^2$	–	0.128E+02	0.873E+01	0.940E+00	0.285E+01
$(e_\sigma)_{k+1}/(e_\sigma)_k^3$	–	0.563E+02	0.146E+02	0.176E+02	0.156E+05
$(e_\varsigma)_{k+1}/(e_\varsigma)_k^3$	–	0.362E+03	0.545E+03	0.420E+03	0.605E+06

Example 12.5 Global Iterations (Algorithm 8.2) Using CPPM with Approximate Elastic Properties for the Local Iterations (Algorithm 12.3).

Question: A strain-controlled, triaxial drained compression loading is to be conducted with $\sigma_1^n = 114$, $\sigma_3^n = 78$, and $p_0^n = 100$. Determine the stresses and strains at the end of the incremental loading with $\Delta\hat{\varepsilon}_1 = 0.05$ and $\Delta\hat{\sigma}_3 = 0.0$. Use Algorithm 8.2 for global iterations with consistent tangent operator and Algorithm 12.3 for local iterations (CPPM with approximate elastic properties) with $\theta = 0.5$. Use the specific triaxial equations developed in Sect. 12.4.2 to perform the calculations. Examine the global convergence behavior of the algorithm. Investigate the effectiveness of using the continuum operator over the consistent operator. Use the values listed in Table 12.1 for the model parameters.

Answer. First, the variables are initialized as (Step 2 of Algorithm 8.2): $\sigma_1^n = \sigma_1^{n+1} = 114$, $\sigma_3^n = \sigma_3^{n+1} = 78$ and $p_0^n = p_0^{n+1} = 100$, $\sigma_{E1}^{n+1} = 114 + 0 = 114$, $\sigma_{E3}^{n+1} = 78 + 0 = 78$,

$$\sigma_{S1}^n = \sigma_{S1}^{n+1} = \sigma_1^n = 114,$$

$$\sigma_{S3}^n = \sigma_{S3}^{n+1} = \sigma_3^n = 78,$$

$$\delta\sigma_1 = \sigma_{E1}^{n+1} - \sigma_{S1}^{n+1} = 0 \quad \text{and} \quad \delta\sigma_3 = \sigma_{E3}^{n+1} - \sigma_{S3}^{n+1} = 0$$

Note that in the above equations, the subscripts "E" and "S" indicate "external force" and "spring force (internal)," respectively. The values of various quantities during the global iteration are summarized in Table 12.6 for the analysis using the consistent operator and in Table 12.7 for analysis using the continuum operator.

Also recall that when the loading is of the mixed type as in the present case (the stress is controlled in direction 3 and the strain is controlled in direction 1), the **D** matrix and the $\delta\boldsymbol{\sigma}$ vector are modified as described in Sect. 8.6. In the present case, the coefficients D_1 and $\delta\sigma_1$ are modified; the modified coefficients are denoted as D_1^* and $\delta\sigma_1^*$ in Tables 12.6 and 12.7.

Table 12.6 Global iterations using consistent tangent operator and with CPPM with approximate elastic properties for local iterations (Example 12.5)

Quantity	Iter. 1	Iter. 2	Iter. 3
D_1	10,177.20	3,422.73	3,526.35
D_2	10,177.20	3,140.53	3,242.24
D_3	4,361.40	3,547.14	3,648.47
D_4	4,361.40	2,038.12	2,100.70
D_5	4,361.40	1,939.05	2,017.29
D_1^*	0.100E+21	0.100E+21	0.100E+21
$\delta\sigma_1^*$	0.500E+19	0.000E+00	0.000E+00
$\delta\varepsilon_1$	0.500E−01	−0.340E−19	0.595E−21
$\delta\varepsilon_3$	−0.150E−01	0.480E−03	−0.815E−05
$\Delta\varepsilon_1$	0.500E−01	0.500E−01	0.500E−01
$\Delta\varepsilon_3$	−0.150E−01	−0.145E−01	−0.145E−01
$\Delta\sigma_1$	46.56	50.01	49.96
$\Delta\sigma_3$	−2.44	0.04	0.00
Δp_0	52.19	54.80	54.76
σ_1^{n+1}	160.56	164.01	163.96
σ_3^{n+1}	75.56	78.04	78.00
p_0^{n+1}	152.19	154.80	154.76
σ_{S1}^{n+1}	160.56	164.01	163.96
σ_{S3}^{n+1}	75.56	78.04	78.00
$\delta\sigma_1$	−46.56	−50.01	−49.96
$\delta\sigma_3$	2.44	−0.04	0.00
e_σ	0.217E−01	0.382E−03	0.113E−06
e_e	−0.731E−01	0.125E−02	0.370E−06
$(e_\sigma)_{k+1}/(e_\sigma)_k$	0.000E+00	0.176E−01	0.297E−03
$(e_\sigma)_{k+1}/(e_\sigma)_k^2$	0.00	0.809E+00	0.776E+00
$(e_\sigma)_{k+1}/(e_\sigma)_k^3$	0.00	0.372E+02	0.203E+04

Table 12.7 Global iterations using continuum tangent operator and using CPPM with approximate elastic properties for local iterations (Example 12.5)

Quantity	Iter. 1	Iter. 2	Iter. 3	Iter. 4	Iter. 5	Iter. 32
D_1	10,177.20	3,772.44	3,770.71	3,769.52	3,768.69	3,766.82
D_2	10,177.20	11,748.13	11,842.44	11,907.91	11,953.36	12,056.73
D_3	4,361.40	4,980.86	4,986.86	4,990.83	4,993.49	4,999.26
D_4	4,361.40	4,980.86	4,986.86	4,990.83	4,993.49	4,999.26
D_5	4,361.40	5,034.41	5,074.48	5,102.19	5,121.39	5,164.90
D_1^*	0.100E+21	0.100E+21	0.100E+21	0.100E+21	0.100E+21	0.100E+21
$\delta\sigma_1^*$	0.500E+19	0.000E+00	0.000E+00	0.000E+00	0.000E+00	0.000E+00
$\delta\varepsilon_1$	0.500E−01	−0.145E−19	−0.100E−19	−0.692E−20	−0.480E−20	−0.256E−24
$\delta\varepsilon_3$	−0.150E−01	0.145E−03	0.100E−03	0.693E−04	0.480E−04	0.256E−08
$\Delta\varepsilon_1$	0.500E−01	0.500E−01	0.500E−01	0.500E−01	0.500E−01	0.500E−01
$\Delta\varepsilon_3$	−0.150E−01	−0.149E−01	−0.148E−01	−0.147E−01	−0.146E−01	−0.145E−01
$\Delta\sigma_1$	46.56	47.60	48.32	48.82	49.16	49.96
$\Delta\sigma_3$	−2.44	−1.70	−1.18	−0.82	−0.57	0.00
Δp_0	52.19	52.97	53.52	53.90	54.16	54.76
σ_1^{n+1}	160.56	161.60	162.32	162.82	163.16	163.96
σ_3^{n+1}	75.56	76.30	76.30	77.18	77.43	78.00
p_0^{n+1}	152.19	152.97	153.52	153.90	154.16	154.76
σ_{S1}^{n+1}	160.56	161.60	162.32	162.82	163.16	163.96
σ_{S3}^{n+1}	75.56	76.30	76.30	77.18	77.43	78.00
$\delta\sigma_1$	−46.56	−47.60	−48.32	−48.82	−49.16	−49.96
$\delta\sigma_3$	2.44	1.70	1.18	0.82	0.57	0.00
e_σ	0.217E−01	0.151E−01	0.105E−01	0.731E−02	0.508E−02	0.273E−06
e_e	−0.731E−01	−0.504E−01	−0.348E−01	−0.241E−01	−0.167E−01	−0.890E−06
$(e_\sigma)_{k+1}/(e_\sigma)_k$	0.000E+00	0.696E+00	0.695E+00	0.695E+00	0.695E+00	0.695E+00
$(e_\sigma)_{k+1}/(e_\sigma)_k^2$	0.0	0.320E+02	0.460E+02	0.661E+02	0.951E+02	0.177E+07
$(e_\sigma)_{k+1}/(e_\sigma)_k^3$	0.0	0.147E+04	0.304E+04	0.629E+04	0.130E+05	0.450E+13

When the consistent operator is used, the results converge in three iterations. The error is too small to reliably evaluate the ratios. However, the results seem to point to a quadratic rate of convergence. When the continuum operator is used, the results unequivocally indicate a linear convergence (Table 12.7). It takes 32 iterations to reduce e_σ to below 1.0×10^{-4}.

12.4.3 Sample Model Simulations

Sample simulations are performed using a uniformly loaded single-element finite element model. The problems analyzed in this section are identical to those in Sects. 12.2.3 and 12.3.3, and the reader is referred to Sect. 12.2.3 for the definition of the problems. The problems are analyzed here using CPPM with approximate elastic properties, summarized in Algorithm 12.3 for the calculation of elasto-plastic stresses. The following control parameters are used: $N_L^{\max} = 100$, $TOL = 1.0 \times 10^{-10}$, $TOL_\sigma = 0.01$, $TOL_\varsigma = 0.01$, $TOL_\phi = 1.0 \times 10^{-5} p_0$, and $\Gamma = 1.0 \times 10^8$ (combined bulk modulus used to simulate undrained behavior as described in Sect. 12.2.3). The midpoint rule ($\theta = 0.5$) is used for the elastic properties. The values listed in Table 12.1 are used for the model parameters.

The results are presented in Figs. 12.9 and 12.10 for the drained and undrained simulations, respectively. The results obtained with a step size $\Delta\varepsilon_1 = 2.5\%$ (which led to a total of 24 and 12 steps in drained and undrained, respectively) and a step size $\Delta\varepsilon_1 = 0.05\%$ (which led to a total of 1,200 and 600 steps in drained and undrained, respectively) are compared with the "exact" solutions. As is the case with the two previous methods, in all cases considered here, the results with a step size $\Delta\varepsilon_1 = 0.05\%$ and the "exact" results are indistinguishable. The results with a step size $\Delta\varepsilon_1 = 2.5\%$ are further discussed below.

As in the case with the cutting-plane algorithm, convergence was achieved in all cases considered here, illustrating the robustness of the algorithm. When the step size is too large ($\Delta\varepsilon = 10\%$), however, small error tolerances (e.g., 1.0×10^{-5} for e_σ) could not be achieved near failure. The tolerance had to be increased to continue the analysis. Except for accuracy, the performance of the method is very similar to the performance of the two-step Euler method discussed in Sect. 12.2.3 and the cutting-plane method discussed in Sect. 12.3.3. While the results seem to be almost as accurate as those obtained with the two-step Euler method in many respects, accuracy of volumetric strain calculated during the drained simulations is not as good with a step size $\Delta\varepsilon_1 = 2.5\%$ (Fig. 12.9).

During the simulation of octahedral plane circular stress path (such as the one shown in Fig. 12.5), the method performed as good as the two-step Euler method and the cutting-plane method. The results using 36 steps and 1,800 steps gave indistinguishable results. The results, however, slightly varied from those obtained using the two-step Euler method. The results are not presented as they do not reveal new information.

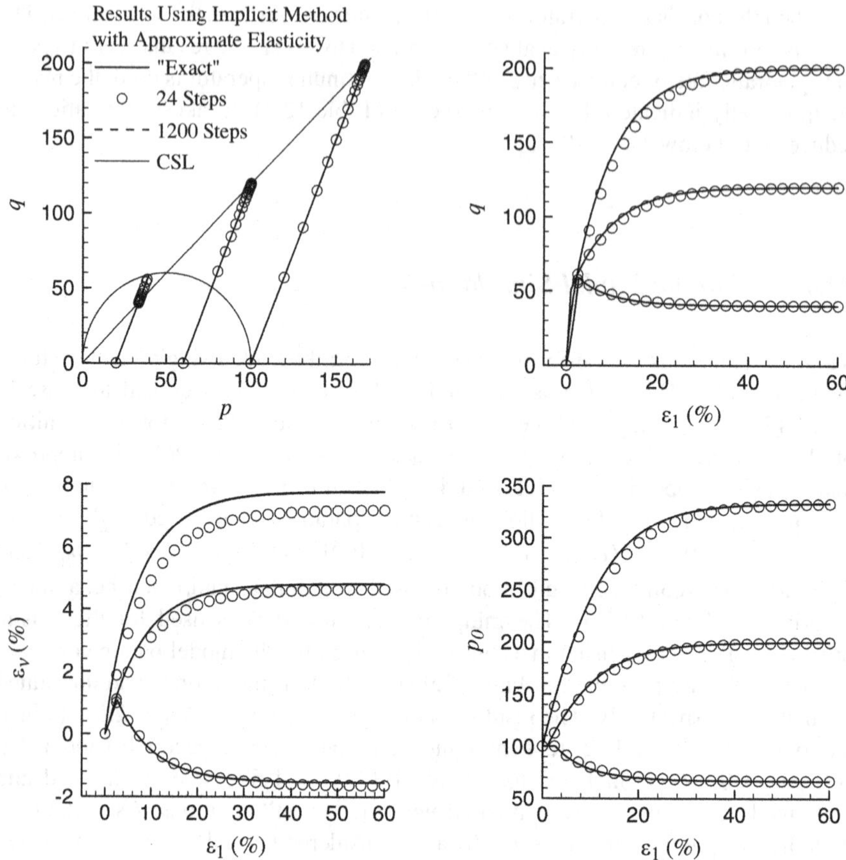

Fig. 12.9 Sample drained triaxial model simulations using CPPM with approximate elastic properties

12.5 The Closest Point Projection (Implicit) Method with Exact Elastic properties: The σ-Space Formulation

12.5.1 Calculation of Elasto-Plastic Stresses

The reader is referred to Sect. 12.4 for a discussion on the use of approximate and exact elastic properties in the application of CPPM to the modified Cam-clay model.

When the plastic strain is zero, the elastic predictor becomes the actual elastic stress. The exact expression for the elastic predictor is presented in Appendix 8.4. It is shown there that

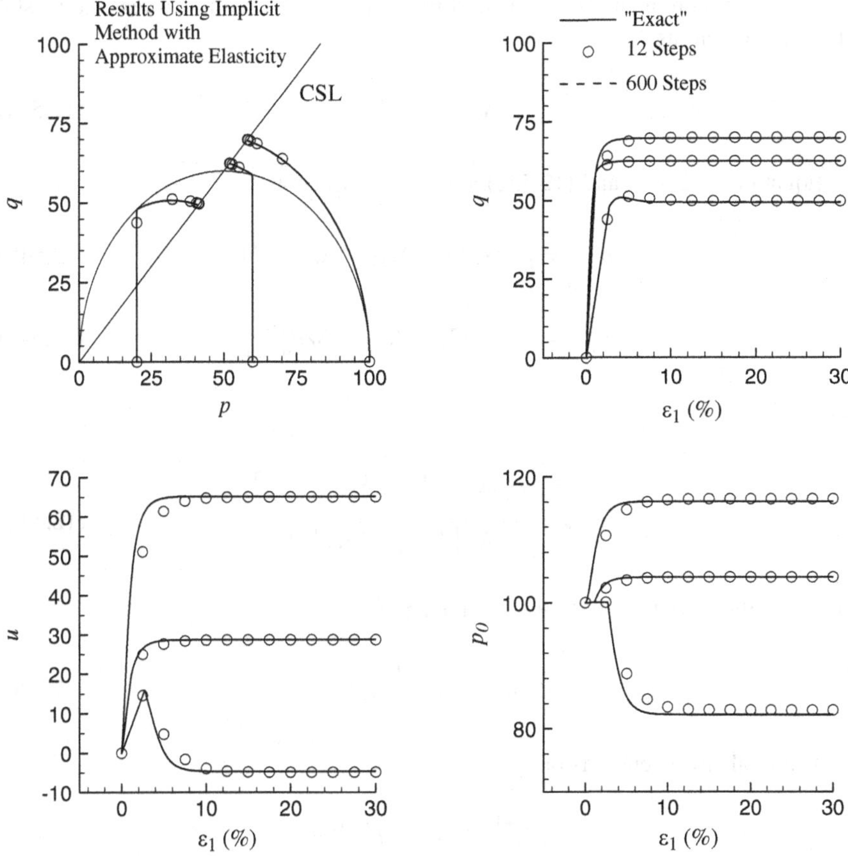

Fig. 12.10 Sample undrained triaxial model simulations using CPPM with approximate elastic properties

$$p^{n+1} = p^n \exp[\bar{K}_0 \Delta\varepsilon_{kk}] \tag{12.53a}$$

$$s_{ij}^{n+1} = s_{ij}^n + 2G_{\text{ave}}\Delta e_{ij} \tag{12.53b}$$

where

$$G_{\text{ave}} = \frac{\bar{G}_0 p^n \left[e^{\bar{K}_0 \Delta\varepsilon_{kk}} - 1\right]}{\bar{K}_0 \Delta\varepsilon_{kk}} \tag{12.53c}$$

The elastic constants $\bar{K}_0$ and $\bar{G}_0$ are defined in (12.2b). $\Delta\varepsilon_{kk}$ and Δe_{ij} are the incremental volumetric and deviatoric strains, respectively.

When the behavior is elasto-plastic, the above equations are modified by subtracting the plastic strain increment from the total strain increment to determine the elastic strain increment. The backward Euler method is used to express the plastic strain increment as

$$\Delta\varepsilon_{ij}^{p} = \Delta\lambda r_{ij}^{n+1} \tag{12.54a}$$

Equations (12.53a) and (12.53b) are then modified as

$$p^{n+1} = p^{n} \exp\left[\bar{K}_0\left\{\Delta\varepsilon_{kk} - \Delta\lambda r_{kk}^{n+1}\right\}\right] \tag{12.54b}$$

$$s_{ij}^{n+1} = s_{ij}^{n} + 2G_{\text{ave}}\left\{\Delta e_{ij} - \Delta\lambda d_{ij}^{n+1}\right\} \tag{12.54c}$$

where

$$G_{\text{ave}} = \frac{\bar{G}_0 p^{n}\left[e^{\bar{K}_0\left\{\Delta\varepsilon_{kk} - \Delta\lambda r_{kk}^{n+1}\right\}} - 1\right]}{\bar{K}_0\left\{\Delta\varepsilon_{kk} - \Delta\lambda r_{kk}^{n+1}\right\}} \tag{12.54d}$$

and d_{ij} is the deviatoric part of r_{ij} defined as

$$d_{ij} = r_{ij} - \frac{1}{3} r_{kk}\delta_{ij} \tag{12.54e}$$

The total stress tensor is obtained as

$$\sigma_{ij}^{n+1} = s_{ij}^{n+1} + p^{n+1}\delta_{ij}$$

For simplicity of notations, let us define

$$\bar{x}_0 = \bar{K}_0\left(\Delta\varepsilon_{kk} - \Delta\lambda r_{kk}^{n+1}\right) \tag{12.55a}$$

$$\Delta y_{ij} = \Delta e_{ij} - \Delta\lambda d_{ij}^{n+1} \tag{12.55b}$$

Then we have

$$\sigma_{ij}^{n+1} = s_{ij}^{n} + 2G_{ave}\Delta y_{ij} + p^{n} e^{\bar{x}_0}\delta_{ij} \tag{12.56a}$$

where

$$G_{ave} = \frac{\bar{G}_0 p^{n}[e^{\bar{x}_0} - 1]}{\bar{x}_0} \tag{12.56b}$$

The hardening rule (12.7a) is integrated as

$$\dot{\varsigma} = \dot{p}_0 = C_1 p_0 \dot{\varepsilon}_v^P = C_1 p_0 r_{kk} \dot{\lambda} \tag{12.57a}$$

$$p_0^{n+1} = p_0^n e^{C_1 \Delta\lambda r_{kk}^{n+1}} \tag{12.57b}$$

Summary of Algorithmic Equations. The algorithmic equations are given by (12.56a), (12.57b), and (12.40c), which are listed below for convenience:

$$\sigma_{ij}^{n+1} = s_{ij}^n + 2G_{ave}\Delta y_{ij} + p^n e^{\bar{x}_0}\delta_{ij} \tag{12.58a}$$

$$p_0^{n+1} = p_0^n e^{C_1 \Delta\lambda r_{kk}^{n+1}} \tag{12.58b}$$

$$\phi^{n+1}(\boldsymbol{\sigma}_{n+1}, p_0^{n+1}) = 0 \tag{12.58c}$$

The Residual Equations and their Linearizations. The residuals $\mathbf{R}_\sigma$, R_ς, and R_ϕ are defined as

$$R_{\sigma_{ij}} = \sigma_{ij}^{n+1} - \left[s_{ij}^n + 2G_{ave}\Delta y_{ij} + p^n e^{\bar{x}_0}\delta_{ij}\right] \tag{12.59a}$$

$$R_\varsigma = p_0^{n+1} - p_0^n e^{C_1 \Delta\lambda r_{kk}^{n+1}} \tag{12.59b}$$

$$R_\phi = \phi^{n+1}(\boldsymbol{\sigma}_{n+1}, \zeta_{n+1}) \tag{12.59c}$$

where $\bar{x}_0$, Δy_{ij}, and G_{ave} are all functions of σ_{ij}^{n+1}, p_0^{n+1}, $\delta\lambda$, and $\Delta\varepsilon_{ij}$, given by (12.55a), (12.55b), and (12.56b), respectively. Linearizing these as done in (12.42), the required expressions for the coefficients **A, B, F, H, E**, ω, β and γ may now be developed as follows:

$$\begin{aligned} R_{\sigma_{ij}} = {} & R_{\sigma_{ij}^1} + \delta_{ip}\delta_{jq}\delta\sigma_{pq} - \left[2\Delta y_{ij}\frac{\partial G_{ave}}{\partial \bar{x}_0} + p^n e^{\bar{x}_0}\delta_{ij}\right]\left[\frac{\partial \bar{x}_0}{\partial \sigma_{pq}^{n+1}}\delta\sigma_{pq} + \frac{\partial \bar{x}_0}{\partial p_0^{n+1}}\delta p_0 + \frac{\partial \bar{x}_0}{\partial \Delta\lambda}\delta\lambda\right] \\ & - 2G_{ave}\left[\frac{\partial \Delta y_{ij}}{\partial \sigma_{pq}^{n+1}}\delta\sigma_{pq} + \frac{\partial \Delta y_{ij}}{\partial p_0^{n+1}}\delta p_0 + \frac{\partial \Delta y_{ij}}{\partial \Delta\lambda}\delta\lambda\right] \end{aligned} \tag{12.60a}$$

$$\begin{aligned} R_\varsigma = {} & R_\varsigma^1 + \delta p_0 - C_1 \Delta\lambda p_0^n e^{C_1 \Delta\lambda r_{kk}^{n+1}}\left[\frac{\partial r_{kk}^{n+1}}{\partial \sigma_{pq}^{n+1}}\delta\sigma_{pq} + \frac{\partial r_{kk}^{n+1}}{\partial p_0^{n+1}}\delta p_0\right] \\ & - C_1 r_{kk}^{n+1} p_0^n e^{C_1 \Delta\lambda r_{kk}^{n+1}}\delta\lambda \end{aligned} \tag{12.60b}$$

$$R_\phi = R_\phi^1 + \frac{\partial \phi}{\partial \sigma_{pq}^{n+1}}\delta\sigma_{pq} + \frac{\partial \phi}{\partial p_0^{n+1}}\delta p_0 \tag{12.60c}$$

Let us define

$$
\begin{aligned}
Z_{ij} &= 2\Delta y_{ij}\frac{\partial G_{ave}}{\partial \bar{x}_0} + p^n e^{\bar{x}_0}\delta_{ij} \\
&= 2\Delta y_{ij}\frac{\bar{G}_0 p^n e^{\bar{x}_0}}{\bar{x}_0} - 2\Delta y_{ij}\frac{\bar{G}_0 p^n (e^{\bar{x}_0}-1)}{\bar{x}_0^2} + p^n e^{\bar{x}_0}\delta_{ij} \\
&= 2\Delta y_{ij}\frac{\bar{G}_0 p^n e^{\bar{x}_0}}{\bar{x}_0} - 2\Delta y_{ij}\frac{G_{ave}}{\bar{x}_0} + p^n e^{\bar{x}_0}\delta_{ij}
\end{aligned}
\tag{12.61}
$$

It then follows that

$$
\frac{\partial \bar{x}_0}{\partial \sigma_{pq}^{n+1}} = -\bar{K}_0\Delta\lambda\frac{\partial r_{kk}^{n+1}}{\partial \sigma_{pq}^{n+1}}; \quad \frac{\partial \bar{x}_0}{\partial p_0^{n+1}} = -\bar{K}_0\Delta\lambda\frac{\partial r_{kk}^{n+1}}{\partial p_0^{n+1}}; \quad \frac{\partial \bar{x}_0}{\partial \Delta\lambda} = -\bar{K}_0 r_{kk}^{n+1} \tag{12.62a}
$$

$$
\frac{\partial \Delta y_{ij}}{\partial \sigma_{pq}^{n+1}} = -\Delta\lambda\frac{\partial d_{ij}^{n+1}}{\partial \sigma_{pq}^{n+1}}; \quad \frac{\partial \Delta y_{ij}}{\partial p_0^{n+1}} = -\Delta\lambda\frac{\partial d_{ij}^{n+1}}{\partial p_0^{n+1}}; \quad \frac{\partial \Delta y_{ij}}{\partial \Delta\lambda} = -d_{ij}^{n+1} \tag{12.62b}
$$

Summary of the Expressions for the Coefficients.
Equations (12.60–12.62) are now arranged to obtain the required coefficients as follows:

$$
A_{ijpq} = \delta_{ip}\delta_{jq} + \bar{K}_0\Delta\lambda Z_{ij}\frac{\partial r_{kk}^{n+1}}{\partial \sigma_{pq}^{n+1}} + 2\Delta\lambda G_{\text{ave}}\frac{\partial d_{ij}^{n+1}}{\partial \sigma_{pq}^{n+1}} \tag{12.63a}
$$

$$
B_{ij} = \bar{K}_0\Delta\lambda Z_{ij}\frac{\partial r_{kk}^{n+1}}{\partial p_0^{n+1}} + 2\Delta\lambda G_{\text{ave}}\frac{\partial d_{ij}^{n+1}}{\partial p_0^{n+1}} \tag{12.63b}
$$

$$
F_{ij} = \bar{K}_0 r_{kk}^{n+1} Z_{ij} + 2G_{\text{ave}} d_{ij}^{n+1} \tag{12.63c}
$$

$$
H_{pq} = -C_1\Delta\lambda p_0^n e^{C_1\Delta\lambda r_{kk}^{n+1}}\frac{\partial r_{kk}^{n+1}}{\partial \sigma_{pq}^{n+1}} \tag{12.63d}
$$

$$
\omega = 1 - C_1\Delta\lambda p_0^n e^{C_1\Delta\lambda r_{kk}^{n+1}}\frac{\partial r_{kk}^{n+1}}{\partial p_0^{n+1}} \tag{12.63e}
$$

$$
\beta = -C_1 r_{kk}^{n+1} p_0^n e^{C_1\Delta\lambda r_{kk}^{n+1}} \tag{12.63f}
$$

$$
E_{pq} = \frac{\partial \phi}{\partial \sigma_{pq}^{n+1}} = n_{pq}^{n+1} \tag{12.63g}
$$

$$
\gamma = \frac{\partial \phi}{\partial p_0^{n+1}} \tag{12.63h}
$$

Surface properties such as $\frac{\partial r_{kk}^{n+1}}{\partial \sigma_{pq}^{n+1}}$ are independent of the integration method employed and are presented in Appendices 7, 8.1, and 8.2.

The Consistent Tangent Operator.
In (12.59), setting $\mathbf{R}_\sigma$, R_ς, and R_ϕ to zero, taking the rate of the equations and arranging the quantities, we have

$$\mathbf{R}_\sigma^{*I} + \mathbf{A}^*\dot{\boldsymbol{\sigma}} + \mathbf{B}^*\dot{\varsigma} + \mathbf{F}^*\dot{\lambda} = \boldsymbol{0}$$

$$R_\varsigma^{*I} + \mathbf{H}^{*^T}\dot{\boldsymbol{\sigma}} + \omega^*\dot{\varsigma} + \beta^*\dot{\lambda} = 0$$

$$R_\phi^{*I} + \mathbf{E}^{*^T}\dot{\boldsymbol{\sigma}} + \gamma^*\dot{\varsigma} = 0$$

It can be easily shown that $\mathbf{A}^* = \mathbf{A}$, $\mathbf{B}^* = \mathbf{B}$, $\mathbf{F}^* = \mathbf{F}$, $\mathbf{H}^* = \mathbf{H}$, $\mathbf{E}^* = \mathbf{E}$, $\omega^* = \omega, \beta^* = \beta, \gamma^* = \gamma$ and

$$R_\varsigma^* = 0, \quad R_\phi^* = 0 \text{ and } R_\sigma^{*1} = \mathbf{C}^*\dot{\boldsymbol{\varepsilon}} \tag{12.64}$$

where

$$C_{ijpq}^* = Z_{ij}\frac{\partial \bar{x}_0}{\partial \varepsilon_{pq}} + 2G_{\text{ave}}\frac{\partial \Delta y_{ij}}{\partial \varepsilon_{pq}}$$

Noting that

$$\frac{\partial \bar{x}_0}{\partial \varepsilon_{pq}} = \bar{K}_0\frac{\partial \varepsilon_{kk}}{\partial \varepsilon_{pq}} = \bar{K}_0\delta_{pq}$$

and

$$\frac{\partial \Delta y_{ij}}{\partial \varepsilon_{pq}} = \frac{\partial \Delta e_{ij}}{\partial \varepsilon_{pq}} = \delta_{ip}\delta_{jq} - \frac{1}{3}\delta_{ij}\delta_{pq}$$

it follows that

$$C_{ijpq}^* = \bar{K}_0 Z_{ij}\delta_{pq} + 2G_{ave}\left[\delta_{ip}\delta_{jq} - \frac{1}{3}\delta_{ij}\delta_{pq}\right] \tag{12.65a}$$

Writing (12.65a) in a form similar to (12.1b), we have,

$$C_{ijpq}^* = \left[\bar{K}_0 Z_{ij} - \frac{2}{3}G_{ave}\delta_{ij}\right]\delta_{pq} + G_{ave}\left[\delta_{ip}\delta_{jq} + \delta_{iq}\delta_{jp}\right] \tag{12.65b}$$

Note that $\mathbf{C}$ must be replaced with $\mathbf{C}^*$ and $\bar{\mathbf{C}}$ must be replaced with $\bar{\mathbf{C}}^* = \mathbf{R}\mathbf{C}^*$. With these equivalences, the previous equations can be applied to the present case as well, as discussed below.

The Implementation Details.
With $\mathbf{C} = \mathbf{C}^*$, the equations in Sect. 12.4.1 remains valid for the present case as well.

Equation (12.48h) is modified as $\bar{\mathbf{C}}^* = \mathbf{R}\mathbf{C}^*$. $\bar{\mathbf{C}}$ is replaced by $\bar{\mathbf{C}}^*$ in (12.48a)–(12.48i).

The details can be summarized as follows:

- Equation (12.47) represents the system equation for iterative quantities $\delta\boldsymbol{\sigma}$, δp_0, and $\delta\lambda$.
- Equations (12.48a–12.48c) remain valid for solving for $\delta\boldsymbol{\sigma}$, δp_0, and $\delta\lambda$.
- Equations (12.48d–12.48i) define the quantities in (12.48a–12.48c).
- Equation (12.49) gives an expression for the trial value for $\Delta\lambda$.
- Equations (12.50a–12.50c) remain valid as update equations.
- Equations (12.51a–12.51c) give expressions for evaluating the errors.
- Equation (12.52) gives an expression for the consistent tangent operator.

Even when the model employs an associated flow rule (i.e., $\mathbf{r} = \mathbf{n}$), we have $\bar{\mathbf{r}} \neq \bar{\mathbf{n}}$ and $\bar{\mathbf{C}} \neq \bar{\mathbf{C}}^T$. Thus $\mathbf{D}$ is non symmetric. The expressions for $\bar{\mathbf{F}}$ and $\bar{\mathbf{B}}$ are developed in Problem 12.8. Note that as in the case of CPPM with approximate elastic properties (10.60a), $\bar{\mathbf{F}} = \mathbf{r}$ in the present case as well. The procedure of calculating the stresses by CPPM with exact elastic properties is summarized in Algorithm 12.4.

Algorithm 12.4. CPPM for Integrating the Constitutive Relations from the Modified Cam-Clay Model with Exact Elastic Properties
Definition of (new) parameters: $N_L^{\max}$ is the maximum number of iterations permitted before the execution is terminated, and N_L is the corresponding counter. TOL is a small, machine-dependent number (it could be set equal to TOL_ϕ). Iflag = 1 if converged and 0 otherwise.

1. Establish $\Delta\boldsymbol{\varepsilon}$, p_0^n, $\boldsymbol{\sigma}_n$, TOL_σ, TOL_ς, TOL_ϕ, TOL, $N_L^{\max}$, and constitutive model parameters.
2. Initialize parameters: $N_L = 0$ and $p_0^{n+1} = p_0^n$.
3. Compute the exact elastic predictor $\boldsymbol{\sigma}_{n+1}^{\text{tr}}$ and $\phi^{\text{tr}}(\boldsymbol{\sigma}_{n+1}^{\text{tr}}, p_0^n)$. Perform loading/unloading check: if $(\phi^{\text{tr}} < TOL)$, set $\boldsymbol{\sigma}_{n+1} = \boldsymbol{\sigma}_{n+1}^{\text{tr}}$ and go to step 12.
4. Set $\boldsymbol{\sigma}_{n+1} = \boldsymbol{\sigma}_{n+1}^{\text{tr}}$ and calculate $\mathbf{r}^{n+1}$.
5. Set $R_\phi^1 = \phi^{\text{tr}}$. Compute the trial value for $\Delta\lambda$. Compute $\boldsymbol{\sigma}_{n+1}^1$ and $p_{0,1}^{n+1}$. Set $\boldsymbol{\sigma}_{n+1} = \boldsymbol{\sigma}_{n+1}^1$ and $p_0^{n+1} = p_{0,1}^{n+1}$.
6. Begin iterations. Compute $R_\phi^1 = \phi(\boldsymbol{\sigma}_{n+1}, p_0^{n+1})$ and $\mathbf{r}^{n+1}$.

(continued)

7. Compute $\boldsymbol{\sigma}^2_{n+1}$ and $p^{n+1}_{0,2}$.
8. Compute $\boldsymbol{R}^1_\sigma = \boldsymbol{\sigma}^1_{n+1} - \boldsymbol{\sigma}^2_{n+1}$ and $R^1_\varsigma = p^{n+1}_{0,1} - p^{n+1}_{0,2}$.
9. Find errors e_σ, e_ς, and e_ϕ and perform convergence check: $e_\sigma \leq TOL_\sigma$, $e_\varsigma \leq TOL_\varsigma$, and $\phi^{i+1} \leq TOL_\phi$. If converged, set iflag $= 1$ and go to step 12.
10. Set $N_L \leftarrow N_L + 1$. If $(N_L > N_L^{\max})$, set iflag $= 0$ and go to step 12.
11. Compute $\delta\lambda$, $\delta\boldsymbol{\sigma}$, and δp_0. Update $\boldsymbol{\sigma}^1_{n+1} \leftarrow \boldsymbol{\sigma}^1_{n+1} + \delta\boldsymbol{\sigma}$, $p^{n+1}_{0,1} \leftarrow p^{n+1}_{0,1} + \delta p_0$, and $\Delta\lambda \leftarrow \Delta\lambda + \delta\lambda$. Set $\boldsymbol{\sigma}_{n+1} = \boldsymbol{\sigma}^1_{n+1}$ and $p^{n+1}_0 = p^{n+1}_{0,1}$. Go to step 6.
12. Return to the main program.

12.5.2 Equations in Triaxial Space

As in the previous cases (Sects. 12.2.3, 12.3.3, and 12.4.3), only the first three components of vectors such as $\boldsymbol{\sigma}$ and $\boldsymbol{\varepsilon}$ are relevant here because the shear components of $\boldsymbol{\sigma}$ and $\boldsymbol{\varepsilon}$ are zero. In addition, because of the assumption of the triaxial condition, the 2nd and 3rd components of $\boldsymbol{\sigma}$ and $\boldsymbol{\varepsilon}$ (and vectors like these), are equal to each other. Among the 36 coefficients in the 6×6 matrices such as $\mathbf{C}$, only the nine components in the 3×3 sub-matrix in the top left hand corner are relevant.

Quantities Needed for the Calculation of Stresses
It was shown in Sect. 12.4.2 that

$$\frac{\partial r_{ij}}{\partial \sigma_{pq}} = 2\delta_{ij}\delta_{pq} + \frac{1}{N^2}\left[\delta_{ip}\delta_{jq} - \frac{1}{3}\delta_{ij}\delta_{pq}\right]$$

$$\frac{\partial r_{kk}}{\partial \sigma_{pq}} = 6\delta_{pq}$$

Also, from the equations presented in Appendix 8.2, we have

$$\frac{\partial r_{ij}}{\partial p_0} = -3\delta_{ij}$$

$$\frac{\partial d_{ij}}{\partial \sigma_{pq}} = \frac{\partial r_{ij}}{\partial \sigma_{pq}} - 2\delta_{pq}\delta_{ij} = \frac{1}{N^2}\left[\delta_{ip}\delta_{jq} - \frac{1}{3}\delta_{ij}\delta_{pq}\right]$$

$$\frac{\partial d_{ij}}{\partial p_0} = 0$$

The relevant deviatoric components are given by

$$s_{11} = \sigma_{11} - p/3; \quad s_{33} = \sigma_{33} - p/3$$

$$d_{11} = r_{11} - r_{kk}/3; \quad d_{33} = r_{33} - r_{kk}/3$$

$$\Delta e_{11} = \Delta\varepsilon_{11} - \Delta\varepsilon_{kk}/3; \quad \Delta e_{33} = \Delta\varepsilon_{33} - \Delta\varepsilon_{kk}/3$$

Then it follows that

$$\Delta y_{11} = \Delta e_{11} - \Delta\lambda d_{11}; \quad \Delta y_{33} = \Delta e_{33} - \Delta\lambda d_{33}$$

After computing the following quantities,

$$\bar{x}_0 = \bar{K}_0(\Delta\varepsilon_{kk} - \Delta\lambda r_{kk}^{n+1}); \quad G_{ave} = \frac{\bar{G}_0 p^n [e^{\bar{x}_0} - 1]}{\bar{x}_0}, \quad \text{and} \quad y_0 = e^{C_1 \Delta\lambda r_{kk}^{n+1}}$$

the first estimates of the stresses and the PIV are obtained from

$$p_1^{n+1} = p^n e^{\bar{x}_0}, \; \sigma_{11}^1 = s_{11}^n + p_1^{n+1} + 2G_{ave} y_{11};$$
$$\sigma_{33}^1 = s_{33}^n + p_1^{n+1} + 2G_{ave} y_{33}, \quad p_{0,1}^{n+1} = p_0^n y_0$$

The other relevant quantities are computed at $(\boldsymbol{\sigma}_{n+1}^1, p_{0,1}^{n+1})$ as

$$Z_{11} = 2\Delta y_{11} \frac{\bar{G}_0 p_1^{n+1}}{\bar{x}_0} - 2\Delta y_{11} \frac{G_{\text{ave}}}{\bar{x}_0} + p_1^{n+1}$$

$$Z_{33} = 2\Delta y_{33} \frac{\bar{G}_0 p_1^{n+1}}{\bar{x}_0} - 2\Delta y_{33} \frac{G_{\text{ave}}}{\bar{x}_0} + p_1^{n+1}$$

$$B_{ij} = \bar{K}_0 \Delta\lambda Z_{ij} \frac{\partial r_{kk}^{n+1}}{\partial p_0^{n+1}} + 2\Delta\lambda G_{\text{ave}} \frac{\partial d_{ij}^{n+1}}{\partial p_0^{n+1}} = -9\bar{K}_0 \Delta\lambda Z_{ij}$$

$$\bar{B}_{ij} = \frac{1}{3} \Delta\lambda \frac{\partial r_{kk}^{n+1}}{\partial p_0^{n+1}} \delta_{ij} = -3\Delta\lambda\delta_{ij}$$

$$H_{ij} = -C_1 \Delta\lambda p_0^n e^{C_1 \Delta\lambda r_{kk}^{n+1}} \frac{\partial r_{kk}^{n+1}}{\partial \sigma_{ij}^{n+1}} = -6C_1 \Delta\lambda p_0^n y_0 \delta_{ij}$$

$$\bar{F}_{ij} = C_{ijpq}^{*-1} F_{pq} = r_{ij}$$

$$\bar{x}_0 = 1/\omega; \quad \omega = 1 - C_1 \Delta\lambda p_0^n e^{C_1 \Delta\lambda r_{kk}^{n+1}} \frac{\partial r_{kk}^{n+1}}{\partial p_0^{n+1}} = 1 + 9C_1 \Delta\lambda p_0^n y_0;$$

$$\beta = -C_1 r_{kk}^{n+1} p_0^n e^{C_1 \Delta\lambda r_{kk}^{n+1}} = -C_1 r_{kk}^{n+1} p_0^n y_0$$

$$\gamma = \frac{\partial \phi}{\partial p_0^{n+1}} = -9p_1^{n+1}$$

$$E_{pq} = n_{pq}^{n+1} = r_{pq}^{n+1}$$

$$\begin{aligned} A_{ijpq} &= \delta_{ip}\delta_{jq} + \bar{K}_0 \Delta\lambda Z_{ij} \frac{\partial r_{kk}^{n+1}}{\partial \sigma_{pq}^{n+1}} + 2\Delta\lambda G_{ave} \frac{\partial d_{ij}^{n+1}}{\partial \sigma_{pq}^{n+1}} \\ &= \delta_{ip}\delta_{jq} - 6\bar{K}_0 \Delta\lambda Z_{ij}\delta_{pq} + 2\Delta\lambda G_{ave} \frac{1}{N^2}\left[\delta_{ip}\delta_{jq} - \frac{1}{3}\delta_{ij}\delta_{pq}\right] \end{aligned}$$

$$\frac{1}{\omega}\mathbf{B}\mathbf{H}^T = x_0^* \begin{bmatrix} Z_{11} & Z_{11} & Z_{11} \\ Z_{33} & Z_{33} & Z_{33} \\ Z_{33} & Z_{33} & Z_{33} \end{bmatrix},$$

where

$$x_0^* = 54 C_1 x_0 p_0^n \bar{K}_0 \Delta\lambda^2 y_0$$

Synthesizing these equations, it follows that

$$\mathbf{R} = \left[\mathbf{A} - \frac{1}{\omega}\mathbf{B}\mathbf{H}^T\right]^{-1}$$

$$\mathbf{A} - \frac{1}{\omega}\mathbf{B}\mathbf{H}^T =$$

$$\begin{bmatrix} 1 & 0 & 0 \\ 0 & 1 & 0 \\ 0 & 0 & 1 \end{bmatrix} + (6\bar{K}_0\Delta\lambda - x_0^*) \begin{bmatrix} Z_{11} & Z_{11} & Z_{11} \\ Z_{33} & Z_{33} & Z_{33} \\ Z_{33} & Z_{33} & Z_{33} \end{bmatrix}$$
$$+ 2\Delta\lambda G_{ave} \begin{bmatrix} 2/(3N^2) & -1/(3N^2) & -1/(3N^2) \\ -1/(3N^2) & 2/(3N^2) & -1/(3N^2) \\ -1/(3N^2) & -1/(3N^2) & 2/(3N^2) \end{bmatrix} = \text{(say)} \begin{bmatrix} T_1 & T_3 & T_3 \\ T_4 & T_2 & T_4 \\ T_4 & T_4 & T_2 \end{bmatrix}$$

where

$$T_1 = 1 + x_3 Z_{11} + \frac{4}{3N^2}\Delta\lambda G_{\text{ave}}$$

$$T_2 = 1 + x_3 Z_{33} + \frac{4}{3N^2}\Delta\lambda G_{\text{ave}}$$

$$T_3 = x_3 Z_{11} - \frac{2}{3N^2}\Delta\lambda G_{ave}$$

$$T_4 = x_3 Z_{33} - \frac{2}{3N^2}\Delta\lambda G_{ave}$$

and

$$x_3 = (6\bar{K}_0\Delta\lambda - \bar{x}_0^*)$$

Let us denote

$$\mathbf{R} = \begin{bmatrix} T_1 & T_3 & T_3 \\ T_4 & T_2 & T_4 \\ T_4 & T_4 & T_2 \end{bmatrix}^{-1} = \begin{bmatrix} S_1 & S_3 & S_3 \\ S_4 & S_2 & S_5 \\ S_4 & S_5 & S_2 \end{bmatrix}$$

and define $\alpha_1 = \frac{T_2}{T_4}$, $\alpha_2 = \frac{T_1}{T_3}$, $Z_1 = 2 - \alpha_2 - \alpha_1\alpha_2$, $Z_2 = 2 - \alpha_2(1 + \alpha_1)$, and $Z_3 = 2 + \alpha_1^2\alpha_2 - \alpha_2 - 2\alpha_1$.

Then it can be shown that $S_1 = -\frac{(1+\alpha_1)}{Z_1 T_3}$; $S_2 = \frac{(\alpha_1\alpha_2 - 1)}{Z_3 T_4}$; $S_3 = \frac{(1-\alpha_1)}{Z_3 T_4}$; $S_4 = \frac{1}{Z_2 T_3}$, and $S_5 = \frac{(1-\alpha_2)}{Z_3 T_4}$

Also, let us define

$$\mathbf{C}^* = \begin{bmatrix} E_1 & E_3 & E_3 \\ E_4 & E_2 & E_5 \\ E_4 & E_5 & E_2 \end{bmatrix}$$

Then, from

$$C^*_{ijpq} = \left[\bar{K}_0 Z_{ij} - \frac{2}{3}G_{ave}\delta_{ij}\right]\delta_{pq} + G_{ave}\left[\delta_{ip}\delta_{jq} + \delta_{iq}\delta_{jp}\right]$$

we have

$$E_1 = Z_{11}\bar{K}_0 - \frac{2}{3}G_{ave} + 2G_{ave}; \quad E_2 = Z_{33}\bar{K}_0 - \frac{2}{3}G_{ave} + 2G_{ave};$$

$$E_3 = Z_{11}\bar{K}_0 - \frac{2}{3}G_{ave}; \quad E_4 = E_5 = Z_{33}\bar{K}_0 - \frac{2}{3}G_{ave}$$

It follows that

$$\bar{\mathbf{C}} = \mathbf{RC} = \begin{bmatrix} \bar{C}_1 & \bar{C}_3 & \bar{C}_3 \\ \bar{C}_4 & \bar{C}_2 & \bar{C}_5 \\ \bar{C}_4 & \bar{C}_5 & \bar{C}_2 \end{bmatrix}$$

where

$$\bar{C}_1 = S_1E_1 + 2S_3E_4; \bar{C}_2 = S_4E_3 + S_2E_2 + S_5E_5; \bar{C}_3 = (E_2 + E_5)S_3 + S_1E_3;$$

$$\bar{C}_4 = (S_2 + S_5)E_4 + S_4E_1; \bar{C}_5 = S_4E_3 + S_2E_5 + S_5E_2$$

$$r_{11} = 6p^{n+1} - 3p_0^{n+1} + 18q^{n+1}/M^2$$

$$r_{33} = 6p^{n+1} - 3p_0^{n+1} - 9q^{n+1}/M^2$$

$$\bar{\mathbf{n}} = \mathbf{n} - \frac{\gamma}{\omega}\mathbf{H} = \mathbf{r} - 9p^{n+1}x_0(6\Delta\lambda C_1 p_0^n y_0 \boldsymbol{\delta})$$

$$\bar{n}_{11} = n_{11} - x_1; \quad \bar{n}_{33} = n_{33} - x_1;$$

where

$$x_1 = 54C_1 p^{n+1} p_0^n x_0 y_0 \Delta\lambda$$

$$\bar{\mathbf{r}} = \bar{\mathbf{F}} - x_0\beta\bar{\mathbf{B}}$$

$$\bar{\mathbf{r}} = \mathbf{r} - x_0\beta\bar{\mathbf{B}} = \mathbf{r} - x_0(-C_1 p_0^n r_{kk}^{n=1} y_0)(-3\Delta\lambda\boldsymbol{\delta})$$

$$\bar{r}_{11} = r_{11} - x_2; \quad \bar{r}_{33} = r_{33} - x_2;$$

where

$$x_2 = 3C_1 p_0^n r_{kk}^{n+1} x_0 y_0 \Delta\lambda$$

$$(\mathbf{RR}_\sigma)_{11} = S_1R_{\sigma_{11}} + 2S_3R_{\sigma_{33}}; \quad (\mathbf{RR}_\sigma)_{33} = S_4R_{\sigma_{11}} + S_2R_{\sigma_{33}} + S_5R_{\sigma_{33}}$$

$$(\bar{\mathbf{C}}\bar{\mathbf{B}})_{11} = \bar{C}_1\bar{B}_{11} + 2\bar{C}_3\bar{B}_{33}; \quad (\bar{\mathbf{C}}\bar{\mathbf{B}})_{33} = \bar{C}_4\bar{B}_{11} + \bar{C}_2\bar{B}_{33} + \bar{C}_5\bar{B}_{33}$$

$$(\bar{\mathbf{C}}\bar{\mathbf{r}})_{11} = C_{r11} = \bar{C}_1\bar{r}_{11} + 2\bar{C}_3\bar{r}_{33}; \quad (\bar{\mathbf{C}}\bar{\mathbf{r}})_{33} = C_{r33} = \bar{C}_4\bar{r}_{11} + \bar{C}_2\bar{r}_{33} + \bar{C}_5\bar{r}_{33}$$

$$\bar{K}_p^* = \bar{\mathbf{n}}^T\bar{\mathbf{C}}\bar{\mathbf{r}} = \bar{n}_{11}(\bar{C}_1\bar{r}_{11} + 2\bar{C}_3\bar{r}_{33}) + 2\bar{n}_{33}(\bar{C}_4\bar{r}_{11} + \bar{C}_2\bar{r}_{33} + \bar{C}_5\bar{r}_{33})$$

$$\bar{K}_p = x_0\gamma\beta$$

$$\bar{K}_p^{**} = \bar{K}_p + K_p^*$$

$$\bar{\mathbf{n}}^T\mathbf{RR}_\sigma = \bar{n}_{11}(S_1R_{\sigma_{11}} + 2S_3R_{\sigma_{33}}) + 2\bar{n}_{33}(S_4R_{\sigma_{11}} + S_2R_{\sigma_{33}} + S_5R_{\sigma_{33}})$$

$$\bar{\mathbf{n}}^T\bar{\mathbf{C}}\bar{\mathbf{B}} = \bar{n}_{11}(\bar{C}_1\bar{B}_{11} + 2\bar{C}_3\bar{B}_{33}) + 2\bar{n}_{33}(\bar{C}_4\bar{B}_{11} + \bar{C}_2\bar{B}_{33} + \bar{C}_5\bar{B}_{33})$$

Quantities Needed for the Calculation of the Consistent Elasto-Plastic Tangent Operator
The details are the same as those in Sect. 12.4.2 and hence will not be repeated here.

The use of these equations to calculate stresses are illustrated in Example 12.6, and to carry out global iterations in Example 12.7. These are also the subjects in Problems 12.9 and 12.10, respectively.

Example 12.6 Local Iterations Using CPPM with Exact Elastic Properties.

Question:. Given $\sigma_1^n = 114, \sigma_3^n = 78, p_0^n = 100, \Delta\varepsilon_1 = 0.01$, and $\Delta\varepsilon_3 = -0.001366$, determine the corresponding stresses and the PIV by CPPM with exact elastic properties. Carry out the iterations a few times and examine the convergence behavior. Use the values listed in Table 12.1 for the model parameters. Assume triaxial conditions.

Answer: The elastic predictor is calculated to be $\sigma_1^{\text{tr}} = 231.14$, $\sigma_3^{\text{tr}} = 108.97$, and $p^{\text{tr}} = 149.69$. Then the following parameters are computed: $N = 0.2309$, $r_{11} = 2125.29$, $r_{33} = -165.43$, $\bar{K}_p = 4.62 \times 10^9$, $\bar{K}_p^* = 67.58 \times 10^9$, $\bar{K}_p^{**} = 72.2 \times 10^9$, $R_\phi = 160231.35$, $\Delta\lambda = R_\phi/\bar{K}_p^{**} = 2.22 \times 10^{-6}$, $\sigma_1^1 = 164.66$, $\sigma_3^1 = 87.58$, $Y_0 = 1.08$, and $p_{0,1} = 107.9$.

With these initial estimates, the iterations begin and the results are presented in Table 12.8, with the main results listed in Table 12.8a, b and the errors listed in Table 12.8c. The results converged in five iterations to within an error of 1.0×10^{-3}. It could be observed that the rate of convergence is between quadratic and cubic. The problem analyzed here is similar to the one analyzed by the cutting-plane algorithm in Example 12.2, where the results converged in five iterations to almost the same level of accuracy as in the present case (compare the results in Table 12.3 with those in Table 12.8c). The convergence behavior of CPPM with approximate elastic properties (Example 12.4) and that with exact elastic properties are very similar as well (compare the results in Table 12.5c with those in Table 12.8c)

Example 12.7 Global Iterations (Algorithm 8.2) Using CPPM with Exact Elastic Properties for the Local Iterations (Algorithm 12.4).

Question: A strain-controlled, triaxial drained compression loading is to be conducted with $\sigma_1^n = 114$, $\sigma_3^n = 78$, and $p_0^n = 100$. Determine the stresses and strains at the end of the incremental loading with $\Delta\hat{\varepsilon}_1 = 0.05$ and $\Delta\hat{\sigma}_3 = 0.0$. Use Algorithm 8.2 for global iterations with consistent tangent operator and Algorithm 12.4 for local iterations. Use the specific triaxial equations developed in Sect. 12.5.2 to perform the calculations. Examine the global convergence behavior of the algorithm. Investigate the effectiveness of using the continuum operator over the consistent operator. Use the values listed in Table 12.1 for the model parameters.

Answer: First, the variables are initialized as (Step 2 of Algorithm 8.2): $\sigma_1^n = \sigma_1^{n+1} = 114$, $\sigma_3^n = \sigma_3^{n+1} = 78$, and $p_0^n = p_0^{n+1} = 100$, $\sigma_{E1}^{n+1} = 114 + 0 = 114$, $\sigma_{E3}^{n+1} = 78 + 0 = 78$,

Table 12.8a Local iterations using CPPM (Algorithm 12.4) with exact elastic properties (Example 12.6)

Quantity	Iter. 1	Iter. 2	Iter. 3	Iter. 4	Iter. 5
p^{n+1}	113.27	97.21	94.62	94.76	94.76
q^{n+1}	77.08	54.24	50.30	50.27	50.28
R_ϕ	0.426E+05	0.607E+04	0.197E+03	−0.230E+00	0.266E−04
r_{11}	1,319.39	927.43	857.55	857.12	857.18
r_{33}	−125.81	−89.62	−85.50	−85.51	−85.52
r_{kk}	1,067.77	748.19	686.54	686.10	686.13
Δy_{11}	0.544E−02	0.350E−02	0.188E−02	0.160E−02	0.159E−02
Δy_{33}	−0.272E−02	−0.175E−02	−0.939E−03	−0.799E−03	−0.797E−03
G_{ave}	3,468.64	3,209.45	3,016.89	2,984.45	2,984.17
p_2^{n+1}	126.81	109.29	96.83	94.78	94.76
σ_1^2	188.54	155.79	132.16	128.32	128.28
σ_3^2	95.94	86.04	79.16	78.01	78.00
$\rho_{0,1}$	104.63	108.96	112.62	113.28	113.29
$R_{\sigma_{11}}$	−23.88	−22.41	−4.02	−0.04	0.00
$R_{\sigma_{33}}$	−8.36	−6.91	−1.31	−0.01	0.00
R_ς	3.27	2.33	0.33	0.00	0.00
γ	−0.102E+04	−0.875E+03	−0.852E+03	−0.853E+03	−0.853E+03
β	−0.213E+07	−0.156E+07	−0.148E+07	−0.148E+07	−0.148E+07
x_0	0.962E+00	0.899E+00	0.851E+00	0.844E+00	0.844E+00
χ_0^*	0.358E−04	0.255E−03	0.568E−03	0.625E−03	0.625E−03
x_1	0.261E+02	0.590E+02	0.848E+02	0.889E+02	0.889E+02
x_2	0.137E+02	0.252E+02	0.342E+02	0.358E+02	0.358E+02
Z_{11}	0.147E+03	0.121E+03	0.103E+03	0.996E+02	0.996E+02
Z_{33}	0.117E+03	0.103E+03	0.940E+02	0.924E+02	0.924E+02
x_3	−0.896E−03	−0.227E−02	−0.324E−02	−0.337E−02	−0.337E−02
T_1	0.132E+01	0.176E+01	0.202E+01	0.205E+01	0.205E+01
T_2	0.130E+01	0.172E+01	0.199E+01	0.202E+01	0.202E+01
T_3	0.353E−01	0.332E−01	−0.965E−02	−0.192E−01	−0.193E−01
T_4	0.850E−02	−0.631E−02	−0.375E−01	−0.435E−01	−0.436E−01

Table 12.8b Local iterations using CPPM (Algorithm 12.4) with exact elastic properties: continuation from Table 12.8a

Quantity	Iter. 1	Iter. 2	Iter. 3	Iter. 4	Iter. 5
S_1	0.756E+00	0.569E+00	0.496E+00	0.489E+00	0.489E+00
S_2	0.771E+00	0.583E+00	0.503E+00	0.495E+00	0.495E+00
S_3	−0.204E−01	−0.110E−01	0.245E−02	0.475E−02	0.477E−02
S_4	−0.492E−02	0.210E−02	0.954E−02	0.108E−01	0.108E−01
S_5	−0.492E−02	0.210E−02	0.954E−02	0.108E−01	0.108E−01
E_1	0.149E+05	0.127E+05	0.112E+05	0.110E+05	0.109E+05
E_2	0.128E+05	0.115E+05	0.106E+05	0.104E+05	0.104E+05
E_3	0.796E+04	0.632E+04	0.517E+04	0.498E+04	0.498E+04
E_4	0.587E+04	0.510E+04	0.457E+04	0.448E+04	0.448E+04
E_5	0.587E+04	0.510E+04	0.457E+04	0.448E+04	0.448E+04
$\overline{C}_1$	0.110E+05	0.714E+04	0.558E+04	0.540E+04	0.540E+04
$\overline{C}_2$	0.980E+04	0.674E+04	0.543E+04	0.527E+04	0.527E+04
$\overline{C}_3$	0.563E+04	0.342E+04	0.260E+04	0.251E+04	0.251E+04
$\overline{C}_4$	0.442E+04	0.301E+04	0.245E+04	0.238E+04	0.238E+04

(continued)

Table 12.8b (continued)

Quantity	Iter. 1	Iter. 2	Iter. 3	Iter. 4	Iter. 5
$\bar{C}_5$	0.442E+04	0.301E+04	0.245E+04	0.238E+04	0.238E+04
$\bar{n}_{11}$	0.129E+04	0.868E+03	0.773E+03	0.768E+03	0.768E+03
$\bar{n}_{33}$	−0.152E+03	−0.149E+03	−0.170E+03	−0.174E+03	−0.174E+03
$\bar{r}_{11}$	0.131E+04	0.902E+03	0.823E+03	0.821E+03	0.821E+03
$\bar{r}_{33}$	−0.139E+03	−0.115E+03	−0.120E+03	−0.121E+03	−0.121E+03
$\bar{K}_p^*$	0.154E+11	0.444E+10	0.270E+10	0.258E+10	0.258E+10
$\bar{K}_p$	0.209E+10	0.122E+10	0.107E+10	0.107E+10	0.107E+10
$\bar{K}_p^{**}$	0.175E+11	0.566E+10	0.377E+10	0.365E+10	0.365E+10
B_{11}	−0.666E−05	−0.180E−04	−0.272E−04	−0.285E−04	−0.286E−04
B_{33}	−0.666E−05	−0.180E−04	−0.272E−04	−0.285E−04	−0.286E−04
$\mathbf{n}^T\mathbf{R}\mathbf{R}_\sigma$	−0.210E+05	−0.974E+04	−0.130E+04	−0.139E+02	−0.386E−03
$\bar{\mathbf{n}}^T\bar{\mathbf{C}}\bar{\mathbf{B}}$	−0.154E+03	−0.150E+03	−0.131E+03	−0.128E+03	−0.128E+03
$\delta\lambda$	0.379E−05	0.306E−05	0.451E−06	0.348E−08	0.471E−13
$\delta\sigma_1$	−31.28	−5.23	0.12	0.01	0.00
$\delta\sigma_3$	−8.45	−1.28	0.15	0.00	0.00
δp_0	3.39	1.66	0.33	0.01	0.00
$\Delta\lambda$	0.601E−05	0.906E−05	0.952E−05	0.952E−05	0.952E−05
σ_1^1	133.37	128.15	128.27	128.28	128.28
σ_3^1	79.13	77.85	78.00	78.00	78.00
$p_{0,1}$	111.29	112.95	113.28	113.29	113.29

Table 12.8c Local iterations using CPPM (Algorithm 12.4) with exact elastic properties: errors

Quantity	Iter. 1	Iter. 2	Iter. 3	Iter. 4	Iter. 5
e_σ	0.129E+00	0.140E+00	0.262E−01	0.279E−03	0.101E−07
e_ς	0.303E−01	0.209E−01	0.294E−02	0.132E−04	0.347E−08
e_ϕ	0.426E+05	0.607E+04	0.197E+03	0.230E+00	0.266E−04
$(e_\sigma)_{k+1}/(e_\sigma)_k$	–	0.109E+01	0.186E+00	0.107E−01	0.363E−04
$(e_\varsigma)_{k+1}/(e_\varsigma)_k$	–	0.691E+00	0.140E+00	0.449E−02	0.263E−03
$(e_\sigma)_{k+1}/(e_\sigma)_k^2$	–	0.839E+01	0.133E+01	0.408E+00	0.130E+00
$(e_\varsigma)_{k+1}/(e_\varsigma)_k^2$	–	0.228E+02	0.670E+01	0.153E+01	0.199E+02
$(e_\sigma)_{k+1}/(e_\sigma)_k^3$	–	0.649E+02	0.945E+01	0.156E+02	0.465E+03
$(e_\varsigma)_{k+1}/(e_\varsigma)_k^3$	–	0.750E+03	0.320E+03	0.519E+03	0.151E+07

$$\sigma_{S1}^n = \sigma_{S1}^{n+1} = \sigma_1^n = 114,$$

$$\sigma_{S3}^n = \sigma_{S3}^{n+1} = \sigma_3^n = 78,$$

$$\delta\sigma_1 = \sigma_{E1}^{n+1} - \sigma_{S1}^{n+1} = 0 \quad \text{and} \quad \delta\sigma_3 = \sigma_{E3}^{n+1} - \sigma_{S3}^{n+1} = 0$$

Note that in the above equations, the subscripts "E" and "S" indicate "external force" and "spring force (internal)," respectively. The values of various quantities during the global iteration are summarized in Table 12.9 for the analysis using the consistent operator and in Table 12.10 for the analysis using the continuum operator.

Table 12.9 Global iterations using consistent operator and using CPPM with exact elastic properties for the local iterations (Example 12.7)

Quantity	Iter. 1	Iter. 2	Iter. 3
D_1	10,177.20	2,452.41	2,538.38
D_2	10,177.20	2,667.21	2,837.60
D_3	4,361.40	2,625.59	2,713.95
D_4	4,361.40	1,603.45	1,670.27
D_5	4,361.40	1,545.63	1,668.35
D_1^*	0.100E+21	0.100E+21	0.100E+21
$\delta\sigma_1^*$	0.500E+19	0.000E+00	0.000E+00
$\delta\varepsilon_1$	0.500E−01	−0.769E−19	0.258E−20
$\delta\varepsilon_3$	−0.150E−01	0.146E−02	−0.475E−04
$\Delta\varepsilon_1$	0.500E−01	0.500E−01	0.500E−01
$\Delta\varepsilon_3$	−0.150E−01	−0.135E−01	−0.136E−01
$\Delta\sigma_1$	37.37	45.19	44.93
$\Delta\sigma_3$	−6.17	0.21	0.00
Δp_0	43.02	48.49	48.31
σ_1^{n+1}	151.37	159.19	158.93
σ_3^{n+1}	71.83	78.21	78.00
p_0^{n+1}	143.02	148.49	148.31
σ_{S1}^{n+1}	151.37	159.19	158.93
σ_{S3}^{n+1}	71.83	78.21	78.00
$\delta\sigma_1$	−37.37	−45.19	−44.93
$\delta\sigma_3$	6.17	−0.21	0.00
e_σ	0.550E−01	0.191E−02	0.203E−05
e_e	−0.185E+00	0.579E−02	0.617E−05
$(e_\sigma)_{k+1}/(e_\sigma)_k$	–	0.347E−01	0.106E−02
$(e_\sigma)_{k+1}/(e_\sigma)_k^2$	–	0.631E+00	0.556E+00
$(e_\sigma)_{k+1}/(e_\sigma)_k^3$	–	0.115E+02	0.292E+03

Also recall that when the loading is of the mixed type as in the present case (the stress is controlled in direction 3 and the strain is controlled in direction 1), the **D** matrix and the $\delta\boldsymbol{\sigma}$ vector are modified as described in Sect. 8.6. In the present case, the coefficients D_1 and $\delta\sigma_1$ are modified; the modified coefficients are denoted as D_1^* and $\delta\sigma_1^*$ in Tables 12.9 and 12.10.

It takes only three iterations to reduce e_σ to below 1.0×10^{-4}, whereas when the continuum operator is used, it takes 22 iterations. The rate of convergence is clearly linear when using the continuum operator (Table 12.10). When using the consistent operator, the results converged too quickly for computing the error ratios accurately (it appears to be quadratic).

12.5.3 Sample Model Simulations

Sample simulations are performed using a uniformly loaded single-element finite element model. The problems analyzed in this section are identical to those in Sects. 12.2.3, 12.3.3, and 12.4.3, and the reader is referred to Sect. 12.2.3 for the definition

Table 12.10 Global iterations using continuum operator and using CPPM with exact elastic properties for the local iterations (Example 12.7)

Quantity	Iter. 1	Iter. 2	Iter. 3	Iter. 4	Iter. 5	Iter. 22
D_1	10,177.20	3,496.02	3,475.45	3,460.41	3,449.43	3,419.68
D_2	10,177.20	11,118.83	11,316.20	11,458.63	11,561.89	11,839.98
D_3	4,361.40	4,634.53	4,625.96	4,617.90	4,611.06	4,588.49
D_4	4,361.40	4,634.53	4,625.96	4,617.90	4,611.06	4,588.49
D_5	4,361.40	4,763.64	4,845.78	4,904.24	4,946.18	5,057.20
D_1^*	0.100E+21	0.100E+21	0.100E+21	0.100E+21	0.100E+21	0.100E+21
$\delta\sigma_1^*$	0.500E+19	0.000E+00	0.000E+00	0.000E+00	0.000E+00	0.000E+00
$\delta\varepsilon_1$	0.500E−01	−0.360E−19	−0.259E−19	−0.187E−19	−0.136E−19	−0.684E−22
$\delta\varepsilon_3$	−0.150E−01	0.388E−03	0.279E−03	0.202E−03	0.147E−03	0.745E−06
$\Delta\varepsilon_1$	0.500E−01	0.500E−01	0.500E−01	0.500E−01	0.500E−01	0.500E−01
$\Delta\varepsilon_3$	−0.150E−01	−0.146E−01	−0.143E−01	−0.141E−01	−0.140E−01	−0.136E−01
$\Delta\sigma_1$	37.37	39.42	40.91	41.99	42.78	44.92
$\Delta\sigma_3$	−6.17	−4.52	−3.31	−2.43	−1.78	−0.01
Δp_0	43.02	44.45	45.49	46.24	46.79	48.30
σ_1^{n+1}	151.37	153.42	154.91	155.99	156.78	158.92
σ_3^{n+1}	71.83	73.48	74.69	75.57	76.22	77.99
p_0^{n+1}	143.02	144.45	145.49	146.24	146.79	148.30
σ_{S1}^{n+1}	151.37	153.42	154.91	155.99	156.78	158.92
σ_{S3}^{n+1}	71.83	73.48	74.69	75.57	76.22	77.99
$\delta\sigma_1$	−37.37	−39.42	−40.91	−41.99	−42.78	−44.92
$\delta\sigma_3$	6.17	4.52	3.31	2.43	1.78	0.01
e_σ	0.550E−01	0.403E−01	0.295E−01	0.216E−01	0.159E−01	0.824E−04
e_e	−0.185E+00	−0.132E+00	−0.949E−01	−0.686E−01	−0.498E−01	−0.251E−03
$(e_\sigma)_{k+1}/(e_\sigma)_k$	–	0.732E+00	0.733E+00	0.733E+00	0.733E+00	0.734E+00
$(e_\sigma)_{k+1}/(e_\sigma)_k^2$	–	0.133E+02	0.182E+02	0.249E+02	0.339E+02	0.654E+04
$(e_\sigma)_{k+1}/(e_\sigma)_k^3$	–	0.242E+03	0.452E+03	0.842E+03	0.157E+04	0.583E+08

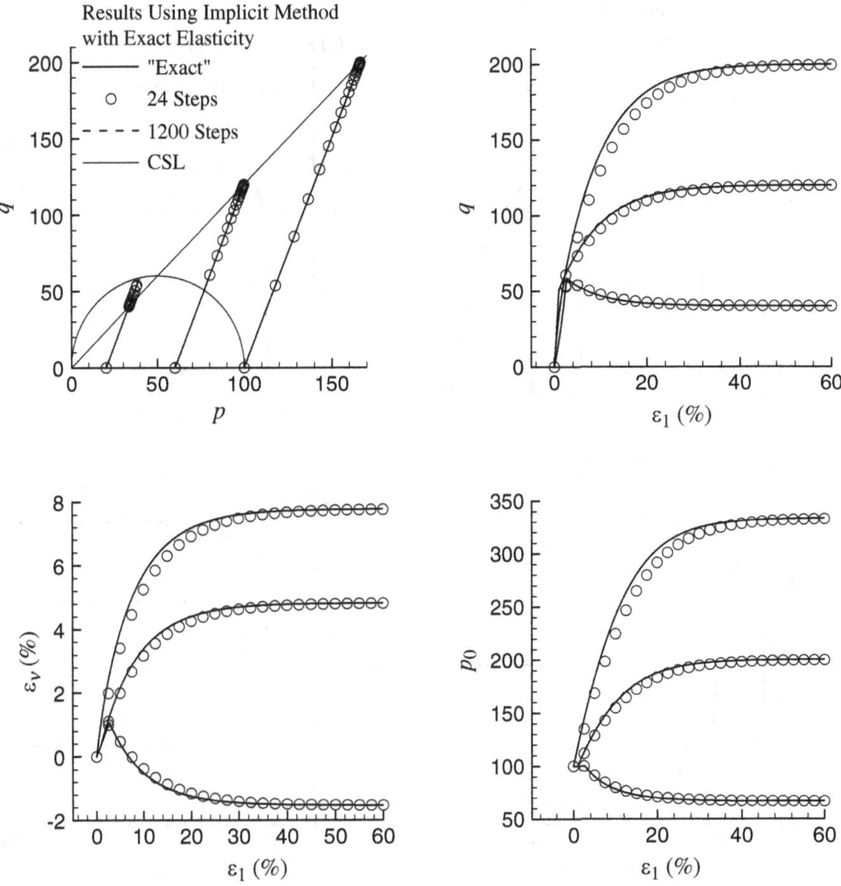

Fig. 12.11 Sample numerical simulations of triaxial drained behavior using CPPM with exact elastic properties

of the problems. The problems are analyzed here using CPPM with exact elastic properties (Algorithm 12.4) with the following control parameters: $N_L^{\max} = 100$, $TOL = 1.0 \times 10^{-10}$, $TOL_\sigma = 0.01$, $TOL_\varsigma = 0.01$, $TOL_\phi = 1.0 \times 10^{-5} p_0$, and $\Gamma = 1.0 \times 10^8$. The values listed in Table 12.1 are used for the model parameters.

The results are presented in Figs. 12.11 and 12.12 for drained and undrained simulations, respectively. The results obtained with a step size $\Delta\varepsilon_1 = 2.5\%$ (which led to a total of 24 and 12 steps in drained and undrained, respectively) and a step size $\Delta\varepsilon_1 = 0.05\%$ (which led to a total of 1,200 and 600 steps in drained and undrained, respectively) are compared with the "exact" solutions. As in previous cases, in all cases considered in the present case, the results with a step size $\Delta\varepsilon_1 = 0.05\%$ and the "exact" results are indistinguishable. The results with a step size $\Delta\varepsilon_1 = 2.5\%$ are further discussed below.

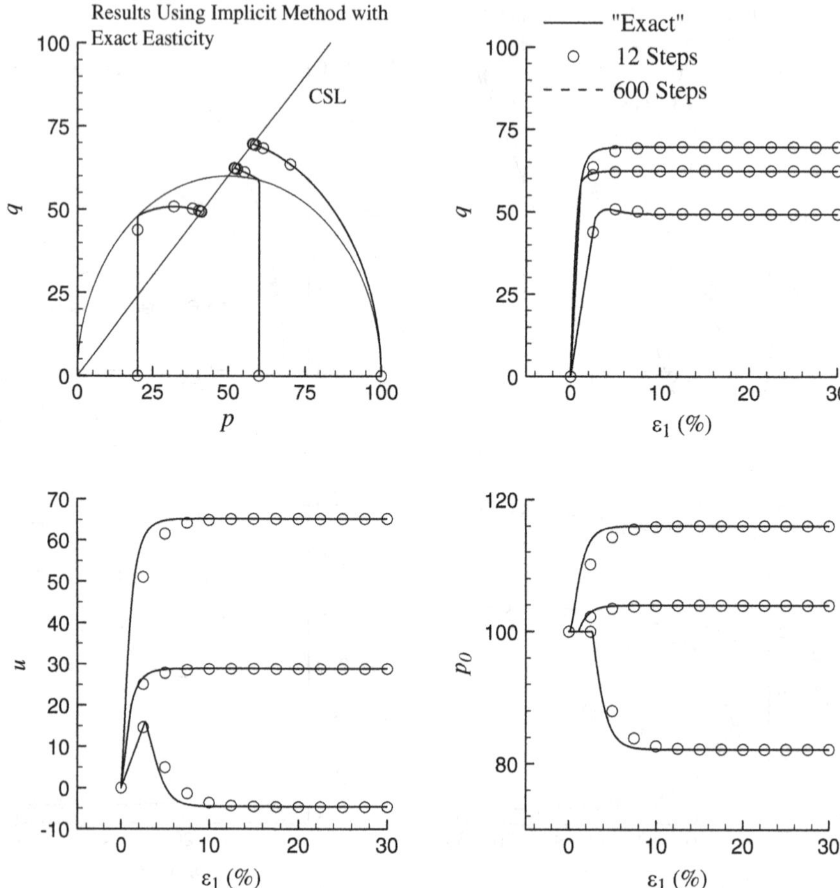

Fig. 12.12 Sample numerical simulations of triaxial undrained behavior using CPPM with exact elastic properties

As in the case with the cutting-plane algorithm and CPPM with approximate elastic properties, convergence was achieved in all cases presented here, illustrating the robustness of the algorithm. As with the case of CPPM that employs approximate elastic properties, when the step size is too large ($\Delta\varepsilon = 10\%$), however, small error tolerances (e.g., 1.0×10^{-5}) for e_σ could not be achieved near failure. The tolerance had to be increased to continue the analysis. The results seem to be almost as accurate as those obtained with the two-step Euler method.

During the simulation of octahedral plane circular stress path (such as the one shown in Fig. 12.5), the method performed as good as the two-step Euler method, the cutting-plane method, and CPPM with approximate elastic properties. The results using 36 steps and 1,800 steps gave indistinguishable results. The results, however, varied from those obtained using the two-step Euler method. The results are not presented as they do not reveal new information.

12.6 Comparison of Performances of the Different σ–Space Methods

12.6.1 General Discussion

Four different methods of implementing the modified Cam-clay model have been presented in the preceding sections. The steps have been developed in the general three-dimensional invariant space and then simplified for triaxial loading conditions for performing the calculations by hand. The convergence behavior of the methods at the local and global levels has been examined. The results indicate that, for the stress paths considered, the methods are robust. As the step size is reduced, the results coincide with the "exact" results.

12.6.2 Accuracy

Except for the two-step Euler method with sub-increments, the remaining three methods are variants of the elastic predictor-plastic corrector method. In the two-step Euler method, the stress rate is integrated as

$$\int_{t}^{t+\Delta t} \dot{\boldsymbol{\sigma}} = \int_{t}^{t+\Delta t} \bar{\mathbf{D}}\dot{\boldsymbol{\varepsilon}} \approx \frac{1}{2}[\bar{\mathbf{D}}_n + \bar{\mathbf{D}}_{n+1}]\Delta\boldsymbol{\varepsilon} \tag{12.66}$$

Hence, the trapezoidal rule is used in an approximate sense (it is approximate because $\bar{\mathbf{D}}_{n+1}$ is computed using only an estimate of $\boldsymbol{\sigma}_{n+1}$). However, the results should be accurate when the step size decreases. The results presented in this chapter confirm this to be the case. However, the consistency is not enforced, and hence in some cases (e.g., models with more complicated hardening rules and yield surfaces), the current stress state may gradually drift off the yield surface.

In the remaining three methods (the cutting-plane, CPPM with approximate elastic properties, and CPPM with exact elastic properties), instead of controlling the step size, the consistency is enforced at the end of the increment. In the cutting-plane method, this is achieved in an explicit manner (i.e., at the beginning of each iteration, quantities computed at the beginning of the step are used), whereas in the other two methods, this is achieved in an implicit manner. The satisfaction of consistency prevents the stress point from drifting off the yield surface, and increases the accuracy.

In addition, the variation of elastic properties during the increment also needs to be accounted for. Consider CPPM with approximate elastic properties. Here, upon convergence, the plastic corrector takes the form

$$\Delta\boldsymbol{\sigma}^{pc} = -\int_{t}^{t+\Delta t} \dot{\lambda}\mathbf{C}\mathbf{r} \approx \Delta\lambda\mathbf{C}_{n+\theta}\mathbf{r}^{n+1} \tag{12.67}$$

It can be seen that in addition to the approximation of **C**, the variation of **r** during the increment is neglected. This is the case with any CPPM (including the implementation using the exact elastic properties). As shown in Fig. 12.13, excellent results are still obtained up to a certain step size and the accuracy deteriorates after that.

As per the use of approximate elastic properties, again for small step sizes, excellent results are obtained. For example, referring to Examples 12.4 and 12.6, for almost the same strain increments ($\Delta\varepsilon_1 = 0.01$ and $\Delta\varepsilon_3 = -0.001418$ in Example 12.4 and $\Delta\varepsilon_1 = 0.01$ and $\Delta\varepsilon_3 = -0.001366$ in Example 12.6), almost the same elastic predictors are calculated by the two methods ($\sigma_1^{tr} = 233.32$ and $\sigma_3^{tr} = 108.69$ in Example 12.4 and $\sigma_1^{tr} = 231.14$ and $\sigma_3^{tr} = 108.97$ in Example 12.6). The advantage of CPPM with approximate elastic properties is that equations need only a minor modification when the elastic model is changed.

12.6.3 Efficiency

The algorithmic details are too dissimilar and complicated to make a step-by-step comparison of the efficiency of computations involved. One method of comparing the efficiencies is to compare the computer CPM time consumption. Here too, the outcome depends on the details of the coding. In the results presented here, which are obtained using a calibration program, the coding is kept as similar as possible. Considering the triaxial drained simulation presented here (e.g., Fig. 12.13), the CPU time requirements for the different methods with different step sizes are compared in Table 12.11. It can be observed that for smaller step sizes, the two-step method is more efficient than the cutting-plane method and the CPPMs, but for larger step sizes, the trend is opposite – the cutting-plane method and CPPMs are much more efficient. The reason is that the when the step size is large, the two-step Euler method divides the increment into very small steps, whereas the CPPMs and the cutting-plane method keep the step size the same and still converges in a reasonable number of iterations.

12.7 The Closest Point Projection (Implicit) Method with Approximate Elastic Properties: The S-Space Formulation

12.7.1 General Description

It is shown in Sect. 10.7.2 (Chap. 10) that the Euler equation for stress (12.40a) can be decomposed into spherical and deviatoric parts. The resulting formulation is known as the *S*-space formulation. When the yield surface is assumed to be a circle

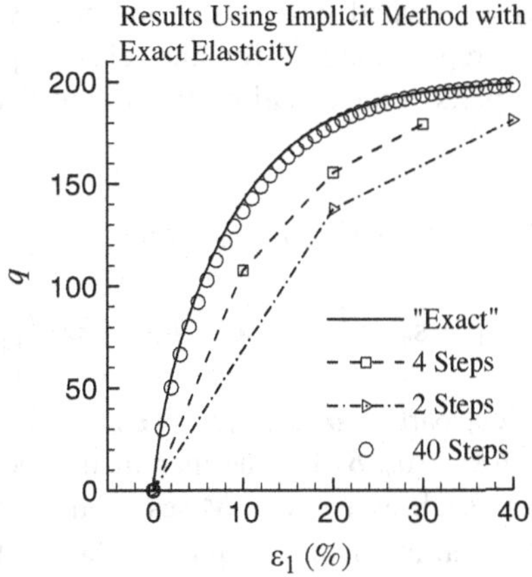

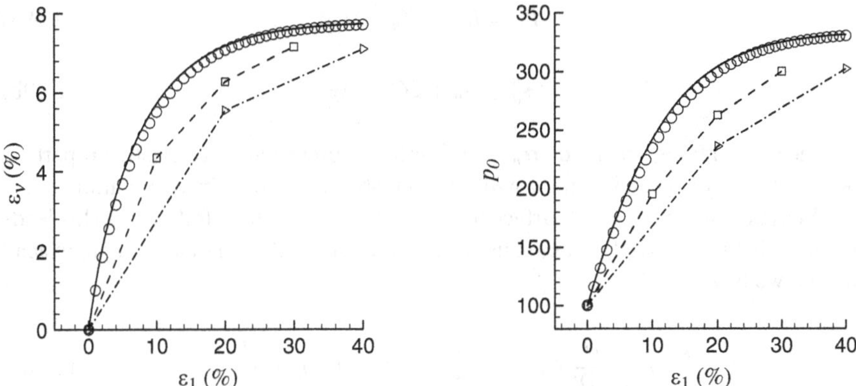

Fig. 12.13 Accuracy of results obtained using CPPM with exact elastic properties as a function of step size (drained tests, normally-consolidated specimens)

Table 12.11 Comparison of efficiency of different integration methods in terms of CPU time

Method	CPU time (seconds) for different step size, $\Delta\varepsilon_1$ (%)			
	0.01	0.05	1.00	2.00
Two-step Euler method	1.931	0.764	0.971	1.832
Cutting plane method	2.219	0.690	0.155	0.142
CPPM with approximate elastic properties	4.382	1.755	0.153	0.090
CPPM with exact elastic properties	4.070	1.277	0.129	0.078

in the octahedral plane, this formulation leads to a more computationally efficient algorithm than that from the σ-space formulation.

The elastic relations are treated in the same manner as in Sect. 12.4.1. Using the approximate Euler representation for the elastic properties (12.39), the spherical and deviatoric parts of the equations (using I instead of p for the spherical part) are

$$I_{n+1} = I_n + 3K_{n+\theta}\Delta\varepsilon_{pp} - 3K_{n+\theta}\Delta\lambda r_{pp}^{n+1} \tag{12.68a}$$

$$\mathbf{s}_{n+1} = \mathbf{s}_n + 2G_{n+\theta}\Delta\mathbf{e} - 2G_{n+\theta}\Delta\lambda\mathbf{r}_{n+1}^{d} \tag{12.68b}$$

where I_{n+1} is the spherical part of $\boldsymbol{\sigma}_{n+1}$, $\mathbf{s}_{n+1} = \boldsymbol{\sigma}_{n+1} - I_{n+1}\boldsymbol{\delta}/3$ is the deviatoric part of $\boldsymbol{\sigma}_{n+1}$ and $\Delta\mathbf{e} = \Delta\boldsymbol{\varepsilon} - \Delta\varepsilon_{pp}\boldsymbol{\delta}/3$ is the incremental deviatoric strain. In pressure-independent models such as the von Mises model considered in Chap. 11, $r_{pp}^{n+1} = 0$, and hence the variation of I is purely elastic. However, in pressure-dependent models such as the modified Cam-clay model, the variation of I is elasto-plastic. Let us define the trial stresses as

$$I_{\theta}^{tr} = I_n + 3K_{n+\theta}\Delta\varepsilon_{pp} \tag{12.69a}$$

$$\mathbf{s}_{\theta}^{tr} = \mathbf{s}_n + 2G_{n+\theta}\Delta\mathbf{e} \tag{12.69b}$$

The trial stresses depend on $\boldsymbol{\sigma}_{n+1}$ for implicit representation of elastic properties, hence the use of the subscript θ with the trial stresses in the above equations.

We will assume the yield surface to be a circle in the octahedral plane. This leads to $\phi_{,\alpha} = 0$. From (12.10), making use of (12.11a) and (12.11b) (and using I_0 instead of p_0), we have

$$r_{ij} = \frac{\partial\phi}{\partial I}\delta_{ij} + \frac{\partial\phi}{\partial J}\frac{s_{ij}}{2J} = (2I - I_0)\delta_{ij} + \frac{1}{N^2}s_{ij} \tag{12.70a}$$

Hence

$$r_{pp}^{n+1} = 6I_{n+1} - 3I_0^{n+1} \tag{12.70b}$$

$$\mathbf{r}_{n+1}^{d} = \frac{1}{N^2}\mathbf{s}_{n+1} \tag{12.70c}$$

Let us first consider the deviatoric part. Combining (12.68b), (12.69b), and (12.70c), we have

$$\begin{aligned} \mathbf{s}_{n+1} &= \mathbf{s}_n + 2G_{n+\theta}\Delta\mathbf{e} - 2G_{n+\theta}\Delta\lambda\mathbf{r}_{n+1}^{d} = \mathbf{s}_{\theta}^{tr} - \frac{2}{N^2}G_{n+\theta}\Delta\lambda\mathbf{s}_{n+1} \\ \mathbf{s}_{n+1} &= \frac{1}{1+\frac{2}{N^2}G_{n+\theta}\Delta\lambda}\mathbf{s}_{\theta}^{tr} = \frac{\mathbf{s}_{\theta}^{tr}}{x_0}; \quad x_0 = 1 + \frac{2}{N^2}G_{n+\theta}\Delta\lambda \end{aligned} \tag{12.71}$$

which represents a radial return in the octahedral plane, just as in the case of the von Mises model ((11.28b) and Fig. 11.1). In terms of the deviatoric invariants, (12.71) becomes

$$J_{n+1} = \frac{J_\theta^{tr}}{x_0}; \quad J_\theta^{tr} = \left(\frac{1}{2} s_{k\ell}^{tr} s_{k\ell}^{tr}\right)^{1/2} \tag{12.72}$$

Backward Euler Equations.
Using invariants for the stresses, the Euler equations (12.40a– 12.40c) are written as (making use of (12.70b))

$$\begin{aligned} I_{n+1} &= I_\theta^{tr} - 3K_{n+\theta}\Delta\lambda r_{pp}^{n+1} = I_\theta^{tr} - 3K_{n+\theta}\Delta\lambda(6I_{n+1} - 3I_0^{n+1}) \\ &= I_\theta^{tr} - 18K_{n+\theta}\Delta\lambda I_{n+1} + 9K_{n+\theta}\Delta\lambda I_0^{n+1} \end{aligned} \tag{12.73a}$$

$$J_{n+1} = \frac{J_\theta^{tr}}{x_0} \tag{12.73b}$$

$$\begin{aligned} I_0^{n+1} &= I_0^n + \Delta\lambda s_{n+1} = I_0^n + C_1\Delta\lambda I_0^{n+1} r_{pp}^{n+1} = I_0^n + C_1\Delta\lambda I_0^{n+1}(6I_{n+1} - 3I_0^{n+1}) \\ &= I_0^n + 6C_1\Delta\lambda I_0^{n+1} I_{n+1} - 3C_1\Delta\lambda (I_0^{n+1})^2 \end{aligned} \tag{12.73c}$$

$$\phi^{n+1}(I_{n+1}, J_{n+1}, I_0^{n+1}) = 0 \tag{12.73d}$$

Because of the nonlinearity of these equations, it is not possible to develop a scalar equation in $\Delta\lambda$ alone as it has been possible in the case of the von Mises model (11.34). As it is done in the σ-space formulations (Sects. 12.4 and 12.5), we use Newton's method to solve for the unknowns $I_{n+1}, J_{n+1}, p_0^{n+1}$, and $\Delta\lambda$; the details are given below.

It is noted that in the case of the σ-space formulation, the Euler equation for the stress contains six equations (12.40a), whereas in the case of the S-space formulation, there are only two equations for the stress, one for I and the other for J. Consequently, the S-space formulation is more computationally efficient than the σ-space formulation. When the rest of the details are the same, the two formulations are fundamentally identical. Hence the accuracy must also be identical, except for numerical differences due to rounding resulting from the differences in the equations.

The Residual Equations and their Linearizations.
The residual equations (the counterparts of (12.41a–12.41c)) are

$$R_I = I_{n+1} - I_\theta^{tr} + 18K_{n+\theta}\Delta\lambda I_{n+1} - 9K_{n+\theta}\Delta\lambda I_0^{n+1} \tag{12.74a}$$

$$R_J = J_{n+1} - \frac{J_\theta^{tr}}{x_0}; \quad x_0 = 1 + \frac{2}{N^2} G_{n+\theta}\Delta\lambda \tag{12.74b}$$

$$R_\varsigma = I_0^{n+1} - I_0^n - 6C_1\Delta\lambda I_0^{n+1} I_{n+1} + 3C_1\Delta\lambda (I_0^{n+1})^2 \tag{12.74c}$$

$$\phi^{n+1}(I_{n+1}, J_{n+1}, I_0^{n+1}) = 0 \tag{12.74d}$$

The linearizations of the above equations (the counterparts of (12.42a–12.42c)) are

$$R_{I0} + A_{11}\delta I + A_{12}\delta J + B_1\delta I_0 + F_1\delta\lambda = 0 \tag{12.75a}$$

$$R_{J0} + A_{21}\delta I + A_{22}\delta J + B_2\delta I_0 + F_2\delta\lambda = 0 \tag{12.75b}$$

$$R_{\varsigma 0} + H_1\delta I + H_2\delta J + \omega\delta I_0 + \beta\delta\lambda = 0 \tag{12.75c}$$

$$R_{\phi 0} + E_1\delta I + E_2\delta J + \gamma\delta I_0 = 0 \tag{12.75d}$$

It follows from (12.69a), (12.69b), and (12.72) that

$$\delta I_\theta^{tr} = 3\frac{\partial K_{n+\theta}}{\partial I_{n+1}}\Delta\varepsilon_{pp}\delta I \tag{12.76a}$$

and $2(J_\theta^{tr})^2 = (s_{k\ell}^n + 2G_{n+\theta}\Delta e_{k\ell})(s_{k\ell}^n + 2G_{n+\theta}\Delta e_{k\ell})$

$$4J_\theta^{tr}\delta J_\theta^{tr} = 4(s_{k\ell}^n + 2G_{n+\theta}\Delta e_{k\ell})\frac{\partial G_{n+\theta}}{\partial I_{n+1}}\delta I\Delta e_{k\ell}$$

$$\delta J_\theta^{tr} = \frac{s_{k\ell}^{tr}\Delta e_{k\ell}}{J_\theta^{tr}}\frac{\partial G_{n+\theta}}{\partial I_{n+1}}\delta I \tag{12.76b}$$

The coefficients of (12.75a–12.75d) are

$$A_{11} = 1 - 3\frac{\partial K_{n+\theta}}{\partial I_{n+1}}\Delta\varepsilon_{pp} + 18\Delta\lambda I_{n+1}\frac{\partial K_{n+\theta}}{\partial I_{n+1}} + 18\Delta\lambda K_{n+\theta} - 9\Delta\lambda I_0^{n+1}\frac{\partial K_{n+\theta}}{\partial I_{n+1}}$$

$$A_{12} = 0$$

$$B_1 = -9\Delta\lambda K_{n+\theta}$$

$$F_1 = 18K_{n+\theta}I_{n+1} - 9K_{n+\theta}I_0^{n+1}$$

$$A_{21} = \frac{2J_\theta^{tr}\Delta\lambda}{x_0^2 N^2}\frac{\partial G_{n+\theta}}{\partial I_{n+1}} - \frac{s_{k\ell}^{tr}\Delta e_{k\ell}}{x_0 J_\theta^{tr}}\frac{\partial G_{n+\theta}}{\partial I_{n+1}}$$

$$A_{22} = 1$$

$$B_2 = 0$$

$$F_2 = \frac{2J_\theta^{tr} G_{n+\theta}}{x_0^2 N^2}$$

$$H_1 = -6C_1 \Delta\lambda I_0^{n+1}$$

$$H_2 = 0$$

$$\omega = 1 - 6C_1 \Delta\lambda I_{n+1} + 6C_1 \Delta\lambda I_0^{n+1}$$

$$\beta = -6C_1 I_0^{n+1} I_{n+1} + 3C_1 (I_0^{n+1})^2$$

$$E_1 = 2I_{n+1} - I_0^{n+1}$$

$$E_2 = \frac{2J_{n+1}}{N^2}$$

$$\gamma = -I_{n+1}$$

The Implementation Details.
Recognizing the coefficients that are zero, (12.75a– 12.75d) are rewritten as

$$R_{I0} + A_{11}\delta I + B_1 \delta I_0 + F_1 \delta\lambda = 0 \tag{12.77a}$$

$$R_{J0} + A_{21}\delta I + \delta J + F_2 \delta\lambda = 0 \tag{12.77b}$$

$$R_{\varsigma 0} + H_1 \delta I + \omega \delta I_0 + \beta \delta\lambda = 0 \tag{12.77c}$$

$$R_{\phi 0} + E_1 \delta I + E_2 \delta J + \gamma \delta I_0 = 0 \tag{12.77d}$$

We will directly manipulate the above equations to solve for the unknowns δI, δJ, δI_0, and λI. This will also later facilitate the determination of an equation for the consistent tangent operator.

Let us find δJ from (12.77b) and substitute into (12.77d)

$$\delta J = -(R_{J0} + A_{21}\delta I + F_2 \delta\lambda) \Rightarrow \tag{12.78a}$$

$$R_{\phi 0} - E_2 R_{J0} + (E_1 - E_2 A_{21})\delta I + \gamma \delta I_0 - E_2 F_2 \delta\lambda = 0 \tag{12.78b}$$

Let us find δI_0 from 12.77c and substitute into (12.77a) and (12.78b)

$$\delta I_0 = -\frac{1}{\omega}(R_{\varsigma 0} + H_1 \delta I + \beta \delta\lambda) \Rightarrow \tag{12.79}$$

$$R_{I0} - \frac{1}{\omega} B_1 R_{\varsigma 0} + \left(A_{11} - \frac{1}{\omega} B_1 H_1 \right) \delta I + \left(F_1 - \frac{1}{\omega} B_1 \beta \right) \delta \lambda = 0$$

$$R_{\phi 0} - E_2 R_{J0} - \frac{\gamma}{\omega} R_{J0} + \left(E_1 - E_2 A_{21} - \frac{\gamma}{\omega} H_1 \right) \delta I - \left(E_2 F_2 + \frac{\gamma}{\omega} \beta \right) \delta \lambda = 0$$

We introduce compact notations and rewrite the above two equations as

$$\bar{R}_{I0} + \bar{A} \delta I + \bar{F} \delta \lambda = 0 \tag{12.80a}$$

$$\bar{R}_{\phi 0} + \bar{E}_1 \delta I + \bar{E}_2 \delta \lambda = 0 \tag{12.80b}$$

where

$$\bar{R}_{I0} = R_{I0} - \frac{1}{\omega} B_1 R_{\varsigma 0}; \quad \bar{A} = A_{11} - \frac{1}{\omega} B_1 H_1; \quad \bar{F} = F_1 - \frac{1}{\omega} B_1 \beta$$

$$\bar{R}_{\phi 0} = R_{\phi 0} - E_2 R_{J0} - \frac{\gamma}{\omega} R_{\varsigma 0}; \quad \bar{E}_1 = E_1 - E_2 A_{21} - \frac{\gamma}{\omega} H_1; \quad \bar{E}_2 = -E_2 F_2 - \frac{\gamma}{\omega} \beta \tag{12.80c}$$

Let us find δI from (12.80a) and substitute into (12.80b) to find $\delta \lambda$ as

$$\delta I = -\frac{1}{\bar{A}} (\bar{R}_{I0} + \bar{F} \delta \lambda) \Rightarrow \tag{12.81a}$$

$$\bar{R}_{\phi 0} - \frac{\bar{E}_1}{\bar{A}} \bar{R}_{I0} + \left(\bar{E}_2 - \frac{\bar{E}_1}{\bar{A}} \bar{F} \right) \delta \lambda = 0$$

$$\delta \lambda = -\frac{R^*_{\phi 0}}{E^*} \tag{12.81b}$$

where

$$R^*_{\phi 0} = \bar{R}_{\phi 0} - \frac{\bar{E}_1}{\bar{A}} \bar{R}_{I0}; E^* = \bar{E}_2 - \frac{\bar{E}_1}{\bar{A}} \bar{F} \tag{12.81c}$$

The procedure then is that we (a) find $\delta \lambda$ from (12.81b), (b) find δI from (12.81a), and (c) find δJ and δI_0 from (12.78a) and 12.79, respectively.

The Consistent Tangent Operator.
Treating $\Delta \varepsilon_{ij}$ as a variable along with I, J, I_0, and $\Delta \lambda$, the Euler equations (12.73a–12.73d) are differentiated to find the rate. Recognizing that $\Delta \varepsilon_{ij}$ appear only in I^{tr}_{θ} and J^{tr}_{θ}, it can be verified that

$$\dot{I}^{tr}_{\theta} = 3K_{n+\theta} \delta_{k\ell} \dot{\varepsilon}_{k\ell} + 3\Delta \varepsilon_{k\ell} \frac{\partial K_{n+\theta}}{\partial I_{n+1}} \dot{I}$$

$$\dot{j}_\theta^{tr} = \frac{G_{n+\theta} s_{k\ell}^{tr}}{J_\theta^{tr}} \dot{e}_{k\ell} + \frac{s_{k\ell}^{tr} \Delta e_{k\ell}}{J_\theta^{tr}} \frac{\partial G_{n+\theta}}{\partial I_{n+1}} \dot{I}$$

As in previous cases (e.g., (12.64)), it can be shown that

$$R_{I0} + A_{11}\dot{I} + B_1 \dot{I}_0 + F_1 \dot{\lambda} = 0 \tag{12.82a}$$

$$R_{J0} + A_{21}\dot{I} + \dot{J} + F_2 \dot{\lambda} = 0 \tag{12.82b}$$

$$R_{\varsigma 0} + H_1 \dot{I} + \omega \dot{I}_0 + \beta \dot{\lambda} = 0 \tag{12.82c}$$

$$R_{\phi 0} + E_1 \dot{I} + E_2 \dot{J} + \gamma \dot{I}_0 = 0 \tag{12.82d}$$

where

$$R_{I0} = -3K\delta_{k\ell}\dot{\varepsilon}_{k\ell}; \quad R_{J0} = -\frac{G_{n+\theta} s_{k\ell}^{tr}}{J_\theta^{tr} x_0} \dot{e}_{k\ell}; \quad R_{\varsigma 0} = 0 \quad \text{and} \quad R_{\phi 0} = 0 \tag{12.82e}$$

It is now left for the reader to show (Problem 12.11) that the algorithmic stress–strain rate equation and the consistent tangent operator $D_{ijk\ell}$ are given by

$$\dot{\sigma}_{ij} = D_{ijk\ell}\dot{\varepsilon}_{k\ell}$$

$$\begin{aligned} D_{ijk\ell} = {} & \left(\frac{K_{n+\theta}}{\bar{A}} - \frac{2}{3} \frac{G_{n+\theta}}{x_0} \right) \delta_{ij}\delta_{k\ell} + \frac{G_{n+\theta}}{x_0} \left(\delta_{ik}\delta_{j\ell} + \delta_{i\ell}\delta_{jk} \right) \\ & + \frac{2\Delta e_{ij} \bar{t}_{k\ell}}{x_0} \frac{\partial G_{n+\theta}}{\partial I_{n+1}} - \frac{2\Delta\lambda s_{ij}^{tr} \bar{t}_{k\ell}}{x_0^2 N^2} \frac{\partial G_{n+\theta}}{\partial I_{n+1}} + \left(\frac{2 s_{ij}^{tr} G_{n+\theta}}{x_0^2 N^2 E^*} + \frac{\bar{F}\delta_{ij}}{3\bar{A}E^*} \right) t_{k\ell} \end{aligned} \tag{12.83a}$$

where

$$t_{k\ell} = \left(\frac{E_2 G_{n+\theta} s_{k\ell}^{tr}}{J_\theta^{tr} x_0} + \frac{3K_{n+\theta}\bar{E}_1 \delta_{k\ell}}{\bar{A}} \right) \tag{12.83b}$$

$$\bar{t}_{k\ell} = \left(\frac{3K_{n+\theta}\delta_{k\ell}}{\bar{A}} + \frac{\bar{F} t_{k\ell}}{\bar{A}E^*} \right) \tag{12.83c}$$

12.7.2 Comparison of Results Obtained Using the σ- and S-Space Formulations

When the steps remain the same (e.g., the type of approximation for the elasticity, etc.), the σ and S-space formulations are fundamentally identical and hence the

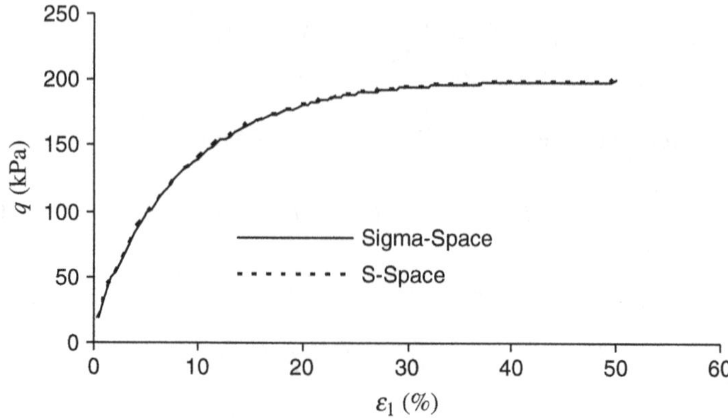

Fig. 12.14 Comparison of model behaviors calculated using the σ and S-space formulations

results obtained with them must be identical. Only one series of analyses is performed to confirm this. The triaxial drained behavior of a normally consolidated specimen (see Fig. 12.9) is simulated using both formulations and the $q - \varepsilon_1$ relations are compared in Fig. 12.14. The initial stress point is on the hydrostatic axis with $p_{init} = p_0 = 100\,\text{kPa}$. The model parameters presented in Table 12.1 are employed. It is seen that the results are identical. The computational efficiencies of the two methods are, however, different from each other. The efficiencies are evaluated for a footing problem in a subsequent section.

12.8 Multi-Element Finite Element Analysis

In this section, we present results from a series of analyses of a footing problem. The problem, along with the boundary conditions and two finite element meshes used to discretize the domain are shown in Fig. 12.15a, b. The analyses are carried out under drained condition with 2D plane strain assumption. Two different meshes are used: a fine mesh consisting of 2,736 three-noded, 2D triangular elements (Fig. 12.15a) and a coarse mesh consisting of 162 eight-noded 2D elements (Fig. 12.15b). During the study, it was noted that the analyses employing these different meshes brought out slightly different convergence difficulties (especially when employing CPPM with approximate elasticity for the integration of stresses).

To simulate a typical rigid footing, the footing is modeled using a set of interconnected rigid bending elements as shown in Fig. 12.15a, b. The column load P is applied as a point load at the mid point of the footing. The foundation soil is assumed to consist of normally consolidated homogeneous clay. Its stress–strain behavior is represented by the modified Cam-clay model with the parameters listed in Table 12.1.

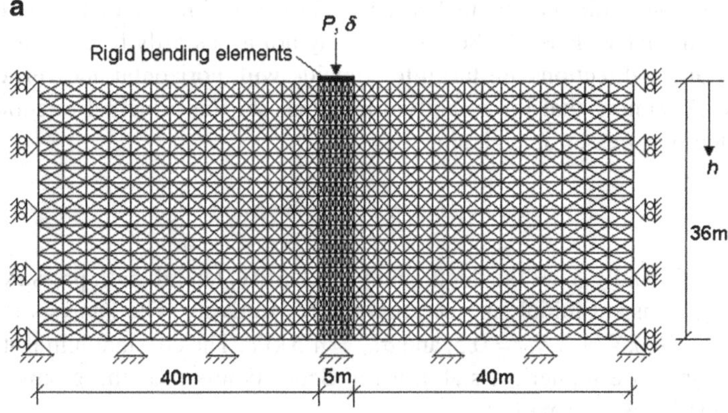

Mesh with 2736 Three-Noded 2D Elements

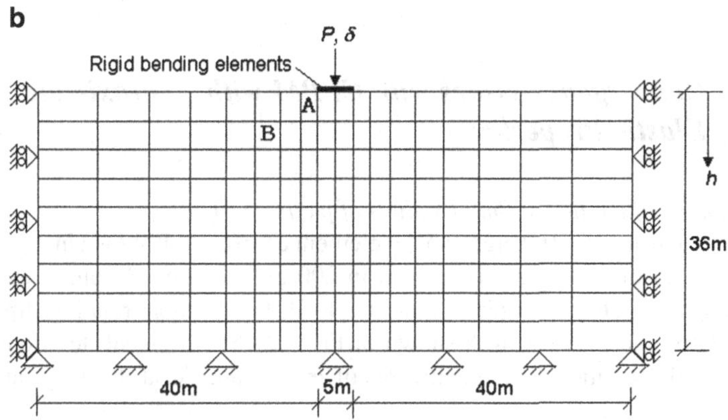

Mesh with 162 Eight-Noded 2D Elements

Fig. 12.15 The footing problem and finite element meshes used for the analyses

It is well known that the failure surface is non-circular in the octahedral plane. The radius of the surface is smaller in triaxial extension than in triaxial compression. This is simulated using M_c and M_e for the slope of the critical state line in compression and extension, respectively. Between these two extremes, the slope of the critical state line M lies in the range $M_e \le M \le M_c$.

For the present analysis, we make the following assumptions: (1) The deformation is plane strain, (2) the failure takes place on the critical state line, and (3) the failure is governed by the Mohr–Coulomb failure criterion. Under these assumptions, it can be shown that at failure, $M = M_{PS} = 0.866$ for $\phi = 30°$. The derivation is presented in Appendix 9.2. (It may be noted that Mohr–Coulomb failure criterion yields $M_c = 1.2$ and $M_e = 0.857$.) The above behavior may be simulated either using a non-circular failure surface such that M assumes a value of 0.866 at failure or using a circular failure surface with a fixed value $M = 0.866$. We choose the latter option in the present analysis. This, however, implies that even the yield surface is circular with $M = 0.866$, which may not necessarily be accurate.

The x- and y-directions are taken to coincide with horizontal and vertical, respectively. Assuming that the x- and y-directions are the principal directions, the initial effective stresses are calculated as

$$\sigma_y = \gamma' h + \sigma_{y0}; \quad \sigma_x = K_0 \sigma_y$$

where h, γ', K_0, and σ_{y0} are the depth, buoyant density, lateral earth pressure coefficient, and a constant initial surface pressure, respectively. The analysis is carried out with $\gamma' = 10\,kN/m^3$, $K_0 = 0.5$, and $\sigma_{y0} = 4.5\,\text{kPa}$. The constant initial stress σ_{y0} is used to increase numerical stability for Gauss points near the ground surface where the stresses are small.

12.8.1 Convergence Issues with CPPM with Approximate Elastic Properties

Convergence Difficulties at Small Values of p_0 and/or I

A run is made using the 162 eight-noded element coarse mesh shown in Fig. 12.15b. A column load of 10,000 kN is applied in 200 global steps. At one stage during the loading indicated by point A in the load-deformation relationship shown in Fig. 12.16, and in element A shown in Fig. 12.15b, the local iterations fail to converge to the solution. The magnitudes of $\boldsymbol{\sigma}_n$, p_0, and $\Delta\varepsilon$ at this stage are

$$\boldsymbol{\sigma}_n = \left(2.501 \times 10^{-5}, 2.606 \times 10^{-5}, 2.815 \times 10^{-5}, 1.760 \times 10^{-5}, 0, 0\right) \text{kPa},$$

$$\Delta\varepsilon = \left(1.844 \times 10^{-4}, -8.792 \times 10^{-6}, 0, 3.074 \times 10^{-4}, 0, 0\right) \quad \text{and}$$

$$p_0^n = 7.371 \times 10^{-5}$$

It is noted that the strain increments are not that large. However, the magnitudes of $\boldsymbol{\sigma}_n$ and p_0^n are very close to zero.

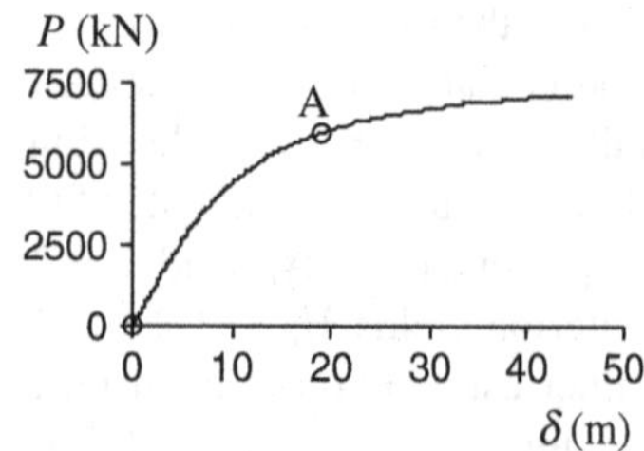

Fig. 12.16 Load-deformation behavior of the footing problem using the 162 eight-noded element coarse mesh

The formulation presented in Sect. 12.4.1 (CPPM with approximate elastic properties) is used with $\theta = 0$ in (12.39d) (explicit elasticity) and $TOL_\sigma = 0.01$, $TOL_\varsigma = 0.01$ and $TOL_\phi = 1.0 \times 10^{-5} p_0^n$ (12.51a–c). While e_ς and R_ϕ become less than the respective tolerances several times during the iteration, e_σ never becomes smaller than TOL_σ as seen in Fig. 12.17a, b within 100 iterations (the same results are plotted in these figures with different vertical scales). Subdividing the load at the global level does not seem to help move the solution forward.

The behavior has been found to stem from the fact that the elastic moduli (12.2a) are pressure dependent and the plastic modulus (12.9a) is both a function of I and p_0. (By expanding (12.9a) with the aid of (12.7b), A7.8, A8.9, and A8.12, one gets $K_p = 9C_1(2I - 3p_0)Ip_0$.) Furthermore, when the size of the yield surface is very small (which occurs when p_0 is very small), small variations in the values of the stresses produce large variations in the key constitutive functions such as K_p and $\mathbf{r}$.

To get through this non convergent point, we will impose lower limits on the values of p_0, K, and G (12.2a) as follows:

$$p_0 \geq p_{0\,\min} \tag{12.84a}$$

$$K \geq \bar{K}_0 p_{\min}; \quad G \geq \bar{G}_0 p_{\min} \tag{12.84b}$$

The limiting yield surface corresponding to $p_{0\,\min}$ is schematically shown in Fig. 12.18. Noting that the behavior corresponds to the case of a shrinking surface (i.e., softening), the plastic volumetric strain increment during the incremental loading from point A to point C must be negative (12.7a). For this to be the case, it is clear that point C must be on the left side of the critical state line, and hence on the left side of the vertical line DE.

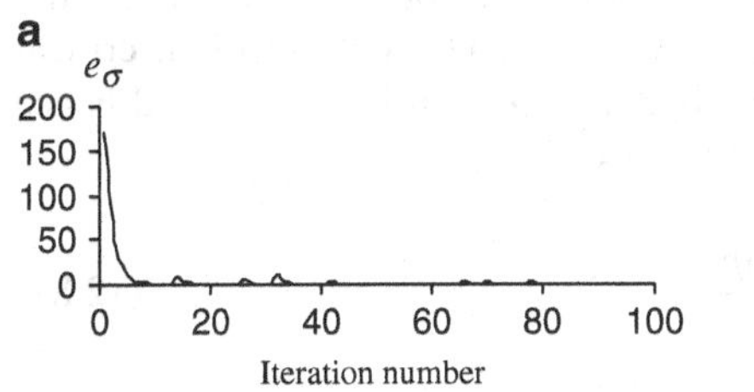

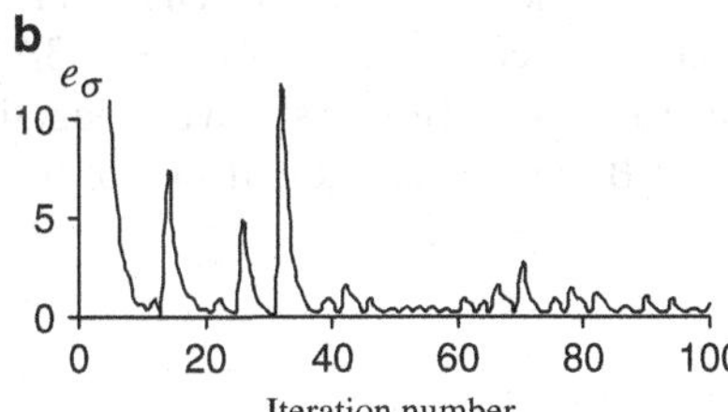

Fig. 12.17 The variation of e_σ during the iteration

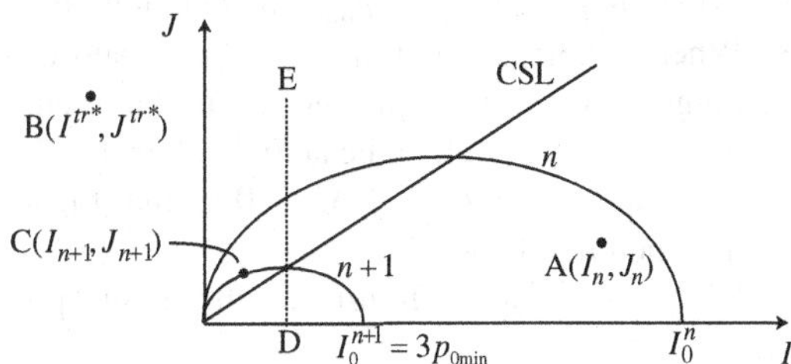

Fig. 12.18 Schematic of the limiting yield surface

To enforce the lower limit on p_0, the strain increment is split up into two parts: $x_0\Delta\boldsymbol{\varepsilon}$ and $(1-x_0)\Delta\boldsymbol{\varepsilon}$, where $0 \leq x_0 \leq 1$. The part $x_0\Delta\boldsymbol{\varepsilon}$ is used to decrease p_0 from p_0^n to $p_0^{n+1} \approx p_{0\,\min}$ and to produce an incremental stress $\Delta\boldsymbol{\sigma}$. The remaining part $(1-x_0)\Delta\boldsymbol{\varepsilon}$ is assumed to cause no change in p_0 and $\Delta\boldsymbol{\sigma}$. As it is difficult to compute the consistent tangent, the continuum tangent is used for this increment for this Gauss point. The value of x_0 is calculated as follows:

Integrating (12.7a) by the backward Euler method and setting a limit for p_0^{n+1}, we have

$$p_0^{n+1} = p_0^n + C_1 p_0^{n+1}\Delta\varepsilon_{kk}^p = p_0^n + C_1 p_0^{n+1}\left[\Delta\varepsilon_{kk} - \Delta\varepsilon_{kk}^e\right] = p_0^n + C_1 p_0^{n+1}\left[\Delta\varepsilon_{kk} - \frac{\Delta I}{3K}\right]$$

$$p_0^{n+1} = p_0^n\left[\frac{1}{1 - C_1\Delta\varepsilon_{kk} + \frac{C_1\Delta I}{3K}}\right] \geq p_{0\,\min}$$

$$\Delta\varepsilon_{kk} \geq \Delta\varepsilon_{kk}^* = \frac{1}{C_1 p_{0\,\min}}\left[p_{0\,\min} - p_0^n + \frac{C_1 p_{0\,\min}\Delta I}{3K}\right] \tag{12.85}$$

Hence the incremental volumetric strain must be larger than $\Delta\varepsilon_{kk}^*$ for p_0^{n+1} to be larger than $p_{0\,\min}$. A value for x_0 is calculated to achieve this as

$$x_0 = \frac{\Delta\varepsilon_{kk}^*}{\Delta\varepsilon_{kk}} \quad \text{for} \quad \Delta\varepsilon_{kk} \neq 0 \tag{12.86}$$

ΔI is an unknown in (12.85). As $0 \leq I_{n+1} \leq 3p_{0\,\min}/2$, we use an average value for I_{n+1} as $I_{n+1} = 3p_{0\,\min}/4$ and calculate an estimate for ΔI as $\Delta I = 3p_{0\,\min}/4 - I_n$.

Let us now examine the character of the plastic corrector $\Delta\boldsymbol{\sigma}^{pc} = -\Delta\lambda\mathbf{Cr}$. It's volumetric component is $\Delta\sigma_{kk}^{pc} = -3\Delta\lambda K r_{kk}$. As $r_{kk}<0$ on the left side of the critical state line, line BC must have a negative slope as shown in Fig. 12.18 and hence point B must be on the left side of line DE; hence,

$$I^{tr} < \frac{3p_{0\,\min}}{2} \tag{12.87}$$

To sum, the modifications to be made for imposing the limits presented in (12.84a) and (12.84b) are as follows:

- When $p<p_{\min}$, set $p = p_{\min}$ and calculate the elastic properties by (12.84b)
- When $I^{tr}<3p_{0\,\min}/2$ (12.87), $p_0>p_{0\,\min}$ and x_0 (computed from (12.86)) lies in the range $0<x_0<1$, multiply the strain increment by x_0. Perform the local iterations for $x_0\Delta\varepsilon$ and establish the actual values for $\Delta\boldsymbol{\sigma}$ and p_0^{n+1} (in view of the fact that ΔI is approximate). Set $\Delta\lambda = 0$ so that the continuum tangent will be used for this increment.
- When $p_0 \leq p_{0\,\min}$, return to the main program after setting $\Delta\boldsymbol{\sigma} = 0$ and $p_0^{n+1} = p_0^n$.

Convergence Difficulties due to p_0 Becoming Negative.
Another common difficulty encountered in our experimentations concerns the value of p_0 becoming negative, leading either to non convergence or convergence to a wrong solution. The two methods that have been found to work with some success are as follows:

- Method 1: At step 11 of Algorithm 12.3, when $p_{0,1}^{n+1}$ is found to be less than a tolerance TOL, set $p_{0,1}^{n+1} = TOL$. A value of $TOL = 1.0 \times 10^{-6} p_a$ has been found to work well.
- Method 2: At step 11 of Algorithm 12.3, when $p_{0,1}^{n+1}$ is found to be less than a tolerance TOL, reset the initial value of $\Delta\lambda$ and start from Step 5. Calculate the initial value of $\Delta\lambda$ as follows: Use the current value of $\Delta\lambda$ if $p_{0,2}^{n+1} \geq TOL$; else, compute $\Delta\lambda$ as (on the basis of the equation in step 7 of Algorithm 12.3)

$$p_{0,2}^{n+1} = p_0^n + \Delta\lambda s^{n+1} = TOL \Rightarrow \Delta\lambda = \frac{TOL - p_0^n}{s^{n+1}} \quad \text{for } s^{n+1} \neq 0. \tag{12.88}$$

If these techniques do not work, control is returned to the main program and the load is subdivided at the global level.

Example 12.8 Using the model parameters listed in Table 12.1, determine the local incremental solution by CPPM with approximate elasticity in the following case:

$$\boldsymbol{\sigma}_n = (10, 10, 10, 0, 0, 0)\,\text{kPa}, \;\; \Delta\boldsymbol{\varepsilon} = (-1.2, -0.8, -1.2, 0, 0, 0) \text{ and } p_0 = 40\,\text{kPa}$$

Solution: In the following, the unit for the stresses and stiffness is kPa. When $p_{0,1}^{n+1}$ becomes less than $TOL = 0.001$, Method 2 (12.88) described above is adopted.

We will use $p_{0\min} = p_{\min} = 1$. Using the unmodified $\Delta\varepsilon$, $I^{tr} = -6,690$. As $I^{tr} < 3p_{0\min}/2 = 1.5$, we will proceed to calculate x_0 as follows:

$$\Delta\varepsilon_{kk}^* = \frac{1}{19.1 \times 1}\left[1.0 - 40.0 + \frac{19.1 \times 1.0 \times (0.75 - 30.0)}{3 \times 700}\right] = -2.0558$$

$$\Delta\varepsilon_{kk} = -3.2$$

$$x_0 = \frac{2.0558}{3.2} = 0.6424$$

The strain increment is reduced accordingly, and the iterations lead to the following solution:

$$I^{tr*} = -4289.25$$

$$\boldsymbol{\sigma}_{n+1} = (-0.009146, 0.02224, -0.009146, 0, 0, 0)$$

$$p_0^{n+1} = 1.0001695$$

It is noted that the calculated value of $p_0^{n+1} \approx p_{0\,\mathrm{min}} = 1.0$ as intended.

When the strain increment is not reduced (i.e., $x_0 = 1$), the solution converges to

$$\boldsymbol{\sigma}_{n+1} = (-0.005905, 0.01435, -0.005905, 0, 0, 0)$$

$$p_0^{n+1} = 0.6471$$

While convergence is achieved, the value of p_0^{n+1} has become less than the limit. If this is allowed to continue, convergence difficulties are sure to be encountered in a subsequent increment (as was the case in the problem stated at the beginning of this section).

It may be pointed out that, for this example, regardless of whether $p_{0,1}^{n+1}$ is prevented from becoming negative or not, the iterations converge to the correct solution. However, for a different initial stress tried, the iterations converged to a wrong solution when $p_{0,1}^{n+1}$ is not adjusted to be positive. The problem tried is as follows:

$$\boldsymbol{\sigma}_n = \left(1.0\times10^{-5}, 1.0\times10^{-5}, 1.0\times10^{-5}, 0, 0, 0\right)\ \text{kPa},$$

$$\Delta\boldsymbol{\varepsilon} = (-1.2, -0.8, -1.2, 0, 0, 0) \quad \text{and} \quad p_0 = 40\ \text{kPa}$$

The converged solution when $p_{0,1}^{n+1}$ is adjusted to remain positive (by Method 2, (12.88)) is as follows:

$$\boldsymbol{\sigma}_{n+1} = (-0.009093, 0.022068, -0.009093, 0, 0, 0) \quad \text{and} \quad p_0^{n+1} = 1.001697,$$

whereas when $p_{0,1}^{n+1}$ is not adjusted to remain positive it is

$$\boldsymbol{\sigma}_{n+1} = (-186.64, -67.46, -186.64, 0, 0, 0) \quad \text{and} \quad p_0^{n+1} = -275.84.$$

There appears to be a solution with the surface completely flipped about the $I = 0$ line.

Divergence at the Global Level.

With the modifications described in the preceding two sections, the analysis of the problem shown in Fig. 12.15b continues until the load is increased to 6,349 kN (where the displacement is 23.5 m). At this point, the global iterations fail to converge within the set limit of 20 iterations. The load is subdivided and the analysis continues until the load becomes 6,876 kN (where the displacement is 33.72 m). At this stage, the local iterations fail to converge in element B shown in Fig. 12.15b. The magnitudes of $\boldsymbol{\sigma}_n$, p_0, and $\Delta\varepsilon$ at this stage are as follows:

$$\boldsymbol{\sigma}_n = (134.49, 44.95, 89.86, 5.39, 0, 0)\ \text{kPa},$$

$$\Delta\boldsymbol{\varepsilon} = (2.1982, -1.7845, 0, 0.2623, 0, 0) \quad \text{and} \quad p_0 = 180.38 \tag{12.89}$$

It is clear that the magnitude of the strain increment is very large. Further investigation reveals that the strain increments are even larger in several other Gauss points and that the behavior is due to divergence at the global level. While measures may be taken to affect convergence at the local level, it is not useful when the solution diverges at the global level. When the load vector is subdivided at the global level in half, the iteration converges locally and globally, and the solution advances.

Noting that the analysis is performed in load control, the iteration is not expected to converge at the global level when the applied load is larger than the collapse load, and that the incremental load vector is to be kept subdivided to advance the solution. The procedure eventually produces the load-displacement curve shown in Fig. 12.16 for the problem presented in Fig. 12.15b.

Before the improvements discussed above are implemented, the convergence behaviors observed during the analyses of the problems in Fig. 12.15a, b are slightly different from each other in the following sense. As pointed out in the preceding sections, in the analysis of the problem in Fig. 12.15b (coarse mesh), convergence difficulties are first encountered at the local level because of p_0 and I becoming very small. This behavior has not been observed during the analysis of the problem in Fig. 12.15a (fine mesh consisting of triangular elements). In both cases, however, the iterative procedure prematurely diverges at the global level at some stage, leading to very large strains and non-convergence at the local level.

Once the corrective measures discussed in this section have been implemented, the overall convergence behaviors of both problems (Fig. 12.15a, b) are very similar. The results from the analysis of the problem in Fig. 12.15b are further discussed in a subsequent section.

Remark. It is clear that the vertical displacement of the footing at failure is unrealistically large, suggesting that the large deformation effects need to be modeled. The theoretical and numerical treatment of the large deformation analysis is beyond the scope of this book. Some results and further comments will be presented on this in Sect. 12.8.5.

12.8.2 Results on Convergence of CPPM with Approximate Elastic Properties

To investigate the convergence behavior systematically and comprehensively, a series of local analyses are performed with each analysis beginning from a specific elastic trial stress point. For convenience, the trial points are chosen to be at the nodes of regular grids. A typical grid chosen in a meridional plane (triaxial compression) and in an octahedral plane are shown in Fig. 12.19a, b, respectively. The runs are made after the improvements discussed in the preceding sections have

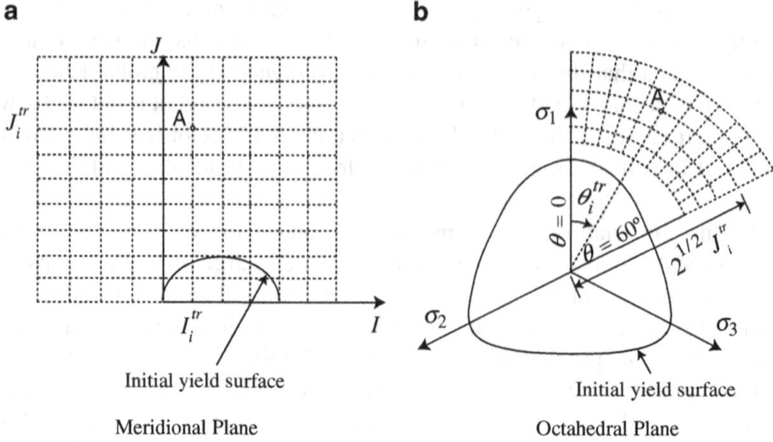

Fig. 12.19 Grids of different elastic trial points

been incorporated (which should affect only the results in the meridional plane presented below). In the following, the unit for the stresses and moduli is kPa.

In the meridional plane (Fig. 12.19a), a 100×100 grid is used. The elastic trial stresses at node i of the grid (point A) are (I_i^{tr}, J_i^{tr}). Runs are made for two different initial sizes for the yield surface: $p_0^n = 2$ and $p_0^n = 10$. The analyses use $p_{0\,\min} = 1$ (and hence the choice of $p_0^n = 2$ is very close to this limit). The tolerances used in the iterations are (12.51) $TOL_\sigma = 0.01$, $TOL_\phi = 0.01$, and $TOL_\phi = 1.0 \times 10^{-5} p_0^n$. The results are presented in the form of contours of the number of iterations required for convergence. The initial stresses chosen for the runs are $\boldsymbol{\sigma}^n = (2, 2, 2, 0, 0, 0)$ for the case with $p_0^n = 2$ and $\boldsymbol{\sigma}^n = (10, 10, 10, 0, 0, 0)$ for the case with $p_0^n = 10$ (hence the soil is assumed to be normally consolidated). Starting from these points, the trial points are reached using suitable (pre-calculated) strain increments. Each point requires different strain increment. For example, the strain increments to reach points A, B, C, and D (Fig. 12.20a) are

$$\Delta\boldsymbol{\varepsilon} = (-0.1927, -0.1927, -0.1860, 0, 0, 0),$$

$$\Delta\boldsymbol{\varepsilon} = (-0.4161, -0.4161, 0.2607, 0, 0, 0),$$

$$\Delta\boldsymbol{\varepsilon} = (-0.1780, -0.1780, 0.4988, 0, 0, 0), \text{ and}$$

$$\Delta\boldsymbol{\varepsilon} = (0.0454, 0.0454, 0.0521, 0, 0, 0), \text{ respectively}$$

The results for $p_0^n = 2$ and $p_0^n = 10$ are presented in Fig. 12.20a, b, respectively. Note that $p^{tr} = I^{tr}/3$ and $q^{tr} = \sqrt{3} J^{tr}$. The following observations are made:

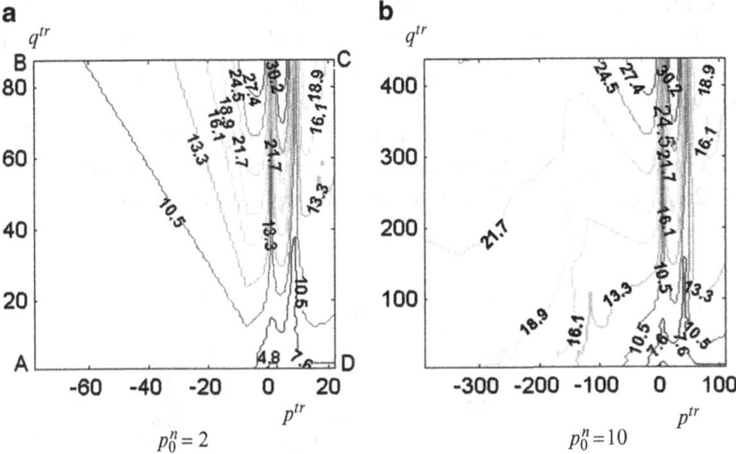

Fig. 12.20 Contours of the number of iterations required for convergence for different elastic trial stress points in the meridional plane

- Noting that the initial sizes of the yield surfaces correspond to $p_0^n = 2$ and $p_0^n = 10$, it is seen that the chosen grids span large areas, posing some of the severest tests for the integration method (with some trial points lying very far from the surface and some points having large negative values of p^{tr}).
- Generally, the number of iterations increases as the trial point moves farther from the initial yield surface. The points directly above the surface seem to take slightly more iterations than those in other parts of the grid.
- Convergence is achieved in all cases considered.

In the octahedral plane (Fig. 12.19b), a 100×100 grid is used. The elastic trial stresses at node i of the grid (point A) are $(\theta_i^{tr}, J_i^{tr})$. Runs are made for two different initial sizes for the yield surface, $p_0^n = 1$ and $p_0^n = 10$, and for each case, runs are made with $n = \frac{M_e}{M_c} = 1$ (circular yield surface) and $n = \frac{M_e}{M_c} = 0.8$ (non-circular yield surface). The octahedral planes are chosen at the same values of p_n; i.e., $p_n = p^{tr}$. The initial stresses chosen for the runs are $\boldsymbol{\sigma}^n = (1, 1, 1, 0, 0, 0)$ for the case with $p_0^n = 1$ and $\boldsymbol{\sigma}^n = (10, 10, 10, 0, 0, 0)$ for the case with $p_0^n = 10$. Starting from these points, the trial points are reached using suitable (pre-calculated) strain increments. For example, the strain increments required to reach the minimum and maximum value of q^{tr} in the grid are

$$\Delta\boldsymbol{\varepsilon} = (-0.0268, -0.0268, 0.0536, 0, 0, 0) \text{ and}$$

$$\Delta\boldsymbol{\varepsilon} = (-1.3403, -1.3403, 2.6805, 0, 0, 0), \text{ respectively.}$$

The ranges selected for the trial stress parameters for the set $(p_n = 1, p_0^n = 1)$ are $0 \leq \theta^{tr} \leq 60°$ and $5 \leq q^{tr} \leq 260$, and for the set $(p_n = 10, p_0^n = 10)$ are

$0 \leq \theta^{tr} \leq 60°$ and $50 \leq q^{tr} \leq 2,600$. The tolerances used in the iterations are (12.51) $TOL_{\sigma} = 0.01$, $TOL_{\varsigma} = 0.01$, and $TOL_{\phi} = 1.0 \times 10^{-5} p_0^n$. The results are presented in the form of contours of the number of iterations required for convergence.

For $n = 1$, the results for $p_0^n = 1$ and $p_0^n = 10$ are presented in Fig. 12.21a, b, respectively. For $n = 0.8$, the results for $p_0^n = 1$ and $p_0^n = 10$ are presented in Fig. 12.22a, b, respectively. The following observations are made:

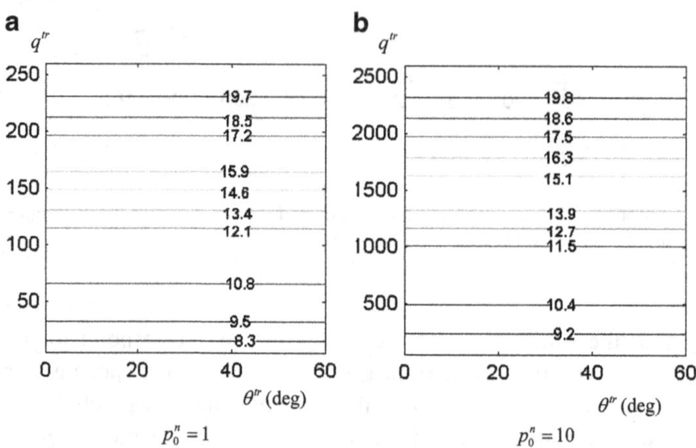

Fig. 12.21 Contours of the number of iterations required for convergence for different elastic trial stress points in the octahedral plane with circular yield surface

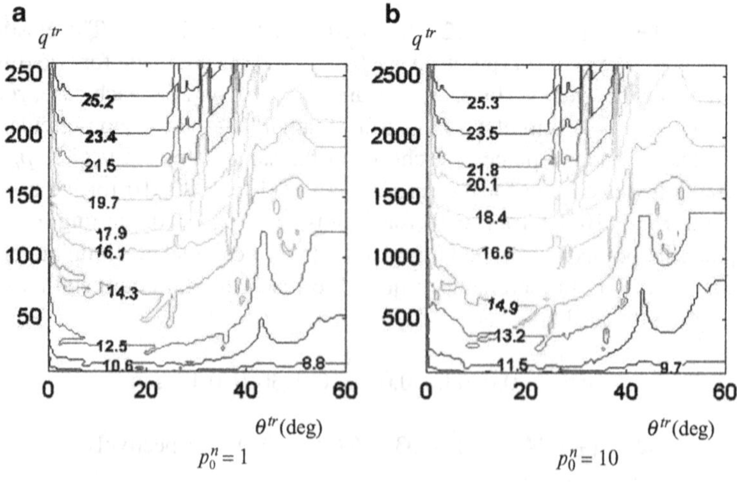

Fig. 12.22 Contours of the number of iterations required for convergence for different elastic trial stress points in the octahedral plane with non-circular yield surface ($n = M_e/M_c = 0.8$)

- Noting that the maximum value of q on the initial yield surface is $M_c p_0^n/2 = 0.866 \times 1/2 = 0.433$ for the case with $p_0^n = 1$, and 4.33 for the case with $p_0^n = 10$, it is seen that the chosen grids span a large range for q^{tr}, posing some of the severest tests for the integration method. Noting that the values of θ^{tr} lie between 0 (triaxial compression) and 60° (triaxial extension), and the yield surface has identical shapes in different quadrants, the range chosen for θ^{tr} covers all possible values.
- For the case with $n = 1$ (circular yield surface), as expected, the number of iterations required for convergence is independent of θ^{tr}. The number of increments increases as the trial point radially moves farther from the initial yield surface.
- For the case with $n = 0.8$ (non-circular yield surface), iterations required for convergence slightly depend on the value of θ^{tr}.
- Convergence is achieved in all cases considered.

12.8.3 Comments on Convergence Issues with the Other Methods

- *CPPM with Exact Elasticity*: The primary difficulties with the use of CPPM with exact elasticity (Sect. 12.5) are that (a) the exponential term in (12.53a) overflows when $\bar{K}_0 \Delta\varepsilon_{kk} \geq e$ (e.g., $e = 20$), (b) the exponential term in (12.54b) overflows when $\bar{K}_0 \left(\Delta\varepsilon_{kk} - \Delta\lambda r_{kk}^{n+1}\right) \geq e$, and (c) the exponential term in (12.58b) overflows when $C_1 \Delta\lambda r_{kk}^{n+1} \geq e$. This may be overcome by sub dividing at the global level as discussed in the preceding section. The problem considered in the preceding section (Fig. 12.15b) reveals that preventing $\Delta\lambda$ becoming negative helps a great deal in preventing the above overflows. That is, whenever $\Delta\lambda \leq TOL$ (e.g., $TOL = 1.0 \times 10^{-20} p_a$), set $\Delta\lambda = TOL$. Note that unlike in the case with approximate elasticity described in the preceding section, (12.58b) ensures that p_0 does not become negative. The methods described in the preceding section (e.g., (12.88)) may be used to impose a lower limit for the value of p_0 if desired.
- *The Cutting-Plane Method:* With this method, the analyses using the meshes shown in Fig. 12.15a, b indicate that the size of the increments must be small (smaller than that for CPPM) and that the rate of convergence is very low near failure. Consequently, as the increment size had to be kept small at the global level to advance the solution, the other problems such as p_0 becoming negative did not surface.

12.8.4 Comparison of Results Obtained Using Different Integration Methods

The results obtained using the fine mesh shown in Fig. 12.15a are considered. The loading is applied in load-controlled mode by increasing P from 0 to 10,000 kN in

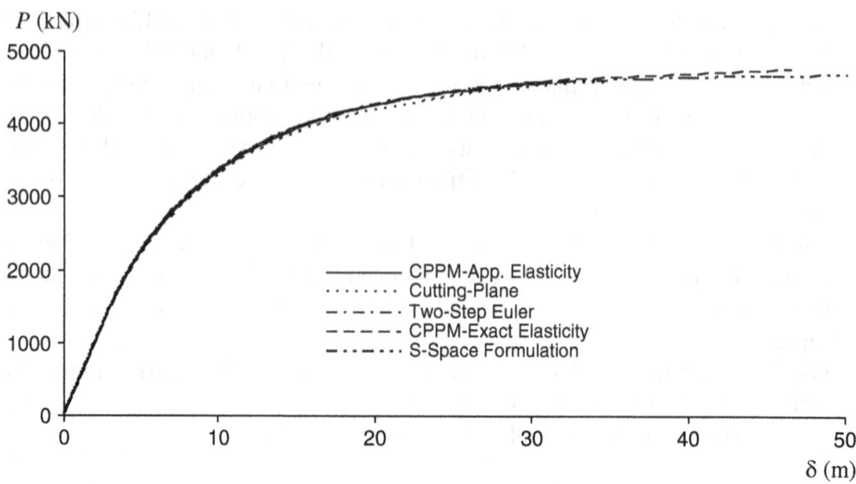

Fig. 12.23 Load-deformation behavior of the footing problem using the 2,736-element fine mesh

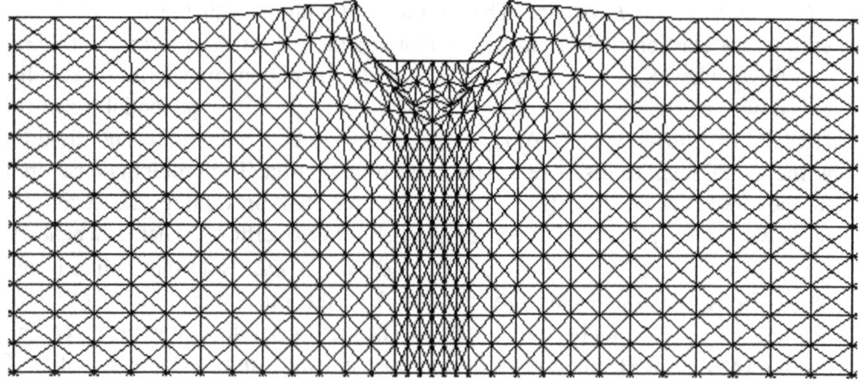

Fig. 12.24 Deformed configuration of the footing problem within a window

200 steps. The load-deformation $(P - \delta)$ curve calculated using the five integration methods presented in this chapter are plotted together in Fig. 12.23. An example of the deformed configuration near failure is shown in Fig. 12.24.

Referring to the $P - \delta$ curves, it may be observed that the results obtained using the different integration methods are virtually identical to each other and the failure occurs at a very large displacement. The analysis continued till the end for all cases, though the load had to be subdivided at the global level at different stages for different methods.

To assess the computational efficiencies of the different integration methods when applied to the footing problem, the clock times taken to run the analyses in a laptop are monitored for the initial portion of the analyses (up to $\delta = 13\,m$).

Table 12.12 Clock times to carry out analysis until $\delta = 13\,m$ with $P = 5,000\,\text{kN}$ applied over NINC increments

Method of integration	t (seconds) for NINC = 200	t (seconds) for NINC = 25
Two-step Euler	146	214
Cutting-plane	196	196
CPPM with approximate elasticity (explicit)	206	117
CPPM with exact elasticity	219	212
S-Space formulation	115	71

The results are compared in Table 12.12 for two series of analyses; one where the load of 5,000 kN is applied over 200 global load steps, and the other where the load of 5,000 kN is applied over 25 global load steps. The time consumptions for subdividing at the global and local levels are included in the times listed.

The S-space formulation is found to be the most efficient. The remaining methods are generally equally competitive, although the two-step Euler method is more efficient than the others when the load step is small, and CPPM with approximate elasticity is more efficient than the others when the load step is large. The cutting-plane method forced immediate sub divisions, hence resulting in the same overall clock time requirement.

12.8.5 Comparison Between the Finite Element and Analytical Bearing Capacities

The bearing capacity from the finite element analysis (Fig. 12.23) is $4,750/(5 \times 1) = 950\,\text{kPa}$. Assuming that plastic flow occurs on the critical state line, the soil is treated as purely frictional as the critical state line passes through the origin. The plane strain value of 0.866 used for M corresponds to a friction angle of 30° (Appendix 9.2). The analytical bearing capacities are calculated as follows:

Terzaghi(s bearing capacity equation (Terzaghi 1943):

$$q_u = qN_q + 0.5B\gamma N_\gamma = 4.5 \times 22.46 + 0.5 \times 5 \times 10 \times 19.13 = 580\,\text{kPa}$$

Meyerhof(s general bearing capacity equation (Mayerhof 1963):

$$q_u = qN_qF_{qd} + 0.5B\gamma N_\gamma = 4.5 \times 18.4 \times 1.026 + 0.5 \times 5 \times 10 \times 22.4 = 645\,\text{kPa}$$

Hence the bearing capacity calculated by the finite element method is roughly 50% higher than the values calculated using the most widely used bearing capacity equations. Both the bearing capacity equations and the finite element method involve many assumptions. In the latter, the effect of large deformation must be considered. A detailed evaluation of the differences is beyond the scope of this chapter. However, one series of analyses is conducted with the large deformation effects accounted for.

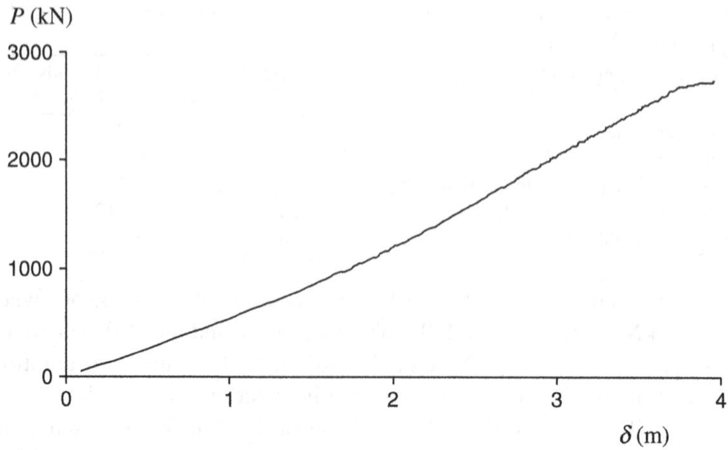

Fig. 12.25 Load-deformation behavior of the footing problem predicted using large-deformation analysis (2,736-element fine mesh, CPPM with exact elasticity)

The analysis is performed using the updated Lagrangian formulation with the Jaumann rate of Cauchy stress (also called the co-rotational rate). The results obtained using the fine mesh (Fig. 12.15a) with CPPM for the local iterations are presented in Fig. 12.25. The analysis is repeated using the Hughes–Winget algorithm (1980), which prevents the development of linear strains during rigid body rotations when the increment step size is large. The results are found to be identical. The following two key observations can be made from the results: (1) The vertical displacement of the footing at failure is much smaller than that obtained using the small-deformation formulation, and (2) the collapse load is smaller than that predicted by the small-deformation formulation. The collapse load is 2,740 kN, which corresponds to a bearing pressure of 548 kPa. This value is very close to the bearing capacity calculated by the Terzaghi(s bearing capacity equation. It is, therefore, clear that in the analysis of "soft" clay such as the one involved in the present analysis, it is necessary to consider the large-deformation effects.

Problems

Solve problems 12.2, 12.3, 12.4, 12.6, and 12.9 with the following parameters: $\sigma_1^n = 90$, $\sigma_3^n = 90$, $p_0^n = 100$, $\Delta\varepsilon_1 = 0.01$, and $\Delta\varepsilon_3 = -0.005$. The modified Cam-clay model parameters of the soil are $\lambda^* = 0.14$, $\kappa = 0.03$, $\nu = 0.3$, and $M_c = 1.2$. The initial void ratio of the soil is 1.1. Assume triaxial conditions. The strain increments correspond to a standard undrained triaxial compression loading.

Solve Problems 12.5, 12.7, and 12.10 with the following parameters: $\sigma_1^n = 90$, $\sigma_3^n = 90$, and $p_0^n = 100$. A strain-controlled, triaxial undrained compression loading is to be conducted with $\Delta\hat{\varepsilon}_1 = 0.05$ and $\Delta\hat{\sigma}_3 = 0.0$ (standard triaxial loading).

Determine the stresses and strains at the end of the incremental loading. Use the procedure of Algorithm 8.2, along with specific triaxial equations developed in Sect. 12.2.2, for global iterations. Use the specific algorithm required in the problem for local iterations. Examine the global convergence behavior of the algorithm. The modified Cam-clay model parameters of the soil are $\lambda^* = 0.14$, $\kappa = 0.03$, $\nu = 0.3$, and $M_c = 1.2$. The initial void ratio of the soil is 1.1.

Problem 12.1 Repeat the analysis presented in Example 8.3 (Chap. 8) by replacing the function in (8.23) by the following.

$$\sigma = \frac{b\varepsilon}{a + \varepsilon}$$

Find the solution for $\bar{f} = 0.5$ with $a = b = 1$.

Problem 12.2 Considering the two-step Euler algorithm, determine the step size ΔT at the end of the first inner iteration in Algorithm 12.1. Use $TOL = 1.0 \times 10^{-3}$ and $\beta_{\min} = 0.1$ and $\beta_{\max} = 0.5$. (The strain increment corresponds to a standard undrained triaxial compression loading.) (**Ans**: max(0.0817,0.1) = 0.1)

Problem 12.3 Modify Algorithm 12.1 (the two-step Euler method) to perform calculations based on the triaxial equations developed in Sect. 12.2.2. Write a computer program to implement your algorithm and obtain accurate solution for Problem 12.2 with $TOL = 1.0 \times 10^{-3}$ and $\beta_{\min} = 0.1$ and $\beta_{\max} = 0.5$. (**Ans**: $\Delta\sigma_1 = 22.745$, $\Delta\sigma_3 = -34.888$, and $p_0^{n+1} = 105.358$).

Problem 12.4 Use the cutting-plane method (Algorithm 12.2), as in Example 12.2, to solve the problem. Perform 5 iterations and examine the convergence behavior. Develop a computer program to perform the required calculations. (**Partial answer**: Results after 5 iterations: $\sigma_1^{n+1} = 112.94$, $\sigma_3^{n+1} = 56.69$, and $p_0^{n+1} = 104.56$).

Problem 12.5 Use the cutting-plane method (Algorithm 12.2) for the local iterations and Algorithm 8.2 for the global iterations. Use the specific triaxial equations developed in Sect. 12.2.2 to perform the calculations. Examine the global convergence behavior of the algorithm. The problem is similar to Example 12.3.

Problem 12.6 Use CPPM with approximate elastic properties (Algorithm 12.3) and $\theta = 0.5$ to solve this problem. Use the triaxial equations developed in Sect. 12.4.2. Perform five iterations, as in Example 12.4, and examine the convergence behavior. Develop a computer program to perform the required calculations. (**Partial answer**: Results after five iterations: $\sigma_1^{n+1} = 113.89$, $\sigma_3^{n+1} = 59.12$, and $p_0^{n+1} = 104.29$).

Problem 12.7 Use CPPM with approximate elastic properties and $\theta = 0.5$ (Algorithm 12.3) for the local iterations and Algorithm 8.2 for the global iterations. Use the specific triaxial equations developed in Sect. 12.4.2 to perform the calculations. Examine the global convergence behavior of the algorithm.

Repeat the analysis using Algorithm 8.2 for global iterations with *continuum tangent* operator and Algorithm 12.3 for local iterations with $\theta = 0.5$, and compare the results with those obtained using consistent operator. The problem is similar to Example 12.5. Develop a computer program to solve this problem.

Problem 12.8 Let the inverse relation be

$$\dot{\varepsilon}_{ij} = C^{*-1}_{ijpq}\dot{\sigma}_{pq},$$

where C^{*}_{ijpq} is given by (12.65b).

(1) Show that $C^{*-1}_{ijpq} = \frac{1}{4G_{ave}}\left(\delta_{ip}\delta_{jq} + \delta_{iq}\delta_{jp}\right) + \left(\frac{1}{3K_0 Z_{kk}}\delta_{ij} - \frac{1}{2G_{ave}Z_{kk}}Z_{ij}\right)\delta_{pq}$
(2) Verify that $C^{*}_{ijpq}C^{*-1}_{pqrs} = \frac{1}{2}\left\{\delta_{ir}\delta_{js} + \delta_{is}\delta_{jr}\right\}$
(3) Show that $\bar{F}_{ij} = C^{*-1}_{ijpq}F_{pq} = r_{ij}$
(4) Show that $\bar{B}_{ij} = C^{*-1}_{ijpq}B_{pq} = \frac{1}{3}\Delta\lambda \frac{\partial r^{n+1}_{kk}}{\partial p^{n+1}_0}\delta_{ij}$
(5) Show that when $\Delta\varepsilon_{ij} \to 0$ and $\Delta\lambda \to 0$, C^{*}_{ijpq} converges to the continuum elastic tangent operator as

$$C^{*}_{ijpq} \to C_{ijpq} = \left[K_n - \frac{2}{3}G_n\right]\delta_{ij}\delta_{pq} + G_n\left[\delta_{ip}\delta_{jq} + \delta_{iq}\delta_{jp}\right]$$

Problem 12.9 Use CPPM with exact elastic properties (Algorithm 12.4) to solve this problem. Perform five iterations, as in Example 12.6, and examine the convergence behavior. Develop a computer program to perform the required calculations. (**Partial answer**: Results after five iterations: $\sigma_1^{n+1} = 113.83$, $\sigma_3^{n+1} = 59.14$, and $p_0^{n+1} = 104.21$).

Problem 12.10 Use CPPM with exact elastic properties (Algorithm 12.4) for the local iterations. Use the specific triaxial equations developed in Sect. 12.5.2 to perform the calculations. Examine the global convergence behavior of the algorithm.

Repeat the analysis using Algorithm 8.2 for global iterations with the *continuum tangent* operator and Algorithm 12.4 for local iterations, and compare the results with those obtained using the consistent operator. The problem is similar to Example 12.7. Develop a computer program to solve this problem.

Problem 12.11 Starting with the (12.82a–12.82e), show that the consistent tangent operator for the S-space formulation considered in Sect. 12.7 is given by (12.83a).

Partial steps:

Show that

$$R^{*}_{\phi 0} = t_{k\ell}\dot{\varepsilon}_{k\ell}; \quad \dot{\lambda} = -\frac{t_{k\ell}\dot{\varepsilon}_{k\ell}}{E^{*}}; \quad \dot{I} = \bar{t}_{k\ell}\dot{\varepsilon}_{k\ell};$$

$$\dot{\sigma}_{ij} = \frac{2G_{n+\theta}\dot{e}_{ij}}{x_0} + \frac{2\Delta e_{ij}}{x_0}\frac{\partial G_{n+\theta}}{\partial I_{n+1}}\dot{I} - \frac{s^{tr}_{ij}}{x_0^2}\left[\left(\frac{2\Delta\lambda}{N^2}\right)\frac{\partial G_{n+\theta}}{\partial I_{n+1}}\dot{I} + \frac{2G_{n+\theta}}{N^2}\dot{\lambda}\right] + \frac{1}{3}\delta_{ij}\dot{I}$$

and then arrive at the required result.

Chapter 13
The Drucker–Prager Model and Its Integration

The Drucker–Prager model was proposed by Drucker and Prager (1952) to describe the stress–strain behavior of pressure-dependent materials such as soil, rock, and concrete. The model uses in the stress space a conical failure surface whose projection in the octahedral plane is a circle and in the meridional plane is a line. In the octahedral plane, the circle is centered at the origin. In the meridional plane, the slope of the line represents the friction angle of the material and the intercept of the line represents the cohesion of the material. The failure surface is, therefore, a generalization of the Mohr–Coulomb failure surface (Chap. 9), represented by a smooth conical surface instead of the irregular hexagonal cone with corners. The details of the Drucker–Prager and Mohr–Coulomb failure surfaces are presented in Chap. 9.

The Drucker–Prager model is a classical model in that it uses a single yield surface and the behavior inside the yield surface is purely elastic. The model employs an associated flow rule and neglects hardening. In the present chapter, however, to increase the utility of the model, we will present an extended version with the following additional features:

- Non-associated flow rule
- Nonlinear isotropic hardening rule
- Non-circular yield, failure, and potential surfaces in the octahedral plane

Simpler versions are obtained as special cases of the above generalized model. For example, a model with an associated flow rule, a non-hardening model (i.e., an elastic perfectly plastic model), etc., are obtained as special cases.

In Chap. 10, we have presented three methods of integrating constitutive laws: (1) The two-step Euler method with sub-stepping, (2) the cutting-plane method, and (3) the closest point projection method (CPPM). Two versions of CPPM, one with approximate elastic properties and the other with exact elastic properties, have been presented. Also, two formulations of implementing CPPM, the σ-space and S-space formulations, have been presented.

The von Mises model is integrated using CPPM in Chap. 11 (with both the σ-space and S-space formulations). The Cam-clay model is integrated using all of

A. Anandarajah, *Computational Methods in Elasticity and Plasticity: Solids and Porous Media*, DOI 10.1007/978-1-4419-6379-6_13,

the methods in Chap. 12. The application of the two-step Euler and cutting-plane methods to different models is fairly straightforward. Hence, in the present chapter, we will only present the details of integrating the Drucker–Prager model using CPPM. The reader is referred to Chaps. 10 and 12 to learn to implement the Drucker–Prager model by one of the other two methods.

When the yield surface is a circle in the octahedral plane, the S-space formulation leads to a computationally efficient algorithm at the local level (requiring the solution of only a scalar equation for $\Delta\lambda$). Such a formulation has been described for the von Mises model in Chap. 11 and for the modified Cam-clay model in Chap. 12. We will describe the formulation for the Drucker–Prager model in the present chapter.

The methods presented in this chapter are as follows:

- CPPM using the σ-space formulation and approximate elastic properties (Sect. 13.2)
- CPPM using the σ-space formulation and exact elastic properties (Sect. 13.3)
- CPPM using the S-space formulation and approximate elastic properties (Sect. 13.4)

We will also discuss different types of convergence difficulties that are generally encountered with the use of CPPM.

The relevant equations associated with the Drucker–Prager model are presented in Sect. 13.1. The reader is referred to other published materials on detailed behavior of the model under various initial and loading conditions.

13.1 The Drucker–Prager Model

Assuming isotropy, the elastic stress–strain relation is expressed as

$$\dot{\boldsymbol{\sigma}} = \mathbf{C}\dot{\boldsymbol{\varepsilon}}^{\mathrm{e}} \tag{13.1a}$$

where

$$C_{ijk\ell} = \left(K - \frac{2}{3}G\right)\delta_{ij}\delta_{k\ell} + G\left(\delta_{ik}\delta_{j\ell} + \delta_{i\ell}\delta_{jk}\right) \tag{13.1b}$$

and G and K are the shear and bulk moduli, respectively. $\boldsymbol{\delta}$ is the Kronecker delta. $\mathbf{C}$ is the elastic stiffness tensor. $\dot{\boldsymbol{\sigma}}$ and $\dot{\boldsymbol{\varepsilon}}^{\mathrm{e}}$ are the rate of stress and rate of elastic strain, respectively.

While in the original model, the elastic properties are assumed to be independent of pressure, we will use pressure-dependent expressions for G and K. For G, we use a modified form of the formula proposed by Richart et al. (1970). K is then evaluated from G and Poisson's ratio v according to the theory of elasticity. The formulas are

$$G = \bar{G}_0(p + p_t)^{n_e}; \quad K = \bar{K}_0(p + p_t)^{n_e} \tag{13.2a}$$

where

$$\bar{G}_0 = G_0 p_a \frac{(2.97 - e_0)^2}{1 + e_0} \frac{1}{p_a^{n_e}} \quad \text{and} \quad \bar{K}_0 = \frac{2(1+v)}{3(1-2v)} \bar{G}_0 \tag{13.2b}$$

and $p_t \geq 0$ is a tension cutoff. When $p_t = 0$ and $n_e = 0.5$, the formula for G becomes identical to the one proposed by Richart et al. (1970) for granular materials. When $n_e = 0$, the model leads to pressure-independent properties G and K. For cohesive soils, a non-zero value may be used for p_t. p_t will be more precisely defined below after giving a definition of the yield surface. v, e_0, G_0, and p_a are Poisson's ratio, the initial void ratio, an elastic model parameter, and atmospheric pressure, respectively.

The plastic constitutive relations are given by the following equations (Chap. 9):

$$\dot{\boldsymbol{\varepsilon}}^{\mathrm{p}} = \dot{\lambda}\mathbf{r}; \quad \mathbf{r} = \frac{\partial \psi}{\partial \boldsymbol{\sigma}}; \quad \dot{\lambda} = \frac{1}{K_p}\mathbf{n}\dot{\boldsymbol{\sigma}}; \quad \mathbf{n} = \frac{\partial \phi}{\partial \boldsymbol{\sigma}} \tag{13.3}$$

$\dot{\boldsymbol{\varepsilon}}^{\mathrm{p}}$ is the plastic strain rate. $\dot{\lambda}$ is the consistency parameter (or the loading index). K_p is the plastic modulus. ϕ and ψ are the yield and potential functions, respectively. The flow rule is associated when $\phi = \psi$, which leads to $\mathbf{r} = \mathbf{n}$. Assuming additivity, the total strain rate is split up into elastic and plastic parts as

$$\dot{\boldsymbol{\varepsilon}} = \dot{\boldsymbol{\varepsilon}}^{\mathrm{e}} + \dot{\boldsymbol{\varepsilon}}^{\mathrm{p}} \tag{13.4}$$

Three stress invariants I, J and θ are defined as (Chap. 2)

$$I = \sigma_{k\ell}\delta_{k\ell} \tag{13.5a}$$

$$s_{ij} = \sigma_{ij} - \frac{1}{3} I \delta_{ij}; \quad J = \left(\frac{1}{2} s_{k\ell} s_{k\ell}\right)^{1/2} \tag{13.5b}$$

$$S = \left(\frac{1}{3} s_{ij} s_{jk} s_{ki}\right)^{1/3} \tag{13.5c}$$

$$-\frac{\pi}{6} \leq \theta = \frac{1}{3} \sin^{-1}\left[\frac{3\sqrt{3}}{2}\left(\frac{S}{J}\right)^3\right] \leq \frac{\pi}{6} \tag{13.5d}$$

θ is the Lode angle. The yield and plastic potential surfaces are given by

$$\phi(I,J,\theta,m) = J - m\alpha I - mk = 0 \tag{13.6a}$$

$$\psi(I,J,\theta,m) = J - m\alpha d_0 I - mk = 0 \tag{13.6b}$$

where

$$\alpha = \alpha_c g(n,\theta); \quad k = k_c g(n,\theta); \quad n = \frac{\alpha_e}{\alpha_c} = \frac{k_e}{k_c} \tag{13.6c}$$

$$g(n,\theta) = \frac{2n}{1+n-(1-n)\sin 3\theta} \tag{13.6d}$$

$$0 \le m \le 1 \tag{13.6e}$$

m is the isotropic hardening parameter in the model. d_0 is the dilation parameter, controlling the ratio between the volumetric and deviatoric plastic strains. The flow rule is associated when $d_0 = 1$ and non-associated when $d_0 \neq 1$. A suitable range for d_0 is

$$0 \le d_0 \le 1 \tag{13.7}$$

α_c, k_c, d_0, and n are model parameters. α and k are, respectively, the slope and intercept of the lines representing the projection of the yield surface in the meridional plane (Chap. 9). For three-dimensional applications where the Lode angle θ varies during the loading, α and k are allowed to vary with θ according to (13.6c). α_c and k_c are the values of α and k, respectively, in triaxial compression. α_e and k_e are the values of α and k, respectively, in triaxial extension. n is the ratio between α_e and α_c as well as that between k_e and k_c. g given by (13.6d) is a smoothing function, allowing α and k to vary smoothly between their values in triaxial compression and extension. In the octahedral plane, the yield and failure surfaces are circular for $n = 1$ and non-circular for $n < 1$. For plane strain applications, we will use circular surfaces with specific values for α and k as follows.

The parameters α and k can be related to the friction angle ϕ and cohesion c of the material according to the Mohr–Coulomb criterion. The relationship depends on the value of the Lode angle. We know that the Lode angle assumes fixed values in triaxial compression and extension. For three-dimensional analyses, non-circular surfaces with the smoothing function given by (13.6d) may then be used. Similarly, at failure, the Lode angle assumes a fixed value in plane strain as well. Then circular failure surfaces with values for α and k calculated from the Mohr–Coulomb failure criterion may be used. The details are presented in Appendix 9.3.

To summarize, in plane strain, assuming the surfaces to be circles in the octahedral plane, we use

$$\alpha = \alpha_{PS} = \frac{\tan\phi}{(9 + 12\tan^2\phi)^{1/2}}; \quad k = k_{PS} = \frac{3c}{(9 + 12\tan^2\phi)^{1/2}} \tag{13.8a}$$

In three dimensions, we use non-circular surfaces in the octahedral plane with

$$\alpha_{c,3D} = \frac{2\sin\phi}{\sqrt{3}(3 - \sin\phi)}; \quad k_{c,3D} = \frac{6c\cos\phi}{\sqrt{3}(3 - \sin\phi)} \tag{13.8b}$$

$$\alpha_{e,3D} = \frac{2\sin\phi}{\sqrt{3}(3 + \sin\phi)}; \quad k_{e,3D} = \frac{6c\cos\phi}{\sqrt{3}(3 + \sin\phi)} \tag{13.8c}$$

where $\alpha_{c,3D}$ and $k_{c,3D}$ are parameters in triaxial compression and $\alpha_{e,3D}$ and $k_{e,3D}$ are parameters in triaxial extension for use in 3D analyses.

The PIV in the model ζ (Chap. 9) is a scalar, given by

$$\varsigma \equiv m \tag{13.9}$$

Geometrically, m modifies the slope and intercept of the yield surface in the meridional plane α and k. As in the case of the von Mises model (Chap. 11), the following nonlinear hyperbolic hardening rule is assumed (Chap. 9):

$$m = m_0 + \frac{h(1 - m_0)\xi^p}{(1 - m_0) + h\xi^p} = m_0 + \frac{c_1\xi^p}{a_2 + b\xi^p} \tag{13.10a}$$

$$b = h; \quad a_2 = 1 - m_0; \quad c_1 = h(1 - m_0) \tag{13.10b}$$

m_0 (initial value of m of a virgin specimen) and h (hardening parameter) are model parameters. Taking the rate of (13.10a), we have

$$\dot{m} = \hat{g}\dot{\xi}_p \tag{13.11a}$$

where

$$\hat{g} = \frac{m(k_f - k_0)^2}{\left[(k_f - k_0) + m\xi^p\right]^2} = \frac{a_1}{(a_2 + b\xi^p)^2} \tag{13.11b}$$

$$a_1 = h(1 - m_0)^2 = ba_2^2 = c_1 a_2 \tag{13.11c}$$

$$\xi^p = \int \left[\frac{1}{2} de^p_{ij} de^p_{ij}\right]^{1/2} \tag{13.11d}$$

$$\dot{\xi}^p = \left[\frac{1}{2}\dot{e}^p_{ij}\dot{e}^p_{ij}\right]^{1/2}; \quad \dot{e}_{ij} = \dot{\varepsilon}_{ij} - \frac{1}{3}\dot{\varepsilon}_{kk}\delta_{ij} \tag{13.11e}$$

From (13.3), (13.11e) becomes

$$\dot{\xi}^{\mathrm{p}} = \left[\frac{1}{2}\dot{e}_{ij}^{\mathrm{p}}\dot{e}_{ij}^{\mathrm{p}}\right] = \left[\frac{1}{2}r_{ij}^{\mathrm{d}}r_{ij}^{\mathrm{d}}\right]^{1/2}\dot{\lambda} = \bar{r}^{\mathrm{d}}\dot{\lambda} \tag{13.12a}$$

where r_{ij}^{d} is the deviatoric part of r_{ij}

$$r_{ij}^{\mathrm{d}} = r_{ij} - \frac{1}{3}r_{pp}\delta_{ij} \tag{13.12b}$$

and $\bar{r}^{\mathrm{d}}$ is the deviatoric invariant of r_{ij}. Combining (13.11a and 13.12a), it follows that

$$\dot{m} = \hat{g}\bar{r}^{\mathrm{d}}\dot{\lambda} = \bar{g}\dot{\lambda}; \quad \bar{g} = \hat{g}\bar{r}^{\mathrm{d}} \tag{13.12c}$$

Thus, the hardening function **s** in the equations presented in Chaps. 9 and 10 is a scalar, given by

$$s = \bar{g} \tag{13.12d}$$

The gradients **r** and **n** can be found using the expression in Appendix 7 once the surface-specific gradients are determined. Taking the gradients of (13.6a and 13.6b), we have

$$\phi_{,I} = -m\alpha; \quad \phi_{,J} = 1; \quad \phi_{,\theta} = -m(\alpha_{\mathrm{c}}I + k_{\mathrm{c}})g_{,\theta}; \quad \phi_{,m} = -(\alpha I + k) \tag{13.13a}$$

$$\psi_{,I} = -m\alpha d_0; \quad \psi_{,J} = 1; \quad \psi_{,\theta} = -m(\alpha_{\mathrm{c}}I + k_{\mathrm{c}})g_{,\theta} \tag{13.13b}$$

From the consistency condition (Chap. 9), it follows that

$$K_{\mathrm{p}} = -\phi_{,\varsigma}s = (\alpha I + k)\bar{g} \tag{13.14a}$$

The yield envelop in the meridional plane intersects the I-axis at $3p_{\mathrm{t0}} \leq 0$, where (from (13.6a))

$$p_{\mathrm{t0}} = \frac{I_{\mathrm{t0}}}{3} = -\frac{k}{3\alpha} = -\frac{k_{\mathrm{c}}}{3\alpha_{\mathrm{c}}} \tag{13.14b}$$

If follows from (13.14a, b) that

$$\begin{aligned} K_{\mathrm{p}} &> 0 \quad \text{for} \quad p > p_{\mathrm{t0}} \\ &= 0 \quad \text{for} \quad p = p_{\mathrm{t0}} \\ &< 0 \quad \text{for} \quad p < p_{\mathrm{t0}} \end{aligned} \tag{13.14c}$$

Table 13.1 The model parameters used in the simulations

Model parameters	α_c	k_c (kPa)	n	m_0	h	ν	G_0	d_0
Set 1 (PS)	0.16	41.5	1	0.2	500	0.3	100	1,0.5
Set 2 (3D)	0.2309	60.0	1,0.8	0.2	500	0.3	100	1,0.5

To make the values of the elastic properties consistent with the value of the plastic modulus, we take the tension cutoff in (13.2a) as

$$p_{\mathrm{t}} = -p_{\mathrm{t0}} \tag{13.14d}$$

Then both the elastic and plastic moduli become zero when $p = p_{\mathrm{t0}}$.

The model parameters used in the simulations presented in this chapter are listed in Table 13.1. The parameters α_{c} and k_{c} have been calculated for $\phi = 30°$ and $c = 50\,\mathrm{kPa}$ using the plane strain and 3D equations (A9.27, A9.28a, and A9.28b) presented in Appendix 9. Note that it is shown in the appendix that the values calculated for the associated ($d_0 = 1$) and non-associated ($d_0 = 0.5$) flow rules are almost the same.

13.2 Application of CPPM to the Drucker–Prager Model: The σ-Space Formulation with Approximate Elastic Properties

The reader is advised to review Chap. 10 for the general details of CPPM and Chap. 12 for an application of CPPM to the modified Cam-clay model (which shares many aspects with the application of CPPM to the Drucker–Prager model presented here).

As the elastic properties are a function of the mean normal pressure (13.2a), the elastic predictor is not a constant in general. The reader is referred to Sects. 12.4 and 12.4.1 on the Cam-clay model for a detailed discussion on this aspect. There are two ways to handle the variable elastic properties:

- Use an Euler approximation (Sect. 12.4.1)
- Integrate the elastic relations exactly (Sect. 12.5.1, Appendix 10)

The equations associated with the former are presented in the present section, and the latter in Sect. 13.3.

13.2.1 *The Backward Euler Equations*

We use the forward Euler approximation for the elastic properties as

$$K = K_n = \bar{K}_0(p_n + p_{\mathrm{t}})^{n_{\mathrm{e}}}; \quad G = G_n = \bar{G}_0(p_n + p_{\mathrm{t}})^{n_{\mathrm{e}}} \tag{13.15}$$

According to this, the elastic predictor remains a constant during the iteration.

An exact relation between m and ξ^{p} is available in the present case (13.10a) and we make use of this to increase the accuracy. We use backward Euler approximation for ξ^{p} as (13.12a)

$$\xi^{\mathrm{p}}_{n+1} = \xi^{\mathrm{p}}_n + \bar{r}^{\mathrm{d}}_{n+1}\Delta\lambda \tag{13.16}$$

This is, however, a secondary relation and will only enter the formulation through the primary equation for m. The primary Euler equations are

$$\boldsymbol{\sigma}_{n+1} = \boldsymbol{\sigma}_n + \mathbf{C}\Delta\boldsymbol{\varepsilon} - \Delta\lambda\mathbf{C}\mathbf{r}^{n+1}(\boldsymbol{\sigma}_{n+1}, m_{n+1}) \tag{13.17a}$$

$$m_{n+1} = m_0 + \frac{h(1-m_0)\xi^{\mathrm{p}}_{n+1}}{(1-m_0) + h\xi^{\mathrm{p}}_{n+1}} \tag{13.17b}$$

$$\phi^{n+1}(\boldsymbol{\sigma}_{n+1}, m_{n+1}) = 0 \tag{13.17c}$$

13.2.2 The Residual Equations and Their Linearizations

The residual equations are given by (10.31a–c)

$$\mathbf{R}_\sigma = \boldsymbol{\sigma}_{n+1} - \left[\boldsymbol{\sigma}_n + \mathbf{C}\Delta\boldsymbol{\varepsilon} - \Delta\lambda\mathbf{C}\mathbf{r}^{n+1}\right] = \mathbf{R}_\sigma(\boldsymbol{\sigma}_{n+1}, m_{n+1}, \Delta\lambda) \tag{13.18a}$$

$$\mathbf{R}_\varsigma = m_{n+1} - m_0 - \frac{c_1(\xi^{\mathrm{p}}_n + \bar{r}^{\mathrm{d}}_{n+1}\Delta\lambda)}{a_2 + b(\xi^{\mathrm{p}}_n + \bar{r}^{\mathrm{d}}_{n+1}\Delta\lambda)} = \mathbf{R}_\varsigma(\boldsymbol{\sigma}_{n+1}, m_{n+1}, \Delta\lambda) \tag{13.18b}$$

$$R_\phi = \phi^{n+1}(\boldsymbol{\sigma}_{n+1}, m_{n+1}) = R_\phi(\boldsymbol{\sigma}_{n+1}, m_{n+1}) \tag{13.18c}$$

where it may be noted that the relation in (13.16) and the definitions in (13.10b) have been used in arriving at (13.18b from 13.17b).

The linearizations of the above equations are (10.34a–c)

$$R^1_{\sigma ij} + A_{ijk\ell}\delta\sigma_{k\ell} + B_{ij}\delta m + F_{ij}\delta\lambda = 0 \tag{13.19a}$$

$$R^1_\varsigma + H_{k\ell}\delta\sigma_{k\ell} + \omega\delta m + \beta\delta\lambda = 0 \tag{13.19b}$$

$$R^1_\phi + E_{k\ell}\delta\sigma_{k\ell} + \gamma\delta m = 0 \tag{13.19c}$$

The coefficients of the above equations are given as (10.36a–h)

$$A_{ijk\ell} = \frac{\partial R_{\sigma ij}}{\partial \sigma_{k\ell}} = \delta_{ik}\delta_{j\ell} - \frac{\partial C_{ijpq}}{\partial \sigma_{k\ell}}\left[\Delta\varepsilon_{pq} - \Delta\lambda r_{pq}^{n+1}\right] + \Delta\lambda C_{ijpq}\frac{\partial r_{pq}^{n+1}}{\partial \sigma_{k\ell}} \tag{13.20a}$$

$$B_{ij} = \frac{\partial R_{\sigma ij}}{\partial m} = \Delta\lambda C_{ijpq}\frac{\partial r_{pq}^{n+1}}{\partial m} \tag{13.20b}$$

$$F_{ij} = \frac{\partial R_{\sigma ij}}{\partial \Delta\lambda} = C_{ijpq} r_{pq}^{n+1} \tag{13.20c}$$

$$\begin{aligned} H_{k\ell} = \frac{\partial R_{\varsigma}}{\partial \sigma_{k\ell}} &= -\frac{1}{\left(a_2 + b\xi_{n+1}^{\mathrm{p}}\right)^2}\left[(a_2 + b\xi_{n+1}^{\mathrm{p}})c_1\Delta\lambda\frac{\partial \bar{r}_{n+1}^{\mathrm{d}}}{\partial \sigma_{k\ell}} - c_1\xi_{n+1}^{\mathrm{p}} b\Delta\lambda\frac{\partial \bar{r}_{n+1}^{\mathrm{d}}}{\partial \sigma_{k\ell}}\right] \\ &= -\frac{a_2}{\left(a_2 + b\xi_{n+1}^{\mathrm{p}}\right)^2}c_1\Delta\lambda\frac{\partial \bar{r}_{n+1}^{\mathrm{d}}}{\partial \sigma_{k\ell}} \end{aligned} \tag{13.20d}$$

$$\omega = \frac{\partial R_{\varsigma}}{\partial m} = 1 - \frac{a_2}{\left(a_2 + b\xi_{n+1}^{\mathrm{p}}\right)^2}c_1\Delta\lambda\frac{\partial \bar{r}_{n+1}^{\mathrm{d}}}{\partial m} \tag{13.20e}$$

$$\beta = \frac{\partial R_{\varsigma}}{\partial \Delta\lambda} = -\frac{a_2}{\left(a_2 + b\xi_{n+1}^{\mathrm{p}}\right)^2}c_1\bar{r}_{n+1}^{\mathrm{d}} \tag{13.20f}$$

$$E_{k\ell} = \frac{\partial R_{\phi}}{\partial \sigma_{k\ell}} = \frac{\partial \phi^{n+1}}{\partial \sigma_{k\ell}} = n_{k\ell}^{n+1} \tag{13.20g}$$

$$\gamma = \frac{\partial R_{\phi}}{\partial m} = \frac{\partial \phi^{n+1}}{\partial m} \tag{13.20h}$$

13.2.3 Evaluation of Coefficients

When explicit Euler representation is used for the elastic properties (13.15),

$$C_{ijpq} = \text{constant} \quad \text{and} \quad \frac{\partial \mathbf{C}}{\partial \boldsymbol{\sigma}_{\mathrm{n+1}}} = 0$$

The other quantities needed for the evaluation of the coefficients in (13.20a–h) are the surface properties $\partial\phi^{n+1}/\partial m, n_{k\ell}^{n+1}, r_{pq}^{n+1}, \partial r_{pq}^{n+1}/\partial\sigma_{k\ell}, \partial r_{pq}^{n+1}/\partial m, \partial\bar{r}_{n+1}^{\mathrm{d}}/\partial\sigma_{k\ell}$, and $\partial\bar{r}_{n+1}^{\mathrm{d}}/\partial m_{n+1}$. These quantities are found as follows:

$\phi_{,m}^{n+1}$ is given by (13.13a).

$n_{k\ell}^{n+1}$ and r_{pq}^{n+1} may be found from the general equation in Appendix 7 (A7.8) using the surface properties given by (13.13a, b).

As done in Appendix 8.2, let us write r_{ij} as (noting that θ is used in the present chapter to denote the Lode angle whereas α is used in the Appendix):

$$r_{ij} = (\psi_{,I})a_{ij} + (\psi_{,J})b_{ij} + (\psi_{,\theta})c_{ij} \tag{13.21a}$$

where from (A7.8), we have

$$a_{ij} = \delta_{ij} \tag{13.21b}$$

$$b_{ij} = \frac{s_{ij}}{2J}$$

$$c_{ij} = \frac{\sqrt{3}}{2J\cos 3\theta}\left[\frac{s_{ir}s_{rj}}{J^2} - \frac{2}{3}\delta_{ij} - \frac{3}{2}\left(\frac{S}{J}\right)^3\frac{s_{ij}}{J}\right] \tag{13.21c}$$

Then, it follows that (A8.25a)

$$\frac{\partial r_{ij}^{n+1}}{\partial \sigma_{pq}^{n+1}} = L_{ijpq}^{1} + L_{ijpq}^{2} + L_{ijpq}^{3}$$

$$L_{ijpq}^{1} = \frac{\partial(\psi_{,I})}{\partial \sigma_{pq}^{n+1}}a_{ij} + (\psi_{,I})L_{ijpq}^{10};\quad L_{ijpq}^{10} = \frac{\partial a_{ij}}{\partial \sigma_{pq}^{n+1}}$$

$$L_{ijpq}^{2} = \frac{\partial(\psi_{,J})}{\partial \sigma_{pq}^{n+1}}b_{ij} + (\psi_{,J})L_{ijpq}^{20};\quad L_{ijpq}^{20} = \frac{\partial b_{ij}}{\partial \sigma_{pq}^{n+1}}$$

$$L_{ijpq}^{3} = \frac{\partial(\psi_{,\theta})}{\partial \sigma_{pq}^{n+1}}c_{ij} + (\psi_{,\theta})L_{ijpq}^{30};\quad L_{ijpq}^{30} = \frac{\partial c_{ij}}{\partial \sigma_{pq}^{n+1}}$$

The expressions for L_{ijpq}^{10}, L_{ijpq}^{20}, and L_{ijpq}^{30} are the same as those in the modified Cam-clay model, given by (A8.25b, A8.26, and A8.27), respectively. From the surface properties listed in (13.13b), we have

$$\psi_{,I} = -m\alpha d_0 \Rightarrow \frac{\partial(\psi_{,I})}{\partial \sigma_{pq}^{n+1}} = -md_0\alpha_c\frac{\partial g}{\partial \sigma_{pq}^{n+1}}$$

where

$$\frac{\partial g}{\partial \sigma_{pq}^{n+1}} = g_{,\theta}\frac{\partial \theta}{\partial \sigma_{pq}^{n+1}}$$

$$\psi_{,J} = 1 \quad \Rightarrow \quad \frac{\partial(\psi_{,J})}{\partial\sigma_{pq}^{n+1}} = 0$$

$$\psi_{,\theta} = -m(\alpha_c I + k_c)g_{,\theta} \Rightarrow \frac{\partial(\psi_{,\theta})}{\partial\sigma_{pq}^{n+1}} = -m\alpha_c g_{,\theta}\delta_{pq} - m(\alpha_c I + k_c)\frac{\partial g_{,\theta}}{\partial\sigma_{pq}^{n+1}}$$

where, from A8.19,

$$g_{,\theta} = -gg^*; \quad g = \frac{2n}{1+n-(1-n)\sin 3\theta}; \quad g^* = -\frac{3(1-n)\cos 3\theta}{1+n-(1-n)\sin 3\theta}$$

$$\frac{\partial g_{,\theta}}{\partial\sigma_{pq}^{n+1}} = -g\frac{\partial g^*}{\partial\sigma_{pq}^{n+1}} - g^*\frac{\partial g}{\partial\sigma_{pq}^{n+1}}$$

Expressions for $\frac{\partial\theta}{\partial\sigma_{pq}^{n+1}}$ and $\frac{\partial g^*}{\partial\sigma_{pq}^{n+1}}$ are given by (A7.7 and A8.21). (Note that g^* here is the same as $\bar{g}$ in A8.21.)

Let us now differentiate (13.21a) with respect to m. Noting that a_{ij}, b_{ij}, and c_{ij} are independent of m, we have

$$\frac{\partial r_{pq}^{n+1}}{\partial m_{n+1}} = \frac{\partial(\psi_{,I})}{\partial m_{n+1}}a_{ij} + \frac{\partial(\psi_{,J})}{\partial m_{n+1}}b_{ij} + \frac{\partial(\psi_{,\theta})}{\partial m_{n+1}}c_{ij}$$

From (13.13b), it follows that

$$\psi_{,I} = -m_{n+1}\alpha d_0 \quad \Rightarrow \quad \frac{\partial(\psi_{,I})}{\partial m_{n+1}} = -\alpha d_0$$

$$\psi_{,J} = 1 \quad \Rightarrow \quad \frac{\partial(\psi_{,J})}{\partial m_{n+1}} = 0$$

$$\psi_{,\theta} = -m_{n+1}(\alpha_c I + k_c)g_{,\theta} \quad \Rightarrow \quad \frac{\partial(\psi_{,\theta})}{\partial m_{n+1}} = -(\alpha_c I + k_c)g_{,\theta}$$

The gradients of $\bar{r}_{n+1}^{\mathrm{d}}$ are now found as

$$\bar{r}_{n+1}^{\mathrm{d}} = \left[\frac{1}{2}r_{ij}^{\mathrm{d}}r_{ij}^{\mathrm{d}}\right]^{1/2} = \left[\frac{1}{2}r_{ij}^{n+1}r_{ij}^{n+1}\right]^{1/2} \qquad \text{(From 13.2a)}$$

$$\frac{\partial\bar{r}_{n+1}^{\mathrm{d}}}{\partial\sigma_{k\ell}^{n+1}} = \frac{1}{2\bar{r}_{n+1}^{\mathrm{d}}}\frac{\partial r_{ij}^{n+1}}{\partial\sigma_{k\ell}^{n+1}}r_{ij}^{n+1}$$

where $\frac{\partial r_{ij}^{n+1}}{\partial \sigma_{k\ell}^{n+1}} r_{ij}^{n+1}$ is evaluated from $\frac{\partial r_{ij}^{n+1}}{\partial \sigma_{k\ell}^{n+1}}$ and r_{ij}^{n+1} using a matrix-vector operation as in Appendix 6 (A6.3d).

$$\frac{\partial \bar{r}_{n+1}^{\mathrm{d}}}{\partial m_{n+1}} = \frac{1}{2\bar{r}_{n+1}^{\mathrm{d}}} \frac{\partial r_{ij}^{n+1}}{\partial m_{n+1}} r_{ij}^{n+1}$$

$$\frac{\partial \bar{r}_{n+1}^{\mathrm{d}}}{\partial \Delta\lambda} = \frac{1}{2\bar{r}_{n+1}^{\mathrm{d}}} \frac{\partial r_{ij}^{n+1}}{\partial \Delta\lambda} r_{ij}^{n+1} = 0$$

The other needed relations are

$$\begin{aligned} C_{ijpq} \frac{\partial r_{pq}^{n+1}}{\partial \sigma_{k\ell}} &= \left[\left(K - \frac{2}{3}G\right)\delta_{ij}\delta_{pq} + G\left(\delta_{ip}\delta_{jq} + \delta_{iq}\delta_{jp}\right)\right] \frac{\partial r_{pq}^{n+1}}{\partial \sigma_{k\ell}} \\ &= \left(K - \frac{2}{3}G\right)\delta_{ij} \frac{\partial r_{pp}^{n+1}}{\partial \sigma_{k\ell}} + 2G \frac{\partial r_{ij}^{n+1}}{\partial \sigma_{k\ell}} \end{aligned}$$

$$\begin{aligned} C_{ijpq} \frac{\partial r_{pq}^{n+1}}{\partial m_{n+1}} &= \left[\left(K - \frac{2}{3}G\right)\delta_{ij}\delta_{pq} + G\left(\delta_{ip}\delta_{jq} + \delta_{iq}\delta_{jp}\right)\right] \frac{\partial r_{pq}^{n+1}}{\partial m_{n+1}} \\ &= \left(K - \frac{2}{3}G\right) \frac{\partial r_{pp}^{n+1}}{\partial m_{n+1}} \delta_{ij} + 2G \frac{\partial r_{ij}^{n+1}}{\partial m_{n+1}} \end{aligned}$$

$$C_{ijpq} r_{pq}^{n+1} = \left(K - \frac{2}{3}G\right)\delta_{ij} r_{pp}^{n+1} + 2G r_{ij}^{n+1}$$

13.2.4 The Implementation Details

The symmetry of the stress and strain tensors allows the tensorial equations (13.19a–c) to be implemented in reduced order matrix-vector form. The details are presented in Appendix 6. The three matrix-vector equations corresponding to (13.19a–c) can be arranged in a single matrix-vector equation system as

$$\begin{bmatrix} \underset{6\times 6}{\mathbf{A}} & \underset{6\times 1}{\mathbf{B}} & \underset{6\times 1}{\mathbf{F}} \\ \underset{1\times 6}{\mathbf{H}^T} & \underset{1\times 1}{\omega} & \underset{1\times 1}{\beta} \\ \underset{1\times 6}{\mathbf{E}^T} & \underset{1\times 1}{\gamma} & \underset{1\times 1}{0} \end{bmatrix} \begin{bmatrix} \underset{6\times 1}{\delta\boldsymbol{\sigma}} \\ \underset{1\times 1}{\delta m} \\ \underset{1\times 1}{\delta\lambda} \end{bmatrix} = \begin{bmatrix} \underset{6\times 1}{-\mathbf{R}_\sigma^1} \\ \underset{1\times 1}{-R_\varsigma^1} \\ \underset{1\times 1}{-R_\phi^1} \end{bmatrix} \tag{13.22}$$

Equation (13.22) can either be directly solved as a system or manipulated to obtain the following equations (10.58a–e):

$$\delta\lambda = \frac{R_\phi^1 - \bar{\mathbf{n}}^T \mathbf{R}\mathbf{R}_\sigma^1 + x_0 R_\varsigma^1 \bar{\mathbf{n}}^T \bar{\mathbf{C}}\bar{\mathbf{B}} - x_0 \gamma R_\varsigma^1}{x_0 \gamma \beta + \mathbf{n}^T \bar{\mathbf{C}}\bar{\mathbf{r}}} \tag{13.23a}$$

$$\delta\sigma = -\mathbf{R}\mathbf{R}_\sigma^1 + x_0 R_\varsigma^1 \bar{\mathbf{C}}\bar{\mathbf{B}} - \bar{\mathbf{C}}\bar{\mathbf{r}}\delta\lambda \tag{13.23b}$$

$$\delta m = -\frac{1}{\gamma}\left[R_\phi^1 + \mathbf{n}^T \delta\boldsymbol{\sigma}\right] \tag{13.23c}$$

where

$$\bar{\mathbf{r}} = \bar{\mathbf{F}} - \frac{\beta}{\omega}\bar{\mathbf{B}} \tag{13.24a}$$

$$\bar{\mathbf{n}} = \mathbf{n} - \frac{\gamma}{\omega}\mathbf{H} \tag{13.24b}$$

$$\bar{\mathbf{F}} = \mathbf{C}^{-1}\mathbf{F} \quad \text{and} \quad \bar{\mathbf{B}} = \mathbf{C}^{-1}\mathbf{B}. \tag{13.24c}$$

$$x_0 = \frac{1}{\omega} \tag{13.24d}$$

$$\bar{\mathbf{C}} = \mathbf{R}\mathbf{C} \tag{13.24e}$$

$$\mathbf{R} = \left[\mathbf{A} - \frac{1}{\omega}\mathbf{B}\mathbf{H}^T\right]^{-1} \tag{13.24f}$$

The method requires an initial value for $\Delta\lambda$, which is calculated from (10.30)

$$\Delta\lambda_1 = \frac{\phi^{\text{tr}}}{\left[\mathbf{n}^T \mathbf{C}\mathbf{r} + K_{\text{p}}\right]^{\text{tr}}} \tag{13.25}$$

We will later show that a more refined initial estimate is required to address convergence difficulties in some zones in the stress space.

The new estimates for $\boldsymbol{\sigma}_{n+1}$, k_{n+1}, and $\Delta\lambda$ are found as (10.37a–c):

$$\boldsymbol{\sigma}_{n+1}^2 = \boldsymbol{\sigma}_{n+1}^1 + \delta\boldsymbol{\sigma} \tag{13.26a}$$

$$m_{n+1}^2 = m_{n+1}^1 + \delta m \tag{13.26b}$$

$$\Delta\lambda_2 = \Delta\lambda_1 + \delta\lambda \tag{13.26c}$$

Note that superscript 2 on these variables does not denote "the square of the variable," but rather the "second estimate."

The following convergence criteria may be used (10.38a–c):

$$e_\sigma = \left\|\mathbf{R}_\sigma^{i+1}\right\| / \left\|\boldsymbol{\sigma}_{n+1}^i\right\| \leq TOL_\sigma \tag{13.27a}$$

$$e_\varsigma = \left\|R_\varsigma^{i+1}\right\| / \left\|m_{n+1}^i\right\| \leq TOL_\varsigma \tag{13.27b}$$

$$R_\phi^{i+1} = \phi^{i+1} \leq TOL_\phi \tag{13.27c}$$

The procedure is summarized in Algorithm 13.1.

The consistent tangent operator is given by (10.66b)

$$\mathbf{D} = \overline{\mathbf{C}} - \frac{(\bar{\mathbf{C}}\bar{\mathbf{r}})(\bar{\mathbf{n}}^T\bar{\mathbf{C}})}{x_0\gamma\beta + \bar{\mathbf{n}}^T\bar{\mathbf{C}}\bar{\mathbf{r}}} \tag{13.28}$$

It was shown in Chap. 10 that when $\Delta\lambda = 0$, the consistent and continuum operators become identical to each other.

Algorithm 13.1. CPPM for Integrating the Constitutive Relations from the Drucker–Prager Model with Explicit Elastic Properties (The σ-Space Formulation)

Definition of (new) parameters: $N_L^{\max}$ is the maximum number of iterations permitted before execution is terminated and N_L is the corresponding counter. TOL is a small, machine-dependent number (it could be set equal to TOL_ϕ). Iflag = 1 if converged and 0 otherwise.

1. Establish $\Delta\boldsymbol{\varepsilon}$, m_n, $\boldsymbol{\sigma}_n$, TOL_σ, TOL_ς, TOL_ϕ, TOL, $N_L^{\max}$, and constitutive model parameters.
2. Initialize parameters as $N_L = 0$ and $m_{n+1} = m_n$.
3. Compute elastic predictor $\boldsymbol{\sigma}_{n+1}^{\mathrm{tr}}$ and $\phi^{\mathrm{tr}}(\boldsymbol{\sigma}_{n+1}^{\mathrm{tr}}, m_n)$. Perform loading/unloading check: if $(\phi^{\mathrm{tr}} < TOL)$, set $\boldsymbol{\sigma}_{n+1} = \boldsymbol{\sigma}_{n+1}^{\mathrm{tr}}$ and go to step 12.
4. Set $\boldsymbol{\sigma}_{n+1} = \boldsymbol{\sigma}_{n+1}^{\mathrm{tr}}$ and calculate $\mathbf{r}_{n+1}$, $\bar{r}_{n+1}^{\mathrm{d}}$, and K_{p}^{n+1}.
5. Set $R_\phi^1 = \phi^{\mathrm{tr}}$. Compute trial value of $\Delta\lambda$. Compute $\xi_{n+1}^{\mathrm{p}} = \xi_n^{\mathrm{p}} + \Delta\lambda\bar{r}_{n+1}^{\mathrm{d}}$ and $\bar{r}_{n+1}^{\mathrm{d}}$. Compute $\boldsymbol{\sigma}_{n+1}^1$ and m_{n+1}^1. Set $\boldsymbol{\sigma}_{n+1} = \boldsymbol{\sigma}_{n+1}^1$ and $m_{n+1} = m_{n+1}^1$.
6. Begin iterations. Compute $R_\phi^1 = \phi(\boldsymbol{\sigma}_{n+1}, m_{n+1})$, $\mathbf{r}_{n+1}$, and $\bar{r}_{n+1}^{\mathrm{d}}$.
7. Compute $\boldsymbol{\sigma}_{n+1}^2$. Compute $\xi_{n+1}^{\mathrm{p}} = \xi_n^{\mathrm{p}} + \Delta\lambda\bar{r}_{n+1}^{\mathrm{d}}$ and $\bar{r}_{n+1}^{\mathrm{d}}$. Compute m_{n+1}^2.
8. Compute $\mathbf{R}_\sigma^1 = \boldsymbol{\sigma}_{n+1}^1 - \boldsymbol{\sigma}_{n+1}^2$ and $R_\varsigma^1 = m_{n+1}^1 - m_{n+1}^2$.
9. Find errors e_σ, e_ς, and e_ϕ and perform convergence check. If converged, set iflag = 1 and go to step 12.
10. Set $N_L \leftarrow N_L + 1$. If $(N_L > N_L^{\max})$, set iflag = 0 and go to step 12.

(continued)

11. Compute $\delta\lambda$, $\delta\boldsymbol{\sigma}$, and δm. Update $\boldsymbol{\sigma}^1_{n+1} \leftarrow \boldsymbol{\sigma}^1_{n+1} + \delta\boldsymbol{\sigma}$, $m^1_{n+1} \leftarrow m^1_{n+1} + \delta m$, and $\Delta\lambda \leftarrow \Delta\lambda + \delta\lambda$. Set $\boldsymbol{\sigma}_{n+1} = \boldsymbol{\sigma}^1_{n+1}$ and $m_{n+1} = m^1_{n+1}$. Go to step 6.
12. Return to the main program

Remark: Noting that $K_{\rm p}$ becomes negative for $p < p_{\rm t0}$ (13.14a–c), the denominator in (13.25) can become negative and sometimes zero. Furthermore, the expressions for the elastic moduli (13.2a) become undefined. We will discuss these issues in detail in Sect. 13.7.2.

13.3 Application of CPPM to the Drucker–Prager Model: The σ-Space Formulation with Exact Elastic Properties

13.3.1 The General Description

When the plastic strain is zero, the elastic predictor becomes the actual elastic stress. The exact expression for the elastic predictor is presented in Appendix 10. It is shown there that (A10.6 and A10.7)

$$p^{\rm tr} = \left[(p_n + p_{\rm t})^{1-n_{\rm e}} + \bar{K}_0(1-n_{\rm e})\Delta\varepsilon_{pp}\right]^{1/(1-n_{\rm e})} - p_{\rm t} \tag{13.29a}$$

$$s^{\rm tr}_{ij} = s^{\rm n}_{ij} + 2G_{\rm ave}\Delta e_{ij} \tag{13.29b}$$

$$G_{\rm ave} = \frac{\bar{G}_0}{\bar{K}_0\Delta\varepsilon_{pp}}(p^{\rm tr} - p_n) \tag{13.29c}$$

where $n_{\rm e} = 0.5$. The quantity within the parenthesis in (13.29a) must be positive; i.e.,

$$p_1 = (p_n + p_{\rm t})^{1-n_{\rm e}} + \bar{K}_0(1-n_{\rm e})\Delta\varepsilon_{pp} \geq 0 \tag{13.29d}$$

When $p_1 < 0$, $p_1^{1/(1-n_{\rm e})}$ becomes undefined when $1/(1-n_{\rm e})$ is not a whole number (e.g., $n_{\rm e} = 0.45 \Rightarrow 1/(1-n_{\rm e}) = 1.818$), and defined, however, representing a wrong functional relation when $1/(1-n_{\rm e})$ is a whole number (e.g., $n_{\rm e} = 0.5 \Rightarrow 1/(1-n_{\rm e}) = 2$).

When the behavior is elasto-plastic, the above equations are modified by using the elastic strain increments instead of the total strain increments. The elastic strains are obtained by subtracting the plastic strains from the total strains. The backward Euler method is used to express the plastic strain increment as (13.3)

$$\Delta\varepsilon^{\rm p}_{ij} = \Delta\lambda r^{n+1}_{ij} \tag{13.30}$$

Let us define

$$x_0 = \Delta\varepsilon_{pp} - \Delta\lambda r^{n+1}_{pp} \tag{13.31a}$$

$$\Delta y_{ij} = \Delta e_{ij} - \Delta\lambda d_{ij}^{n+1} \tag{13.31b}$$

where d_{ij} is the deviatoric part of r_{ij} defined as

$$d_{ij} = r_{ij} - \frac{1}{3} r_{kk}\delta_{ij} \tag{13.31c}$$

Equations (13.29a–c) are then modified as

$$p_{n+1} = \left[(p_n + p_t)^{1-n_e} + \bar{K}_0(1 - n_e)x_0 \right]^{1/(1-n_e)} - p_t \tag{13.32a}$$

$$s_{ij}^{n+1} = s_{ij}^{n} + 2G_{ave}\Delta y_{ij} \tag{13.32b}$$

$$G_{ave} = \frac{\bar{G}_0}{\bar{K}_0 x_0}(p_{n+1} - p_n) \tag{13.32c}$$

Note that the restriction in (13.29d) applies for the term within the parenthesis in (13.32a). The total stress tensor is obtained as

$$\sigma_{ij}^{n+1} = s_{ij}^{n+1} + p_{n+1}\delta_{ij} \tag{13.33a}$$

13.3.2 The Backward Euler Equations, Residuals, and Linearizations

When the hardening rule remains the same as in Sect. 13.2, only the equations associated with σ_{ij}^{n+1} change and the remaining equations are the same as those in Sect. 13.2. The residual equations (13.18b, c), the linearized equations (13.19b, c), and the coefficients $H_{k\ell}$, ω, β, $E_{k\ell}$, and γ remain the same. In the following, we will present the equations associated with σ_{ij}^{n+1} (i.e., the counterparts of (13.18a, 13.19a, 13.20a, b, c)).

The residual $\mathbf{R}_\sigma$ is defined as

$$R_{\sigma ij} = \sigma_{ij}^{n+1} - \left[s_{ij}^{n+1} + p_{n+1}\delta_{ij} \right] \tag{13.34a}$$

Linearizing this, we have

$$R_{\sigma ij}^{1} + A_{ijk\ell}\delta\sigma_{k\ell} + B_{ij}\delta m + F_{ij}\delta\lambda = 0 \tag{13.34b}$$

13.3.3 Evaluation of Coefficients and Implementation Details

The coefficients are given by

$$A_{ijk\ell} = \delta_{ik}\delta_{j\ell} - \frac{\partial s_{ij}}{\partial \sigma_{k\ell}^{n+1}} - \frac{\partial p_{n+1}}{\partial \sigma_{k\ell}^{n+1}}\delta_{ij}$$

$$B_{ij} = -\frac{\partial s_{ij}}{\partial m_{n+1}} - \frac{\partial p_{n+1}}{\partial m_{m+1}}\delta_{ij}$$

$$F_{ij} = -\frac{\partial s_{ij}}{\partial \Delta\lambda} - \frac{\partial p_{n+1}}{\partial \Delta\lambda}\delta_{ij}$$

where

$$\frac{\partial s_{ij}}{\partial \sigma_{k\ell}^{n+1}} = 2\Delta y_{ij}\frac{\partial G_{\text{ave}}}{\partial \sigma_{k\ell}^{n+1}} + 2G_{\text{ave}}\frac{\partial \Delta y_{ij}}{\partial \sigma_{k\ell}^{n+1}}; \quad \frac{\partial \Delta y_{ij}}{\partial \sigma_{k\ell}^{n+1}} = -\Delta\lambda\frac{\partial d_{ij}^{n+1}}{\partial \sigma_{k\ell}^{n+1}}$$

$$\frac{\partial s_{ij}}{\partial m_{n+1}} = 2\Delta y_{ij}\frac{\partial G_{\text{ave}}}{\partial m_{n+1}} + 2G_{\text{ave}}\frac{\partial \Delta y_{ij}}{\partial m_{n+1}}; \quad \frac{\partial \Delta y_{ij}}{\partial m_{n+1}} = -\Delta\lambda\frac{\partial d_{ij}^{n+1}}{\partial m_{n+1}}$$

$$\frac{\partial s_{ij}}{\partial \Delta\lambda} = 2\Delta y_{ij}\frac{\partial G_{\text{ave}}}{\partial \Delta\lambda} + 2G_{\text{ave}}\frac{\partial \Delta y_{ij}}{\partial \Delta\lambda}; \quad \frac{\partial \Delta y_{ij}}{\partial \Delta\lambda} = -\Delta\lambda\frac{\partial d_{ij}^{n+1}}{\partial \Delta\lambda}$$

$$\frac{\partial G_{\text{ave}}}{\partial \sigma_{k\ell}^{n+1}} = \frac{\bar{G}_0}{\bar{K}_0 x_0}\frac{\partial p_{n+1}}{\partial \sigma_{k\ell}^{n+1}} - \frac{\bar{G}_0}{\bar{K}_0 x_0^2}(p_{n+1} - p_n)\frac{\partial x_0}{\partial \sigma_{k\ell}^{n+1}}$$

$$\frac{\partial G_{\text{ave}}}{\partial m_{n+1}} = \frac{\bar{G}_0}{\bar{K}_0 x_0}\frac{\partial p_{n+1}}{\partial m_{n+1}} - \frac{\bar{G}_0}{\bar{K}_0 x_0^2}(p_{n+1} - p_n)\frac{\partial x_0}{\partial m_{n+1}}$$

$$\frac{\partial G_{\text{ave}}}{\partial \Delta\lambda} = \frac{\bar{G}_0}{\bar{K}_0 x_0}\frac{\partial p_{n+1}}{\partial \Delta\lambda} - \frac{\bar{G}_0}{\bar{K}_0 x_0^2}(p_{n+1} - p_n)\frac{\partial x_0}{\partial \Delta\lambda}$$

$$\frac{\partial x_0}{\partial \sigma_{k\ell}^{n+1}} = -\Delta\lambda\frac{\partial r_{pp}^{n+1}}{\partial \sigma_{k\ell}^{n+1}}$$

$$\frac{\partial x_0}{\partial m_{n+1}} = -\Delta\lambda\frac{\partial r_{pp}^{n+1}}{\partial m_{n+1}}$$

$$\frac{\partial x_0}{\partial \Delta\lambda} = -r_{pp}^{n+1}$$

Let us define

$$p_0 = (p_n + p_t)^{1-n_e} + \bar{K}_0(1 - n_e)x_0$$

Then we have

$$\frac{\partial p_{n+1}}{\partial \sigma_{k\ell}^{n+1}} = \frac{1}{1-n_e} p_0^{n_e/(1-n_e)} \bar{K}_0(1-n_e)\frac{\partial x_0}{\partial \sigma_{k\ell}^{n+1}} = \bar{K}_0 p_0^{n_e/(1-n_e)} \frac{\partial x_0}{\partial \sigma_{k\ell}^{n+1}}$$

$$\frac{\partial p_{n+1}}{\partial m_{n+1}} = \frac{1}{1-n_e} p_0^{n_e/(1-n_e)} \bar{K}_0(1-n_e)\frac{\partial x_0}{\partial m_{n+1}} = \bar{K}_0 p_0^{n_e/(1-n_e)} \frac{\partial x_0}{\partial m_{n+1}}$$

$$\frac{\partial p_{n+1}}{\partial \Delta\lambda} = \frac{1}{1-n_e} p_0^{n_e/(1-n_e)} \bar{K}_0(1-n_e)\frac{\partial x_0}{\partial \Delta\lambda} = \bar{K}_0 p_0^{n_e/(1-n_e)} \frac{\partial x_0}{\partial \Delta\lambda}$$

The above equations may now be synthesized to evaluate the required coefficients.

The system (13.22) remains valid for the present case as well. As in Sect. 13.2, (13.22) can either be solved as a system or manipulated to obtain separate equations for $\delta\boldsymbol{\sigma}$, δm, and $\delta\lambda$. One change from the equations in Sect. 13.2 is that in place of the standard elastic stiffness tensor **C**, here we will need to use a modified elastic tensor $\mathbf{C}^*$ as explained below.

13.3.4 The Consistent Tangent Operator

Let us now consider the evaluation of the consistent tangent operator. In (13.18) (replacing (13.18a) with (13.34a)), let us set $\mathbf{R}_\sigma$, R_ς, and R_ϕ to zero, take the rate of the equations, and arrange the terms as

$$\mathbf{R}_\sigma^{*1} + \mathbf{A}^*\dot{\boldsymbol{\sigma}} + \mathbf{B}^*\dot{\varsigma} + \mathbf{F}^*\dot{\lambda} = 0$$

$$R_\varsigma^{*1} + \mathbf{H}^{*T}\dot{\boldsymbol{\sigma}} + \omega^*\dot{\varsigma} + \beta^*\dot{\lambda} = 0$$

$$R_\phi^{*1} + \mathbf{E}^{*T}\dot{\boldsymbol{\sigma}} + \gamma^*\dot{\varsigma} = 0$$

It can be easily shown that $\mathbf{A}^* = \mathbf{A}$, $\mathbf{B}^* = \mathbf{B}$, $\mathbf{F}^* = \mathbf{F}$, $\mathbf{H}^* = \mathbf{H}$, $\mathbf{E}^* = \mathbf{E}$ $\omega^* = \omega$, $\beta^* = \beta$, $\gamma^* = \gamma$, and

$$R_\varsigma^* = 0, \quad R_\phi^* = 0 \quad \text{and} \quad R_\sigma^{*1} = \mathbf{C}^*\dot{\boldsymbol{\varepsilon}},$$

where

$$\begin{aligned} C_{ijpq}^* &= -\frac{\partial R_{\sigma ij}}{\partial \varepsilon_{pq}} \\ &= \frac{\partial s_{ij}^{n+1}}{\partial \varepsilon_{pq}} + \delta_{ij}\frac{\partial p_{n+1}}{\partial \varepsilon_{pq}} \end{aligned}$$

Note that

$$\frac{\partial s_{ij}^{n+1}}{\partial \varepsilon_{pq}} = 2G_{\text{ave}} \frac{\partial \Delta y_{ij}}{\partial \varepsilon_{pq}} + 2\Delta y_{ij} \frac{\partial G_{\text{ave}}}{\partial \varepsilon_{pq}}$$

$$\frac{\partial \Delta y_{ij}}{\partial \varepsilon_{pq}} = \frac{\partial \Delta e_{ij}}{\partial \varepsilon_{pq}} = \delta_{ip}\delta_{jq} - \frac{1}{3}\delta_{ij}\delta_{pq}$$

$$\frac{\partial p_{n+1}}{\partial \varepsilon_{pq}} = p_0^{n_e/(1-n_e)} \bar{K}_0 \frac{\partial x_0}{\partial \varepsilon_{pq}}$$

$$\frac{\partial x_0}{\partial \varepsilon_{pq}} = \delta_{pq}$$

$$\frac{\partial G_{\text{ave}}}{\partial \varepsilon_{pq}} = -\frac{G_{\text{ave}}}{x_0} \frac{\partial x_0}{\partial \varepsilon_{pq}} + \frac{\bar{G}_0}{\bar{K}_0 x_0} p_0^{n_e/(1-n_e)} \bar{K}_0 \frac{\partial x_0}{\partial \varepsilon_{pq}} = \frac{1}{x_0}\left(\bar{G}_0 p_0^{n_e/(1-n_e)} - G_{\text{ave}}\right)\delta_{pq}$$

It then follows that

$$\begin{aligned} C^*_{ijpq} &= 2G_{\text{ave}}\left[\delta_{ip}\delta_{jq} - \frac{1}{3}\delta_{ij}\delta_{pq}\right] + 2\Delta y_{ij}\frac{1}{x_0}\left(\bar{G}_0 p_0^{n_e/(1-n_e)} - G_{\text{ave}}\right)\delta_{pq} \\ &\quad + p_0^{n_e/(1-n_e)}\bar{K}_0\delta_{ij}\delta_{pq} \\ &= \left[\bar{K}_0 p_0^{n_e/(1-n_e)}\delta_{ij} + \frac{2\Delta y_{ij}}{x_0}\left(\bar{G}_0 p_0^{n_e/(1-n_e)} - G_{\text{ave}}\right) - \frac{2}{3}G_{\text{ave}}\delta_{ij}\right]\delta_{pq} \\ &\quad + G_{\text{ave}}\left(\delta_{ip}\delta_{jq} + \delta_{iq}\delta_{jp}\right) \end{aligned} \tag{13.35}$$

It may be verified that when $\Delta\lambda = 0$, we have $\mathbf{C}^* = \mathbf{C}$.

The general structure of the algorithm for calculating the stresses by CPPM with elastic properties is the same as in Algorithm 13.1. The equations in the present section are to be used in calculating the stresses (Steps 3, 5, and 7), residual $\boldsymbol{R}_\sigma^1$ (Step 8), and the corrections (Step 11).

13.4 Application of CPPM to the Drucker–Prager Model: The S-Space Formulation with Approximate Elastic Properties

13.4.1 The General Description

It is shown in Sect. 10.7.2 (Chap. 10) that the Euler equation for stress (13.17a) can be decomposed into spherical and deviatoric parts. The resulting formulation is called the *S*-space formulation. When the *yield surface is assumed to be a circle* in the octahedral plane, this formulation leads to a more computationally efficient algorithm than that from the σ-space formulation. We have already seen this for the von Mises model (Sect. 11.3) and the modified Cam-clay model (Sect. 12.7).

As in Sect. 13.2, we will use the forward Euler approximation for the elastic properties as

$$K = K_n = \bar{K}_0(p_n + p_t)^{n_e}; \quad G = G_n = \bar{G}_0(p_n + p_t)^{n_e} \tag{13.36}$$

According to this, the elastic predictor remains a constant during the iteration.

The Euler equations for the elasto-plastic stresses (using I instead of p for the spherical part) are

$$I_{n+1} = I_n + 3K\Delta\varepsilon_{pp} - 3K\Delta\lambda r_{pp}^{n+1} \tag{13.37a}$$

$$\mathbf{s}_{n+1} = \mathbf{s}_n + 2G\Delta\mathbf{e} - 2G\Delta\lambda\mathbf{r}_{n+1}^{\mathrm{d}} \tag{13.37b}$$

where I_{n+1} is the spherical part of $\boldsymbol{\sigma}_{n+1}$, $\mathbf{s}_{n+1} = \boldsymbol{\sigma}_{n+1} - I_{n+1}\boldsymbol{\delta}/3$ is the deviatoric part of σ_{n+1}, and $\Delta\mathbf{e} = \Delta\boldsymbol{\varepsilon} - \Delta\varepsilon_{pp}\boldsymbol{\delta}/3$ is the incremental deviatoric strain. In pressure-independent models such as the von Mises model considered in Chap. 11, $r_{pp}^{n+1} = 0$, and hence the variation of I is purely elastic. However, in pressure-dependent models such as the modified Cam-clay model presented in Chap. 12 and the Drucker–Prager considered here, the variation of I is elasto-plastic. Let us define the trial stresses as

$$I^{\mathrm{tr}} = I_n + 3K\Delta\varepsilon_{pp} \tag{13.38a}$$

$$\mathbf{s}^{\mathrm{tr}} = \mathbf{s}_n + 2G\Delta\mathbf{e}; \quad J^{\mathrm{tr}} = \left(\frac{1}{2}s_{k\ell}^{\mathrm{tr}}s_{k\ell}^{\mathrm{tr}}\right)^{1/2} \tag{13.38b}$$

The yield surface is assumed to be a circle in the octahedral plane. This leads to $\phi_{,\theta} = 0$. From (A7.8), making use of (13.13b), we have

$$r_{ij} = \frac{\partial\psi}{\partial I}\delta_{ij} + \frac{\partial\psi}{\partial J}\frac{s_{ij}}{2J} = -m\alpha d_0\delta_{ij} + \frac{s_{ij}}{2J} \tag{13.39a}$$

$$n_{ij} = \frac{\partial\phi}{\partial I}\delta_{ij} + \frac{\partial\phi}{\partial J}\frac{s_{ij}}{2J} = -m\alpha\delta_{ij} + \frac{s_{ij}}{2J} \tag{13.39b}$$

Hence, it follows that

$$r_{pp}^{n+1} = -3m\alpha d_0 \tag{13.40a}$$

$$\mathbf{r}^{\mathrm{d}}_{n+1} = \frac{1}{2J_{n+1}}\mathbf{s}_{n+1} \tag{13.40b}$$

$$\bar{r}^{\mathrm{d}}_{n+1} = \frac{1}{2} \tag{13.40c}$$

Let us first consider the deviatoric part. Combining (13.37b, 13.38b, and 13.40b), we have

$$\mathbf{s}_{n+1} = \mathbf{s}_n + 2G\Delta\mathbf{e} - 2G\Delta\lambda\mathbf{r}^{\mathrm{d}}_{n+1} = \mathbf{s}^{\mathrm{tr}} - \frac{1}{J_{n+1}}G\Delta\lambda\mathbf{s}_{n+1} \tag{13.41a}$$

$$\mathbf{s}_{n+1} = \frac{1}{1+\frac{G\Delta\lambda}{J_{n+1}}}\mathbf{s}^{\mathrm{tr}} = \frac{\mathbf{s}^{\mathrm{tr}}}{x_0}; \quad x_0 = 1 + \frac{G\Delta\lambda}{J_{n+1}} \tag{13.41b}$$

which represents a radial return in the octahedral plane, just as in the case of the von Mises model (11.28b and Fig. 11.1). By squaring both sides of the first part of (13.41b), we obtain

$$J^2_{n+1}\left(\frac{J_{n+1}+G\Delta\lambda}{J_{n+1}}\right)^2 = (J^{\mathrm{tr}})^2$$

$$J_{n+1} = J^{\mathrm{tr}} - G\Delta\lambda \tag{13.41c}$$

13.4.2 The Backward Euler Equations

Using invariants for the stresses and the exact representation for the hardening rule, the backward Euler equations are (from 13.37a, 13.38a, 13.40a, 13.51c, 13.17b, and 13.17c)

$$I_{n+1} = I^{\mathrm{tr}} - 3K\Delta\lambda r^{n+1}_{pp} = I^{\mathrm{tr}} + 3K\Delta\lambda(3m\alpha d_0) = I^{\mathrm{tr}} + (9K\alpha d_0)m\Delta\lambda \tag{13.42a}$$

$$J_{n+1} = J^{\mathrm{tr}} - \Delta\lambda G \tag{13.42a}$$

$$m_{n+1} = m_0 + \frac{h(1-m_0)\xi^{\mathrm{p}}_{n+1}}{(1-m_0)+h\xi^{\mathrm{p}}_{n+1}} = m_0 + \frac{c_1\xi^{\mathrm{p}}_{n+1}}{a_2+b\xi^{\mathrm{p}}_{n+1}} \tag{13.42c}$$

$$\phi^{n+1}(\boldsymbol{\sigma}_{n+1}, m_{n+1}) = 0 \tag{13.42d}$$

As before (13.16), we will use backward Euler approximation for ξ^{p} as (using (13.40c))

$$\xi^{\mathrm{p}}_{n+1} = \xi^{\mathrm{p}}_{n} + \bar{r}^{\mathrm{d}}_{n+1}\Delta\lambda = \xi^{\mathrm{p}}_{n} + \Delta\lambda/2 \tag{13.43}$$

13.4.3 Solution of Equations

As in the case of the von Mises model, it is possible to combine the Euler equations to arrive at a scalar equation for $\Delta\lambda$. We pursue this avenue here. By expanding (13.42c) with the aid of (13.43 and 13.40c), we get

$$\begin{aligned} m_{n+1} &= m_0 + \frac{c_1\xi^{\mathrm{p}}_{n+1}}{a_2 + b\xi^{\mathrm{p}}_{n+1}} = m_0 + \frac{c_1(\xi^{\mathrm{p}}_{n} + \Delta\lambda/2)}{a_2 + b(\xi^{\mathrm{p}}_{n} + \Delta\lambda/2)} \\ &= \frac{m_0a_2 + m_0b\xi^{\mathrm{p}}_{n} + m_0b\Delta\lambda/2 + c_1(\xi^{\mathrm{p}}_{n} + \Delta\lambda/2)}{a_2 + b(\xi^{\mathrm{p}}_{n} + \Delta\lambda/2)} = \frac{\alpha_{11} + \alpha_{12}\Delta\lambda}{\alpha_{13} + \alpha_{14}\Delta\lambda} \end{aligned} \tag{13.44a}$$

where

$$\alpha_{11} = m_0a_2 + m_0b\xi^{\mathrm{p}}_{n} + c_1\xi^{\mathrm{p}}_{n} \tag{13.44b}$$

$$\alpha_{12} = (m_0b + c_1)/2 \tag{13.44c}$$

$$\alpha_{13} = a_2 + b\xi^{\mathrm{p}}_{n} \tag{13.44d}$$

$$\alpha_{14} = b/2 \tag{13.44e}$$

With the aid of (13.44a), let us expand the expression for I_{n+1} (13.42c) as

$$\begin{aligned} I_{n+1} &= I^{\mathrm{tr}} + (9K\alpha d_0)m\Delta\lambda \\ &= I^{\mathrm{tr}} + (9K\alpha d_0)\left[\frac{\alpha_{11} + \alpha_{12}\Delta\lambda}{\alpha_{13} + \alpha_{14}\Delta\lambda}\right]\Delta\lambda \\ &= \frac{I^{\mathrm{tr}}(\alpha_{13} + \alpha_{14}\Delta\lambda) + (9K\alpha d_0)\alpha_{11}\Delta\lambda + (9K\alpha d_0)\alpha_{12}\Delta\lambda^2}{\alpha_{13} + \alpha_{14}\Delta\lambda} \\ &= \frac{\beta_{11}\Delta\lambda^2 + \beta_{12}\Delta\lambda + \beta_{13}}{\alpha_{13} + \alpha_{14}\Delta\lambda} \end{aligned} \tag{13.45a}$$

where

$$\beta_{11} = 9K\alpha d_0\alpha_{12} \tag{13.45b}$$

$$\beta_{12} = 9K\alpha d_0\alpha_{11} + I^{\text{tr}}\alpha_{14} \tag{13.45c}$$

$$\beta_{13} = I^{\text{tr}}\alpha_{13} \tag{13.45d}$$

Let us now substitute J_{n+1}, I_{n+1}, and m_{n+1} from (13.42a, 13.45a, and 13.44a), respectively into the equation of the yield surface given by (13.6a) as

$$J_{n+1} - \alpha m_{n+1} I_{n+1} - k m_{n+1} = 0$$

$$J^{\text{tr}} - \Delta\lambda G - \alpha\left[\frac{\alpha_{11} + \alpha_{12}\Delta\lambda}{\alpha_{13} + \alpha_{14}\Delta\lambda}\right]\left[\frac{\beta_{11}\Delta\lambda^2 + \beta_{12}\Delta\lambda + \beta_{13}}{\alpha_{13} + \alpha_{14}\Delta\lambda}\right] - k\left[\frac{\alpha_{11} + \alpha_{12}\Delta\lambda}{\alpha_{13} + \alpha_{14}\Delta\lambda}\right] = 0$$

$$\begin{aligned}&(J^{\text{tr}} - \Delta\lambda G)\left(\alpha_{13}^2 + 2\alpha_{13}\alpha_{14}\Delta\lambda + \alpha_{14}^2\Delta\lambda^2\right) - \alpha(\alpha_{11} + \alpha_{12}\Delta\lambda)\left(\beta_{11}\Delta\lambda^2 + \beta_{12}\Delta\lambda + \beta_{13}\right)\\ &\quad - k(\alpha_{11} + \alpha_{12}\Delta\lambda)(\alpha_{13} + \alpha_{14}\Delta\lambda) = 0\end{aligned}$$

$$B_1\Delta\lambda^3 + B_2\Delta\lambda^2 + B_3\Delta\lambda + B_4 = 0 \tag{13.46a}$$

where

$$B_1 = -\alpha_{14}^2 G - \alpha\alpha_{12}\beta_{11} \tag{13.46b}$$

$$B_2 = -2\alpha_{13}\alpha_{14}G + \alpha_{14}^2 J^{\text{tr}} - \alpha\alpha_{11}\beta_{11} - \alpha\alpha_{12}\beta_{12} - k\alpha_{12}\alpha_{14} \tag{13.46c}$$

$$B_3 = 2\alpha_{13}\alpha_{14}J^{\text{tr}} - \alpha_{13}^2 G - \alpha\alpha_{11}\beta_{12} - \alpha\alpha_{12}\beta_{13} - k\alpha_{11}\alpha_{14} - k\alpha_{12}\alpha_{13} \tag{13.46d}$$

$$B_4 = \alpha_{13}^2 J^{\text{tr}} - \alpha\alpha_{11}\beta_{13} - k\alpha_{11}\alpha_{13} \tag{13.46e}$$

Equation (13.46a) can be solved for $\Delta\lambda$. The Euler equations (13.42a, b, c) and (13.41b) can then be used to find J_{n+1}, I_{n+1}, m_{n+1}, and $\mathbf{s}_{n+1}$, respectively. The stress tensor is then assembled as

$$\boldsymbol{\sigma}_{n+1} = \mathbf{s}_{n+1} + \frac{1}{3}I_{n+1}\boldsymbol{\delta} \tag{13.47}$$

13.4.4 The Consistent Tangent Operator

Taking the rate of (13.47), we have (leaving the subscripts and superscripts $n+1$ out)

$$\dot{\boldsymbol{\sigma}} = \dot{\mathbf{s}} + \frac{1}{3}\dot{I}\boldsymbol{\delta}$$

From (13.41a), it follows that

$$\dot{\mathbf{s}} = \dot{\mathbf{s}}^{\mathrm{tr}} - \frac{1}{J}G\Delta\lambda\dot{\mathbf{s}} + \frac{1}{J^2}G\Delta\lambda\mathbf{s}\dot{J} - \frac{1}{J}G\mathbf{s}\dot{\lambda} \tag{13.48}$$

From (13.42a and 13.41c), we have

$$\dot{I}_{n+1} = \dot{I}^{\mathrm{tr}} + (9K\alpha d_0)\dot{m}\Delta\lambda + (9K\alpha d_0)m\dot{\lambda} \tag{13.49a}$$

$$\dot{J}_{n+1} = \dot{J}^{\mathrm{tr}} - G\dot{\lambda} \tag{13.49b}$$

Differentiating (13.38a and 13.38b), we get

$$\dot{I}^{\mathrm{tr}} = 3K\dot{\varepsilon}_{pp} \tag{13.50a}$$

$$\dot{\mathbf{s}}^{\mathrm{tr}} = 2G\dot{\mathbf{e}}; \tag{13.50b}$$

$$2(J^{\mathrm{tr}})^2 = s_{k\ell}^{\mathrm{tr}}s_{k\ell}^{\mathrm{tr}} \quad \Rightarrow \quad \dot{J}^{\mathrm{tr}} = \frac{Gs_{k\ell}^{\mathrm{tr}}\dot{e}_{k\ell}}{J^{\mathrm{tr}}} \tag{13.50c}$$

The rate of the hardening variable is (from 13.11a and 13.11b)

$$\dot{m} = \hat{g}\dot{\xi}_{\mathrm{p}}; \quad \hat{g} = \frac{a_1}{(a_2 + b\xi^{\mathrm{p}})^2} \tag{13.51}$$

We will take the rate of (13.46a) to find $\dot{\lambda}$. But let us first evaluate the rate of the associated constants. From (13.44b–e), we have

$$\dot{\alpha}_{11} = \dot{\alpha}_{12} = \dot{\alpha}_{13} = \dot{\alpha}_{14} = 0 \tag{13.52a}$$

From (13.45b–d), we get

$$\dot{\beta}_{11} = 0; \quad \dot{\beta}_{12} = \alpha_{14}\dot{I}^{\mathrm{tr}}; \quad \dot{\beta}_{13} = \alpha_{13}\dot{I}^{\mathrm{tr}} \tag{13.52b}$$

Equations (13.46b–e) lead to

$$\dot{B}_1 = 0 \tag{13.52c}$$

$$\dot{B}_2 = \left[\alpha_{14}^2\dot{J}^{\mathrm{tr}} - \alpha\alpha_{12}\dot{\beta}_{12}\right] = \left[\alpha_{14}^2\frac{Gs_{k\ell}^{\mathrm{tr}}}{J^{\mathrm{tr}}} - \alpha\alpha_{12}\alpha_{14}(3K)\delta_{k\ell}\right]\dot{\varepsilon}_{k\ell}$$

$$\dot{B}_2 = f_{k\ell}\dot{\varepsilon}_{k\ell}; \quad f_{k\ell} = \alpha_{14}^2\frac{Gs_{k\ell}^{\mathrm{tr}}}{J^{\mathrm{tr}}} - \alpha\alpha_{12}\alpha_{14}(3K)\delta_{k\ell} \tag{13.52d}$$

$$\dot{B}_3 = 2\alpha_{13}\alpha_{14}\dot{J}^{\text{tr}} - \alpha\alpha_{11}\dot{\beta}_{12} - \alpha\alpha_{12}\dot{\beta}_{13} = \left[2\alpha_{13}\alpha_{14}\frac{Gs_{k\ell}^{\text{tr}}}{J^{\text{tr}}} - \alpha(\alpha_{11}\alpha_{14} + \alpha_{12}\alpha_{13})(3K)\delta_{k\ell}\right]\dot{\varepsilon}_{k\ell}$$

$$\dot{B}_3 = g_{k\ell}\dot{\varepsilon}_{k\ell}; \quad g_{k\ell} = 2\alpha_{13}\alpha_{14}\frac{Gs_{k\ell}^{\text{tr}}}{J^{\text{tr}}} - \alpha(\alpha_{11}\alpha_{14} + \alpha_{12}\alpha_{13})(3K)\delta_{k\ell} \tag{13.52e}$$

$$\dot{B}_4 = \alpha_{13}^2\dot{J}^{\text{tr}} - \alpha\alpha_{11}\dot{\beta}_{13} = \left[\alpha_{13}^2\frac{Gs_{k\ell}^{\text{tr}}}{J^{\text{tr}}} - \alpha\alpha_{11}\alpha_{13}(3K)\delta_{k\ell}\right]\dot{\varepsilon}_{k\ell}$$

$$\dot{B}_4 = h_{k\ell}\dot{\varepsilon}_{k\ell}; \quad h_{k\ell} = \alpha_{13}^2\frac{Gs_{k\ell}^{\text{tr}}}{J^{\text{tr}}} - \alpha\alpha_{11}\alpha_{13}(3K)\delta_{k\ell} \tag{13.52f}$$

Now take the rate of (13.46a) to find $\dot{\lambda}$ as

$$(3B_1\Delta\lambda^2 + 2B_2\Delta\lambda + B_3)\dot{\lambda} + (\dot{B}_1\Delta\lambda^3 + \dot{B}_2\Delta\lambda^2 + \dot{B}_3\Delta\lambda + \dot{B}_4) = 0$$

$$\dot{\lambda} = -\frac{\dot{B}_2\Delta\lambda^2 + \dot{B}_3\Delta\lambda + \dot{B}_4}{3B_1\Delta\lambda^2 + 2B_2\Delta\lambda + B_3} = -\frac{f_{k\ell}\Delta\lambda^2 + g_{k\ell}\Delta\lambda + h_{k\ell}}{3B_1\Delta\lambda^2 + 2B_2\Delta\lambda + B_3}\dot{\varepsilon}_{k\ell}$$

$$\dot{\lambda} = \bar{f}_{k\ell}\dot{\varepsilon}_{k\ell}; \quad \bar{f}_{k\ell} = -\frac{f_{k\ell}\Delta\lambda^2 + g_{k\ell}\Delta\lambda + h_{k\ell}}{3B_1\Delta\lambda^2 + 2B_2\Delta\lambda + B_3} \tag{13.53}$$

Let's assemble the equations to evaluate $\dot{\mathbf{s}}$ (13.48) and $\dot{I}$ (13.49a) as

$$\dot{\mathbf{s}} = \dot{\mathbf{s}}^{\text{tr}} - \frac{1}{J}G\Delta\lambda\dot{\mathbf{s}} + \frac{1}{J^2}G\Delta\lambda\mathbf{s}\dot{J} - \frac{1}{J}G\mathbf{s}\dot{\lambda}$$

$$\dot{s}_{ij}\left(1 + \frac{1}{J}G\Delta\lambda\right) = 2G\left\{\frac{1}{2}\left(\delta_{ik}\delta_{j\ell} + \delta_{i\ell}\delta_{jk}\right) - \frac{1}{3}\delta_{ij}\delta_{k\ell}\right\}\dot{\varepsilon}_{k\ell}$$
$$+ \frac{1}{J^2}G\Delta\lambda s_{ij}\left(\frac{Gs_{k\ell}^{\text{tr}}}{J^{\text{tr}}} - G\bar{f}_{k\ell}\right)\dot{\varepsilon}_{k\ell} - \frac{1}{J}Gs_{ij}\bar{f}_{k\ell}\dot{\varepsilon}_{k\ell}$$

$$\dot{s}_{ij} = \left\{\frac{G}{x_0}\left(\delta_{ik}\delta_{j\ell} + \delta_{i\ell}\delta_{jk}\right) - \frac{2G}{3}\delta_{ij}\delta_{k\ell}\right\}\dot{\varepsilon}_{k\ell}$$
$$+ \left(\frac{G^2\Delta\lambda s_{ij}s_{k\ell}^{\text{tr}}}{x_0J^2J^{\text{tr}}} - \frac{G^2\Delta\lambda s_{ij}\bar{f}_{k\ell}}{x_0J^2} - \frac{Gs_{ij}\bar{f}_{k\ell}}{x_0J}\right)\dot{\varepsilon}_{k\ell}$$

$$\dot{I} = 3K\delta_{k\ell}\dot{\varepsilon}_{k\ell} + (9K\alpha d_0)\left(\hat{g}\bar{r}^{\text{d}}\bar{f}_{k\ell}\dot{\varepsilon}_{k\ell}\right)\Delta\lambda + (9K\alpha d_0)m(\bar{f}_{k\ell}\dot{\varepsilon}_{k\ell})$$

Putting $\dot{\mathbf{s}}$ and $\dot{I}$ together to evaluate the total stress rate, the consistent tangent operator $D_{ijk\ell}$ at point $n + 1$ is obtained as

$$\dot{\sigma}_{ij} = D_{ijk\ell}\dot{\varepsilon}_{k\ell}$$

$$\begin{aligned} D_{ijk\ell} = & \left(K - \frac{2}{3}\frac{G}{x_0}\right)\delta_{ij}\delta_{k\ell} + \frac{G}{x_0}\left(\delta_{ik}\delta_{j\ell} + \delta_{i\ell}\delta_{jk}\right) \\ & - \left(\frac{Gs_{ij}^{n+1}\bar{f}_{k\ell}}{J_{n+1}} - \frac{G^2\Delta\lambda s_{ij}^{n+1}s_{k\ell}^{\text{tr}}}{x_0 J_{n+1}^2 J^{\text{tr}}} - 3m\alpha K d_0 \delta_{ij}\bar{f}_{k\ell} - \frac{3}{2}\alpha K d_0 \Delta\lambda\hat{g}\delta_{ij}\bar{f}_{k\ell}\right) \end{aligned} \quad (13.54)$$

13.5 Sample Calculation of Incremental Quantities

13.5.1 Problem Definition and Calculation of Trial Stresses

In this section, we first present some sample calculations for finding the incremental stress and PIV for a given incremental strain. The nonlinear isotropic hardening law presented in this chapter (13.10a) is considered. The model parameters listed in Table 13.1 are employed. The values of relevant quantities at the initial point n and the incremental strain are listed below. The unit is kPa for stresses, yield stresses, and moduli. We will work in the principal stress space. We use properties Set 2 listed in Table 13.1 with $n = 1$ and $d_0 = 1$. The incremental stress $\Delta\sigma$ and PIV Δm (or σ_{n+1} and m_{n+1}) are to be found.

$$\boldsymbol{\sigma}_n = \{100 \quad 50 \quad 75\}$$

$$m_n = 0.3$$

$$e_0 = 0.9$$

$$\Delta\boldsymbol{\varepsilon} = \{0.03 \quad -0.028 \quad 0.01\}$$

We will perform the calculations first using the S-space formulation and then using the σ-space formulation. We use approximate elastic properties (Sects. 13.2 and 13.4). For both of these methods, the initial sets of calculations are identical. These calculations are first performed using (say) the steps in Sects. 13.2 and 13.4. The initial and final yield surfaces, and the initial, trial, and final stresses are marked in octahedral and meridional planes in Fig. 13.1 (the explanations are provided below).

$$I_n = 225, \quad p_n = \frac{I_n}{3} = 75$$

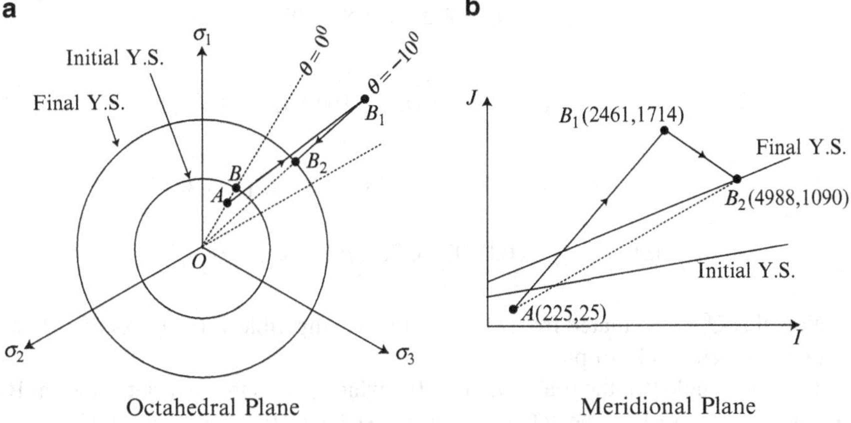

Fig. 13.1 Sketch of yield surfaces and stress points in stress space

$$p_t = -p_{t0} = \frac{k_c}{3\alpha_c} = \frac{60}{3 \times 0.2309} = 86.62$$

$$G = G_n = 28668.91, \quad K = K_n = 62115.97 \qquad \text{(From 13.15)}$$

$$J_n = 25, \quad \mathbf{s}_n = \{25 \quad -25 \quad 0\}, \quad \phi^n = -8.588$$

To mark the point representing the initial state in the octahedral plane, we calculate the Lode angle θ (13.5d). The value is 0°. The initial point then plots at Point A. In the octahedral plane, $OA = |\mathbf{s}_n| = \sqrt{2}J_n = 35.36$. In the meridional plane, the coordinates of Point A are $(225, 25)$. In the octahedral plane, noting that $OB = \sqrt{2}m_n(\alpha I_n + k) = 47.49$ (using 13.6b), it is seen that the initial point is inside the surface. Similarly, Point A plots inside the initial yield surface in the meridional plane as well. Note that θ is calculated only for the purpose of marking the point; it is not required by the algorithm.

As $\phi^n < 0$, the initial point is inside the surface (as we have already noted above). We now calculate the trial stress to determine if the strain increment is elastic or elasto-plastic.

$$\Delta\varepsilon_{pp} = 0.012$$

$$\Delta\mathbf{e} = \Delta\boldsymbol{\varepsilon} - \frac{1}{3}\Delta\varepsilon_{pp}\boldsymbol{\delta} = \{0.026 \quad -0.032 \quad 0.006\}$$

$$\mathbf{s}^{tr} = \mathbf{s}^n + 2G\Delta\mathbf{e} = \{1515.78 \quad -1859.81 \quad 344.03\}$$

$$p^{\text{tr}} = p_n + K\Delta\varepsilon_{pp} = 820.39$$

$$\boldsymbol{\sigma}^{\text{tr}} = \mathbf{s}^{\text{tr}} + \frac{1}{3}p^{\text{tr}}\boldsymbol{\delta} = \{\,2336.18 \quad -1039.42 \quad 1164.42\,\}$$

$$I^{\text{tr}} = 2461.18, \quad J^{\text{tr}} = 1713.89, \quad \phi^{\text{tr}} = 1525.38$$

$$(\xi^{\text{p}})^{\text{tr}} = \xi_n^{\text{p}} = 0.00022857; \quad m^{\text{tr}} = m_n = 0.3$$

Note that ξ_n^{p} is calculated from (13.10a) to be compatible with m_n. As $\phi^{\text{tr}}>0$, the strain increment is elasto-plastic.

The Lode angle θ at the trial point is $-10°$ which places the trial state at Point B_1 as shown in Fig. 13.1. In the octahedral plane, we have $OB_1 = \sqrt{2}J^{\text{tr}} = 2423.81$. In the meridional plane, the coordinates of Point B_1 are (2,461.18, 1,713.89).

13.5.2 *The Plastic Corrector by the S-Space Formulation (Sect. 13.4)*

Here the procedure presented in Sect. 13.4 is followed. We now calculate a value for $\Delta\lambda$ which will bring the trial stress point back to the surface with the expansion of the surface properly accounted for. This entails the solution of the cubic equation 13.46a. Now let us calculate the coefficients of (13.46a) as follows:

$$\alpha_{11} = 0.2743, \alpha_{12} = 250, \alpha_{13} = 0.9143, \alpha_{14} = 250$$

$$\beta_{11} = 3.2276 \times 10^7, \beta_{12} = 6.507 \times 10^5, \beta_{13} = 2250.22$$

$$B_1 = -3.655 \times 10^9, B_2 = 5.065 \times 10^7, B_3 = 5.705 \times 10^5, B_4 = 1275.1$$

Taking the real positive solution among the three possible solutions, we have

$$\Delta\lambda = 0.021765$$

The remaining quantities are calculated as follows:

$$\xi_{n+1}^{\text{p}} = \xi_n^{\text{p}} + \Delta\lambda/2 = 0.011111$$

$$m_{n+1} = m_0 + \frac{c_1\xi_{n+1}^{\text{p}}}{a_2 + b\xi_{n+1}^{\text{p}}} = 0.8993$$

$$J_{n+1} = J^{\text{tr}} - G\Delta\lambda = 1089.92$$

$$x_0 = 1 + \frac{G\Delta\lambda}{J_{n+1}} = 1.5725$$

$$I_{n+1} = I^{\mathrm{tr}} - 3K\Delta\lambda r_{pp}^{n+1} = I^{\mathrm{tr}} + 3K\Delta\lambda(3m_{n+1}\alpha d_0) = 4988.16$$

$$\mathbf{s}_{n+1} = \frac{\mathbf{s}^{\mathrm{tr}}}{x_0} = \{963.94 \quad -1182.72 \quad 218.77\}$$

$$\boldsymbol{\sigma}_{n+1} = \mathbf{s}_{n+1} + \frac{1}{3}I_{n+1}\boldsymbol{\delta} = \{2626.65 \quad 480.01 \quad 1881.49\}$$

The Lode angle θ at the final point $n + 1$ is $-10°$. This places the point on the final surface at Point B_2. In the octahedral plane, we have $OB_2 = \sqrt{2}J_{n+1} = 1541.15$. Note that Point B_2 is on the radial line OB_1, confirming that the plastic corrector represents a radial return from the trial stress point. The point in the meridional plane is shown in Fig. 13.1b.

To summarize, the required solution using the S-space formulation is as follows:

$$\Delta\boldsymbol{\sigma} = \{2526.65 \quad 430.01 \quad 1806.49\}$$

$$\Delta\lambda = 0.021765$$

$$\xi_{n+1}^{\mathrm{p}} = 0.011111$$

$$m_{n+1} = 0.8993$$

13.5.3 The Plastic Corrector by the σ-Space Formulation (Sect. 13.2)

Here the procedure presented in Sect. 13.2 is followed. We need a starting value for $\Delta\lambda$, which is calculated from (13.25) as follows:

$$\hat{g} = \frac{a_1}{(a_2 + b\xi_n^{\mathrm{p}})^2} = 382.81$$

$$K_{\mathrm{p}} = \hat{g}(\alpha I_n + k)/2 = 120276.63$$

$$\phi_{,I} = -0.06928, \quad \phi_{,J} = 1$$

$$r_{ij} = n_{ij} = -m_n\alpha d_0\delta_{ij} + \frac{s_{ij}^{\mathrm{tr}}}{2J^{\mathrm{tr}}} = \{0.3729 \quad -0.6118 \quad 0.0311\}$$

$$n_{ij}C_{ijk\ell}r_{k\ell} = \left(K - \frac{2}{3}G\right)n_{pp}r_{pp} + 2G(n_{ij}r_{ij}) = 9Km_n^2\alpha^2 d_0 + G = 31352.32$$

$$\Delta\lambda_{tr} = \frac{\phi^{\text{tr}}}{\left[\mathbf{n}^T\mathbf{Cr} + K_{\text{p}}\right]^{\text{tr}}} = 0.01$$

The stresses and PIV are recalculated using this value of $\Delta\lambda$ in (13.17a, 13.16, and 13.17b) (Step 5, Algorithm 13.1) as

$$\boldsymbol{\sigma}_{n+1} = \boldsymbol{\sigma}_n + \mathbf{C}\Delta\boldsymbol{\varepsilon} - \Delta\lambda\mathbf{C}\mathbf{r}_{n+1} = \{2210.9 \quad -596.6 \quad 1236.4\}$$

$$\xi^{\text{p}}_{n+1} = \xi^{\text{p}}_n + \Delta\lambda\bar{r}^{\text{d}}_{n+1} = 0.0052585$$

$$m_{n+1} = 0.8134$$

At this point, the iterations (Steps 6 through 11 in Algorithm 13.1) begin and continue until convergence. The key results during each iteration are listed in Table 13.2. The final value of $\Delta\lambda$ (which is not listed in the table) is 0.021765.

To summarize, the required solution using the σ-space formulation is as follows:

$$\Delta\boldsymbol{\sigma} = \{2526.65 \quad 430.01 \quad 1806.49\}$$

$$\Delta\lambda = 0.021765$$

$$\xi^{\text{p}}_{n+1} = 0.011111$$

$$m_{n+1} = 0.8993$$

During the iterations, the coefficient matrix in (13.22) is to be found and (13.22) solved to determine the iterative changes to the stress, PIV and $\Delta\lambda$: $\delta\boldsymbol{\sigma}, \delta m$, and $\delta\lambda$. For the sake of completeness, the equation is presented below for the second iteration.

Table 13.2 Results during σ-space iteration

Iter. no.	$\boldsymbol{\sigma}^1_{n+1}$			m^1	$\boldsymbol{\sigma}^2_{n+1}$			m^2	e_s	e_ς	e_ϕ
	σ_{11}	σ_{22}	σ_{33}		σ_{11}	σ_{22}	σ_{33}				
1	2,211	−597	1,236	0.813	2,433	−374	1,459	0.813	0.15	0	840
2	2,614	464	1,868	0.972	2,694	543	1,947	0.899	0.04	0.08	77
3	2,627	479	1,881	0.899	2,626	479	1,881	0.899	0.80×10^{-4}	0.40×10^{-6}	0.69
4	2,627	480	1,881	0.899	2,627	480	1,881	0.899	0.15×10^{-7}	0.29×10^{-7}	0.14×10^{-4}

$$\begin{bmatrix} 1.1571 & 0.0835 & -0.2406 & -934.0 & -16466.1 \\ 0.0835 & 1.0444 & -0.1279 & -934.0 & -72930.9 \\ -0.2406 & -0.1279 & 1.3685 & -934.0 & -36066.6 \\ 0 & 0 & 0 & 1 & -3.98 \\ 0.2178 & -0.7669 & -0.1241 & -1202.2 & 0 \end{bmatrix} \begin{bmatrix} \delta\sigma_{11} \\ \delta\sigma_{22} \\ \delta\sigma_{33} \\ \delta m \\ \delta\lambda \end{bmatrix} = \begin{bmatrix} -79.4 \\ -79.4 \\ -79.4 \\ 0.07273 \\ -76.65 \end{bmatrix} \tag{13.55}$$

The solution is as follows:

$$\delta\boldsymbol{\sigma} = \{12.42 \quad 15.13 \quad 13.36\}$$

$$\delta m = -0.0725$$

$$\delta\lambda = 4.806 \times 10^{-5}$$

The examination of the right-hand side reveals that $\mathbf{R}_\sigma = (-79.4 \quad -79.4 \quad -79.4)$, whose deviatoric component is zero. It is left for the reader to provide an explanation for this (Problem 13.5). It is noted that the results converged in four iterations.

13.5.4 Comparison of Results and Discussion

By comparing the results from the two methods presented above, the following observations can be made:

1. The solutions from the S-space formulation and the σ-space formulation are identical. These two methods are fundamentally the same and the results are expected to be the same.
2. The σ-space formulation is seen to involve significantly more computations than the S-space formulation; hence, whenever possible, the S-space formulation is preferred over the σ-space formulation.

13.6 Simulation of Stress–Strain Relations

In this section, we simulate the stress–strain behavior of a uniformly loaded specimen. To achieve this, we use a single-element finite element model shown in Fig. 13.2. The mesh consists of one 8-noded brick element (Chap. 5).

The consolidated-drained behavior of the one-element model in triaxial compression is shown in Fig. 13.3. The properties used in the analyses are listed in Table 13.1 as Set 2 (3D). The value of n does not have an influence on the response

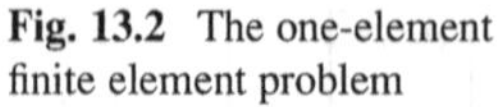

Fig. 13.2 The one-element finite element problem

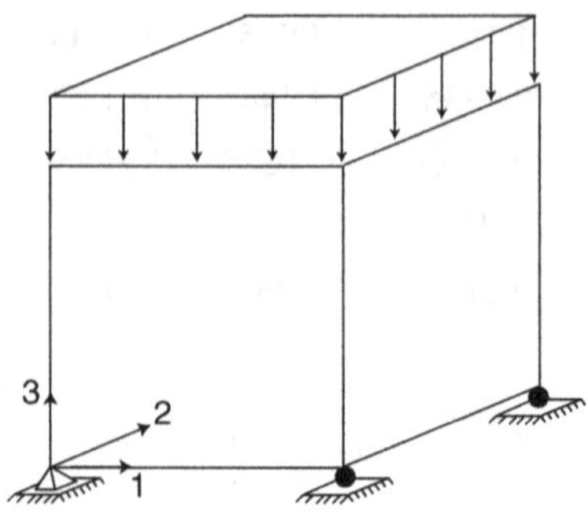

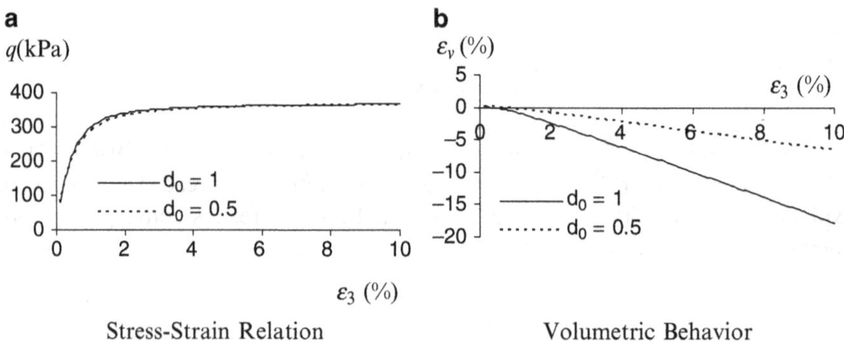

Fig. 13.3 The response of the one-element problem: effect of d_0

as the stress path lies only in triaxial compression. The initial effective confining pressure is $\sigma_1 = \sigma_2 = \sigma_3 = 100\,\text{kPa}$ and the initial value of the hardening variable is 0.3 (i.e., $m_n = 0.3$). The $q - \varepsilon_3$ relation and $\varepsilon_v - \varepsilon_3$ relation are shown in Fig. 13.3a, b, respectively, where $q = \sigma_3 - \sigma_1$ and ε_v is the volumetric strain. Note that direction 3 is the axial direction. A smooth stress–strain relation with a definite failure stress is obtained. The stress–strain relation (i.e., $q - \varepsilon_3$ relation) seems almost insensitive to the value of d_0. The volumetric strain keeps increasing indefinitely, and the rate of increase is controlled by the dilation parameter d_0.

The local integration is performed using the S-space formulation. The analyses are then repeated using (a) σ-space formulation of CPPM with approximate elastic properties, (b) σ-space formulation of CPPM with exact elastic properties, (c) the cutting-plane algorithm (not described in this chapter), and (d) the two-step Euler method with sub-stepping (not described in this chapter). Except for the cutting-plane method, the results obtained with the other methods coincide for NINC = 100 (for $\hat{\varepsilon}_3 = 20\%$) and NINC = 5, where NINC is the number of increments used for $\hat{\varepsilon}_3 = 20\%$ (the analysis is carried out in strain-controlled mode). The results from the cutting-plane method are also the same for NINC = 100, but for NINC = 5, the algorithm automatically subdivided the prescribed strain at the global level to 20 increments; the results then are the same as those from the other methods.

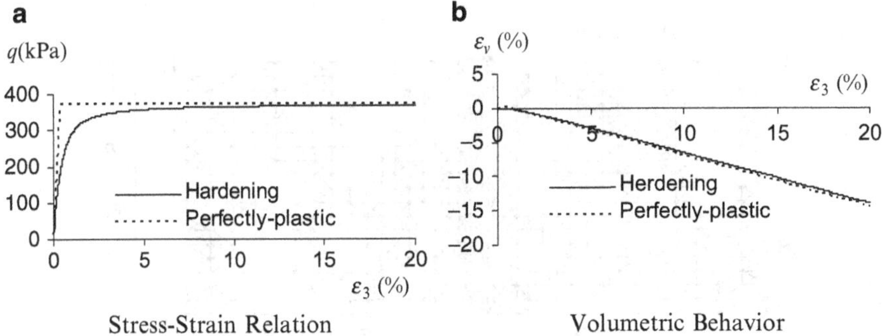

Stress-Strain Relation

Volumetric Behavior

Fig. 13.4 The response of the one-element problem: effect of hardening

The influence of employing hardening on the response of the one-element problem is shown in Fig. 13.4. The analysis is performed with $d_0 = 0.5$. It is seen that the use of hardening has a significant influence on the stress–strain behavior, but almost negligible influence on the volumetric behavior.

13.7 Multi-element Finite Element Analyses

13.7.1 Problem Definition

A multi-element finite element problem (2D plane strain), representing a typical boundary-value problem encountered in geotechnical engineering (the shallow footing problem shown in Fig. 13.5), is considered in this section. The loading is applied through a rigid plate, simulated using plane-frame bending elements. The mesh consists of 2,736 three-noded 2D triangular elements. The foundation soil is assumed to consist of a homogeneous material whose stress–strain behavior is represented by the Drucker–Prager model presented in this chapter. The plane-strain model parameter set listed in Table 13.1 (Set 1) with yield and potential surfaces whose projections in the octahedral plane are circles is used (i.e., $n = 1$).

The x- and y-directions are taken to coincide with horizontal and vertical directions, respectively. Assuming that the x- and y-directions are the principal directions, the initial effective stresses are calculated as

$$\sigma_y = \gamma' h + \sigma_{y0}; \quad \sigma_x = K_0 \sigma_y$$

where h, γ', K_0, and σ_{y0} are the depth, buoyant density, lateral earth pressure coefficient, and a constant surface pressure, respectively. The analysis is carried out with $\gamma' = 10\,\text{kN/m}^3$, $K_0 = 0.5$, and $\sigma_{y0} = 4.5\,\text{kPa}$. The constant initial stress σ_{y0} is used to increase numerical stability at small stresses.

A vertical load P of 30,000 kN is applied in load-controlled mode in an initial global increment of 100. Upon non-convergence at the global level, the load is to be

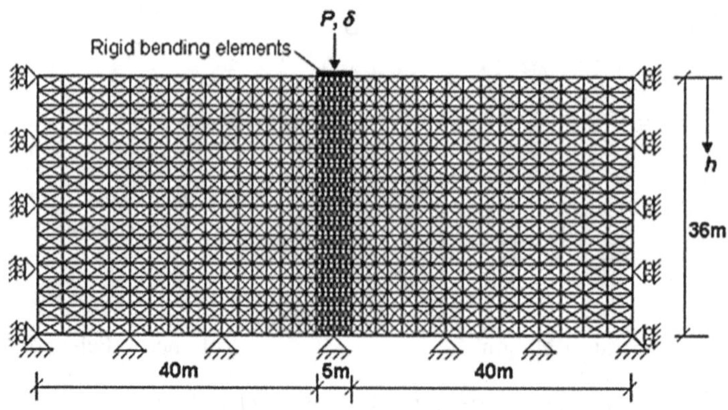

Fig. 13.5 The footing problem

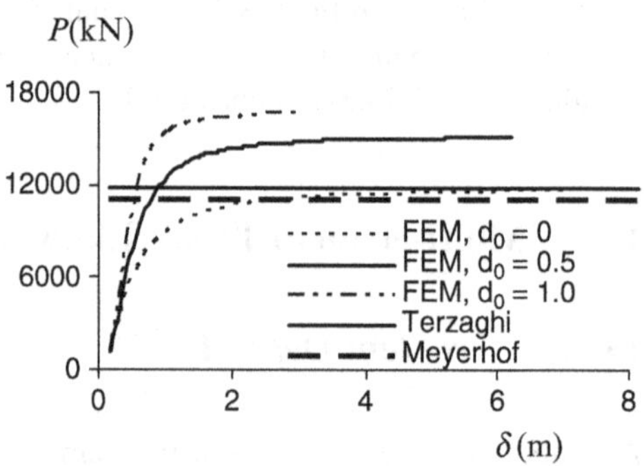

Fig. 13.6 The load-displacement behavior

subdivided in half. However, the problem of non-convergence first began appearing at the local level, which is addressed in detail below. Once this problem is addressed, the global iterations proceeded as intended, resulting in the load–deformation curves shown in Fig. 13.6. The different load–deformation curves correspond to different values of the dilation parameter d_0. The deformed configuration near failure is shown in Fig. 13.7 within a window around the footing. We will return to the discussion of the results after addressing the convergence problem in the following.

13.7.2 Convergence Issues with CPPM with Approximate Elastic Properties

13.7.2.1 Convergence Difficulty Near the Apex and the Apex Region Strategy to Overcome the Difficulty

The case with $d_0 = 0.5$ is considered, for which, the collapse load is 15,100 kN. The local iterations first fail to converge in the element on the surface and immediately

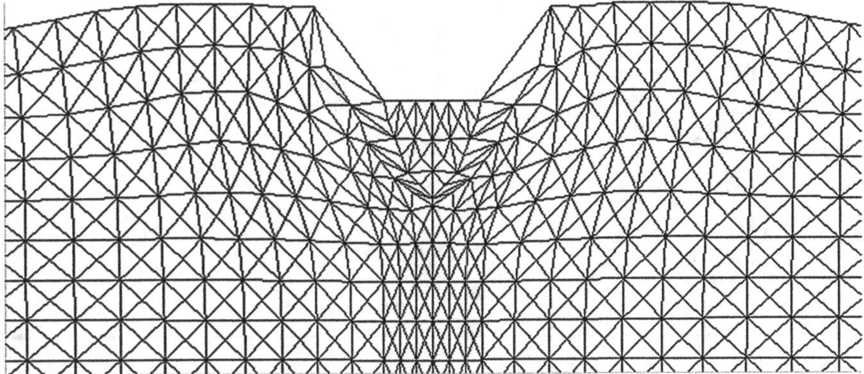

Fig. 13.7. Deformed configuration at failure within a window

left of the footing when $P = 12,300\,kN$ and $\delta = 0.8261\,m$. This occurs during the second global iteration. The procedure does not seem to be diverging. The values of $\boldsymbol{\sigma}_n$, m_n, and $\Delta\boldsymbol{\varepsilon}$ are

$$\boldsymbol{\sigma}_n = (-82.9264, -79.3504, -80.4405, 1.91303, 0, 0)\,\text{kPa},$$

$$\Delta\boldsymbol{\varepsilon} = \left(-4.776 \times 10^{-3}, 1.282 \times 10^{-3}, 0, 9.647 \times 10^{-4}, 0, 0\right) \quad \text{and} \quad m_n = 0.994039$$

The analysis is performed with $n = 1$ in Table 13.1 (Set 1). Hence, the yield surface is a circle in octahedral plane and $\alpha = \alpha_c$ and $k = k_c$. The tension cut-off in (13.2) is taken to be the same as the point of intersection of the yield surface with the I-axis; i.e., $p_t = -p_{t0} = 86.46\,\text{kPa}$ (13.14b). The trial stresses I^{tr} and J^{tr} are $I^{tr} = -363.4\,\text{kPa}$ and $J^{tr} = 64.1\,kPa$, respectively.

By examining the above numbers, it may be seen that (a) the initial stress point n is very near the corner (intersection of the yield surface and the I-axis) where the values of the elastic and plastic moduli (13.14a and 13.14c) are zero, and (b) the trial stress point lies to the left of the cut-off $3p_{t0}$ (13.14b) in the $I - J$ space, and that the plastic corrector from such a point has the potential for ending up at the corner or left of it. The non-convergence observed has to do with these issues. A method is needed for dealing with these issues.

Here, we first describe the overall strategy proposed by Bicanic and Pearce (1996), and follow up with some additional specific strategies. Referring to the $I - J$ stress space shown in Fig. 13.8, the zone outside the yield surface is divided into *regular*, *transition*, and *apex zones*. When a trial stress point lies in either the regular or transition zones, there exists a corresponding closest point projection on the yield surface in the sense of the assumption in (13.17a). Points D and G are examples of such points, and they are projected onto points E and I, respectively. No such projection exists for trial points in the apex zone. Point A is an example of such a point. Such points are projected onto the apex (point B). However, the

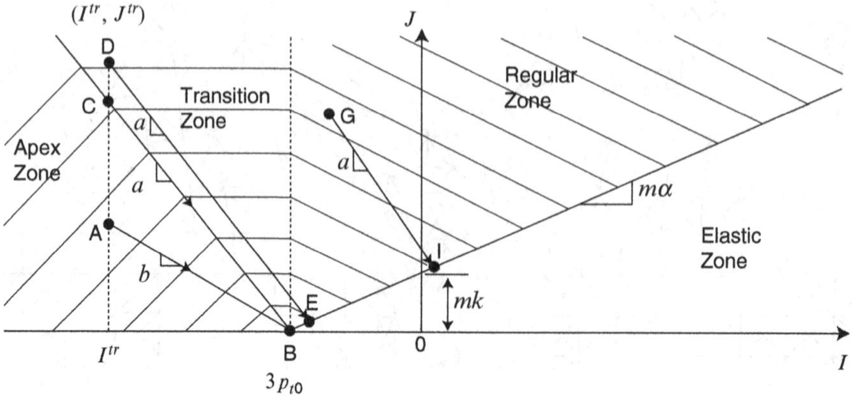

Fig. 13.8 Plastic correctors from trial points in different zones

gradient of the potential surface **r** is undefined at the apex. At the apex, $J = 0$ and $s_{ij} = 0$ and hence **r** is undefined (13.39a).

Specific issues to be addressed are the following:

- Determine the boundary CB between the apex and transition zones
- Determine the incremental solution for trial stress points lying in the apex zone
- Determine the incremental solution for the trial stress points lying in the transition zone

To establish the boundary between the apex and transition zones, let us expand the plastic corrector (13.17a) as

$$\Delta\sigma_{ij}^{\text{pc}} = -\Delta\lambda C_{ijk\ell} r_{k\ell} = K r_{kk} \Delta\lambda\delta_{ij} + 2G\Delta\lambda r_{ij}^{\text{d}} \tag{13.56a}$$

where r_{ij}^{d} is the deviatoric part of r_{ij} (13.12b). The absolute value of the slope of the plastic corrector in the $I - J$ space is determined as

$$a = \left|\frac{2G\bar{r}^{\text{d}}}{3K r_{kk}}\right| \tag{13.56b}$$

where $\bar{r}^{\text{d}}$ is given by (13.12a). For a circular surface, **r** is given by (13.39a), which is repeated here:

$$r_{ij} = -d_0 m\alpha\delta_{ij} + \frac{s_{ij}}{2J} \tag{13.57}$$

By combining (13.56b and 13.57)

$$a = \frac{2G(1/2)}{3K(3d_0 m\alpha)} = \frac{G}{9K d_0 m\alpha} \tag{13.58}$$

Hence, the line separating the transition and apex zones BC has a slope a given by (13.58). The trial points C, D, and G are returned to points B, E, and I, respectively, on the surface at the slope a.

Let us define the slope of the line connecting the trial point (say, point A) and the apex (point B) as

$$b = \frac{J^{\mathrm{tr}}}{|3p_{\mathrm{t}0} - I^{\mathrm{tr}}|} \tag{13.59}$$

Introducing a tolerance TOL_1 to account for the numerical nature of the problem, we will consider a trial point to be

$$\text{in the transition zone if } b > (1 + TOL_1)a \tag{13.60a}$$

$$\text{and in the apex zone if } b \leq (1 + TOL_1)a \tag{13.60b}$$

We will use $TOL_1 = 0.01$(i.e., 1%) in the following.

For a non-hardening (elastic perfectly plastic) model, we have $m = 1$. Then a value for a can be easily calculated from (13.58) and the criteria in (13.60a, b) can be readily implemented. For a hardening material, however, $m_{n+1} \neq m_n$ and hence a cannot be determined a priori. A set of nonlinear equations is to be solved as described below.

Consider a trial point such as point A that is in the apex zone. Point A is to be projected onto point B; i.e., vector AB is to be the plastic corrector. *Let us assume that the volumetric part of* $\mathbf{r}$ *at point B is still given by the volumetric part in (13.57)*; i.e.,

$$r_{kk} = -3d_0 m\alpha \tag{13.61}$$

Hence it follows that

$$3p_{\mathrm{t}0} - I^{\mathrm{tr}} = -\Delta\lambda(3Kr_{kk})_{n+1} = 9Kd_0 m_{n+1}\alpha\Delta\lambda \Rightarrow \Delta\lambda = \frac{3p_{\mathrm{t}0} - I^{\mathrm{tr}}}{9Kd_0 m_{n+1}\alpha}$$

Defining

$$a_1 = \frac{3p_{\mathrm{t}0} - I^{\mathrm{tr}}}{9Kd_0\alpha}, \quad \text{we have} \quad \Delta\lambda = \frac{a_1}{m_{n+1}} \tag{13.62}$$

Note that in the above, α is assumed to be independent of θ (i.e., the surface is circle). From (13.44a)

$$m_{n+1} = \frac{\alpha_{11} + \alpha_{12}\Delta\lambda}{\alpha_{13} + \alpha_{14}\Delta\lambda} \tag{13.63}$$

where the constants such as α_{11} are given by (13.44b–e).

Combining (13.62 and 13.63), we arrive at a quadratic equation for m_{n+1}:

$$\alpha_{13}m_{n+1}^2 + (\alpha_{14}a_1 - \alpha_{11})m_{n+1} - \alpha_{12}a_1 = 0 \tag{13.64}$$

Among the two solutions, the relevant solution is chosen on the basis of $m_0 \le m_{n+1} \le 1$

When the surface is non-circular, Bicanic and Pearce (1996) use the extension parameters α_e and k_e in the above analyses. As $\alpha_e \le \alpha_c$, the corresponding apex zone is the largest (from 13.62), encompassing all possible trial points that are likely to end up in the apex zone.

Noting that the values of the elastic and plastic moduli are zero at point B, as suggested by Crisfield (1987), a small elastic stiffness is used for the tangent stiffness needed for the global analysis.

For example, consider the following problem (the unit for the stresses and stiffness is kPa):

$$\boldsymbol{\sigma}_n = (10, 10, 10, 0, 0, 0) \tag{13.65a}$$

$$\Delta\boldsymbol{\varepsilon} = \left(-2.674 \times 10^{-3}, 0.9483 \times 10^{-3}, -2.674 \times 10^{-3}, 0, 0, 0\right) \tag{13.65b}$$

and

$$m_n = 0.2 \tag{13.65c}$$

The calculated parameters are $K = 47730, G = 22029, p_{t0} = 85.41, I^{tr} = -600$, $J^{tr} = 92.14$, $m_{n+1} = 0.8318$, $a = 0.7706$, and $b = 0.2681$. As $b<(1 + TOL_1)a = 1.01 \times 0.7706 = 0.7783$, by the criteria given by (13.60), the trial stress point is determined to be in the apex zone. The apex zone strategy is thus used for the solution. The calculated incremental stress is $\Delta\boldsymbol{\sigma} = (-94.42, -94.42, -94.42, 0, 0, 0)$. A flag is set to indicate that the increment is elastic for the purpose of calculating the tangent stiffness needed for the global analysis.

With these modifications, the analysis continues with no further convergence difficulties at the local level. Hence for the case considered in this section, where the yield and potential surfaces are assumed to be circles in the octahedral plane, the behaviors from the transition and regular zones are no different from each other (based only on the analysis of the footing problem considered here). However, as described below, this is not the case when the surface is non-circular in the octahedral plane.

13.7.2.2 Convergence Issues Associated with Non-circular Surfaces in the Octahedral Plane

The finite element analysis described in the preceding section is repeated with one change to the values of the model parameters: the value of $n = k_e/k_c = \alpha_e/\alpha_c$

(Table 13.1) is changed from 1 to 0.8. This is done for the purpose of investigating potential convergence difficulties associated with non-circular yield and failure surfaces.

The local iterations fail to converge within 100 iterations at a Gauss point where (using kPa for the unit of stresses and moduli)

$$\boldsymbol{\sigma}_n = (-35.7, 4.77, -24.4, 22.0, 0, 0) \tag{13.66a}$$

$$\Delta\boldsymbol{\varepsilon} = (-0.0205, 0.0101, 0, 0.0323, 0, 0) \tag{13.66b}$$

and

$$m_n = 0.9895 \tag{13.66c}$$

The initial point is on the surface. The invariants of σ_n are $I_n = -55.3$, $J_n = 30.4$, and $\theta_n = 8.62°$. Noting that $3p_{t0} = -259.4$, the initial stress point is between the apex and the $I = 0$ line. The values of the moduli are $K = 40302$ and $G = 18601$. The trial stresses are $I^{tr} = -1,306.7$, $J^{tr} = 863.78$, and $\theta^{tr} = -7.05°$. The values of a and b (13.58 and 13.59) are $a = 0.8085$ and $b = 0.8248$. Thus $b > 1.01a$ (13.60a) and the trial stress point is in the transition zone. Hence, the convergence failure is not caused by the trial stress point being in the apex zone.

The convergence failure in this case is due to the high curvature of the yield and potential surfaces. Such a convergence failure was observed by Bicanic and Pearce (1996). Below, we describe two methods for overcoming this: (1) A method proposed by Bicanic and Pearce (1996) who obtain improved initial estimates to use in the iterations by first projecting the trial point onto an auxiliary circular surface and (2) another method developed here which is on the basis of sub-stepping to establish improved initial estimates, and to progressively establish the final solution.

13.7.2.3 Obtain Improved Initial Estimate: Auxiliary Surface Projection Strategy

Bicanic and Pearce (1996) advocate the use of an initial state other than the elastic trial state – a state from which the final solution can be reached. They found that an improved initial state can be obtained by the closest point projection of the trial stress onto a circular surface whose size is defined by k_e and α_e (13.8c). In the case of the Drucker–Prager model, such a projection can be obtained analytically by following the procedure in Sect. 13.4.

That is, in Algorithm 13.1, we skip steps 4 and 5. Instead, assuming that the yield and potential surfaces are circles in the octahedral plane, we use the procedure in Sect. 13.4 and obtain the values of $\Delta\lambda$, m_{n+1}, and $\boldsymbol{\sigma}_{n+1}$. Then we set $\boldsymbol{\sigma}^1_{n+1} = \boldsymbol{\sigma}_{n+1}$ and $m^1_{n+1} = m_{n+1}$, and go to step 6. We follow the rest of the steps as in Algorithm 13.1. (It may be noted that Lode angle $\theta^1 = \theta^{tr}$, but the final value θ_{n+1} is in general different from θ^{tr}).

The proposal is on the basis of the premise that the final solution is likely to be very close to that based on the circular surface, hence leading to convergence during subsequent iterations. Bicanic and Pearce (1996) found that the above procedure improves the convergence behavior for both the ideal plasticity and softening plasticity, but the non-convergence still exists for softening plasticity. One could use a "line search" strategy to locate the solution (Dutko et al. 1993; Armero and Pérez-Foguet 2002; Pérez-Foguet and Armero 2002). However, Bicanic and Pearce (1996) point out that such a strategy is not only computationally expensive, but also cannot completely overcome the convergence problem in cases of high curvature. To this end, Bicanic and Pearce (1996) employ an under-relaxation scaling of the stress update. The reader is referred to their paper for further details. Alternatively, the sub-stepping technique presented in the next section also can be used.

Now returning to the earlier problem (13.66a–c), let us apply the auxiliary surface projection strategy as the initial state for the local iterations. With $k_e = nk_c = 33.2$ and $\alpha_e = n\alpha_c = 0.128$, and using the method of Sect. 13.4, we get

$$\Delta\lambda = 0.04631; \quad m_{n+1} = 0.9912; \quad \boldsymbol{\sigma}_{n+1} = (-82.11, -78.93, -80.03, 1.67, 0, 0)$$

$$I_{n+1} = -241.06; \quad J_{n+1} = 2.32 \quad \text{and} \quad \alpha_{n+1} = -7.05^\circ.$$

Using these as initial estimates, the Newton-based iterations (steps 6–11, Algorithm 13.1) converges in five iterations, leading to

$$\Delta\lambda = 0.04345, \quad m_{n+1} = 0.9911, \quad \boldsymbol{\sigma}_{n+1} = (-55.01, -29.27, -47.68, 13.59, 0, 0),$$

$$I_{n+1} = -131.96, \quad J_{n+1} = 18.99, \quad \alpha_{n+1} = 9.61^\circ. \tag{13.67}$$

It is noted that the final solution is reasonably close to the initial estimates in terms of the values of $\Delta\lambda$ (0.04631 vs. 0.04345) and m_{n+1} (0.9912 vs. 0.9911). However, during the iteration, the Lode angle changes from –7.05° to 9.61°, indicating that the solution for the stress is not too close to the initial estimate in the octahedral plane.

13.7.2.4 Obtain Improved Initial Estimate: Sub-stepping Strategy

Let us define a sub-increment as

$$\Delta\boldsymbol{\varepsilon}^* = x_0\Delta\boldsymbol{\varepsilon}; \quad 0 < x_0 \leq 1 \tag{13.68}$$

where $\Delta\boldsymbol{\varepsilon}$ is the actual strain increment coming from the main program and x_0 is a linear scale factor. Let the solution for $\Delta\boldsymbol{\varepsilon}^*$ be

$$\mathbf{y} = \{\Delta\lambda, \boldsymbol{\sigma}_{n+1}, m_{n+1}\}$$

It is clear that $\mathbf{y}$ is a function of x_0 i.e.,

$$\mathbf{y} = \mathbf{y}(x_0)$$

Depending on the nonlinearity of constitutive functions between points n and $n+1$ and the magnitude and direction of $\Delta\boldsymbol{\varepsilon}$, $\mathbf{y}(x_0)$ could be a highly nonlinear function of x_0 between points n and $n+1$, resulting in the failure of Newton's method to establish the solution in a single step. However, for a well-constructed constitutive model, regardless of the magnitude and direction of $\Delta\boldsymbol{\varepsilon}$, a solution exists when $x_0 = 1$. Also, Newton's method leads to the actual solution when the "starting set" is close to the "final set." It follows from these observations that one could establish a multi-level iteration technique to arrive at the final solution. A good success has been achieved with the following method for the cases tried thus far.

Let the strain increment $\Delta\boldsymbol{\varepsilon}$ be divided into sub-increments as $\Delta\boldsymbol{\varepsilon} = \Delta\boldsymbol{\varepsilon}_1 + \Delta\boldsymbol{\varepsilon}_2 + \cdots + \Delta\boldsymbol{\varepsilon}_m$. Starting from the set $(\boldsymbol{\sigma}_n, m_n)$, let the solution for $\Delta\boldsymbol{\varepsilon}_1^* = \Delta\boldsymbol{\varepsilon}_1$ be $(\boldsymbol{\sigma}_{n+1}, m_{n+1}, \Delta\lambda)_1$, for $\Delta\boldsymbol{\varepsilon}_2^* = \Delta\boldsymbol{\varepsilon}_1 + \Delta\boldsymbol{\varepsilon}_2$ be $(\boldsymbol{\sigma}_{n+1}, m_{n+1}, \Delta\lambda)_2$, ... and for $\Delta\boldsymbol{\varepsilon}_m^* = \Delta\boldsymbol{\varepsilon}_1 + \Delta\boldsymbol{\varepsilon}_2 + \cdots + \Delta\boldsymbol{\varepsilon}_m$ be $(\boldsymbol{\sigma}_{n+1}, m_{n+1}, \Delta\lambda)_m$.

It is clear that the required solution is $(\boldsymbol{\sigma}_{n+1}, m_{n+1}, \Delta\lambda) = (\boldsymbol{\sigma}_{n+1}, m_{n+1}, \Delta\lambda)_m$.

During the iteration for $\Delta\boldsymbol{\varepsilon}_1^*$, we use$(\sigma_n, m_n)$ as the starting set. We follow Algorithm 13.1 in its entirety for this iteration. $\Delta\boldsymbol{\varepsilon}_1^*$ must be small enough for convergence to be achieved. During the iteration for $\Delta\boldsymbol{\varepsilon}_2^*$, we use $(\boldsymbol{\sigma}_{n+1}, m_{n+1}, \Delta\lambda)_1$ as the starting set. We skip steps 1–5 in Algorithm 13.1 and go directly into step 6. Here, the starting set $(\boldsymbol{\sigma}_{n+1}, m_{n+1}, \Delta\lambda)_1$ and the solution $(\boldsymbol{\sigma}_{n+1}, m_{n+1}, \Delta\lambda)_2$ must be as near as needed for the convergence to be achieved. The procedure is continued until the solution is found.

It remains to develop a rational and efficient method of finding suitable sizes for $\Delta\boldsymbol{\varepsilon}_1^*, \Delta\boldsymbol{\varepsilon}_2^*, \ldots \Delta\boldsymbol{\varepsilon}_m^*$ that lead to convergence. The procedure that has been tried here is as follows:

- Start with the original strain increment $\Delta\boldsymbol{\varepsilon}$ and perform the local iterations. Upon non-convergence within a specified number of iterations (say, 100), divide the current strain increment in half. Continue this until convergence is achieved. The strain increment at this stage is $\Delta\boldsymbol{\varepsilon}_1$.
- Set the starting estimate for $\Delta\boldsymbol{\varepsilon}_2$ as $\Delta\boldsymbol{\varepsilon}_2 = \Delta\boldsymbol{\varepsilon}_1$. Upon non-convergence within a specified number of iterations (say, 100), divide the current value of $\Delta\boldsymbol{\varepsilon}_2$ in half; i.e., $\Delta\boldsymbol{\varepsilon}_2^* = \Delta\boldsymbol{\varepsilon}_1 + \Delta\boldsymbol{\varepsilon}_1/2$. Continue this until convergence is achieved. This establishes the value for $\Delta\boldsymbol{\varepsilon}_2$.
- Set the starting estimate for $\Delta\boldsymbol{\varepsilon}_3$ as $\Delta\boldsymbol{\varepsilon}_3 = \Delta\boldsymbol{\varepsilon}_2$. Continue the process as above until the end.

This procedure is applied to the problem stated earlier (13.66). The convergence is achieved in four iterations when

$\Delta\boldsymbol{\varepsilon}_1^* = \Delta\boldsymbol{\varepsilon}_1 = \Delta\boldsymbol{\varepsilon}/2 = (-0.01024, 0.005065, 0, 0.016135, 0, 0)$. The solution is $(\boldsymbol{\sigma}_{n+1}, m_{n+1}, \Delta\lambda)_1$, where $\boldsymbol{\sigma}_{n+1} = (-45.7, -12.5, -36.3, 17.6, 0, 0)$, $m_{n+1} = 0.9904$, and $\Delta\lambda = 0.02173$. The invariants of the stress are $I_{n+1} = -94.5$, $J_{n+1} = 24.6$, and $\alpha_{n+1} = 9.71°$.

Now setting $\Delta\boldsymbol{\varepsilon}_2 = \Delta\boldsymbol{\varepsilon}_1 = \Delta\boldsymbol{\varepsilon}/2$, the second strain increment is $\Delta\boldsymbol{\varepsilon}_2^* = \Delta\boldsymbol{\varepsilon}_1 + \Delta\boldsymbol{\varepsilon}_2 = \Delta\boldsymbol{\varepsilon} = (-0.0205, 0.0101, 0, 0.0323, 0, 0)$ (i.e., the actual full increment). Using $(\boldsymbol{\sigma}_{n+1}, m_{n+1}, \Delta\lambda)_1$ as the starting set, the iterations converge in five iterations to the same final solution as obtained by the "Auxiliary Surface Projection Strategy" (13.67).

Remarks: The Sub-stepping Strategy

- The method of obtaining the improved starting set by the sub-stepping strategy presented in this section could be used, not only to address the convergence issues associated with non-circular surfaces, but in general when non-convergence is encountered as a result of the elastic trial state being too far from the final solution.
- The method presented in this section is different from the traditional method of sub-stepping. Upon convergence, the present method is the same as the backward Euler method and hence a consistent tangent operator exists. The sub-stepping strategy is simply a way of getting improved initial estimates. In the traditional method, the strain increment is subdivided into smaller steps and the closest point projection is performed separately for each sub-increment. The process yields a series of consistency parameters $\Delta\lambda_1$, $\Delta\lambda_2$, etc. There is no direct way of calculating the consistent tangent operator. One could use the continuum operator for this global increment; however, the rate of convergence may no longer be quadratic. Alternatively, one could use special formulations such as the one proposed by Perez-Foguet et al. (2001) to calculate a consistent operator. For further details of this method, the reader is referred to the above publication.
- The method presented in this section could also be used with CMMP with exact elastic properties.
- An efficient method for determining the size of sub-steps needs to be developed before the sub-stepping method can serve a useful purpose in practical calculations.

Remark: Other Methods of Obtaining Improved Starting Sets

Many other ways of obtaining an improved starting set are conceivable. One such method that has been found to work for the case involving a circular yield and potential surfaces is as follows: Whenever the trial stress point falls in the transition zone (Fig. 13.8), the point (e.g., point D) is projected into the regular zone by simply shifting the point horizontally in the $I-J$ space to a point just right of the apex, where $I = 3p_{t0} + TOL$. TOL is a suitable small tolerance. This method has been found to be as effective as the "auxiliary surface projection strategy."

13.7.3 Results on Convergence of CPPM with Approximate Elastic Properties

To investigate the convergence behavior systematically and comprehensively, a series of local analyses are performed with each analysis beginning from

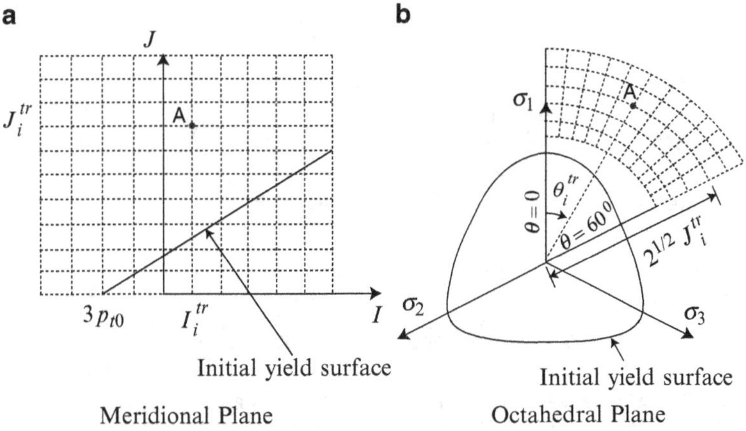

Fig. 13.9 Grids of different elastic trial points

a specific elastic trial stress point. For convenience, the trial points are chosen to be at the nodes of a regular grid. A typical grid chosen in a meridional plane (triaxial compression) and an octahedral plane are shown in Fig. 13.9a and b, respectively.

In the meridional plane (Fig. 13.9a), a 150 × 150 grid is used. The elastic trial stresses at node i of the grid (point A) are $(I_i^{\mathrm{tr}}, J_i^{\mathrm{tr}})$. Runs are made using the model parameter Set 1 listed in Table 13.1 with $n = k_e/k_c = \alpha_e/\alpha_c = 1$ and $d_0 = 0.5$, and the initial parameters $m_n = 0.4$ and $\sigma^n = (10, 10, 10, 0, 0, 0)$. The tolerances used in the iterations are (13.27) $TOL_\sigma = 0.01$, $TOL_\varsigma = 0.01$, and $TOL_\phi = 0.0001$. The results are presented in the form of contours of the number of iterations required for convergence. In the stress space, starting from the initial stress point, the elastic trial points are reached using suitable (pre-calculated) strain increments. Each point requires a specific strain increment.

Not shown here are the results obtained for the case when both the apex region and auxiliary surface strategies are not employed. When these strategies are not employed, the iterations fail to converge in the apex region within 100 iterations, and take fairly large numbers of iterations to converge in the transition zone near the border between the apex and transition zones. No convergence difficulties are encountered in the remainder of the transition zone, and in the regular zone.

The effectiveness of using the apex region and auxiliary surface projection strategies are illustrated in Fig. 13.10a and b. Note that $p^{\mathrm{tr}} = I^{\mathrm{tr}}/3$ and $q^{\mathrm{tr}} = \sqrt{3}J^{\mathrm{tr}}$. Both cases employ the apex region strategy, which correctly identifies the trial points in the apex region and projects the trial points to the apex, as described in Sect. 13.7.2. No iterations are involved for the trial points in the apex region (left of line AB in Fig. 13.10a, b).

As seen in Fig. 13.10a, when no special procedures are used, while convergence is achieved for points in the transition and regular zones, the numbers of iterations

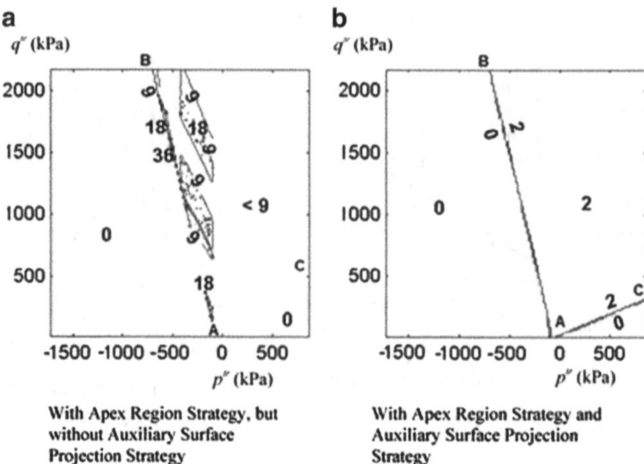

Fig. 13.10 Contours of number of iterations required for convergence for different elastic trial stress points in the triaxial compression meridional plane

are fairly high for the points in the transition zone near the boundary between the apex and transition zones (line AB). It is seen from Fig. 13.10b that the use of the auxiliary surface projection strategy basically eliminates the sluggishness in the rate of convergence. It takes only 2–3 iterations for the points in the transition and regular zones. Note that the region below line AC represents the elastic domain, where there are no iterations involved.

In the octahedral plane (Fig. 13.9b), a 100×100 grid is used. The elastic trial stresses at node i of the grid (point A) are $(\theta_i^{tr}, J_i^{tr})$. As the cases of $n = k_e/k_c = \alpha_e/\alpha_c \neq 1$ (non-circular surfaces) represent the more severe test cases than that of $n = 1$, only the results for cases with $n = 0.8$ are presented and discussed. Results are presented for the cases where the elastic trial stress points lie on the following two octahedral planes: one chosen at $p^{tr} = 10\,\text{kPa}$ (well to the right on the apex and hence in the regular zone, Fig. 13.8) and another chosen at $p^{tr} = -200\,\text{kPa}$ (well to the left of the apex, and hence in the apex and transition zones). The results for $p^{tr} = 10\,\text{kPa}$ and $p^{tr} = -200\,\text{kPa}$ are presented in Figs. 13.11 and 13.12, respectively. In all cases, the apex zone strategy is employed. In each figure, the results obtained with and without the use of the auxiliary surface projection strategy are compared.

Referring first to Fig. 13.11, at $p^{tr} = 10\,\text{kPa}$, the size of the elastic domain (where there are no iterations involved and hence the number of iterations required is zero) is too small compared to the range of trial stresses considered. Hence, the elastic domain is not reflected in the results presented in Fig. 13.11. Once the trial stress points are outside the elastic domain, they are in the regular zone, and convergence is seen to have been achieved in all cases in at most 5–6 iterations. The use of the auxiliary surface projection strategy does not appear to have any discernable effect.

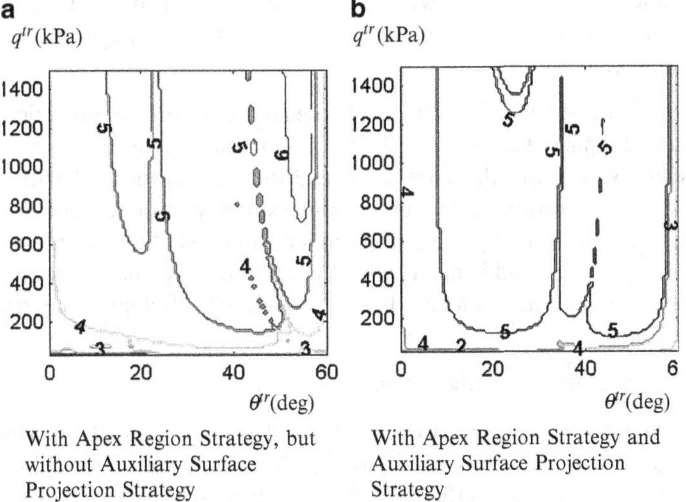

With Apex Region Strategy, but without Auxiliary Surface Projection Strategy

With Apex Region Strategy and Auxiliary Surface Projection Strategy

Fig. 13.11 Contours of number of iterations required for convergence for different elastic trial stress points in the octahedral plane at $p^{tr} = 10\,\text{kPa}$

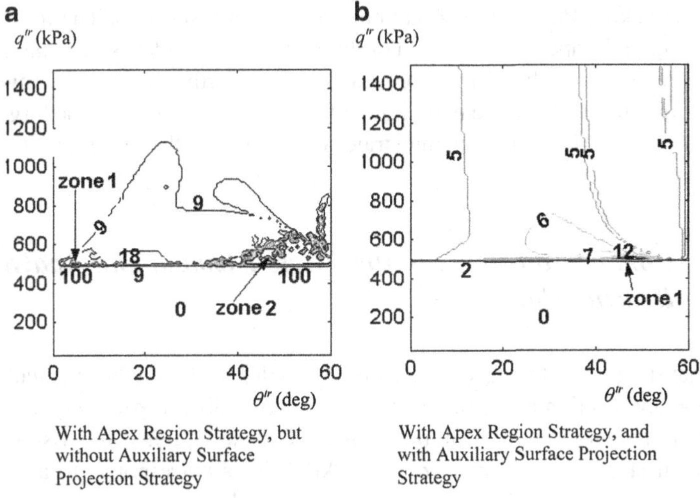

With Apex Region Strategy, but without Auxiliary Surface Projection Strategy

With Apex Region Strategy, and with Auxiliary Surface Projection Strategy

Fig. 13.12 Contours of number of iterations required for convergence for different elastic trial stress points in the octahedral plane at $p^{tr} = -200\,\text{kPa}$

Now referring to Fig. 13.12a, it is seen that there are pockets of zones (zone 1 and zone 2 indicated in the figure), where non-convergence prevails in 100 iterations. The use of the auxiliary surface, as shown in Fig. 13.12b, eliminates the zones of non-convergence. The horizontal band below $q^{tr} = 500\,\text{kPa}$ represents the apex zone, where the apex zone strategy projects the trial points to the apex without the use of iterations. In Fig. 13.12b, while there appears to be a zone (zone 1) where the number of iterations required for convergence is relatively large, convergence is achieved nevertheless.

Remark: As pointed out earlier, the use of the auxiliary surface projection strategy and the sub-stepping strategy described in this section produces similar outcome in terms of overcoming the convergence difficulties associated with the non-circular surfaces. However, the sub-stepping strategy is more general in that it can address the non-convergence arising as a result of any factor (not just from the non-circular surfaces) and hence is useful in other more complex models as well. In the present case of the Drucker–Prager model, both strategies may be simultaneously employed.

Remarks: Convergence of Other Integration Methods

- In CPPM with exact elastic properties (Sect. 13.3), one difficulty concerns, for a given strain increment, the quantity (p_1) within the parenthesis in the expression for p^{tr} (13.29a) or p_{n+1} (13.32a) becoming negative. Noting that p_{n+1} could never be less than $p_{\text{t}} = 3p_{\text{t0}}$, a strategy that seeks to use an improved starting set could be employed to address this. The sub-stepping strategy described in this section can be used to address further convergence issues.
- The S-space formulation presented in this chapter (Sect. 13.4) is only applicable to the Drucker–Prager model that employs circular surfaces. The trial stresses ending up in the apex zone is still a problem; this can be addressed using the apex zone strategy described in the present section. Furthermore, the convergence difficulties arising from the magnitude of strain increment being too large can be addressed using the sub-stepping strategy proposed in the present section.

13.7.4 Comparison Between the Finite Element and Analytical Bearing Capacities

Recall that the parameters α_{c} and k_{c} listed in Tables 13.1 have been calculated for a cohesion and friction angle of $c = 50\,\text{kPa}$ and $\phi = 30°$, respectively. As pointed out earlier, to avoid numerical problems, the Gauss points are assigned with small initial stresses $(\sigma_1 = \sigma_2 = \sigma_3 = 4.5\,\text{kPa})$. This corresponds to an overburden pressure $q = 4.5\,\text{kPa}$ or a footing depth of 0.45 m at a submerged density $\gamma\prime = 10\,kN/m^3$. Accounting for this, the analytical bearing capacities are calculated as follows:

By using Terzaghi's bearing capacity equation (Terzaghi 1943)

$$q_u = cN_{\text{c}} + qN_q + 0.5B\gamma N_\gamma = 50 \times 37.16 + 4.5 \times 22.46 + 0.5 \times 5 \times 10 \times 19.13 = 2,363\,\text{kPa}$$

By using Mayerhof's general bearing capacity equation (Mayerhof 1963)

$$q_u = cN_cF_{cd} + qN_qF_{qd} + 0.5B\gamma N_\gamma$$
$$= 50 \times 30.14 \times 1.04 + 4.5 \times 18.4 \times 1.026 + 0.5 \times 5 \times 10 \times 22.4 = 2{,}213\,\text{kPa}$$

The corresponding collapse loads ($Q_u = q_u(\text{kPa}) \times (5\,\text{m})$) are marked in Fig. 13.6. The results presented in Fig. 13.6 indicate that the collapse load predicted by the finite element method is a function of the dilation parameter d_0. It is seen that the range of collapse loads predicted by the finite element method for $0 \leq d_0 \leq 1$ is within about 40% of those predicted by the analytical methods. The bearing capacity value predicted by the finite element method for $d_0 = 0$ (which corresponds to zero dilation) is very close to the analytical bearing capacity values.

Problems

Problem 13.1 Determine under what conditions, if any, the consistent operator given by (13.28) involved in the σ-space formulation will be symmetric.

Problem 13.2 Verify that in general the consistent tangent operator **D** given by (13.54) for the S-space formulation is non-symmetric.

Problem 13.3 Show that when $\Delta\lambda \rightarrow 0$ and $\Delta\boldsymbol{\varepsilon} \rightarrow 0$, the consistent operator given by (13.54) becomes equal to the continuum operator.

Problem 13.4 Considering the nonlinear isotropic hardening rule (13.10a) considered in this chapter and using the material properties listed in Table 13.1 (Set 2 with $n = 1$ and $d_0 = 1$) and the initial state variables

$$\boldsymbol{\sigma}_n = \{100 \quad 50 \quad 75\}\,\text{kPa}, m_n = 0.3, \text{and } e_0 = 0.9$$

determine the incremental stress and PIV for the following incremental strain:

$$\Delta\boldsymbol{\varepsilon} = \{0.03 \quad -0.028 \quad 0.01\}$$

Use the S-space formulation presented in Sect. 13.4. Refer to the example presented in Sect. 13.5. Sketch the initial and final stresses on the octahedral plane and mark the initial, trial, and final stress points on it.

Problem 13.5 An example of the system matrix given by (13.22) is given in (13.55). The examination of the right-hand side of this equation reveals that $\mathbf{R}_\sigma = (-79.4 \quad -79.4 \quad -79.4)$, whose deviatoric component is zero. Explain why the deviatoric component is zero in this case.

Problem 13.6 In Problem 8.3, you are asked to develop a computer program based on Algorithm 8.2 and the coding guide in Appendix 3 to calibrate a constitutive model. In that program, you called a subroutine called "MISES" where a nonlinear elastic model is implemented (a) to calculate the incremental stress for a given incremental strain, and (b) to calculate the consistent or continuum operator needed for the global iteration.

Modify this program to implement the Drucker–Prager model presented in the present chapter by the two formulations (the σ- and S-space formulations with approximate elastic properties) and perform the following tasks:

a. Using the properties in Table 13.1, reproduce the results in Fig. 13.4 to verify your code.
b. Investigate the influence of the load step size on the convergence behavior and accuracy of results. (You may make one run with a large number of steps and use the results as "exact" in the accuracy evaluations.)
c. Modify your code to include an option of using the continuum operator along with the consistent operator that you already have. Investigate the convergence behavior using both the continuum and consistent operators.

Problem 13.7 Use model parameter Set 1 listed in Table 13.1 with $d_0 = 0.5$ in solving this problem. By using CPPM with approximate (explicit) elastic properties, determine if the elastic trial stress points in the following cases are in the apex, transition, regular, or elastic zones:

Case 1:

$$\boldsymbol{\sigma}_n = (10, 10, 10, 0, 0, 0);$$

$$\Delta\boldsymbol{\varepsilon} = \left(-6.05 \times 10^{-3}, 7.71 \times 10^{-3}, -6.05 \times 10^{-3}, 0, 0, 0\right); \quad m_n = 0.2$$

Case 2:

$$\boldsymbol{\sigma}_n = (10, 10, 10, 0, 0, 0);$$

$$\Delta\boldsymbol{\varepsilon} = \left(-20.7 \times 10^{-4}, -2.6 \times 10^{-4}, -20.7 \times 10^{-4}, 0, 0, 0\right); \quad m_n = 0.2$$

Determine the incremental stress for Case 2.

Chapter 14
The Sliding-Rolling Granular Material Model and Its Integration

The stress–strain behavior of cohesive soil such as clay and that of cohesionless soil or granular materials such as sand share many common characteristics (e.g. pressure-dependency, critical state failure, and shear-induced dilation). However, some aspects of the stress–strain behavior of granular materials are very different from those of cohesive soils. For example, during an undrained (constant-volume) loading, granular materials may experience liquefaction (loss of effective mean normal pressure and/or loss of stiffness and strength), whereas cohesive soils do not. For this reason, a proper mathematical representation of the stress–strain behavior of granular materials requires a specific constitutive law. For instance, the Cam-clay model described in Chap. 12 is not suitable for granular materials. While the Drucker–Prager model may be used to represent some aspects (e.g. pressure-dependency), it is incapable of representing the others (e.g. density-dependent strain softening or liquefaction-induced softening).

Many models have been presented in the past to describe the stress–strain behavior of granular materials. One such model is treated in this chapter. The focus here is on computational issues and thus the finer details of the constitutive model will not be presented; the reader is referred to published materials. The model is described in detail in Anandarajah (2008a). The microstructural basis of the model may be found in Anandarajah (2004, 2008b). An application of the model to the modeling of soil liquefaction is presented in Anandarajah (2008c). The model is known as the sliding-rolling model. The basic structure of the model was derived from a microstructural theory known as the sliding-rolling theory. In this theory, the plastic deformation is considered to result from the grain-to-grain sliding and rolling. The model is capable of simulating many features that the Drucker–Prager model is incapable of simulating. For instance, the model can simulate the effects of fabric anisotropy, density-dependent strain-softening, complex stress ratio-dependent dilation and compaction, critical state failure, plastic behavior during stress reversal, and liquefaction during undrained loading.

If you recall, the Cam-clay and Drucker–Prager models presented in this book employ the isotropic hardening behavior. The hardening variable ς is a scalar. In these models, the rate of change of the yield surface size depends on the rate of

A. Anandarajah, *Computational Methods in Elasticity and Plasticity: Solids and Porous Media*, DOI 10.1007/978-1-4419-6379-6_14,

change of the scalar ς. The yield surface's center of similarity or the orientation does not vary during loading. Both the isotropic and kinematic hardening behaviors are (individually) considered for the von Mises model. The back stress, which is a second order tensor, is used to model the kinematic hardening behavior.

In addition to a scalar hardening variable, the sliding-rolling model employs two 2nd order tensorial variables α_{ij}^{η} and δ_{ij}^{a}. As in the case of the von Mises model, α_{ij}^{η} represents the current center of the yield surface in the octahedral plane. δ_{ij}^{a} represents the current center of the failure surface (which in turn models plastic material anisotropy). Hence, the model embodies not only the isotropic hardening behavior, but also a combination of the kinematic and rotational hardening behaviors.

The basic principles of integrating constitutive models are presented in Chap. 10. The specific details pertaining to the sliding-rolling model are summarized in the present chapter. Only one method of integration (the closest point projection method or CPPM) is presented. The elastic law is integrated both approximately and exactly. Detailed analyses are presented only for the method employing approximate elastic properties.

14.1 The Sliding-Rolling Granular Model

14.1.1 Definition of Yield and Failure Surfaces and Hardening Variables

The cross sections of the yield and failure surfaces in the meridional and octahedral planes are shown schematically in Fig. 14.1a, b, respectively. The yield surface is a cone $\phi_1 = 0$ with a cap $\phi_2 = 0$ placed at $p = p_0 = I_0/3$, where p is the mean normal pressure. The shape of the yield surface is circular in the octahedral plane. The surface $\phi_2 = 0$ is planar. According to the sliding-rolling theory, $\phi_1 = 0$ is

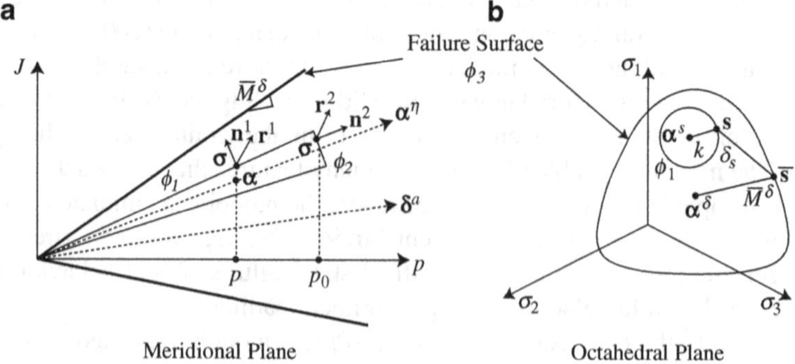

Fig. 14.1 Schematic of yield and failure surfaces

associated with interparticle sliding (mechanism 1) and $\phi_2 = 0$ is associated with interparticle rolling (mechanism 2). At the corner between the two surfaces, the plastic deformation comes from both mechanisms; hence the theory is a multi-mechanism model. For the sake of simplicity, the primary focus in this book is limited to the plastic behavior associated with mechanism 1. For completeness, however, some details concerning the behavior associated with mechanism 2 are also presented. The reader is referred to Anandarajah (2008a) for further details concerning mechanism 2.

The yield surface is allowed to move around in the octahedral plane through a kinematic hardening rule. The change in the size of the yield surface in the octahedral plane (i.e. its radius k), however, is neglected. p_0 is allowed to change through a scalar hardening rule. The yield surface generally stays inside an outer surface $\phi_3 = 0$, known as the virtual critical state failure surface (simply known as the "failure surface" hereafter). While at the critical state failure, the yield surface becomes inside of and tangential to the failure surface; at intermediate stages, the surface is allowed to go outside. This allows the strain-softening behavior exhibited by dense granular materials to be simulated.

Let σ_{ij} be the current stress tensor. α_{ij} is defined as a tensor directed along the axis of the yield cone, having the same units as the stress tensor. The following quantities are defined:

$$\alpha_{ij}^{\eta} = \frac{\alpha_{ij}^{S}}{I}; \quad \alpha_{ij}^{S} = \alpha_{ij} - \frac{1}{3}\alpha_{kk}\delta_{ij}; \quad I = \sigma_{kk} \tag{14.1}$$

I and α_{kk} are the traces of σ_{ij} and α_{ij}, respectively. α_{ij}^{S} is the deviatoric part of α_{ij}, which represents the location of the center of the yield surface in the octahedral plane at a given I. α_{ij}^{η} equals α_{ij}^{S} normalized with I. Hence α_{ij}^{η} is a dimensionless tensor, simply representing the direction of the axis of the cone $\phi_1 = 0$ in the stress space. δ_{ij} is the Kronecker delta; $\delta_{ij} = 1$ for $i = j$ and $\delta_{ij} = 0$ for $i \neq j$. Hence, the δ_{ij} represents the direction of the space diagonal or the p–axis, where $p = I/3$.

The axis of the failure surface $\phi_3 = 0$ is taken along the direction of a second order tensor δ_{ij}^{a}, defined as

$$\delta_{ij}^{\mathrm{a}}\delta_{ij}^{\mathrm{a}} = 3 \tag{14.2a}$$

$$\delta_{ij}^{\mathrm{a}} = \delta_{ij} \text{ for isotropic state, and} \tag{14.2b}$$

$$\delta_{ij}^{\mathrm{a}} \neq \delta_{ij} \text{ for anisotropic state} \tag{14.2c}$$

Hence, δ_{ij}^{a} is a dimensionless tensor representing the direction of $\phi_3 = 0$ in the stress space. The center of ϕ_3 in the octahedral plane at a given I is then given by

$$\alpha_{ij}^{\delta} = \frac{I}{\delta_{kk}^{\mathrm{a}}}\delta_{ij}^{\mathrm{ad}} \tag{14.2d}$$

where $\delta_{ij}^{\mathrm{ad}}$ is the deviatoric part of δ_{ij}^{a} defined as

$$\delta_{ij}^{\mathrm{ad}} = \delta_{ij}^{\mathrm{a}} - \frac{1}{3}\delta_{kk}^{\mathrm{a}}\delta_{ij} \tag{14.2e}$$

With the aid of the above equations and Fig. 14.1, the geometrical meaning of the *hardening variables* $\boldsymbol{\alpha}^{\eta}$, $\boldsymbol{\delta}^{\mathrm{a}}$, and p_0 may now be understood.

The equations for ϕ_1 and ϕ_2 are in the form

$$\phi_1(\boldsymbol{\sigma}, \boldsymbol{\alpha}^{\eta}, p_o) = 0 \tag{14.3a}$$

$$\phi_2(\boldsymbol{\sigma}, \boldsymbol{\alpha}^{\eta}, p_o) = 0 \tag{14.3b}$$

It may be noted that both ϕ_1 and ϕ_2 are independent of $\boldsymbol{\delta}^{\mathrm{a}}$. As we will see later, it is not necessary to explicitly define the mathematical form of ϕ_3. However, its size and location in the octahedral plane must be defined. At a given value of I, the location is given by α_{ij}^{δ} (14.2d).

As shown in Fig. 14.1, the parameters representing the radii of ϕ_1 and ϕ_3 in the octahedral plane are k and $\bar{M}^{\delta}$ respectively. Note that $\bar{M}^{\delta}$ is defined with respect to $\boldsymbol{\delta}^{\mathrm{a}}$. k is assumed to be independent of the Lode angle θ (Chap. 2). However, to simulate the three-dimensional behavior realistically, $\bar{M}^{\delta}$ is allowed to vary as a function of the Lode angle. The Lode angle to be used here is defined with respect to the center of ϕ_3, instead of the origin of the deviatoric stress tensor $\mathbf{s}$, where

$$s_{ij} = \sigma_{ij} - \frac{1}{3}I\delta_{ij} \tag{14.3c}$$

The equations of ϕ_1 and ϕ_2, and their gradients with respect to the state variables $\boldsymbol{\sigma}$, $\boldsymbol{\alpha}^{\eta}$, $\boldsymbol{\delta}^{\mathrm{a}}$, and p_0 are given in Appendix 12.1.

While the parameter k is kept a constant, the parameter $\bar{M}^{\delta}$ is allowed to vary as a function of the void ratio to simulate the effect of density. $\bar{M}^{\delta}$ varies about its value on the critical state line $M^{c\delta}$. $\bar{M}^{\delta} > M^{c\delta}$ when the material is dense. $\bar{M}^{\delta} < M^{c\delta}$ when the material is loose. $\bar{M}^{\delta} = M^{c\delta}$ when the material is on the critical state. The effect of void ratio is introduced through a variable I_s known as the state index. The details are presented in Appendices 12.2. It may be noted from Appendix 12.2 that the definition of I_s requires the introduction of two lines in the $e - p$ space, known as the critical state line (in the $e - p$ space) and the upper reference line. The definitions of these lines require four model parameters Γ_s, Γ_U, λ^*, and n_p. As shown in Appendix 12.3, $M^{c\delta}$ (which is defined with respect to $\boldsymbol{\delta}^{\mathrm{a}}$) is a function of the interparticle friction angle ϕ_{μ}. The slope of the critical state line in the $p - q$ space M_{CSL} (which is defined with respect to $\boldsymbol{\delta}$) depends not only on ϕ_{μ}, but also on another parameter A_{CSL}. The parameter A_{CSL} defines the rotation of ϕ_3 at the critical state. As described in Appendix 12.3, the relationship between $\bar{M}^{\delta}$ and $M^{c\delta}$ requires an additional model parameter c_{α}. When $\boldsymbol{\delta}^{\mathrm{a}} = \boldsymbol{\delta}$, we have $\bar{M}^{\delta} = M^{c\delta}$ for $c_{\alpha} = 0$ and

$\bar{M}^{\delta} \neq M^{c\delta}$ for $c_{\alpha}>0$. Whether $\bar{M}^{\delta}>M^{c\delta}$ or $\bar{M}^{\delta}<M^{c\delta}$ depends on the state index I_s. For $I_s>1$ (dense soil), we have $\bar{M}^{\delta}>M^{c\delta}$ and for $I_s<1$ (loose soil), we have $\bar{M}^{\delta}>M^{c\delta}$.

14.1.2 Constitutive Laws and Hardening Rules

With the assumption of additivity, the total strain rate is split up into elastic and plastic strain rates as

$$\dot{\boldsymbol{\varepsilon}} = \dot{\boldsymbol{\varepsilon}}^{\mathrm{e}} + \dot{\boldsymbol{\varepsilon}}^{\mathrm{p}} \tag{14.4}$$

Assuming elastic isotropy, the elastic stress–strain relation is

$$\dot{\boldsymbol{\sigma}} = \mathbf{C}\dot{\boldsymbol{\varepsilon}}^{\mathrm{e}} \tag{14.5a}$$

where

$$C_{ijk\ell} = \left(K - \frac{2}{3}G\right)\delta_{ij}\delta_{k\ell} + G\left(\delta_{ik}\delta_{j\ell} + \delta_{i\ell}\delta_{jk}\right) \tag{14.5b}$$

G and K are the shear and bulk moduli, respectively. $\mathbf{C}$ is the elastic stiffness tensor. $\dot{\boldsymbol{\sigma}}$ and $\dot{\boldsymbol{\varepsilon}}^{\mathrm{e}}$ are the rate of stress and elastic strain, respectively. G and K are pressure dependent. On the basis of the swelling relation in the $e - p$ space $e = e^{*} - \kappa(p/p_{\mathrm{a}})^{0.5}$, an equation is derived for K. G is then related to K in terms of Poisson's ratio v. The resulting equations are as follows:

$$K = K(p) = \bar{K}_0 p^{0.5}; \quad \bar{K}_0 = 2p_a^{0.5}\frac{1 + e_0}{\kappa} \tag{14.6a}$$

$$G = G(p) = \bar{G}_0 p^{0.5}; \quad \bar{G}_0 = \bar{K}_0\frac{3(1 - 2v)}{2(1 + v)} \tag{14.6b}$$

If a strain increment activates both surfaces ϕ_1 and ϕ_2, the total plastic strain rate depends on contributions from each surface. Hence in general

$$\dot{\boldsymbol{\varepsilon}}^{p} = \sum_{k=1}^{2}\dot{\lambda}_k\mathbf{r}^k; \quad \dot{\lambda}_k = \frac{1}{K_{\mathrm{p}}^k}\mathbf{n}^k{:}\dot{\boldsymbol{\sigma}} > 0 \tag{14.7}$$

where $\mathbf{r}^1$ and $\mathbf{r}^2$ are the directions of the plastic strain rate associated with ϕ_1 and ϕ_2, $\mathbf{n}^1$ and $\mathbf{n}^2$ (Appendix 12.1) are tensors normal to ϕ_1 and ϕ_2, K_{p}^1 and K_{p}^2 are the plastic moduli associated with mechanisms 1 and 2, and $\dot{\lambda}_1$ and $\dot{\lambda}_2$ are the plastic consistency parameters associated with mechanisms 1 and 2.

Generally, surfaces known as plastic potential surfaces are used to define directions such as $\mathbf{r}^1$ and $\mathbf{r}^2$. While this is the approach taken in defining $\mathbf{r}^2$, $\mathbf{r}^1$ is defined as a linear combination of $\mathbf{n}^1$ and $\boldsymbol{\sigma}$ as

$$\mathbf{r}^1 = \mathbf{n}^1 + d\frac{\boldsymbol{\sigma}}{I}$$

The above equation is on the basis of the sliding-rolling theory. The parameter d is one of the key parameters in the model. d controls the ratio between the rates of volumetric to deviatoric plastic strains. The sliding-rolling theory leads to an equation for d in terms of ϕ_μ. In addition, just as $\bar{M}^\delta$ is a function of the void ratio, d is a function of the void ratio as well. An approach similar to that used for $\bar{M}^\delta$ is used to account for the void ratio dependency of d. This introduces another model parameter c_d into the formulation. The details are summarized in Appendix 12.4. For $I_s > 1$ (dense soil), as c_d increases, d decreases and the vector $\mathbf{r}^1$ rotates towards $\mathbf{n}^1$. The volumetric strain becomes more dilative. For $I_s < 1$ (loose soil), the trend is opposite.

A modified ellipse is used for the plastic potential to define $\mathbf{r}^2$. No additional model parameters are introduced in the definition of the ellipse. The details concerning the definition of $\mathbf{r}^1$ and $\mathbf{r}^2$ are presented in Appendix 12.4.

The hardening rules are as follows:

$$\dot{\boldsymbol{\alpha}}^\eta = \xi_1\dot{\lambda}_1\bar{\mathbf{h}}; \quad \xi_1 = \xi_1(\boldsymbol{\sigma},\boldsymbol{\alpha}^\eta,\boldsymbol{\delta}^{\mathrm{a}},p_0) \quad \text{and} \quad \bar{\mathbf{h}} = \bar{\mathbf{h}}(\boldsymbol{\sigma},\boldsymbol{\alpha}^\eta,\boldsymbol{\delta}^{\mathrm{a}},p_0) \tag{14.8a}$$

$$\dot{\boldsymbol{\delta}}^{\mathrm{a}} = \bar{\lambda}^*\bar{\mathbf{r}}\sum_{k=1}^{2}\dot{\lambda}_k; \quad \bar{\lambda}^* = \bar{\lambda}^*(\boldsymbol{\sigma},\boldsymbol{\alpha}^\eta,\boldsymbol{\delta}^{\mathrm{a}},p_0) \quad \text{and} \quad \bar{\mathbf{r}} = \bar{\mathbf{r}}(\boldsymbol{\sigma},\boldsymbol{\alpha}^\eta,\boldsymbol{\delta}^{\mathrm{a}},p_0) \tag{14.8b}$$

$$\dot{p}_o = \sum_{k=1}^{2} g_k\dot{\lambda}_k; \quad g_k = g_k(\boldsymbol{\sigma},\boldsymbol{\alpha}^\eta,\boldsymbol{\delta}^{\mathrm{a}},p_0) \tag{14.8c}$$

It is seen that $\bar{\mathbf{h}}$ and $\bar{\mathbf{r}}$ are the directions of $\dot{\boldsymbol{\alpha}}^\eta$ and $\dot{\boldsymbol{\delta}}^{\mathrm{a}}$, respectively. $\dot{\boldsymbol{\alpha}}^\eta$ depends only on plastic deformation due to mechanism 1. $\dot{\boldsymbol{\delta}}^{\mathrm{a}}$ and $\dot{p}_o$ depend on plastic deformation due to both mechanisms. The function ξ_1 introduces one model parameter c_ℓ. During loading, the function $\bar{\lambda}^*$ is defined completely in terms of A_{CSL}. During stress reversal, however, an additional parameter λ_{UL} is used to control $\dot{\boldsymbol{\delta}}^{\mathrm{a}}$. The functions g_1 and g_2 depend only on the model parameters already defined; namely λ^*, κ, and n_p. See Appendix 12.2 for the definition of λ^* and n_p. κ is the slope of the swelling line in the $e - \ln p$ space. The details of the hardening functions are summarized in Appendix 12.5.

As seen in Fig. 14.1, the radius of the failure surface $\bar{M}^\delta$ is allowed to vary as a function of the Lode angle defined with respect to the center of ϕ_3 (called the anisotropic Lode angle θ^{A}). The parameter M^δ (which appears in the definition of $\mathbf{r}^2$) is also allowed to vary with θ^{A}. The parameter M_{CSL}, however, is expressed as a function of the Lode angle defined with respect to the origin of the stress axes in the

Table 14.1a Model parameters in triaxial compression

Set no.	λ^*	κ	Γ_U	Γ_S	n_p	ϕ_μ	c_d	c_α	c_ℓ	A_{CSL}	λ_{UL}
1	0.03	0.007	0.888	0.76	0.4	23^0	0.4	0.31	1.6	1.4	80
2	0.03	0.007	0.888	0.752	0.4	23^0	0.4	0.31	10.0	1.4	80

Table 14.1b Ratio between model parameter values in triaxial extension and compression

Set no.	$\frac{M^e_{\text{CSL}}}{M^c_{\text{CSL}}} = \frac{M^{\delta e}}{M^{\delta c}} = \frac{\bar{M}^{\delta e}}{\bar{M}^{\delta c}}$	$\frac{\Gamma^e_S}{\Gamma^c_S}$	$\frac{c^e_d}{c^c_d}$	$\frac{c^e_\alpha}{c^c_\alpha}$	$\frac{c^e_\ell}{c^c_\ell}$
1	0.8	0.9474	1.00	0.001	0.20
2	0.8	0.9474	1.00	0.001	0.20

octahedral plane (called the isotropic Lode angle θ^I). In matching the laboratory experimental stress–strain data with model simulations, it was found necessary to permit some of the model parameters also vary with the Lode angles. In the present case, the parameters c_d, c_α, c_ℓ, and Γ_S are allowed to vary with θ^I. The details are summarized in Appendix 12.6.

The results presented in this chapter are obtained using the model parameter values listed in Table 14.1a, b. Poisson's ratio is assumed to be 0.3. The value of the parameter k is fixed at 0.05.

14.1.3 Consistency Condition and Plastic Modulus

The gradients of the yield surfaces with respect to stress are

$$\mathbf{n}^1 = \frac{\partial \phi_1}{\partial \boldsymbol{\sigma}}; \quad \mathbf{n}^2 = \frac{\partial \phi_2}{\partial \boldsymbol{\sigma}} \tag{14.9a}$$

Let us define

$$T_1 = \frac{\partial \phi_1}{\partial p_o}; \quad T_2 = \frac{\partial \phi_2}{\partial p_o} \tag{14.9b}$$

$$\mathbf{P}_1 = \frac{\partial \phi_1}{\partial \boldsymbol{\alpha}^\eta}; \quad \mathbf{P}_2 = \frac{\partial \phi_2}{\partial \boldsymbol{\alpha}^\eta} \tag{14.9c}$$

The expressions for the above gradients are given in Appendix 12.1. It may be noted that $T_1 = 0$.

Applying the consistency condition to the yield surfaces, we have

$$\dot{\phi}_k = \frac{\partial \phi_k}{\partial \boldsymbol{\sigma}} \dot{\boldsymbol{\sigma}} + \frac{\partial \phi_k}{\partial \boldsymbol{\alpha}^\eta} \dot{\boldsymbol{\alpha}}^\eta + \frac{\partial \phi_k}{\partial p_o} \dot{p}_o = 0; \quad k = 1, 2 \tag{14.10}$$

Note that ϕ_1 and ϕ_2 are independent of $\boldsymbol{\delta}^{\mathrm{a}}$. The influence of $\boldsymbol{\delta}^{\mathrm{a}}$ comes only through the evolution of $\boldsymbol{\alpha}^{\eta}$. Combining (14.7), (14.8a), and (14.8c) and using the notations in (14.9a–c), we arrive at

$$[\mathbf{A}]\left[\dot{\boldsymbol{\lambda}}\right] = \begin{bmatrix} A_{11} & A_{12} \\ A_{21} & A_{22} \end{bmatrix} \begin{bmatrix} \dot{\lambda}_1 \\ \dot{\lambda}_2 \end{bmatrix} = \begin{bmatrix} K_p^1 + \xi_1 \mathbf{P}_1.\bar{\mathbf{h}} & 0 \\ T_2 g_1 + \xi_1 \mathbf{P}_2.\bar{\mathbf{h}} & K_{\mathrm{p}}^2 + T_2 g_2 \end{bmatrix} \begin{bmatrix} \dot{\lambda}_1 \\ \dot{\lambda}_2 \end{bmatrix} = 0 \tag{14.11a}$$

$$\text{For } \dot{\boldsymbol{\lambda}} \neq \mathbf{0} \Rightarrow |\mathbf{A}| = A_{11}A_{22} - A_{12}A_{21} = 0 \tag{14.11b}$$

Equation (14.11b) is satisfied when the diagonal elements of $\mathbf{A}$ are zero as

$$A_{11} = 0 \Rightarrow K_{\mathrm{p}}^1 = -\xi_1 \mathbf{P}_1.\bar{\mathbf{h}} \tag{14.12a}$$

$$A_{22} = 0 \Rightarrow K_{\mathrm{p}}^2 = -T_2 g_2 \tag{14.12b}$$

It may be noted that the expression for K_{p}^1 when only Mechanism 1 is active is the same as the one given by (14.12a). Similarly, the expression for K_{p}^2 when only Mechanism 2 is active is the same as the one given by (14.12b). These properties ensure that, regardless of whether Mechanism 1, Mechanism 2 or both are active, the stress point will continue to lie on the surfaces. The stress point will not lag behind the movement of the surfaces, ending up inside the surfaces, or move faster, ending up outside the surface.

14.1.4 Combined Elasto-Plastic Law and Continuum Tangent Operator

Combining (14.4), (14.5a) and (14.7), we have

$$\dot{\boldsymbol{\sigma}} = \mathbf{C}\dot{\boldsymbol{\varepsilon}}^{\mathrm{e}} = \mathbf{C}[\dot{\boldsymbol{\varepsilon}} - \dot{\boldsymbol{\varepsilon}}^{\mathrm{p}}] = \mathbf{C}\left[\dot{\boldsymbol{\varepsilon}} - \sum \dot{\lambda}_k\, \mathbf{r}^k\right] \tag{14.13}$$

Multiplying the above equation by $\mathbf{n}^{\ell}$, it follows that

$$\mathbf{n}^{\ell}\dot{\boldsymbol{\sigma}} = \mathbf{n}^{\ell}\mathbf{C}\left[\dot{\boldsymbol{\varepsilon}} - \sum \dot{\lambda}_k\, \mathbf{r}^k\right] = \mathbf{n}^{\ell}\mathbf{C}\dot{\boldsymbol{\varepsilon}} - \sum \dot{\lambda}_k \mathbf{n}^{\ell}\mathbf{C}\mathbf{r}^k \tag{14.14a}$$

$$\dot{\lambda}_{\ell} K_{\mathrm{p}}^{\ell} = \mathbf{n}^{\ell}\mathbf{C}\dot{\boldsymbol{\varepsilon}} - \sum \dot{\lambda}_k\, \mathbf{n}^{\ell}\mathbf{C}\mathbf{r}^k \tag{14.14b}$$

$$[\mathbf{B}]\left[\dot{\boldsymbol{\lambda}}\right] = \begin{bmatrix} K_p^1 + \mathbf{n}^1\mathbf{C}\mathbf{r}^1 & \mathbf{n}^1\mathbf{C}\mathbf{r}^2 \\ \mathbf{n}^2\mathbf{C}\mathbf{r}^1 & K_{\mathrm{p}}^2 + \mathbf{n}^2\mathbf{C}\mathbf{r}^2 \end{bmatrix} \begin{bmatrix} \dot{\lambda}_1 \\ \dot{\lambda}_2 \end{bmatrix} = \begin{bmatrix} \mathbf{n}^1\mathbf{C}\ \dot{\boldsymbol{\varepsilon}} \\ \mathbf{n}^2\mathbf{C}\ \dot{\boldsymbol{\varepsilon}} \end{bmatrix} = \begin{bmatrix} Q_1 \\ Q_2 \end{bmatrix} = [\mathbf{Q}] \tag{14.14c}$$

$$\dot{\lambda}_k = \sum B_{km}^{-1} Q_m = \sum B_{km}^{-1} [\mathbf{n}^m \mathbf{C} \dot{\boldsymbol{\varepsilon}}] \tag{14.14d}$$

Let us define

$$\mathbf{A}_m = \mathbf{n}^m \mathbf{C} \quad \text{and} \quad \mathbf{E}_k = \mathbf{C}\mathbf{r}^k \tag{14.15}$$

Substituting (14.14d) into (14.13), and simplifying the resulting equation using (14.15), we have

$$\dot{\boldsymbol{\sigma}} = \bar{\mathbf{D}} \dot{\boldsymbol{\varepsilon}} \tag{14.16a}$$

$$\bar{\mathbf{D}} = \mathbf{C} - \sum_{m=1}^{2} \sum_{k=1}^{2} B_{km}^{-1} [\mathbf{A}_m \mathbf{E}_k] \tag{14.16b}$$

Equation (14.16a) is the overall elasto-plastic stress–strain rate relation. Equation (14.16b) presents an expression for is the continuum elasto-plastic tangent operator.

Uniqueness of solution is a concern in any multi-mechanism model. This is treated in detail in Anandarajah (2008a).

14.2 Application of CPPM to Sliding-Rolling Model: The σ-Space Formulation with Approximate Elastic Properties

Chapter 10 presents the general details of CPPM. In addition, Chaps. 12 and 13 present the details of implementation for pressure-dependent models and are useful to review. As the elastic properties are a function of the mean normal pressure (14.6a and b), the elastic predictor is not a constant in general; it varies during the iteration. The reader is referred to Sects. 12.4 and 12.4.1 on the Cam-clay model for a discussion on this aspect. There are two ways to handle the variable elastic properties:

- Use an Euler approximation (Sects. 12.4.1 and 13.2)
- Integrate the elastic relations exactly (Sects. 12.5.1 and 13.3, Appendices 8.4 and 10)

The presentation in this section is restricted to the former. The key equations associated with CPPM with exact elastic properties will be briefly presented in Sect. 14.4.

14.2.1 The Backward Euler Equations

We will use the forward Euler (explicit) approximation for the elastic properties as

$$K = K_n = K(\boldsymbol{\sigma}_n); \quad G = G_n = G(\boldsymbol{\sigma}_n) \tag{14.17}$$

According to this, the elastic predictor remains a constant during the iteration. The primary Euler equations are

$$\boldsymbol{\sigma}_{n+1} = \boldsymbol{\sigma}_n + \mathbf{C}\Delta\boldsymbol{\varepsilon} - \mathbf{C}\{\Delta\lambda_1 \mathbf{r}^1_{n+1} + \Delta\lambda_1 \mathbf{r}^1_{n+1}\} \tag{14.18a}$$

$$\boldsymbol{\alpha}^{\eta}_{n+1} = \boldsymbol{\alpha}^{\eta}_n + \xi_1 \Delta\lambda_1 \bar{\mathbf{h}}_{n+1} \tag{14.18b}$$

$$\boldsymbol{\delta}^{\mathrm{a}}_{n+1} = \frac{\sqrt{3}}{(\mathbf{y}_{n+1} \cdot \mathbf{y}_{n+1})^{1/2}} \mathbf{y}_{n+1}, \quad \text{where} \quad \mathbf{y}_{n+1} = \boldsymbol{\delta}^{\mathrm{a}}_n + \lambda^* \{\Delta\lambda_1 + \Delta\lambda_2\} \bar{\mathbf{r}}_{n+1} \tag{14.18c}$$

$$p_0^{n+1} = p_0^n + \{g_1^{n+1}\Delta\lambda_1 + g_2^{n+1}\Delta\lambda_2\} \tag{14.18d}$$

$$\phi_1^{n+1}(\boldsymbol{\sigma}_{n+1}, \boldsymbol{\alpha}^{\eta}_{n+1}, p_0^{n+1}) = 0 \quad \text{if} \quad \phi_1^{\mathrm{tr}} > 0 \tag{14.18e}$$

$$\phi_2^{n+1}(\boldsymbol{\sigma}_{n+1}, \boldsymbol{\alpha}^{\eta}_{n+1}, p_0^{n+1}) = 0 \quad \text{if} \quad \phi_2^{\mathrm{tr}} > 0 \tag{14.18f}$$

Noting that $\boldsymbol{\sigma}$, $\boldsymbol{\alpha}^{\eta}$, and $\boldsymbol{\delta}^{\mathrm{a}}$ are symmetric second order tensors, there are 20 variables (if one mechanism is activated) or 21 variables (if both mechanisms are activated) in the above equations.

14.2.2 The Residual Equations and Their Linearizations

The residual equations are given by (10.31a–c)

$$\begin{aligned}\mathbf{R}_{\sigma} &= \boldsymbol{\sigma}_{n+1} - [\boldsymbol{\sigma}_n + \mathbf{C}\Delta\boldsymbol{\varepsilon} - \mathbf{C}\{\Delta\lambda_1 \mathbf{r}^1_{n+1} + \Delta\lambda_2 \mathbf{r}^2_{n+1}\}] \\ &= \mathbf{R}_{\sigma}(\boldsymbol{\sigma}_{n+1}, \boldsymbol{\alpha}^{\eta}_{n+1}, \boldsymbol{\delta}^{\mathrm{a}}_{n+1}, p_0^{n+1}) = 0\end{aligned} \tag{14.19a}$$

$$\mathbf{R}_{\alpha^{\eta}} = \boldsymbol{\alpha}^{\eta}_{n+1} - [\boldsymbol{\alpha}^{\eta}_n + \xi_1 \Delta\lambda_1 \bar{\mathbf{h}}_{n+1}] = \mathbf{R}_{\alpha^{\eta}}(\boldsymbol{\sigma}_{n+1}, \boldsymbol{\alpha}^{\eta}_{n+1}, \boldsymbol{\delta}^{\mathrm{a}}_{n+1}, p_0^{n+1}) = 0 \tag{14.19b}$$

$$\mathbf{R}_{\delta^{\mathrm{a}}} = \boldsymbol{\delta}^{\mathrm{a}}_{n+1} - \frac{\sqrt{3}}{(\mathbf{y}_{n+1} \cdot \mathbf{y}_{n+1})^{1/2}} \mathbf{y}_{n+1} = \mathbf{R}_{\delta^{\mathrm{a}}}(\boldsymbol{\sigma}_{n+1}, \boldsymbol{\alpha}^{\eta}_{n+1}, \boldsymbol{\delta}^{\mathrm{a}}_{n+1}, p_0^{n+1}) = 0 \tag{14.19c}$$

$$R_{p_0} = p_0^{n+1} - \left[p_0^n + \left\{g_1^{n+1}\Delta\lambda_1 + g_2^{n+1}\Delta\lambda_2\right\}\right] = R_{p_0}\left(\boldsymbol{\sigma}_{n+1}, \boldsymbol{\alpha}_{n+1}^{\eta}, \boldsymbol{\delta}_{n+1}^{\mathrm{a}}, p_0^{n+1}\right) = 0 \tag{14.19d}$$

$$R_{\phi_1} = \phi_1^{n+1}\left(\boldsymbol{\sigma}_{n+1}, \boldsymbol{\alpha}_{n+1}^{\eta}, p_0^{n+1}\right) = 0 \quad \text{if} \quad \phi_1^{\mathrm{tr}} > 0 \tag{14.19e}$$

$$R_{\phi_2} = \phi_2^{n+1}\left(\boldsymbol{\sigma}_{n+1}, \boldsymbol{\alpha}_{n+1}^{\eta}, p_0^{n+1}\right) = 0 \quad \text{if} \quad \phi_2^{\mathrm{tr}} > 0 \tag{14.19f}$$

Linearizing the above equations (as in 10.34a–c) and presenting them in a matrix-vector form (for the case where $\phi_1^{\mathrm{tr}} > 0$ and $\phi_2^{\mathrm{tr}} > 0$), we have

$$\underset{21\times 21}{\begin{bmatrix} A_{11} & A_{12} & A_{13} & A_{14} & A_{15} & A_{16} \\ A_{21} & A_{22} & A_{23} & A_{24} & A_{25} & A_{26} \\ A_{31} & A_{32} & A_{33} & A_{34} & A_{35} & A_{36} \\ A_{41} & A_{42} & A_{43} & A_{44} & A_{45} & A_{46} \\ A_{51} & A_{52} & A_{53} & A_{54} & A_{55} & A_{56} \\ A_{61} & A_{62} & A_{63} & A_{64} & A_{65} & A_{66} \end{bmatrix}} \underset{21\times 1}{\begin{bmatrix} \underset{6\times 1}{\delta\bar{\boldsymbol{\sigma}}} \\ \underset{6\times 1}{\delta\bar{\boldsymbol{\alpha}}^{\eta}} \\ \underset{6\times 1}{\delta\bar{\boldsymbol{\delta}}^{\mathrm{a}}} \\ \underset{1\times 1}{\delta p_0} \\ \underset{1\times 1}{\delta\lambda_1} \\ \underset{1\times 1}{\delta\lambda_2} \end{bmatrix}} = \underset{21\times 1}{\begin{bmatrix} \underset{6\times 1}{-\mathbf{R}_{\sigma}^0} \\ \underset{6\times 1}{-\mathbf{R}_{\alpha^{\eta}}^0} \\ \underset{6\times 1}{-\mathbf{R}_{\delta^{\mathrm{a}}}^0} \\ \underset{1\times 1}{-R_{p_0}^0} \\ \underset{1\times 1}{-R_{\phi_1}^0} \\ \underset{1\times 1}{-R_{\phi_2}^0} \end{bmatrix}} \tag{14.20}$$

In the above, it may be noted that $\delta\bar{\boldsymbol{\sigma}}$, $\delta\bar{\boldsymbol{\alpha}}^{\eta}$, and $\delta\bar{\boldsymbol{\delta}}^{\mathrm{a}}$ are vector forms of the second order tensors $\delta\boldsymbol{\sigma}$, $\delta\boldsymbol{\alpha}^{\eta}$, and $\delta\boldsymbol{\delta}^{\mathrm{a}}$.

14.2.3 The Implementation Details

Equation (14.20) may be directly solved to find the iterative changes. The variables may then be updated as (leaving the over bar for generality)

$$\boldsymbol{\sigma}_{n+1}^2 = \boldsymbol{\sigma}_{n+1}^1 + \delta\boldsymbol{\sigma} \tag{14.21a}$$

$$\boldsymbol{\alpha}_{n+1}^{\eta,2} = \boldsymbol{\alpha}_{n+1}^{\eta,1} + \delta\boldsymbol{\alpha}^{\eta} \tag{14.21b}$$

$$\boldsymbol{\delta}_{n+1}^{\mathrm{a},2} = \boldsymbol{\delta}_{n+1}^{\mathrm{a},1} + \delta\boldsymbol{\delta}^{\mathrm{a}} \tag{14.21c}$$

$$p_0^{(n+1),2} = p_0^{(n+1),1} + \delta p_0 \tag{14.21d}$$

$$\Delta\lambda_1^2 = \Delta\lambda_1^1 + \delta\lambda_1 \tag{14.21e}$$

$$\Delta\lambda_2^2 = \Delta\lambda_2^1 + \delta\lambda_2 \tag{14.21f}$$

Note that superscript 2 on these variables does not denote "square of the variable", but rather the "second estimate."

The following convergence criteria (as in 10.38a–c) may be used:

$$e_\sigma = \left\|\mathbf{R}_\sigma^{i+1}\right\| / \left\|\boldsymbol{\sigma}_{n+1}^{i}\right\| \le TOL_\sigma \tag{14.22a}$$

$$e_{\alpha^\eta} = \left\|\mathbf{R}_{\alpha^\eta}^{i+1}\right\| \le TOL_{\alpha^\eta} \tag{14.22b}$$

$$e_{\delta^{\mathrm{a}}} = \left\|\mathbf{R}_{\delta^{\mathrm{a}}}^{i+1}\right\| / \left\|\boldsymbol{\delta}_{n+1}^{a,i}\right\| \le TOL_{\delta^{\mathrm{a}}} \tag{14.22c}$$

$$e_{p_0} = \left\|R_{p_0}^{i+1}\right\| / \left\|p_0^{(n+1),i}\right\| \le TOL_{p_0} \tag{14.22d}$$

$$R_{\phi_1}^{i+1} = \phi_1^{i+1} \le TOL_\phi \tag{14.22e}$$

$$R_{\phi_2}^{i+1} = \phi_2^{i+1} \le TOL_\phi \tag{14.22f}$$

Finally, the possibility that the mechanisms chosen based on ϕ_1^{tr} and ϕ_2^{tr} could change during the iteration must be considered (Simo and Hughes 1998). To address this, an outer iteration loop (referred to here as the "mode" loop) is introduced. When either $\Delta\lambda_1$ or $\Delta\lambda_2$ become negative, the corresponding mechanism is dropped and the iteration is restarted from the beginning. The algorithm has worked well for small strain increments. However, when the strain increment is large, $\Delta\lambda_1$ or $\Delta\lambda_2$ could become negative during the iteration even when the corresponding mechanism is not necessarily inactive. Sufficient number of iterations must be performed within the inner loop to establish the signs of $\Delta\lambda_1$ and $\Delta\lambda_2$.

The consistent elasto-plastic tangent operator is found by taking the rate of the residual (14.19) with the strain increment $\Delta\boldsymbol{\varepsilon}$ treated as a variable. When this is done, one obtains an equation that is similar to (14.20) as follows: (a) The 21 × 21 coefficient matrix remains the same, (b) the 21 × 1 vector on the left-hand side contains rates (e.g. $\dot{\bar{\boldsymbol{\sigma}}}$), and (c) in the vector on the right-hand side, $-\mathbf{R}_\sigma^0 = \mathbf{C}\dot{\boldsymbol{\varepsilon}}$ and all other terms are zero. Let us write the resulting equation as

$$\begin{bmatrix} \underset{6\times 6}{\mathbf{C}_{11}} & \underset{15\times 6}{\mathbf{C}_{12}} \\ \underset{6\times 15}{\mathbf{C}_{21}} & \underset{15\times 15}{\mathbf{C}_{22}} \end{bmatrix} \begin{bmatrix} \underset{6\times 1}{\dot{\bar{\boldsymbol{\sigma}}}} \\ \underset{15\times 1}{\dot{\bar{\mathbf{X}}}} \end{bmatrix} = \begin{bmatrix} \underset{6\times 1}{\mathbf{C}\dot{\boldsymbol{\varepsilon}}} \\ \underset{15\times 1}{\mathbf{0}} \end{bmatrix} \tag{14.23}$$

This equation is manipulated to obtain (leaving the over bar for generality)

$$\dot{\boldsymbol{\sigma}} = \mathbf{D}\dot{\boldsymbol{\varepsilon}} \tag{14.24a}$$

$$\mathbf{D} = \mathbf{R}\mathbf{C} \tag{14.24b}$$

$$\mathbf{R} = \left[\mathbf{C}_{11} - \mathbf{C}_{12}\mathbf{C}_{22}^{-1}\mathbf{C}_{21}\right]^{-1} \tag{14.24c}$$

D is the consistent tangent operator.

Algorithm 14.1. CPPM for Integrating the Constitutive Relations from the Sliding-Rolling Granular Material Model with Explicit Elastic Properties by the $\boldsymbol{\sigma}$–Space Formulation.
Definition of (new) parameters: $N_M^{\max}$ and $N_L^{\max}$ are the maximum number of mode changes and Newton's iterations, respectively, permitted before execution is terminated, and N_M and N_L are the corresponding counters. N_{LM} is the number of Newton's iterations performed before checking for possible mode changes. TOL is a small, machine-dependent number. Iflag = 1 if converged, 0 otherwise. $\mathbf{x} = \{\boldsymbol{\alpha}^\eta, \boldsymbol{\delta}^a, p_0\}$ and $\mathbf{R}_x = \{\mathbf{R}_{\alpha^\eta}, \mathbf{R}_{\delta^a}, R_{p_0}\}$. e = void ratio.

1. Establish $\Delta\boldsymbol{\varepsilon}$, $\mathbf{x}_n$, e_n, $\boldsymbol{\sigma}_n$, TOL_{α^η}, TOL_{δ^a}, TOL_{p_0}, TOL_σ, TOL_ϕ, TOL, $N_L^{\max}$, $N_M^{\max}$ and constitutive model parameters.
2. Initialize parameters as $N_L = N_M = 0$, $e_{n+1} = e_n - \Delta\varepsilon_{kk}/(1+e_n)$, and $\mathbf{x}_{n+1} = \mathbf{x}_n$. Set loading index $L_i = 0$ for $i = 1-2$.
3. Compute the elastic predictor $\boldsymbol{\sigma}_{n+1}^{\text{tr}}$ and $\phi_1^{\text{tr}}(\boldsymbol{\sigma}_{n+1}^{\text{tr}}, \mathbf{x}_{n+1})$ and $\phi_2^{\text{tr}}(\boldsymbol{\sigma}_{n+1}^{\text{tr}}, \mathbf{x}_{n+1})$.
4. Perform loading/unloading check: if $([\phi_1^{\text{tr}} < TOL]$ and $\phi_2^{\text{tr}} < TOL$), set $\boldsymbol{\sigma}_{n+1} = \boldsymbol{\sigma}_{n+1}^{\text{tr}}$ and Iflag = 1, go to step 16. If not, set $L_i = 1$ if $\phi_i^{\text{tr}} > 0$ for $i = 1-2$.
5. Begin iterations to establish modes.
6. Compute $\Delta\lambda_i^{\text{tr}}$ by (14.27) if $L_i = 1$ for $i = 1-2$.
7. Using $\Delta\lambda_i^{\text{tr}}$, compute $\mathbf{x}_{n+1}^1$ and $\boldsymbol{\sigma}_{n+1}^1$. Set $\Delta\lambda_i^1 = \Delta\lambda_i^{\text{tr}}$ if $L_i = 1$ for $i = 1-2$. Set $\mathbf{x}_{n+1} = \mathbf{x}_{n+1}^1$, $\boldsymbol{\sigma}_{n+1} = \boldsymbol{\sigma}_{n+1}^1$, and $\Delta\lambda_i = \Delta\lambda_i^1$ if $L_i = 1$ for $i = 1-2$.
8. Begin Newton's iterations.
9. Compute ϕ_i if $L_i = 1$ for $i = 1-2$. Compute $\mathbf{x}_{n+1}^2$ and $\boldsymbol{\sigma}_{n+1}^2$ by (14.18).
10. Compute residuals $\mathbf{R}_\sigma = \boldsymbol{\sigma}_{n+1}^1 - \boldsymbol{\sigma}_{n+1}^2$, $\mathbf{R}_x = \mathbf{x}_{n+1}^1 - \mathbf{x}_{n+1}^2$ by (14.19).
11. Find errors e_σ, e_{α^η}, e_{δ^a} and e_{p_0}, and R_{ϕ_i} (if $L_i = 1$) for $i = 1-2$. Perform convergence check. If converged, set Iflag = 1 and go to step 16.
12. Set $N_L \leftarrow N_L + 1$. If $(N_L > N_L^{\max})$, set Iflag = 0 and go to step 16.
13. Determine corrections $\delta\mathbf{x}$, $\delta\boldsymbol{\sigma}$ and $\delta\lambda_i$ (if $L_i = 1$) for $i = 1-2$ (14.20). Evaluate $\Delta\lambda_i$ if $L_i = 1$ for $i = 1-2$ (14.21e–f).
14. Check possible change in modes if $N_L > N_{LM}$: if $\Delta\lambda_i \leq 0$ set $L_i = 0$ for $i = 1-2$. If there is a change in modes: $N_M \leftarrow N_M + 1$. If $(N_M > N_M^{\max})$, set Iflag = 0 and go to step 16. Otherwise, go to step 5.
15. Evaluate $\mathbf{x}_{n+1}^1 \leftarrow \mathbf{x}_{n+1}^1 + \delta\mathbf{x}$ and $\boldsymbol{\sigma}_{n+1}^1 \leftarrow \boldsymbol{\sigma}_{n+1}^1 + \delta\boldsymbol{\sigma}$. Go to step 8.
16. Return to the main program.

By following a procedure similar to that in (10.22) of Chap. 10 (which leads to 10.24 and 10.30), the trial values of $\Delta\lambda_1$ and $\Delta\lambda_2$ may be found as follows. Linearizing the yield functions about the trial stress point i, it follows that

$$\phi_k^{i+1} = \phi_k^i + \frac{\partial \phi_k^i}{\partial \boldsymbol{\sigma}} \delta\boldsymbol{\sigma} + \frac{\partial \phi_k^i}{\partial \boldsymbol{\alpha}^\eta} \delta\boldsymbol{\alpha}^\eta + \frac{\partial \phi_k^i}{\partial p_o} \delta p_o = 0 \tag{14.25}$$

From the plastic corrector and the evolution rules for the hardening variables (14.8), we have

$$\delta\boldsymbol{\sigma} = -\mathbf{C}\{\Delta\lambda_1^{\text{tr}}\mathbf{r}_{\text{tr}}^1 + \Delta\lambda_2^{\text{tr}}\mathbf{r}_{\text{tr}}^2\} \tag{14.26a}$$

$$\delta\boldsymbol{\alpha} = \xi_1^{\text{tr}}\Delta\lambda_1^{\text{tr}}\bar{\mathbf{h}}_{\text{tr}} \tag{14.26b}$$

$$\delta p_0 = \{g_1^{\text{tr}}\Delta\lambda_1^{\text{tr}} + g_2^{\text{tr}}\Delta\lambda_2^{\text{tr}}\} \tag{14.26c}$$

Substituting these into (14.25) and making use of the compact notations in (14.9a–c), we get

$$\begin{bmatrix} \mathbf{n}_{\text{tr}}^1\mathbf{C}\mathbf{r}_{\text{tr}}^1 - \xi_1^{\text{tr}}(\mathbf{P}_1^{\text{tr}}.\bar{\mathbf{h}}_{\text{tr}}) & \mathbf{n}_{\text{tr}}^1\mathbf{C}\mathbf{r}_{\text{tr}}^2 \\ \mathbf{n}_{\text{tr}}^2\mathbf{C}\mathbf{r}_{\text{tr}}^1 - \xi_1^{\text{tr}}(\mathbf{P}_2^{\text{tr}}.\bar{\mathbf{h}}_{\text{tr}}) - T_2^{\text{tr}}g_1^{\text{tr}} & \mathbf{n}_{\text{tr}}^2\mathbf{C}\mathbf{r}_{\text{tr}}^2 - T_2^{\text{tr}}g_2^{\text{tr}} \end{bmatrix} \begin{bmatrix} \Delta\lambda_1^{\text{tr}} \\ \Delta\lambda_2^{\text{tr}} \end{bmatrix} = \begin{bmatrix} \phi_1^{\text{tr}} \\ \phi_2^{\text{tr}} \end{bmatrix} \tag{14.27}$$

The procedure for calculating the stresses is summarized in Algorithm 14.1.

14.3 Sample Calculations

14.3.1 Triaxial Drained Compression Behavior

The model behavior is presented under triaxial drained compression loading. The one-element finite element model shown in Fig. 14.2 is used for the simulation. The model consists of an 8-noded brick element (Chap. 5). The initial stresses applied on the specimen are $\sigma_{11} = \sigma_{22} = \sigma_{33} = 80\,\text{kPa}$. It is understood that for a triaxial loading, $\sigma_{12} = \sigma_{23} = \sigma_{13} = 0$. The initial values of the hardening variables are $\boldsymbol{\alpha}^\eta = \mathbf{0}$, $\boldsymbol{\delta}^{\text{a}} = \boldsymbol{\delta}$ and $p_0 = 88{,}714\,\text{kPa}$. The test, therefore, starts from the p-axis with isotropic material properties. The axis of the yield cone is aligned with the p-axis and the planar cap is placed at 88,714 kPa. The initial void ratio is 0.637. When a swelling line is run through the point ($e_0 = 0.637, p = 80\,\text{kPa}$) in the $e-p$ space, it intersects the critical state line at $p_0 = 88{,}714$ kPa.

The combination ($e_0 = 0.637, p = 80\,\text{kPa}$) places the initial point in the "dense" zone ($I_s = 1.75 > 1$). Hence, during the triaxial compression loading, a net

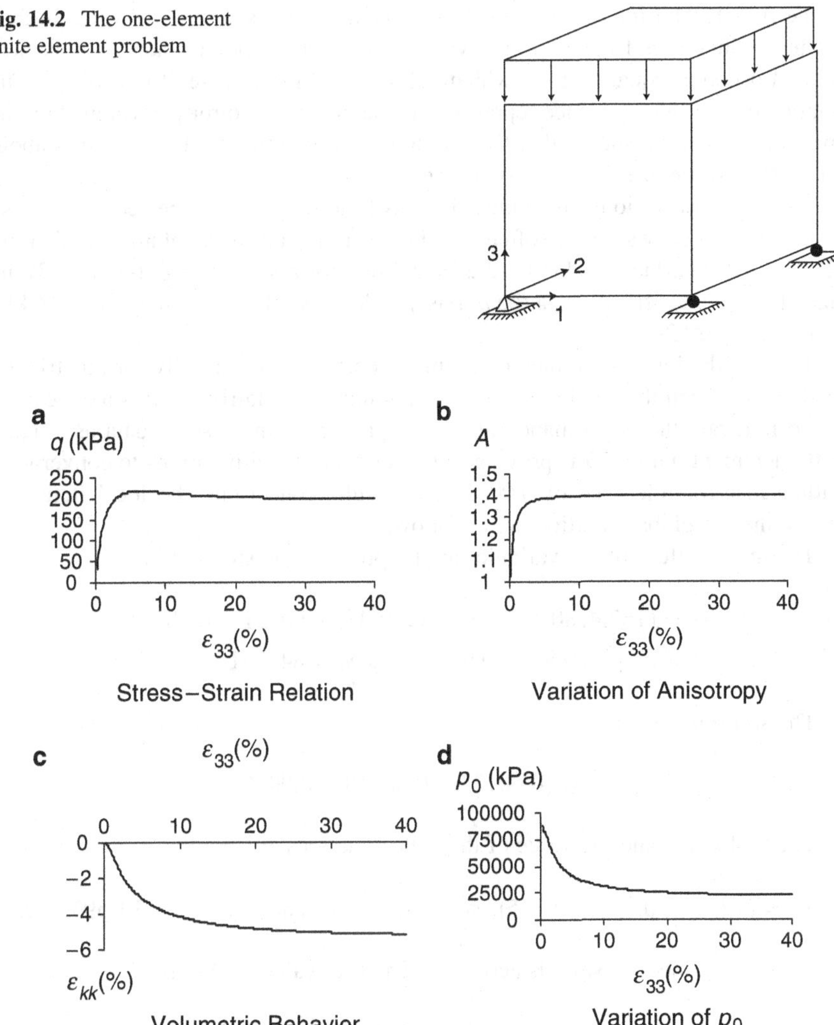

Fig. 14.2 The one-element finite element problem

Fig. 14.3 The model behavior under triaxial drained compression

volumetric dilation and strain softening are expected. The simulations are performed using the model parameters Set 1 presented in Table 14.1a, b.

The loading is applied in strain-controlled mode; i.e. keeping σ_{11} and σ_{22} constant, ε_{33} is increased from 0 to $\hat{\varepsilon}$. The maximum (uniform) strain increment that the integration method is able to handle is about 0.125%. The results shown in Fig. 14.3 are obtained when $\hat{\varepsilon} = 40\%$ is applied in 320 uniform increments. Note that $q = \sigma_{33} - \sigma_{11}$ is the deviatoric stress.

The rotation of the failure surface is quantified in terms of the anisotropic index A, defined as $A = \delta^{\mathrm{a}}_{33}/\delta^{\mathrm{a}}_{11}$. When the rotation is zero, we have $\delta^{\mathrm{a}}_{ij} = \delta_{ij}$ and hence $A = 1$. The initial value of A is unity in the present case. As the surface rotates, A increases from its initial value. As the stress point reaches the critical state, A becomes equal to A_{CSL}, which is 1.4 in the present case (Table 14.1a). The rotation of the failure surface represents the degree of anisotropy developed by the specimen. It can be shown that the rotation corresponding to $A_{CSL} = 1.4$ is about 26% of the slope of the critical state line.

The key observations are summarized as follows: (1) As expected, the stress–strain relation shows strain softening (though mild) and net volumetric dilation, (2) the failure surface gradually rotates and reaches a saturation point, and (3) the size of the yield surface along the p-axis, p_0, decreases from a value of 88,714 kPa to a value of 22,200 kPa.

To sum, the integration method seems to perform satisfactorily for this triaxial loading, although the need to keep the strain increment fairly small is a concern.

To illustrate the performance of the integration method, some data associated with increment number 5 is provided. It takes four global iterations to converge to within an error tolerance of 0.0001. The results concerning the local iterations during the 4th global iteration are as follows.

The initial values of the variables are (expressing the stress in kPa)

$$\begin{aligned}&\boldsymbol{\sigma} = (167.4, 80, 80),\ \boldsymbol{\alpha}^{\eta} = (0.1641, -0.0821, -0.0821),\\ &\boldsymbol{\delta}^{a} = (1.106, 0.9423, 0.9423)\ \text{and}\ p_0 = 84,631.\end{aligned}$$

The strain increment is

$$\Delta\boldsymbol{\varepsilon} = (0.00125, -0.00108, -0.00108)$$

The trial stress and the value of the yield functions are

$$\boldsymbol{\sigma}^{tr} = (192.7, -0.125, -0.125),\ \phi_1^{tr} = 81.7 > 0 \quad \text{and} \quad \phi_2^{tr} = -130,593 < 0.$$

Hence only mechanism 1 is activated. The trial value of $\Delta\lambda_1$ is

$$\Delta\lambda_1^{tr} = 0.002643.$$

The estimates of variables using this value of $\Delta\lambda_1$ are

$$\begin{aligned}&\phi_1 = 5.06,\ \boldsymbol{\sigma} = (170.2, 88.5, 88.5),\ \boldsymbol{\alpha}^{\eta} = (0.1272, -0.0636, -0.0636),\\ &\boldsymbol{\delta}^{a} = (0.1811, 1.218, 1.218)\ \text{and}\ p_0 = 82,710\end{aligned}$$

The values of errors during the Newton iterations are listed in Table 14.2. It is seen that the errors very quickly decrease. The final values of the variables are

$$\begin{aligned}&\boldsymbol{\sigma} = (180.7, 80, 80),\ \boldsymbol{\alpha}^{\eta} = (0.1825, -0.0913, -0.0913),\\ &\boldsymbol{\delta}^{a} = (1.1252, 0.9311, 0.9311),\ \text{and}\ p_0 = 84,819.\end{aligned}$$

Table 14.2 Change in the values of errors during the iteration

Iteration no.	R_{ϕ_1}	e_σ	e_{α^η}	e_{δ^a}	e_{p_0}
1	5.06	0.0258	0.00128	0.3053	0.009
2	0.841	0.0113	0.00092	0.0519	0.0062
3	0.240	4.51×10^{-5}	5.8×10^{-5}	0.0016	0.00036
4	0.0058	1.02×10^{-7}	1.27×10^{-7}	2.19×10^{-8}	1.74×10^{-5}
5	1.44×10^{-5}	1.04×10^{-12}	4.69×10^{-14}	9.28×10^{-15}	5.53×10^{-8}

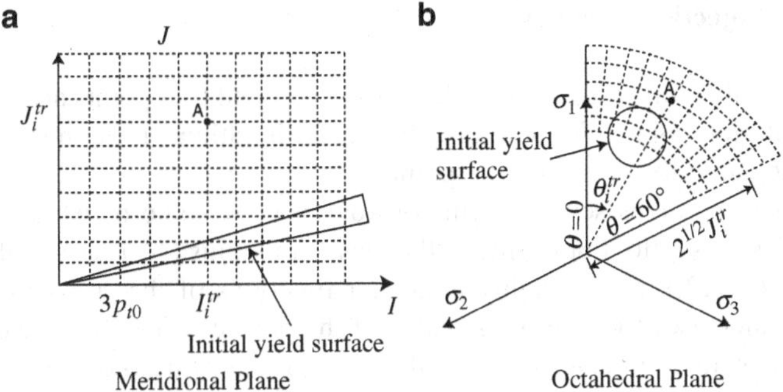

Fig. 14.4 Grids of different elastic trial stress points

14.3.2 Further Results on Convergence of CPPM with Approximate Elastic Properties

To investigate the convergence behavior systematically and comprehensively, a series of local analyses are performed, with each analysis beginning from a specific elastic trial stress point. For convenience, the trial points are chosen to be at the nodes of a regular grid. A typical grid chosen in a meridional plane (in triaxial compression) and one in an octahedral plane are shown in Fig. 14.4a, b, respectively.

In the meridional plane (Fig. 14.4a), a 100×100 grid is used. The elastic trial stresses at node i of the grid (point A) are (I_i^{tr}, J_i^{tr}). The initial values chosen for the variables are (using kPa as the unit for stress and noting that the shear components of the tensors are zero)

$$\boldsymbol{\sigma} = (10, 10, 10);\ \boldsymbol{\alpha}^\eta = (0, 0, 0);\ \boldsymbol{\delta}^a = (1, 1, 1) \text{ and } p_0 = 93,469.$$

The test, therefore, starts from the isotropic stress axis with isotropic material properties. The axis of the yield cone is aligned with the p–axis and the planar cap is placed at $p_0 = 93,469$. The initial void ratio is 0.637. The combination $(e_0 = 0.637, p = 10)$ places the initial point in the "dense" zone ($I_s = 1.87 > 1$). Starting from this initial state, a suitable strain increment is applied to get to the trial point A. The Newton's iterations are then performed to find the plastic corrector. The tolerances used in the local iterations (14.22) are $TOL_\sigma = 0.01$, $TOL_{\phi_1} = 0.0001$, $TOL_{\alpha^\eta} = 0.01$, $TOL_{\delta^a} = 0.01$, and $TOL_{p_0} = 0.01$.

It is found that, for some trial points, convergence occurs usually within about 4–5 iterations, and for the remaining trial points, convergence does not occur within 100 iterations. In the following section, we will develop and implement an auxiliary surface projection strategy in an attempt to improve the convergence behavior.

14.3.2.1 Obtain Improved Initial Estimate: Auxiliary Surface Projection Strategy

A closer examination indicated that the non-convergent trial points are too far from the final solution points. Furthermore, during the iterations, mean normal pressure becomes negative for some starting points.

The model involves several forms of nonlinearity, some of which are on the basis of the stress ratio. For example, the plastic modulus (14.12a) and the dilation parameter d (A12.11a) are highly nonlinear functions of the stress ratio. As the stress ratio increases from zero, the value of the plastic modulus decreases from a positive value to eventually zero on the critical state line. For loose sand, this variation is monotonic. However, for dense sand, the plastic modulus could first change from positive to negative and then increase to zero. Similar nonlinear behavior exists for the variation of the dilation parameter d. These forms of nonlinearity are the reasons for the observed non-convergence.

To overcome this difficulty, the auxiliary surface projection strategy proposed by Bicanic and Pearce (1996) is tried. The effectiveness of the method has been examined for the Drucker–Prager model in Chap. 13. It has been shown that the convergence behavior improves significantly. In that model, the trial point is projected onto a circular yield surface (even when the yield surface is non-circular). The projected point is used as the initial state for the subsequent Newton's iterations that establish the plastic corrector. As the Drucker–Prager model is simple, involving only isotropic hardening, it was possible to obtain the solution (the stress, plastic consistency parameter, and hardening variable) analytically. This is, however, difficult in the case of the sliding-rolling model as it involves complex isotropic, kinematic, and rotational hardening rules. Hence an approximate strategy is used, where the current stress $\boldsymbol{\sigma}$ is projected to obtain a modified stress $\boldsymbol{\sigma}^*$, keeping the hardening variables unchanged. The consistency parameter and the hardening variables are then found using the modified stress $\boldsymbol{\sigma}^*$. The method is applied not only to project the trial stress $\boldsymbol{\sigma}^{tr}$, but any current stress $\boldsymbol{\sigma}$.

Referring to Fig. 14.5a, the current stress point A is projected onto the yield surface by dropping a perpendicular to point B. In the octahedral plane (Fig. 14.5b), point A is projected to point B by connecting A to the yield surface center. Referring to Fig. 14.5a, we have

$$\frac{I^* - I}{J - J^*} = \frac{J^*}{I^*} \Rightarrow (I^*)^2 - I^*I - JJ^* - (J^*)^2 = 0 \qquad (14.28)$$

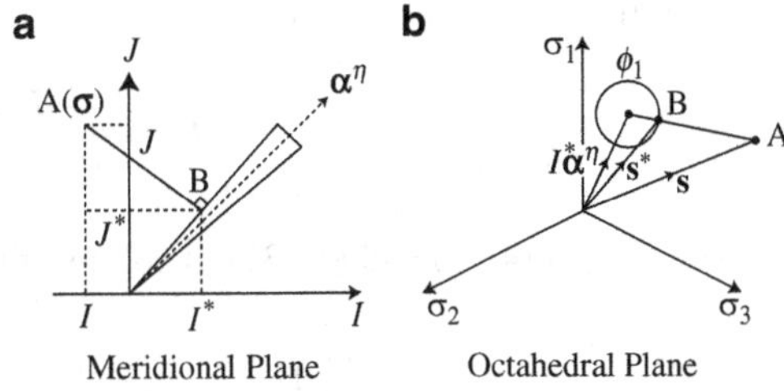

Fig. 14.5 The projected stress $\boldsymbol{\sigma}^*$

We express $\mathbf{s}^*$ as

$$\mathbf{s}^* = x_0 I^* \boldsymbol{\alpha}^\eta + (1 - x_0)\mathbf{s}; \quad 0 \leq x_0 \leq 1 \tag{14.29}$$

Then the projected stress is

$$\boldsymbol{\sigma}^* = \mathbf{s}^* + \frac{1}{3} I^* \boldsymbol{\delta} \tag{14.30}$$

$\boldsymbol{\sigma}^*$ can be calculated if x_0 is known. x_0 is found by utilizing the fact that $\boldsymbol{\sigma}^*$ is on the yield surface (on the basis of the current values of the hardening variables). The yield function is (Appendix 12.1)

$$\phi_1 = f_J^* - k^* I^* = 0 \tag{14.31}$$

where

$$\begin{aligned} f_J^* &= \left[\frac{1}{2}(\mathbf{s}^* - I^*\boldsymbol{\alpha}^\eta)(\mathbf{s}^* - I^*\boldsymbol{\alpha}^\eta)\right]^{1/2} \\ &= (1 - x_0)\left[\frac{1}{2}(\mathbf{s} - I^*\boldsymbol{\alpha}^\eta)(\mathbf{s} - I^*\boldsymbol{\alpha}^\eta)\right]^{1/2} \\ &= (1 - x_0)\left[\frac{1}{2}\left\{2J^2 + 2(I^*)^2\alpha_J^2 - 4I^* y_0^2\right\}\right]^{1/2} \end{aligned} \tag{14.32}$$

and

$$y_0^2 = \frac{1}{2}\boldsymbol{\alpha}^\eta : \mathbf{s}.$$

In finding the value of x_0, the following approximation helps simplify the equation:

$$\mathbf{s}^* \approx I^*\boldsymbol{\alpha}^\eta \Rightarrow J^* = I^*\alpha_J; \ \alpha_J = \left(\frac{1}{2}\boldsymbol{\alpha}^\eta : \boldsymbol{\alpha}^\eta\right)^{1/2} \tag{14.33}$$

Then from (14.28), we have

$$I^* = \frac{I + J\alpha_J}{1 + \alpha_J^2} \tag{14.34}$$

Now from (14.31) and (14.32), it follows that

$$(1 - x_0)\left[\frac{1}{2}\left\{2J^2 + 2(I^*)^2\alpha_J^2 - 4I^*y_0^2\right\}\right]^{1/2} = k^*I^*$$

$$x_0 = 1 - \frac{\sqrt{2}k^*I^*}{\left\{2J^2 + 2(I^*)^2\alpha_J^2 - 4I^*y_0^2\right\}} \tag{14.35}$$

Using this value of x_0, we calculate $\mathbf{s}^*$ from 14.29). Using this value of $\mathbf{s}^*$ and the value of I^* from (14.34), we evaluate $\boldsymbol{\sigma}^*$ from (14.30).

Depending on how small the current value of I and α_J are, I^* may be calculated to be negative. To handle such situations, a lower limit is imposed as $I^* \geq I^*_{\text{min}}$. A suitable value is used for I^*_{min}; e.g. $I^*_{\text{min}} = 0.01\,p_{\text{a}}$, where p_{a} is the atmospheric pressure.

Algorithm 14.1 is modified as follows: In step (6), instead of using $\boldsymbol{\sigma}_{n+1} = \boldsymbol{\sigma}^{\text{tr}}$ in evaluating the parameters needed for calculating $\Delta\lambda_1^{\text{tr}}$ (such as $\mathbf{r}^1$), use $\boldsymbol{\sigma}_{n+1} = \boldsymbol{\sigma}^*$. Follow the same approach in step (9).

It is found that the above projection method significantly improves the convergence behavior. Convergence within a reasonable number of iterations is now possible for the majority of the trial points. However, for some trial points, nonconvergence is still encountered. For these trial points, the hardening variables appear to change drastically, leading eventually to divergence. To overcome this difficulty, in the following section, we will develop and implement a scaling strategy that regulates the variation of the hardening variables during the iteration.

Remarks.

- The projection strategy is necessary only when the stress ratio at the current stress point is large. The algorithm may be modified so that the projection strategy is activated only when the stress ratio at the current point is larger than a limit.
- When the projection strategy is employed, during the calculation of stresses, the coefficient matrix in (14.20) must be modified. It is left for the reader as an exercise. No such modification is needed in calculating the consistent tangent operator.
- The use of the projection strategy is likely to slow down the rate of convergence somewhat.

14.3.2.2 Restrict the Newton's Step Size

The projection method presented above controls only $\boldsymbol{\sigma}$. However, depending on the strain increment, the hardening variables also could get too far from the final solution, causing non-convergence of Newton's iterations. One way to overcome this problem is to place a limit on the variation of the hardening variables during each iteration. In step (15) of Algorithm 14.1, the hardening variables are updated as

$$\mathbf{x}_{n+1}^{k+1} = \mathbf{x}_{n+1}^{k} + \delta\mathbf{x} \tag{14.36}$$

where k denotes iteration number. When the problem is treated as one of minimization of a scalar function, the methods of nonlinear optimization could be employed to find the final solution (Fletcher 1987). In such approaches, a line search is employed in certain direction to minimize the scalar function. In fact, $\delta\mathbf{x}$ in (14.36) is one such viable search direction (and there are many others). In such a case, (14.36) becomes

$$\mathbf{x}_{n+1}^{k+1} = \mathbf{x}_{n+1}^{k} + \alpha\delta\mathbf{x} \tag{14.37}$$

where α is a scalar parameter to be determined by using a suitable line search strategy. A value is found for α such that $\mathbf{x}_{n+1}^{k+1}$ reduces the value of the scalar function as much as possible.

The same strategy may be used in the present case, but with a different objective. The objective here is to find a value of α such that the variations of the hardening variables do not exceed specified limits. As the total incremental changes in the hardening variables are proportional to $\Delta\lambda_1$ and $\Delta\lambda_2$, the iterative changes of the hardening variables $\delta\mathbf{x}$ are approximately proportional to the iterative changes $\delta\lambda_1$ and $\delta\lambda_2$. Hence, finding a value of α in (14.37) is tantamount to finding a suitable scale factor β by which $\delta\lambda_1$ and $\delta\lambda_2$ are scaled back as

$$\delta\lambda_1 \leftarrow \beta\delta\lambda_1 \quad \text{and} \quad \delta\lambda_2 \leftarrow \beta\delta\lambda_2 \tag{14.38}$$

Let us say that we wish to control the rotation of the yield and failure surfaces. In addition, we wish to ensure that p_0 does not become less than a limit. Essentially, we wish to avoid p_0 becoming negative.

Let us define

$$x_1 = \alpha_J^{k+1}; \quad \alpha_J^{k+1} = \left(\frac{1}{2}\boldsymbol{\alpha}^{\eta} : \boldsymbol{\alpha}^{\eta}\right)_{k+1}^{1/2}$$

$$x_2 = \frac{(\delta_J^{\mathrm{a}})^{k+1}}{(\delta_{kk}^{\mathrm{a}})^{k+1}}; \quad (\delta_J^{\mathrm{a}})^{k+1} = \left(\frac{1}{2}\boldsymbol{\delta}^{sa} : \boldsymbol{\delta}^{sa}\right)_{k+1}^{1/2}; \quad \boldsymbol{\delta}^{sa} = \boldsymbol{\delta}^{\mathrm{a}} - \frac{1}{3}\delta_{kk}^{\mathrm{a}}\boldsymbol{\delta}$$

$$x_3 = p_0^{k+1}$$

Let the limit on x_i be x_i^{lim}. Then we calculate a scale factor for each variable x_i as

$$\beta_i = \frac{x_i^{\mathrm{lim}} - x_i^k}{\delta x_i} \quad \text{for} \quad |\delta x_i| > \varepsilon; \quad \text{and} \quad \beta_i = 1 \quad \text{for} \quad |\delta x_i| \le \varepsilon$$

where ε is a suitable small number. When $\beta_i \le \varepsilon_2$ for some suitable value of ε_2 (e.g. 1.0×10^{-10}), the local iteration is declared non-convergent and the control is returned to the main program. Otherwise, a value for β to be used in (14.38) is obtained as

$$\beta = \min(\beta_i)$$

The strategy has been implemented with $x_1^{\lim} = x_2^{\lim} = 0.1$ and $x_3^{\lim} = 1\,\text{kPa}$. For the trial points in the meridional plane (Fig. 14.4a), convergence is achieved for all but two points. Among the 10,000 trial points considered, the trial points $(p, q) = (6.667, 58.197)\,\text{kPa}$ and $(p, q) = (8.0, 58.197)\,\text{kPa}$ lead to non-convergence within 100 iterations. These rare instances of non-convergence are handled using the sub-stepping strategy described in the next section.

14.3.2.3 Obtain Improved Initial Estimate: Sub-Stepping Strategy

A sub-stepping strategy has been introduced in Chap. 13; the reader is referred to this chapter for details (see (13.68) and the description there under). When this sub-stepping strategy is incorporated along with the projection strategy and scaling strategy described in the preceding sections, the convergence problem previously encountered at a few trial points disappeared.

For example, consider the analysis that begins from the trial point $(p, q) = (8.0, 58.197)\,\text{kPa}$ in the meridional plane (one of the two points that lead to non-convergence). The initial state variables for this point are (expressing the stress in kPa and noting that the shear components of the tensors are zero)

$$\boldsymbol{\sigma} = (10, 10, 10),\ \boldsymbol{\alpha}^{\eta} = (0, 0, 0),\ \boldsymbol{\delta}^{\mathrm{a}} = (1, 1, 1) \quad \text{and} \quad p_0 = 93{,}470.$$

The strain increment is:

$$\Delta\boldsymbol{\varepsilon} = (0.002797, -0.001466, -0.001466)$$

The strain is first divided in half. The iterations converge to the following solution:

$$\boldsymbol{\sigma} = (17.84, 12.91, 12.91),\ \boldsymbol{\alpha}^{\eta} = (0.06369, -0.03184, -0.03184),$$
$$\boldsymbol{\delta}^{\mathrm{a}} = (1.0212, 0.9892, 0.9892),\ \Delta\lambda_1 = 0.002041\ \text{(only mechanism 1 is activated)},$$
$$e = 0.6371 \quad \text{and} \quad p_0 = 92{,}739.$$

Using these as the initial estimate, the application of the full strain increment successfully converges to the solution

$$\boldsymbol{\sigma} = (29.32, 19.25, 19.25),\ \boldsymbol{\alpha}^{\eta} = (0.08711, -0.04356, -0.04356),$$

$$\boldsymbol{\delta}^{a} = (1.04298, 0.9778, 0.9778),\ \Delta\lambda_1 = 0.004059\ \text{(only mechanism 1 is activated)},$$

$$e = 0.63722 \quad \text{and} \quad p_0 = 91,565.$$

Remarks.

- The strategies described in the preceding sections effectively deal with trial/current stress points with positive mean pressure. When the strain increments involved very large negative volumetric component, the trial point may fall in the apex region. These cases are not treated in this chapter.
- The strategy of restricting the Newton's step size may be systematically implemented using a suitable line search algorithm (Fletcher 1980, 1987).

14.3.2.4 Number of Iterations Required for Convergence

The strategies described in the preceding sections (auxiliary surface projection strategy, scaling strategy, and sub-stepping strategy) have been implemented. The procedure is then used to investigate the number of iterations required for convergence for trial points chosen at the grid points shown in Fig. 14.4a, b. The results are presented in Fig. 14.6a for the trial points in the meridional plane (Fig. 14.4a) and in Fig. 14.6b for the trial points in the octahedral plane (Fig. 14.4b). It may be observed that convergence is achieved in all cases in fewer than 10 iterations.

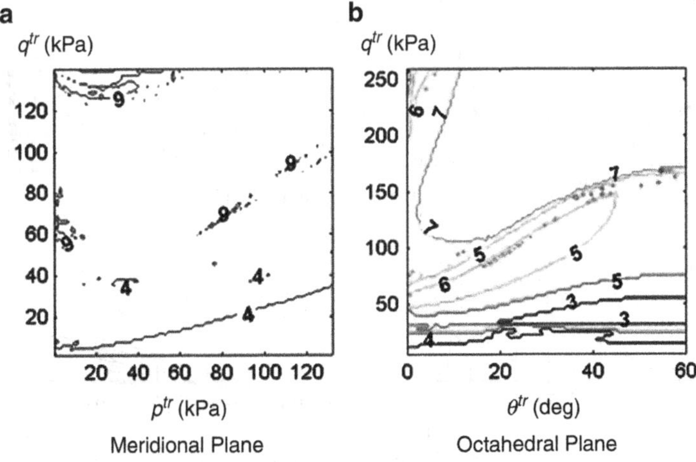

Fig. 14.6 Contours of number of iterations required for convergence for different elastic trial stress points

14.3.3 Application of the Sliding-Rolling Model to Earthquake Analysis Using CPPM with Approximate Elastic Properties

One of the main uses of sophisticated constitutive models such as the sliding-rolling model is to numerically model the earthquake behavior of structures involving liquefiable granular materials. As a prerequisite, the model must be able to capture the stress–strain behavior of uniformly loaded laboratory specimens. Using the model parameters Set 2 listed in Table 14.1a, b, the undrained cyclic behaviors of triaxial specimens are simulated. The initial state variables are $\boldsymbol{\sigma} = (100, 100, 100)$ kPa, $\boldsymbol{\alpha}^{\eta} = (0, 0, 0)$, $\boldsymbol{\delta}^{a} = (1, 1, 1)$, and $e = 0.675$. The model behaviors under constant-amplitude stress loading and strain loading are presented in Fig. 14.7a, b respectively. It may be seen that the specimens experience liquefaction in both cases. The mean normal pressure becomes close to zero during the loading.

The model is then used to analyze the response of a level ground subjected to an earthquake base acceleration (Fig. 14.8). The soil deposit is assumed to consist of a homogeneous granular material of 10m thickness and to rest on bedrock. The constitutive behavior is modeled using the sliding-rolling model with the parameters Set 2 listed in Table 14.1a, b.

An eight-element mesh involving 8-noded 2D elements is used (Fig. 14.8a). The analysis is performed under the plane strain condition, with the left and right vertical sides "slaved" to simulate the free-field condition. The horizontal and vertical directions are assumed to be the principal directions. The vertical effective stress is calculated as $\sigma'_{33} = \gamma' \times \text{depth}$ with $\gamma' = 9.55\,\text{kN/m}^3$. The horizontal

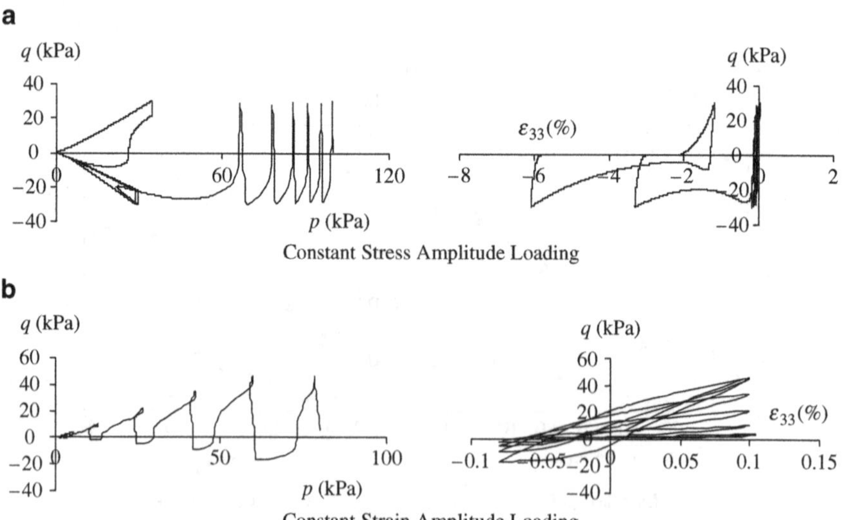

Fig. 14.7 Theoretical triaxial undrained stress–strain behavior

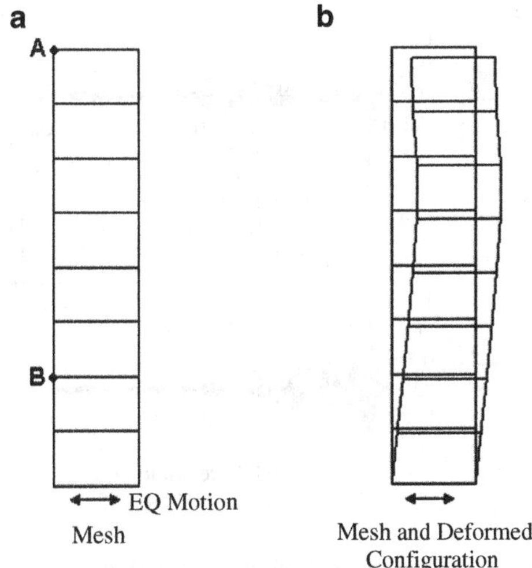

Fig. 14.8 Mesh and deformed configuration: the earthquake problem

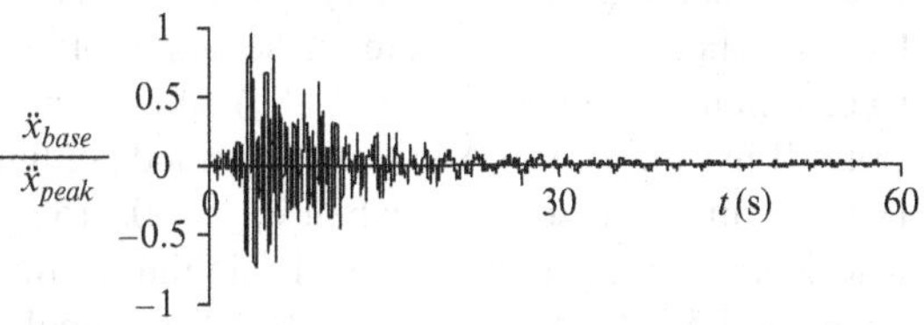

Fig. 14.9 Horizontal base acceleration: 1994 Northridge record

stresses are calculated as $\sigma'_{11} = \sigma'_{22} = K_0\sigma'_{33}$ with $K_0 = 0.55$. The initial void ratio is assumed to be 0.675. A uniform vertical surcharge of 4.5 kPa is applied at the ground level to avoid numerical problems caused by the stresses being too small at Gauss points near the ground surface.

The analysis is carried out under fully coupled conditions (Chaps. 6 and 7). This requires the following additional properties:

Normalized permeability tensor $\mathbf{k}/\gamma_w = (0.0003, 0.0003, 0.0003)\ \mathrm{m^4/(kN\,s)}$
Bulk modulus of solids $K_s = 2.5 \times 10^7$ kPa
Bulk modulus of water $K_w = 1.0 \times 10^6$ kPa
Porosity $n = 0.40$

One of the ground accelerations recorded during the 1994 Northridge earthquake is scaled (Fig. 14.9) and used as the base motion in the present analysis. The time-history has 3,000 points recorded at a time interval of 0.02 s. This interval is used as the time step for the dynamic analysis as well. The Rayleigh damping model is used to apply a small damping. The parameters used for the Rayleigh damping model are (see Chap. 5 for details): $\omega_1 = \omega_2 = 10\,\mathrm{rad/s}$ and $\beta = 0.05$. The Rayleigh damping supplements the energy dissipation modeled by the sliding-rolling model, making the finite element analysis more stable.

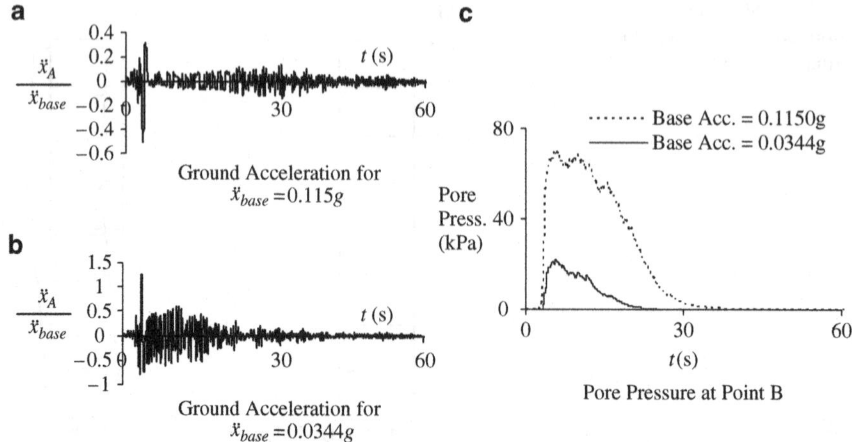

Fig. 14.10 Results of earthquake analysis

The analysis is repeated for two different values of the base acceleration amplitudes, one small ($\ddot{x}_{\text{base}} = 0.0334\,g$) and the other moderately large ($\ddot{x}_{\text{base}} = 0.115\,g$). The calculated responses are presented in Fig. 14.10 in terms of the ground accelerations (point A in Fig. 14.8a) and the pore pressures at a depth of 7.5 m (point B in Fig. 14.8a). A typical deformed configuration is shown in Fig. 14.8b at a certain time. It is seen that when $\ddot{x}_{\text{base}} = 0.115\,g$, the soil experiences liquefaction at a depth of 7.5 m. Note that the initial vertical effective stress at a depth of 7.5 m is $7.5 \times 9.55 + 4.5 = 76\,\text{kPa}$. The calculated peak pore pressure buildup (Fig. 14.10c) is close to this value. As a result of liquefaction, and the consequence softening, the soil looses the ability to transmit dynamic motion. This is evident from the calculated ground acceleration shown in Fig. 14.10a. When the base acceleration is low ($\ddot{x}_{\text{base}} = 0.0334\,g$), the pore pressure buildup is small. The deposit is able to transmit the motion. The calculated peak ground acceleration is slightly higher than the peak base acceleration. In summary, the results indicate that (a) the sliding-rolling theory is capable of representing the liquefaction behavior of soil deposits during an earthquake loading, and (b) CPPM with approximate elastic properties provides an effective means of integrating the sliding-rolling model.

14.4 Application of CPPM to Sliding-Rolling Model: The σ-Space Formulation with Exact Elastic Properties

The details of the steps are similar to those in Sect. 13.3. When the plastic strain is zero, the elastic predictor becomes the actual elastic stress. The exact expression for the elastic predictor is presented in Appendix 10. With the tension cut-off taken as zero (i.e. $p_t = 0$), we have

$$p^{\text{tr}} = \left[p_n^{1-n_e} + \bar{K}_0(1 - n_e)\Delta\varepsilon_{pp}\right]^{1/(1-n_e)} \tag{14.39a}$$

$$s_{ij}^{\text{tr}} = s_{ij}^{n} + 2G_{\text{ave}}\Delta e_{ij} \tag{14.39b}$$

$$G_{\text{ave}} = \frac{\bar{G}_0}{\bar{K}_0\Delta\varepsilon_{pp}}\left(p^{\text{tr}} - p_n\right) \tag{14.39c}$$

where $n_e = 0.5$. The quantity within the parenthesis in (14.39a) must be positive; i.e.

$$p_1 = p_n^{1-n_e} + \bar{K}_0(1 - n_e)\Delta\varepsilon_{pp} \geq 0 \tag{14.39d}$$

When $p_1 < 0$, $p_1^{1/(1-n_e)}$ represents a wrong functional relation.

When the behavior is elasto-plastic, the above equations are modified by using the elastic strain increments instead of the total strain increments. The elastic strains are obtained by subtracting the plastic strains from the total strains. The backward Euler method is used to express the plastic strain increment as (14.7)

$$\Delta\boldsymbol{\varepsilon}^{\text{p}} = \sum \Delta\lambda_k \mathbf{r}^{k,n+1} \tag{14.40}$$

Let us define

$$x_0 = \Delta\varepsilon_{pp} - \sum \Delta\lambda_k r_{pp}^{k,n+1} \tag{14.41a}$$

$$\Delta y_{ij} = \Delta e_{ij} - \sum \Delta\lambda_k \mathbf{d}^{k,n+1} \tag{14.41b}$$

where d_{ij} is the deviatoric part of r_{ij} defined as

$$d_{ij} = r_{ij} - \frac{1}{3} r_{kk}\delta_{ij} \tag{14.41c}$$

Equations 14.39a–c are then modified as

$$p^{\text{tr}} = \left[p_n^{1-n_e} + \bar{K}_0(1 - n_e)x_0\right]^{1/(1-n_e)} \tag{14.42a}$$

$$s_{ij}^{n+1} = s_{ij}^{n} + 2G_{\text{ave}}\Delta y_{ij} \tag{14.42b}$$

$$G_{\text{ave}} = \frac{\bar{G}_0}{\bar{K}_0 x_0}\left(p_{n+1} - p_n\right) \tag{14.42c}$$

Note that the restriction in (14.39d) applies to the term within the parenthesis in (14.42a). The total stress tensor is obtained as

$$\sigma_{ij}^{n+1} = s_{ij}^{n+1} + p_{n+1}\delta_{ij} \tag{14.43}$$

The residual $\mathbf{R}_\sigma$ is defined as

$$R_{\sigma ij} = \sigma_{ij}^{n+1} - \left[s_{ij}^{n+1} + p_{n+1}\delta_{ij} \right] \tag{14.44}$$

The expressions for the remaining residuals given by (14.19b–f) are unchanged. Hence, only the first row of (14.20) is modified. The rest of the details are the same as those in Sect. 14.2.3 with one difference: The matrix $\mathbf{C}$ in the term $\mathbf{C}\dot{\boldsymbol{\varepsilon}}$ on the right-hand side of (14.23) becomes $\mathbf{C}^*$ defined in Sect. 13.3.4.

Problems

Problem 14.1 It is common to use the Drucker–Prager model described in Chap. 13 in numerical modeling of geotechnical problems. However, the Drucker–Prager model has some limitations and the Sliding-rolling model presented in this chapter is intended to address those. With this background, answer the following questions:

(a) List three features of the sliding-rolling model, which are fundamentally different from those of the Drucker–Prager model.
(b) What are the specific features of the sliding-rolling model that permit the description of the liquefaction behavior of granular materials?
(c) Can the sliding-rolling model describe Bauschinger effect? Explain.
(d) Can the sliding-rolling model describe the directional dependence of the stress–strain behavior of anisotropic granular materials? Explain.

Problem 14.2 Referring to Fig. 14.3, sketch the expected model behavior in the four graphs for the following initial void ratios: $e = 0.6$ and 0.7. How does the model parameter c_α control these relationships?

Problem 14.3 The local iterations associated with the sliding-rolling model are significantly more computationally intensive than those associated with, say, the Cam-clay model or the Drucker–Prager model presented in the previous chapters. List three reasons for this.

Appendices

Appendix 1. The Virtual Work Principle

In this section, we provide a formal proof of the virtual work principle. We will make use of the notations presented in Fig. 5.2a. First, let us recognize the fact that in solid mechanics problems, there are two types of variable sets involved:

- Kinematic variables (deformation related): $(\mathbf{u}, \boldsymbol{\varepsilon})$
- Static variables (load related): $(\boldsymbol{\sigma}, \hat{\mathbf{t}}, \mathbf{b})$

If these quantities are actual solution of a problem, then $\boldsymbol{\sigma}$ and $\mathbf{b}$ will be related by the equilibrium equations

$$\sigma_{ij,j} + b_i = 0 \text{ in } \Omega \tag{A1.1}$$

The stress will be related to strain by a relationship such as (not necessarily an elastic)

$$\dot{\sigma}_{ij} = g(\dot{\varepsilon}_{ij}) \tag{A1.2}$$

and to the boundary traction by

$$t_i = \sigma_{ij} n_j \tag{A1.3}$$

The strains will be related to the deformation $\mathbf{u}$ by the strain-displacement relation (assuming small strains and deformations)

$$\varepsilon_{ij} = \frac{1}{2}(u_{i,j} + u_{j,i})$$

We are often interested in approximate solutions. In the process of seeking an approximate solution, we start with functions that may not satisfy all of the above relations exactly and simultaneously. In this context, let us define a special set of static and kinematic variables as follows:

- Kinematically admissible set $(\bar{\mathbf{u}}, \bar{\boldsymbol{\varepsilon}})$: One where $\bar{\mathbf{u}}$ is a continuous displacement field and $\bar{\boldsymbol{\varepsilon}}$ is a consistent strain field related to $\bar{\mathbf{u}}$ by the strain-displacement relation

$$\bar{\varepsilon}_{ij} = \frac{1}{2}(\bar{u}_{i,j} + \bar{u}_{j,i}) \tag{A1.4}$$

- Statically admissible set $(\bar{\boldsymbol{\sigma}}, \bar{\mathbf{t}}, \mathbf{b})$: One which satisfies the equilibrium equation as

$$\bar{\sigma}_{ij,j} + b_i = 0 \quad \text{in} \quad \Omega \tag{A1.5}$$

and the $(\bar{\boldsymbol{\sigma}}, \bar{\mathbf{t}})$ relation

$$\bar{t}_i = \bar{\sigma}_{ij} n_j \quad \text{on the boundary } \Gamma \tag{A1.6}$$

Again, it is reminded that the two sets $(\bar{\mathbf{u}}, \bar{\boldsymbol{\varepsilon}})$ and $(\bar{\boldsymbol{\sigma}}, \bar{\mathbf{t}}, \mathbf{b})$ are not necessarily related at this point. The displacement $\bar{\mathbf{u}}$ is referred to as the *virtual (not actual) displacement*. We will now prove the virtual work principle starting with the external virtual work defined in terms of the kinematically and statically admissible sets as

$$\begin{aligned}
\bar{W}^{e} &= \int_{\Gamma} \bar{t}_i \bar{u}_i \mathrm{d}s + \int_{\Omega} b_i \bar{u}_i \mathrm{d}v \\
&= \int_{\Gamma} \bar{\sigma}_{ij} n_j \bar{u}_i \mathrm{d}s + \int_{\Omega} b_i \bar{u}_i \mathrm{d}v \quad \text{(From A 1.6)} \\
&= \int_{\Gamma} (\bar{\sigma}_{ij} \bar{u}_i)_{,j} \mathrm{d}v + \int_{\Omega} b_i \bar{u}_i \mathrm{d}v \quad \left(\text{using Gauss theorem} \int_{\Gamma} (..)_j n_j ds = \int_{\Omega} (..)_{j,j} dv, \text{ Chap. 2}\right) \\
&= \int_{\Gamma} (\bar{\sigma}_{ij,j} + b_i) \bar{u}_i \, \mathrm{d}v + \int_{\Omega} \bar{\sigma}_{ij} \bar{u}_{i,j} \, \mathrm{d}v \\
&= 0 + \int_{\Omega} \bar{\sigma}_{ij} \bar{u}_{i,j} \, \mathrm{d}v \text{ (from A 1.5)} \\
&= \int_{\Omega} \bar{\sigma}_{ij} \left[\frac{1}{2}(\bar{u}_{i,j} + \bar{u}_{j,i}) + \frac{1}{2}(\bar{u}_{i,j} - \bar{u}_{j,i}) \right] \mathrm{d}v \\
&= \int_{\Omega} \bar{\sigma}_{ij} (\bar{\varepsilon}_{ij} + \bar{\omega}_{ij}) \mathrm{d}v \quad \text{(From A 1.4)}
\end{aligned} \tag{A1.7}$$

$\bar{\omega}_{ij} = \frac{1}{2}(\bar{u}_{i,j} - \bar{u}_{j,i})$ is the antisymmetric displacement gradient, with the property $\bar{\omega}_{ij} = -\bar{\omega}_{ji}$ and $\bar{\omega}_{ij} = 0$ for $i = j$. The product of a symmetric tensor like $\bar{\sigma}_{ij}$ with an antisymmetric tensor like $\bar{\omega}_{ij}$ is zero (as the product of the diagonal terms is zero because of the fact that the diagonal elements of $\bar{\omega}_{ij}$ are zero, and the product of the remaining terms can be grouped and shown to be zero as, for instance, $\bar{\sigma}_{12}\bar{\omega}_{12} + \bar{\sigma}_{21}\bar{\omega}_{21} = \bar{\sigma}_{12}\bar{\omega}_{12} - \bar{\sigma}_{12}\bar{\omega}_{12} = 0$). Thus, we have

$$\bar{W}^{\mathrm{e}} = \int_{\Omega} \bar{\sigma}_{ij}\bar{\varepsilon}_{ij}\mathrm{d}v = \bar{U} \tag{A1.8}$$

Let us define

$$\bar{U} = \int_{\Omega} \bar{\sigma}_{ij}\bar{\varepsilon}_{ij}\mathrm{d}v \tag{A1.9}$$

$\bar{U}$ is known as the *internal virtual work*. The virtual work principle states that External virtual work = Internal virtual work

$$\bar{W}^{\mathrm{e}} = \bar{U}$$

$$\int_{\Gamma} \bar{t}_i\bar{u}_i\mathrm{d}s + \int_{\Omega} b_i\bar{u}_i\mathrm{d}v = \int_{\Omega} \bar{\sigma}_{ij}\bar{\varepsilon}_{ij}\mathrm{d}v \tag{A1.10}$$

Appendix 2. General Stress–Strain Relation for an Elastic Material

There is a special relationship that all elastic materials, linear or nonlinear, must obey, and we derive that relationship here with the aid of the laws of thermodynamics. The first law concerns the conservation of energy. Referring to the system shown in Fig. 5.2a, let us define

- $\dot{W}$ = rate of work done by the external forces on the system
- $\dot{U}$ = rate of internal strain energy stored
- $\dot{V}$ = rate of internal kinetic energy stored
- $\dot{W}_1$ = rate of other forms of energy (e.g., chemical, thermal, electromagnetic, etc.) supplied to the system
- $\dot{W}_2$ = rate of other forms of energy (e.g., chemical, thermal, electromagnetic, etc.) withdrawn from the system

The first law of thermodynamics states that the energy cannot be destroyed. Thus, for a purely elastic material, the energy balance reads

$$\dot{W} + \dot{W}_1 - \dot{W}_2 = \dot{U} + \dot{V} \tag{A2.1}$$

(For an inelastic material, a term reflecting internal energy dissipation must be added to the equation.) Neglecting $\dot{V}$, $\dot{W}_1$and $\dot{W}_2$, we have for an elastic material

$$\dot{W} = \dot{U} \tag{A2.2}$$

Considering an infinitesimal time δt, (A2.2) can be written as

$$\frac{\delta W}{\delta t} = \frac{\delta U}{\delta t}$$

For a nonzero δt, we have

$$\delta W = \delta U \tag{A2.3}$$

Now, we make use of the virtual work expression given by (A1.10). Let the variational displacement $\delta \mathbf{u}$ be our virtual displacement, where $\mathbf{u}$ is the actual displacement. Let $\delta \mathbf{u}$ be applied over a time period δt and $\delta \boldsymbol{\varepsilon}$ be the corresponding virtual strain. Let the static admissible set be the actual stresses and tractions. Then by (A1.8), we have

$$\delta \bar{W} = \delta W = \int_{\Omega} \sigma_{ij} \delta \varepsilon_{ij} \mathrm{d}v \tag{A2.4}$$

From (A2.3) and (A2.4), we have

$$\delta U = \int_{\Omega} \sigma_{ij} \delta \varepsilon_{ij} \mathrm{d}v \tag{A2.5}$$

Equation (A2.5) gives a quantitative expression for calculating the change in internal energy of the system in terms of stresses and strains during an infinitesimal change. By differentiating (A2.5), we obtain the relation that we are interested in:

$$\sigma_{ij} = \frac{\partial U}{\partial \varepsilon_{ij}} \tag{A2.6}$$

Equation (A2.6) states that for an elastic material, the stresses can be derived from a scalar function – the internal energy function U (also called the potential function). For the special case of the linear elastic material, U is a quadratic function of strain as

$$U = \boldsymbol{\varepsilon}^{\mathrm{T}}\mathbf{D}\boldsymbol{\varepsilon} \tag{A2.7}$$

Then from (A2.6), one recovers the generalized Hooke's law

$$\boldsymbol{\sigma} = \mathbf{D}\boldsymbol{\varepsilon} \tag{A2.8}$$

where $\mathbf{D}$ is a constant (i.e., independent of stress or strain) fourth order tensor, known as the elastic stiffness tensor.

Appendix 3. Computer Coding Guidelines for Algorithm 8.2: Global Iteration Strategy for Model Calibration

Definition of variables:

NGMAX: Maximum number of global iterations permitted ($N_{\mathrm{G}}^{\max}$)
ERMAX1: TOL_1
ERMAX2: TOL_2
NPROP: Number of model parameters
NPIV: Number of PIVs
NUM: Number of increments that load is to be divided into
IELAST = 1: Find elastic stiffness, = 0: Find elasto-plastic stiffness
ISTIFF = 1: Find stiffness, = 0: Do not find stiffness
ISTRES = 1: Find incremental stress, = 0: Do not find incremental stress

Definition of arrays:

SIGB(6): $\boldsymbol{\sigma}_n$
SIGF(6): $\boldsymbol{\sigma}_{n+1}$
SIGSB(6): $\boldsymbol{\sigma}_n^{\mathrm{s}}$
SIGSF(6): $\boldsymbol{\sigma}_{n+1}^{\mathrm{s}}$
SIGE(6): $\boldsymbol{\sigma}_{n+1}^{\mathrm{e}}$
EPB(6): $\boldsymbol{\varepsilon}_n$
DSIG(6): $\Delta\boldsymbol{\sigma}$
DEP(6): $\Delta\boldsymbol{\varepsilon}$
PROP(NPROP): E, v, etc.
STOR(2*NPIV + 1): 1 to NPIV = ξ_n
STOR(2*NPIV + 1): NPIV + 1 to 2*NPIV= ξ_{n+1}
STOR(2*NPIV + 1): 2*NPIV + 1 = ALOD
ICOD(k) = 1 strain control, = 0 for stress control; $k = 1-6$
V(6): Either $\boldsymbol{\sigma}^{\mathrm{e}}$ or $\boldsymbol{\varepsilon}^{\mathrm{e}}$ depending on whether ICOD(k) = 0 or 1, respectively
DV(6): Either $\Delta\boldsymbol{\sigma}^{\mathrm{e}} = \boldsymbol{\sigma}^{\mathrm{e}}/\mathrm{NUM}$ or $\Delta\boldsymbol{\varepsilon}^{\mathrm{e}} = \boldsymbol{\varepsilon}^{\mathrm{e}}/\mathrm{NUM}$ depending on whether ICOD(k) = 0 or 1, respectively

R(6): Temporary array used for incremental load
R0(6): Residual stress $\boldsymbol{\sigma}^{e}_{n+1} - \boldsymbol{\sigma}^{s}_{n+1}$
C(6,6): $\mathbf{D}^{t}_{n+1}$

Input Data

Read in or type in:

```
NPROP
NPIV
NGMAX
ERMAX1 and ERMAX2
(PROP(I), I = 1,NPROP)
(SIGB(I), I = 1,6)
(STOR(I), I = 1,NPIV)
((ICOD(I),V(I)), I = 1,6), NUM
XXX = 1.0E + 20*(Atmospheric Pressure) (a large value)
```

Initialize

```
EPB(I) = 0; SIGSB(I) = SIGB(I); SIGE(I) = SIGB(I) for I = 1,6
DV(I) = [V(I)-SIGB(I)]/NUM if ICOD(I) = 0
      = [V(I)-EPB(I)]/NUM if ICOD(I) = 1 for I = 1,6
```

Start Analysis

Start load step loop

```
FOR INC = 1,NUM
Set DEP(I) = 0; DSIG(I) = 0;
   IF[ICOD(I) = 0] then SIGE(I) = SIGE(I) + DV(I) for I = 1,6
```

Start iteration loop

```
   FOR ITNO = 1,NGMAX
C Call material model module and calculate tangent stiffness C(6,6)
   ISTIFF = 1; ISTRE = 0
C Calculate elastic stiffness for the first iteration and elasto-plastic stiffness for
C the subsequent iterations
      IF[ITNO = 1] then IELAST = 1; else IELAST = 0
C The name MISES is arbitrary; you may call it ELAST or DRUCKER depending
      on the constitutive model that you are implementing
      CALL MISES(C,PROP,STOR,SIGB,EPB,DSIG,DEP,NPROP,NPIV, IELAST,
         ISTIFF,ISTRE)
C Modify C(6,6) and R(6) for strain-controlled loading (refer to Sect. 8.6)
      FOR I = 1,6
         IF(ICOD(I) = 1) C(I,I) = XXX
         IF(ITNO = 1) THEN
            IF(ICOD(I) = 0) R(I) = DV(I)
            IF(ICOD(I) = 1) R(I) = DV(I)*XXX
      ELSE
         IF(ICOD(I) = 0) R(I) = R0(I)
         IF(ICOD(I) = 1) R(I) = 0
      ENDIF
      NEXT I
```

```
C Solve $\mathbf{D}^{t}_{n+1}\delta\boldsymbol{\varepsilon} = \delta\boldsymbol{\sigma}$
      CALL SOLVE(C,R)
C (It is assumed that R contains stress $\delta\boldsymbol{\sigma}$ on entering, and strain $\delta\boldsymbol{\varepsilon}$ on returning)
C Define incremental strain from iterative strain
      DEP(I) = DEP(I) + R(I) for I = 1,6
C Call material model module and calculate incremental stress
      ISTIFF = 0; ISTRE = 1; IELAST = 0
      CALL MISES(C,PROP,STOR,SIGB,EPB,DSIG,DEP,NPROP,NPIV, IELAST,
         ISTIFF,ISTRE)
C Update stresses and spring forces and define residual load
      FOR I = 1,6
      SIGF(I) = SIGB(I) + DSIG(I)
      SIGSF(I) = SIGSB(I) + DSIG(I)
      R0(I) = SIGE(I)-SIGSF(I)
      NEXT I
C Calculate errors: ERSIG: error based on stresses, EREN: error based on energy
      ERSIG = 0.0
      EREN = 0.0
      ERRT = 0.0
      FOR I = 1,6
         ERRT = ERRT + SIGSF(I)**2
         IF(ICOD(I) = 0) THEN
            ERSIG = ERSIG + R0(I)**2
            EREN = EREN + R0(I)*DEP(I)
      ENDIF
      NEXT I
      ERSIG = SQRT(ERSIG)/SQRT(ERRT)
      EREN = ABS(EREN)/SQRT(ERRT)
C Check for convergence
      IF(ERSIG < ERMAX1 AND EREN < ERMAX2) Exit iteration loop
C End of iteration loop
      NEXT ITNO
C Update stresses, spring forces and strains
      FOR I = 1,6
         SIGB(I) = SIGF(I)
         SIGSB(I) = SIGSF(I)
         EPB(I) = EPB(I) + DEP(I)
      NEXT I
C Update PIVs
      STOR(I) = STOR(I + NPIV) for I = 1,NPIV
C Print results and continue with the next load step
      NEXT INC
C STOP
```

Appendix 4. Convexity of a Function

Referring to Fig. A4.1, the function $\phi(\sigma)$is convex

$$\bar{\phi} \le \phi^* \tag{A4.1a}$$

where

$$\bar{\sigma} = \beta\sigma_1 + (1-\beta)\sigma_2; \quad \beta \in [0,1] \tag{A4.1b}$$

$$\phi^* = \beta\phi_1 + (1-\beta)\phi_2 \tag{A4.1c}$$

When the function $\phi(\sigma)$ is smooth, it is convex if and only if

$$\phi_2 - \phi_1 \ge (\sigma_2 - \sigma_1)\phi'(\sigma_1) \tag{A4.2}$$

where $\phi'(\sigma_1)$ is the slope of the function at σ_1. The idea can be easily generalized for a smooth, multi-variable function.

Appendix 5. Use of the Elastic Predictor for the Determination of Loading/Unloading Event

The relevant proof may be found in Simo and Hughes (1998), and is summarized here in a slightly different form. From the definition of a multi-variable, smooth, convex function, presented in Appendix 4, we have

$$\phi^{\mathrm{tr}}_{n+1}(\boldsymbol{\sigma}^{\mathrm{tr}}_{n+1}, \zeta_n) - \phi_{n+1}(\boldsymbol{\sigma}_{n+1}, \zeta_{n+1}) \ge -\Delta\boldsymbol{\sigma}^{\mathrm{pc}}\left[\frac{\partial\phi}{\partial\boldsymbol{\sigma}_{n+1}}\right] - \Delta\zeta\left[\frac{\partial\phi}{\partial\zeta_{n+1}}\right] \tag{A5.1}$$

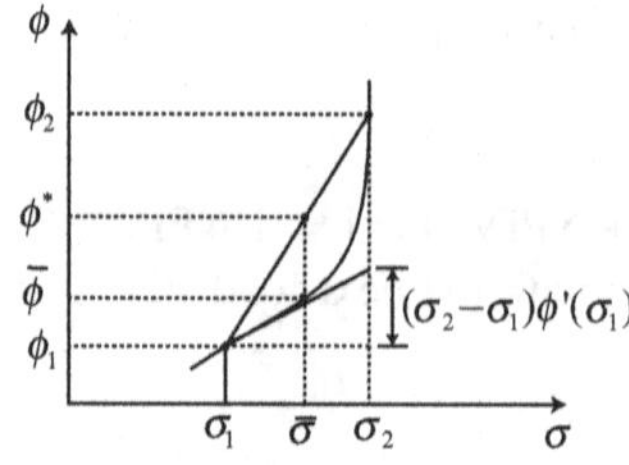

Fig. A4.1 Schematic of a one-dimensional smooth convex function

where $\Delta\boldsymbol{\sigma}^{\mathrm{pc}}$ is the plastic corrector. Using the backward Euler integration method (assuming that the elastic stiffness tensor is a constant), we have

$$\Delta\boldsymbol{\sigma}^{\mathrm{pc}} = -\int_{t}^{t+\Delta t} \dot{\lambda}\mathbf{C}\mathbf{r} = -\Delta\lambda\mathbf{C}\mathbf{r}_{n+1} \qquad \text{(from 10.5c)}$$

and $\Delta\boldsymbol{\zeta} = \int_{t}^{t+\Delta t} \dot{\lambda}\mathbf{s} = \Delta\lambda\mathbf{s}_{n+1}$(from 10.2c). Here it is noted that $\Delta\lambda \geq 0$, with $\Delta\lambda = 0$ for elastic behavior and $\Delta\lambda > 0$ for elasto-plastic behavior.

Substituting these into (A5.1), and noting that

$\mathbf{n}_{n+1} = \dfrac{\partial\phi}{\partial\boldsymbol{\sigma}_{n+1}}$ and $K_{\mathrm{p}} = -(\partial_{\zeta}\phi)^{T}\mathbf{s}$ (10.2e), we have

$$\phi^{\mathrm{tr}}_{n+1}\left(\boldsymbol{\sigma}^{\mathrm{tr}}_{n+1}, \zeta_{n}\right) - \phi_{n+1}\left(\boldsymbol{\sigma}_{n+1}, \zeta_{n+1}\right) \geq \Delta\lambda\left[\mathbf{n}^{T}_{n+1}\mathbf{C}\mathbf{r}_{n+1} + K^{n+1}_{\mathrm{p}}\right] \qquad \text{(A5.2)}$$

As $\Delta\lambda \geq 0$, the term on the right hand side is non-negative when the quantity within the bracket is non-negative. For a model employing associated flow rule, we have $\mathbf{n}_{n+1} = \mathbf{r}_{n+1}$. As the elastic stiffness tensor is positive definite, the first term is positive because $\mathbf{n}^{T}_{n+1}\mathbf{C}\mathbf{r}_{n+1} = \mathbf{r}^{T}_{n+1}\mathbf{C}\mathbf{r}_{n+1} > 0$. When the model does not permit strain softening, the plastic modulus $K^{n+1}_{\mathrm{p}} \geq 0$. Thus, the quantity within the bracket is non-negative. For other cases (i.e., models using non-associated flow rule and/or models permitting strain softening), the term on the right hand side is still non-negative so far as the terms within the bracket add up to a non-negative number.

When the right hand side is non-negative, (A5.2) becomes

$$\phi^{\mathrm{tr}}_{n+1}\left(\boldsymbol{\sigma}^{\mathrm{tr}}_{n+1}, \zeta_{n}\right) - \phi_{n+1}\left(\boldsymbol{\sigma}_{n+1}, \zeta_{n+1}\right) \geq 0$$

or

$$\phi^{\mathrm{tr}}_{n+1}\left(\boldsymbol{\sigma}^{\mathrm{tr}}_{n+1}, \zeta_{n}\right) \geq \phi_{n+1}\left(\boldsymbol{\sigma}_{n+1}, \zeta_{n+1}\right) \qquad \text{(A5.3)}$$

Now consider the following two cases:

1. Case 1: $\phi^{\mathrm{tr}}_{n+1}\left(\boldsymbol{\sigma}^{\mathrm{tr}}_{n+1}, \zeta_{n}\right) < 0$. It follows from (A5.3) that $\phi_{n+1}\left(\boldsymbol{\sigma}_{n+1}, \zeta_{n+1}\right) < 0$. Hence the step is elastic and $\Delta\lambda = 0$.
2. Case 2: $\phi^{\mathrm{tr}}_{n+1}\left(\boldsymbol{\sigma}^{\mathrm{tr}}_{n+1}, \zeta_{n}\right) > 0$. The stress point is outside the current yield surface, and thus must cause plastic deformation. Hence, the step is elasto-plastic, $\phi_{n+1}\left(\boldsymbol{\sigma}_{n+1}, \zeta_{n+1}\right) = 0$ and $\Delta\lambda > 0$.

In conclusion, we have

$$\phi^{\mathrm{tr}}_{n+1}\left(\boldsymbol{\sigma}^{\mathrm{tr}}_{n+1}, \zeta_{n}\right) > 0 \quad \Rightarrow \quad \text{Elasto} - \text{Plastic and } \Delta\lambda > 0 \qquad \text{(A5.4a)}$$

$$\phi^{\mathrm{tr}}_{n+1}\left(\boldsymbol{\sigma}^{\mathrm{tr}}_{n+1}, \zeta_{n}\right) < 0 \quad \Rightarrow \quad \text{Elastic and } \Delta\lambda = 0 \qquad \text{(A5.4b)}$$

Appendix 6. Simplification of Tensorial Operations Based on Symmetries

Using Voigt's notation, let us place the elements of the stress and strain tensors in vectors as

$$\bar{\boldsymbol{\sigma}} = \{\sigma_{11} \quad \sigma_{22} \quad \sigma_{33} \quad \sigma_{12} \quad \sigma_{23} \quad \sigma_{13}\} \tag{A6.1a}$$

$$\bar{\boldsymbol{\varepsilon}} = \{\varepsilon_{11} \quad \varepsilon_{22} \quad \varepsilon_{33} \quad \gamma_{12} \quad \gamma_{23} \quad \gamma_{13}\} \tag{A6.1b}$$

where $\gamma_{12} = 2\varepsilon_{12}$, $\gamma_{23} = 2\varepsilon_{23}$, and $\gamma_{13} = 2\varepsilon_{13}$. The shear components in the stress vector are the same as the corresponding shear components in the stress tensor, whereas the shear components in the strain vector are twice the corresponding shear components in the strain tensor. With this difference in mind, in the following, we will refer to second order tensors as either a "stress-like" quantity or a "strain-like" quantity. We use similar representations for the incremental stresses and strains. We now present a few tensorial operations that could be simplified into reduced order matrix-vector operations.

Operation 1

$$\beta = A_{ij}B_{ij}$$

where $A_{ij} = A_{ji}$ and $B_{ij} = B_{ji}$. Then A_{ij} and B_{ij} can be placed in vectors like in (A6.1a) and (A6.1b). The required operations can then be performed as

$$\beta = \bar{\mathbf{A}}.\bar{\mathbf{B}}$$

where $\bar{\mathbf{A}} = \{A_{11}\ A_{22}\ A_{33}\ A_{12}\ A_{23}\ A_{13}\}$
and $\bar{\mathbf{B}} = \{B_{11} \quad B_{22} \quad B_{33} \quad 2B_{12} \quad 2B_{23} \quad 2B_{13}\}$

$\bar{\mathbf{A}}$ is a stress-like quantity, and $\bar{\mathbf{B}}$ is a strain-like quantity.

Operation 2

$$K_{ij} = L_{ijk\ell}M_{k\ell} \tag{A6.2a}$$

where

$$M_{k\ell} = M_{\ell k} \tag{A6.2b}$$

and

$$L_{ijk\ell} = L_{jik\ell} \Rightarrow K_{ij} = K_{ji} \tag{A6.2c}$$

Then K_{ij} and M_{ij} can be placed in vectors like in (A6.1a) and (A6.1b). The required operations can then be performed as

$$\bar{\mathbf{K}}_{6\times 1} = \underset{6\times 6}{\bar{\mathbf{L}}}\ \underset{6\times 1}{\bar{\mathbf{M}}} \tag{A6.3a}$$

When $\bar{\mathbf{M}}$ is a stress-like quantity (A6.1a), we have

$$\begin{bmatrix} K_{11} \\ K_{22} \\ K_{33} \\ K_{12} \\ K_{23} \\ K_{12} \end{bmatrix} = \begin{bmatrix} L_{1111} & L_{1122} & L_{1133} & L_{1112}+L_{1121} & L_{1123}+L_{1132} & L_{1113}+L_{1131} \\ L_{2211} & L_{2222} & L_{2233} & L_{2212}+L_{2221} & L_{2223}+L_{2232} & L_{2213}+L_{2231} \\ L_{3311} & L_{3322} & L_{3333} & L_{3312}+L_{3321} & L_{3323}+L_{1132} & L_{3313}+L_{3331} \\ L_{1211} & L_{1222} & L_{1233} & L_{1212}+L_{1221} & L_{1223}+L_{1232} & L_{1213}+L_{1231} \\ L_{2311} & L_{2322} & L_{2333} & L_{2312}+L_{2321} & L_{2323}+L_{2332} & L_{2313}+L_{2331} \\ L_{1211} & L_{1222} & L_{1233} & L_{1212}+L_{1221} & L_{1223}+L_{1232} & L_{1213}+L_{1231} \end{bmatrix} \begin{bmatrix} M_{11} \\ M_{22} \\ M_{33} \\ M_{12} \\ M_{23} \\ M_{13} \end{bmatrix} \tag{A6.3b}$$

When M_{ij} is a strain-like quantity (A6.1b), we have

$$\begin{bmatrix} K_{11} \\ K_{22} \\ K_{33} \\ K_{12} \\ K_{23} \\ K_{12} \end{bmatrix} = \begin{bmatrix} L_{1111} & L_{1122} & L_{1133} & (L_{1112}+L_{1121})/2 & (L_{1123}+L_{1132})/2 & (L_{1113}+L_{1131})/2 \\ L_{2211} & L_{2222} & L_{2233} & (L_{2212}+L_{2221})/2 & (L_{2223}+L_{2232})/2 & (L_{2213}+L_{2231})/2 \\ L_{3311} & L_{3322} & L_{3333} & (L_{3312}+L_{3321})/2 & (L_{3323}+L_{1132})/2 & (L_{3313}+L_{3331})/2 \\ L_{1211} & L_{1222} & L_{1233} & (L_{1212}+L_{1221})/2 & (L_{1223}+L_{1232})/2 & (L_{1213}+L_{1231})/2 \\ L_{2311} & L_{2322} & L_{2333} & (L_{2312}+L_{2321})/2 & (L_{2323}+L_{2332})/2 & (L_{2313}+L_{2331})/2 \\ L_{1211} & L_{1222} & L_{1233} & (L_{1212}+L_{1221})/2 & (L_{1223}+L_{1232})/2 & (L_{1213}+L_{1231})/2 \end{bmatrix} \times \begin{bmatrix} M_{11} \\ M_{22} \\ M_{33} \\ 2M_{12} \\ 2M_{23} \\ 2M_{13} \end{bmatrix} \tag{A6.3c}$$

Also when $L_{ijk\ell} = L_{ji\ell k}$, the terms on the last three columns simplify as $L_{1112} + L_{1121} = 2L_{1112}$, etc.

Following a similar approach, the operation

$$K_{k\ell} = L_{ijk\ell} M_{ij} \tag{A6.3d}$$

can be easily performed. It is left for the reader as an exercise.

Appendix 7. General Expression for the Gradient of Yield/Potential Function with Respect to Stresses

The yield and plastic potentials typically take the same functional form, and thus, the equations derived in this section can be applied to either one.

We consider yield surfaces that are given by the following function:

$$\phi(I, J, \alpha, \zeta) = 0 \tag{A7.1}$$

By differentiating the yield function with respect to $\boldsymbol{\sigma}$, we have

$$r_{ij} = \frac{\partial \phi}{\partial I}\frac{\partial I}{\partial \sigma_{ij}} + \frac{\partial \phi}{\partial J}\frac{\partial J}{\partial \sigma_{ij}} + \frac{\partial \phi}{\partial \alpha}\frac{\partial \alpha}{\partial \sigma_{ij}} \tag{A7.2}$$

The stress invariants I, J and α are defined as

$$I = \sigma_{k\ell}\delta_{k\ell}$$

$$s_{ij} = \sigma_{ij} - \frac{1}{3}I\delta_{ij}; \quad J = \left(\frac{1}{2}s_{k\ell}s_{k\ell}\right)^{1/2}$$

$$S = \left(\frac{1}{3}s_{ij}s_{jk}s_{ki}\right)^{1/3}$$

$$-\frac{\pi}{6} \leq \alpha = \frac{1}{3}\sin^{-1}\left[\frac{3\sqrt{3}}{2}\left(\frac{S}{J}\right)^3\right] \leq \frac{\pi}{6}$$

Differentiating these quantities with respect to the stress tensor, we have

$$\frac{\partial I}{\partial \sigma_{ij}} = \delta_{ij} \tag{A7.3}$$

$$\frac{\partial s_{ij}}{\partial \sigma_{k\ell}} = \delta_{ik}\delta_{j\ell} - \frac{1}{3}\delta_{ij}\delta_{k\ell} \tag{A7.4}$$

From $2J^2 = s_{pq}s_{pq}$, we have

$$4J\frac{\partial J}{\partial \sigma_{ij}} = \left(\delta_{pi}\delta_{qj} - \frac{1}{3}\delta_{pq}\delta_{ij}\right)s_{pq} + \ldots$$

Noting that $s_{kk} = 0$, the above equation simplifies to

$$4J\frac{\partial J}{\partial \sigma_{ij}} = 2s_{ij}$$

$$\frac{\partial J}{\partial \sigma_{ij}} = \frac{s_{ij}}{2J} \tag{A7.5}$$

Similarly, noting that $3S^3 = s_{pq}s_{qr}s_{rp}$, we have

$$9S^2 \frac{\partial S}{\partial \sigma_{ij}} = (\delta_{pi}\delta_{qj} - \frac{1}{3}\delta_{pq}\delta_{ij})s_{qr}s_{rp} + \ldots$$
$$= (s_{jr}s_{ri} - \frac{1}{3}s_{pr}s_{rp}\delta_{ij}) + \ldots$$
$$= 3s_{ir}s_{rj} - \frac{6J^2}{3}\delta_{ij}$$

$$\frac{\partial S}{\partial \sigma_{ij}} = \frac{1}{3S^2}s_{ir}s_{rj} - \frac{2J^2}{9}\delta_{ij} \tag{A7.6}$$

From

$$\sin 3\alpha = \frac{3\sqrt{3}}{2}\frac{S^3}{J^3}$$

we have

$$3\cos 3\alpha \frac{\partial \alpha}{\partial \sigma_{ij}} = \frac{3\sqrt{3}}{2}\frac{3S^2}{J^3}\frac{\partial S}{\partial \sigma_{ij}} - \frac{3\sqrt{3}}{2}\frac{3S^3}{J^4}\frac{\partial J}{\partial \sigma_{ij}}$$

$$= \frac{3\sqrt{3}}{2}\frac{3S^2}{J^3}\left[\frac{1}{3S^2}s_{ir}s_{rj} - \frac{2J^2}{9}\delta_{ij}\right] - \frac{3\sqrt{3}}{2}\frac{3S^3}{J^4}\frac{s_{ij}}{2J}$$

(using (A7.5) and (A7.6))

$$\frac{\partial \alpha}{\partial \sigma_{ij}} = \frac{\sqrt{3}}{2J\cos 3\alpha}\left[\frac{s_{ir}s_{rj}}{J^2} - \frac{2}{3}\delta_{ij} - \frac{3}{2}\left(\frac{S}{J}\right)^3\frac{s_{ij}}{J}\right] \tag{A7.7}$$

Synthesizing the above equations, we arrive at

$$r_{ij} = \frac{\partial \phi}{\partial I}\delta_{ij} + \frac{\partial \phi}{\partial J}\frac{s_{ij}}{2J} + \frac{\partial \phi}{\partial \alpha}\frac{\sqrt{3}}{2J\cos 3\alpha}\left[\frac{s_{ir}s_{rj}}{J^2} - \frac{2}{3}\delta_{ij} - \frac{3}{2}\left(\frac{S}{J}\right)^3\frac{s_{ij}}{J}\right] \tag{A7.8}$$

Equation (A7.8) is quite general and can be used with any surface that is represented in a functional form given by (A7.1). Depending on the specific shape of the surface, $\partial_I\phi$, $\partial_J\phi$, and $\partial_\alpha\phi$ will vary.

Appendix 8. Equations Related to the Cam-Clay Model

Appendix 8.1. First Gradients of Yield Function with Respect to State Variables

The Cam-clay ellipse is given in a 3-invariant space as

$$\phi(I, J, \alpha, p_0) = I^2 + \left(\frac{J}{N}\right)^2 - 3p_0 I = 0 \tag{A8.1}$$

where $N = \frac{M}{3\sqrt{3}}$; $M = g(n, \alpha)M_c$; $g(n, \alpha) = \dfrac{2n}{1 + n - (1 - n)\sin 3\alpha}$;
and

$$n = \frac{M_e}{M_c} \tag{A8.2}$$

Differentiating (A8.1), we have

$$\frac{\partial \phi}{\partial I} = 2I - 3p_0 \tag{A8.3}$$

$$\frac{\partial \phi}{\partial J} = \frac{2J}{N^2} \tag{A8.4}$$

$\dfrac{\partial \phi}{\partial \alpha} = -\dfrac{2J^2}{N^3}\dfrac{\partial N}{\partial \alpha} = -\dfrac{J}{N}\dfrac{\partial \phi}{\partial J}\dfrac{\partial N}{\partial \alpha}$ (using (A8.4))

From (A8.2), we have

$$\frac{\partial N}{\partial \alpha} = \frac{M_c}{3\sqrt{3}}\frac{\partial g}{\partial \alpha} = \frac{M_c}{3\sqrt{3}}\left[\frac{(2n)(1-n)(3\cos 3\alpha)}{\{1 + n - (1-n)\sin 3\alpha\}^2}\right]$$

$$= \frac{3N(1-n)\cos 3\alpha}{1 + n - (1-n)\sin 3\alpha} \tag{A8.5}$$

Thus, it follows that

$$\frac{\partial \phi}{\partial \alpha} = J\bar{g}\frac{\partial \phi}{\partial J};$$

where

$$\bar{g} = -\frac{3(1-n)\cos 3\alpha}{1 + n - (1-n)\sin 3\alpha} \tag{A8.6}$$

The gradient of ϕ with respect to I_0 is given by

$$\frac{\partial \phi}{\partial I_0} = -I \tag{A8.7}$$

Equations (A8.3), (A8.4), (A8.6), and (A8.7) furnish the key gradients $\partial_I \phi$, $\partial_J \phi$, $\partial_\alpha \phi$, and $\partial_{I_0} \phi$, respectively.

Appendix 8.2. Second Gradients of Yield Function with Respect to State Variables

Repeating (A7.8), we have

$$r_{ij} = \frac{\partial \phi}{\partial I}\delta_{ij} + \frac{\partial \phi}{\partial J}\frac{s_{ij}}{2J} + \frac{\partial \phi}{\partial \alpha}\frac{\sqrt{3}}{2J\cos 3\alpha}\left[\frac{s_{ir}s_{rj}}{J^2} - \frac{2}{3}\delta_{ij} - \frac{3}{2}\left(\frac{S}{J}\right)^3\frac{s_{ij}}{J}\right] \tag{A8.8}$$

where from (A8.3), (A8.4), (A8.6), and (A8.7),

$$\frac{\partial \phi}{\partial I} = 2I - 3p_0 \tag{A8.9}$$

$$\frac{\partial \phi}{\partial J} = \frac{2J}{N^2} \tag{A8.10}$$

$$\frac{\partial \phi}{\partial \alpha} = J\bar{g}\frac{\partial \phi}{\partial J};$$

where

$$\bar{g} = -\frac{3(1-n)\cos 3\alpha}{1+n-(1-n)\sin 3\alpha} \tag{A8.11}$$

$$\frac{\partial \phi}{\partial I_0} = -I \tag{A8.12}$$

The 2nd term in (A8.8) is to be excluded for $J = 0$, and the 3rd term for $J = 0$ and/or $\alpha = \pi/6$ or $-\pi/6$.

From (A7.3)–(A7.7)

$$\frac{\partial I}{\partial \sigma_{ij}} = \delta_{ij} \tag{A8.13}$$

$$\frac{\partial s_{ij}}{\partial \sigma_{k\ell}} = \delta_{ik}\delta_{j\ell} - \frac{1}{3}\delta_{ij}\delta_{k\ell} \tag{A8.14}$$

$$\frac{\partial J}{\partial \sigma_{ij}} = \frac{s_{ij}}{2J} \tag{A8.15}$$

$$\frac{\partial S}{\partial \sigma_{ij}} = \frac{1}{3S^2} s_{ir}s_{rj} - \frac{2J^2}{9}\delta_{ij} \tag{A8.16}$$

$$\frac{\partial \alpha}{\partial \sigma_{ij}} = \frac{\sqrt{3}}{2J\cos 3\alpha}\left[\frac{s_{ir}s_{rj}}{J^2} - \frac{2}{3}\delta_{ij} - \frac{3}{2}\left(\frac{S}{J}\right)^3 \frac{s_{ij}}{J}\right] \tag{A8.17}$$

With the aid of equations in Appendix 8.1,

$$N = \frac{M}{3\sqrt{3}} = \frac{g(n,\alpha)M_c}{3\sqrt{3}},$$

where

$$g(n,\alpha) = \frac{2n}{1 + n - (1-n)\sin 3\alpha} \tag{A8.18}$$

Differentiating g with respect to α, we have

$$\frac{\partial g}{\partial \alpha} = \frac{(2n)(1-n)(3\cos 3\alpha)}{\{1+n-(1-n)\sin 3\alpha\}^2} = \frac{3(1-n)g^2\cos 3\alpha}{2n} \tag{A8.19}$$

Differentiating N with respect to σ_{pq} and making use of (A8.19), we have

$$\begin{aligned}\frac{\partial N}{\partial \sigma_{pq}} &= \frac{M_c}{3\sqrt{3}}\frac{\partial g(n,\alpha)}{\partial \sigma_{pq}} = \frac{M_c}{3\sqrt{3}}\frac{\partial g(n,\alpha)}{\partial \alpha}\frac{\partial \alpha}{\partial \sigma_{pq}}\\ &= \frac{M_c}{3\sqrt{3}}\frac{3(1-n)g^2\cos 3\alpha}{2n}\frac{\partial \alpha}{\partial \sigma_{pq}}\\ &= \frac{M_c}{2\sqrt{3}}\frac{1-n}{n}g^2\cos 3\alpha\frac{\partial \alpha}{\partial \sigma_{pq}}\\ &= \frac{3N}{2}\frac{1-n}{n}g\cos 3\alpha\frac{\partial \alpha}{\partial \sigma_{pq}}\end{aligned} \tag{A8.20}$$

Expressing $\bar{g}$ given by (A8.11) in terms of g, we have

$$\bar{g} = -\frac{3(1-n)\cos 3\alpha}{1+n-(1-n)\sin 3\alpha} = -\frac{3(1-n)}{2n} g \cos 3\alpha$$

Differentiating $\bar{g}$, we get

$$\begin{aligned}
\frac{\partial \bar{g}}{\partial \sigma_{pq}} &= \frac{9(1-n)}{2n} g \sin 3\alpha \frac{\partial \alpha}{\partial \sigma_{pq}} - \frac{3(1-n)}{2n} \cos 3\alpha \frac{\partial g}{\partial \sigma_{pq}} \\
&= \frac{9(1-n)}{2n} g \sin 3\alpha \frac{\partial \alpha}{\partial \sigma_{pq}} - \frac{3(1-n)}{2n} \cos 3\alpha \frac{\partial g}{\partial \alpha} \frac{\partial \alpha}{\partial \sigma_{pq}} \\
&= \frac{9(1-n)}{2n} g \sin 3\alpha \frac{\partial \alpha}{\partial \sigma_{pq}} \\
&\quad - \frac{3(1-n)}{2n} \cos 3\alpha \left\{ \frac{3(1-n) g^2 \cos 3\alpha}{2n} \right\} \frac{\partial \alpha}{\partial \sigma_{pq}} \quad \text{(using (A8.19))} \\
&= \left[\frac{9(1-n)}{2n} g \sin 3\alpha - \frac{9(1-n)^2}{4n^2} g^2 \cos^2 3\alpha \right] \frac{\partial \alpha}{\partial \sigma_{pq}}
\end{aligned} \tag{A8.21}$$

By differentiation of (A8.9) to (A8.12), we arrive at

$$\frac{\partial [\partial_I \phi]}{\partial \sigma_{pq}} = 2\delta_{pq} \tag{A8.22}$$

$$\begin{aligned}
\frac{\partial [\partial_J \phi]}{\partial \sigma_{pq}} &= \frac{2}{N^2} \frac{\partial J}{\partial \sigma_{pq}} - \frac{4J}{N^3} \frac{\partial N}{\partial \sigma_{pq}} \\
&= \frac{2}{N^2} \frac{s_{pq}}{2J} - \frac{4J}{N^3} \frac{3N}{2} \frac{1-n}{n} g \cos 3\alpha \frac{\partial \alpha}{\partial \sigma_{pq}} \quad \text{(using (A8.15) and (A8.20))} \\
&= \frac{1}{N^2} \frac{s_{pq}}{J} - \frac{6J}{N^2} \frac{1-n}{n} g \cos 3\alpha \frac{\partial \alpha}{\partial \sigma_{pq}}
\end{aligned} \tag{A8.23}$$

$$\begin{aligned}
\frac{\partial [\partial_\alpha \phi]}{\partial \sigma_{pq}} &= J\bar{g} \frac{\partial [\partial_J \phi]}{\partial \sigma_{pq}} + \bar{g} [\partial_J \phi] \frac{\partial J}{\partial \sigma_{pq}} + J [\partial_J \phi] \frac{\partial \bar{g}}{\partial \sigma_{pq}} \\
&= J\bar{g} \frac{\partial [\partial_J \phi]}{\partial \sigma_{pq}} + \bar{g} [\partial_J \phi] \frac{s_{pq}}{2J} + J [\partial_J \phi] \left[\frac{9(1-n)}{2n} g \sin 3\alpha - \frac{9(1-n)^2}{4n^2} g^2 \cos^2 3\alpha \right] \\
&\quad \times \frac{\partial \alpha}{\partial \sigma_{pq}} \quad \text{(using (A8.15) and (A8.21))}
\end{aligned} \tag{A8.24}$$

Now, let us differentiate (8.8). Let us denote the terms associated with $\partial_I\phi$, $\partial_J\phi$, and $\partial_\alpha\phi$ by L^1_{ijpq}, L^2_{ijpq}, and L^3_{ijpq}, respectively. First, let us write r_{ij} as

$$r_{ij} = (\partial_I\phi)a_{ij} + (\partial_J\phi)b_{ij} + (\partial_\alpha\phi)c_{ij}$$

where

$$a_{ij} = \frac{\partial I}{\partial \sigma_{ij}} = \delta_{ij}$$

$$b_{ij} = \frac{\partial J}{\partial \sigma_{ij}} = \frac{s_{ij}}{2J}$$

$$c_{ij} = \frac{\partial \alpha}{\partial \sigma_{ij}} = \frac{\sqrt{3}}{2J\cos 3\alpha}\left[\frac{s_{ir}s_{rj}}{J^2} - \frac{2}{3}\delta_{ij} - \frac{3}{2}\left(\frac{S}{J}\right)^3\frac{s_{ij}}{J}\right]$$

It then follows:

$$\frac{\partial r_{ij}}{\partial \sigma_{pq}} = L^1_{ijpq} + L^2_{ijpq} + L^3_{ijpq} \tag{A8.25a}$$

$$L^1_{ijpq} = \frac{\partial(\partial_I\phi)}{\partial \sigma_{pq}}a_{ij} + (\partial_I\phi)L^{10}_{ijpq}; \quad L^{10}_{ijpq} = \frac{\partial a_{ij}}{\partial \sigma_{pq}}$$

$$L^2_{ijpq} = \frac{\partial(\partial_J\phi)}{\partial \sigma_{pq}}b_{ij} + (\partial_J\phi)L^{20}_{ijpq}; \quad L^{20}_{ijpq} = \frac{\partial b_{ij}}{\partial \sigma_{pq}}$$

$$L^3_{ijpq} = \frac{\partial(\partial_\alpha\phi)}{\partial \sigma_{pq}}c_{ij} + (\partial_\alpha\phi)L^{30}_{ijpq}; \quad L^{30}_{ijpq} = \frac{\partial c_{ij}}{\partial \sigma_{pq}}$$

From the equations presented above, we have

$$L^{10}_{ijpq} = 0 \tag{A8.25b}$$

$$\begin{aligned} L^{20}_{ijpq} &= \frac{1}{2J}\frac{\partial s_{ij}}{\partial \sigma_{pq}} - \frac{1}{2J^2}\frac{\partial J}{\partial \sigma_{pq}}s_{ij} \\ &= \frac{1}{2J}\left[\delta_{ip}\delta_{jq} - \frac{1}{3}\delta_{ij}\delta_{pq}\right] - \frac{1}{2J^2}\frac{s_{pq}}{2J}s_{ij} \quad \text{(using A8.14 and A8.15)} \end{aligned} \tag{A8.26}$$

Let us write c_{ij} as

$$c_{ij} = \frac{\sqrt{3}}{2\cos 3\alpha}\bar{A}_{ij} = a_0\bar{A}_{ij}$$

where $\bar{A}_{ij} = \frac{s_{ir}s_{rj}}{J^3} - \frac{2}{3}\delta_{ij}\frac{1}{J} - \frac{3}{2}\frac{S^3}{J^5}s_{ij}$ and $a = \frac{\sqrt{3}}{2\cos 3\alpha}$

Then we have

$$\begin{aligned} L^{30}_{ijpq} &= \frac{\partial c_{ij}}{\partial \sigma_{pq}} = \frac{3\sqrt{3}\sin 3\alpha}{2\cos^2 3\alpha}\frac{\partial \alpha}{\partial \sigma_{pq}}\bar{A}_{ij} \\ &+ \frac{a_0}{J^3}\frac{\partial s_{ir}}{\partial \sigma_{pq}}s_{rj} + \frac{a_0}{J^3}\frac{\partial s_{rj}}{\partial \sigma_{pq}}s_{ir} + a_0\left[\frac{15}{2}\frac{S^3}{J^6}s_{ij} + \frac{2}{3}\frac{1}{J^2}\delta_{ij} - \frac{3}{J^4}s_{ir}s_{rj}\right]\frac{\partial J}{\partial \sigma_{pq}} \\ &- \frac{9a_0}{2}\frac{S^2}{J^5}s_{ij}\frac{\partial S}{\partial \sigma_{pq}} - \frac{3}{2}\frac{S^3}{J^5}\frac{\partial s_{ij}}{\partial \sigma_{pq}} \end{aligned} \tag{A8.27}$$

With the aid of (A8.14), (A8.15), and (A8.16), and noting that

$$\begin{aligned} &\frac{\partial s_{ir}}{\partial \sigma_{pq}}s_{rj} = s_{pj}\delta_{iq} - \frac{1}{3}s_{ij}\delta_{pq}; \quad \frac{\partial s_{rj}}{\partial \sigma_{pq}}s_{ir} = s_{pi}\delta_{jq} - \frac{1}{3}s_{ij}\delta_{pq}; \\ &\frac{\sqrt{3}}{2\cos 3\alpha}\bar{A}_{ij} = \frac{\partial \alpha}{\partial \sigma_{pq}} \end{aligned} \tag{A8.28}$$

L^{30}_{ijpq} can be evaluated from (A8.27).

Using the expressions for $\frac{\partial[\partial_I \phi]}{\partial \sigma_{pq}}$, $\frac{\partial[\partial_J \phi]}{\partial \sigma_{pq}}$ and $\frac{\partial[\partial_\alpha \phi]}{\partial \sigma_{pq}}$ from (A8.22), (A8.23), and (A8.24), L^1_{ijpq}, L^2_{ijpq}, and L^3_{ijpq} can now be evaluated. $\frac{\partial r_{ij}}{\partial \sigma_{pq}}$ is then calculated from A8.25a.

Also note that from (A8.8)

$$r_{kk} = 3\frac{\partial \phi}{\partial I} \tag{A8.29a}$$

$$\frac{\partial r_{kk}}{\partial \sigma_{pq}} = 3\frac{\partial[\partial_I \phi]}{\partial \sigma_{pq}} = 6\delta_{pq} \tag{A8.29b}$$

Furthermore, from (A8.8), we have

$$\begin{aligned} \frac{\partial r_{ij}}{\partial p_0} &= \frac{\partial[\partial_I \Phi]}{\partial p_0}\delta_{ij} + \frac{\partial[\partial_J \Phi]}{\partial p_0}\frac{s_{ij}}{2J} + \frac{\partial[\partial_\alpha \Phi]}{\partial p_0}\frac{\sqrt{3}}{2J\cos 3\alpha}\left[\frac{s_{ir}s_{rj}}{J^2} - \frac{2}{3}\delta_{ij} - \frac{3}{2}\left(\frac{S}{J}\right)^3\frac{s_{ij}}{J}\right] \\ &= -3\delta_{ij} \end{aligned} \tag{A8.29c}$$

Defining the deviatoric part of r_{ij} as

$$d_{ij} = r_{ij} - \frac{1}{3} r_{kk} \delta_{ij} \tag{A8.29d}$$

it follows that

$$\frac{\partial d_{ij}}{\partial \sigma_{pq}} = \frac{\partial r_{ij}}{\partial \sigma_{pq}} - 2\delta_{pq}\delta_{ij} \tag{A8.29e}$$

$$\frac{\partial d_{ij}}{\partial p_0} = -3\delta_{ij} + 3\delta_{ij} = 0 \tag{A8.29f}$$

Appendix 8.3. Equations Associated with the Hardening Rule

The Cam-clay model employs the following isotropic hardening rule:

$$\dot{\varsigma} = \dot{p}_0 = C_1 p_0 \dot{\varepsilon}_v^P = C_1 p_0 r_{kk} \dot{\lambda} = s\dot{\lambda} \tag{A8.30a}$$

where

$$C_1 = \frac{1 + e_0}{\lambda^* - \kappa} \quad \text{is a constant and} \quad s = C_1 p_0 r_{kk}. \tag{A8.30b}$$

The relevant gradients of s are

$$\frac{\partial s}{\partial \sigma_{pq}} = C_1 p_0 \frac{\partial r_{kk}}{\partial \sigma_{pq}} = 3C_1 p_0 \frac{\partial [\partial_I \phi]}{\partial \sigma_{pq}} = 6C_1 p_0 \delta_{pq} \quad \text{(using A8.29b)} \tag{A8.31}$$

$$\frac{\partial s}{\partial \varsigma} = \frac{\partial s}{\partial p_0} = C_1 r_{kk} + C_1 p_0 \frac{\partial r_{kk}}{\partial p_0} = 3C_1 [\partial_I \phi] - 9C_1 p_0 \quad \text{(usingA8.29a} - \text{c)} \tag{A8.32}$$

Appendix 8.4. Exact Elastic Predictor for the Cam-Clay Elastic Relations

Splitting the elastic constitutive relations into volumetric and deviatoric parts, we have

$$\dot{p} = K\dot{\varepsilon}_{kk} \tag{A8.32a}$$

and

$$\dot{s}_{ij} = 2G\dot{e}_{ij} \tag{A8.32b}$$

where

$$\dot{s}_{ij} = \dot{\sigma}_{ij} - \dot{p}\delta_{ij} \tag{A8.33a}$$

and

$$\dot{e}_{ij} = \dot{\varepsilon}_{ij} - \frac{1}{3}\dot{\varepsilon}_{kk}\delta_{ij} \tag{A8.33b}$$

Let us write the K vs. p and G vs. p relationships as

$$K = \bar{K}_0 p \quad \text{and} \quad G = \bar{G}_0 p \tag{A8.34a}$$

where

$$\bar{K}_0 = \frac{(1+e_0)}{\kappa} \quad \text{and} \quad \bar{G}_0 = \frac{3(1-2\nu)}{2(1+\nu)}\bar{K}_0 \tag{A8.34b}$$

Let us define

$$\delta\varepsilon_{ij} = x\Delta\varepsilon_{ij}, \quad x \in [0,1] \tag{A8.35}$$

with $x = 0$ at the beginning of the increment $x = 1$ at the end of the increment. $\Delta\varepsilon_{ij}$ is the total strain increment, which is a constant.

Combining (A8.32a) and (A8.34a) and integrating the rate equations, we have

$$\int_{p^n}^{p} \frac{\dot{p}}{p} = \bar{K}_0 \int_{0}^{\delta\varepsilon_{kk}} \dot{\varepsilon}_{kk} \tag{A8.36a}$$

$$p = p^n \exp[\bar{K}_0 \delta\varepsilon_{kk}] \quad \text{or} \quad p = p^n \exp[\bar{K}_0 x \Delta\varepsilon_{kk}] \tag{A8.36b}$$

Noting that $\delta e_{ij} = x\Delta e_{ij}$, it follows that

$$\dot{e}_{ij} = \dot{x}\Delta e_{ij} \tag{A8.37a}$$

Equation (A.8.32b) is written as

$$\dot{s}_{ij} = 2G\dot{e}_{ij} = 2\bar{G}_0 p \Delta e_{ij} \dot{x} = 2\bar{G}_0 p^n e^{x\bar{K}\Delta\varepsilon_{kk}} \Delta e_{ij} \dot{x} \quad \text{(using A8.36b)} \tag{A8.37b}$$

Integrating this equation, we get

$$\int_{s_{ij}^n}^{s_{ij}} \dot{s}_{ij} = \int_0^x 2\bar{G}_0 p^n e^{x\bar{K}_0\Delta\varepsilon_{kk}} \Delta e_{ij} \dot{x}$$

$$s_{ij} = s_{ij}^n + \frac{2\bar{G}_0 p^n \left[e^{x\bar{K}_0\Delta\varepsilon_{kk}} - 1 \right]}{\bar{K}_0 \Delta\varepsilon_{kk}} \Delta e_{ij} \tag{A8.38}$$

The stresses at the end of the increment are obtained from (A8.36b) and (A8.38) for $x = 1$ as

$$p^{n+1} = p^n \exp[\bar{K}_0 \Delta\varepsilon_{kk}] \tag{A8.39a}$$

$$s_{ij}^{n+1} = s_{ij}^n + \frac{2\bar{G}_0 p^n \left[e^{\bar{K}_0\Delta\varepsilon_{kk}} - 1 \right]}{\bar{K}_0 \Delta\varepsilon_{kk}} \Delta e_{ij} = s_{ij}^n + 2G_{\text{ave}} \Delta e_{ij} \tag{A8.39b}$$

Note that when $\Delta\varepsilon_{kk} = 0$, (A8.39b) is not valid since G_{ave} goes to ∞. Applying L'Hospital's rule, it can be shown that

$$G_{\text{ave}} \Rightarrow \bar{G}_0 p^n = G_n \quad \text{as} \quad \Delta\varepsilon_{kk} \Rightarrow 0 \tag{A8.40}$$

This must be recognized and dealt with in computer implementations. For example, it could be handled as follows:

$$G_{\text{ave}} = \frac{\bar{G}_0 p^n \left[e^{\bar{K}_0\Delta\varepsilon_{kk}} - 1 \right]}{\bar{K}_0 \Delta\varepsilon_{kk}} \quad \text{for} \quad \Delta\varepsilon_{kk} \geq TOL \tag{A8.41a}$$

and

$$G_{\text{ave}} = \frac{\bar{G}_0 p^n \left[e^{\bar{K}_0(TOL)} - 1 \right]}{\bar{K}_0 (TOL)} \quad \text{for} \quad \Delta\varepsilon_{kk} < TOL \tag{A8.41b}$$

where TOL is a suitable small number

The stress tensor at the end of the increment is obtained as

$$\sigma_{ij}^{n+1} = s_{ij}^{n+1} + p^{n+1} \delta_{ij} \tag{A8.42}$$

Appendix 9. Parameters for Plane Strain Analysis

In this section, we derive expressions for the model parameters that control the strength of materials that is suitable for use in two-dimensional plane strain analyses. Drucker and Prager (1952) had done this for the Drucker–Prager model. We will summarize their work. In addition, we extend their work to the von Mises (Chap. 11) and Cam-clay (Chap. 12) models. We derive the equations on the basis of the Tresca failure criterion (Chap. 9) for the von Mises model, and the Mohr–Coulomb failure criterion (Chap. 9) for the Cam-clay and Drucker–Prager models.

In plane strain, we have

$$\dot{\varepsilon}_{13} = \dot{\varepsilon}_{23} = \dot{\varepsilon}_{33} = 0 \tag{A9.1}$$

It is shown in Chap. 9 that when the material experiences flow, the rate of change of stress is zero; i.e., $\dot{\boldsymbol{\sigma}} = 0$. This implies that the rate of change of elastic strain is zero; i.e., $\dot{\varepsilon}^{\mathrm{e}}_{ij} = 0$. It then follows that

$$\dot{\varepsilon}_{ij} = \dot{\varepsilon}^{\mathrm{p}}_{ij} \tag{A9.2}$$

where $\varepsilon^{\mathrm{p}}_{ij}$ is the plastic strain. Hence (A9.1) becomes

$$\dot{\varepsilon}^{\mathrm{p}}_{13} = \dot{\varepsilon}^{\mathrm{p}}_{23} = \dot{\varepsilon}^{\mathrm{p}}_{33} = 0 \tag{A9.3}$$

The plastic strain rate is expressed as (Chap. 9)

$$\dot{\varepsilon}^{\mathrm{p}}_{ij} = \dot{\lambda} r_{ij} \tag{A9.4}$$

where $\dot{\lambda}$ is the consistency parameter (or loading index) and r_{ij} is the direction of plastic strain rate. r_{ij} depends on the specific constitutive model employed. Combining (A9.3) and (A9.4), we get

$$r_{13} = r_{23} = r_{33} = 0 \tag{A9.5}$$

Referring to Fig. A9.1a, for the metal forging problem (Chap. 11) or the bearing capacity problem (Chaps. 12 and 13), let us assume that the failure takes place by plastic flow at points along a failure surface such as the one shown. The Mohr circle for the stresses acting on an element on the failure surface (Point C, Fig. A9.1b) is shown in Fig. A9.1c. Denoting the distance to the center as x_0 and the radius as r_0, it is seen that

$$x_0 = \frac{1}{2}(\sigma_{11} + \sigma_{22}) \tag{A9.6a}$$

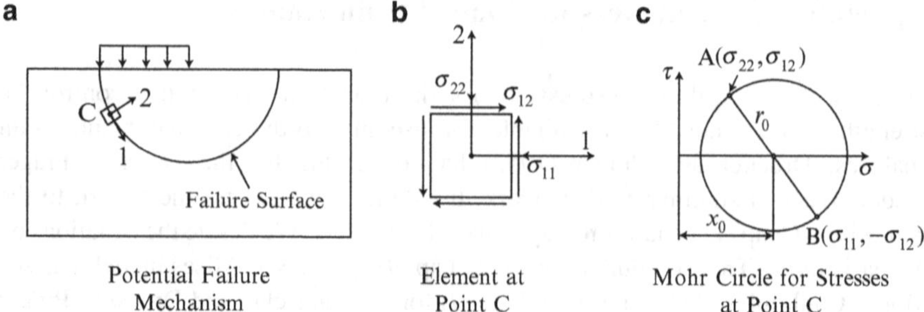

Fig. A9.1 Potential failure mechanism for the metal forging/bearing capacity problem

$$r_0 = \left[\left(\frac{\sigma_{11} - \sigma_{22}}{2}\right)^2 + \sigma_{12}^2\right]^{1/2} \tag{A9.6b}$$

We now consider three models (the von Mises model, the Cam-clay model, and the Drucker-Prager model) and develop equations for the failure parameters for use in plane strain analyses.

Appendix 9.1 The von Mises Model (Chap. 11)

For the von Mises model, we have (11.8b)

$$r_{ij} = 2s_{ij} \tag{A9.7}$$

Hence from (A9.5), we have

$$s_{13} = s_{23} = s_{33} = 0 \tag{A9.8}$$

As $s_{ij} = \sigma_{ij}$ for $i \neq j$, it follows that $\sigma_{13} = \sigma_{23} = 0$. From the third condition in (A9.8), we get

$$s_{33} = \sigma_{33} - \frac{I}{3} = \sigma_{33} - \frac{1}{3}(\sigma_{11} + \sigma_{22} + \sigma_{33}) = 0$$

$$\sigma_{33} = \frac{1}{2}(\sigma_{11} + \sigma_{22})$$

$$I = 3\sigma_{33} = \frac{3}{2}(\sigma_{11} + \sigma_{22}) \tag{A9.9a}$$

Similarly, noting (A9.8), it follows that

$$
\begin{aligned}
2J^2 = s_{ij}s_{ij} &= s_{11}^2 + s_{22}^2 + 2s_{12}^2 \\
&= \left(\sigma_{11} - \frac{1}{2}(\sigma_{11} + \sigma_{22})\right)^2 + \left(\sigma_{22} - \frac{1}{2}(\sigma_{11} + \sigma_{22})\right)^2 + 2\sigma_{12}^2 \\
&= 2\left[\left(\frac{\sigma_{11} - \sigma_{22}}{2}\right)^2 + \sigma_{12}^2\right]
\end{aligned}
\tag{A9.9b}
$$

$$
J = \left[\left(\frac{\sigma_{11} - \sigma_{22}}{2}\right)^2 + \sigma_{12}^2\right]^{1/2}
$$

With the aid of (A9.6a) and (A9.6b), (A9.9a) and (A9.9b) become (Fig. A9.1b, c):

$$
I = 3x_0 \tag{A9.10a}
$$

$$
J = r_0 \tag{A9.10b}
$$

Hence the maximum shear stress at point C is J. Let us match this with the maximum shear stress in a uniaxial specimen at failure as

$$
J = \frac{\sigma_f}{2} \tag{A9.11}
$$

This is tantamount to using the Tresca failure criterion (see the remark below). From the equation of the von Mises failure surface (11.3), we have

$$
s_{ij}s_{ij} = \frac{2}{3}k^2 \quad \Rightarrow \quad J = \frac{k}{\sqrt{3}} \tag{A9.12}
$$

Combining (A9.10a), (A9.11), and (A9.12):

$$
k = \frac{\sqrt{3}}{2}\sigma_{\mathrm{f}} = 0.866\sigma_{\mathrm{f}} \tag{A9.13}
$$

Noting that for 3D analyses, (9.58b) implies that $k = \sigma_{\mathrm{f}}$, we summarize the results as follows:

$$
\text{Plane strain: } k = k_{\mathrm{PS}} = \frac{\sqrt{3}}{2}\sigma_{\mathrm{f}} = 0.866\sigma_{\mathrm{f}} \tag{A9.14a}
$$

$$
\text{Three dimensions: } k = k_{\mathrm{3D}} = \sigma_{\mathrm{f}} \tag{A9.14b}
$$

Remarks

1. Matching the Tresca and von Mises Criteria

The Tresca failure criterion is on the basis of the assumption that failure occurs when the maximum shear stress reaches a limit. The criterion is analyzed in Chap. 9. Referring to Fig. 9.27b, the side of the failure surface in sector OAB is given by (9.69)

$$y = -\frac{1}{\sqrt{3}}x + \sqrt{\frac{2}{3}}k$$

Considering the state of stresses in Appendix 9.1 (refer to (A9.9b)) and defining $\bar{q} = \sigma_{11} - \sigma_{22}$, we have

$$s_{11} = \bar{q}/2, \quad s_{22} = -\bar{q}/2 \quad \text{and} \quad s_{12} = \sigma_{12}$$

$$S^3 = \left(\frac{1}{3}s_{ij}s_{jk}s_{ki}\right) = \frac{1}{3}\left(s_{11}^3 + 3s_{11}s_{12}^2 + 3s_{22}s_{12}^2 + s_{22}^3\right) = 0$$

From the equation for the Lode angle in Chapter 2, it follows that the Lode angle $\alpha = 0$. Let the size of the failure surface at $\alpha = 0$ be ℓ. Then $x = \ell\cos 60^\circ = \ell/2$; $y = \ell\sin 60^\circ = \ell\sqrt{3}/2$. Substituting these into the equation of the failure surface, we have

$$\frac{\ell\sqrt{3}}{2} = -\frac{1}{\sqrt{3}}\frac{\ell}{2} + \sqrt{\frac{2}{3}}k \Rightarrow \ell = \frac{k}{\sqrt{2}}$$

Referring to Fig. 9.30, we have

$$OA = \sqrt{\frac{2}{3}}k$$

$$\frac{\ell}{OA} = \frac{1}{\sqrt{2}}\sqrt{\frac{3}{2}} = 0.8666 \Rightarrow \ell = 0.8666(OA)$$

This agrees with (A9.14a).

2. Geotechnical Engineering Applications

In geotechnical engineering applications, the von Mises model is sometimes used to analyze cohesive soils with the assumption of zero friction. In this case, the radius of the Mohr circle (Fig. A9.1c) is the cohesion c. Using $J = c$ ((A9.10b)), (A9.12) leads to $k = c\sqrt{3}$. Hence we have the following:

$$\text{Plane strain: } k = k_{PS} = c\sqrt{3} = 1.732c \tag{A9.15a}$$

$$\text{Three dimensions: } k = k_{3\mathrm{D}} = 2c \tag{A9.15b}$$

Appendix 9.2 The Modified Cam-Clay Model (Chap. 12)

We will only consider the "ultimate" failure, which always occur on the critical state line. The failure points lie on the critical state line, where during flow, $r_{kk} = 0$. By the Mohr-Coulomb failure criterion, the shear strength can be represented by

$$\tau_\mathrm{f} = \sigma \tan \phi \tag{A9.16}$$

For the modified Cam-clay model, assuming a circular failure surface in the octahedral plane (12.10), we have

$$r_{ij} = \frac{\partial \phi}{\partial I}\delta_{ij} + \frac{\partial \phi}{\partial J}\frac{s_{ij}}{2J}$$

When flow occurs, $r_{kk} = 0$ and hence (making use of A8.10)

$$r_{ij} = \frac{\partial \phi}{\partial J}\frac{s_{ij}}{2J} = \frac{s_{ij}}{N^2}$$

For plane strain, as in the case with the von Mises model, (A9.5) becomes

$$s_{13} = s_{23} = s_{33} = 0$$

which leads to identical expressions for I and J as in (A9.10a) and (A9.10b); i.e.,

$$I = 3x_0 \tag{A9.16a}$$

$$J = r_0 \tag{A9.16b}$$

Now we use the Mohr-Coulomb criterion to find the plane strain parameters. During the footing failure, slip occurs along the failure surface and hence the plane parallel to the failure surface (the plane 2) becomes the failure plane. The failure envelop must pass through Point A in Fig. A9.1c. As cohesion is zero for the case considered here, the failure envelop also must pass through the origin in Fig. A9.1c. Hence, it is easy to see

$$r_0 = x_0 \sin \phi \tag{A9.17}$$

Combining (A9.16a), (A9.16b), and (A9.17), we have

$$N = \left(\frac{J}{I}\right)_{\text{CSL}} = \frac{r_0}{3x_0} = \frac{\sin\phi}{3}$$

and

$$M = 3\sqrt{3}N = \sqrt{3}\sin\phi$$

Summarizing the results, we have the following:

$$\text{Plane strain}: M_{\text{PS}} = \sqrt{3}\sin\phi \qquad \text{(A9.18a)}$$

$$\text{Three dimensions}: M_{\text{c,3D}} = \frac{6\sin\phi}{3-\sin\phi}; \quad M_{\text{e,3D}} = \frac{6\sin\phi}{3+\sin\phi} \qquad \text{(A9.18b)}$$

where $M_{c,\text{3D}}$ and $M_{e,3D}$ values of M to be used in triaxial compression and extension respectively in 3D analyses (according to the Mohr-Coulomb criterion).

For example, for $\phi = 30°$, $M_{PS} = 0.866$, $M_{c,3D} = 1.2$, and $M_{e,3D} = 0.857 = 0.714M_{c,3D}$.

Remarks

Equation (A9.18a) can also be derived in a different way as follows. Considering the state of stresses in Appendix 9.1 and defining $\bar{q} = \sigma_{11} - \sigma_{22}$, we have

$$s_{11} = \bar{q}/2, \quad s_{22} = -\bar{q}/2 \text{ and } s_{12} = \sigma_{12}$$

$$S^3 = \left(\frac{1}{3}s_{ij}s_{jk}s_{ki}\right) = \frac{1}{3}\left(s_{11}^3 + 3s_{11}s_{12}^2 + 3s_{22}s_{12}^2 + s_{22}^3\right) = 0$$

From the equation for the Lode angle in Chap. 2, it follows that the Lode angle $\alpha = 0$. Consider (9.77) for the Mohr-Coulomb failure surface in the Octahedral plane (Fig. 9.30)

$$y = -\frac{\sqrt{3}}{(1+2k_A)}x + \frac{\sqrt{6}(1-k_A)p + 2\sqrt{6}c\sqrt{k_A}}{1+2k_A}$$

Let the size of the failure surface at $\alpha = 0$ be ℓ. Then $x = \ell\cos 60° = \ell/2$; $y = \ell\sin 60° = \ell\sqrt{3}/2$. Substituting these into the above equation, we have

$$\frac{\ell\sqrt{3}}{2} = -\frac{\sqrt{3}}{(1+2k_A)}\frac{\ell}{2} + \frac{\sqrt{6}(1-k_A)p + 2\sqrt{6}c\sqrt{k_A}}{1+2k_A}$$

$$\ell\sqrt{3}\frac{1+k_A}{1+2k_A} = \frac{\sqrt{6}(1-k_A)p + 2\sqrt{6}c\sqrt{k_A}}{1+2k_A} \Rightarrow \ell = \frac{1}{\sqrt{3}}\frac{\sqrt{6}(1-k_A)p + 2\sqrt{6}c\sqrt{k_A}}{1+k_A}$$

Referring to Fig. 9.30, we have from (9.78)

$$OA = \frac{\sqrt{6}(1 - k_A)p + 2\sqrt{6}c\sqrt{k_A}}{1 + 2k_A}$$

$$\frac{M_{PS}}{M_c} = \frac{\ell}{OA} = \frac{1}{\sqrt{3}}\frac{1 + 2k_A}{1 + k_A} = \frac{1}{\sqrt{3}}\frac{1 + 2\frac{1-\sin\phi}{1+\sin\phi}}{1 + \frac{1-\sin\phi}{1+\sin\phi}} = \frac{3 - \sin\phi}{2\sqrt{3}}$$

$$M_{PS} = \frac{3 - \sin\phi}{2\sqrt{3}}\frac{6\sin\phi}{3 - \sin\phi} = \sqrt{3}\sin\phi$$

The above equation agrees with (A9.18a). This suggests that if the three-dimensional failure surface in the octahedral plane can match the Mohr-Coulomb failure surface (i.e., the irregular hexagonal cone shown in Fig. 9.30), then the three-dimensional failure criterion may be directly used in two-dimensional analyses.

Appendix 9.3 The Drucker–Prager Model (Chap. 13)

The Drucker–Prager model may be used for a general $c - \phi$ material. By the Mohr–Coulomb failure criterion, the shear strength can be represented by

$$\tau_f = c + \sigma\tan\phi \tag{A9.19}$$

For the Drucker−Prager model model, assuming a circular failure surface in the octahedral plane (13.39a)

$$r_{ij} = \frac{\partial\psi}{\partial I}\delta_{ij} + \frac{\partial\psi}{\partial J}\frac{s_{ij}}{2J}$$

Noting that the hardening parameter $m = 1$ at failure, (13.39a) becomes

$$r_{ij} = -d_0\alpha\delta_{ij} + \frac{s_{ij}}{2J}$$

For plane strain, (A9.5) becomes

$$s_{13} = s_{23} = 0$$

$$s_{33} = 2d_0\alpha J \tag{A9.20}$$

Using $I = \sigma_{11} + \sigma_{22} + \sigma_{33} \Rightarrow \sigma_{33} = I - (\sigma_{11} + \sigma_{22})$

Equation (A9.20) is expressed as

$$s_{33} = \sigma_{33} - \frac{I}{3} = I - \sigma_{11} - \sigma_{22} - \frac{I}{3} = 2d_0\alpha J$$

$$I = 3\left[\frac{1}{2}(\sigma_{11} + \sigma_{22}) + \mathrm{d}_0\alpha J\right] = 3x_0 + 3d_0\alpha J \tag{A9.21}$$

where x_0 is defined in (A9.6a). Similarly, J is expressed as

$$2J^2 = s_{11}^2 + s_{22}^2 + s_{33}^2 + 2s_{12}^2$$

$$2J^2 = \left[\sigma_{11} - \frac{\sigma_{11} + \sigma_{22}}{2} - d_0\alpha J\right]^2 + \left[\sigma_{22} - \frac{\sigma_{11} + \sigma_{22}}{2} - d_0\alpha J\right]^2 + 4d_0^2\alpha^2J^2 + 2\sigma_{12}^2$$

$$2J^2 = \left[\frac{\sigma_{11} - \sigma_{22}}{2} - d_0\alpha J\right]^2 + \left[-\frac{\sigma_{11} - \sigma_{22}}{2} - d_0\alpha J\right]^2 + 4d_0^2\alpha^2J^2 + 2\sigma_{12}^2$$

$$J^2 = \left[\left(\frac{\sigma_{11} - \sigma_{22}}{2}\right)^2 + \sigma_{12}^2\right] + 3d_0^2\alpha^2J^2$$

$$J = \frac{r_0}{\left(1 - 3d_0^2\alpha^2\right)^{1/2}} \tag{A9.22}$$

where r_0 is defined by (A9.6b). At failure, the stress point must lie on the Drucker–Prager failure surface, which is given by

$$J = \alpha I + k$$

Substituting for I from (A9.21), we have

$$J = \alpha(3x_0 + 3d_0\alpha J) + k$$

$$J\left(1 - 3d_0\alpha^2\right) = 3\alpha x_0 + k$$

Substituting for J from (A9.22), we have

$$\frac{r_0}{\left(1 - 3d_0^2\alpha^2\right)^{1/2}}\left(1 - 3d_0\alpha^2\right) = 3\alpha x_0 + k$$

$$r_0 = \frac{3\alpha\left(1 - 3d_0^2\alpha^2\right)^{1/2}}{\left(1 - 3d_0\alpha^2\right)}x_0 + \frac{k\left(1 - 3d_0^2\alpha^2\right)^{1/2}}{\left(1 - 3d_0\alpha^2\right)} \tag{A9.23}$$

According to the Mohr-Coulomb criterion, it is easy to show that for a $c - \phi$ material

$$r_0 = x_0 \sin\phi + c\cos\phi \tag{A9.24}$$

Comparing (A9.23) and (A9.24), it follows that

$$\frac{3\alpha\left(1 - 3d_0^2\alpha^2\right)^{1/2}}{(1 - 3d_0\alpha^2)} = \sin\phi \tag{A9.25a}$$

$$\frac{k\left(1 - 3d_0^2\alpha^2\right)^{1/2}}{(1 - 3d_0\alpha^2)} = c\cos\phi \tag{A9.25b}$$

Case A: Associated Flow Rule $(d_0 = 1)$

For this case, (A9.25a) becomes

$$\frac{3\alpha}{(1 - 3d_0\alpha^2)^{1/2}} = \sin\phi$$

$$9\alpha^2 = \left(1 - 3d_0\alpha^2\right)\sin^2\phi$$

$$\alpha = \frac{\sin\phi}{\sqrt{3}(3 + \sin^2\phi)^{1/2}}$$

From the relation $\sin^2\phi = \frac{\tan^2\phi}{1+\tan^2\phi}$, it is easy to show that the above equation becomes

$$\alpha = \frac{\tan\phi}{\left(9 + 12\tan^2\phi\right)^{1/2}} \tag{A9.26a}$$

From (A9.25b), we have

$$\frac{k}{\left(1 - 3\alpha^2\right)^{1/2}} = c\cos\phi$$

Combining this with (A9.26a), it is easy to show

$$k = \frac{3c}{\left(9 + 12\tan^2\phi\right)^{1/2}} \tag{A9.26b}$$

Equations (A9.26a) and (A9.26b) are the original relations derived by Drucker and Prager (1952).

Summarizing the results, we have

Plane strain:

$$\alpha_{PS} = \frac{\tan\phi}{(9 + 12\tan^2\phi)^{1/2}}; \quad k_{PS} = \frac{3c}{(9 + 12\tan^2\phi)^{1/2}} \tag{A9.27}$$

Three dimensions:

$$\alpha_{c,3D} = \frac{2\sin\phi}{\sqrt{3}(3 - \sin\phi)}; \quad k_{c,3D} = \frac{6c\cos\phi}{\sqrt{3}(3 - \sin\phi)} \tag{A9.28a}$$

$$\alpha_{e,3D} = \frac{2\sin\phi}{\sqrt{3}(3 + \sin\phi)}; \quad k_{e,3D} = \frac{6c\cos\phi}{\sqrt{3}(3 + \sin\phi)} \tag{A9.28b}$$

where $\alpha_{c,3D}$ and $k_{c,3D}$ are parameters in triaxial compression and $\alpha_{e,3D}$ and $k_{e,3D}$ are parameters in triaxial extension for use in 3D analyses.

For example, for $\phi = 30°$, $\alpha_{PS} = 0.16$, $k_{PS} = 0.83c$, $\alpha_{c,3D} = 0.23$, $k_{c,3D} = 1.2c$, $\alpha_{e,3D} = 0.16$, and $k_{c,3D} = 186c$.

Case B: Non-Associated Flow Rule ($d_0 \neq 1$)

For this case, squaring both sides of (A9.25a), one gets the following equation for α:

$$a_1\alpha^4 + a_2\alpha^2 + a_3 = 0 \tag{A9.29}$$

where

$$a_1 = 9d_0^2(3 + \sin^2\phi); \quad a_2 = -(9 + 6d_0\sin^2\phi) \text{ and } a_3 = \sin^2\phi;$$

Equation (A9.29) can be solved to evaluate α for a given value of ϕ. From (A9.25b), we have

$$k = \frac{(1 - 3d_0\alpha^2)}{(1 - 3d_0^2\alpha^2)^{1/2}} c\cos\phi \tag{A9.30}$$

Once α is found from (A9.29), the value is substituted into (A9.30) to find the value for k.

For example, consider the case for $\phi = 30°$.

For $d_0 = 1$: $a_1 = 29.25$, $a_2 = -10.5$, and $a_3 = 0.25 \Rightarrow \alpha = 0.1601$, which agrees with that calculated earlier using the analytical (A9.27).

For $d_0 = 0.5$: $a_1 = 7.31$, $a_2 = -9.75$, and $a_3 = 0.25 \Rightarrow \alpha = 0.1617$, which is very close to that calculated with $d_0 = 1$

Appendix 10. Exact Elastic Predictor for the Exponential Elastic Relation

Splitting elastic constitutive relations into volumetric and deviatoric parts, we have

$$\dot{p} = K\dot{\varepsilon}_{kk} \tag{A10.1a}$$

and

$$\dot{s}_{ij} = 2G\dot{e}_{ij} \tag{A10.1b}$$

where

$$\dot{s}_{ij} = \dot{\sigma}_{ij} - \dot{p}\delta_{ij} \tag{A10.2a}$$

and

$$\dot{e}_{ij} = \dot{\varepsilon}_{ij} - \frac{1}{3}\dot{\varepsilon}_{kk}\delta_{ij} \tag{A10.2b}$$

Let us write the G vs. p and K vs. p relationships as

$$G = \bar{G}_0(p + p_{\mathrm{t}})^{n_e}; \quad K = \bar{K}_0(p + p_{\mathrm{t}})^{n_e} \tag{A10.3a}$$

where

$$\bar{G}_0 = G_0 p_{\mathrm{a}} \frac{(2.97 - e_0)^2}{1 + e_0} \quad \text{and} \quad \bar{K}_0 = \frac{2(1 + v)}{3(1 - 2v)}\bar{G}_0 \tag{A10.3b}$$

p_t is the tension cutoff.

Let us define

$$\delta\varepsilon_{ij} = x\Delta\varepsilon_{ij}, \quad x \in [0, 1] \tag{A10.4}$$

with $x = 0$ at the beginning of the increment and $x = 1$ at the end of the increment. $\Delta\varepsilon_{ij}$ is the total strain increment, which is a constant.

Combining (A10.1a) and (A10.3a) and integrating the rate equations, we have

$$\int_{p^n}^{p} \frac{\dot{p}}{(p + p_{\mathrm{t}})^{n_e}} = \bar{K}_0 \int_0^{\delta\varepsilon_{kk}} \dot{\varepsilon}_{kk}$$

$$(p + p_{\mathrm{t}})^{1-n_e} - (p_n + p_{\mathrm{t}})^{1-n_e} = \bar{K}_0(1 - n_e)\delta\varepsilon_{pp}$$

$$p = \left[(p_n + p_{\mathrm{t}})^{1-n_e} + \bar{K}_0(1 - n_e)\delta\varepsilon_{pp}\right]^{1/(1-n_e)} - p_{\mathrm{t}}$$

$$p = \left[(p_n + p_t)^{1-n_e} + \bar{K}_0(1 - n_e)x\Delta\varepsilon_{pp} \right]^{1/(1-n_e)} - p_t \quad \text{(A10.5)}$$

When $x = 1$, the above equation gives the elastic predictor ("trial" stress) as

$$p^{tr} = \left[(p_n + p_t)^{1-n_e} + \bar{K}_0(1 - n_e)\Delta\varepsilon_{pp} \right]^{1/(1-n_e)} - p_t \quad \text{(A10.6)}$$

Noting that $\delta e_{ij} = x\Delta e_{ij}$, we have

$$\dot{e}_{ij} = \dot{x}\Delta e_{ij}$$

Using the above equation and (A10.5), the deviatoric stress–strain rate equation becomes

$$\dot{s}_{ij} = 2G\dot{e}_{ij} = 2\bar{G}_0(p + p_t)^{n_e}\Delta e_{ij}\dot{x}$$

$$\dot{s}_{ij} = 2G\dot{e}_{ij} = 2\bar{G}_0\left[(p_n + p_t)^{1-n_e} + \bar{K}_0(1 - n_e)x\Delta\varepsilon_{pp} \right]^{n_e/(1-n_e)} \Delta e_{ij}\dot{x}$$

Integrating the above equation, we have

$$\int_{s_{ij}^n}^{s_{ij}} \dot{s}_{ij} = \int_0^1 2\bar{G}_0\left[(p_n + p_t)^{1-n_e} + \bar{K}_0(1 - n_e)x\Delta\varepsilon_{pp} \right]^{n_e/(1-n_e)} \Delta e_{ij}\dot{x}$$

The elastic predictor is computed by letting $x = 1$ as

$$s_{ij}^{tr} = s_{ij}^n + 2G_{ave}\Delta e_{ij},$$

where

$$G_{ave} = \frac{\bar{G}_0}{\bar{K}_0\Delta\varepsilon_{pp}}(p^{tr} - p_n) \quad \text{(A10.7)}$$

The stress tensor is obtained as

$$\sigma_{ij}^{tr} = s_{ij}^{tr} + p^{tr}\delta_{ij} \quad \text{(A10.8)}$$

Note that when $\Delta\varepsilon_{pp} = 0$, (A10.7) is not valid as G_{ave} goes to ∞. Let us find the limit for the case of $n_e = 0.5$.

$$p^{tr} = \left[(p_n + p_t)^{0.5} + 0.5\bar{K}_0\Delta\varepsilon_{pp}\right]^2 - p_t$$
$$= p_n + p_t + \bar{K}_0(p_n + p_t)^{0.5}\Delta\varepsilon_{pp} + 0.25\bar{K}_0^2\Delta\varepsilon_{pp}^2 - p_t$$

As $\Delta\varepsilon_{pp} \to 0$, we have

$$p^{tr} \to p_n + \bar{K}_0(p_n + p_t)^{0.5}\Delta\varepsilon_{pp}$$

$$G_{ave} = \frac{\bar{G}_0}{\bar{K}_0\Delta\varepsilon_{pp}}\left[\bar{K}_0(p_n + p_t)^{0.5}\Delta\varepsilon_{pp}\right] = \bar{G}_0(p_n + p_t)^{0.5} = G_n$$

This must be recognized and dealt with in computer implementations. For example, it could be handled as follows:

$$G_{ave} = \frac{\bar{G}_0}{\bar{K}_0\Delta\varepsilon_{pp}}(p^{tr} - p_n) \quad \text{for} \quad \Delta\varepsilon_{kk} \geq TOL$$

and

$$G_{ave} = \frac{\bar{G}_0}{\bar{K}_0(TOL)}(\bar{p}^{tr} - p_n) \quad \text{for} \quad \Delta\varepsilon_{kk} < TOL$$

where $\bar{p}^{tr} = \left[(p_n + p_t)^{1-n_e} + \bar{K}_0(1 - n_e)(TOL)\right]^{1/(1-n_e)} - p_t$ and TOL is a suitable small number.

Appendix 11. Computer Coding Guidelines for Computing Incremental Quantities for Von Mises Model Using the *S*–Space Formulation (Sect. 11.3.2)

C The formulation presented in Sect. 11.3.2 uses nonlinear hardening and the "Exact" C method of integrating the hardening variable.
C

C Definition of Variables:
C Control parameters
C NPROP: Number of model parameters (=5)
C NPIV: Number of PIVs (=6)
C IELAST = 1: Find elastic stiffness, = 0: Find elasto-plastic stiffness
C ISTIFF = 1: Find stiffness, = 0: Do not find stiffness
C ISTRES = 1: Find incremental stress, = 0: Do not find incremental stress

```
C   Property array (brought in from the main program)
C   PROP(1): E; PROP(2): ν; PROP(3): m; PROP(4): k_f; PROP(5): k_0
C   State variable array (brought in from the main program)
C   STOR(1): Δλ; STOR(2): L (loading index); STOR(3):k_n; STOR(4):k_{n+1};
C   STOR(5): ξ^p_n; STOR(6): ξ^p_{n+1}
C   Local variables
C   Properties
C   XM: m; XKF: k_f; XK0: k_0; XE: E; XNU: ν; XK: K_{n+1}; XG: G_{n+1}
C   Hardening variable
C   YSB: k_n; YSF: k_{n+1}; ZETAB: ξ^p_n; ZETAF: ξ^p_{n+1}
C   Stresses
C   SIGB(6): σ_n; SIGBX(3,3): σ^n_ij; SIGF(6): σ_{n+1}; SIGFX(3,3): σ^{n+1}_ij
C   DSIG(6): Δσ; DSIGX(3,3): Δσ_ij
C   SB(3,3): s^n_ij; SF(3,3): s^{n+1}_ij; STRI(3,3): s^tr_ij; RJB: J_n; RJF: J_{n+1}; RJTRI: J^tr
C   RIB: I_n; RIF: I_{n+1}; RITRI: I^tr
C   Strains
C   DEP(6): Δε; DEPX(3,3): Δε_ij
C   Other
C   DEL(3,3): δ_ij
C   DLAM: Δλ
C   C(6,6): D^t_{n+1}
C==================================================================
```

C ISTIFF = 0 and ISTRE = 1: Begin Calculations to find incremental
C stresses and the PIV

```
      Initialize XM, XKF, XK0, XE, XNU
      Initialize YSB, YSF = YSB, ZETAB, ZETAF = ZETAB
      Initialize SIGB(6), SIGBX(3,3)
      Initialize DEP(6), DEPX(3,3) (note: DEPX(1,2) = DEP(4)/2, etc.)
      Initialize DEL(3,3)
      XK = XE/(3.0*(1–2*XNU)), XG = XE/(2.0*(1 + XNU))
      Calculate: RIB, SB(3,3)
      Calculate RITRI, STRI(3,3), RJTRI
      Set SIGFX(3,3) = STRI(3,3) + RITRI*DEL(3,3)/3.0 and define SIGF(6)
      FTRI = 2.0*RJTRI**2–2*YSF**2/3
C Decide if the strain increment causes elastic or elasto-plastic response
      IF(FTRI.LE.0.0) THEN
C Elastic
      DSIG(6) = SIGF(6)-SIGB(6)
      STOR(3) = 0.0
      RETURN TO THE MAIN PROGRAM
      ENDIF
C Elasto-Plastic
      Calculate B1, B2 and B3
      Solve for DLAM = Δλ (choose positive of the roots)
```

```
      Calculate X0 = x0
      SF(3,3) = STRI(3,3)/X0
      SIGFX(3,3) = SF(3,3) + RITRI*DEL(3,3)/3
      Calculate DSIG(6)
      ZETAF = ZETAB + 2*DLAM*RJTRI/X0
      Calculate constants C1, A2 and B
      YSF = XK0 + C1*ZETAF/(A2 + B*ZETAF)
      STOR(1) = DLAM
      STOR(4) = YSF
      STOR(6) = ZETAF
      RETURN TO THE MAIN PROGRAM
C=========================================================
C ISTIFF = 1 and ISTRE = 0: Begin Calculations to find the tangent
C stiffness operator
C Determine if elastic or elasto-plastic stiffness is required
      IF(IELAST = 1 OR STOR(2) ≤ 0) THEN
C Find elastic stiffness
      Calculate elastic stiffness C(6,6)
      RETURN TO THE MAIN PROGRAM
      ENDIF
C
C Find elasto-plastic consistent tangent operator
      Initialize XM, XKF, XK0, XE, XNU
      Initialize YSF, ZETAF
      Initialize DLAM
      Initialize SIGB(6), SIGBX(3,3)
      Initialize DEP(6), DEPX(3,3) (note: DEPX(1,2) = DEP(4)/2, etc.)
      Initialize DEL(3,3)
      XK = XE/(3.0*(1–2*XNU)), XG = XE/(2.0*(1 + XNU))
      Calculate: RIB, SB(3,3)
      Calculate RITRI, STRI(3,3), RJTRI
      X0 = 1.0 + 4.0*XG*DLAM
C

Calculate B1, B2, B3, B1BAR, B2BAR, B3BAR, F1BAR
C Calculate consistent operator
C The following code is in Fortran; modify this to suit the language of your choice
C Define an array II(6) as
      DATA II/11,22,33,12,23,13/
      X1 = XK-2.0*XG/(3.0*X0)
      X2 = XG/X0
      F1 = 4.0*XG*F1BAR/(X0**2)
C
      DO 100 M = 1,6
      I = II(M)/10
      J = MOD(II(M),10)
```

```
      DO 100 N = 1,6
      K = II(N)/10
      L = MOD(II(N),10)
      C(M,N) = X1*DEL(I,J)*DEL(K,L)
      C(M,N) = C(M,N) + X2*(DEL(I,K)*DEL(J,L) + DEL(I,L)*DEL(J,K))
      C(M,N) = C(M,N)-F1*STRI(I,J)*STRI(K,L)
100 CONTINUE
C RETURN TO THE MAIN PROGRAM
C=========================================================
```

Appendix 12. Equations Related to the Sliding-Rolling Model

Appendix 12.1 The Surface Properties

Referring to Fig. 14.1, the equation of ϕ_1 is given by

$$\phi_1 = f_J - k^* I = 0 \tag{A12.1}$$

where

$$f_J = \left(\frac{1}{2} f_{k\ell} f_{k\ell}\right)^{1/2}; \quad f_{k\ell} = s_{k\ell} - I\alpha^{\eta}_{k\ell} \tag{A12.2a}$$

$$k^* = \frac{k}{3\sqrt{3}}(1 + 27\eta_{\alpha}^2)^{1/2}; \quad \eta_{\alpha}\left(\frac{1}{2}\alpha^{\eta}_{k\ell}\alpha^{\eta}_{k\ell}\right)^{1/2} \tag{A12.2b}$$

The relevant gradients of ϕ_1 are as follows:

$$\mathbf{n}^1 = \frac{\partial \phi_1}{\partial \boldsymbol{\sigma}} = \frac{\mathbf{f}}{2f_J} - \left[\frac{\mathbf{f} : \boldsymbol{\alpha}^{\eta}}{2f_J} + k^*\right]\boldsymbol{\delta} \tag{A12.3a}$$

$$\mathbf{P}_1 = \frac{\partial \phi_1}{\partial \boldsymbol{\alpha}^{\eta}} = -\left(\frac{I}{2f_J}\right)\mathbf{f} - \frac{3\sqrt{3}kI}{2(1 + 27\eta_{\alpha}^2)^{1/2}}\boldsymbol{\alpha}^{\eta} \tag{A12.3b}$$

$$\frac{\partial \phi_1}{\partial \boldsymbol{\delta}^{\mathrm{a}}} = \mathbf{0} \tag{A12.3c}$$

$$T_1 = \frac{\partial \phi_1}{\partial p_{\mathrm{o}}} = 0 \tag{A12.3d}$$

The equation of ϕ_2 is

$$\phi_2 = 9J\eta_\alpha + \frac{I}{3} - 27p_o(\eta_\alpha)^2 - p_o = 0 \tag{A12.4}$$

J is the deviatoric stress invariant (Chap. 2). The relevant gradients of ϕ_2 are as follows:

$$\mathbf{n}^2 = \frac{\partial \phi_2}{\partial \boldsymbol{\sigma}} = \frac{9\eta_\alpha}{2J}\mathbf{s} + \frac{1}{3}\boldsymbol{\delta} \tag{A12.5a}$$

$$\mathbf{P}_2 = \frac{\partial \phi_2}{\partial \boldsymbol{\alpha}^\eta} = \left[\frac{9J}{2\eta_\alpha} - 27p_o\right]\boldsymbol{\alpha}^\eta \tag{A12.5b}$$

$$\frac{\partial \phi_2}{\partial \boldsymbol{\delta}^a} = \mathbf{0} \tag{A12.5c}$$

$$T_2 = \frac{\partial \phi_2}{\partial p_o} = -27(\eta_\alpha)^2 - 1 \tag{A12.5d}$$

Appendix 12.2 The Definition of State Index

Ishihara (1993) defines a parameter known as the state index on the basis of two lines in the $e - p$ space, one of which is the standard critical state line (in the $e - p$ space) and the other known as the "upper reference line." The upper reference line represents the loosest possible states for the material. In the sliding-rolling model, the critical state line is given by

$$e = \Gamma_S - \lambda^* \left(\frac{p}{p_a}\right)^{n_p} \tag{A12.6a}$$

The upper reference line is given by

$$e = \Gamma_U - \lambda^* \left(\frac{p}{p_a}\right)^{n_p} \tag{A12.6b}$$

In the above equations, e is the current void ratio, p_a is the atmospheric pressure, and Γ_S, Γ_U, λ^*, and n_p are model parameters. It is noted that the two lines are taken to be parallel to each other in the $e - \ln p$ space. The state index is defined as

$$I_s = \frac{e_U - e}{e_U - e_S} \tag{A12.7}$$

where e_U and e_S are the values of the void ratio on the upper reference and critical state lines respectively at the current value of p. It then follows that $I_s = 0$ on the upper reference line, $I_s = 1$ on the critical state line, $0 < I_s < 1$ ("loose") in the region between the critical state and upper reference lines, and $I_s > 1$ ("dense") in the region below the critical state line.

Appendix 12.3 The Critical State Line in the $p - q$ *Space*

Denoting the slope of the critical state line with respect to the $\boldsymbol{\delta}^a$ axis as $M^{c\delta}$, it was deduced from the sliding-rolling theory that (Anandarajah 2008a)

$$M^{c\delta} = \frac{3(1 - S^{c\delta})}{1 + 2S^{c\delta}} \tag{A12.8a}$$

where

$$S^{c\delta} = \tan(45 - \phi_\mu) \tag{A12.8b}$$

ϕ_μ is the interparticle friction angle. The actual slope of the critical state line in the $p - q$ space (i.e., with respect to the $\boldsymbol{\delta}$–axis) depends on the orientation of $\boldsymbol{\delta}^a$ at the critical state. An index (called the anisotropic index) A_{CSL} is used for this purpose. Hence, the slope of the critical state line in the $p - q$ space M_{CSL} is a function of two model parameters $\phi_\mu$and A_{CSL}. It was shown (Anandarajah 2008a) that

$$M_{CSL} = M^{c\delta} + \frac{3(A_{CSL} - 1)}{(A_{CSL} + 2)} \tag{A12.8c}$$

It may be noted that the second term in the above equation is zero when $A_{CSL} = 1$.

Following Wood et al. (1994), $\bar{M}^\delta$ is related to $M^{c\delta}$ by

$$\bar{M}^\delta = M^{c\delta} f_A \exp\{c_\alpha (I_s - 1)\} \tag{A12.9a}$$

where

$$f_A = \exp(\delta^a_{J,CSL} - \delta^a_J); \quad \delta^a_{J,CSL} = \frac{A_{CSL} - 1}{(2 + A^2_{CSL})^{1/2}}; \quad \delta^a_J = \left(\frac{1}{2}\delta^{ad}_{ij}\delta^{ad}_{ij}\right)^{1/2} \tag{A12.9b}$$

When $I_s = 1$ (critical state) and $\delta_J^a = \delta_{J,CSL}^a$ (isotropy), we have $\bar{M}^\delta = M^{c\delta}$. When $I_s > 1$ (dense soil) and $\delta_J^a = \delta_{J,CSL}^a$, we have $\bar{M}^\delta > M^{c\delta}$. The peak strength is primarily controlled by the parameter c_α.

Appendix 12.4 The Plastic Strain Rate Directions

The direction $\mathbf{r}^1$ is defined as a linear combination of $\mathbf{n}^1$ and $\boldsymbol{\sigma}$ as

$$\mathbf{r}^1 = \mathbf{n}^1 + d\frac{\boldsymbol{\sigma}}{I} \tag{A12.10}$$

The parameter d depends on the stress ratio and void ratio. Similarly to (A12.9a), the following function is employed to account for the effect of void ratio:

$$d = \frac{d^c}{f_A}\exp\{-c_d(I_s - 1)I_s^4\} \tag{A12.11a}$$

where f_A is given by (A12.9b).

As the stress ratio is increased from zero, the plastic volumetric behavior of a granular material changes from contractive to dilative at a certain stress ratio. The radial line drawn at this stress ratio in the $p - q$ space is known as the phase transformation line (Ishihara et al. 1975; Habib and Luong 1978). The slope of the phase transformation line is a function of the initial density, and hence a function of I_s.

When $I_s = 1$ and $\delta_J^a = \delta_{J,CSL}^a$, we have from (A12.11a) that $d = d^c$. In this case, the phase transformation line coincides with the critical state line. When $I_s > 1$ and $\delta_J^a = \delta_{J,CSL}^a$, we have $d < d^c$. This lowers the slope of the phase transformation line from the slope of the critical state line. The behavior is controlled by the parameter c_d.

From the sliding-rolling theory, the parameter d^c is expressed as

$$d^c = d_0\left(1 - \frac{q}{pM_{CSL}}\right) + d_{CSL}\left(\frac{q}{pM_{CSL}}\right)^{0.95}; \quad d_0 = \frac{2\sqrt{3}\tan\phi_\mu}{3 + \tan\phi_\mu}; \quad d_{CSL} = \frac{M_{CSL}}{\sqrt{3}}; \tag{A12.11b}$$

M_{CSL} is the actual slope of the critical state line in the $p - q$ space, given by (A12.8c).

The flow direction $\mathbf{r}^2$ is taken to be normal to a modified ellipse. The ellipse is defined such that it is normal to δ_{ij}^a and its normal has a zero volumetric component on the CSL. The expression for $\mathbf{r}^2$ is

$$\mathbf{r}^2 = \frac{1}{IN^{c\delta}}\left(\frac{(N^\delta)^2\eta_1}{(N^\delta)^2+\eta_1^2}\right)\mathbf{s}^\delta + (N^\delta)^2\left(\frac{(N^\delta)^2-\eta_1^2}{(N^\delta)^2+\eta_1^2}\right)\boldsymbol{\delta}^a \quad \text{(A12.12a)}$$

$$\mathbf{s}^\delta = \mathbf{s} - \boldsymbol{\alpha}^\delta; \quad \eta_1 = \frac{1}{I}\left(\frac{1}{2}\mathbf{s}^\delta : \mathbf{s}^\delta\right)^{1/2}; \quad N^\delta = \frac{M^\delta}{3\sqrt{3}} \quad \text{(A12.12b)}$$

It is easily verified that on the δ^a_{ij}-axis, $\mathbf{s}^\delta = 0$, and hence $\mathbf{r}^2$ and $\boldsymbol{\delta}^a$ have the same direction. When $N^\delta = \eta_1$, $\mathbf{r}^2$ and $\mathbf{s}^\delta$ have the same direction (and hence $\mathbf{r}^2$ is normal to δ_{ij}).

It is pointed out that $M^\delta \neq \bar{M}^\delta$. The value of M^δ is chosen so as to ensure unique solution for points at the intersection of ϕ_1 and ϕ_2. The following equation is used for M^δ:

$$M^\delta = M^{c\delta} f_A \exp\{-c_s(I_s - 1)\} \quad \text{(A12.13)}$$

A value of $c_s = 0.75$ ensures a unique solution (Anandarajah 2008a).

Appendix 12.5 The Hardening Functions

Kinematic Hardening

Following Mroz et al. (1979), the image stress $\bar{\mathbf{s}}$ (Fig. 14.1) is found using a proportionality rule as

$$\frac{\bar{\mathbf{s}} - \alpha^\delta}{\bar{M}^\delta} = \frac{\mathbf{s} - \boldsymbol{\alpha}^S}{k^*} \rightarrow \bar{\mathbf{s}} = \boldsymbol{\alpha}^\delta + \frac{\bar{M}^\delta}{k^*}\left(\mathbf{s} - \alpha^S\right) \quad \text{(A12.14a)}$$

The direction $\bar{\mathrm{h}}$ is taken as

$$\bar{\mathbf{h}} = [(\bar{\mathbf{s}} - \mathbf{s})\eta_s + \mathbf{s}(1-\eta_s)]/p; \quad \eta_s = \frac{\delta_s}{\delta_{s_0}} \quad \text{(A12.14b)}$$

δ_s is the scalar distance between the current stress point ($\mathbf{s}$) and the image stress point ($\bar{\mathbf{s}}$) as shown in Fig. 14.1. δ_{s_0} is the maximum value of δ_s. Noting that the size of ϕ_3 varies with the density and Lode angle, δ_{s_0} is computed as $\delta_{s_0} = 2I\bar{N}^\delta$, where $\bar{N}^\delta$ is calculated at the current density and Lode angle.

The hardening function ξ_1 is defined as

$$\xi_1 = c_\ell \bar{r}_v^2 \left(\frac{p}{p_a}\right)^{1/2}\left(1 + \frac{f_p}{(q/p)^2}\right)\left(\frac{g_{r1}}{g_{r2}}\right) \quad \text{(A12.15a)}$$

$$f_p = \exp\left[-c_p\left(\frac{p_a}{p}\right)^2\right]; \quad g_{r1} = (\mathbf{r}^1 : \mathbf{r}^1)^{1/2}; \quad \bar{g}_{r2} = (\bar{\mathbf{r}}^2 : \bar{\mathbf{r}}^2)^{1/2} \tag{A12.15b}$$

where $\bar{r}_v^2$ is the volumetric part of $\bar{\mathbf{r}}^2$, which is defined by (A12.12a) using $\bar{M}^\delta$ instead of M^δ. It is noted that as the sign of $\bar{r}_v^2$ changes from positive to negative, $\dot{\boldsymbol{\alpha}}^\eta$ also changes from positive to negative. c_ℓ is a model parameter.

p_a, g_{r1}, and $\bar{g}_{r2}$ are introduced for dimensional consistency. Note that without the factor f_p, ξ_1 goes to infinity when the stress path crosses the p–axis. While this is the intended effect for monotonic and drained cyclic loading, this prevents complete liquefaction for undrained loading. The role of f_p is to essentially remove the term involving η as the stress path approaches the origin. $c_p > 0$ is a constant. It may be noted that as $p \to 0, f_p \to 0$ and as $p \to \infty, f_p \to 1$. However, as p increases from zero, the above transition must occur quickly. A value of $c_p = 10^{-4}$ has been found to produce the intended results. Note that for $p/p_{\mathrm{a}} = 0.005$, 0.01, 0.05, and 0.1, $f_p = 0.0183$, 0.37, 0.96, and 0.99, respectively.

Rotational Hardening

The direction of $\dot{\boldsymbol{\delta}}^{\mathrm{a}}$ is taken as

$$\bar{\mathbf{r}} = \mathbf{r}^{2,\mathrm{ad}} = \mathbf{r}^2 - \frac{1}{3} r_{kk}^{2,\mathrm{a}} \boldsymbol{\delta}^{\mathrm{a}} \tag{A12.16a}$$

where

$$r_{kk}^{2,\mathrm{a}} = r_{k\ell}^2 \delta_{k\ell}^{\mathrm{a}} \tag{A12.16b}$$

Noting that $\mathbf{r}^{2,ad}$ is normal to the $\boldsymbol{\delta}^a$-axis, it follows that $\dot{\boldsymbol{\delta}}^a = 0$ when the loading takes place along the $\boldsymbol{\delta}^a$-axis. This is because $\mathbf{r}^2$ and $\boldsymbol{\delta}^a$ are parallel to each other on the $\boldsymbol{\delta}^a$ axis. The hardening function is expressed as

$$\lambda^* = \frac{1}{2}[\lambda_{\mathrm{L}} + \lambda_{\mathrm{UL}}] + \frac{1}{2} x[\lambda_{\mathrm{L}} - \lambda_{\mathrm{UL}}](x\eta_\beta)^{1/5} \tag{A12.17a}$$

where

$$\lambda_{\mathrm{L}} = c_0\left(1 - \frac{\eta^{\mathrm{a}}}{\eta_{\max}^{\mathrm{a}}}\right) f_3; \quad \eta^{\mathrm{a}} = \frac{\delta_J^{\mathrm{a}}}{\delta_{kk}^{\mathrm{a}}};$$
$$c_0 = -81\ \ln(0.05)\frac{(A_{\mathrm{CSL}} - 1)}{(A_{\mathrm{CSL}} + 2)M_{\mathrm{CSL}}^3} \tag{A12.17b}$$

$$\eta_\beta = \frac{\beta}{\beta_{\mathrm{CSL}}}; \quad \beta = \frac{(\mathbf{s} - \boldsymbol{\alpha}^S) : \boldsymbol{\alpha}^\delta}{I^2 N^{c\delta}(1 + \eta_\alpha^2)^{1/2}} \tag{A12.17c}$$

$$x = 1 \quad \text{for } \eta_\beta > 0 \quad \text{and} \quad x = -1 \quad \text{for } \eta_\beta \le 0. \tag{A12.17d}$$

The constant c_0 controls the rate of change of δ_{ij}^{a} during loading and λ_{UL} controls the rate of change during stress reversal. Discrete element analyses on random assemblies of spherical particles (Anandarajah 2008b) indicated that A increases from 1.0 to A_{CSL} as the stress ratio increases from zero to a value near M_{CSL} (say, $0.95M_{CSL}$). On this basis, the expression given by (A12.17b) was developed for c_0 as a function of A_{CSL}. Hence, instead of c_0, A_{CSL} is used as the input parameter for the model. Thus, the evolution of anisotropy is controlled by A_{CSL} during loading, and λ_{UL} during unloading. The function f_3 as well as the rest of the details is the same as those in Anandarajah (2008a), and the reader is referred to this publication.

Isotropic Hardening

Consider a case where only mechanism 2 is active. Following Schofield and Wroth (1968), it may be shown that

$$g_2 = g_I r_v^2 \tag{A12.18a}$$

where

$$g_I = \frac{(1 + e_0)p_0}{\left[\lambda n_p \left(\frac{p_o}{p_a}\right)^{n_p} - \kappa n_e \left(\frac{p_o}{p_a}\right)^{n_e}\right]}; \quad n_e = 0.5 \tag{A12.18b}$$

r_v^2 is the volumetric part of r_{ij}^2. When both mechanisms are active, (A12.18a) is generalized to obtain an expression for g_1 as

$$g_1 = g_I r_v^1 \tag{A12.18c}$$

Appendix 12.6 The Lode-Angle Dependence

The dependence of model parameters on the third stress invariant can be conveniently considered using the Lode angle (Anandarajah 1994, 2008a). Let us define a Lode angle $\theta(\sigma_{ij}, a_{ij})$ in a generalized sense as

$$-\frac{\pi}{6} \le \theta(\sigma_{ij}, a_{ij}) = \frac{1}{3}\sin^{-1}\left[\frac{3\sqrt{3}}{2}\left(\frac{J_S^*}{J^*}\right)^3\right] \le \frac{\pi}{6} \tag{A12.19a}$$

$$J_S^* = \left(\frac{1}{2}S_{ij}^*S_{ij}^*\right)^{1/2}; \quad J^* = \left(\frac{1}{3}S_{ij}^*S_{jk}^*S_{ki}^*\right)^{1/3};$$
$$S_{ij}^* = \sigma_{ij}^* - \frac{1}{3}I^*\delta_{ij}; \quad I^* = \sigma_{ij}^*\delta_{ij}; \quad \sigma_{ij}^* = \sigma_{ij} - a_{ij} \tag{A12.19b}$$

Let us define isotropic and anisotropic Lode angles as

$$\theta^I = \theta(\sigma_{ij}, a_{ij} = o_{ij}) \quad \text{and} \quad \theta^A = \theta(\sigma_{ij}, a_{ij} = \alpha_{ij}^\delta) \tag{A12.20}$$

where o_{ij} is a null tensor. The value of a given parameter y at any value of the Lode angle is then defined using a smoothing function (Anandarajah 1994, 2008a) as

$$y = g(\theta, n)y_c; \quad g(\theta, n) = \frac{2n}{1 + n - (1 - n)\sin 3\theta}; \quad n = \frac{y_e}{y_c} \tag{A12.21}$$

where y_c and y_e are values of y at $\theta = \pi/6$ and $\theta = -\pi/6$, respectively. The parameters M_{CSL}, M^δ, and $\bar{M}^\delta$ are expressed as

$$M_{CSL} = g(\theta^I, n)M_{CSL}^C; \quad M^\delta = g(\theta^A, n)M^{c\delta}; \quad \bar{M}^\delta = g(\theta^A, n)\bar{M}^{c\delta};$$
$$n = M_{CSL}^e/M_{CSL}^c = M^{e\delta}/M^{c\delta} = \bar{M}^{e\delta}/\bar{M}^{c\delta}; \tag{A12.22}$$

In the simulation of cyclic behavior, it was found necessary to allow some of the other model parameters also vary with θ^I. However, this must be done without introducing discontinuity as θ^I changes from $\pi/6$ to $-\pi/6$ as the stress path crosses the p-axis. The following approach is taken:

$$y = y_\theta + (\bar{y} - y_\theta)f_s; \quad \bar{y} = (y_c + y_e)/2; \quad y_\theta = g(\theta^I, n)y_c; \quad n = y_e/y_c$$
$$f_s = \exp\left(-\alpha_1 \frac{q}{pM_{CSL}}\right) \tag{A12.23}$$

f_s is another smoothing function, whose purpose is to allow y to change across the p-axis smoothly but rapidly. A value of $\alpha_1 = 10.0$ was found to serve the purpose. This was found to be necessary for the parameters c_d, c_α, c_ℓ, and Γ_S. Data from triaxial compression and extension tests are necessary to determine the parameters.

References

Ahmadi, G. and Farshad, M. (1974). On the continuum theory of solid-fluid mixtures – A superimposed model of equipresent constituents. *Indian Journal of Technology*, 12: 195.

Alanso, E.E., Gens, A. and Josa, A. (1990). A constitutive model for partially saturated soils. *Geotechnique*, 40: 405–430.

Anandarajah, A. (1990). Time-domain radiation boundary for analysis of plane Love-wave propagation problems. *International Journal of Numerical Methods in Engineering*, 29: 1049–1063.

Anandarajah, A. (1993a). VELACS Project: elasto-plastic finite element prediction of the liquefaction behavior of centrifuge models nos. 1, 3 and 4a. Proceedings of the International Conference on Verification of Numerical Procedures for the Analysis of Soil Liquefaction Problems, Davis, CA, Oct. 17–20. (Eds. K. Arulanandan and R. F. Scott), pp. 1075–1104.

Anandarajah, A. (1993b). Dynamic analysis of axially-loaded footings in time domain. *Soils and Foundations, Japanese Society of Soil Mechanics and Foundation Engineering*, 33(1): 40–54.

Anandarajah, A. (1994). Procedures for elasto-plastic liquefaction modeling of sand. *Journal of Engineering Mechanics Division, ASCE*, 120(7): 1563–1589.

Anandarajah, A. (2000). Fully-coupled analysis of a single pile founded in liquefiable sand. In ASCE Geotechnical Special Publication Number 110: Computer Simulation of Earthquake Effects. (Eds. K. Arulanandan, A. Anandarajah and X. S. Li), pp. 117–131.

Anandarajah, A. (2004). Sliding and rolling constitutive theory for granular materials. *Journal of Engineering Mechanics, ASCE*, 130(6): 665–681.

Anandarajah, A. (2008a). Multi-mechanism anisotropic model for granular materials. *International Journal of Plasticity*, 24(5): 804–846 (Online version in 2007, and print version in 2008).

Anandarajah, A. (2008b). The critical state of granular materials based on the sliding-rolling theory. *Journal of Geotechnical and Geoenvironmental Engineering, ASCE*, 134(1): 125–135.

Anandarajah, A. (2008c). Modeling liquefaction by multimechnism model. *Journal of Geotechnical and Geoenvironmental Engineering, ASCE*, 134(7): 949–959.

Anandarajah, A. and Chen, J. (1997). Van der Waals attractive force between clay particles in water and contaminants. *Soils and Foundations, Japanese Society of Soil Mechanics and Foundation Engineering*, 37(2): 27–37.

Anandarajah, A. and Dafalias, Y.F. (1986). Bounding surface plasticity, part 3: Application to anisotropic soils. *Journal of Engineering Mechanics Division, ASCE*, 112(12): 1292–1318.

Anandarajah, A. and Lu, N. (1992). Numerical study of the electrical double-layer repulsion between non-parallel clay particles of finite length. *International Journal for Numerical and Analytical Methods in Geomechanics*, 15(10): 683–703.

Anandarajah, A., Rashidi, H. and Arulanandan, K. (1995). Elasto-plastic finite element analyses of earthquake pile-soil-structure interaction problems tested in a centrifuge. *Computers and Geotechnics*, 17: 301–325.

Anderson, J.C., Leaver, K.D., Rawlings, R.D. and Alexander, J.M. (1990). Materials Science (4th Edition). Chapman and Hall, London.

Archer, J.S. (1965). Consistent matrix formulations for structural analysis using finite element techniques. *AIAA Journal*. 3(10): 1910–1918.

Armero, F. and Pérez-Foguet, A. (2002). On the formulation of closest-point projection algorithm for elasto-plasticity. Part I: The variational structure. *International Journal of Numerical Methods in Engineering*, 53(1): 297–329.

Armstrong, P.J. and Frederick, C.O. (1966). A mathematical representation of the multiaxial Bauschinger effect. *G.E.G.B. Report RD/B/N 731*.

Atkin, R.J. and Craine, R.E. (1976a). Continuum theories of mixtures: basic theory and historical development. *Quarterly Journal of Mechanics and Applied Mathematics*, 29: 209–244.

Atkin, R.J. and Craine, R.E. (1976b). Continuum theories of mixtures: applications. *Journal of the Institute of Mathematics and its Applications*, 17: 153–207.

Atkinson, J.H. and Bransby, P.L. (1978). The Mechanics of Soils: An Introduction to Critical State Soil Mechanics. McGraw-Hill, Maidenhead.

Axelsson, O. (1976). A class of iterative methods for finite element equations. *CMAME*, 9(2): 123–137.

Axelsson, O. (1996). Iterative Solution Methods. Cambridge University Press, New York, 654 pages.

Bathe, K-J. (1982). Finite Element Procedures in Engineering Analysis. Prentice-Hall, Inc., Englewood Cliffs, NJ.

Bear, J. and Bachmat, Y. (1986). Macroscopic modeling of transport phenomena in porous media: 2. Application to mass momentum and energy transport. *Transport in Porous Media*, 1: 241–269.

Bicanic, N. and Pearce, C.J. (1996). Computational aspects of softening plasticity model for plain concrete. *Mechanics of Cohesive and Frictional Materials*, 1: 75–94.

Biot, M.A. (1941). General theory of three-dimensional consolidation. *Journal of Applied Physics*, 12: 155–164.

Biot, M.A. (1955). Theory of elasticity and consolidation for a porous anisotropic solid. *Journal of Applied Physics*, 26: 182–185.

Biot, M.A. (1956). Theory of propagation of elastic waves in a fluid-saturated porous solid. I: Low frequency range. *Journal of the Acoustic Society of America*, 28: 168–191.

Biot, M.A. and Willis, P.A. (1957). Elastic coefficients of the theory of consolidation. *Journal of Applied Mechanics*, 24: 594–601.

Bishop, A.W. (1959). The principle of effective stress. *Teknisk Ukeblad*, 39: 859–863.

Boresi, A.P., Schmidt, R.J. and Sidebottom, O.M. (1993). Advanced Mechanics of Materials. Wiley, New York, 811 pages.

Bowen, R.M. (1975). Theory of mixtures. In Continuum Physics. (Ed. A. C. Eringen), Academic, New York, Vol. 3, pp. 1–127.

Bowen, R.M. (1982). Compressible porous media models by use of the theory of mixtures. *International Journal of Engineering Science*, 20(6): 697–735.

Borja, R.I. (1991). Cam-clay plasticity, Part II: Implicit integration of constitutive equation based on a nonlinear elastic stress predictor. *Computer Methods in Applied Mechanics and Engineering*, 88: 225–240.

Borja, R.I. and Lee, S.R. (1990). Cam-clay plasticity, Part I: Implicit integration of elasto-plastic constitutive relations. *Computer Methods in Applied Mechanics and Engineering*, 78(1): 49–72.

Bridgeman, P.W. (1952). Studies in large plastic flow and fracture. Metallurgy and Metallurgical Engineering Series, McGraw-Hill, New York.

Bushnell, D. (1977). A strategy for the solution of problems involving large deflections, plasticity and creep. *International Journal of Numerical Methods in Engineering*, 11: 683–708.

Butkov, E. (1968). Mathematical Physics. Addison-Wesley, Reading, MA, 735 pages.

Chaboche, J.L. (1977). Viscoplastic constitutive equations for the description of cyclic and anisotropic behavior of metals. *Bulletin De l' Academie Plolonaise Des Sciences, Sévie Sc. Et Techn.* 25(1): 33.

Chaboche, J.L. (1986). Time independent constitutive theories for cyclic plasticity. *International Journal of Plasticity*, 2(2): 149.

Clough, R.W. and Penzien, J. (1975). Dynamics of Structures. McGraw Hill, New York, 634 pages.

Cook, R.D. (1981). Concepts and Applications of Finite Element Analysis. Wiley, New York, 537 pages.

Coleman, B.D. (1964). Thermodynamics of materials with memory. *Archive for Rational Mechanics and Analysis*, 17: 1.

Coulomb, C.A. (1773). Sur une application des regles de maxims et minims a qulques problems de statique relatifs à l'architecture. *Mémoires de Mathématique et de Physique, Académie des Sciences, Paris*, 7: 343.

Coussy, O. (1995). Mechanics of Porous Media. Wiley, Chichester.

Crisfield, M.A. (1987). Plasticity computations using Mohr-Coulomb yield criterion. *Engineering Computation*, 4: 300–308.

Crisfield, M.A. (1991). Nonlinear Finite Element Analysis of Solids and Structures: Vol. 1: Essentials, Wiley, New York.

Dafalias, Y.F. (1986). Bounding surface plasticity. I: Theory. *Journal of Engineering Mechanics, ASCE*, 112(EM12): 1242–1262.

Dafalias, Y.F. and Herrmann, L.R. (1986). Bounding surface plasticity. II: Application to isotropic cohesive soils. *Journal of Engineering Mechanics, ASCE*, 112(EM12): 1263–1291.

Dafalias, Y.F. and Popov, E.P. (1974). A model of nonlinearly hardening materials for complex loading. *Proceedings of the 7th U.S. National Congress of Applied Mechanics, (Abstract)*, Boulder, CO, p. 146.

Dafalias, Y.F. and Popov, E.P. (1975). A model for nonlinearly hardening materials for complex loading. *Acta Mechanica*, 21: 173–192.

Darcy, H.J. (1856). Les Fontaines Publiques de la ville de Dijon. V. Dalmont, Paris.

De Boer, R. (1996). Highlights in the historical development of the porous media theory. *Applied Mechanics Review*, 49: 201–262.

De Borst, R. (1986). Nonlinear Analysis of Frictional Materials. Ph.D. Thesis. Inst. TNO for Building Materials and Structures. Delft.

De Borst, R. and Heeres, O.M. (2002). A unified approach to the implicit integration of standard, non-standard and visco plastic models. *International Journal for Numerical and Analytical Methods in Geomechanics*, 26: 1059–1070.

De Saint-Venant, B. (1870). Memoire sur l'établissement des equations différentielles mouvements intérieurs opérés dans les corps solids ductiles au delà des limites ou l'élasticité pourrait les remener a leur premier état. *C. R. Académie des Sciences, Paris*, 70: 473.

De Souza Neto, E.A., Peric, D. and Owen, D.R.J. (1994). A model for elasto-plastic damage at finite strains: algorithmic issues and applications. *Engineering Computation*, 11: 257–281.

Dieter, G.E. (1986). Mechanical Metallurgy. (3rd Edition). McGraw Hill, New York.

Drucker, D.C. (1950). Some implications of work hardening and ideal plasticity. *Quarterly Journal of Applied Mechanics*, 7: 411.

Drucker, D.C. (1951). A more fundamental approach to plastic stress-strain relations. *Proceedings of the 1st U.S. National Congress on Applied Mechanics. ASME*, 487–491.

Drucker, D.C. and Prager, W. (1952). Soil mechanics and plasticity analysis or limit design. *Quarterly of Applied Mathematics*, 10: 157.

Drew, D.A. (1971). Averaged field equation for two-phase media. *Studies in Applied Mechanics*, 50: 133–166.

Duff, I.S. and van der Vorst, H.A. (1999). Developments and trends in the parallel solution of linear systems. *Parallel Computing*, 25(13–14): 1931–1970.

Dunne, F. and Petrinic, N. (2005). Introduction to Computational Plasticity. Oxford University Press. 241 pages.

Dutko, M., Peric, D. and Owen, D.R.J. (1993). Universal anisotropic yield criterion based on superquadratic functional representation: Part I: Algorithmic issues and accuracy analysis. *Computational Methods in Applied Mechanics and Engineering*, 109: 73–93.

Eringen, A.C. (1967). Mechanics of Continua. Wiley, New York, 502 pages.

Fisher, R.A. (1948). The fracture of liquids. *Journal of Applied Physics*, 19: 1062–1067.

Fletcher, R. (1987). Practical Methods of Optimization (2nd Edition). Wiley, Chichester, 436 pages.

Fredlund, D.G. and Rahardjo, H. (1993). Soil Mechanics for Unsaturated Soils. Wiley, New York.

Gawin, D. and Schrefler, B.A. (1996). Thermo-hydro-mechanical analysis of partially saturated porous materials. *Engineering Computations*, 13(7): 113–143.

Gear, C.W. (1971). *Numerical Initial Value Problems in Ordinary Differential Equations*. Prentice Hall. 253 pages.

Ghaboussi, J. and Wilson, E.L. (1972). Variational formulation of dynamics of fluid saturated porous elastic solids. *Journal of Engineering Mechanics, ASCE*, 98(EM4): 947–963.

Gray, W.G. and Hassanizadeh, M. (1989). Averaging theorems and averaged equations for transport of interface properties in multi-phase systems. *International Journal of Multi-Phase Flow*, 15: 81–95.

Gray, W.G. and Hassanizadeh, M. (1990a). Paradoxes and realities in unsaturated flow theory. *Water Resources Research*, 27(8): 1847–1854.

Gray, W.G. and Hassanizadeh, M. (1990b). Unsaturated flow theory including interfacial phenomena. *Water Resources Research*, 27(8): 1855–1863.

Green, A.E. and Naghdi, P.M. (1965). A dynamical theory of interacting continua. *International Journal of Engineering Science*, 3: 231–241.

Green, D.W., Winandy, J.E. and Kretschmann, D.E. (1999). Mechanical properties of Wood. In Wood Handbook: Wood as an Engineering Material. Gen. Tech. Rep. FPL-GTR-113. Forest Service, Forest Products Laboratory, U. S. Department of Agriculture, Madison, WI, 463 pages.

Griffiths, D.V. and Smith, I.M. (1991). Numerical Methods for Engineers. CRC Press, Boca Raton, FL.

Guan, Y. and Fredlund, D. (1997). Use of tensile strength of water for the direct measurement of high soil suction. *Canadian Geotechnical Journal*, 34(4): 604–614.

Gurtin, M.E., Oliver, M.L. and Williams, W.O. (1972). On balance of forces for mixtures. *Quarterly of Applied Mathematics*, 30: 527–530.

Guymon, G.L., Scott, V.H. and Herrmann, L.R. (1970). A general numerical solution of the two-dimensional differential-convection equation by the finite element method. *Water Research*, 6: 1611–1615.

Habib, P. and Luong, M.P. (1978). Sols pulvérulents sous chargement cyclique. Matériaux et Structures Sous Chargement Cyclique, Association Amicale des Ingénieurs Anciens Elèves de l'Ecole Nationale des Ponts et Chaussées (Palaiseau, 28–29, Sept.), pp. 49–79.

Hairer, E., Norsett, S.P. and Wanner, G. (1987). Solving Ordinary Differential Equations. Springer-Verlag.

Hashash, Y.M.A. and Whittle, A.J. (1992). Integration of the modified Cam-clay model in nonlinear finite element analysis. *Computers and Geotechnics*, 14: 59–83.

Hassanizadeh, M. and Gray, W.G. (1979a). General conservation equations for multi-phase systems: 1 Averaging procedure. *Advances in Water Resources*, 2: 131–144.

Hassanizadeh, M. and Gray, W.G. (1979b). General conservation equations for multi-phase systems: 1 Mass, momenta, energy and entropy equations. *Advances in Water Resources*, 2: 191–203.

Hassanizadeh, M. and Gray, W.G. (1980). General conservation equations for multi-phase systems: 3 Constitutive theory for porous media flow. *Advances in Water Resources*, 3: 25–40.

Hassanizadeh, M. and Gray, W.G. (1990). Mechanics and thermodynamics of multi-phase flow in porous media including inter-phase transport. *Advances in Water Resources*, 13(4): 169–186.

Herrmann, L.R. (1965). Elasticity equations for incompressible or nearly incompressible materials by a variational theorem. *Journal of American Institute of Aeronautics and Astronautics*, 3: 1896.

Herrmann, L.R., Kaliakin, V., Shen, C.K., Mish, K.D. and Zhu, Z-Y. (1987). Numerical implementation of plasticity model for cohesive soil. *Journal of Engineering Mechanics Division, ASCE*, 113(4): 500–519.

Hilber, H.M., Hughes, T.J.R. and Taylor, R.L. (1977). Improved numerical dissipation for time integration algorithms in structural mechanics. *International Journal of Earthquake Engineering and Structural Dynamics* 5: 283–292

Hill, R. (1948). A variational principle of maximum plastic work in classical plasticity. *Quarterly Journal of Mechanics and Applied Mathematics*, 1: 18.

Hill, R. (1950). The Mathematical Theory of Plasticity. Oxford University Press, New York.

Hinton, E., Rock, A. and Zienkiewicz, O.C. (1976). A note on mass lumping is related process in the finite element method. *International Journal of Earthquake Engineering and Structural Dynamics*, 4: 245–249.

Huber, M.T. (1904). Czasopismo Techniczne, Lemberg, Austria, Vol. 22, 181.

Hughes, T.J.R. and Winget, J. (1980). Finite rotation effects in numerical integration of rate constitutive equations arising in large deformation analysis. *International Journal for Numerical Methods in Engineering*, 15: 1862–1867.

Il'iushin, A.A. (1961). On the postulate of plasticity. *Prikl. Mat. Mekh.*, 25: 503.

Ishihara, K. (1993). Liquefaction and flow failure during earthquakes. 33rd Rankine Lecture. *Geotechnique*, 43(3): 351–415.

Ishihara, K., Tatsuoka, F. and Yasuda, S. (1975). Undrained deformation and liquefaction of sand under cyclic stresses. *Soils and Foundations*, 15(1): 29–44.

Ishii, M. (1975). Thermo-Fluid Dynamic Theory of Two-Phase Flow, Eyrolles, Paris.

Jeremic, B. and Sture, S. (1997). Implicit integrations in elastoplastic geotechnics. *Mechanics of Cohesive-Frictional Materials*, 2(2): 165–183.

Kalaydjian, F. (1987). A macroscopic description of multiphase flow involving spacetime and evolution of fluid/fluid interfaces. *Transport in Porous Media*, 2: 537–552.

Kaliakin, V.N. (2001). Introduction to Approximate Solution Techniques, Numerical Modeling and Finite Element Methods. CRC Press, Newark.

Kassimali, A. (1999). Matrix Analysis of Structures. Brooks/Cole Publishing Company, New York, 592 pages.

Key, S.W., Stone, C.M. and Krieg, R.D. (1980). A solution strategy for the quasi-static, large-deformation inelastic response of axisymmetric solids. *Proceedings of the Europe/U.S. Workshop on Nonlinear Finite Element Analysis in Structural Mechanics*, Bochum.

Koiter, W.T. (1953). Stress-strain relations, uniqueness and variational theorems for elasto-plastic materials with singular yield surfaces. *Quarterly of Applied Mathematics*, 11: 350.

Kojic, M. and Bathe, K.J. (1987). The effective stress function algorithm for thermo-elasto-plasticity and creep. *Int J Num Meth Engng*, 24: 1509–1532.

Kojic, M. and Bathe, K.J. (2005). Inelastic Analysis of Solids and Structures. Springer. 414 pages.

Kopal, Z. (1955). Numerical Analysis. Wiley, New York.

Krieg, R.D. (1975). A practical two-surface plasticity theory. *Journal of Applied Mechanics*, 42: 641–646.

Krieg, R.D. and Krieg, D.B. (1977). Accuracies of numerical solution methods for the elastic perfectly plastic model. *Transactions of ASME Journal of Pressure Vessel Technology*, 99: 510–515.

Lempriere, B.M. (1968). Poisson's ratio in orthotropic materials. *AIAA Journal*, 6(11): 2226.

Levy, M. (1870). Memoire sur des equations generales des mouvements interieurs des corps solides ductiles au dela limites ou l'élasticite pourrait les ramener a leur premier état. *C. R. Académie des Sciences, Paris*, 70: 1323.

Levy, M. (1871). Extrait du memoire sur les equations generales des mouvements interieurs des corps solides ductiles au dela limites ou l'élasticite pourrait les ramener a leur premier état. *Journal de Mathématiques Pures et Appliquées*, 16: 369.

Lewis, R.W. and Schrefler, B.A. (1998). The Finite Element Method in the Deformation and Consolidation of Porous Media. Wiley, New York.

Li, X.S. (2004). Modeling the hysteresis response for arbitrary wetting/drying paths. *Computers and Geotechnics*, 32: 133–137.

Likos, W.J. and Lu, N. (2004). Hysteresis of capillary stress in unsaturated granular soil. *Journal of Engineering Mechanics, ASCE*, 130(6): 646–655.

Loret, B. and Prevost, J.H. (1986). Accurate numerical solutions for Drucker-Prager elasto-plastic models. *Computer Methods in Applied Mechanics and Engineering*, 54(3): 259–277.

Lu, N. (2008). Is metric suction a stress variable? *Journal of Geotechnical and Geoenvironmental Engineering, ASCE*, 134(7): 899–905.

Lu, N. and Likos, W.J. (2004). Unsaturated Soil Mechanics. Wiley.

Lubliner, J. (1990). Plasticity Theory. Macmillan, New York.

Luenberger, D.G. (1984). Linear and Nonlinear Programming. (2nd Edition). Addison-Wesley Publishing Company, 491 pages.

Malkus, D.S. and Hughes, T.J.R. (1978). Mixed finite element method – reduced and selective integration techniques: a unification concepts. *Computer Methods in Applied Mechanics and Engineering*, 15: 63–81.

Mallett, R.H. and Schmit, L.A. (1967). Nonlinear structural analysis by energy search. *Journal of the Structural Division*, 93(ST3): 221–234.

Manzari, M.T. and Nour, M.A. (1997). On Implicit Integration of Bounding Surface Plasticity Models, *Computers and Structures*, 63(3): 385–395.

Manzari, M.T. and Prachathananukit, R. (2001). On integration of a cyclic plasticity model for sand. *International Journal of Numerical and Analytical Methods in Geomechanics*, 25(6): 525–549.

Marle, C.M. (1982). On macroscopic equations governing multiphase flow with diffusion and chemical reactions in porous media. *International Journal of Engineering Science*, 20: 643–662.

Marques, J.M.M.C. (1984). Stress computation in elasto-plasticity. *Engineering Computations*, 1: 42–51.

Martin, J.B. (1975). Plasticity: Fundamentals and General Results. MIT Press, Cambridge, MA.

Mei, C.C. and Foda, M.A. (1982). Boundary layer theory of waves in poro-elastic sea bed. In: Soil Mechanics–Transient and Cyclic Loads, Eds. G.N. Pande and O.C. Zienkiewicz. Jonh Wiley and Sons, pp. 17–36.

Meroi, E.A., Schrefler, B.A. and Zienkiewicz, O.C. (1995). Large strain static and dynamic semisaturated soil behavior. *International Journal of Numerical and Analytical Methods in Geomechanics*, 19: 81–106.

Mase, G.E. (1970). Continuum Mechanics. Schaum's Outline Series. McGraw Hill, New York, 221 pages.

Meyerhof, G.G. (1963). Some recent research on the bearing capacity of foundations. *Canadian Geotechnical Journal*, 1(1): 16–26.

Mondkar, D.P. and Powell, G.H. (1977). Evaluation of state determination calculation in nonlinear analysis. In *4th International Conference on Structural Mechanics Reaction Technology*, San Francisco.

Morland, L.W. (1972). A simple constitutive theory for a fluid-saturated porous solids. *Journal of Geophysical Research*, 77: 890–900.

Mroz, Z. (1967). On the description of anisotropic work hardening. *Journal of Mechanics and Physics of Solids*, 15: 163–175.

Mroz, Z., Norris, V.A. and Zienkiewicz, O.C. (1979). Application of an anisotropic hardening model in the analysis of elasto-plastic deformation of soils. *Geotechnique*, 29(1): 1–34.

Muraleetharan, K.K. and Wei, C. (1999). Dynamic behavior of unsaturated porous media: governing equations using the theory of mixtures with interfaces. *International Journal for Numerical and Analytical Methods in Geomechanics*, 23: 1579–1608.

Muraleetharan, K.K., Liu, C., Wei, C-F., Kibbey, T.C.G., and Chen, L. (2009). An elastoplastic framework for coupling hydraulic and mechanical behavior of unsaturated soils. *International Journal of Plasticity*, 25, 473–490.

Nayak, G.C. and Zienkiewicz, O.C. (1972). Elasto-plastic stress analysis, a generalization for various constitutive relations including strain softening. *International Journal for Numerical Methods in Engineering*, 5: 113–135.

Nguyen, Q.S. (1977). On the elasto-plastic initial boundary value problem and its numerical integration. *International Journal for Numerical Methods in Engineering*, 11: 817–832.

Nigmatulin, R.I. (1979). Spatial averaging in the mechanics of heterogeneous and dispersed systems. *International Journal of Multi-Phase Flow*, 5: 353–385.

Nyssen, C. (1981). An efficient and accurate iterative method, allowing large increment steps, to solve elasto-plastic problems. *Computers & Structures*, 13: 63–71.

Ortiz, M. and Simo, J.C. (1986). Analysis of a new class of integration algorithms for elastoplastic constitutive relations. *International Journal for Numerical Methods in Engineering*, 23: 353–366.

Ortiz, M. and Popov, E.P. (1985). Accuracy and stability of integration algorithms for elastoplastic constitutive equations. *International Journal for Numerical Methods in Engineering*, 21: 1561–1576.

Pérez-Foguet, A. and Armero, F. (2002). On the formulation of closest-point project algorithms in elasto-plasticity: Part II: Globally convergent scheme. *International Journal for Numerical Methods in Engineering*, 55: 331–374.

Pérez-Foguet, A., Rodriguez-Ferran, A. and Huerta, A. (2001). Consistent tangent matrices for substepping schemes. *Computational Methods in Applied Mechanics and Engineering*, 190: 4627–4647.

Pietruszczak, S. and Pande, G.N. (1996). Constitutive relations for partially saturated soils containing gas inclusions. *Journal of Geotechnical Engineering, ASCE*, 122(1): 50–59.

Prager, W. (1955). The theory of plasticity: a survey of recent achievements. *Proceedings of the Institution of Mechanical Engineers*, 169: 41.

Prandtl, L. (1921). Uber die eindringungsfestigkeit (harte) plastischer baustoffe und die festigkeit von schneiden, *Zeitschrift fur angewandte Mathematik und Mechanik*, 1(1): 15–20.

Prandtl, L. (1924). Spannungsverteilung in plastischen Koerpern. In *Proceedings of the 1st International Congress on Applied Mechanics*. Delft, p. 43.

Prevost, J.H. (1978). Plasticity theory for soil stress-strain behavior. *Proc. ASCE, Engineering Mechanics*, 104: 1177–1196.

Prevost, J.H. (1980). Mechnics of continuous porous media. *International Journal of Engineering Science*, 18(5): 787–800.

Reddy, J.N. (1993). An Introduction to the Finite Element Method. McGraw Hill, New York.

Rees, D.W.A. (2000). Mechanics of Solids and Structures. Imperial College Press, London, 734 pages.

Reuss, A. (1930). Beruecksichtigung der elastischen Formaenderungen in der Plastizitaetstheorie. *Z. Angen. Math. Mech*, 10: 266.

Roscoe, K.H. and Burland, J.B. (1968). On the generalized stress-strain behavior of wet clay. In Engineering Plasticity. (Eds. J. Heyman and F.A. Leckie), Cambridge University Press, Cambridge, England, pp. 535–609.

Runesson, K. and Samuelsson, A. (1985). Aspects of numerical techniques in small deformation plasticity. In NUMETA 85, Numerical Methods in Engineering: Theory and Applications. (Eds. J. Middleton and G.N. Pande), A.A. Balkema, Rotterdam, Vol. 1, pp. 337–348.

Sandhu, R.S. and Wilson, E.L. (1969). Finite element analysis of flow in saturated porous elastic media. *Journal of Engineering Mechanics, ASCE*, 95: 641–652.

Schiffman, R.L. (1970). Stress components of a porous medium. *Journal of Geophysical Research*, 75: 4035–4038.

Schofield, A. and Wroth, P. (1968). The Critical State Soil Mechanics. McGraw Hill, London.

Schrefler, B.A. and Simoni, L. (1995). Numerical solutions of thermo-hydro-mechanical problems. In Modern Issues in Non-Saturated Soils. (Eds. A. Gens, P. Jouanna and B.A. Schrefler), Springer, Berlin, pp. 213–276.

Shreyer, H.L., Kulak, R.F. and Kramer, J.M. (1979). Accurate numerical solution for elasto-plastic models. *Transactions ASME Journal of Pressure Vessel Technology*, 101: 226–234.

Selvadurai, A.P.S. (1996a). Mechanics of Porous Media. Kluwer, Dordrecht, The Netherlands.

Selvadurai, A.P.S. (1996b). Thermally-induced pore pressure generation in a nearly-saturated cementitious material. In Unsaturated Soils. (Ed. T. Schanz), Springer, Berlin, Vol. II: 15–28.

Sekiguchi, H., Ohta, K. (1977). Induced anisotropy and time dependency in clays. In Constitutive Equations of Soils. *Proceedings 9th International Conference on Soil Mechanics and Foundation Engineering, Special Session 9, Tokyo*, ISSMFE, pp. 229–238.

Shield, R. and Ziegler, H. (1958). On Prager's hardening rule. *Zeitschrift für angewandte Mathematik und Physik*, 9a: 260.

Simácek, P., Kaliakin, V.N. and Pipes, R.B. (1993). Pathologies associated with the numerical analysis of hyper-anisotropic materials. *International Journal for Numerical Methods in Engineering*, 36: 3487–3508.

Simo, J.C. and Govindjee, S. (1988). Exact closed-form solution of the return mapping algorithm for plane stress elasto-viscoplasticity. *Engineering Computations*, 3: 254–258.

Simo, J.C. and Taylor, R.L. (1985). Consistent tangent operators for rate-independent elastoplasticity. *Computer Methods in Applied Mechanics and Engineering*, 48: 101–118.

Simo, J.C. and Taylor, R.L. (1986). Return mapping algorithm for plane stress elastoplasticity. *International Journal for Numerical Methods in Engineering*, 22: 649–670.

Simo, J.C. and Ortiz, M. (1985). A unified approach to finite deformation elasto-plasticity based on the use of hyperelastic constitutive equations. *Computer Methods in Applied Mechanics and Engineering*, 49: 221–245.

Simo, J.C., Ju, J.W., Pister, K.S. and Taylor, R.L. (1988). An assessment of the cap model: consistent return algorithms and rate dependent extension. *Journal of Engineering Mechanics, ASCE*, 114: 191–218.

Simo, J.C. and Hughes, T.J.R. (1998). Computational Inelasticity. Springer, New York, 392 pages.

Skempton, A.W. (1960). Effective stress in soils, concrete and rocks. Proceedings of a Conference on Pore Pressure and Suction in Soils. Butterworths, London, pp. 4–16.

Slattery, J.M. (1981). Momentum, Energy and Mass Transfer in Continua. (2nd Edition). McGraw Hill, New York.

Sloan, S.W. (1987). Substepping schemes for the numerical integration of elastoplastic stress-strain relations. *Int J Num Meth Engng*, 24: 893–911.

Soboyejo, W. (2003). Mechanical Properties of Engineered Materials. Marcel Dekker, Inc, New York.

Stewart, G.W. (1973). Introduction to Matrix Computations. Academic, New York, 441 pages.

Stroud, A.H. and Secrest, D. (1966). Gaussian Quadrature Formulas. Prentice-Hall, Englewood Cliffs, NJ.

Surana, K.S. (1978). Lumped mass matrices with non-zero inertia for general shell and axisymmetric shell elements. International Journal of Numerical Methods in Engineering, 12(11): 1635–1650.

Taylor, G.I. (1947). A connection between criterion of yield and the strain ratio relationship in plastic solids. *Proceedings of the Royal Society*, A191: 441.

Taylor, D.W. (1948). Fundamentals of Soil Mechanics. Wiley, New York.

Tresca, H. (1864). Sur l'écoulement des corps solides soumis á de fortes pressions. *C. R. Académie des Sciences, Paris*, 59: 754.

Terzaghi, K. (1925). Erdbaumechanik auf bodenphysikalischer Grundlage. Leipzig, Deuticke.

Terzaghi, K. (1936). The shearing resistance of saturated soils. *Proceedings of the 1st ICSMFE*, 1: 54–56.

Terzaghi, K. (1943). Theoretical Soil Mechanics. Wiley, New York.

Truesdell, C. (1965). The Elements of Continuum Mechanics. Springer, New York.

Truesdell, C. and Toupin, R. (1960). The classical field theories. In Handbuch der physic. (Ed. S. Flugge), Springer, Berlin, Vol. III/1.

Valanis, K.C. (1971). A theory of viscoplasticity without a yield surface, Part I: General theory. *Archives of Mechanics*, 23: 517.

Valanis, K.C. and Reed, H.E. (1982). A new endochronic plasticity models for soils. In Soil Mechanics – Transient and Cyclic Loads. (Eds. G.N. Pande and O.C. Zienkiewicz), Wiley, New York.

von Mises, R. (1913). Mechanik der festen Körper im plastisch deformablen Zustland. *Nachr. Ges. Wiss. Göttingen*, 1: 582–592.

von Mises, R. (1928). Mechnik der plastischen Formanderung von Kristallen. *Z. Angen. Math. Mech*, 8: 161.

Voyiadjis, G.Z., and Song, C.R. (2006). The Coupled Theory of Mixtures in Geomechanics with Applications, Springer, Heidelberg, ISBN: 3540-25130–8, 438 p.

Wang, X., Wang, L.B. and Xu, L.M. (2004). Formulation of the return mapping algorithm for elasto-platsic soil models. *Computers and Geotechnics*, 31: 315–338.

Whitaker, S. (1986). Flow in porous media II: The governing equations for immiscible two-phase flow. *Transport in Porous Media*, 1: 105–126.

Whitaker, S. (1999). The Method of Volume Averaging. Kluwer, Dordrecht/Boston/London, 219 pages.

Wilkins, M.L. (1964). Calculation of elasto-plastic flow. In Methods of Computational Physics 3, Academic, New York.

Williams, W.O. (1973). On the theory of mixtures. *Archives for Rational Mechanics and Analysis*, 51: 239–260.

Wood, D.M., Belkheir, K. and Liu, D.F. (1994). Strain softening and state parameter for sand modeling. *Geotechnique*, 44(2): 335–339.

Wroth, C.P. and Houlsby, G.T. (1985). Soil mechanics: property characterization and analysis procedures. *Proceedings of the 11th International Conference on Soil Mechanics and Foundation Engineering, San Francisco*, 1: 1–55.

Yoder, P.J. and Whirley, B.G. (1984). On the numerical integration of elasto-plastic models. *Journal of Applied Mechanics, ASME*, 51: 283–288.

Ziegler, H. (1959). A modification of Prager's hardening rule. *Quarterly of Applied Mathematics*, 17: 55.

Zienkiewicz, O.C. (1977). The Finite Element Method. McGraw Hill, New York, 787 pages.

Zienkiewicz, O.C. (1982). Basic formulation of static and dynamic behavior of soils and other porous materials. In Numerical Methods in Geomechanics. (Ed. J.B. Martins), D. Reidel, Boston and London.

Zienkiewicz, O.C. and Shiomi, T. (1984). Dynamic behavior of saturated porous media: the generalized Biot formulation and its solution. *International Journal for Numerical and Analytical Methods in Geomechanics*, 8: 71–96.

Zienkiewicz, O.C., Chan, A.H.C., Pastor, M., Paul, D.K. and Shiomi, T. (1990a). Static and dynamic behavior of geomaterials: a rational approach to quantitative solutions, Part 1, Fully saturated problems. *Proceedings of the Royal Society London*, A429: 285–309.

Zienkiewicz, O.C., Xie, Y.M., Schrefler, B.A., Ledesma, A. and Bicanic, N. (1990b). Static and dynamic behavior of geomaterials: a rational approach to quantitative solutions, Part 2, Semi-saturated problems. *Proceedings of the Royal Society London*, A429: 310–323.

Zienkiewicz, O.C. and Taylor, R.L. (1991). The Finite Element Method. McGraw-Hill, Oxford, 807 pages.

Zienkiewicz, O.C., Chan, A.H.C., Pastor, M., Schrefler, B.A. and Shiomi, T. (1999). Computational Geomechanics with Special Reference to Earthquake Engineering. Wiley, New York.

Index

R

S